# Springer Series in Statistics

# Springer Series in Statistics

*Alho/Spencer:* Statistical Demography and Forecasting.
*Andersen/Borgan/Gill/Keiding:* Statistical Models Based on Counting Processes.
*Atkinson/Riani:* Robust Diagnostic Regression Analysis.
*Atkinson/Riani/Ceriloi:* Exploring Multivariate Data with the Forward Search.
*Berger:* Statistical Decision Theory and Bayesian Analysis, 2nd edition.
*Borg/Groenen:* Modern Multidimensional Scaling: Theory and Applications, 2nd edition.
*Brockwell/Davis:* Time Series: Theory and Methods, 2nd edition.
*Bucklew:* Introduction to Rare Event Simulation.
*Cappé/Moulines/Rydén:* Inference in Hidden Markov Models.
*Chan/Tong: Chaos:* A Statistical Perspective.
*Chen/Shao/Ibrahim:* Monte Carlo Methods in Bayesian Computation.
*Coles:* An Introduction to Statistical Modeling of Extreme Values.
*Devroye/Lugosi:* Combinatorial Methods in Density Estimation.
*Diggle/Ribeiro:* Model-based Geostatistics
*Efromovich:* Nonparametric Curve Estimation: Methods, Theory, and Applications.
*Eggermont/LaRiccia:* Maximum Penalized Likelihood Estimation, Volume I: Density Estimation.
*Fahrmeir/Tutz:* Multivariate Statistical Modeling Based on Generalized Linear Models, 2nd edition.
*Fan/Yao:* Nonlinear Time Series: Nonparametric and Parametric Methods.
*Ferraty/Vieu:* Nonparametric Functional Data Analysis: Theory and Practice.
*Fienberg/Hoaglin:* Selected Papers of Frederick Mosteller
*Frühwirth-Schnatter:* Finite Mixture and Markov Switching Models.
*Ghosh/Ramamoorthi:* Bayesian Nonparametrics
*Glaz/Naus/Wallenstein:* Scan Statistics.
*Good:* Permutation Tests: Parametric and Bootstrap Tests of Hypotheses, 3rd edition.
*Gouriéroux:* ARCH Models and Financial Applications.
*Gu:* Smoothing Spline ANOVA Models.
*Györfi/Kohler/Krzyżak/Walk:* A Distribution-Free Theory of Nonparametric Regression.
*Haberman:* Advanced Statistics, Volume I: Description of Populations.
*Hall:* The Bootstrap and Edgeworth Expansion.
*Härdle:* Smoothing Techniques: With Implementation in S.
*Harrell:* Regression Modeling Strategies: With Applications to Linear Models, Logistic Regression, and Survival Analysis.
*Hart:* Nonparametric Smoothing and Lack-of-Fit Tests.
*Hastie/Tibshirani/Friedman:* The Elements of Statistical Learning: Data Mining, Inference, and Prediction.
*Hedayat/Sloane/Stufken:* Orthogonal Arrays: Theory and Applications.

*(Continued after index)*

Sylvia Frühwirth-Schnatter

# Finite Mixture and Markov Switching Models

Springer

Sylvia Frühwirth-Schnatter
IFAS—Institut für Angewandte Statistik
Johannes Kepler Universität
4040 Linz
Austria
Sylvia.fruehwirth-schnatter@jku.at

Library of Congress Control Number: 2006923106

ISBN-10: 0-387-32909-9 e-ISBN 0-387-35768-8
ISBN-13: 978-0387-32909-3

Printed on acid-free paper.



Printed in the United States of America. (MVY)

9 8 7 6 5 4 3 2 1

springer.com

To my husband

*Rudi Frühwirth,*

to our sons

*Felix, Matthias, and Stephan,*

and to my parents

*Elfriede and Karl Schnatter*

# Preface

Modelling based on finite mixture distributions is a rapidly developing area with the range of applications exploding. Finite mixture models are nowadays applied in such diverse areas as biometrics, genetics, medicine, and marketing whereas Markov switching models are applied especially in economics and finance. There exist various features of finite mixture distributions that render them useful in statistical modelling. First, finite mixture distributions arise in a natural way as marginal distribution for statistical models involving discrete latent variables such as clustering or latent class models. On the other hand, we find that statistical models which are based on finite mixture distributions capture many specific properties of real data such as multimodality, skewness, kurtosis, and unobserved heterogeneity. Their extension to Markov mixture models is able to deal with many features of practical time series, for example, spurious long-range dependence and conditional heteroscedasticity.

Finite mixture models provide a straightforward, but very flexible extension of classical statistical models. The price paid for this flexibility is that inference for these models is somewhat of a challenge. Although the specific models discussed in this book are very different, they share common features as far as inference is concerned, namely a discrete latent structure that causes certain fundamental difficulties in estimation, the need to decide on the unknown number of groups, states, and clusters, and great similarities in the algorithms used for practical estimation.

In the beginning, my intention was to write the book entirely from a Bayesian viewpoint, which has been the only way of statistical thinking that was to able to satisfy my own intellectual needs. I was introduced into the Bayesian approach as a student during a course on reliability theory read by Reinhold Viertl in the winter term 1981/82 at the University of Technology. I became a practical Bayesian a few months later when I had the incredible luck to start my scientific career on a project using Bayesian methods for flood design in hydrology (Kirnbauer et al., 1987).

However, the more this book project progressed, the clearer it became that a lot would be said about finite mixture and Markov switching models, about their mathematical formulation, their properties, and their applications, that would have been said with the very same words by any non-Bayesian. Therefore I decided to put the whole project on a broader basis as far as statistical inference is concerned.

I hope that by reading this book many frequent users of statistical models will become familiar with the finite mixture and Markov switching modelling approach, and by using the software developed especially for this book may succeed in pursuing this approach also in practice.

I am grateful to several researchers who raised my interest in the models and methods discussed in the book. Dieter Gutknecht introduced me to Kalman filtering and the application of (switching) state space models in hydrology in 1984 and the wonderful hours we spent discussing our ideas encouraged me to follow a scientific career. Sylvia Kaufmann drew my attention to Markov switching models and their usefulness in empirical economics and finance in 1994, which was the starting point for a rewarding friendship and cooperation. In 1995 Thomas Otter introduced me to the world of Bayesian methods in marketing research and I owe him and my former PhD student Regina Tüchler wonderful and exciting experiences with using finite mixture models for this line of research.

This book project was started when I was a member of the Statistics Department of the Vienna University of Economics and Business Administration and I would like to thank Helmut Strasser for his continuous support and encouragement. I am indebted to colleagues at the Department of Applied Statistics of the Johannes Kepler University in Linz for their enduring patience with my difficulty reconciling my duties as department head and bringing this project to an end. I am particularly grateful to Helga Wagner, who provided useful comments and help with proofreading the book. I am grateful to several anonymous publisher's referees for many helpful suggestions for improving the presentation of the material and to John Kimmel of Springer for his support.

Finally, I am greatly indebted to my husband Rudi Frühwirth for his love and his continuous understanding and support for my research activities throughout the years.

Linz and Vienna, Austria
February 2005

*Sylvia Frühwirth-Schnatter*

# Contents

# 1

# Finite Mixture Modeling

## 1.1 Introduction

Many statistical models involve finite mixture distribution in some way or other. The following illuminating statistical problem where a finite mixture distribution arises in a natural way was first noted by Feller (1943). Consider a population made up of $K$ subgroups, mixed at random in proportion to the relative group sizes $\eta_1, \dots, \eta_K$. Assume that interest lies in some random feature $Y$ which is heterogeneous across and homogeneous within the subgroups. Due to heterogeneity, $Y$ has a different probability distribution in each group, usually assumed to arise from the same parametric family $p(y|\boldsymbol{\theta})$, however, with the parameter $\boldsymbol{\theta}$ differing across the groups. The groups may be labeled through a discrete indicator variable $S$ taking values in the set $\{1, \dots, K\}$.

When sampling randomly from such a population, we may record not only $Y$, but also the group indicator $S$. The probability of sampling from the group labeled $S$ is equal to $\eta_S$, whereas conditional on knowing $S$, $Y$ is a random variable following the distribution $p(y|\boldsymbol{\theta}_S)$ with $\boldsymbol{\theta}_S$ being the parameter in group $S$. The joint density $p(y, S)$ is given by

$$p(y, S) = p(y|S)p(S) = p(y|\boldsymbol{\theta}_S)\eta_S.$$

A finite mixture distribution arises if it is not possible to record the group indicator $S$; what we observe is only the random variable $Y$. The marginal density $p(y)$ is obviously given by the following mixture density,

$$p(y) = \sum_{S=1}^{K} p(y, S) = \eta_1 p(y|\boldsymbol{\theta}_1) + \cdots + \eta_K p(y|\boldsymbol{\theta}_K). \tag{1.1}$$

The Fishery Data analyzed in Titterington et al. (1985), a data set of the length of 256 snappers, is an interesting example from biology, showing how a mixture distribution arises when unobserved heterogeneity is present in a population for which a particular random characteristic is observed. Similar

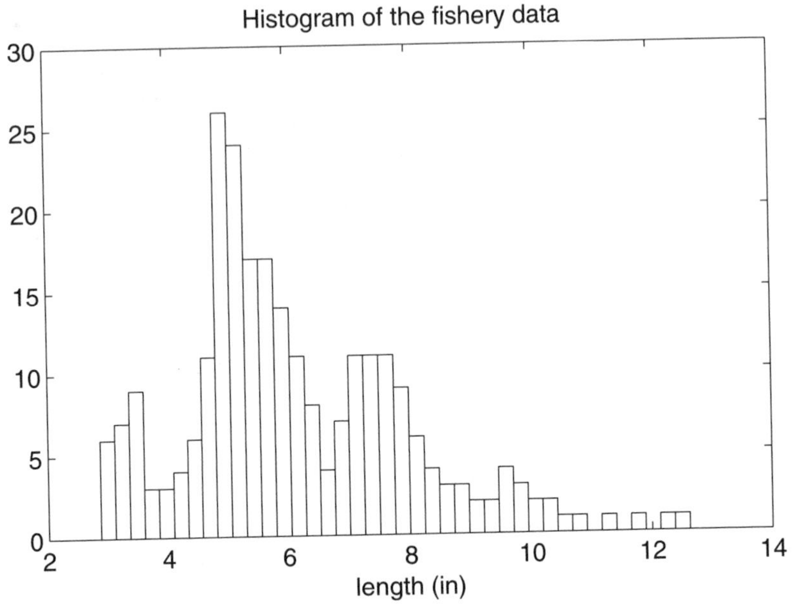


**Fig. 1.1.** FISHERY DATA, empirical distribution of the observations

data are found in several other areas such as marketing (Rossi et al., 2005, Chapter 5) or public health (Spiegelhalter et al., 2003). The histogram of the FISHERY DATA displayed in Figure 1.1 shows various modes, a possible explanation being that the fish belong to different age groups, and thus have different expected length. As age is hard to measure, no observations concerning the age group a fish belongs to are available, and one has to cope with unobserved heterogeneity.

More generally, heterogeneity in a sample occurs whenever the mean of a random characteristic $Y$ is different among the observed subjects. A common way to deal with heterogeneity is to assume that $Y$ follows a normal distribution with subject specific mean $\mu_i^s$, $Y \sim \mathcal{N}\left(\mu_i^s, \sigma^2\right)$. If the deviation of $\mu_i^s$ from a common mean $\mu$ could be explained by an observed factor $z_i$,

$$\mu_i^s = \beta_1 + z_i\beta_2,$$

then observed heterogeneity is present, and the unknown parameters could be estimated from a regression type model:

$$Y = \beta_1 + z_i\beta_2 + \varepsilon_i, \qquad \varepsilon_i \sim \mathcal{N}\left(0, \sigma^2\right). \tag{1.2}$$

If the factor $z_i$, however, is unobserved, then we have to cope with unobserved heterogeneity. For the FISHERY DATA, for instance, the unobserved factor $z_i$ could be the age of a fish. If the differences in the mean are caused by an unobserved categorical variable $z_i$ with $K$ categories, then the distribution $p(\mu_i^s)$ of $\mu_i^s$ will be discrete:

$$\mu_i^s = \begin{cases} \mu_1, & S_i = 1, \\ \vdots & \vdots \\ \mu_K, & S_i = K. \end{cases} \tag{1.3}$$

Here we coded the categorical variable $z_i$ through an indicator $S_i$ taking the values $1, \ldots, K$. Let $\eta_1, \ldots, \eta_K$ denote the discrete distribution of $z_i$ in the population. If the data are a random sample from the population under investigation, then the marginal distribution of $Y$ given in (1.2) is a finite mixture of $K$ normal distributions with equal variances:

$$Y \sim \eta_1 \mathcal{N}\left(\mu_1, \sigma^2\right) + \cdots + \eta_K \mathcal{N}\left(\mu_K, \sigma^2\right). \tag{1.4}$$

Thus a finite mixture model captures unobserved heterogeneity caused by an unobserved categorical variable.

Many other statistical issues have a similar finite mixture structure in the sense that the marginal distribution of a random variable of interest has a mixture density as in (1.1). Such applied finite mixture problems were studied in rather different scientific communities, such as unsupervised clustering in neural network applications, latent class analysis in the social sciences, and regime switching models in economics, just to mention a few. One of the major goals of this book is to alert the reader to the many different statistical problems where such finite mixture distributions arise. A second goal of the book is to show that many common features are shared by these methods. Some of these topics have a vast literature of their own, such as unsupervised clustering or latent class analysis, with a host of applications in many areas.

## 1.2 Finite Mixture Distributions

### 1.2.1 Basic Definitions

Consider a random variable or random vector $\mathbf{Y}$, taking values in a sample space $\mathcal{Y} \subset \Re^r$, which may be discrete or continuous, univariate or multivariate. In what follows, the probability distribution of a random variable $\mathbf{Y}$ is generally characterized by its probability density function. It is understood that probability density functions are defined with respect to an appropriate measure on $\Re^r$ which is either the Lebesgue measure, or a counting measure, or a combination of the two, depending on the context.

$\mathbf{Y}$ is said to arise from a finite mixture distribution, if the probability density function $p(\mathbf{y})$ of this distribution takes the form of a mixture density for all $\mathbf{y} \in \mathcal{Y}$:

$$p(\mathbf{y}) = \eta_1 p_1(\mathbf{y}) + \cdots + \eta_K p_K(\mathbf{y}),$$

where $p_k(\mathbf{y})$ is a probability density function for all $k = 1, \dots, K$. A single density $p_k(\mathbf{y})$ is referred to as the component density. $K$ is called the number of components. The parameters $\eta_1, \dots, \eta_K$ are called the weights; the vector $\boldsymbol{\eta} = (\eta_1, \dots, \eta_K)$ is called the weight distribution. $\boldsymbol{\eta}$ takes a value in the unit simplex $\mathcal{E}_K$ which is a subspace of $(\Re^+)^K$, defined by following constraint,

$$\eta_k \geq 0, \qquad \eta_1 + \dots + \eta_K = 1. \tag{1.5}$$

In most applications one assumes that all component densities arise from the same parametric distribution family $\mathcal{T}(\boldsymbol{\theta})$ with density $p(\mathbf{y}|\boldsymbol{\theta})$, indexed by a parameter $\boldsymbol{\theta} \in \Theta$, although this need not be the case:

$$p(\mathbf{y}|\boldsymbol{\vartheta}) = \eta_1 p(\mathbf{y}|\boldsymbol{\theta}_1) + \dots + \eta_K p(\mathbf{y}|\boldsymbol{\theta}_K). \tag{1.6}$$

The mixture density function $p(\mathbf{y}|\boldsymbol{\vartheta})$ is indexed by the parameter $\boldsymbol{\vartheta} = (\boldsymbol{\theta}_1, \dots, \boldsymbol{\theta}_K, \boldsymbol{\eta})$ taking values in the parameter space $\Theta_K = \Theta^K \times \mathcal{E}_K$. Unless stated otherwise, we assume that the finite mixture distribution is unconstrained in the sense that no constraints are imposed on the component parameters $\boldsymbol{\theta}_1, \dots, \boldsymbol{\theta}_K$ and that the weight distribution $\boldsymbol{\eta}$ is unconstrained apart from the natural constraint (1.5).

A random variable $\mathbf{Y}$ with density (1.6) is said to arise from a finite mixture of $\mathcal{T}(\boldsymbol{\theta})$ distributions, abbreviated by

$$\mathbf{Y} \sim \eta_1 \mathcal{T}(\boldsymbol{\theta}_1) + \dots + \eta_K \mathcal{T}(\boldsymbol{\theta}_K).$$

This book is mainly concerned with finite mixtures from well-known distribution families, an important example being finite mixtures of multivariate normal distributions $\mathcal{N}_r(\boldsymbol{\mu}, \boldsymbol{\Sigma})$ with density $f_N(\mathbf{y}; \boldsymbol{\mu}, \boldsymbol{\Sigma})$, where the mixture density is given by

$$p(\mathbf{y}|\boldsymbol{\vartheta}) = \eta_1 f_N(\mathbf{y}; \boldsymbol{\mu}_1, \boldsymbol{\Sigma}_1) + \dots + \eta_K f_N(\mathbf{y}; \boldsymbol{\mu}_K, \boldsymbol{\Sigma}_K), \tag{1.7}$$

which reduces for $r = 1$ to a mixture of univariate normal distributions $\mathcal{N}(\mu, \sigma^2)$:

$$p(y|\boldsymbol{\vartheta}) = \eta_1 f_N(y; \mu_1, \sigma_1^2) + \dots + \eta_K f_N(y; \mu_K, \sigma_K^2). \tag{1.8}$$

Historically seen, a mixture of two univariate normal densities with different means and different variances is the oldest known application of a finite mixture distribution (Pearson, 1894).

Another popular example is a mixture of Poisson distributions $\mathcal{P}(\mu)$ with density $f_P(y; \mu)$, where the following mixture density results,

$$p(y|\boldsymbol{\vartheta}) = \eta_1 f_P(y; \mu_1) + \dots + \eta_K f_P(y; \mu_K); \tag{1.9}$$

see Feller (1943) for an early treatment of this distribution. Yet another example is a finite mixture of exponential distributions $\mathcal{E}(\lambda)$ with the mixture density being given as

$$p(y|\boldsymbol{\vartheta}) = \eta_1 f_E(y; \lambda_1) + \cdots + \eta_K f_E(y; \lambda_K), \tag{1.10}$$

where $f_E(y; \lambda_k)$ is the density of an exponential distribution parameterized as in Appendix A.1.4.

These examples are special cases of mixtures from the general exponential family (Barndorff-Nielsen, 1978), where the component densities in (1.6) take the following form,

$$p(\mathbf{y}|\boldsymbol{\theta}_k) = \exp\left\{\phi(\boldsymbol{\theta}_k)' u(\mathbf{y}) + c(\mathbf{y}) - g(\boldsymbol{\theta}_k)\right\}, \tag{1.11}$$

where $\phi : \Theta \rightarrow \Theta$, $u : \mathcal{Y} \rightarrow \Theta$, $c : \mathcal{Y} \rightarrow \Re$, and $g(\boldsymbol{\theta}_k)$ defines the normalizing constant:

$$\exp\{g(\boldsymbol{\theta}_k)\} = \int \exp\left\{\phi(\boldsymbol{\theta}_k)' u(\mathbf{y}) + c(\mathbf{y})\right\} d\mathbf{y}.$$

Shaked (1980) gives a general mathematical treatment of mixtures from the exponential family.

### 1.2.2 Some Descriptive Features of Finite Mixture Distributions

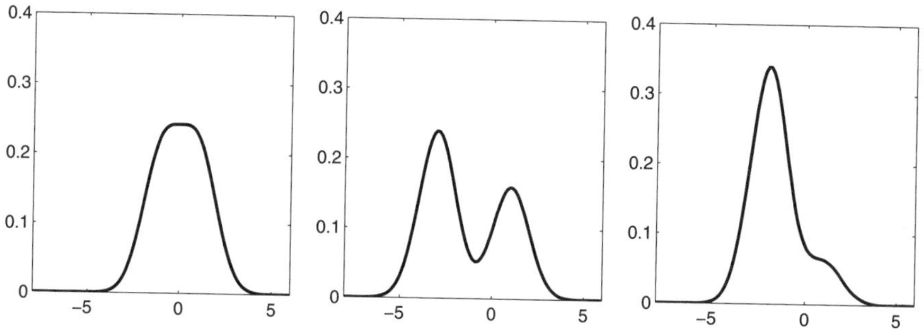


**Fig. 1.2.** Densities of various two-component mixtures of univariate normal distributions, with equal variance $\sigma_1^2 = \sigma_2^2 = 1$, but different weights and different means (left-hand side: $\mu_1 = -1$, $\mu_2 = 1$, $\eta_1 = 0.5$; middle: $\mu_1 = -3$, $\mu_2 = 1$, $\eta_1 = 0.6$; right-hand side: $\mu_1 = -2$, $\mu_2 = 1$, $\eta_1 = 0.85$)

The most striking property of a mixture density is that the shape of the density is extremely flexible. Figure 1.2, showing the density of various two-component mixtures of univariate normal distributions, demonstrates that a mixture density may be multimodal, or even if it is unimodal, may exhibit considerable skewness or additional humps. For this reason, finite mixture distributions offer a flexible way to describe rather unsmooth data structures such as the FISHERY DATA studied in Subsection 1.1, by summarizing the characteristics of the data in terms of the number, the location, and the spread of the mixture components.

Finite mixture models are often used for the purpose of clustering, meaning that one wants to find homogeneous groups among the data. This approach assumes the existence of $K$ hidden groups as in Subsection 1.1 and hopes to reconstruct them from fitting a $K$ component mixture density. If heterogeneity among the groups is large, the within-group distribution is normal and the groups are balanced, then in fact all groups will generate clusters of data around a mode corresponding to the mean of the hidden group; see, for instance, the middle plot in Figure 1.2. The other examples in Figure 1.2 show that mixture densities are not necessarily multimodal, and that a two-component mixture may generate only a single cluster

Some families, such as mixtures of exponential distributions, always generate unimodal densities, as is easily verified from the mixture density (1.10), because the mode of this density lies at zero and the derivative of $p(y|\boldsymbol{\vartheta})$ with respect to $y$ is negative on $\Re^+$:

$$\frac{\partial p(y|\boldsymbol{\vartheta})}{\partial y} = -\eta_1 \lambda_1 f_E(y;\lambda_1) - \cdots - \eta_K \lambda_K f_E(y;\lambda_K) < 0.$$

For mixtures of normal distributions the situation is much more complex, but in particular for applications in clustering it is important to understand how many modes exist in the mixture density. This does not really answer the question of whether a population with a unimodal normal mixture density should be described as being a homogeneous one arising from a nonnormal distribution or as consisting of two homogeneous normal groups that are not well separated.

Conditions for the number of modes in a mixture of univariate and multivariate normal densities have been derived recently in Ray and Lindsay (2005), generalizing several earlier works on univariate (Robertson and Fryer, 1969; Behboodian, 1970; see also Titterington et al., 1985, pp.160), and multivariate mixtures (Carreira-Perpiñán and Williams, 2003). Ray and Lindsay (2005) show that the number of modes in a $K$-component normal mixture density reduces to studying the mixture density along a $(K-1)$-dimensional manifold of $\Re^r$ which they call the ridgeline surface. For $K=2$ this manifold reduces to the following path from $\boldsymbol{\mu}_1$ to $\boldsymbol{\mu}_2$, indexed by $\alpha \in [0,1]$,

$$\mathbf{y}^\star(\alpha) = ((1-\alpha)\boldsymbol{\Sigma}_1^{-1} + \alpha\boldsymbol{\Sigma}_2^{-1})^{-1}((1-\alpha)\boldsymbol{\Sigma}_1^{-1}\boldsymbol{\mu}_1 + \alpha\boldsymbol{\Sigma}_2^{-1}\boldsymbol{\mu}_2),$$

which depends on the component parameters, but is independent of the weights $(\eta_1, \eta_2)$. The values of the two-component mixture density along this path,

$$h(\alpha) = \eta_1 f_N(\mathbf{y}^\star(\alpha);\boldsymbol{\mu}_1,\boldsymbol{\Sigma}_1) + \eta_2 f_N(\mathbf{y}^\star(\alpha);\boldsymbol{\mu}_2,\boldsymbol{\Sigma}_2),$$

is called the elevation plot. Ray and Lindsay (2005, Theorems 1 and 2 which are formulated for general $K$), give a rigorous proof that for $K=2$ all critical points of the mixture density $p(\mathbf{y}|\boldsymbol{\vartheta})$ lie on $\mathbf{y}^\star(\alpha)$ and that every critical point of $h(\alpha)$ gives rise to a local maximum of $p(\mathbf{y}|\boldsymbol{\vartheta})$, iif it is a local maximum of $h(\alpha)$. Thus the number of local maxima of $h(\alpha)$ over the unit interval is equal to the number of modes of the mixture density.

It is easy to verify that every critical point of $h(\alpha)$ satisfies the following equation,

$$\Phi(\alpha) = \frac{(1-\alpha) f_N(\mathbf{y}^\star(\alpha); \boldsymbol{\mu}_2, \boldsymbol{\Sigma}_2)}{\alpha f_N(\mathbf{y}^\star(\alpha); \boldsymbol{\mu}_1, \boldsymbol{\Sigma}_1) + (1-\alpha) f_N(\mathbf{y}^\star(\alpha); \boldsymbol{\mu}_2, \boldsymbol{\Sigma}_2)} = \eta_1.$$

Because $\Phi(0) = 1$, $\Phi(1) = 0$, and $\Phi(\alpha) \in [0,1]$, the function $\Phi(\alpha)$ which depends only on the component parameters, but not on the weight distribution, is oscillating between 0 and 1. For any $\eta_1$, there will be an uneven number $2M - 1 \geq 1$ of intersections of $\Phi(\alpha)$ with $\eta_1$, and $M$ defines the actual number of modes of $p(\mathbf{y}|\boldsymbol{\vartheta})$. Independent of $\eta_1$, there exists a maximum number $L^\star$ of such intersections, and $M^\star = (L^\star + 1)/2$ defines an upper limit for the number of modes of $p(\mathbf{y}|\boldsymbol{\vartheta})$ for a certain set of component parameters $\boldsymbol{\mu}_1, \boldsymbol{\mu}_2, \boldsymbol{\Sigma}_1$, and $\boldsymbol{\Sigma}_2$. The actual number of modes could be smaller for extreme values of $\eta_1$ being close to 0 or 1. $M^\star = L/2 + 1$, where $L$ is the number of sign changes of $\Phi(\alpha)$ on $[0,1]$, or equivalently, the number of local minima and maxima of $\Phi(\alpha)$ on $[0,1]$. Ray and Lindsay (2005, Theorem 3) show that the critical points of $\Phi(\alpha)$ are located at the zeros of the following function $q(\alpha)$,

$$q(\alpha) = 1 - \alpha(1-\alpha)(\boldsymbol{\mu}_2 - \boldsymbol{\mu}_1)' \mathbf{D}(\alpha)(\boldsymbol{\mu}_2 - \boldsymbol{\mu}_1), \tag{1.12}$$

where

$$\begin{aligned}\mathbf{D}(\alpha) &= \boldsymbol{\Sigma}_1^{-1}\mathbf{C}(\alpha)^{-1}\boldsymbol{\Sigma}_2^{-1}\mathbf{C}(\alpha)^{-1}\boldsymbol{\Sigma}_2^{-1}\mathbf{C}(\alpha)^{-1}\boldsymbol{\Sigma}_1^{-1},\\ \mathbf{C}(\alpha) &= (1-\alpha)\boldsymbol{\Sigma}_1^{-1} + \alpha\boldsymbol{\Sigma}_2^{-1}.\end{aligned}$$

Because $q(0) = q(1) = 1$, the number $L$ of sign changes is even. If (1.12) does not have any zeros in [0,1] for a specific component parameter, then $L = 0$ and the corresponding mixture density is unimodal, independent of $\eta_1$.

### Modes in a Homoscedastic Mixture of Two Normals

This leads to an explicit solution for homoscedastic mixtures with $\boldsymbol{\Sigma}_1 = \boldsymbol{\Sigma}_2 = \boldsymbol{\Sigma}$. In this case $q(\alpha)$ reduces to

$$q(\alpha) = 1 - \alpha(1-\alpha) d_M(\boldsymbol{\mu}_2; \boldsymbol{\mu}_1, \boldsymbol{\Sigma})^2, \tag{1.13}$$

where $d_M(\boldsymbol{\mu}_2; \boldsymbol{\mu}_1, \boldsymbol{\Sigma})$ is the Mahalanobis distance of the component mean $\boldsymbol{\mu}_2$ from the component given by $(\boldsymbol{\mu}_1, \boldsymbol{\Sigma})$ which is defined in (1.17).

It follows that any two-component homoscedastic mixture of normal distributions, irrespective of the dimension $r$ and the weights $\eta_1$ and $\eta_2$, has at most two modes, because $q(\alpha)$ is quadratic. A two-component homoscedastic mixture of normal distributions is unimodal, iff $q(\alpha)$, defined in (1.13), does not have any zeros in $[0,1]$ which is the case, iff the Mahalanobis distance is smaller than 2: $d_M(\boldsymbol{\mu}_2;\boldsymbol{\mu}_1,\boldsymbol{\Sigma}) < 2$. If $d_M(\boldsymbol{\mu}_2;\boldsymbol{\mu}_1,\boldsymbol{\Sigma}) \geq 2$, then two modes are possible, depending on the intersection of $\Phi(\alpha)$ with $\eta_1$. The zeros of (1.13) are given by

$$\alpha_{1,2} = \frac{1 \pm \sqrt{d_M(\boldsymbol{\mu}_2;\boldsymbol{\mu}_1,\boldsymbol{\Sigma})^2 - 4}}{2}, \tag{1.14}$$

thus the mixture density is bimodal if $\eta_1$ lies in the interior of $[\Phi(\alpha_1), \Phi(\alpha_2)]$. Straightforward calculations show that the boundaries of this interval are entirely determined by the Mahalanobis distance and are given by

$$\Phi(\alpha_i) = \frac{1-\alpha_i}{1-\alpha_i+\alpha_i \exp(-d_M(\boldsymbol{\mu}_2;\boldsymbol{\mu}_1,\boldsymbol{\Sigma})^2(\alpha_i - \frac{1}{2}))}.$$

The larger the Mahalanobis distance between the means, the closer $\eta_1$ may move toward 0 or 1 without losing bimodality. The closer the Mahalanobis distance moves toward 2, the closer to 1/2 the weight $\eta_1$ must be to obtain a bimodal density. If $\eta_1 = \eta_2 = 1/2$, then bimodality holds for $d_M(\boldsymbol{\mu}_2;\boldsymbol{\mu}_1,\boldsymbol{\Sigma}) > 2$.

### Modes in Heteroscedastic Mixtures of Two Normals

Furthermore, Ray and Lindsay (2005) give explicit results for scaled mixtures of normals, where $\boldsymbol{\Sigma}_2 = \omega\boldsymbol{\Sigma}_1$, which reduces to a heteroscedastic mixture in the univariate case. In this case $q(\alpha)$ reduces to

$$q(\alpha) = 1 - d_M(\boldsymbol{\mu}_2;\boldsymbol{\mu}_1,\boldsymbol{\Sigma}_1)^2 \frac{\omega\alpha(1-\alpha)}{((1-\alpha)\omega+\alpha)^3}, \tag{1.15}$$

where $d_M(\boldsymbol{\mu}_2;\boldsymbol{\mu}_1,\boldsymbol{\Sigma}_1)$ is the Mahalanobis distance of the component mean $\boldsymbol{\mu}_2$ from the component given by $(\boldsymbol{\mu}_1,\boldsymbol{\Sigma}_1)$ which is defined in (1.17). Ray and Lindsay (2005) show that $q(\alpha)$ does not have any zeros, and the corresponding mixture is unimodal, iff

$$d_M(\boldsymbol{\mu}_2;\boldsymbol{\mu}_1,\boldsymbol{\Sigma}_1) \leq \left(\frac{2(\omega^2-\omega+1)^{3/2} - (2\omega^3 - 3\omega^2 - 3\omega + 2)}{\omega}\right)^{1/2}. \tag{1.16}$$

This reduces to the same conditions as given by Robertson and Fryer (1969) for univariate heteroscedastic mixtures of normals. If the Mahalanobis distance $d_M(\boldsymbol{\mu}_2;\boldsymbol{\mu}_1,\boldsymbol{\Sigma}_1)$ is larger than this bound, then $q(\alpha)$ has two zeros $\alpha_{1,2}$ in [0,1], and the mixture is bimodal if $\eta_1$ lies in the interior of $[\Phi(\alpha_1), \Phi(\alpha_2)]$.

For more general mixtures, Ray and Lindsay (2005) suggest plotting the function $\Phi(\alpha)$ over $\alpha$, which they call the $\Phi$-plot, and determining the number $M$ of modes from the number $2M-1$ of intersections of this plot with $\eta_1$.

### 1.2.3 Diagnosing Similarity of Mixture Components

It is often of interest to diagnose similarity of mixture components and to see how much they overlap. Leisch (2004) uses the Kullback–Leibler distance,

$$I(p(\mathbf{y}|\boldsymbol{\theta}_k), p(\mathbf{y}|\boldsymbol{\theta}_l)) = \int p(\mathbf{y}|\boldsymbol{\theta}_k) \log(p(\mathbf{y}|\boldsymbol{\theta}_k)/p(\mathbf{y}|\boldsymbol{\theta}_l)) d\mathbf{y}$$

to diagnose which components overlap in a mixture model. This distance is not symmetric and may be substituted by the symmetrized Kullback–Leibler distance

$$J(p(\mathbf{y}|\boldsymbol{\theta}_k), p(\mathbf{y}|\boldsymbol{\theta}_l)) = I(p(\mathbf{y}|\boldsymbol{\theta}_k), p(\mathbf{y}|\boldsymbol{\theta}_l)) + I(p(\mathbf{y}|\boldsymbol{\theta}_l), p(\mathbf{y}|\boldsymbol{\theta}_k)).$$

For multivariate mixtures of normals this is equal to

$$\begin{aligned} &J(p(\mathbf{y}|\boldsymbol{\theta}_k), p(\mathbf{y}|\boldsymbol{\theta}_l)) \\ &= \frac{1}{2}(\boldsymbol{\mu}_k - \boldsymbol{\mu}_l)'(\boldsymbol{\Sigma}_k^{-1} + \boldsymbol{\Sigma}_l^{-1})(\boldsymbol{\mu}_k - \boldsymbol{\mu}_l) + \frac{1}{2}\text{tr}\left((\boldsymbol{\Sigma}_k\boldsymbol{\Sigma}_l^{-1} + \boldsymbol{\Sigma}_k^{-1}\boldsymbol{\Sigma}_l)\right) - r, \end{aligned}$$

where $r = \dim(\mathbf{y})$. For $\boldsymbol{\Sigma}_k = \boldsymbol{\Sigma}_l = \boldsymbol{\Sigma}$ this is related to the squared Mahalanobis distance

$$d_M(\boldsymbol{\mu}_k; \boldsymbol{\mu}_l, \boldsymbol{\Sigma})^2 = (\boldsymbol{\mu}_k - \boldsymbol{\mu}_l)' \boldsymbol{\Sigma}^{-1}(\boldsymbol{\mu}_k - \boldsymbol{\mu}_l),$$

which is defined for an arbitrary $\mathbf{y} \in \Re^r$ as

$$d_M(\mathbf{y}; \boldsymbol{\mu}_k, \boldsymbol{\Sigma}_k) = ((\mathbf{y} - \boldsymbol{\mu}_k)' \boldsymbol{\Sigma}_k^{-1}(\mathbf{y} - \boldsymbol{\mu}_k))^{1/2}. \tag{1.17}$$

Scott and Szewczyk (2001) introduced a symmetric similarity measure that lies between 0 and 1:

$$S(p(\mathbf{y}|\boldsymbol{\theta}_k), p(\mathbf{y}|\boldsymbol{\theta}_l)) = \frac{\int p(\mathbf{y}|\boldsymbol{\theta}_k)p(\mathbf{y}|\boldsymbol{\theta}_l)d\mathbf{y}}{(\int p(\mathbf{y}|\boldsymbol{\theta}_k)^2 d\mathbf{y})^{1/2}(\int p(\mathbf{y}|\boldsymbol{\theta}_l)^2 d\mathbf{y})^{1/2}}, \tag{1.18}$$

which is equal to 1, iff $p(\mathbf{y}|\boldsymbol{\theta}_k) = p(\mathbf{y}|\boldsymbol{\theta}_l)$ almost surely. Note that

$$S(p(\mathbf{y}|\boldsymbol{\theta}_k), p(\mathbf{y}|\boldsymbol{\theta}_l)) \leq 1$$

by the Cauchy–Schwarz inequality. For multivariate mixtures of normals this reduces to

$$\begin{aligned} &S(p(\mathbf{y}|\boldsymbol{\theta}_k), p(\mathbf{y}|\boldsymbol{\theta}_l)) \\ &= \exp\left(-\frac{1}{2}(\boldsymbol{\mu}_k - \boldsymbol{\mu}_l)'(\boldsymbol{\Sigma}_k + \boldsymbol{\Sigma}_l)^{-1}(\boldsymbol{\mu}_k - \boldsymbol{\mu}_l)\right) \frac{2^{r/2}\sqrt[4]{|\boldsymbol{\Sigma}_k^{-1}||\boldsymbol{\Sigma}_l^{-1}|}}{\sqrt{|\boldsymbol{\Sigma}_k^{-1} + \boldsymbol{\Sigma}_l^{-1}|}}. \end{aligned}$$

Ray and Lindsay (2005) suggest applying the methods they developed for diagnosing the number of modes for a two-component mixture to any pair of components in a mixture of multivariate normal distributions to study how close they are, and if they constitute a unimodal cluster in the data. The weight distribution for this analysis is obtained from the relative weights $\eta_k/(\eta_k + \eta_l)$ and $\eta_l/(\eta_k + \eta_l)$.

#### The Point Processes Representation

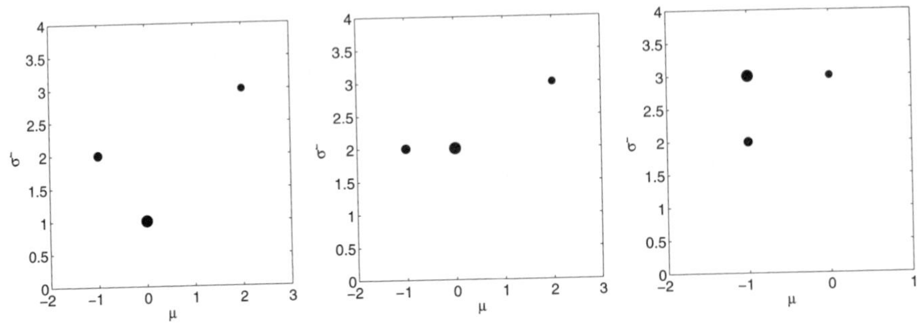


**Fig. 1.3.** Point process representation of various mixtures of three univariate normal densities with the size of the point being proportional to the weight of the corresponding component

For fixed $K$, and a fixed parametric family, any finite mixture distribution has a representation as a marked point process, a viewpoint introduced by Stephens (2000a). A finite mixture distribution may be seen as a distribution of the points $\{\boldsymbol{\theta}_1, \ldots, \boldsymbol{\theta}_K\}$ over the space $\Theta$, with each point $\boldsymbol{\theta}_k$ having an associated mark $\eta_k$, with the marks being constrained to sum to unity; see Figure 1.3 for the point process representation of various mixtures of three univariate normal distributions. The point process representation is a graphical summary of the components of the mixture distributions and turns out to be rather useful when dealing with the issue of identification in Section 1.3.

### 1.2.4 Moments of a Finite Mixture Distribution

A nice feature of mixture distributions is that moments are quite easily available. To determine the expectation $\mathrm{E}(H(\mathbf{Y})|\boldsymbol{\vartheta})$ of a function $H(\mathbf{Y})$ of $\mathbf{Y}$ with respect to the mixture density (1.6), consider first the expectation $\mathrm{E}(H(\mathbf{Y})|\boldsymbol{\theta}_k)$ of $H(\mathbf{Y})$ with respect to the component density $p(\mathbf{y}|\boldsymbol{\theta}_k)$:

$$\mathrm{E}(H(\mathbf{Y})|\boldsymbol{\theta}_k) = \int_{\mathcal{Y}} H(\mathbf{y}) p(\mathbf{y}|\boldsymbol{\theta}_k)\, d\mathbf{y}.$$

If $\mathrm{E}(H(\mathbf{Y})|\boldsymbol{\theta}_k)$ exists for all $k = 1, \ldots, K$, then $\mathrm{E}(H(\mathbf{Y})|\boldsymbol{\vartheta})$ is obviously given by

$$\mathrm{E}(H(\mathbf{Y})|\boldsymbol{\vartheta}) = \sum_{k=1}^{K} \mathrm{E}(H(\mathbf{Y})|\boldsymbol{\theta}_k)\eta_k. \tag{1.19}$$

For a univariate random variable $Y$, for instance, the mean $\mu$ and the variance $\sigma^2$ of the distribution generated by $p(y|\boldsymbol{\vartheta})$ are obtained with $H(Y) = Y$ and $H(Y) = (Y - \mu)^2$, respectively:

$$\mu = \mathrm{E}(Y|\boldsymbol{\vartheta}) = \sum_{k=1}^{K} \mu_k \eta_k, \tag{1.20}$$

$$\sigma^2 = \mathrm{Var}(Y|\boldsymbol{\vartheta}) = \sum_{k=1}^{K} (\mu_k^2 + \sigma_k^2)\eta_k - \mu^2, \tag{1.21}$$

provided that the component moments $\mu_k = \mathrm{E}(Y|\boldsymbol{\theta}_k)$ and $\sigma_k^2 = \mathrm{Var}(Y|\boldsymbol{\theta}_k)$ exist. Higher-order moments around zero are easily obtained from the corresponding higher-order moments of the component densities:

$$\mathrm{E}(Y^m|\boldsymbol{\vartheta}) = \sum_{k=1}^{K} \mathrm{E}(Y^m|\boldsymbol{\theta}_k)\eta_k.$$

Higher-order moments around the mean result with $H(Y) = (Y - \mu)^m$ and the help of the binomial formula:

$$\begin{aligned} \mathrm{E}((Y-\mu)^m|\boldsymbol{\vartheta}) &= \sum_{k=1}^{K} \mathrm{E}((Y - \mu_k + \mu_k - \mu)^m|\boldsymbol{\theta}_k)\eta_k \\ &= \sum_{k=1}^{K}\sum_{n=0}^{m} \binom{m}{n} (\mu_k - \mu)^{m-n}\mathrm{E}((Y-\mu_k)^n|\boldsymbol{\theta}_k)\eta_k. \end{aligned} \tag{1.22}$$

Teicher (1960) proves that a finite mixture of nonempty, distinct normal distributions cannot be normal. Therefore finite mixtures of normal distributions are apt to capture skewness and excess kurtosis. The following higher-order moments of a mixture of normal distributions follow immediately from (1.22).

$$\mathrm{E}((Y-\mu)^3|\boldsymbol{\vartheta}) = \sum_{k=1}^{K} \eta_k \left((\mu_k-\mu)^2 + 3\sigma_k^2\right)(\mu_k - \mu),$$

$$\mathrm{E}((Y-\mu)^4|\boldsymbol{\vartheta}) = \sum_{k=1}^{K} \eta_k \left((\mu_k-\mu)^4 + 6(\mu_k-\mu)^2\sigma_k^2 + 3\sigma_k^4\right).$$

It is evident that skewness is present, iff at least two component means are different.

### 1.2.5 Statistical Modeling Based on Finite Mixture Distributions

Mixture modeling is a rapidly developing area, with the range of applications exploding. There exist various features of finite mixture distributions that render them useful in statistical modeling. First, as shown in Subsection 1.1 finite mixture distributions arise in a natural way as marginal distributions for statistical models involving discrete latent variables. On the other hand, from the data-oriented perspective, it turns out that statistical models that are based on finite mixture distributions are able to capture many specific features of real data, such as multimodality, skewness, and kurtosis.

### The Standard Finite Mixture Model and Extension Discussed in the Book

If the empirical distribution of data $\mathbf{y}_1, \dots, \mathbf{y}_N$ exhibits multimodality, skewness, or excess kurtosis, it may be assumed that they are independent realizations of a random variable $\mathbf{Y}$ from a finite mixture distribution. This is the standard finite mixture model considered in the monographs by Everitt and Hand (1981), Titterington et al. (1985), McLachlan and Basford (1988), and McLachlan and Peel (2000).

Finite mixture models provide a straightforward, but very flexible extension of classical statistical models. Although the extension appears quite simple, their estimation results in complex computational problems. The price to be paid for the high flexibility of finite mixture models is that their inference is somewhat of a challenge, as discussed in great length throughout Chapter 2 to Chapter 5.

A particularly important special case, discussed in much detail in Chapter 6, is mixtures of normal distribution. Chapter 7 demonstrates that this model is helpful for practical data analysis both of univariate and multivariate data, for instance, for unsupervised clustering and density estimation. Chapter 9 is devoted to a thorough discussion of finite mixture modeling of non-Gaussian data.

As in Subsection 1.1, any standard finite mixture model may be regarded as a hierarchical latent variable model, where the distribution of the observations $\mathbf{y} = (\mathbf{y}_1, \dots, \mathbf{y}_N)$ depends on hidden discrete indicator variables $\mathbf{S} = (S_1, \dots, S_N)$. On a first layer of the model, the joint sampling distribution of $\mathbf{y} = (\mathbf{y}_1, \dots, \mathbf{y}_N)$ is specified conditional on the whole sequence of indicators $\mathbf{S} = (S_1, \dots, S_N)$:

$$p(\mathbf{y}|\mathbf{S}, \boldsymbol{\vartheta}) = \prod_{i=1}^{N} p(\mathbf{y}_i|S_i, \boldsymbol{\vartheta}) = \prod_{i=1}^{N} p(\mathbf{y}_i|\boldsymbol{\theta}_{S_i}). \tag{1.23}$$

A second layer of the model specifies the joint distribution $p(\mathbf{S}|\boldsymbol{\vartheta})$ of the indicators which is a discrete distribution over the lattice

$$\mathcal{S}_K = \{(S_1, \dots, S_N) : S_i \in \{1, \dots, K\}, i = 1, \dots, N\}.$$

In the standard finite mixture model it is assumed that the indicators $S_1, \dots, S_N$ are independent, and their joint distribution reads:

$$p(\mathbf{S}|\boldsymbol{\eta}) = \prod_{i=1}^{N} p(S_i|\boldsymbol{\eta}), \tag{1.24}$$

where $\Pr(S_i = k|\boldsymbol{\eta}) = \eta_k$. Because $S_i \sim \text{MulNom}(1, \boldsymbol{\eta})$, Titterington (1990) proposed the name "hidden multinomial model" for the standard finite mixture model.

One modification of $p(\mathbf{S}|\boldsymbol{\eta})$ is to assume a uniform distribution over $\mathcal{S}_K$,

$$p(\mathbf{S}|\boldsymbol{\eta}) = \frac{1}{K^N},$$

which is a prior often used in classification, as discussed in Chapter 7. This model is equivalent to a hidden multinomial model where the weights in the underlying mixture distribution are equal to $\eta_k = 1/K$.

This hierarchical formulation provides the basis for the formulation of more complex finite mixture models. These extensions concern both the joint distribution $p(\mathbf{S}|\boldsymbol{\vartheta})$ of the indicators as well as the joint distribution $p(\mathbf{y}|\mathbf{S}, \boldsymbol{\vartheta})$.

To include covariates in finite mixture modeling, the mean of the density $p(\mathbf{y}_i|\boldsymbol{\theta}_k)$, which is equal to $\mu_k$ for a standard finite mixture model, may be replaced by $\mathbf{x}_i\boldsymbol{\beta}_k$, where $\mathbf{x}_i$ is a row vector of exogenous regressors, and $\boldsymbol{\beta}_k$ is a vector of unknown parameters. This leads to the switching regression model discussed in Chapter 8.

To deal with time series data, finite mixture models are needed that are able to capture autocorrelation. One may allow for autocorrelation through appropriate specification of the distribution of $p(\mathbf{y}|\mathbf{S}, \boldsymbol{\vartheta})$. Additional flexibility is achieved by assuming that $S_i$ follow a Markov chain. The resulting extension is called the finite Markov mixture model and is discussed in detail in Chapter 10. Estimation of the Markov mixture model in Chapter 11 reveals many similarities with the standard finite mixture model. Chapter 12 demonstrates that this model class is rather useful in nonlinear time series analysis, because it is able to deal with many stylized features of practical time series such as spurious long-range dependence and conditional heteroscedasticity. Various monographs deal with specific aspects of time series analysis based on finite Markov mixture models, such as hidden Markov models for discrete-valued time series (MacDonald and Zucchini, 1997) and Markov switching vector autoregressive time series (Krolzig, 1997). Application of Markov switching models in bioinformatics is reviewed in Koski (2001) and applications in financial economics in Bhar and Hamori (2004).

The monograph on state space models with regime switching by Kim and Nelson (1999) shows that state space models may be successfully combined with Markov switching models in analyzing time series in finance and economics; actually this model is much older and goes back to engineering applications in the 1970s. Such switching state space models are the last extension of the standard finite mixture model that is discussed in this book in Chapter 13.

### Extensions Not Discussed in the Book

There many extensions that would deserve investigation, but are beyond the scope of this book. These concern in particular more flexible approaches for modeling the distribution $p(\mathbf{S})$ of the indicators $S = (S_1, \ldots, S_N)$, especially finite mixture modeling of spatial data, where the domain of $\mathbf{S}$ is a two-dimensional lattice. If a Markov random field model (Besag, 1974) is assumed as a generating mechanism for $\mathbf{S}$, the hidden Markov random field model results. This model is popular in statistical image analysis, where the elements of $\mathbf{S}$ are colors or levels of gray, and $\mathbf{y}_i$ is an observed, noise-corrupted image; see the monograph of Li and Gray (2000) for a review. We refer to Qian and Titterington (1991) and Rydén and Titterington (1998) for a Bayesian analysis of these models, and to Green and Richardson (2002) for an interesting application in disease mapping.

A further extension is nonparametric mixture modeling, where it is first assumed that $\mathbf{y}_i$ is drawn from a parametric distribution family $\mathcal{T}(\boldsymbol{\theta}_i^s)$ with density $p(\mathbf{y}_i|\boldsymbol{\theta}_i^s)$, indexed by a random parameter $\boldsymbol{\theta}_i^s \in \Theta$ drawn from a distribution $W(\boldsymbol{\theta})$. Marginally, the distribution of $\mathbf{y}_i$ is a mixture distribution over densities $p(\mathbf{y}_i|\boldsymbol{\theta})$:

$$p(\mathbf{y}_i) = \int_{\Theta} p(\mathbf{y}_i|\boldsymbol{\theta})W(d\boldsymbol{\theta}).$$

If $W(\boldsymbol{\theta})$ were a continuous distribution, perhaps depending on an unknown hyperparameter, the standard random-effects model results, which does not allow for any grouping of the observations. Grouping of the data is achieved by introducing discreteness into the distribution $W(\boldsymbol{\theta})$.

The nonparametric mixture approach which is outlined in the monograph by Böhning (2000) assumes that $W(\boldsymbol{\theta})$ is a discrete distribution over $K \leq N$ atoms $\{\boldsymbol{\theta}_1, \dots, \boldsymbol{\theta}_K\}$, with the position of the atoms, their probability mass and their number being unknown. In the nonparametric ML approach these quantities are estimated from the data.

In the Bayesian nonparametric mixture approach $W(\boldsymbol{\theta})$ is modeled as a random distribution being drawn from a Dirichlet process $\mathcal{DP}(\alpha, W_0)$; see Ferguson (1973) for the definition and properties of this process. This Dirichlet process prior gives positive probability to ties, and as noted by Green and Richardson (2001), specifies implicitly a joint distribution $p(\mathbf{S}|\alpha, W_0)$ on the set of allocations $\mathbf{S} = (S_1, \dots, S_N)$. Green and Richardson (2001) relate this approach to the more explicit hidden multinomial model, where $S_i \sim \text{MulNom}(1, \boldsymbol{\eta})$, and the following priors are used in a Bayesian analysis, $\boldsymbol{\eta}$ follows a Dirichlet distribution, $\boldsymbol{\eta} \sim \mathcal{D}(e_0, \dots, e_0)$, and $\boldsymbol{\theta}_k \sim W_0$. They show convergence of the multinomial approach to the Dirichlet process prior approach, if with increasing $K$ the parameter $e_0$ of the Dirichlet prior on the weight distribution $\boldsymbol{\eta}$ goes to 0 in such a way that $Ke_0$ goes to $\alpha$.

## 1.3 Identifiability of a Finite Mixture Distribution

Statistical models based on finite mixture distributions have mathematical features that make them an interesting objective for a rigorous mathematical treatment as exemplified in Lindsay (1995). The present monograph deals mostly with the practical aspects of finite mixture models, in particular with describing the features they imply on the data and how they are estimated. Nevertheless we have to touch upon the more formal issue of identifiability of a mixture distribution, which is essential for parameter estimation, whenever a mixture distribution appears as part of a statistical model.

A parametric family of distributions, indexed by a parameter $\boldsymbol{\vartheta} \in \Theta$, which is defined over a sample space $\mathcal{Y}$, is said to be identifiable if any two parameters $\boldsymbol{\vartheta}$ and $\boldsymbol{\vartheta}^\star$ in $\Theta$ define the same probability law on $\mathcal{Y}$, iff $\boldsymbol{\vartheta}$ and $\boldsymbol{\vartheta}^\star$ are identical (see, e.g., Rothenberg, 1971). In terms of the corresponding probability densities $p(\mathbf{y}|\boldsymbol{\vartheta})$ and $p(\mathbf{y}|\boldsymbol{\vartheta}^\star)$ this means that if the densities are identical for almost every $\mathbf{y} \in \mathcal{Y}$, then the parameters $\boldsymbol{\vartheta}$ and $\boldsymbol{\vartheta}^\star$ need to be identical:

$$p(\mathbf{y}|\boldsymbol{\vartheta}) = p(\mathbf{y}|\boldsymbol{\vartheta}^\star) \quad \text{for almost all } \mathbf{y} \in \mathcal{Y} \rightarrow \boldsymbol{\vartheta} = \boldsymbol{\vartheta}^\star. \tag{1.25}$$

If for any two parameters $\boldsymbol{\vartheta}$ and $\boldsymbol{\vartheta}^\star$ in $\Theta$, which are distinct, the densities $p(\mathbf{y}|\boldsymbol{\vartheta})$ and $p(\mathbf{y}|\boldsymbol{\vartheta}^\star)$ are identical for almost every $\mathbf{y} \in \mathcal{Y}$, then this family of distributions is not identifiable. Any subset $\mathcal{U}(\boldsymbol{\vartheta})$ of $\Theta$, defined as

$$\mathcal{U}(\boldsymbol{\vartheta}) = \{\boldsymbol{\vartheta}^\star \in \Theta : p(\mathbf{y}|\boldsymbol{\vartheta}^\star) = p(\mathbf{y}|\boldsymbol{\vartheta}), \text{for almost every } \mathbf{y} \in \mathcal{Y}\},$$

which contains more than one point in $\Theta$ is called a nonidentifiability set.

A single normal distribution, indexed by $\boldsymbol{\vartheta} = (\mu, \sigma^2)$, for instance, is clearly identifiable, whereas for mixtures of probability distributions, the issue of identifiability is considerably more involved. Identifiability problems for finite mixture distributions are studied in Teicher (1963), Yakowitz and Spragins (1968), Chandra (1977), Redner and Walker (1984), and Crawford (1994).

In general, for a finite mixture distribution one has to distinguish among three types of nonidentifiability. Nonidentifiability due to invariance to relabeling the components of the mixture distribution and nonidentifiability due to potential overfitting, discussed in Subsection 1.3.1 and Subsection 1.3.2, respectively, may be ruled out through formal identifiability constraints, as explained in Subsection 1.3.3. The last type of nonidentifiability is a generic property of a certain class of mixture distributions and is investigated in Subsection 1.3.4.

### 1.3.1 Nonidentifiability Due to Invariance to Relabeling the Components

First of all, nonidentifiability of a finite mixture distribution is caused by the invariance of a mixture distribution to relabeling the components, as first noted by Redner and Walker (1984).

Consider a mixture of two normal distributions, $p(y|\boldsymbol{\vartheta}) = \eta_1 f_N(y; \mu_1, \sigma_1^2) + \eta_2 f_N(y; \mu_2, \sigma_2^2)$, where $\boldsymbol{\theta}_k = (\mu_k, \sigma_k^2) \in \Theta = \Re \times \Re^+$ and $\boldsymbol{\vartheta} = (\boldsymbol{\theta}_1, \boldsymbol{\theta}_2, \eta_1, \eta_2) \in \Theta_2 = \Theta^2 \times \mathcal{E}_2$. Take an arbitrary parameter $\boldsymbol{\vartheta} \in \Theta_2$ with $\boldsymbol{\theta}_1 \neq \boldsymbol{\theta}_2$, and define the parameter $\boldsymbol{\vartheta}^\star = (\boldsymbol{\theta}_2, \boldsymbol{\theta}_1, \eta_2, \eta_1)$, which is obtained by interchanging the order of the components. Then the distribution induced by $\boldsymbol{\vartheta}$ and $\boldsymbol{\vartheta}^\star$ is the same although evidently the two parameters are distinct:

$$\begin{aligned} p(y|\boldsymbol{\vartheta}^\star) &= \eta_2 f_N(y;\mu_2,\sigma_2^2) + \eta_1 f_N(y;\mu_1,\sigma_1^2) = \\ &\eta_1 f_N(y;\mu_1,\sigma_1^2) + \eta_2 f_N(y;\mu_2,\sigma_2^2) = p(y|\boldsymbol{\vartheta}). \end{aligned} \tag{1.26}$$

Because of this invariance, a mixture of two normal distributions is not identifiable in the strict sense defined above.

For the general finite mixture distribution with $K$ components defined in (1.6), there exist $s = 1, \dots, K!$ equivalent ways of arranging the components. Each of them may be described by a permutation $\rho_s : \{1, \dots, K\} \rightarrow \{1, \dots, K\}$, where the value $\rho_s(k)$ is assigned to each value $k \in \{1, \dots, K\}$. Let $\boldsymbol{\vartheta} = (\boldsymbol{\theta}_1, \dots, \boldsymbol{\theta}_K, \eta_1, \dots, \eta_K)$ be an arbitrary point in the parameter space $\Theta_K = \Theta^K \times \mathcal{E}_K$, and define the following subset $\mathcal{U}^P(\boldsymbol{\vartheta}) \subset \Theta_K$.

$$\begin{aligned} \mathcal{U}^P(\boldsymbol{\vartheta}) = \bigcup_{s=1}^{K!} \{ &\boldsymbol{\vartheta}^\star \in \Theta_K : \\ &\boldsymbol{\vartheta}^\star = (\boldsymbol{\theta}_{\rho_s(1)}, \dots, \boldsymbol{\theta}_{\rho_s(K)}, \eta_{\rho_s(1)}, \dots, \eta_{\rho_s(K)})\}. \end{aligned} \tag{1.27}$$

Any point $\boldsymbol{\vartheta}^\star \in \mathcal{U}^P(\boldsymbol{\vartheta})$ generates the same mixture distribution as $\boldsymbol{\vartheta}$, which is easily seen by rearranging the components of the mixture density (1.6) according to the permutation $\rho_s$ used in the definition of $\boldsymbol{\vartheta}^\star$:

$$\begin{aligned} p(\mathbf{y}|\boldsymbol{\vartheta}) &= \eta_1 p(\mathbf{y}|\boldsymbol{\theta}_1) + \cdots + \eta_K p(\mathbf{y}|\boldsymbol{\theta}_K) \\ &= \eta_{\rho_s(1)} p(\mathbf{y}|\boldsymbol{\theta}_{\rho_s(1)}) + \cdots + \eta_{\rho_s(K)} p(\mathbf{y}|\boldsymbol{\theta}_{\rho_s(K)}) = p(\mathbf{y}|\boldsymbol{\vartheta}^\star). \end{aligned}$$

The set $\mathcal{U}^P(\boldsymbol{\vartheta})$ contains $K!$ distinct parameters $\boldsymbol{\vartheta}^\star$, iff all $K$ component parameters $\boldsymbol{\theta}_1, \dots, \boldsymbol{\theta}_K$ are distinct points in $\Theta$. The set contains only $K!/L!$ distinct parameters $\boldsymbol{\vartheta}^\star$, if $L$ among the $K$ component parameters $\boldsymbol{\theta}_1, \dots, \boldsymbol{\theta}_K$ of $\boldsymbol{\vartheta}$ are identical. Thus for each $\boldsymbol{\vartheta} \in \Theta_K$, where at least two component parameters $\boldsymbol{\theta}_k$ and $\boldsymbol{\theta}_l$ differ in at least one element, the set $\mathcal{U}^P(\boldsymbol{\vartheta})$ is a nonidentifiability set in $\Theta_K$.

In a certain sense, this is not a severe identifiability problem, as all parameters $\boldsymbol{\vartheta}^\star \in \mathcal{U}^P(\boldsymbol{\vartheta})$ are related to each other, and in fact only differ in the way the components are arranged. Nevertheless this identifiability plays a certain role in later chapters that deal with parameter estimation.

In the point process representation of a finite mixture distribution, considered earlier in Subsection 1.2.3, we find that all parameters $\boldsymbol{\vartheta}^\star \in \mathcal{U}^P(\boldsymbol{\vartheta})$ are matched into the same points. Thus the point process representation of a finite mixture distribution provides a way of looking at the component parameters of a finite mixture distribution, which is insensitive to the precise labeling of the components.

### 1.3.2 Nonidentifiability Due to Potential Overfitting

A further identifiability problem, noted by Crawford (1994), is nonidentifiability due to potential overfitting. Consider a finite mixture distribution with $K$ components, defined as in (1.6), where $\boldsymbol{\vartheta} = (\boldsymbol{\theta}_1, \ldots, \boldsymbol{\theta}_K, \eta_1, \ldots, \eta_K) \in \Theta_K = \Theta^K \times \mathcal{E}_K$. Next consider a finite mixture distribution from the same parametric family, however, with $K-1$ rather than $K$ components. Crawford (1994) showed that any mixture with $K-1$ components defines a nonidentifiability subset in the larger parameter space $\Theta_K$, which corresponds to mixtures with $K$ components, where either one component is empty or two components are equal.

Consider, for instance, a mixture of two normal distributions with arbitrary parameter $\boldsymbol{\vartheta}_2 = (\mu_1, \sigma_1^2, \mu_2, \sigma_2^2, \eta_1, \eta_2) \in \Theta_2$. Any mixture of two normal distributions may be written as a mixture of three normal distributions by adding a third component with weight $\eta_3 = 0$:

$$\begin{aligned} p(y|\boldsymbol{\vartheta}) &= \eta_1 f_N(y; \mu_1, \sigma_1^2) + \eta_2 f_N(y; \mu_2, \sigma_2^2) \\ &= \eta_1 f_N(y; \mu_1, \sigma_1^2) + \eta_2 f_N(y; \mu_2, \sigma_2^2) + 0 \times f_N(y; \mu_3, \sigma_3^2). \end{aligned}$$

In the parameter space $\Theta_3$, the parameter $\boldsymbol{\vartheta} = (\mu_1, \sigma_1^2, \mu_2, \sigma_2^2, \mu_3, \sigma_3^2, \eta_1, \eta_2, 0)$ corresponding to this mixture lies in a nonidentifiability set, as the density $p(y|\boldsymbol{\vartheta})$ is the same for arbitrary values $\mu_3$ and $\sigma_3^2$. The same nonidentifiability set results if a mixture of three normal distributions is generated by splitting one component of a mixture of two normal distributions:

$$\begin{aligned} p(y|\boldsymbol{\vartheta}) &= \eta_1 f_N(y; \mu_1, \sigma_1^2) + \eta_2 f_N(y; \mu_2, \sigma_2^2) \\ &= \eta_1 f_N(y; \mu_1, \sigma_1^2) + (\eta_2 - \eta_3) f_N(y; \mu_2, \sigma_2^2) + \eta_3 f_N(y; \mu_2, \sigma_2^2). \end{aligned}$$

Again in $\Theta_3$, the parameter $\boldsymbol{\vartheta} = (\mu_1, \sigma_1^2, \mu_2, \sigma_2^2, \mu_2, \sigma_2^2, \eta_1, \eta_2 - \eta_3, \eta_3)$ lies in a nonidentifiability set, as the density $p(y|\boldsymbol{\vartheta})$ is the same for arbitrary values of $\eta_3$ with $0 \leq \eta_3 \leq \eta_2$. Furthermore this is the same nonidentifiability set as above.

In general, let a mixture with $K-1$ components be generated by an arbitrary parameter $\boldsymbol{\vartheta}_{K-1} \in \Theta_{K-1} = \Theta^{K-1} \times \mathcal{E}_{K-1}$, given by

$$\boldsymbol{\vartheta}_{K-1} = (\boldsymbol{\theta}_1^{K-1}, \ldots, \boldsymbol{\theta}_{K-1}^{K-1}, \eta_1^{K-1}, \ldots, \eta_{K-1}^{K-1}).$$

Define the following subset $\mathcal{U}^Z(\boldsymbol{\vartheta}_{K-1})$ of $\Theta_K = \Theta^K \times \mathcal{E}_K$:

$$\begin{aligned} \mathcal{U}^Z(\boldsymbol{\vartheta}_{K-1}) = \bigcup_{k=1}^{K} \bigcup_{s=1}^{(K-1)!} & \{\boldsymbol{\vartheta} \in \Theta_K : \eta_k = 0, \boldsymbol{\theta}_k \in \Theta, \\ & (\boldsymbol{\theta}_1, \ldots, \boldsymbol{\theta}_{k-1}, \boldsymbol{\theta}_{k+1}, \ldots, \boldsymbol{\theta}_K) = (\boldsymbol{\theta}_{\rho_s(1)}^{K-1}, \ldots, \boldsymbol{\theta}_{\rho_s(K-1)}^{K-1}), \\ & (\eta_1, \ldots, \eta_{k-1}, \eta_{k+1}, \ldots, \eta_K) = (\eta_{\rho_s(1)}^{K-1}, \ldots, \eta_{\rho_s(K-1)}^{K-1})\Big\}, \end{aligned} \tag{1.28}$$

which contains mixtures with $K$ components, where one component is empty, whereas the remaining $K-1$ components are defined by $\boldsymbol{\vartheta}_{K-1}$. The set

$\mathcal{U}^Z(\boldsymbol{\vartheta}_{K-1})$ is a nonidentifiability set in the parameter space $\Theta_K$ because it contains infinitely many parameters $\boldsymbol{\vartheta}$, each generating the same mixture distribution:

$$p(\mathbf{y}|\boldsymbol{\vartheta}) = \sum_{j=1,j\neq k}^{K} \eta_j p(\mathbf{y}|\boldsymbol{\theta}_j) = \sum_{j=1}^{K-1} \eta_j^{K-1} p(\mathbf{y}|\boldsymbol{\theta}_j^{K-1}). \tag{1.29}$$

The set $\mathcal{U}^Z(\boldsymbol{\vartheta}_{K-1})$ is part of an even larger nonidentifiability set. Define the following subset $\mathcal{U}^E(\boldsymbol{\vartheta}_{K-1})$ of $\Theta_K$.

$$\begin{aligned}\mathcal{U}^E(\boldsymbol{\vartheta}_{K-1}) = \bigcup_{k=1}^{K-1} \bigcup_{s=1}^{(K-1)!} \Big\{ & \boldsymbol{\vartheta} \in \Theta_K : \boldsymbol{\theta}_K = \boldsymbol{\theta}_{\rho_s(k)}^{K-1}, \eta_k + \eta_K = \eta_{\rho_s(k)}^{K-1}, \\ (\boldsymbol{\theta}_1, \ldots, \boldsymbol{\theta}_{K-1}) &= (\boldsymbol{\theta}_{\rho_s(1)}^{K-1}, \ldots, \boldsymbol{\theta}_{\rho_s(K-1)}^{K-1}), (\eta_1, \ldots, \eta_{k-1}, \eta_{k+1}, \ldots, \eta_{K-1}) \\ &= (\eta_{\rho_s(1)}^{K-1}, \ldots, \eta_{\rho_s(k-1)}^{K-1}, \eta_{\rho_s(k+1)}^{K-1}, \ldots, \eta_{\rho_s(K-1)}^{K-1}) \Big\},\end{aligned} \tag{1.30}$$

which contains mixtures with $K$ components, where two component densities are equal and are obtained from the mixture defined by $\boldsymbol{\vartheta}_{K-1}$ by splitting one component into two identical ones. Again, $\mathcal{U}^E(\boldsymbol{\vartheta}_{K-1})$ is a nonidentifiability set in $\Theta_K$, which contains infinitely many parameters $\boldsymbol{\vartheta}$, inducing the following mixture distribution,

$$p(\mathbf{y}|\boldsymbol{\vartheta}) = \sum_{j=1,j\neq k}^{K-1} \eta_j p(\mathbf{y}|\boldsymbol{\theta}_j) + (\eta_k + \eta_K) p(\mathbf{y}|\boldsymbol{\theta}_k) = \sum_{j=1}^{K-1} \eta_j^{K-1} p(\mathbf{y}|\boldsymbol{\theta}_j^{K-1}).$$

As this mixture distribution is the same as for all parameters in $\mathcal{U}^Z(\boldsymbol{\vartheta}_{K-1})$, the union $\mathcal{U}^Z(\boldsymbol{\vartheta}_{K-1}) \cup \mathcal{U}^E(\boldsymbol{\vartheta}_{K-1})$ is an even larger nonidentifiability subset of $\Theta_K$. Note that all parameters in subset $\mathcal{U}^Z(\boldsymbol{\vartheta}_{K-1}) \cup \mathcal{U}^E(\boldsymbol{\vartheta}_{K-1})$ are related by relabeling, however, there are no longer $K!$, but only $(K-1)!$ different ways of relabeling.

The subset $\mathcal{U}^Z(\boldsymbol{\vartheta}_{K-1}) \cup \mathcal{U}^E(\boldsymbol{\vartheta}_{K-1})$ plays a prominent role in Chapter 4, dealing with mixtures with an unknown number of components, because they correspond to that part of the parameter space $\Theta_K$ which deals with the case that a mixture with $K$ components is overfitting the number of components. Therefore this kind of nonidentifiability is called nonidentifiability due to potential overfitting.

Again, it is interesting to study this kind of nonidentifiability in the point process representation of the finite mixture distribution. All parameters in $\mathcal{U}^E(\boldsymbol{\vartheta}_{K-1})$ are mapped into the $K-1$ points $\{\boldsymbol{\theta}_k^{K-1}, k = 1, \ldots, K-1\}$. For all parameters in $\mathcal{U}^Z(\boldsymbol{\vartheta}_{K-1})$, the component with $\eta_k = 0$ disappears by definition, whereas the remaining components are mapped again into the $K-1$ points $\{\boldsymbol{\theta}_k^{K-1}, k = 1, \ldots, K-1\}$. Thus all points in $\mathcal{U}^Z(\boldsymbol{\vartheta}_{K-1}) \cup \mathcal{U}^E(\boldsymbol{\vartheta}_{K-1})$ have the same point process representation as the underlying $(K-1)$-component finite mixture distribution.

Provided that $K > 2$, nonidentifiability due to potential overfitting may become even more complicated. For each $L = 2, \dots, K-1$, we may consider a finite mixture distribution with $K - L$ components, defined by an arbitrary parameter $\boldsymbol{\vartheta}_{K-L} = (\boldsymbol{\theta}_1^{K-L}, \dots, \boldsymbol{\theta}_{K-L}^{K-L}, \eta_1^{K-L}, \dots, \eta_{K-L}^{K-L}) \in \Theta_{K-L}$. For each $L = 2, \dots, K-1$, the parameter $\boldsymbol{\vartheta}_{K-L} \in \Theta_{K-L}$ defines a nonidentifiability subset in $\Theta_K$, which corresponds to mixtures of $K$ components where either $L$ components are empty or $L+1$ component densities are equal. The corresponding nonidentifiability sets are obtained by modifying the definitions in (1.28) and (1.30), that were given for $L = 1$, in an obvious way.

### 1.3.3 Formal Identifiability Constraints

Identifiability may be achieved in a formal manner by imposing constraints on the parameter space in such a way that no different parameters generate the same distribution and condition (1.25) is fulfilled. Loosely speaking, for finite mixture distributions formal identifiability is achieved by constraining the parameter space $\Theta_K$ is such a way that the density (1.6) is assumed to be a mixture of $K$ distinct, nonempty components.

First of all, a positivity constraint on the weights avoids nonidentifiability due to empty components. If the parameter space $\Theta_K$ is restricted to that subset of $\Theta^K \times \mathcal{E}_K$ where the condition

$$\eta_k > 0, \qquad k = 1, \dots, K \tag{1.31}$$

is fulfilled, then the nonidentifiability set $\mathcal{U}^E(\boldsymbol{\vartheta}_{K-1})$ defined in (1.28) is empty.

Second, an inequality condition on the component parameters avoids nonidentifiability due to equal components. For mixtures with a univariate component parameter $\theta_k$ this condition evidently reads: $\theta_k \neq \theta_{k'}$, $\forall k \neq k', k, k' = 1, \dots, K$. For mixtures with a multivariate component parameter one could require that *all* elements of $\boldsymbol{\theta}_k$ differ from those of $\boldsymbol{\theta}_{k'}$:

$$\theta_{k,j} \neq \theta_{k',j}, \qquad \forall j = 1, \dots, d, \tag{1.32}$$

$\forall k, k' = 1, \dots, K, k \neq k'$. This strong constraint rules out many interesting mixtures with multivariate component parameters that might occur in practice; consider, for instance, a mixture of three normal distributions, where $\sigma_1^2 \neq \sigma_2^2$, but $\sigma_2^2 = \sigma_3^2$. For a mixture of multivariate normal distributions it is even more unrealistic to assume that *all* elements of the variance–covariance matrices $\boldsymbol{\Sigma}_1, \dots, \boldsymbol{\Sigma}_K$ are different in all components.

Finite mixtures are identifiable under a much weaker inequality constraint requiring only that any two parameters $\boldsymbol{\theta}_k$ and $\boldsymbol{\theta}'_k$ differ in *at least one* element which need not be the same for all components or, more formally, $\forall k, k' = 1, \dots, K, k \neq k'$,

$$\exists j(k,k') \in \{1, \dots, d\} : \theta_{k,j(k,k')} \neq \theta_{l,j(k,k')}. \tag{1.33}$$

In the point process representation this condition simply means that the points corresponding to the $K$ component parameters are distinct. If the parameter space $\Theta_K$ is restricted to that subset of $\Theta^K \times \mathcal{E}_K$, where condition (1.33) is fulfilled, then the nonidentifiability set $\mathcal{U}^Z(\boldsymbol{\vartheta}_{K-1})$ defined in (1.30) is empty.

Furthermore, both constraints (1.32) or (1.33) force a unique labeling. The stronger constraint (1.32) imposes a strict order constraint on any of the $d$ elements $\theta_{k,j}, j = 1, \ldots, d$:

$$\theta_{1,j} < \cdots < \theta_{K,j}. \tag{1.34}$$

Under constraint (1.34) invariance to relabeling disappears, because in the restricted parameter space the set $\mathcal{U}^P(\boldsymbol{\vartheta})$ defined in (1.27) contains just the point $\boldsymbol{\vartheta}$, and is no longer an nonidentifiability set.

Consider now mixtures with multivariate component parameters, where only the weaker constraint (1.33) holds. If the same element $\theta_{k,j}$ of $\boldsymbol{\theta}_k$ is different for all components, then this particular element could be used to force a unique labeling as in (1.34). It easy to verify in the point process representation, if such an element exists at all, by considering the projection of the points onto the various axes $\theta_j$ of $\boldsymbol{\theta}$.

The situation is more complicated if such an element does not exist. Consider, for instance, a mixture of three normal distributions, where $\mu_1 = \mu_2$, $\sigma_1^2 \neq \sigma_2^2$, $\mu_2 \neq \mu_3$, $\sigma_2^2 = \sigma_3^2$, where neither $\mu_k$ nor $\sigma_k^2$ fulfills constraint (1.34). Note that a strict order constraint on a single element $\theta_{k,j}$ of $\boldsymbol{\theta}_k$ corresponds to $K - 1$ strict inequalities which have to be fulfilled by the pairs $\theta_{k,j}$ and $\theta_{k+1,j}$ for all $k = 1, \ldots, K-1$. It is possible to substitute some of these inequalities by a constraint on a different element of $\boldsymbol{\theta}$. For the mixture of three normal distributions mentioned above, one could use one inequality involving $(\mu_2, \mu_3)$ and another involving $(\sigma_1^2, \sigma_2^2)$ to describe the differences between the component parameters.

Assume that such a constraint has been formulated, defining a certain subset $\mathcal{R}$ of the unconstrained space $\Theta^K \times \mathcal{E}_K$. This constraint forces a unique labeling, iff the set $\mathcal{U}^P(\boldsymbol{\vartheta})$ defined in (1.27) contains just the single point $\boldsymbol{\vartheta}$ for all $\boldsymbol{\vartheta} \in \mathcal{R}$. Naturally, the identification of a valid constraint in higher dimensions may be somewhat of a challenge.

### Difficulties with Commonly Used Constraints

Various formal identifiability constraints used in the literature actually fail to achieve identifiability, an example being the following constraint on the weights,

$$0 < \eta_1 < \cdots < \eta_K, \tag{1.35}$$

which is applied, for example, in Aitkin and Rubin (1985) and Lenk and DeSarbo (2000). This constraint rules out empty components and induces a unique labeling, if the weights are actually different, but does not rule out nonidentifiability due to potentially equal parameters. Assume the parameter space $\Theta_K$ is restricted to that subset of $\Theta^K \times \mathcal{E}_K$, where condition (1.35) is fulfilled. Then the nonidentifiability set $\mathcal{U}^E(\boldsymbol{\vartheta}_{K-1})$ defined in (1.30) still contains infinitely many parameters.

For finite mixtures with multivariate parameters, it is common practice to put an order constraint as in (1.34) on an arbitrary element $\theta_{k,j}$ of $\boldsymbol{\theta}_k$ without checking if this constraint actually holds. However, invariance to relabeling is not ruled out if (1.34) is violated for any two components, say $\theta_{k,j} = \theta_{k+1,j}$ for some $k$ between 1 and $K-1$ for a certain way of arranging the components. If $\boldsymbol{\theta}_k$ is distinct from $\boldsymbol{\theta}_{k+1}$ for any other element, then the set $\mathcal{U}^P(\boldsymbol{\vartheta})$ defined in (1.27) contains the point $\boldsymbol{\vartheta} = (\boldsymbol{\theta}_1, \ldots, \boldsymbol{\theta}_k, \boldsymbol{\theta}_{k+1}, \ldots, \boldsymbol{\theta}_K, \eta_1, \ldots, \eta_k, \eta_{k+1}, \ldots, \eta_K)$ as well as the point $\boldsymbol{\vartheta}^\star = (\boldsymbol{\theta}_1, \ldots, \boldsymbol{\theta}_{k+1}, \boldsymbol{\theta}_k, \ldots, \boldsymbol{\theta}_K, \eta_1, \ldots, \eta_{k+1}, \eta_k, \ldots, \eta_K)$, causing $\mathcal{U}^P(\boldsymbol{\vartheta})$ to be a nonidentifiability set. Therefore order constraints have to be selected carefully to rule out invariance to relabeling the components.

### 1.3.4 Generic Identifiability

Finite mixture distributions may remain unidentifiable, even if a formal identifiability constraint rules out any of the nonidentifiability problems described above. A well-known example of a nonidentifiable family, mentioned already in Teicher (1961), is finite mixtures of uniform distributions. Consider, for instance, the following mixture distributions taken from Everitt and Hand (1981).

$$\begin{aligned} Y &\sim \frac{1}{2}\mathcal{U}[-2,1] + \frac{1}{2}\mathcal{U}[-1,2], \\ Y &\sim \frac{1}{3}\mathcal{U}[-1,1] + \frac{2}{3}\mathcal{U}[-2,2], \end{aligned}$$

which both have density:

$$p(y) = \begin{cases} 1/6, & -2 \leq y < -1, \\ 1/3, & -1 \leq y \leq 1, \\ 1/6, & 1 < y < 2. \end{cases}$$

A second example is finite mixtures of binomial distributions (Teicher, 1961); consider, for instance, a mixture of two binomial distributions as in Titterington et al. (1985, p.35):

$$Y \sim \eta_1 \text{BiNom}(2, \pi_1) + (1 - \eta_1)\text{BiNom}(2, \pi_2).$$

The density of the corresponding mixture distribution is different from zero only for $y = 0, 1, 2$ and takes the values

$$\begin{aligned} \Pr(Y = 0|\boldsymbol{\vartheta}) &= \eta_1(1 - \pi_1)^2 + (1 - \eta_1)(1 - \pi_2)^2, \\ \Pr(Y = 1|\boldsymbol{\vartheta}) &= 2\eta_1\pi_1(1 - \pi_1) + 2(1 - \eta_1)\pi_2(1 - \pi_2), \end{aligned}$$

and $\Pr(Y = 2|\boldsymbol{\vartheta}) = 1 - \Pr(Y = 0|\boldsymbol{\vartheta}) - \Pr(Y = 1|\boldsymbol{\vartheta})$. Evidently the mixture distribution is the same for any parameter $\boldsymbol{\vartheta} = (\pi_1, \pi_2, \eta_1)$ fulfilling these two equations and therefore unidentifiable.

Generic identifiability of a certain family of finite mixtures of distributions $\mathcal{T}(\boldsymbol{\theta})$ with density $p(\mathbf{y}|\boldsymbol{\theta})$ indexed by a parameter $\boldsymbol{\theta} \in \Theta$ is a class property that has been defined in Yakowitz and Spragins (1968) in the following way. Consider two arbitrary members of this class,

$$Y_1 \sim \eta_1 \mathcal{T}(\boldsymbol{\theta}_1) + \cdots + \eta_K \mathcal{T}(\boldsymbol{\theta}_K), \qquad Y_2 \sim \eta_1^\star \mathcal{T}(\boldsymbol{\theta}_1^\star) + \cdots + \eta_{K^\star}^\star \mathcal{T}(\boldsymbol{\theta}_{K^\star}^\star).$$

Assume that all weights are positive and that for each mixture the component parameters are distinct in the weak sense defined in (1.33). The class of finite mixtures of distributions $\mathcal{T}(\boldsymbol{\theta})$ is said to be generically identifiable, if equality of the corresponding mixture density functions,

$$\sum_{k=1}^{K} \eta_k p(\mathbf{y}|\boldsymbol{\theta}_k) = \sum_{l=1}^{K^\star} \eta_l^\star p(\mathbf{y}|\boldsymbol{\theta}_l^\star),$$

for almost every $\mathbf{y} \in \mathcal{Y}$, implies that $K = K^\star$ and that the two mixtures are equivalent apart from arranging the components. Yakowitz and Spragins (1968, p.210) prove that a family of finite mixture distributions is identifiable, iff the members $\mathcal{T}(\boldsymbol{\theta})$ of the underlying distribution family are linearly independent over the field of real numbers.

Utilizing Titterington et al. (1985, Corollary 3.11) it is often easier to verify identifiability by showing that some transform $G(z; \boldsymbol{\theta})$ of $\mathcal{T}(\boldsymbol{\theta})$ such as the characteristic function or the moment-generating function is linearly independent. Consider, as an example, a mixture of normal distributions with homoscedastic variance:

$$Y \sim \eta_1 \mathcal{N}\left(\mu_1, \sigma^2\right) + \cdots + \eta_K \mathcal{N}\left(\mu_K, \sigma^2\right).$$

Identifiability follows from Titterington et al. (1985, Corollary 3.11) with $G(z; \boldsymbol{\theta})$ being equal to the characteristic function:

$$G(z; \boldsymbol{\theta}) = \mathrm{e}^{\mathrm{i} z \mu_k} \mathrm{e}^{-\sigma^2 z^2/2},$$

because for arbitrary $K$

$$\sum_{k=1}^{K} \eta_k G(z; \boldsymbol{\theta}_k) = \left( \sum_{k=1}^{K} \eta_k \mathrm{e}^{\mathrm{i} z \mu_k} \right) \mathrm{e}^{-\sigma^2 z^2/2} = 0,$$

is possible for all $z \in \Re$, iff $\eta_1 = \cdots = \eta_K = 0$.

Using a slightly different sufficient condition on transforms $G(z; \boldsymbol{\theta})$ of $\mathcal{T}(\boldsymbol{\theta})$, Teicher (1963) proves that many mixtures of univariate continuous densities, especially univariate mixtures of normals, mixtures of exponential and Gamma distributions are generically identifiable. These results are extended by Yakowitz and Spragins (1968) to various multivariate families such as multivariate mixtures of normals. Discrete mixtures need not be identifiable, as demonstrated for a mixture of two binomial distributions. Whereas mixtures of Poisson distributions (Feller, 1943; Teicher, 1960) as well as mixtures of negative binomial distributions (Yakowitz and Spragins, 1968) are identifiable, mixtures of BiNom $(n, \pi)$-distributions are not identifiable, if $n < 2K-1$ (Teicher, 1963).

Another useful result on the identifiability of mixtures appears in Teicher (1967) where it is shown that mixtures of $r$-fold product densities defined for $\mathbf{y} \in \mathcal{Y}^r$ as

$$p(\mathbf{y}|\boldsymbol{\theta}) = \prod_{j=1}^{r} p(\mathbf{y}_j|\boldsymbol{\theta}_j),$$

with $\boldsymbol{\theta} = (\boldsymbol{\theta}_1, \dots, \boldsymbol{\theta}_r) \in \Theta^r$ are identifiable if the chosen parametric family of densities $p(\mathbf{y}|\boldsymbol{\theta}), \boldsymbol{\theta} \in \Theta$, defined on $\mathcal{Y}$, is identifiable.

# 2

# Statistical Inference for a Finite Mixture Model with Known Number of Components

## 2.1 Introduction

Assume that $N$ observations $\mathbf{y} = (\mathbf{y}_1, \dots, \mathbf{y}_N)$, drawn randomly from a finite mixture of $\mathcal{T}(\boldsymbol{\theta})$ distributions with density $p(\mathbf{y}|\boldsymbol{\theta})$ indexed by a parameter $\boldsymbol{\theta} \in \Theta$, are available, which should be used to make inferences about the underlying mixture structure. For the resulting finite mixture model that reads

$$p(\mathbf{y}_i|\boldsymbol{\vartheta}) = \sum_{k=1}^{K} \eta_k p(\mathbf{y}_i|\boldsymbol{\theta}_k), \tag{2.1}$$

three kinds of statistical inference problem have to be considered. First, modeling of data by a finite mixture model requires some specification of $K$, the number of components. Statistical inference for finite mixtures with an unknown number of components is a very delicate issue that is postponed until Chapter 4. For the rest of this and the next chapter it is assumed that the number of components is known. Second, the component parameters $\boldsymbol{\theta}_1, \dots, \boldsymbol{\theta}_K$ and the weight distribution $\boldsymbol{\eta} = (\eta_1, \dots, \eta_K)$ may be unknown and should be estimated from the data. In what follows, we denote all distinct parameters appearing in the mixture model (2.1) by $\boldsymbol{\vartheta} = (\boldsymbol{\theta}_1, \dots, \boldsymbol{\theta}_K, \boldsymbol{\eta})$. In order to obtain a vector of distinct parameters one of the redundant weights $\eta_1, \dots, \eta_K$ has to be omitted, because each $\eta_k$ is completely determined given the remaining weights. It is, however, usually not necessary to be explicit about which parameter this should be. The final problem is allocation, by assigning each observation $\mathbf{y}_i$ to a certain component and by making inference on the hidden discrete indicators $\mathbf{S} = (S_1, \dots, S_N)$.

As an introduction to statistical inference problems for finite mixtures, we start in Section 2.2 with allocation of each observation under the assumption that the component parameters and the weight distribution are known, a problem that allows us to recall Bayes' rule. Section 2.3 deals with estimating the parameter $\boldsymbol{\vartheta}$, when the allocations are known, and provides complete-data maximum likelihood estimation as well as an introduction into complete-data

Bayesian inference. Section 2.4 deals with parameter estimation when the allocations are unknown, using methods of moments and maximum likelihood estimation. The Bayesian approach to this most interesting inference problem is briefly introduced; a full discussion is given in Chapter 3.

## 2.2 Classification for Known Component Parameters

Assume that the finite mixture distribution (2.1) is known exactly, with precise values assigned the number $K$ of components, the component parameters $\boldsymbol{\theta}_1, \ldots, \boldsymbol{\theta}_K$, and the weight distribution $\boldsymbol{\eta}$, and the only challenge is to classify a set of $N$ observations $\{\mathbf{y}_1, \ldots, \mathbf{y}_N\}$ into each component. This classification problem is a common and very old problem in statistics; for a review see Cormack (1971), McLachlan and Basford (1988), Everitt et al. (2001), and Press (2003, Chapter 10) for a discussion from a Bayesian point of view.

### 2.2.1 Bayes' Rule for Classifying a Single Observation

Exact knowledge of the component parameters leads to an inference problem, which is easily solved by Bayes' rule (Bayes, 1763). Introduce, as in Subsection 1.2.5, a discrete indicator $S_i$, taking values in $\{1, \ldots, K\}$, which associates each observation $\mathbf{y}_i$ with a certain component in (2.1). Classification of a single observation $\mathbf{y}_i$ aims at deriving the conditional probability $\Pr(S_i = k|\mathbf{y}_i, \boldsymbol{\vartheta})$ of the event $\{S_i = k\}$, having observed the event $\{\mathbf{Y} = \mathbf{y}_i\}$. Bayes' rule shows how to compute this probability for each $k = 1, \ldots, K$ for observations from a discrete mixture distribution:

$$\Pr(S_i = k|\mathbf{y}_i, \boldsymbol{\vartheta}) = \frac{\Pr(\mathbf{Y} = \mathbf{y}_i|S_i = k, \boldsymbol{\vartheta})\Pr(S_i = k|\boldsymbol{\vartheta})}{\sum_{j=1}^{K} \Pr(\mathbf{Y} = \mathbf{y}_i|S_i = j, \boldsymbol{\vartheta})\Pr(S_i = j|\boldsymbol{\vartheta})}. \tag{2.2}$$

$\Pr(S_i = k|\boldsymbol{\vartheta})$ is the prior probability that observation $\mathbf{y}_i$ falls into class $k$, which is equal to the class size: $\Pr(S_i = k|\boldsymbol{\vartheta}) = \eta_k$. For a discrete mixture, $\Pr(\mathbf{Y} = \mathbf{y}_i|S_i = k, \boldsymbol{\vartheta})$ is easily obtained from the component-specific probability density function: $p(\mathbf{y}_i|\boldsymbol{\theta}_k)$.

It is convenient to rewrite Bayes' rule (2.2) in the following way,

$$\Pr(S_i = k|\mathbf{y}_i, \boldsymbol{\vartheta}) = \frac{p(\mathbf{y}_i|\boldsymbol{\theta}_k)\eta_k}{\sum_{j=1}^{K} p(\mathbf{y}_i|\boldsymbol{\theta}_j)\eta_j}, \tag{2.3}$$

as this result also holds if we are dealing with observations from continuous rather than a discrete mixture distribution.

The denominator in (2.3) remains the same, whatever the value of $k$, and is equal to the sum of the numerators over all $k$. For this reason, Bayes' rule is usually formulated up to proportionality:

$$\Pr(S_i = k|\mathbf{y}_i, \boldsymbol{\vartheta}) \propto p(\mathbf{y}_i|\boldsymbol{\theta}_k)\eta_k. \tag{2.4}$$

The right-hand side is evaluated for each $k = 1, \dots, K$, with the resulting values being normalized, to obtain a proper posterior distribution.

**Table 2.1.** Data from a mixture of two Poisson distributions with $\eta_1 = \eta_2 = 0.5$, $\mu_1 = 1$, and two different values of $\mu_2$; probability $\Pr(S_i = 1|y_i, \eta_1, \eta_2, \mu_1, \mu_2)$ of correct classification of data from the first component

| | $y_i = 0$ | $y_i = 1$ | $y_i = 2$ | $y_i = 3$ | $y_i = 4$ |
|---|---|---|---|---|---|
| $\mu_2 = 5$ | 0.9820 | 0.9161 | 0.6859 | 0.3040 | 0.0803 |
| $\mu_2 = 25$ | 1.0000 | 1.0000 | 1.0000 | 1.0000 | 1.0000 |

A common classification rule, also called the naïve Bayes' classifier, assigns each observation to the class with the highest posterior probability (Anderson, 1984, Chapter 6), because this minimizes the expected misclassification risk, see also Subsection 7.1.7. How well this classifier works depends on the difference between the parameters in the various mixture components, as the following example exemplifies.

Consider count data, assumed to arise from a mixture of two Poisson distributions with $\mu_1, \mu_2, \eta_1$, and $\eta_2$ being known. Because Bayes' rule (2.4) yields

$$\Pr(S_i = k|y_i, \mu_1, \mu_2, \eta_1, \eta_2) \propto \eta_k \mu_k^{y_i} e^{-\mu_k},$$

for $k = 1$ and $k = 2$, $y_i$ is assigned to class 1, iff

$$y_i < \frac{\mu_2 - \mu_1 + \log \eta_1 - \log(1 - \eta_1)}{\log \mu_2 - \log \mu_1}.$$

The difference between $\mu_1$ and $\mu_2$ will strongly influence the discriminative power of this classification rule. Consider, for instance, $\eta_1 = \eta_2 = 0.5$ and $\mu_1 = 1$. Observations from the first component take, with probability 0.9963, values between 0 and 4. As demonstrated in Table 2.1, the misclassification risk is rather high for a mixture with $\mu_2 = 5$, as observations between 2 and 4 are likely to arise from both components, whereas the misclassification risk is zero for a mixture with $\mu_2 = 25$.

### 2.2.2 The Bayes' Classifier for a Whole Data Set

What difference does it make to classify all observations $\mathbf{y} = (\mathbf{y}_1, \dots, \mathbf{y}_N)$ jointly, rather than individually as in Subsection 2.2.1? Joint classification is identical to individual classification if all component parameters are known; the situation, however, is different under unknown component parameters, which is known as the clustering problem (Everitt et al., 2001) and is studied in detail in Section 7.1.

Let $\mathbf{S} = (S_1, \dots, S_N)$ be the sequence of all allocations. Joint classification aims at deriving the probability of the event $\{S_1 = k_1, \dots, S_N = k_N\}$ for all possible allocations $(k_1, \dots, k_N)$ of the $N$ observations into $K$ classes, having observed $\mathbf{y} = (\mathbf{y}_1, \dots, \mathbf{y}_N)$. The density $p(\mathbf{S}|\boldsymbol{\vartheta}, \mathbf{y})$ of this distribution is obtained from Bayes' rule as in Subsection 2.2.1,

$$p(\mathbf{S}|\boldsymbol{\vartheta}, \mathbf{y}) \propto p(\mathbf{y}|\mathbf{S}, \boldsymbol{\vartheta}) p(\mathbf{S}|\boldsymbol{\vartheta}). \tag{2.5}$$

In (2.5), $p(\mathbf{y}|\mathbf{S}, \boldsymbol{\vartheta})$ is the density of the sampling distribution of the whole sequence $(\mathbf{y}_1, \dots, \mathbf{y}_N)$, if the allocations $\mathbf{S}$ are known. Under the assumption that the data are sampled independently, this density reads:

$$p(\mathbf{y}|\mathbf{S}, \boldsymbol{\vartheta}) = p(\mathbf{y}|\mathbf{S}, \boldsymbol{\theta}_1, \dots, \boldsymbol{\theta}_K) = \prod_{i=1}^{N} p(\mathbf{y}_i|\boldsymbol{\theta}_{S_i}). \tag{2.6}$$

In (2.5), $p(\mathbf{S}|\boldsymbol{\vartheta})$ is the probability density of the joint distribution of the sequence $\mathbf{S} = (S_1, \dots, S_N)$ of all unobserved allocations, before having observed the data. To specify this prior distribution, it is common to assume that the allocations, like the data, are independent a priori:

$$p(\mathbf{S}|\boldsymbol{\vartheta}) = \prod_{i=1}^{N} p(S_i|\boldsymbol{\vartheta}).$$

It is important to note that Bayes' rule (2.5) combines the information contained in the likelihood $p(\mathbf{y}|\mathbf{S}, \boldsymbol{\vartheta})$ with the prior distribution $p(\mathbf{S}|\boldsymbol{\vartheta})$, in order to derive the posterior distribution $p(\mathbf{S}|\boldsymbol{\vartheta}, \mathbf{y})$.

For known component parameters $\boldsymbol{\vartheta}$ the joint posterior density $p(\mathbf{S}|\boldsymbol{\vartheta}, \mathbf{y})$ simplifies in the following way.

$$p(\mathbf{S}|\boldsymbol{\vartheta}, \mathbf{y}) = \prod_{i=1}^{N} p(S_i|\boldsymbol{\vartheta}, \mathbf{y}_i),$$

where $p(S_i|\boldsymbol{\vartheta}, \mathbf{y}_i)$ is the density of the individual posterior classification distribution $\Pr(S_i = k|\boldsymbol{\vartheta}, \mathbf{y}_i)$, given in (2.4). As mentioned above, joint allocation of all observations may be carried out independently for each individual observation $\mathbf{y}_i$, if the component parameters are known.

A general way of assessing the quality of the classification rule based on Bayes' rule (2.4) is to consider the entropy $\text{EN}(\boldsymbol{\vartheta}|\mathbf{y})$ of $p(\mathbf{S}|\mathbf{y}, \boldsymbol{\vartheta})$ (Celeux and Soromenho, 1996) which is defined as

$$\text{EN}(\boldsymbol{\vartheta}|\mathbf{y}) = \text{E}(-\sum_{i=1}^{N} \sum_{k=1}^{K} I_{\{S_i=k\}} \log \Pr(S_i = k|\mathbf{y}_i, \boldsymbol{\vartheta})),$$

where the expectation is with respect to the classification distribution $p(\mathbf{S}|\mathbf{y}, \boldsymbol{\vartheta})$; therefore:

$$\mathrm{EN}(\boldsymbol{\vartheta}|\mathbf{y}) = -\sum_{i=1}^{N}\sum_{k=1}^{K} \Pr(S_i = k|\mathbf{y}_i, \boldsymbol{\vartheta}) \log \Pr(S_i = k|\mathbf{y}_i, \boldsymbol{\vartheta}) \geq 0. \tag{2.7}$$

For a fixed value of $\boldsymbol{\vartheta}$, the entropy is a measure of how well the data $\mathbf{y} = (\mathbf{y}_1, \dots, \mathbf{y}_N)$ are classified given a mixture distribution defined by $\boldsymbol{\vartheta}$. The entropy is 0 for a perfect classification, where for all observations $\Pr(S_i = k_i|\mathbf{y}_i, \boldsymbol{\vartheta}) = 1$ for a certain value of $k_i$, otherwise the entropy may be considerably larger. It is equal to $N \log K$ for a uniform distribution, where $\Pr(S_i = k_i|\mathbf{y}_i, \boldsymbol{\vartheta}) = 1/K$.

In the Poisson example considered above, $\mathrm{EN}(\boldsymbol{\vartheta}|\mathbf{y})/(N \log 2)$ is, for large $N$, equal to 0.38 for $\mu_2 = 5$ and equal to 0 for $\mu_2 = 25$.

## 2.3 Parameter Estimation for Known Allocation

In this section, attention is shifted toward estimating the component parameters $\boldsymbol{\theta}_1, \dots, \boldsymbol{\theta}_K$ and the weight distribution $\boldsymbol{\eta}$ from data $\mathbf{y} = (\mathbf{y}_1, \dots, \mathbf{y}_N)$ drawn randomly from the finite mixture distribution (2.1), under the assumption that the allocations $\mathbf{S} = (S_1, \dots, S_N)$ are observed as well. There exist $K!$ different ways of connecting the data with the various components of the mixture distribution, but once the labeling scheme has been fixed, parameter estimation could be based on the complete or "fully categorized" (Titterington et al., 1985) data $(\mathbf{y}, \mathbf{S})$ using standard methods of statistical inference, such as maximum likelihood estimation or Bayesian estimation, which is the preferred approach throughout this book. Subsequently, $\boldsymbol{\vartheta} = (\boldsymbol{\theta}_1, \dots, \boldsymbol{\theta}_K, \boldsymbol{\eta})$ denotes all unknown parameters.

### 2.3.1 The Complete-Data Likelihood Function

For known allocations, both maximum likelihood as well as Bayesian estimation are based on the complete-data likelihood function which is equal to the sampling distribution $p(\mathbf{y}, \mathbf{S}|\boldsymbol{\vartheta})$ of the complete data $(\mathbf{y}, \mathbf{S})$, regarded as a function of the unknown parameter $\boldsymbol{\vartheta}$.

To specify $p(\mathbf{y}, \mathbf{S}|\boldsymbol{\vartheta})$ we exploit the hierarchical latent variable representation of a finite mixture model given in Subsection 1.2.5 in (1.23) and (1.24):

$$p(\mathbf{y}, \mathbf{S}|\boldsymbol{\vartheta}) = p(\mathbf{y}|\mathbf{S}, \boldsymbol{\vartheta})p(\mathbf{S}|\boldsymbol{\vartheta}) = \prod_{i=1}^{N} p(\mathbf{y}_i|S_i, \boldsymbol{\vartheta})p(S_i|\boldsymbol{\vartheta}).$$

Because $p(\mathbf{y}_i|S_i = k, \boldsymbol{\vartheta}) = p(\mathbf{y}_i|\boldsymbol{\theta}_k)$ and $\Pr(S_i = k|\boldsymbol{\vartheta}) = \eta_k$ the complete-data likelihood function reads:

$$p(\mathbf{y}, \mathbf{S}|\boldsymbol{\vartheta}) = \prod_{i=1}^{N} \prod_{k=1}^{K} \left( p(\mathbf{y}_i|\boldsymbol{\theta}_k)\eta_k \right)^{I_{\{S_i=k\}}} \tag{2.8}$$
$$= \prod_{k=1}^{K} \left( \prod_{i:S_i=k} p(\mathbf{y}_i|\boldsymbol{\theta}_k) \right) \left( \prod_{k=1}^{K} \eta_k^{N_k(\mathbf{S})} \right),$$

with $N_k(\mathbf{S}) = \#\{S_i = k\}$ counting the number of observations allocated to component $k$. When regarded as a function of $\boldsymbol{\vartheta} = (\boldsymbol{\theta}_1, \ldots, \boldsymbol{\theta}_K, \boldsymbol{\eta})$, the complete-data likelihood function exhibits a rather convenient structure that highly facilitates parameter estimation. It reduces to the product of $K+1$ factors, with the first $K$ factors corresponding to a certain component parameter $\boldsymbol{\theta}_k$, whereas the last factor depends only on the weight distribution $\boldsymbol{\eta}$.

### 2.3.2 Complete-Data Maximum Likelihood Estimation

In complete-data maximum likelihood (ML) estimation, the logarithm of the complete-data likelihood function, $\log p(\mathbf{y}, \mathbf{S}|\boldsymbol{\vartheta})$, is maximized with respect to $\boldsymbol{\vartheta}$. Due to the factorization discussed earlier, estimation reduces to $K+1$ independent estimation problems. Each component parameter $\boldsymbol{\theta}_k$ is estimated for all $k = 1, \ldots, K$ from the observations in group $k$ ($S_i = k$), only, whereas estimation of $\boldsymbol{\eta}$ is based on the number $N_1(\mathbf{S}), \ldots, N_K(\mathbf{S})$ of allocations to each group. Elementary analysis yields the following complete-data ML estimator for the weights for $k = 1, \ldots, K$,

$$\hat{\eta}_k = \frac{N_k(\mathbf{S})}{N} = \frac{\#\{S_i = k\}}{N}. \tag{2.9}$$

The precise estimator $\hat{\boldsymbol{\theta}}_k$ of $\boldsymbol{\theta}_k$ depends on the parametric family chosen as the component density. For univariate mixtures of normals, defined in (1.8), for instance, the complete-data ML estimators of $\mu_k$ and $\sigma_k^2$ are equal to the sample mean $\overline{y}_k(\mathbf{S})$ and the sample variance $s_{y,k}^2(\mathbf{S})$ in group $k$:

$$\hat{\mu}_k = \overline{y}_k(\mathbf{S}) = \frac{1}{N_k(\mathbf{S})} \sum_{i:S_i=k} y_i,$$
$$\hat{\sigma}_k^2 = s_{y,k}^2(\mathbf{S}) = \frac{1}{N_k(\mathbf{S})} \sum_{i:S_i=k} (y_i - \overline{y}_k(\mathbf{S}))^2. \tag{2.10}$$

It is known that the ML estimator $\hat{\boldsymbol{\theta}}_k$ is consistent and asymptotically normal, provided that certain regularity conditions hold (Lehmann, 1983; Casella and Berger, 2002):

$$\sqrt{N_k(\mathbf{S})}(\hat{\boldsymbol{\theta}}_k - \boldsymbol{\theta}_k^{\text{true}}) \to_d \mathcal{N}\left(0, \mathcal{I}(\boldsymbol{\theta}_k^{\text{true}})^{-1}\right), \tag{2.11}$$

where $\mathcal{I}(\boldsymbol{\theta}_k)$ is the expected Fisher information matrix defined as

$$\mathcal{I}(\boldsymbol{\theta}_k) = -\int_{\mathcal{Y}} \frac{\partial^2}{\partial \boldsymbol{\theta}_k^2} \log p(\mathbf{y}_i|\boldsymbol{\theta}_k) p(\mathbf{y}_i|\boldsymbol{\theta}_k) d\mathbf{y}_i, \tag{2.12}$$

with $p(\mathbf{y}_i|\boldsymbol{\theta}_k)$ being the density of component $k$. Approximate confidence intervals for the unknown parameter $\boldsymbol{\theta}_k$ are derived from (2.11), after substituting $\boldsymbol{\theta}_k^{\text{true}}$ by $\hat{\boldsymbol{\theta}}_k$. For a full discussion of ML estimation we refer to standard textbooks such as Lehmann (1983) and Casella and Berger (2002).

Although complete-data ML estimation is straightforward for most mixtures from the exponential family, it may fail if some of the group sizes $N_k(\mathbf{S})$ are too small, in which case the complete-data ML estimator may not exist, may be degenerate, or may lie on the boundary of the parameter space. If, for instance, a certain group $k$ is empty, the ML estimator $\hat{\eta}_k$ lies on the boundary of the parameter space, violating an important regularity condition for ML estimation. The ML estimator $\hat{\boldsymbol{\theta}}_k$ may lie on the boundary of the parameter space, even if the group size $N_k(\mathbf{S})$ is positive. Consider, for instance, the complete-data ML estimator $\hat{\sigma}_k^2$ for a univariate mixture of normal distributions given in (2.10) if the group $k$ contains only one or two identical observations. Finally, even if all group sizes $N_k(\mathbf{S})$ are large enough to obtain an ML estimator in the interior of the parameter space, standard errors and confidence intervals based on the asymptotic normal distribution (2.11) may be inaccurate, unless $N_k(\mathbf{S})$ is quite large.

#### Complete-Data ML Estimation for a Mixture of Poisson Distributions

For a mixture of Poisson distributions, defined earlier in (1.9), it is easy to verify that the complete-data ML estimator of $\mu_k$ is equal to the sample mean in group $k$, $\hat{\mu}_k = \overline{y}_k(\mathbf{S})$. The complete-data ML estimator lies in the interior of the parameter space, iff each group is nonempty ($N_k(\mathbf{S}) > 0$) and contains at least one nonzero observation. The expected Fisher information is equal to $\mathcal{I}(\mu_k) = 1/\mu_k$. Asymptotic 95% confidence intervals for $\mu_k$, based on (2.11), are equal to $\overline{y}_k(\mathbf{S}) \pm 1.96\sqrt{\overline{y}_k(\mathbf{S})/N_k(\mathbf{S})}$, where the expected Fisher information has been evaluated at $\mu_k = \overline{y}_k(\mathbf{S})$. The effective coverage probability may be considerably smaller, if the sample size $N$ and the true values of $\mu_k$ and $\eta_k$ are very small; see Table 2.2 below, where asymptotic 95% confidence intervals are compared with Bayesian confidence intervals.

### 2.3.3 Complete-Data Bayesian Estimation of the Component Parameters

In Bayesian estimation the complete-data likelihood $p(\mathbf{y}, \mathbf{S}|\boldsymbol{\vartheta})$, regarded as a function of $\boldsymbol{\vartheta}$ as for maximum likelihood estimation, is combined with a prior distribution $p(\boldsymbol{\vartheta})$ on the parameter $\boldsymbol{\vartheta}$ to obtain the complete-data posterior distribution $p(\boldsymbol{\vartheta}|\mathbf{y}, \mathbf{S})$ using Bayes' theorem. The elements of Bayesian inference are laid out in many excellent texts such as the classical monographs by Zellner (1971) and Box and Tiao (1973). Further introductions useful for the Bayesian inference problems considered in this book are Antelman (1997), Koop (2003), and Press (2003).

To give a very short introduction to Bayes' theorem, consider a mixture of two Poisson distributions, where $\boldsymbol{\vartheta} = (\mu_1, \mu_2, \eta_1)$ takes just the two values $\boldsymbol{\vartheta} = (1, 2, .5)$ and $\boldsymbol{\vartheta} = (1, 3, .2)$. Given the complete data $(\mathbf{S}, \mathbf{y})$, Bayes' rule, which was discussed in Subsection 2.2.1 in the context of classification, yields the following posterior probabilities for the two values of $\boldsymbol{\vartheta}$,

$$\begin{aligned}\Pr(\boldsymbol{\vartheta} = (1, 2, .5)|\mathbf{S}, \mathbf{y}) &\propto p(\mathbf{y}, \mathbf{S}|\boldsymbol{\vartheta} = (1, 2, .5))\Pr(\boldsymbol{\vartheta} = (1, 2, .5)),\\ \Pr(\boldsymbol{\vartheta} = (1, 3, .2)|\mathbf{S}, \mathbf{y}) &\propto p(\mathbf{y}, \mathbf{S}|\boldsymbol{\vartheta} = (1, 3, .2))\Pr(\boldsymbol{\vartheta} = (1, 3, .2)).\end{aligned}$$

Using probability densities rather than probabilities of events, Bayes' rule may be rewritten as

$$p(\boldsymbol{\vartheta}|\mathbf{S}, \mathbf{y}) \propto p(\mathbf{y}, \mathbf{S}|\boldsymbol{\vartheta})p(\boldsymbol{\vartheta}). \tag{2.13}$$

If $\boldsymbol{\vartheta}$ takes values in a continuous parameter space, rather than in a discrete one, formula (2.13), which is now called Bayes' theorem, still holds and yields the complete-data posterior density $p(\boldsymbol{\vartheta}|\mathbf{S}, \mathbf{y})$. Bayes' theorem (2.13) combines the information about $\boldsymbol{\vartheta}$, contained in the complete-data likelihood $p(\mathbf{y}, \mathbf{S}|\boldsymbol{\vartheta})$, with the prior information contained in the prior distribution $p(\boldsymbol{\vartheta})$.

The complete-data likelihood function factors into $K + 1$ products, and a similar structure is assumed for the prior density $p(\boldsymbol{\vartheta})$:

$$p(\boldsymbol{\vartheta}) = p(\boldsymbol{\eta}) \prod_{k=1}^{K} p(\boldsymbol{\theta}_k). \tag{2.14}$$

Under this prior, the complete-data posterior density $p(\boldsymbol{\vartheta}|\mathbf{S}, \mathbf{y})$ of a finite mixture model factors in the same convenient way:

$$p(\boldsymbol{\vartheta}|\mathbf{S}, \mathbf{y}) = \prod_{k=1}^{K} p(\boldsymbol{\theta}_k|\mathbf{y}, \mathbf{S})p(\boldsymbol{\eta}|\mathbf{S}), \tag{2.15}$$

where

$$p(\boldsymbol{\theta}_k|\mathbf{y}, \mathbf{S}) \propto \prod_{i:S_i=k} p(\mathbf{y}_i|\boldsymbol{\theta}_k)p(\boldsymbol{\theta}_k) \tag{2.16}$$

$$p(\boldsymbol{\eta}|\mathbf{S}) \propto \prod_{k=1}^{K} \eta_k^{N_k(\mathbf{S})} p(\boldsymbol{\eta}). \tag{2.17}$$

Hence, complete-data Bayesian estimation may be carried independently for each component parameter $\boldsymbol{\theta}_k$ and for the weight distribution $\boldsymbol{\eta}$. The structure of the complete-data posterior $p(\boldsymbol{\eta}|\mathbf{S})$ is discussed in much detail in Subsection 2.3.4; we focus for the rest of this subsection on the component parameter $\boldsymbol{\theta}_k$.

**Table 2.2.** Complete-data estimation of $\mu_k$ for a mixture of Poisson distributions comparing confidence intervals for $\mu_k$ obtained from the Bayesian and the ML approach

| | 95% Confidence Interval (data set of size $N = 100$) | | |
|---|---|---|---|
| | $\eta_k^{\text{true}} = 0.05, \mu_k^{\text{true}} = 0.5$ | $\eta_k^{\text{true}} = 0.1, \mu_k^{\text{true}} = 1$ | $\eta_k^{\text{true}} = 0.5, \mu_k^{\text{true}} = 1$ |
| ML | [−0.110, 0.682] | [0.444, 1.422] | [0.826, 1.419] |
| Bayes | [0.059, 0.917] | [0.535, 1.524] | [0.854, 1.450] |
| | Coverage rate for replications over 1000 data sets | | |
| | $\eta_k^{\text{true}} = 0.05, \mu_k^{\text{true}} = 0.5$ | $\eta_k^{\text{true}} = 0.1, \mu_k^{\text{true}} = 1$ | $\eta_k^{\text{true}} = 0.5, \mu_k^{\text{true}} = 1$ |
| ML | 0.813 | 0.925 | 0.949 |
| Bayes | 0.959 | 0.949 | 0.952 |

## Complete-Data Estimation of the Component Parameters

A closed solution for the complete-data posterior distribution $p(\boldsymbol{\theta}_k|\mathbf{y}, \mathbf{S})$ of the component parameters $\boldsymbol{\theta}_k$ results for many finite mixture distributions, such as univariate and multivariate mixtures of normals, mixtures of exponentials and mixtures of binomial distributions, if the prior is chosen appropriately.

For a mixture of Poisson distributions, for instance, the complete-data likelihood $p(\mathbf{y}, \mathbf{S}|\boldsymbol{\vartheta})$, after dropping factors independent of $\boldsymbol{\vartheta} = (\mu_1, \dots, \mu_K, \boldsymbol{\eta})$, reads:

$$p(\mathbf{y}, \mathbf{S}|\mu_1, \dots, \mu_K, \boldsymbol{\eta}) \propto \prod_{k=1}^{K} \mu_k^{N_k(\mathbf{S})\overline{y}_k(\mathbf{S})} \mathrm{e}^{-N_k(\mathbf{S})\mu_k} \left( \prod_{k=1}^{K} \eta_k^{N_k(\mathbf{S})} \right),$$

where $N_k(\mathbf{S}) = \#\{S_i = k\}$. Regarded as a function of $\mu_k$, the $k$th factor is the density of a Gamma distribution; see Subsection A.1.6. Under the flat prior $p(\boldsymbol{\vartheta}) \propto$ constant, the complete-data posterior density takes the form:

$$p(\mu_1, \dots, \mu_K, \boldsymbol{\eta}|\mathbf{S}, \mathbf{y}) = \prod_{k=1}^{K} p(\mu_k|\mathbf{S}, \mathbf{y}) p(\boldsymbol{\eta}|\mathbf{S}),$$

where $p(\mu_k|\mathbf{S}, \mathbf{y})$ is the density of the $\mathcal{G}\left(N_k(\mathbf{S})\overline{y}_k(\mathbf{S}) + 1, N_k(\mathbf{S})\right)$-distribution. The inclusion of an arbitrary prior $p(\boldsymbol{\vartheta})$ would destroy this closed solution with the exceptions of a specific family of prior distributions, called natural conjugate priors (Press, 2003, Chapter 5) which exist for many problems in Bayesian statistics involving exponential families (Diaconis and Ylvisaker, 1979).

For complete data from a mixture of Poisson distributions, for instance, the natural conjugate prior distribution for $\mu_k$ is a $\mathcal{G}\left(a_{0,k}, b_{0,k}\right)$-distribution. It is easy to verify that the complete-data posterior $p(\mu_k|\mathbf{S}, \mathbf{y})$ is equal to the $\mathcal{G}\left(a_k(\mathbf{S}), b_k(\mathbf{S})\right)$-distribution, where

$$a_k(\mathbf{S}) = a_{0,k} + N_k(\mathbf{S})\overline{y}_k(\mathbf{S}), \qquad b_k(\mathbf{S}) = b_{0,k} + N_k(\mathbf{S}). \tag{2.18}$$

### The Normalizing Constant

In (2.16), the right-hand side is equal to the nonnormalized complete-data posterior, from which the complete-data posterior density $p(\boldsymbol{\theta}_k|\mathbf{y}, \mathbf{S})$ is obtained by dividing by the normalizing constant:

$$p(\boldsymbol{\theta}_k|\mathbf{y}, \mathbf{S}) = \frac{\prod_{i:S_i=k} p(\mathbf{y}_i|\boldsymbol{\theta}_k)p(\boldsymbol{\theta}_k)}{\int \prod_{i:S_i=k} p(\mathbf{y}_i|\boldsymbol{\theta}_k)p(\boldsymbol{\theta}_k)d\boldsymbol{\theta}_k}.$$

For conjugate problems, the normalizing constant is not really needed for Bayesian estimation, but it sometimes appears in a different context, in which case it is easily evaluated by dividing the nonnormalized by the normalized posterior for arbitrary $\boldsymbol{\theta}_k$:

$$\int \prod_{i:S_i=k} p(\mathbf{y}_i|\boldsymbol{\theta}_k)p(\boldsymbol{\theta}_k)d\boldsymbol{\theta}_k = \frac{\prod_{i:S_i=k} p(\mathbf{y}_i|\boldsymbol{\theta}_k)p(\boldsymbol{\theta}_k)}{p(\boldsymbol{\theta}_k|\mathbf{y}, \mathbf{S})}. \tag{2.19}$$

### Comparing the Bayesian and the ML Approach

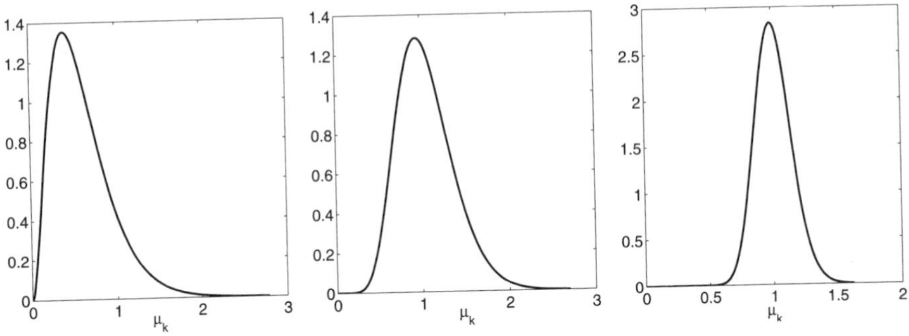


**Fig. 2.1.** Complete-data posterior density of the component mean $\mu_k$ for 100 observations arising from a mixture of Poisson distributions, where $\eta_k^{\text{true}} = 0.05$ and $\mu_k^{\text{true}} = 0.5$ (left-hand side); $\eta_k^{\text{true}} = 0.1$ and $\mu_k^{\text{true}} = 1$ (middle) ; $\eta_k^{\text{true}} = 0.5$ and $\mu_k^{\text{true}} = 1$ (right-hand side)

The complete-data posterior distribution $p(\boldsymbol{\theta}_k|\mathbf{S}, \mathbf{y})$ could be used to draw inferences on the unknown parameter $\boldsymbol{\theta}_k$. For illustration, we return to the component mean $\mu_k$ in a mixture of Poisson distributions. The posterior density is given by $\mu_k|\mathbf{y}, \mathbf{S} \sim \mathcal{G}(a_k(\mathbf{S}), b_k(\mathbf{S}))$, with $a_k(\mathbf{S})$ and $b_k(\mathbf{S})$ being defined by (2.18). The posterior density is plotted in Figure 2.1 for three artificial data sets. As it turns out, the posterior density centers around the true value of $\mu_k$, and the mode or any other location parameter of the posterior density could be used as an estimator of $\mu_k$; see Press (2003, Chapter 8) for more details on Bayesian point estimation.

Depending on the shape parameter $a_k(\mathbf{S})$, the posterior $p(\mu_k|\mathbf{y}, \mathbf{S})$ is rather skewed for the first data set whereas it is close to a normal distribution for the third one. As the posterior shape parameter $a_k(\mathbf{S})$ increases, the $\mathcal{G}(a_k(\mathbf{S}), b_k(\mathbf{S}))$-distribution converges to a normal distribution with mean and variance given by

$$\frac{a_k(\mathbf{S})}{b_k(\mathbf{S})} = \frac{a_{0,k} + N_k(\mathbf{S})\overline{y}_k(\mathbf{S})}{b_{0,k} + N_k(\mathbf{S})} \approx \overline{y}_k(\mathbf{S}) + o(1/N_k(\mathbf{S})), \tag{2.20}$$

$$\frac{a_k(\mathbf{S})}{b_k(\mathbf{S})^2} = \frac{a_{0,k} + N_k(\mathbf{S})\overline{y}_k(\mathbf{S})}{(b_{0,k} + N_k(\mathbf{S}))^2} \approx \frac{\overline{y}_k(\mathbf{S})}{N_k(\mathbf{S})} + o(1/N_k(\mathbf{S})), \tag{2.21}$$

where $o(1/N_k(\mathbf{S})) \to 0$ as $N_k(\mathbf{S}) \to \infty$. Such asymptotic normality of the posterior distribution holds for many problems in Bayesian inference; see, for example, Press (2003, Chapter 7).

To compare the Bayesian approach with the ML approach discussed in Subsection 2.3.2, we consider the problem of interval estimation for $\mu_k$. Within a Bayesian approach, a 95% credibility interval for $\mu_k$ is obtained from the 0.025 and 0.975 percentile of the $\mathcal{G}(a_k(\mathbf{S}), b_k(\mathbf{S}))$ posterior distribution. Table 2.2 compares this credibility interval with the approximate 95% confidence interval obtained by ML estimation for artificial data sets of size $N = 100$ for different values of $\mu_k^{\text{true}}$ and $\eta_k^{\text{true}}$. Partly these intervals agree; partly they are substantially different. As the shape parameter $a_k(\mathbf{S})$ increases, the posterior moments given in (2.20) and (2.21) converge to the ML estimator and the inverse of the expected Fisher information evaluated at the ML estimator, and the difference between the Bayesian and the ML interval estimator disappears.

The expected value of $a_k(\mathbf{S}) = N_k(\mathbf{S})\overline{y}_k(\mathbf{S}) + a_{0,k}$ is equal to $N\eta_k\mu_k + a_{0,k}$. Thus there is little difference between the Bayesian and the ML interval estimator, whenever $N$ is large. The smaller $N$, $\eta_k$, or $\mu_k$, the larger the difference is between the ML and the Bayesian approach. Whenever $a_k(\mathbf{S})$ is small, the Gamma posterior density of $\mu_k$ automatically accounts for departure from asymptotic normality (see again Figure 2.1), whereas asymptotic ML theory is not able to do so. Table 2.2, which compares the effective coverage probability of 95% Bayesian and ML intervals over 1000 data sets simulated for the three different parameter settings, shows that in cases where $a_k(\mathbf{S})$ is too small for asymptotic theory to hold, the effective coverage probability of the Bayesian credibility interval is much closer to the nominal value than the effective coverage probability of asymptotic confidence intervals based on ML estimation.

### 2.3.4 Complete-Data Bayesian Estimation of the Weights

For complete-data Bayesian estimation of the weights $\boldsymbol{\eta} = (\eta_1, \ldots, \eta_K)$, the complete-data likelihood $p(\mathbf{S}|\boldsymbol{\eta})$ is combined with a prior distribution $p(\boldsymbol{\eta})$, to obtain the posterior

$$p(\boldsymbol{\eta}|\mathbf{S}) \propto \prod_{k=1}^{K} \eta_k^{N_k(\mathbf{S})} p(\boldsymbol{\eta}),$$

where $N_k(\mathbf{S}) = \#\{S_i = k\}$ counts the number of observations in group $k$. Due to the constraint $\sum_k \eta_k = 1$ the group sizes are not independent. The complete-data likelihood, when regarded as a function of $\boldsymbol{\eta} = (\eta_1, \ldots, \eta_K)$, is the density of a Dirichlet distribution; see Subsection A.1.3 for more details on this distribution family. The conjugate prior distribution family is again the Dirichlet distribution (Bernardo and Girón, 1988), $\boldsymbol{\eta} \sim \mathcal{D}(e_{0,1}, \ldots, e_{0,K})$, where

$$p(\boldsymbol{\eta}) \propto \prod_{k=1}^{K} \eta_k^{e_{0,k}-1},$$

leading to the following posterior distribution,

$$p(\boldsymbol{\eta}|\mathbf{S}) \propto \prod_{k=1}^{K} \eta_k^{N_k(\mathbf{S})+e_{0,k}-1}.$$

This is the density of a Dirichlet distribution, $\boldsymbol{\eta}|\mathbf{S} \sim \mathcal{D}(e_1(\mathbf{S}), \ldots, e_K(\mathbf{S}))$, where:

$$e_k(\mathbf{S}) = e_{0,k} + N_k(\mathbf{S}), \qquad k = 1, \ldots, K. \tag{2.22}$$

The posterior mean of the unknown weight distribution is given by

$$\mathrm{E}(\eta_k|\mathbf{S}) = \frac{e_{0,k} + N_k(\mathbf{S})}{\sum_{j=1}^{K} e_{0,j} + N}, \qquad k = 1, \ldots, K, \tag{2.23}$$

whereas the posterior mode is equal to:

$$\eta_k^\star = \frac{e_{0,k} + N_k(\mathbf{S}) - 1}{\sum_{j=1}^{K} e_{0,j} + N - K}, \qquad k = 1, \ldots, K.$$

#### Choosing the Prior Distribution

When dealing with data from a mixture distribution, considerably more attention must be addressed to choosing the prior than is necessary in a Bayesian analysis of more conventional statistical models. Complete-data estimation of the weight distribution $\boldsymbol{\eta}$ is closely related to Bayesian analysis of observed binary or multinomial data (Congdon, 2005). For $K = 2$, for instance, where the Dirichlet distribution reduces to a Beta distribution, one of the weights, say $\eta_1$, is estimated from the "binary" data $\mathbf{S} = (S_1, \ldots, S_N)$. Prior distributions, that are common for a Bayesian analysis of observed binary data, may be applied, such as the uniform prior $\eta_1 \sim \mathcal{B}(1, 1)$, Jeffreys' prior

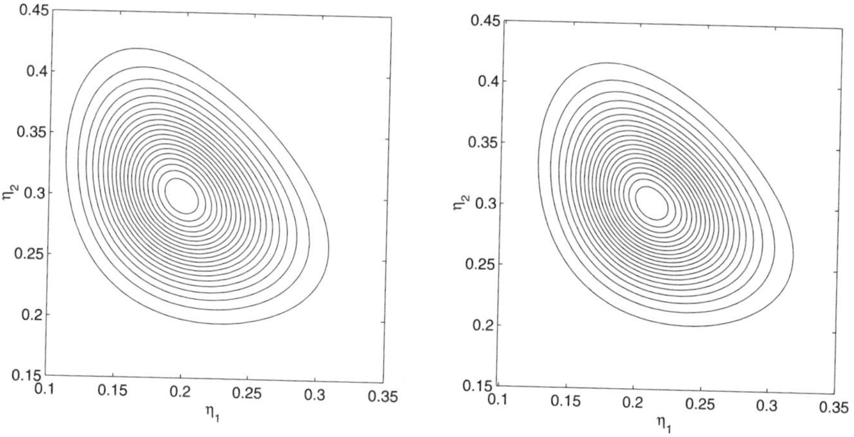


**Fig. 2.2.** Synthetic data of size $N = 100$ from a mixture with three components; contours of the posterior density $p(\eta_1, \eta_2|\mathbf{y})$ for two different priors; $e_{0,k} \equiv 0.5$ (left-hand side) and $e_{0,k} \equiv 4$ (right-hand side)

$\eta_1 \sim \mathcal{B}\left(\frac{1}{2}, \frac{1}{2}\right)$ (Box and Tiao, 1973, p.59), or a prior that is uniform in the natural parameter of the exponential family representation (Gelman et al., 2004); that is, $\eta_1/\eta_2 \propto$ constant, which corresponds to the improper prior $\eta_1 \sim \mathcal{B}(0,0)$. Dealing with "multinomial" data for $K > 2$, one could use the prior $\boldsymbol{\eta} \sim \mathcal{D}(1, \ldots, 1)$, which is uniform over the unit simplex $\mathcal{E}_K$, the prior $\boldsymbol{\eta} \sim \mathcal{D}\left(\frac{1}{2}, \ldots, \frac{1}{2}\right)$, or the improper prior $\boldsymbol{\eta} \sim \mathcal{D}(0, \ldots, 0)$ (Bernardo and Girón, 1988).

Bayesian analysis of observed binary or multinomial data is insensitive to choosing the prior only if none of the observed categories is rare. Consider, for illustration, 100 observations from a finite mixture distribution with three components where $N_1(\mathbf{S}) = 20, N_2(\mathbf{S}) = 30$, and $N_3(\mathbf{S}) = 50$. For $K = 3$ there are two free parameters, for instance $\eta_1$ and $\eta_2$. Figure 2.2 shows the posterior density $p(\eta_1, \eta_2|\mathbf{y})$ under two different priors, which hardly have an effect on the posterior.

Bayesian analysis of observed binary or multinomial data, however, tends to be sensitive to specific prior choices, if some of the observed categories are rare. An improper prior should be avoided for data drawn from a finite mixture distribution, as the posterior distribution $p(\boldsymbol{\eta}|\mathbf{S})$ obtained from such a prior need not be proper. When drawing data from a finite mixture distribution it easily happens that one of the categories, say $k$, is not observed in the sample, in particular if the corresponding weight $\eta_k$ is close to zero (*rare categories*). This event will occur in a sample of size $N$ with probability $(1-\eta_k)^N$, which is in fact rather likely for small weights $\eta_k$. For $N = 50$, for instance, categories with probabilities $\eta_k = 0.01/0.02/0.03$ will not be observed with probabilities as high as 0.605/0.364/0.218, respectively. If one of the categories, say $k$, is not observed in the sample, then $e_k = e_{0,k}$ and an improper prior leads to an improper posterior distribution. On the other hand, any proper prior will be highly influential for small groups, as the following discussion demonstrates.

### Estimation for Empty Components

If a category $k$ is not observed in the sample, then $N_k(\mathbf{S}) = 0$, and a nonregular problem results for ML estimation. If $\eta_k$ is among the free parameters, then the mode of the likelihood function will lie on the boundary of the parameter space as the maximum is attained at $\hat{\eta}_k = 0$. If $\eta_k$ is not among the free parameters, then $\sum_{j \neq k} \hat{\eta}_j = 1$ and the ML estimator lies in the nonidentifiability set corresponding to a reduced mixture with $K - 1$ categories. In both cases the likelihood function is nonregular, and asymptotic confidence intervals for $\eta_k$ are not available from standard asymptotic theory.

In a Bayesian context, under the $\mathcal{D}(e_{0,1}, \ldots, e_{0,K})$-prior, the marginal posterior of $\eta_k$ is easily obtained from the joint posterior (2.22) (see Subsection A.1.3):

$$\eta_k|\mathbf{S} \sim \mathcal{B}\left(e_{0,k}, \sum_{j=1, j\neq k}^{K} e_{0,j} + N\right). \tag{2.24}$$

The variance of this distribution is approximately equal to

$$\mathrm{Var}(\eta_k|\mathbf{S}) \approx e_{0,k}/(\sum_{j=1}^{K} e_{0,j} + N)^2.$$

Thus even if category $k$ is not directly observed, the total number of observations in the other categories is highly informative about $\eta_k$. Confidence intervals for $\eta_k$ are available from density (2.24). They are shrunken toward the posterior mean $\mathrm{E}(\eta_k|\mathbf{S})$, given by (2.23) with rate $o(1/N)$ rather than the regular rate $o(1/\sqrt{N})$.

As for other Bayesian inference problems based on nonregular likelihoods, the prior parameter $e_{0,k}$ will have a substantial effect on the shape of the posterior density if category $k$ is not observed. For illustration we consider two synthetic data sets of $N = 100$ observations from a two-component mixture distribution with unknown weights. Assume $N_1(\mathbf{S}) = 40, N_2(\mathbf{S}) = 60$ for data set 1, and $N_1(\mathbf{S}) = 0, N_2(\mathbf{S}) = 100$ for data set 2, with category 1 being never observed. Figure 2.3 shows the posterior density of $\eta_1$ for both data sets for various prior parameters $e_{0,k}$. Whereas for data set 1 the posterior density of $\eta_1$ is hardly affected by the prior, for data set 2, where category 1 is never observed, the posterior density of $\eta_1$ is extremely influenced by the prior. The posterior of $\eta_1$ is strongly pulled toward 0 and unbounded at 0 for $e_{0,k} = 0.5$, or more generally for any prior with $0 < e_{0,k} < 1$. The posterior is bounded with the mode lying at 0 for $e_{0,k} = 1$. If $e_{0,k} > 1$, then the posterior has

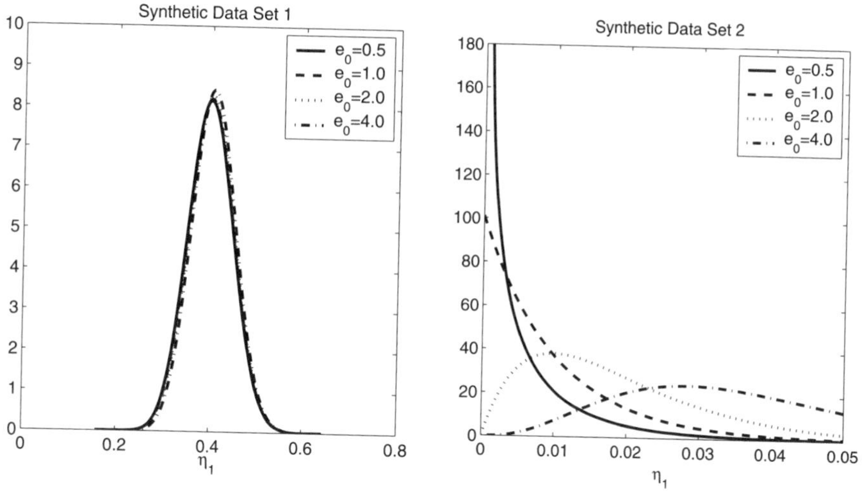


**Fig. 2.3.** Synthetic data 1 and 2: posterior densities for $\eta_1$ under different priors with $e_{0,k} \equiv e_0$; $e_0 = 0.5$ (full line), $e_0 = 1$ (dashed line), $e_0 = 2$ (dotted line), $e_0 = 4$ (dash dotted line); data set 1 (left-hand side), data set 2 (right-hand side)

a mode inside the unit interval, and is completely bounded away from 0 for $e_{0,k} = 4$.

The influence of the prior is also evident from Table 2.3 reporting 95%-confidence regions for $\eta_1$ for both data sets.

**Table 2.3.** Synthetic data set 1 and 2: 95% confidence region for $\eta_1$ based on various priors under $K = 2$

| Prior Parameter | 95% Confidence Region for $\eta_1$ | | | |
|---|---|---|---|---|
| $e_{0,k} \equiv e_0$ | Data Set 1 | | Data Set 2 | |
| | Lower | Upper | Lower | Upper |
| 0.5 | 0.313 | 0.493 | 0 | 0.019 |
| 1 | 0.312 | 0.493 | 0 | 0.029 |
| 2 | 0.317 | 0.498 | 0.0007 | 0.044 |
| 4 | 0.324 | 0.493 | 0.0063 | 0.071 |

The influence of the prior on the posterior density of the weight distribution under an unobserved category is even more striking in higher dimensions. For illustration we reconsider data set 1 introduced above, however, this time we assume that the data arise from a mixture distribution with three components, therefore $N_1(\mathbf{S}) = 40$, $N_2(\mathbf{S}) = 60$, and $N_3(\mathbf{S}) = 0$. There are two free parameters, and one may choose parameterizations excluding or including the weight of the unobserved category, such as $(\eta_1, \eta_2)$ and $(\eta_2, \eta_3)$, respectively. Figure 2.4 shows the contours of the corresponding bivariate posterior densities $p(\eta_1, \eta_2|\mathbf{y})$ and $p(\eta_2, \eta_3|\mathbf{y})$ for two different priors. In contrast to Figure 2.2, these are highly nonregular posterior densities. To study the influence of the prior, it is helpful to distinguish the following cases when estimating the weight $\eta_k$ of an unobserved category.

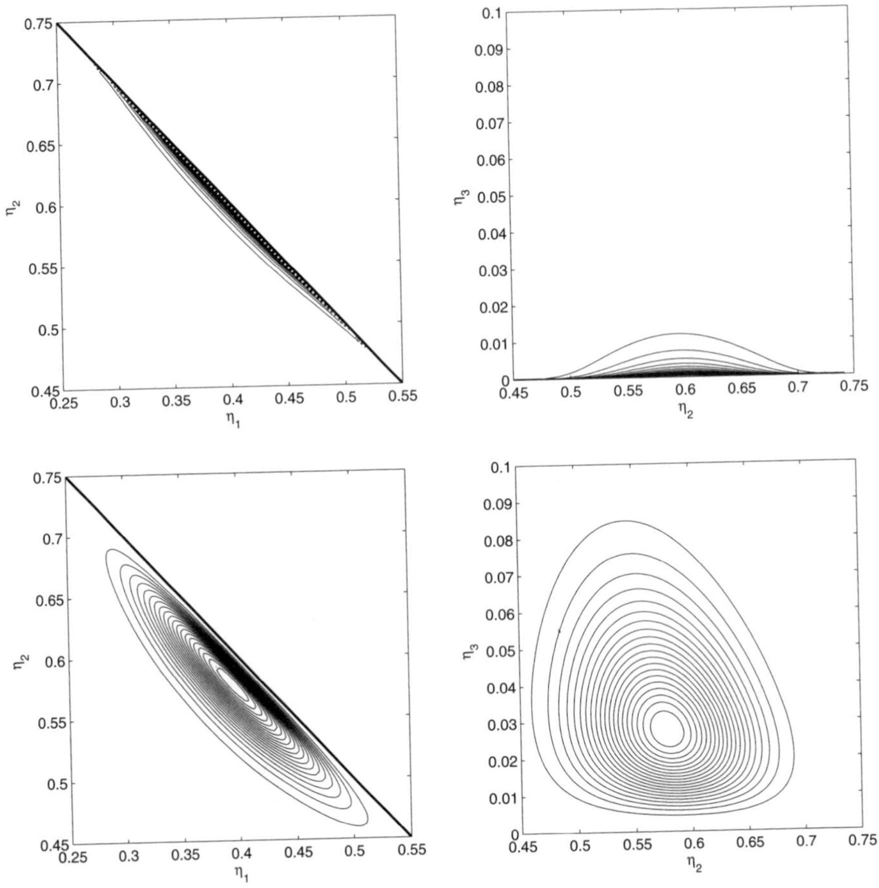


**Fig. 2.4.** Synthetic data, $K = 3$, contours of the posterior densities of the bivariate posterior distribution $p(\eta_1, \eta_2|\mathbf{y})$ (left-hand side) and $p(\eta_2, \eta_3|\mathbf{y})$ (right-hand side) for various priors; $e_{0,k} \equiv 0.5$ (top) and $e_{0,k} \equiv 4$ (bottom)

If $0 < e_{0,k} < 1$, then if $\eta_k$ is among the free parameters, the posterior is unbounded in $\eta_k = 0$, because of the term $1/(\eta_k^{1-e_{0,k}})$ appearing in the joint posterior. If $\eta_k$ is not among the free parameters, $\eta_k$ measures how far the free parameters are from perfect linear dependence: $\eta_k = 1 - \sum_{j \neq k} \eta_j$. As the marginal density of $p(\eta_k|\mathbf{S})$, given in (2.24), is unbounded at $\eta_k = 0$

with most of the mass close to 0, the free parameters are nearly linearly dependent and the posterior $p(\boldsymbol{\eta}|\mathbf{S})$ concentrates over the unit simplex $\mathcal{E}_{K-1}$, which corresponds to the reduced mixture, where only $K-1$ categories are present; see the top of Figure 2.4.

If $e_{0,k} > 1$, then the mode of the posterior $p(\boldsymbol{\eta}|\mathbf{S})$ lies inside the parameter space whatever choice is made for the free parameters. If $\eta_k$ is among the free parameters, then the posterior $p(\boldsymbol{\eta}|\mathbf{S})$ is pulled away from the boundary of the parameter space, where $\eta_k = 0$. If $\eta_k$ is not among the free parameters, then the posterior $p(\boldsymbol{\eta}|\mathbf{S})$ is pulled away from the subspace corresponding to the reduced model.

The bottom of Figure 2.4 illustrates for $e_{0,k} = 4$ how the posterior of the weight distribution is forced away from the reduced model. What looks like an undue influence, is in fact a big blessing when testing for the number of components in a finite mixture model. When allowing the posterior density of a model with $K$ components to concentrate over the subspace corresponding to a model with $K-1$ components when one of the categories is unobserved, then there is no way to distinguish between the two models, because their predictive power in terms of the complete-data likelihood will be the same. Only by bounding the posterior away from a model with $K-1$ components by putting an appropriate prior on $\eta_k$, will it be possible to distinguish the models.

## 2.4 Parameter Estimation When the Allocations Are Unknown

In this section we assume that in the mixture distribution (2.1) the true number $K$ of distinct components and the parametric distribution family the component densities arise from, are known, whereas the component parameters $\boldsymbol{\theta}_1, \ldots, \boldsymbol{\theta}_K$, the weight distribution $\boldsymbol{\eta}$ and the allocations $\mathbf{S}$ are unknown. In this case estimation of the parameters of the mixture distribution is not straightforward. Even for mixtures of standard distribution such as the univariate normal distribution no explicit estimation method exists and in general some numerical method is required for practical estimation.

In the early days of mixture modeling (Pearson, 1894), the method of moments was the most widely applied estimation technique; it is briefly reviewed in Subsection 2.4.1. With the advent of modern computer technology attention turned to methods based on the mixture likelihood function, which is defined in Subsection 2.4.2, namely maximum likelihood estimation (see Subsection 2.4.4), and Bayesian estimation which is briefly summarized in Subsection 2.4.5, whereas a full account appears in Chapter 3. For a thorough discussion of non-Bayesian parameter estimation for finite mixtures we refer the reader to the excellent monographs by Everitt and Hand (1981) and Titterington et al. (1985), and, published more recently with emphasis on maximum likelihood estimation based on the EM algorithm, by McLachlan and Peel (2000).

### 2.4.1 Method of Moments

The basic idea of the method of moment estimator is to select a set of moments $\mathrm{E}(H_j(\mathbf{Y})|\boldsymbol{\vartheta})$ of the random variable $\mathbf{Y}$, and to determine the parameters $\boldsymbol{\vartheta}$ of the mixture model in such a way that the theoretical moments $\mathrm{E}(H_j(\mathbf{Y})|\boldsymbol{\vartheta})$ match the empirical counterpart $\overline{H}_j$, given by the sample average of $H_j(\cdot)$ over the observed values $\mathbf{y}_1, \ldots, \mathbf{y}_N$:

$$\overline{H}_j = \frac{1}{N} \sum_{i=1}^{N} H_j(\mathbf{y}_i).$$

Matching the sample moments $\overline{H}_j$ to the corresponding theoretical moments of the finite mixture distribution derived in Subsection 1.2.4 yields:

$$\sum_{k=1}^{K} \eta_k \mathrm{E}(H_j(\mathbf{Y})|\boldsymbol{\theta}_k) = \overline{H}_j. \tag{2.25}$$

This typically is a nonlinear equation in the unknown parameter $\boldsymbol{\vartheta} = (\boldsymbol{\theta}_1, \ldots, \boldsymbol{\theta}_K, \eta_1, \ldots, \eta_K)$, even if the component-specific moments $\mathrm{E}(H_j(\mathbf{Y})|\boldsymbol{\theta}_k)$ are available in closed form. The system of equations generated by (2.25) may be solved with respect to $\boldsymbol{\vartheta}$, if the number of functions $H_j(\mathbf{Y})$ is equal to the number of distinct parameters in $\boldsymbol{\vartheta}$, and linear dependence among the equations is avoided. The method of moments estimator is a (not necessarily unique) solution to this system of equations. The larger the number of components $K$, however, the more equations will be needed.

For univariate data, the method of moments is typically applied by considering $H_j(Y) = Y^j$ for $j = 1, \ldots, \dim(\boldsymbol{\vartheta})$; another suggestion has been made by Lindsay (1989) for univariate mixtures of normals. Quandt and Ramsey (1978) use the moment-generating function $H_j(Y) = \exp(jY)$ with different values $j$ for fitting univariate mixtures of normals and switching regression models.

Although the method of moments is the oldest estimation method known for mixtures, dating back to the problem of estimating the five parameters of a mixture of two normal distributions (Pearson, 1894), some potential pitfalls have to be mentioned, in particular numerical problems with solving (2.25) and loss of efficiency in comparison to other estimators; see the small simulation experiment in Subsection 2.4.7, where different estimators are compared, and Day (1969) who shows by means of a simulation study that the method of moments estimator may be inefficient compared to maximum likelihood estimation for bivariate mixtures of two normal distributions. Lindsay and Basak (1993) illustrate for multivariate mixtures of normals that a computationally attractive estimator may result if the moment equations (2.25) are designed carefully.

**Methods of Moments for a Poisson Mixture Distribution**

As discussed by Everitt (1985), the method of moments for data from a Poisson mixture distribution could be based on the factorial moments $H_j(Y) = Y!/(Y-j)!$ for $j = 1, \dots, 2K-1$. Because $\mathrm{E}(H_j(Y)|\mu_k) = \mu_k^j$, the corresponding system of equations reads:

$$\sum_{k=1}^{K} \eta_k \mu_k^j = v_j,$$

where $v_j$ are the sample factorial moments:

$$v_j = \frac{1}{N} \sum_{i:y_i \geq j} y_i(y_i - 1) \cdots (y_i - (j-1)).$$

Note that $v_1 = \overline{y}$. For a mixture of two Poisson distributions, the following set of equations results.

$$\begin{aligned}
\eta_1(\mu_1 - \mu_2) + \mu_2 &= \overline{y}, \\
\eta_1(\mu_1^2 - \mu_2^2) + \mu_2^2 &= v_2, \\
\eta_1(\mu_1^3 - \mu_2^3) + \mu_2^3 &= v_3.
\end{aligned}$$

The method of moments estimator $\hat{\mu}_1^{MM}$ and $\hat{\mu}_2^{MM}$ is given as the roots of the equation $\mu^2 - b\mu + c = 0$, where

$$b = \frac{v_3 - \overline{y}v_2}{v_2 - \overline{y}^2}, \qquad c = \overline{y}b - v_2,$$

whereas the moment estimator $\hat{\eta}_1^{MM}$ reads:

$$\hat{\eta}_1^{MM} = \frac{\hat{\mu}_2^{MM} - \overline{y}}{\hat{\mu}_2^{MM} - \hat{\mu}_1^{MM}}.$$

### 2.4.2 The Mixture Likelihood Function

In this subsection we derive the mixture likelihood $p(\mathbf{y}|\boldsymbol{\vartheta})$ of $\boldsymbol{\vartheta}$ given $N$ random observations $\mathbf{y} = (\mathbf{y}_1, \dots, \mathbf{y}_N)$ from the mixture distribution (2.1). We assume that no information concerning the allocation of $\mathbf{y}_i$ to a certain component is available, and define the mixture likelihood function $p(\mathbf{y}|\boldsymbol{\vartheta})$ as the joint distribution of $\mathbf{y}_1, \dots, \mathbf{y}_N$ under $\boldsymbol{\vartheta}$. The mixture likelihood function takes the form:

$$p(\mathbf{y}|\boldsymbol{\vartheta}) = \prod_{i=1}^{N} p(\mathbf{y}_i|\boldsymbol{\vartheta}) = \prod_{i=1}^{N} \left( \sum_{k=1}^{K} \eta_k p(\mathbf{y}_i|\boldsymbol{\theta}_k) \right), \tag{2.26}$$

which may be expanded into a sum of $K^N$ individual terms. For Poisson mixtures with $K$ components, for instance, the mixture likelihood reads:

$$p(\mathbf{y}|\boldsymbol{\vartheta}) = \prod_{i=1}^{N} \frac{1}{\Gamma(y_i+1)} \left( \sum_{k=1}^{K} \eta_k \mu_k^{y_i} \mathrm{e}^{-\mu_k} \right). \qquad (2.27)$$

Titterington et al. (1985) called $p(\mathbf{y}|\boldsymbol{\vartheta})$ the uncategorized likelihood, as no information concerning the component to which observation $\mathbf{y}_i$ belongs is incorporated, in contrast to the complete-data likelihood function $p(\mathbf{y}, \mathbf{S}|\boldsymbol{\vartheta})$ defined earlier in (2.8).

The mixture likelihood function is the basis both for maximum likelihood estimation as well as Bayesian estimation. An important difference between the mixture likelihood and the complete-data likelihood function defined earlier in (2.8) lies in the mathematical structure of these functions. Whereas the complete-data likelihood may be decomposed into $K+1$ independent factors, no such decomposition is possible for the mixture likelihood making parameter estimation much more difficult.

Hathaway (1986) noted that the mixture likelihood and the complete-data likelihood are related. The contribution $p(\mathbf{y}_i|\boldsymbol{\vartheta})$ of the $i$th observation to the mixture likelihood in (2.26) is equal to the normalizing constant in the Bayes' classifier for this observation; see (2.3). By rewriting the Bayes' classifier as $p(\mathbf{y}_i|\boldsymbol{\theta}_k)\eta_k = \Pr(S_i = k|\mathbf{y}_i, \boldsymbol{\vartheta})p(\mathbf{y}_i|\boldsymbol{\vartheta})$, and substituting this into (2.8) the following results,

$$p(\mathbf{y}, \mathbf{S}|\boldsymbol{\vartheta}) = \prod_{i=1}^{N} p(\mathbf{y}_i|\boldsymbol{\vartheta}) \prod_{k=1}^{K} \Pr(S_i = k|\mathbf{y}_i, \boldsymbol{\vartheta})^{I_{\{S_i=k\}}}.$$

Taking the log of both sides yields:

$$\log p(\mathbf{y}|\boldsymbol{\vartheta}) = \log p(\mathbf{y}, \mathbf{S}|\boldsymbol{\vartheta}) - \sum_{i=1}^{N} \sum_{k=1}^{K} I_{\{S_i=k\}} \log \Pr(S_i = k|\mathbf{y}_i, \boldsymbol{\vartheta}). \quad (2.28)$$

The second term is a measure of loss of information in the mixture likelihood function compared to the complete-data likelihood function which is zero when the mixture model enables perfect classification; see Subsection 2.2.2.

Finally, the mixture likelihood, when considered as a function of the unknown parameters $\boldsymbol{\vartheta}$, is quite different from the regular likelihood functions common to many statistical models; see Subsection 2.4.3 for some illustration.

### 2.4.3 A Helicopter Tour of the Mixture Likelihood Surface for Two Examples

As a consequence of invariance to relabeling the components of a mixture model, the mixture likelihood function usually has $K!$ different, but equivalent

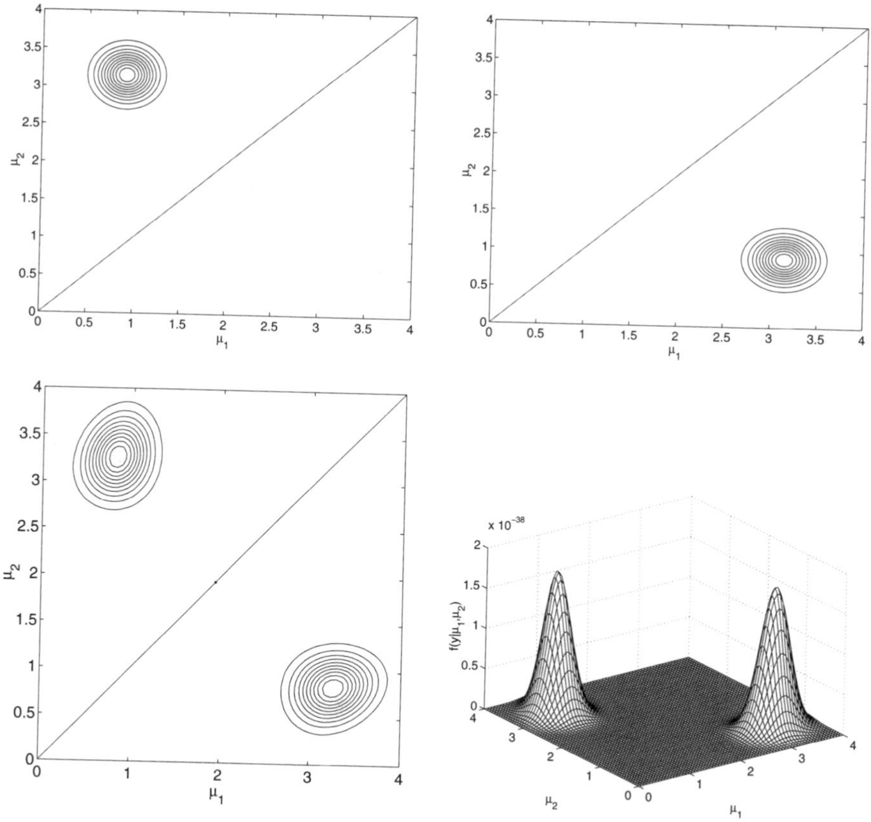


**Fig. 2.5.** HIDDEN AGE GROUPS — Synthetic Data Set 1; top: contours of the complete-data likelihood under labeling 1 (left-hand side) and labeling 2 (right-hand side); bottom: contours and surface of the mixture likelihood

modes corresponding to all different ways of labeling. For illustration, we study the surface of the likelihood function in more detail for synthetic data. Assume that data come from a population with two unobserved categories. We may, for instance, observe a random feature $Y$ in a population with two age groups. Depending on the unobserved age, the expected value of $Y$ is equal to $\mu_y$ for the younger and $\mu_e$ for the elder age group. Consider now, as a model for these data, a mixture of two normal distributions, where to keep the discussion simple, we assume that both components have equal weights, $\eta_1 = \eta_2 = 0.5$, and equal variances, $\sigma_1^2 = \sigma_2^2 = 1$:

$$p(y) = 0.5 f_N(y; \mu_1, 1) + 0.5 f_N(y; \mu_2, 1). \tag{2.29}$$

One of these components will model the distribution of $Y$ for the younger age group, whereas the other component will model the distribution of $Y$ for

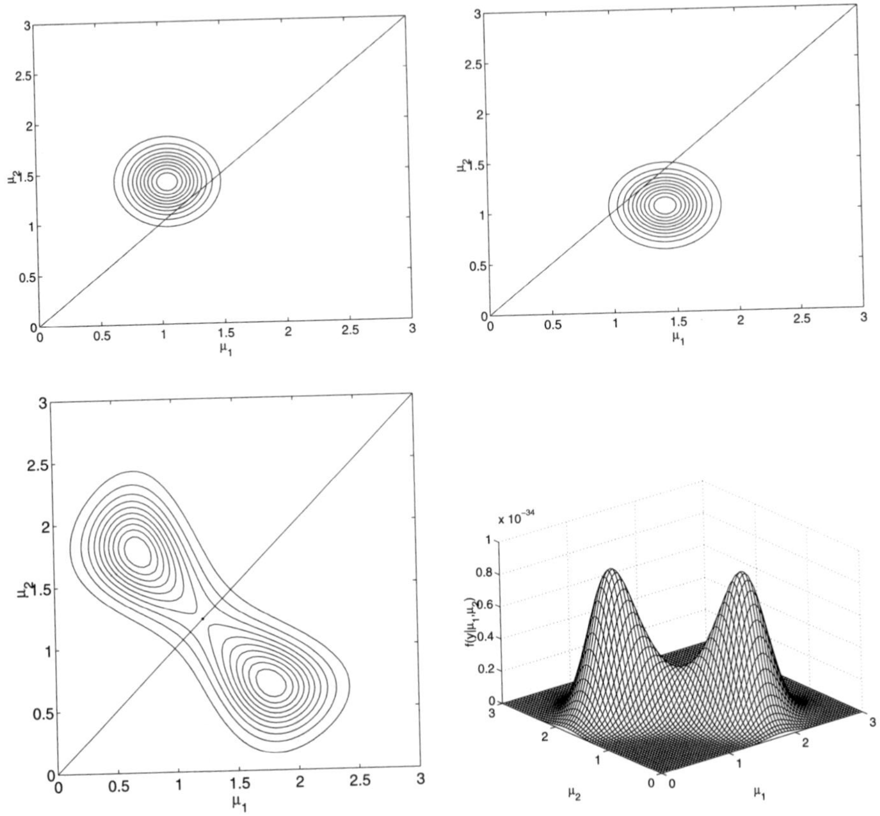


**Fig. 2.6.** HIDDEN AGE GROUPS — Synthetic Data Set 2; top: contours of the complete-data likelihood under labeling 1 (left-hand side) and labeling 2 (right-hand side); bottom: contours and surface of the mixture likelihood

the elder age group. Note, however, that the model is indifferent as to what component corresponds to what age group.

To interpret the parameters of the components correctly, we have to assign a labeling that associates a certain age group with a certain component. Let $S_i$ be equal to the component indicator. For two categories, there are two ways to assign a labeling:

| | Younger Age Group | Elder Age Group |
|---|---|---|
| Labeling 1 | $S_i = 1$ | $S_i = 2$ |
| Labeling 2 | $S_i = 2$ | $S_i = 1$ |

For labeling 1, the younger age group is associated with the first component, thus the expected value of $Y$ in this group is equal to $\mu_1$, whereas for labeling 2, the younger age group is associated with the second component, thus the expected value of $Y$ in this group is equal to $\mu_2$.

For illustration we consider two synthetic data sets of $N = 100$ observations simulated from (2.29). For both data sets the expected value of $Y$ for the younger age group is equal to $\mu_y = 1$, whereas the expected value of $Y$ for the elder age group varies over the two data sets. For synthetic data set 1 with $\mu_e = 3$ the expected value of $Y$ differs significantly between the age groups; for synthetic data set 2 with $\mu_e = 1.5$ the difference is small compared to the variance.

First we assume that the data are categorized. If for each observation $y_i$ the age category were observed, we may select one of the two possible labelings, and determine the complete-data likelihood $p(\mathbf{y}|\mathbf{S}, \mu_1, \mu_2)$. Note, however, that depending on the labeling we selected, the complete-data likelihood would be concentrated over different regions of the parameter space. The tops of Figure 2.5 and Figure 2.6 show the contours of the complete-data likelihood defined in (2.8) for the various synthetic data sets under the two ways of labeling. For synthetic data set 1, labeling 1 leads to a likelihood concentrated over the region $\mu_1 < \mu_2$, whereas for labeling 2, the likelihood is concentrated over the region $\mu_1 > \mu_2$. For this data set, parameters fulfilling the constraint $\mu_1 < \mu_2$ may be clearly associated with labeling 1. For the other data set, however, parameters fulfilling the constraint $\mu_1 < \mu_2$ may not be associated uniquely with labeling 1, as also under labeling 2 parameters fulfilling this constraint have considerable likelihood.

If the data are uncategorized, one would consider the mixture likelihood $p(\mathbf{y}|\boldsymbol{\vartheta})$ defined in (2.26) for parameter estimation which, however, reflects the arbitrariness of the labeling of the hidden categories. Any two parameters $\boldsymbol{\vartheta} = (\mu_1, \mu_2)$ and $\boldsymbol{\vartheta}^\star = (\mu_2, \mu_1)$ generate the same functional value for the likelihood. This may, but need not, cause multimodality of the likelihood function. The bottoms of Figure 2.5 and Figure 2.6 show the contours and the surfaces of the mixture likelihood $p(\mathbf{y}|\boldsymbol{\vartheta})$ for both synthetic data sets. For synthetic data set 1, the mixture likelihood function in Figure 2.5 has two well-separated modes, one concentrated over the region $\mu_1 < \mu_2$, the other concentrated over the region $\mu_1 > \mu_2$. In comparison to the complete-data likelihood displayed in the top of the same figure, it is clear that the two modes correspond to the two ways of labeling the hidden groups. For synthetic data set 2, the mixture likelihood function in Figure 2.6 again has two modes, which are, however, not well separated. Again, in comparison to the complete-data likelihood it becomes clear that this lack of separation reflects the fact that parameter values around the unidentifiability set $\mathcal{U}^E(\mu), \mu \in \Re$, which corresponds to the line $\mu_1 = \mu_2$ in this particular example, have considerable likelihood under both ways of labeling.

### Formal Identifiability Versus Unique Labeling in the Mixture Likelihood Function

Formal identifiability constraints were introduced in Subsection 1.3.3. It is a common misunderstanding that an *arbitrary* formal identifiability constraint

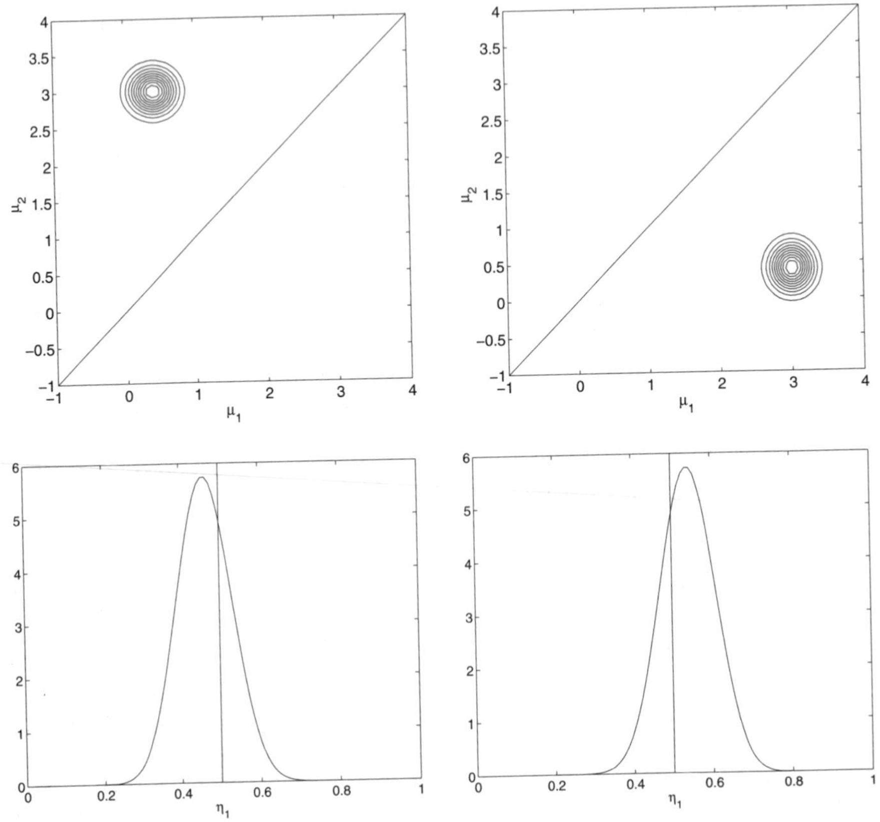


**Fig. 2.7.** HIDDEN AGE GROUPS — Synthetic Data Set 3; complete-data likelihood $p(\mathbf{y}|\mu_1, \mu_2, \eta_1, \mathbf{S})$ under labeling 1 (left-hand side) and labeling 2 (right-hand side); top: contours of $p(\mathbf{y}|\mu_1, \mu_2, \eta_1, \mathbf{S})$ for (arbitrary) $\eta_1$ fixed; bottom: profile of $p(\mathbf{y}|\mu_1, \mu_2, \eta_1, \mathbf{S})$ for (arbitrary) $\mu_1$ and $\mu_2$ fixed

leads to a unique labeling. As pointed out by Celeux (1998), Celeux et al. (2000), Stephens (2000b), and Frühwirth-Schnatter (2001b), this is not necessarily the case.

Consider, for illustration, the mixture likelihood function of synthetic data set 3, generated from the following mixture model,

$$p(y) = \eta_1 f_N(y; \mu_1, 1) + (1 - \eta_1) f_N(y; \mu_2, 1), \tag{2.30}$$

where $\mu_1 = 1$, $\mu_2 = 3$, and $\eta_1 = 0.45$. Thus the first group is slightly smaller than the second group. For this model, one could use the constraint $\eta_1 < \eta_2$, which is equivalent to $\eta_1 < 0.5$, or the constraint $\mu_1 < \mu_2$ to achieve formal identifiability, but only the second constraint will induce a unique labeling.

Figure 2.7 shows contours of the complete-data likelihood $p(\mathbf{y}|\mu_1, \mu_2, \eta_1, \mathbf{S})$ over $(\mu_1, \mu_2)$ for (arbitrary) $\eta_1$ fixed and a profile over $\eta_1$ for (arbitrary)

$(\mu_1, \mu_2)$ fixed. Under labeling 1, the complete-data likelihood is concentrated over the region $\mu_1 < \mu_2$. Although the complete-data likelihood has a lot of mass over $\eta_1 < 0.5$, considerable likelihood is also given to the region $\eta_1 > 0.5$. Under labeling 2, the complete-data likelihood is concentrated over $\mu_1 > \mu_2$. This time the likelihood has a lot of mass over $\eta_1 > 0.5$, however, considerable likelihood is also given to $\eta_1 < 0.5$. As a consequence, parameters fulfilling the constraint $\mu_1 < \mu_2$ are very likely to occur only under labeling 1, whereas parameters fulfilling the constraint $\eta_1 < 0.5$ may occur under both ways of labeling.

This has consequences for the ability of any of the two constraints to force a unique labeling. As the constraint $\mu_1 < \mu_2$ is likely to occur only under labeling 1, the constraint $\mu_1 < \mu_2$ is able to identify parameters from a single modal region of the mixture likelihood as demonstrated by Figure 2.8. Parameters fulfilling the constraint $\eta_1 < 0.5$, however, are likely under both ways of labeling. Therefore, if we consider the mixture likelihood under the constraint $\eta_1 < 0.5$, we obtain a function with two modes, and parameters with high likelihood may come from both modal regions.

### 2.4.4 Maximum Likelihood Estimation

With the availability of powerful computers and elaborate numerical algorithms, maximum likelihood (ML) estimation became the preferred approach to parameter estimation for finite mixture models for many decades. Redner and Walker (1984) provide a concise and excellent review of ML estimation for finite mixture models. ML estimation was used for a univariate mixture of two normal distributions with $\sigma_1^2 = \sigma_2^2$ as early as Rao (1948). Further pioneering work for ML estimation was done by Hasselblad (1966) for univariate mixtures of normals, Hasselblad (1969) for general mixtures from the exponential family, and Day (1969) and Wolfe (1970) for multivariate mixtures of normals.

In these early papers, the ML estimator $\hat{\boldsymbol{\vartheta}}$ is obtained by maximizing the mixture likelihood $p(\mathbf{y}|\boldsymbol{\vartheta})$ with respect to $\boldsymbol{\vartheta}$ using some direct method such as Newton's method (Hasselblad, 1966) or a gradient method (Quandt, 1972). An iterative scheme for maximizing the likelihood function, developed by Hasselblad (1966, 1969), is nothing but an early variant of the EM algorithm, introduced later by Dempster et al. (1977) which is the most commonly applied method to find the ML estimator for a finite mixture model nowadays.

#### The EM Algorithm

The expectation-maximization (EM) algorithm was introduced for general latent variable models in the seminal paper by Dempster et al. (1977), who also mentioned applications to finite mixture models. The use of the EM algorithm for the estimation of mixture models has been studied in detail in

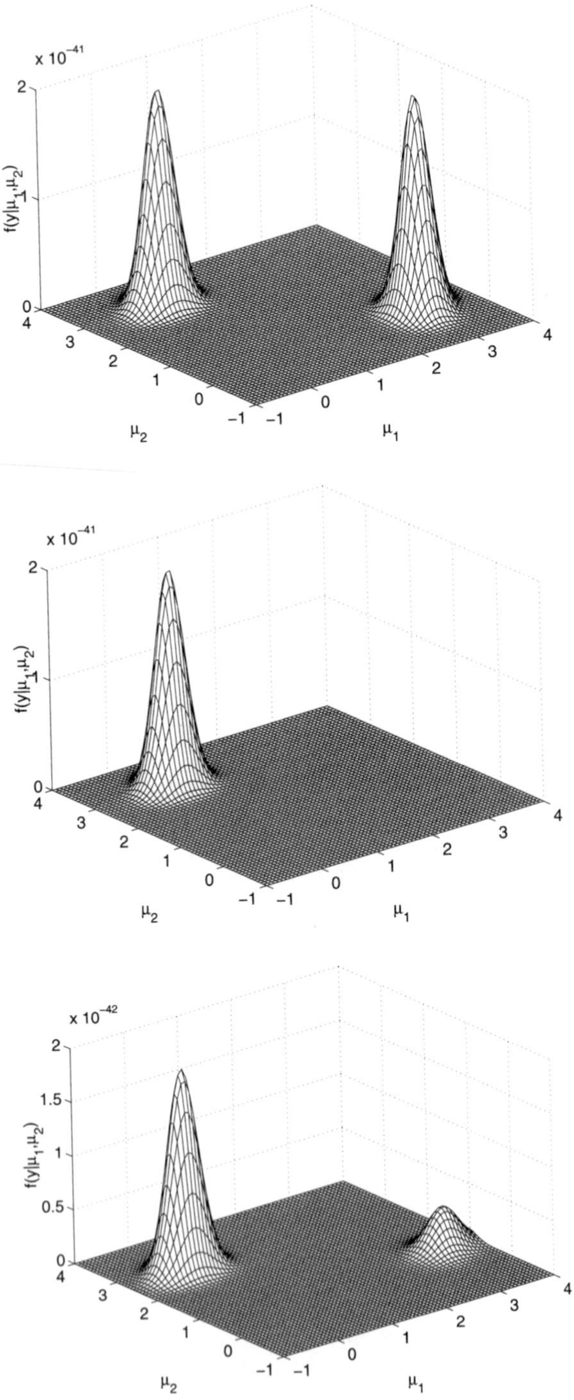


**Fig. 2.8.** Hidden age groups — Synthetic Data Set 3; marginalized mixture likelihood $p(\mathbf{y}|\mu_1, \mu_2)$; top: unconstrained; middle: under the constraint $\mu_1 < \mu_2$; bottom: under the constraint $\eta_1 < \eta_2$

Redner and Walker (1984). Meng (1997) provides a very inspiring general-level tutorial on the EM algorithm in the context of finite mixtures of Poisson distributions, whereas the monograph of McLachlan and Peel (2000) gives full details for a wide range of finite mixture models.

To implement the EM algorithm for a finite mixture model, the log of complete-data likelihood function $p(\mathbf{y}, \mathbf{S}|\boldsymbol{\vartheta})$, defined earlier in (2.8), is written as

$$\log p(\mathbf{y}, \mathbf{S}|\boldsymbol{\vartheta}) = \sum_{i=1}^{N} \sum_{k=1}^{K} D_{ik} \log(\eta_k p(\mathbf{y}_i|\boldsymbol{\theta}_k)), \tag{2.31}$$

where $D_{ik}$ is a 0/1 coding of the allocations $S_i$: $D_{ik} = 1$, iff $S_i = k$. Starting from $\hat{\boldsymbol{\vartheta}}^{(0)}$, the EM algorithm iterates between two steps: an E-step, where the conditional expectation of $\log p(\mathbf{y}, \mathbf{S}|\boldsymbol{\vartheta})$, given the current data and given the current parameter is computed, and an M-step in which parameters that maximize the expected complete-data log likelihood function, obtained from the E-step are determined. Under fairly mild regularity conditions, the EM algorithm converges to a local maximum of the mixture likelihood function (Dempster et al., 1977; Wu, 1983). For mixture models, the E-step leads for $m \geq 1$ to the following estimator of $D_{ik}$,

$$\hat{D}_{ik}^{(m)} = \frac{\hat{\eta}_k^{(m-1)} p(\mathbf{y}_i|\hat{\boldsymbol{\theta}}_k^{(m-1)})}{\sum_{j=1}^{K} \hat{\eta}_j^{(m-1)} p(\mathbf{y}_i|\hat{\boldsymbol{\theta}}_j^{(m-1)})}, \tag{2.32}$$

and the M-step involves maximizing

$$\sum_{i=1}^{N} \sum_{k=1}^{K} \hat{D}_{ik}^{(m)} \log(\eta_k p(\mathbf{y}_i|\boldsymbol{\theta}_k))$$

with respect to all unknown components in $\boldsymbol{\vartheta} = (\boldsymbol{\theta}_1, \ldots, \boldsymbol{\theta}_K, \boldsymbol{\eta})$, leading to a new estimate $\hat{\boldsymbol{\vartheta}}^{(m)}$. It is easy to verify that for an arbitrary mixture

$$\hat{\eta}_k^{(m)} = \frac{n_k}{N}, \qquad n_k = \sum_{i=1}^{N} \hat{D}_{ik}^{(m)}, \tag{2.33}$$

whereas the estimator of the component parameters $\boldsymbol{\theta}_k$ of course depends on the distribution family underlying the mixture. For mixtures of Poisson distributions, for instance, the estimator of the component mean $\mu_k$ reads:

$$\hat{\mu}_k^{(m)} = \frac{1}{n_k} \sum_{i=1}^{N} \hat{D}_{ik}^{(m)} y_i.$$

A disadvantage of the EM algorithm compared to direct maximization of the likelihood function is much slower convergence. Following Redner and

Walker (1984), who recommended combining the EM algorithm with Newton's method, several authors used hybrid algorithms for mixture estimation; see Aitkin and Aitkin (1996) for a review.

### Asymptotic Properties of the ML estimator

In simulation studies the ML estimator generally leads to smaller mean squared errors than moment estimators; see, for instance, Tan and Chang (1972) and the small simulation study in Subsection 2.4.7.

A theoretical underpinning for the ML estimator is provided by asymptotic theory. Consider the likelihood equation

$$\frac{\partial}{\partial \boldsymbol{\vartheta}} \log p(\mathbf{y}|\boldsymbol{\vartheta}) = 0, \tag{2.34}$$

where $\boldsymbol{\vartheta}$ denotes the parameter $(\boldsymbol{\theta}_1, \ldots, \boldsymbol{\theta}_K, \eta_1, \ldots, \eta_{K-1})$ with the redundant weight $\eta_K$ being omitted. Let $\boldsymbol{\vartheta}^{\text{true}}$ be the true value of $\boldsymbol{\vartheta}$. Define for an arbitrary parameter $\boldsymbol{\vartheta}$ the expected Fisher information matrix as

$$\mathcal{I}(\boldsymbol{\vartheta}) = \int_{\mathcal{Y}} \frac{\partial}{\partial \boldsymbol{\vartheta}} \log p(\mathbf{y}|\boldsymbol{\vartheta}) (\frac{\partial}{\partial \boldsymbol{\vartheta}} \log p(\mathbf{y}|\boldsymbol{\vartheta}))' p(\mathbf{y}|\boldsymbol{\vartheta}) d\mathbf{y}, \tag{2.35}$$

and assume that $\mathcal{I}(\boldsymbol{\vartheta}^{\text{true}})$ is well defined and positive definite. Then Redner and Walker (1984, p.211) prove under certain boundedness conditions on the partial derivatives of $p(\mathbf{y}|\boldsymbol{\vartheta})$ with respect to the components of $\boldsymbol{\vartheta}$, that in any sufficiently small neighborhood of $\boldsymbol{\vartheta}^{\text{true}}$ for a sufficiently large sample size $N$ there exists a unique solution $\hat{\boldsymbol{\vartheta}}$ of the likelihood equation (2.34) in that neighborhood, which locally maximizes the log likelihood function. Furthermore, this ML estimator is consistent, efficient, and asymptotically normal:

$$\sqrt{N}(\hat{\boldsymbol{\vartheta}} - \boldsymbol{\vartheta}^{\text{true}}) \rightarrow_d \mathcal{N}\left(0, \mathcal{I}(\boldsymbol{\vartheta}^{\text{true}})^{-1}\right).$$

### Practical Difficulties with ML Estimation

ML estimation may encounter various practical difficulties. First, it may be difficult to find the global maximum of the likelihood numerically. Several studies report convergence failures particularly when the sample size is small or the components are not well separated; see, for instance, Finch et al. (1989). Recently, more attention has been paid to choosing starting values that increase the chance of convergence (Karlis and Xekalaki, 2003; Biernacki et al., 2003).

Second, for the important special case of mixtures of normal distributions, the mixture likelihood is unbounded, as discussed later in Subsection 6.1.2. In this case the ML estimator as a global maximizer of the likelihood function does not exist, however, it usually exists as a local maximizer. The practical difficulty is to identify this local maximum and avoid spurious modes in the course of maximizing the log likelihood function; see McLachlan and Peel (2000, Section 3.10).

Third, as for any incomplete data problems, the provision of standard errors is not straightforward within ML estimation of finite mixture models, in particular when using the EM algorithm, although various papers (Louis, 1982; Meng and Rubin, 1991) show how to obtain approximate standard errors from the EM algorithm. A further problem noted in several papers is singularity of the matrix of second partial derivatives of the log likelihood function.

Finally, as McLachlan and Peel (2000, p.68) warn, "In particular for mixture models, it is well known that the sample size $N$ has to be very large, before asymptotic theory of maximum likelihood applies." The regularity conditions are often violated, including cases of great practical concern, among them small data sets, mixtures with small component weights, and overfitting mixtures with too many components.

### 2.4.5 Bayesian Parameter Estimation

Let $\mathbf{y} = \{\mathbf{y}_1, \dots, \mathbf{y}_N\}$ be $N$ randomly selected, uncategorized observations from the mixture distribution (2.1). Let $\boldsymbol{\vartheta} = (\boldsymbol{\theta}_1, \dots, \boldsymbol{\theta}_K, \boldsymbol{\eta})$ denote all unknown parameters appearing in the mixture model. As in Subsection 2.3.3, where Bayesian parameter estimation for known allocations has been discussed, one has to assume that a prior distribution $p(\boldsymbol{\vartheta})$ on $\boldsymbol{\vartheta}$ is available. Using Bayes' theorem, the mixture likelihood $p(\mathbf{y}|\boldsymbol{\vartheta})$ defined in (2.26) is combined with the prior $p(\boldsymbol{\vartheta})$ in a similar way as was done for the complete-data likelihood function in Subsection 2.3.3 to obtain the posterior density $p(\boldsymbol{\vartheta}|\mathbf{y})$:

$$p(\boldsymbol{\vartheta}|\mathbf{y}) \propto p(\mathbf{y}|\boldsymbol{\vartheta})p(\boldsymbol{\vartheta}). \tag{2.36}$$

There are various reasons why one might be interested in adopting a Bayesian approach for finite mixture models. The inclusion of a proper prior within a Bayesian approach will generally introduce a smoothing effect on the mixture likelihood function and reduce the risk of obtaining spurious modes in cases where the EM algorithm leads to degenerate solutions. This is shown in particular in Section 6.1 for finite mixtures of normal distributions. Second, as the whole posterior distribution $p(\boldsymbol{\vartheta}|\mathbf{y})$ is available, it is much easier to address the issue of parameter uncertainty. Finally, Bayesian estimation does not rely on asymptotic normality, and yields valid inference also in cases where regularity conditions are violated, such as small data sets and mixtures with small component weights; see again Subsection 2.3.3.

In contrast to Subsection 2.3.3, however, where the allocations were assumed to be known, practical Bayesian estimation is more involved when the allocations are unknown. Unfortunately, for the mixture likelihood (2.26) no natural conjugate prior is available, meaning that whatever prior $p(\boldsymbol{\vartheta})$ one chooses, the posterior density obtained from (2.36) does not belong to any tractable distribution family. For this reason, Bayesian estimation of even simple mixture problems proved to be extremely difficult prior to the advent of Markov chain Monte Carlo methods. An illuminating insight into the difficulties one had to face appears in Bernardo and Girón (1988) for a mixture model where only the weights distribution $\boldsymbol{\eta}$ is unknown.

This situation changed only rather recently in the early 1990s with the widespread availability of Markov chain Monte Carlo (MCMC) methods and their application to Bayesian estimation of finite mixture models. Like the EM algorithm, practical Bayesian estimation using MCMC methods is based on the work of Dempster et al. (1977) who realized that a finite mixture model may always be expressed in terms of an incomplete data problem by introducing the allocations as missing data. As shown by several pioneering papers (West, 1992; Smith and Roberts, 1993; Diebolt and Robert, 1994; Escobar and West, 1995; Mengersen and Robert, 1996; Raftery, 1996b), it is surprisingly straightforward to sample from the posterior density (2.36) using MCMC techniques such as data augmentation (Tanner and Wong, 1987) and Gibbs sampling (Gelfand and Smith, 1990). A very detailed account of Bayesian inference for finite mixture models is given in Chapter 3.

### 2.4.6 Distance-Based Methods

Distance-based methods consider that value $\boldsymbol{\vartheta}$ as an estimator for an unknown parameter that minimizes a suitably defined distance $\delta(\hat{F}_N, F_{\boldsymbol{\vartheta}})$ between the empirical distribution $\hat{F}_N$ and the distribution function $F_{\boldsymbol{\vartheta}}$ of the mixture distribution $p(\mathbf{y}|\boldsymbol{\vartheta})$. Titterington et al. (1985, Section 4.5) provides a comprehensive review of properties of minimum distance estimators for finite mixture models under a wide range of distance functions. The ML estimator results if $\delta(\cdot)$ is equal to the Kullback–Leibler distance (Kullback and Leibler, 1951). Another commonly used distance is the Hellinger distance (Beran, 1977) which is applied to finite mixture models by Lindsay (1994), Cutler and Cordero-Braña (1996), and, in the context of mixtures of Poisson distributions, by Karlis and Xekalaki (1998, 2001).

Further variants of distance-based methods in finite mixture modeling are penalized minimum distance estimation (Chen and Kalbfleisch, 1996) and Bayesian distance-based estimation (Mengersen and Robert, 1996; Celeux et al., 2000; Sahu and Cheng, 2003; Hurn et al., 2003).

### 2.4.7 Comparing Various Estimation Methods

For illustration, a simulation experiment is carried out in order to compare various estimation methods for different sample sizes ranging from small ($N = 20$) over medium ($N = 200$) to large ($N = 2000$). 100 data sets each consisting of $N$ observations were simulated from a two-component Poisson mixture model with $\eta_1^{\text{true}} = 0.3$, $\mu_1^{\text{true}} = 1$, and $\mu_2^{\text{true}} = 5$. The unknown parameter $\boldsymbol{\vartheta} =$

**Table 2.4.** Performance of different estimation methods, namely method of moments (MM), maximum likelihood (ML), and Bayesian estimation (Bayes), for 100 data sets simulated from a mixture of two Poisson distributions with $\eta_1^{\text{true}} = 0.3$, $\mu_1^{\text{true}} = 1$, and $\mu_2^{\text{true}} = 5$ for different sample sizes $N$

| | Bias | | | MSE | | |
|---|---|---|---|---|---|---|
| N=20 | $\mu_1$ | $\mu_2$ | $\eta_1$ | $\mu_1$ | $\mu_2$ | $\eta_1$ |
| MM | −0.85773 | −0.2835 | −0.12096 | 1.6361 | 0.39866 | 0.04378 |
| ML | −0.055929 | −0.07033 | 0.0010964 | 0.18881 | 0.29199 | 0.0085535 |
| Bayes | 0.31284 | −0.14187 | 0.082573 | 0.14742 | 0.24411 | 0.0070211 |
| | Bias | | | MSE | | |
| N=200 | $\mu_1$ | $\mu_2$ | $\eta_1$ | $\mu_1$ | $\mu_2$ | $\eta_1$ |
| MM | −0.13395 | −0.026314 | −0.026936 | 0.12128 | 0.058986 | 0.0035519 |
| ML | −0.017062 | 0.015664 | −0.0032702 | 0.035213 | 0.031482 | 0.0017613 |
| Bayes | 0.051202 | 0.068292 | 0.019925 | 0.033 | 0.035072 | 0.0014094 |
| | Bias | | | MSE | | |
| N=2000 | $\mu_1$ | $\mu_2$ | $\eta_1$ | $\mu_1$ | $\mu_2$ | $\eta_1$ |
| MM | −0.018328 | −0.017891 | −0.001898 | 0.014131 | 0.0062658 | 0.0005169 |
| ML | −0.003916 | −0.0084198 | −0.0017287 | 0.0031607 | 0.0038663 | 0.00019152 |
| Bayes | 0.0077498 | −0.0063272 | 0.00052239 | 0.0032099 | 0.0040859 | 0.00018001 |

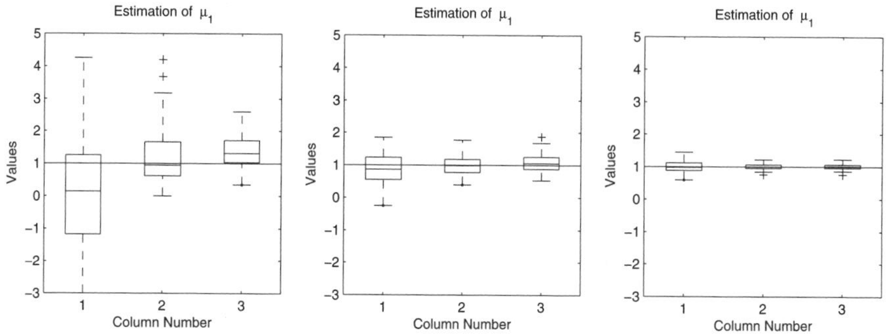


**Fig. 2.9.** Distribution of the estimator $\hat{\mu}_1$ for $N$ observations arising from a mixture of two Poisson distributions with $\eta_1^{\text{true}} = 0.3$, $\mu_1^{\text{true}} = 1$, and $\mu_2^{\text{true}} = 5$ (1 ... MM, 2 ... ML, 3 ... Bayes); $N = 20$ (left-hand side), $N = 200$ (middle), $N = 2000$ (right-hand side)

$(\mu_1, \mu_2, \eta_1)$ is estimated for each of the 100 data sets based on the method of moments, the ML estimator using the EM algorithm, and Bayesian estimation, using the priors $\mu_k \sim \mathcal{G}(0.5, 0.5/\overline{y})$ and $\boldsymbol{\eta} \sim \mathcal{D}(4, 4)$. For $N = 20$, the EM algorithm fails three times whereas the method of moments fails once. For each of the 100 data sets, Bayesian estimation is carried out using 2000 draws from a random permutation Gibbs sampler (Frühwirth-Schnatter, 2001b), after a burn-in of 500 draws. The mean of the posterior draws is used as an estimator of $\boldsymbol{\vartheta}$, after the model has been identified using unsupervised clustering. These computational issues are explained in full detail in Chapter 3.

Table 2.4 evaluates the estimation error through the average bias $\hat{\mu}_k - \mu_k^{\mathrm{true}}$ for $k = 1, 2$ and $\hat{\eta}_1 - \eta_1^{\mathrm{true}}$ and the average mean squared error (MSE) $(\hat{\mu}_k - \mu_k^{\mathrm{true}})^2$ for $k = 1, 2$ and $(\hat{\eta}_1 - \eta_1^{\mathrm{true}})^2$ over all replications. The method of moments has the largest bias and the largest MSE for all samples sizes $N$. For $N = 20$, the Bayes estimator has a larger bias, but is more efficient in terms of the MSE than the ML estimator. As expected, there is little difference between ML estimation and Bayesian estimation for $N = 2000$. For further illustration, Figure 2.9 shows the distribution of the various estimators $\hat{\mu}_1$ over all replications for the different sample sizes.

# 3

# Practical Bayesian Inference for a Finite Mixture Model with Known Number of Components

## 3.1 Introduction

Assume as in Chapter 2 that $N$ observations $\mathbf{y} = (\mathbf{y}_1, \dots, \mathbf{y}_N)$, drawn randomly from a finite mixture of $\mathcal{T}(\boldsymbol{\theta})$ distributions with density $p(\mathbf{y}|\boldsymbol{\theta})$ indexed by a parameter $\boldsymbol{\theta} \in \Theta$, are available, which should be used to make inferences about the underlying mixture structure. In this chapter we outline in detail Bayesian inference for the standard finite mixture model,

$$p(\mathbf{y}_i|\boldsymbol{\vartheta}) = \sum_{k=1}^{K} \eta_k p(\mathbf{y}_i|\boldsymbol{\theta}_k), \tag{3.1}$$

when the number of components is known.

If $\boldsymbol{\vartheta} = (\boldsymbol{\theta}_1, \dots, \boldsymbol{\theta}_K, \boldsymbol{\eta})$ are unknown parameters that need to be estimated from the data then, as noted earlier, from a Bayesian perspective all information contained in the data $\mathbf{y}$ about $\boldsymbol{\vartheta}$ is summarized in terms of the posterior density $p(\boldsymbol{\vartheta}|\mathbf{y})$, which is derived using Bayes' theorem:

$$p(\boldsymbol{\vartheta}|\mathbf{y}) \propto p(\mathbf{y}|\boldsymbol{\vartheta})p(\boldsymbol{\vartheta}). \tag{3.2}$$

By Bayes' theorem, the data-dependent mixture likelihood function $p(\mathbf{y}|\boldsymbol{\vartheta})$, defined earlier, is combined with a prior density $p(\boldsymbol{\vartheta})$ in order to obtain the mixture posterior density $p(\boldsymbol{\vartheta}|\mathbf{y})$. For Bayesian estimation, we have to assume that such a prior distribution $p(\boldsymbol{\vartheta})$ is available. For finite mixture models it is not possible to choose an improper prior such as $p(\boldsymbol{\vartheta}) \propto$ constant, because this leads to an improper mixture posterior density $p(\boldsymbol{\vartheta}|\mathbf{y})$. This problem and choosing proper priors are discussed in Section 3.2.

Within a Bayesian analysis of a finite mixture model we are interested in the entire mixture posterior density $p(\boldsymbol{\vartheta}|\mathbf{y})$, which to a large extent is dominated by the mixture likelihood function $p(\mathbf{y}|\boldsymbol{\vartheta})$, and, as discussed in Section 3.3, inherits all of its properties, in particular the invariance to relabeling the mixture components.

In Section 3.4 we discuss inference for the group indicators $\mathbf{S}$ without parameter estimation which is interesting in its own right and provides an opportunity to introduce recent concepts of computational Bayesian statistics such as Gibbs sampling and the Metropolis–Hastings algorithm. Gibbs sampling, together with data augmentation, is also useful for drawing Bayesian inference about the parameters of a mixture model (reviewed in detail in Section 3.5) and is the most commonly used approach for obtaining draws from the mixture posterior $p(\boldsymbol{\vartheta}|\mathbf{y})$; other sampling-based approaches such as the Metropolis–Hastings algorithm are briefly discussed in Section 3.6. Finally, it is discussed in Section 3.7 how draws from the mixture posterior density $p(\boldsymbol{\vartheta}|\mathbf{y})$ could be used within a Bayesian approach to obtain inference on quantities of interest such as the unknown component parameters.

## 3.2 Choosing the Prior for the Parameters of a Mixture Model

### 3.2.1 Objective and Subjective Priors

For Bayesian estimation of a finite mixture model a prior $p(\boldsymbol{\vartheta})$ has to be selected for the unknown parameters $\boldsymbol{\vartheta} = (\boldsymbol{\theta}_1, \dots, \boldsymbol{\theta}_K, \boldsymbol{\eta})$. As in Press (2003, Chapter 5), one may distinguish between objective and subjective priors.

Objective priors should reflect the notion of having no prior information, however, there exists no general agreement about how knowing little about a parameter $\boldsymbol{\vartheta}$ should be expressed in terms of a probability distribution $p(\boldsymbol{\vartheta})$. Very often improper priors, which are not integrable over the parameter space, are used to express complete ignorance, in the hope that the data are informative enough to turn the improper prior $p(\boldsymbol{\vartheta})$ into a proper posterior distribution $p(\boldsymbol{\vartheta}|\mathbf{y})$. The choice of objective priors is particularly difficult for finite mixture models, as common improper priors will lead to improper posteriors; see Subsection 3.2.2.

Subjective priors bring prior knowledge into the analysis, and offer the advantage of being proper. For finite mixture models, such priors are usually obtained by choosing priors that are conjugate for the complete-data likelihood function; see Subsection 3.2.3. It is common to assume that the parameters $\boldsymbol{\theta}_1, \dots, \boldsymbol{\theta}_K$ are independent of the weight distribution $\boldsymbol{\eta}$:

$$p(\boldsymbol{\vartheta}) = p(\boldsymbol{\theta}_1, \dots, \boldsymbol{\theta}_K) p(\boldsymbol{\eta}). \tag{3.3}$$

For finite mixture models, the standard prior for the weight distribution $\boldsymbol{\eta}$ is the $\mathcal{D}(e_0, \dots, e_0)$-distribution, which arises from the same prior distribution family as for complete-data Bayesian inference considered in Subsection 2.3.4, however, the hyperparameters of the prior are assumed to be the same, in order to obtain an invariant prior. The precise prior on the component parameters $\boldsymbol{\theta}_1, \dots, \boldsymbol{\theta}_K$ depends on the distribution family underlying the mixture distribution.

It is not always easy to assess the parameters of a subjective prior, also called hyperparameters. Results from a Bayesian analysis of finite mixture models using subjective prior information is often highly dependent on particular choices of hyperparameters. To reduce this sensitivity, it is common practice in the context of finite mixture modeling to use hierarchical priors where the hyperparameter is equipped with a prior of its own; see Subsections 3.2.4.

In any case, for a Bayesian analysis of finite mixture models the prior distribution has to be selected with some care.

### 3.2.2 Improper Priors May Cause Improper Mixture Posteriors

Assume that in (3.3), complete ignorance about $\boldsymbol{\theta}_1, \ldots, \boldsymbol{\theta}_K$ is expressed in terms of the product of independent improper priors:

$$p(\boldsymbol{\theta}_1, \ldots, \boldsymbol{\theta}_K) \propto \prod_{k=1}^{K} p^\star(\boldsymbol{\theta}_k), \tag{3.4}$$

with $\int p^\star(\boldsymbol{\theta}_k) d\boldsymbol{\theta}_k = \infty$. Roeder and Wasserman (1997b) show that the mixture posterior $p(\boldsymbol{\vartheta}|\mathbf{y})$ is improper under prior (3.4), by rewriting the mixture likelihood $p(\mathbf{y}|\boldsymbol{\vartheta})$ as a sum over complete-data likelihoods:

$$p(\mathbf{y}|\boldsymbol{\vartheta}) = \sum_{\mathbf{S} \in \mathcal{S}_K} p(\mathbf{y}|\mathbf{S}, \boldsymbol{\theta}_1, \ldots, \boldsymbol{\theta}_K) p(\mathbf{S}|\boldsymbol{\eta}), \tag{3.5}$$

where summation runs over all $K^N$ possible classifications $\mathbf{S}$. Under prior (3.4), the mixture posterior is proportional to

$$p(\boldsymbol{\vartheta}|\mathbf{y}) \propto \sum_{\mathbf{S} \in \mathcal{S}_K} p(\mathbf{y}|\mathbf{S}, \boldsymbol{\theta}_1, \ldots, \boldsymbol{\theta}_K) \prod_{k=1}^{K} p^\star(\boldsymbol{\theta}_k) p(\mathbf{S}|\boldsymbol{\eta}) p(\boldsymbol{\eta}), \tag{3.6}$$

and is proper, if the integral over the right-hand side is finite. The normalizing constant turns out to be

$$\sum_{\mathbf{S} \in \mathcal{S}_K} c_1(\mathbf{S}) c_2(\mathbf{S}), \tag{3.7}$$

$$c_1(\mathbf{S}) = \prod_{k=1}^{K} \int \left( \prod_{i:S_i=k} p(\mathbf{y}_i|\boldsymbol{\theta}_k) \right) p^\star(\boldsymbol{\theta}_k) d\boldsymbol{\theta}_k,$$

$$c_2(\mathbf{S}) = \int p(\mathbf{S}|\boldsymbol{\eta}) p(\boldsymbol{\eta}) d\boldsymbol{\eta}.$$

To obtain a proper posterior distribution, $c_1(\mathbf{S})$ and $c_2(\mathbf{S})$ have to be finite for *all* classifications $\mathbf{S}$. Note that the hidden multinomial prior on $\mathbf{S}$ assigns positive probability to partitions $\mathbf{S}$, where one component, say $j$, is empty.

In this case, the complete-data likelihood does not contain any information about $\boldsymbol{\theta}_j$ and $c_1(\mathbf{S})$ is not finite under the improper prior (3.4) because

$$\int \left( \prod_{i:S_i=j} p(\mathbf{y}_i|\boldsymbol{\theta}_j) \right) p^\star(\boldsymbol{\theta}_j) d\boldsymbol{\theta}_j = \int p^\star(\boldsymbol{\theta}_j) d\boldsymbol{\theta}_j = \infty.$$

To obtain proper posterior distributions under the prior (3.4), Wasserman (2000) modifies the prior distribution $p(\mathbf{S})$ of the allocations $\mathbf{S}$, by restricting $\mathcal{S}_K$ to allocations with nonempty components.

### 3.2.3 Conditionally Conjugate Priors

Whereas it is not possible to choose simple conjugate priors for the mixture likelihood $p(\mathbf{y}|\boldsymbol{\vartheta})$, a conjugate analysis is possible for the complete-data likelihood $p(\mathbf{y}, \mathbf{S}|\boldsymbol{\vartheta})$, if the component densities in the mixture come from the exponential family as in (1.11),

$$p(\mathbf{y}_i|\boldsymbol{\theta}_k) = \exp\left\{ \phi(\boldsymbol{\theta}_k)' u(\mathbf{y}_i) - g(\boldsymbol{\theta}_k) + c(\mathbf{y}_i) \right\};$$

see also Subsection 2.3.3. To formulate a joint prior for $\boldsymbol{\theta}_1, \ldots, \boldsymbol{\theta}_K$, the component parameters are assumed to be independent a priori, given a hyperparameter $\boldsymbol{\delta}$:

$$p(\boldsymbol{\theta}_1, \ldots, \boldsymbol{\theta}_K|\boldsymbol{\delta}) = \prod_{k=1}^{K} p(\boldsymbol{\theta}_k|\boldsymbol{\delta}). \tag{3.8}$$

If for each component the prior $p(\boldsymbol{\theta}_k|\boldsymbol{\delta})$ takes the form

$$p(\boldsymbol{\theta}_k|\boldsymbol{\delta}) \propto \exp\left\{ \phi(\boldsymbol{\theta}_k)' a_0 - g(\boldsymbol{\theta}_k) b_0 \right\}, \tag{3.9}$$

with hyperparameter $\boldsymbol{\delta} = (a_0, b_0)$, then the conditional posterior $p(\boldsymbol{\theta}_k|\mathbf{S}, \mathbf{y})$ is given by:

$$p(\boldsymbol{\theta}_k|\mathbf{S}, \mathbf{y}) \propto \exp\left\{ \phi(\boldsymbol{\theta}_k)' a_k - g(\boldsymbol{\theta}_k) b_k \right\}, \tag{3.10}$$

which is again a density from the chosen exponential family with

$$a_k = a_0 + \sum_{i:S_i=k} u(\mathbf{y}_i), \qquad b_k = b_0 + N_k(\mathbf{S}),$$

where $N_k(\mathbf{S}) = \#\{S_i = k\}$. For mixtures of Poisson distributions, for instance, Bayesian inference for the complete data problem, already studied in Subsection 2.3.3, leads to the conditionally conjugate prior $\mu_k \sim \mathcal{G}(a_0, b_0)$, where $a_0$ as well as $b_0$ have to be positive to obtain a proper posterior distribution.

### 3.2.4 Hierarchical Priors and Partially Proper Priors

For practical Bayesian inference, prior (3.8) is assessed by choosing the hyperparameter $\boldsymbol{\delta}$. Prior (3.8) acts as a kind of shrinkage prior, pulling all component parameters $\boldsymbol{\theta}_k$ toward a common center defined by $\mathrm{E}(\boldsymbol{\theta}_k|\boldsymbol{\delta})$, where both the center of the prior as well as the amount of shrinkage may crucially depend on $\boldsymbol{\delta}$. For illustration, consider a mixture of Poisson distributions, and rewrite the conditionally conjugate $\mathcal{G}(a_0, b_0)$-prior introduced in Subsection 3.2.3 as $\mu_k \sim (a_0/b_0)W_k$, $W_k \sim \mathcal{G}(a_0, a_0)$. Evidently, this prior induces shrinkage of $\mu_k$ toward the prior mean $\mathrm{E}(\mu_k) = a_0/b_0,$, with shrinkage being more pronounced the larger $a_0$.

In particular for mixtures with small components, the posterior distribution may be sensitive to specific choices of $\boldsymbol{\delta}$. To reduce sensitivity to specific choices of $\boldsymbol{\delta}$, it is common practice to use hierarchical priors, which treat $\boldsymbol{\delta}$ as an unknown quantity with a prior $p(\boldsymbol{\delta})$:

$$p(\boldsymbol{\theta}_1, \ldots, \boldsymbol{\theta}_K, \boldsymbol{\delta}) = p(\boldsymbol{\delta}) \prod_{k=1}^{K} p(\boldsymbol{\theta}_k|\boldsymbol{\delta}). \tag{3.11}$$

As a result, $\boldsymbol{\theta}_1, \ldots, \boldsymbol{\theta}_K$ are dependent a priori:

$$p(\boldsymbol{\theta}_1, \ldots, \boldsymbol{\theta}_K) = \int p(\boldsymbol{\theta}_1, \ldots, \boldsymbol{\theta}_K|\boldsymbol{\delta})p(\boldsymbol{\delta})d\boldsymbol{\delta} \neq \prod_{k=1}^{K} p(\boldsymbol{\theta}_k).$$

Such priors have been applied to finite mixtures of normal distributions in Mengersen and Robert (1996), Richardson and Green (1997), and Roeder and Wasserman (1997b).

Partially proper priors (Roeder and Wasserman, 1997b) are hierarchical priors where the prior $p(\boldsymbol{\delta})$ of the hyperparameter $\boldsymbol{\delta}$ is improper. Although, marginally, the prior $p(\boldsymbol{\theta}_k)$ is improper, the posterior distribution is proper.

#### A Hierarchical Prior for Poisson Mixtures

For a mixture of Poisson distributions, a hierarchical prior is obtained by assuming that $b_0$ is a random parameter with a prior of its own:

$$\mu_k|b_0 \sim \mathcal{G}(a_0, b_0), \qquad b_0 \sim \mathcal{G}(g_0, G_0). \tag{3.12}$$

Then the component means $\mu_1, \ldots, \mu_K$ are dependent a priori, and the joint prior $p(\mu_1, \ldots, \mu_K)$, where $b_0$ is integrated out, is available in closed form, if the $\mathcal{G}(g_0, G_0)$-prior is proper:

$$p(\mu_1, \ldots, \mu_K) = \int p(\mu_1, \ldots, \mu_K | b_0) p(b_0) db_0 \tag{3.13}$$
$$= \frac{G_0^{g_0} \Gamma(g_0 + K a_0) \left( \prod_{k=1}^{K} \mu_k \right)^{a_0 - 1}}{\Gamma(a_0)^K \Gamma(g_0) \left( G_0 + \sum_{k=1}^{K} \mu_k \right)^{g_0 + K a_0}}.$$

A partially proper prior results if the $\mathcal{G}(g_0, G_0)$-prior is improper, for example, if $g_0 = 0.5$ and $G_0 = 0$.

### 3.2.5 Other Priors

Reference priors were suggested by Bernardo (1979) as prior distributions having a minimal effect on the final inference, relative to the data. The derivation of such a reference prior, however, is less than obvious for mixture models. Reference priors depend on the asymptotic behavior of the relevant posterior distributions. Although several papers have established the limiting properties of maximum likelihood estimators in finite mixture models (see Subsection 2.4.4), the derivation of reference priors for general finite mixture models still seems infeasible.

Some investigations appear in Bernardo and Girón (1988) for a mixture model where only the weight distribution $\boldsymbol{\eta}$ is unknown. For a mixture of two known densities, the reference prior for $\eta_1$ is virtually Jeffrey's $\mathcal{B}\left(\frac{1}{2}, \frac{1}{2}\right)$-prior, when the two densities are well separated, whereas the uniform $\mathcal{B}(1, 1)$ would approximate the reference prior when the two densities are very close. For a mixture of more than two known densities Bernardo and Girón (1988) suggest that a Dirichlet distribution with parameters ranging in the interval $[\frac{1}{2}, 1]$ is a reasonable approximation to the reference prior.

### 3.2.6 Invariant Prior Distributions

Because the components in a mixture density may be arbitrarily arranged, it is usual to choose priors that reflect this information, by being invariant to relabeling the components. Consider all $s = 1, \ldots, K!$ different permutations $\rho_s$:$\{1, \ldots, K\} \rightarrow \{1, \ldots, K\}$, where the value $\rho_s(k)$ is assigned to each value $k \in \{1, \ldots, K\}$. Let $\boldsymbol{\vartheta} = (\boldsymbol{\theta}_1, \ldots, \boldsymbol{\theta}_K, \eta_1, \ldots, \eta_K)$ be an arbitrary parameter in $\Theta_K = \Theta^K \times \mathcal{E}_K$, and define for each permutation $\rho_s$ the parameter $\tilde{\boldsymbol{\vartheta}}_s$ by

$$\tilde{\boldsymbol{\vartheta}}_s = (\boldsymbol{\theta}_{\rho_s(1)}, \ldots, \boldsymbol{\theta}_{\rho_s(K)}, \eta_{\rho_s(1)}, \ldots, \eta_{\rho_s(K)}). \tag{3.14}$$

A prior density $p(\boldsymbol{\vartheta})$ is invariant to relabeling the components of the mixture model, if the following identity holds for all $\boldsymbol{\vartheta} \in \Theta_K$, for any of the $K!$ permutations $\rho_s(\cdot)$;

$$p(\tilde{\boldsymbol{\vartheta}}_s) = p(\boldsymbol{\vartheta}). \tag{3.15}$$

Any of the prior distributions discussed so far in this section is invariant by construction.

Nonsymmetric priors have been applied in the hope that this eliminates all modes of the mixture likelihood function but one and Bayesian inference leads to a unimodal posterior distribution. Because this is not necessarily the case (see, for instance, Chib, 1995), the recommendation is to use an invariant prior unless there is a structural asymmetry in the mixture distribution. One example is Bayesian outlier modeling based on finite mixture, where it is sensible to choose priors that are not invariant, because the outlier component is much smaller than the other components by definition; see Section 7.2 for more detail.

## 3.3 Some Properties of the Mixture Posterior Density

### 3.3.1 Invariance of the Posterior Distribution

The mixture posterior density $p(\boldsymbol{\vartheta}|\mathbf{y})$ defined in (3.2) is to a large extent dominated by the mixture likelihood function $p(\mathbf{y}|\boldsymbol{\vartheta})$, which is invariant to relabeling the components of the mixture distribution. Under an invariant prior, the mixture posterior distribution inherits the invariance of the mixture likelihood to relabeling the components of the mixture, and the following identity holds for all $\boldsymbol{\vartheta} \in \Theta_K$, for any of the $K!$ permutations $\rho_s(\cdot)$;

$$p(\tilde{\boldsymbol{\vartheta}}_s|\mathbf{y}) = p(\boldsymbol{\vartheta}|\mathbf{y}). \tag{3.16}$$

It is quite illuminating to study the behavior of the posterior density as $N$ increases. The following considerations are purely heuristic, without providing a formal proof.

Let $\boldsymbol{\vartheta}^{\text{true}} = (\boldsymbol{\theta}_1^{\text{true}}, \dots, \boldsymbol{\theta}_K^{\text{true}}, \eta_1^{\text{true}}, \dots, \eta_K^{\text{true}})$ denote the true value of $\boldsymbol{\vartheta}$. Assume that $\boldsymbol{\vartheta}^{\text{true}}$ fulfills the formal identifiability constraints of Subsection 1.3.3, with $\eta_k^{\text{true}} > 0$, and $\boldsymbol{\theta}_k^{\text{true}} \neq \boldsymbol{\theta}_l^{\text{true}}$, for all $k \neq l$, where in a multiparameter setting not all components of all parameters need to be different. Let $\mathcal{U}^P(\boldsymbol{\vartheta}^{\text{true}})$ be the set defined in (1.27). Due to the formal identifiability constraints the mixture model is not overfitting and the set $\mathcal{U}^P(\boldsymbol{\vartheta}^{\text{true}})$ contains $K!$ distinct points, obtained from relabeling all components of $\boldsymbol{\vartheta}^{\text{true}}$ through all possible permutations of $\{1, \dots, K\}$.

Then with increasing number of observations, the posterior density has $K!$ equivalent modes and becomes proportional to an invariant mixture of asymptotic normal distributions, with the modes lying in the set $\mathcal{U}^P(\boldsymbol{\vartheta}^{\text{true}})$:

$$p(\boldsymbol{\vartheta}|\mathbf{y}) \approx \frac{1}{K!} \sum_{s=1}^{K!} f_N(\tilde{\boldsymbol{\vartheta}}_s; \boldsymbol{\vartheta}^{\text{true}}, \mathcal{I}(\boldsymbol{\vartheta}^{\text{true}})).$$

### 3.3.2 Invariance of Seemingly Component-Specific Functionals

The invariance property of the mixture posterior density $p(\boldsymbol{\vartheta}|\mathbf{y})$, discussed in the previous subsection, causes state independence of many functionals derived from the posterior distribution, which are seemingly component specific, like the posterior mean $\mathrm{E}(\boldsymbol{\theta}_k|\mathbf{y})$.

#### Marginal Distributions of Component-Specific Parameters

Consider, as an example the marginal distribution of the component parameter $\boldsymbol{\theta}_k$, which is defined in the usual way as

$$p(\boldsymbol{\theta}_k|\mathbf{y}) = \int_{\Theta^{K-1}\times\mathcal{E}_K} p(\boldsymbol{\vartheta}|\mathbf{y})d(\boldsymbol{\theta}_1,\ldots,\boldsymbol{\theta}_{k-1},\boldsymbol{\theta}_{k+1},\ldots,\boldsymbol{\theta}_K,\eta_1,\ldots,\eta_K).$$

Consider an arbitrary permutation $\rho_s(1),\ldots,\rho_s(K)$ of $\{1,\ldots,K\}$, which is different from the identity, to transform the parameter in this integration. The Jacobian of the transformation being 1, the area of integration being unchanged, one obtains:

$$p(\boldsymbol{\theta}_k|\mathbf{y}) = \int_{\Theta^{K-1}\times\mathcal{E}_K} p(\tilde{\boldsymbol{\vartheta}}_s|\mathbf{y}) \\ d(\boldsymbol{\theta}_{\rho_s(1)},\ldots,\boldsymbol{\theta}_{\rho_s(k-1)},\boldsymbol{\theta}_{\rho_s(k+1)},\ldots,\boldsymbol{\theta}_{\rho_s(K)},\eta_{\rho_s(1)},\ldots,\eta_{\rho_s(K)}).$$

By the invariance property (3.16) this is equal to:

$$p(\boldsymbol{\theta}_k|\mathbf{y}) = \int_{\Theta^{K-1}\times\mathcal{E}_K} p(\boldsymbol{\vartheta}|\mathbf{y}) \\ d(\boldsymbol{\theta}_{\rho_s(1)},\ldots,\boldsymbol{\theta}_{\rho_s(k-1)},\boldsymbol{\theta}_{\rho_s(k+1)},\ldots,\boldsymbol{\theta}_{\rho_s(K)},\eta_{\rho_s(1)},\ldots,\eta_{\rho_s(K)}).$$

Marginalization is with respect to all unknown parameters except $\boldsymbol{\theta}_{\rho_s(k)}$, therefore $p(\boldsymbol{\theta}_k|\mathbf{y}) = p(\boldsymbol{\theta}_{\rho_s(k)}|\mathbf{y})$. Because this holds all permutations $s = 1,\ldots,K!$, the seemingly component-specific marginal posterior densities $p(\boldsymbol{\theta}_k|\mathbf{y})$ are actually state-independent and the same for all $k \neq k'$:

$$p(\boldsymbol{\theta}_k|\mathbf{y}) = p(\boldsymbol{\theta}_{k'}|\mathbf{y}). \tag{3.17}$$

It could be proven in a similar way that the marginal posterior density of the component weight $\eta_k$ is state-independent:

$$p(\eta_k|\mathbf{y}) = p(\eta_{k'}|\mathbf{y}), \tag{3.18}$$

for all $k \neq k'$. State independence holds for other marginal densities, such as the marginal distribution of any two parameters from different components where $k \neq k'$ and $\rho_s$ arbitrary:

$$p(\boldsymbol{\theta}_k,\boldsymbol{\theta}_{k'}|\mathbf{y}) = p(\boldsymbol{\theta}_{\rho_s(k)},\boldsymbol{\theta}_{\rho_s(k')}|\mathbf{y}). \tag{3.19}$$

As this relation holds in particular for $\rho_s(k) = k'$ and $\rho_s(k') = k$, the posterior in (3.19) is symmetric:

$$p(\boldsymbol{\theta}_k, \boldsymbol{\theta}_{k'}|\mathbf{y}) = p(\boldsymbol{\theta}_{k'}, \boldsymbol{\theta}_k|\mathbf{y}). \tag{3.20}$$

For a mixture of univariate normal distributions, for instance, we obtain $\forall k, k' = 1, \dots, K, k \neq k'$:

$$\begin{aligned}
&p(\mu_k|\mathbf{y}) = p(\mu_{k'}|\mathbf{y}), \qquad p(\sigma_k^2|\mathbf{y}) = p(\sigma_{k'}^2|\mathbf{y}),\\
&p(\mu_k, \sigma_k^2|\mathbf{y}) = p(\mu_{k'}, \sigma_{k'}^2|\mathbf{y}),\\
&p(\mu_k, \mu_{k'}|\mathbf{y}) = p(\mu_1, \mu_2|\mathbf{y}) = p(\mu_2, \mu_1|\mathbf{y}),\\
&p(\sigma_k^2, \sigma_{k'}^2|\mathbf{y}) = p(\sigma_1^2, \sigma_2^2|\mathbf{y}) = p(\sigma_2^2, \sigma_1^2|\mathbf{y}).
\end{aligned}$$

#### The Posterior Mean

The posterior mean is a commonly used point estimator, which is optimal with respect to a quadratic loss function; see, for instance, Zellner (1971) and Berger (1985). From the mixture posterior distribution, the following result may be derived,

$$\mathrm{E}(\tilde{\boldsymbol{\vartheta}}_s|\mathbf{y}) = \mathrm{E}(\boldsymbol{\vartheta}|\mathbf{y}), \tag{3.21}$$

where the parameter $\tilde{\boldsymbol{\vartheta}}_s$ has been defined for each permutation $\rho_s$ in (3.14). Identity (3.21) follows from the invariance property (3.16).

As (3.21) holds for all permutations, it follows that the seemingly component-specific posterior mean of $\boldsymbol{\theta}_k$ and $\eta_k$ is actually state-independent:

$$\mathrm{E}(\boldsymbol{\theta}_k|\mathbf{y}) = \mathrm{E}(\boldsymbol{\theta}_{k'}|\mathbf{y}), \qquad \mathrm{E}(\eta_k|\mathbf{y}) = \mathrm{E}(\eta_{k'}|\mathbf{y}),$$

for any $k \neq k'$. Consequently, the mean $\mathrm{E}(\boldsymbol{\vartheta}|\mathbf{y})$ of the mixture posterior is not a sensible point estimator for the component parameters and the weight distribution. More sensible point estimators are discussed in Subsection 3.7.6.

### 3.3.3 The Marginal Posterior Distribution of the Allocations

We now turn to the posterior density $p(\mathbf{S}|\mathbf{y})$ of the allocations $\mathbf{S}$, which is of importance when dealing with Bayesian clustering in Section 7.1. $p(\mathbf{S}|\mathbf{y})$ is a discrete distribution over the lattice

$$\mathcal{S}_K = \{(S_1, \dots, S_N) : S_i \in \{1, \dots, K\}, i = 1, \dots, N\}. \tag{3.22}$$

As noted by Chen and Liu (1996) and Casella et al. (2000), for many mixture models it is possible to derive an explicit form for the marginal posterior $p(\mathbf{S}|\mathbf{y})$ of the indicators $\mathbf{S}$, where dependence on the parameter $\boldsymbol{\vartheta}$ is integrated out. By Bayes' theorem, the marginal posterior $p(\mathbf{S}|\mathbf{y})$ is given by

$$p(\mathbf{S}|\mathbf{y}) \propto p(\mathbf{y}|\mathbf{S})p(\mathbf{S}), \tag{3.23}$$

where the integrated likelihood $p(\mathbf{y}|\mathbf{S})$ and the integrated prior $p(\mathbf{S})$ are equal to

$$p(\mathbf{y}|\mathbf{S}) = \int p(\mathbf{y}|\mathbf{S}, \boldsymbol{\theta}_1, \ldots, \boldsymbol{\theta}_K)p(\boldsymbol{\theta}_1, \ldots, \boldsymbol{\theta}_K)d(\boldsymbol{\theta}_1, \ldots, \boldsymbol{\theta}_K),$$
$$p(\mathbf{S}) = \int p(\mathbf{S}|\boldsymbol{\eta})p(\boldsymbol{\eta})d\boldsymbol{\eta}.$$

Assume that the prior $p(\boldsymbol{\vartheta})$ takes exactly the same form as (3.3) and (3.8). Then:

$$p(\mathbf{y}|\mathbf{S}) = \prod_{k=1}^{K} \int \prod_{i:S_i=k} p(\mathbf{y}_i|\boldsymbol{\theta}_k)p(\boldsymbol{\theta}_k)d\boldsymbol{\theta}_k,$$
$$p(\mathbf{S}) = \int p(\mathbf{S}|\boldsymbol{\eta})p(\boldsymbol{\eta})d\boldsymbol{\eta}.$$

Under the conditionally conjugate prior $\boldsymbol{\eta} \sim \mathcal{D}(e_0, \ldots, e_0)$ we obtain:

$$p(\mathbf{S}) = \frac{\Gamma(Ke_0)\prod_{k=1}^{K}\Gamma(N_k(\mathbf{S}) + e_0)}{\Gamma(N + Ke_0)\Gamma(e_0)^K}, \tag{3.24}$$

where $N_k(\mathbf{S}) = \#\{S_i = k\}$. If the component densities in the mixture come from the exponential family as in (1.11), then under a conditionally conjugate prior $p(\boldsymbol{\theta}_k)$, the integrated likelihood $p(\mathbf{y}|\mathbf{S})$ is the product of the normalizing constants of each nonnormalized complete-data posterior, which are easily derived from (2.19):

$$p(\mathbf{y}|\mathbf{S}) = \prod_{k=1}^{K} \left( \frac{p(\boldsymbol{\theta}_k)}{p(\boldsymbol{\theta}_k|\mathbf{y}, \mathbf{S})} \prod_{i:S_i=k} p(\mathbf{y}_i|\boldsymbol{\theta}_k) \right). \tag{3.25}$$

For a mixture of Poisson distributions, for instance, this yields:

$$p(\mathbf{y}|\mathbf{S}) = \prod_{i=1}^{N} \frac{1}{\Gamma(y_i + 1)} \frac{b_0^{Ka_0}}{\Gamma(a_{0,k})^K} \prod_{k=1}^{K} \frac{\Gamma(a_k(\mathbf{S}))}{b_k(\mathbf{S})^{a_k(\mathbf{S})}},$$

where $a_k(\mathbf{S})$ and $b_k(\mathbf{S})$ are the posterior moments of the complete-data posterior densities given in (2.18):

$$a_k(\mathbf{S}) = a_0 + N_k(\mathbf{S})\overline{y}_k(\mathbf{S}),$$
$$b_k(\mathbf{S}) = b_0 + N_k(\mathbf{S}).$$

### 3.3.4 Invariance of the Posterior Distribution of the Allocations

State invariance occurs also for the seemingly component dependent allocations $\mathbf{S}$. $p(\mathbf{S}|\mathbf{y})$ is a marginal density obtained from integrating the joint posterior $p(\mathbf{S}, \boldsymbol{\vartheta}|\mathbf{y})$ with respect to $\boldsymbol{\vartheta}$:

$$p(\mathbf{S}|\mathbf{y}) = \int_{\Theta_K} p(\mathbf{S}, \boldsymbol{\vartheta}|\mathbf{y}) d\boldsymbol{\vartheta}.$$

Because this holds for any $\mathbf{S}$, it also holds for $\tilde{\mathbf{S}}_s = (\rho_s(S_1), \ldots, \rho_s(S_N))$ for an arbitrary permutation. When using the same permutation for transforming the parameter $\boldsymbol{\vartheta}$ in this integration, we obtain:

$$p(\tilde{\mathbf{S}}_s|\mathbf{y}) = \int_{\Theta_K} p(\tilde{\mathbf{S}}_s, \tilde{\boldsymbol{\vartheta}}_s|\mathbf{y}) d\tilde{\boldsymbol{\vartheta}}_s = \int_{\Theta_K} p(\mathbf{S}, \boldsymbol{\vartheta}|\mathbf{y}) d\tilde{\boldsymbol{\vartheta}}_s = p(\mathbf{S}|\mathbf{y}),$$

because the joint posterior is invariant to relabeling, and the order of integration may be rearranged arbitrarily. Therefore, for an arbitrary permutation $\rho_s(\cdot)$ of $\{1, \ldots, K\}$, the posterior density $p(\mathbf{S}|\mathbf{y})$ is invariant to relabeling:

$$p(S_1, \ldots, S_N|\mathbf{y}) = p(\rho_s(S_1), \ldots, \rho_s(S_N)|\mathbf{y}). \tag{3.26}$$

It follows that any two sequences $\mathbf{S}$ and $\mathbf{S}'$ that imply the same partition of the data obtain the same posterior probability. Consider, as a simple example, $N = 3$ and $K = 2$; then there are only four different partitions, each of which has the same posterior probability:

$$\begin{aligned} p(1,1,1|\mathbf{y}) &= p(2,2,2|\mathbf{y}), \qquad & p(2,1,1|\mathbf{y}) &= p(1,2,2|\mathbf{y}), \\ p(1,2,1|\mathbf{y}) &= p(2,1,2|\mathbf{y}), \qquad & p(1,1,2|\mathbf{y}) &= p(2,2,1|\mathbf{y}). \end{aligned}$$

#### The Marginal Posterior of a Single Allocation

When a finite mixture model is fitted to data with the aim of performing posterior clustering, one would hope to infer how likely the event $\{S_i = k\}$ is in light of the data. A natural candidate appears to be the posterior probability $\Pr(S_i = k|\mathbf{y})$. Somewhat surprisingly, it turns out that this marginal posterior probability is state-independent and equal to $1/K$, regardless of the data:

$$\Pr(S_i = k|\mathbf{y}) = \frac{1}{K}. \tag{3.27}$$

This follows from (3.26), by integrating both sides with respect to the indicators $(S_1, \ldots, S_{i-1}, S_{i+1}, \ldots, S_N)$, which yields that the seemingly component-specific posterior probability $\Pr(S_i = k|\mathbf{y})$ is actually state invariant:

$$\Pr(S_i = k|\mathbf{y}) = \Pr(\rho_s(S_i) = k|\mathbf{y}) = \Pr(S_i = \rho_s^{-1}(k)|\mathbf{y}).$$

As this holds for all permutations, (3.27) follows immediately.

## 3.4 Classification Without Parameter Estimation

One of the most the challenging inference problems in finite mixture modeling, commonly known as the clustering problem, is classifying observations from a mixture distribution into $K$ groups without knowing the component parameters. This interesting issue is studied in detail in Section 7.1; some aspects, however, are addressed at this point because they provide a good opportunity to introduce two important MCMC technique, namely Gibbs sampling and the Metropolis–Hastings algorithm, which are of relevance not only for classification, but also for Bayesian parameter estimation.

Bayesian clustering without parameter estimation is based on the marginal posterior distribution $p(\mathbf{S}|\mathbf{y})$ of the hidden allocation vector $\mathbf{S}$, where the mixture parameter $\boldsymbol{\vartheta}$ is integrated out, which is known up to a normalizing constant explicitly for mixtures from the exponential family; see again Subsection 3.3.3. $p(\mathbf{S}|\mathbf{y})$ is a discrete distribution over the lattice $\mathcal{S}_K$, defined in (3.22), which increases rapidly with the number of observations and the number of components. For $N = 10$ and $K = 3$, for instance, there are 59,049 different allocations $\mathbf{S}$, whereas for $N = 100$ and $K = 3$ the number of different allocations is of the order $5 \cdot 10^{47}$. For a very small data set from a mixture with very few components it would be possible to determine $p(\mathbf{S}|\mathbf{y})$ for all $K^N$ possible allocations, and to find the allocation with the highest posterior probability $p(\mathbf{S}|\mathbf{y})$. With increasing sample size and increasing number of components, however, this is infeasible, and some search strategy has to be implemented to find an optimal allocation. Exploring the space $\mathcal{S}_K$, however, is in general quite a challenge.

Common search strategies that are applied in a Bayesian context are based on sampling allocations $\mathbf{S}^{(1)}, \ldots, \mathbf{S}^{(M)}$ from the marginal posterior distribution $p(\mathbf{S}|\mathbf{y})$, which are then used for further inference, as explained in Subsection 7.1.7. Direct sampling of $\mathbf{S}$ from $p(\mathbf{S}|\mathbf{y})$ is not simple, as unconditionally the allocations $S_1, \ldots, S_N$ are correlated. Chen and Liu (1996) showed how sampling of the allocation through Markov chain Monte Carlo methods is feasible. An MCMC sampler starts from some preliminary classification $\mathbf{S}^{(0)}$. During sweep $m$, $m \geq 1$, of the MCMC sampler, the allocation $S_i$ of each observation $\mathbf{y}_i$ is resampled in an appropriate manner, and the updated allocations are then stored as $\mathbf{S}^{(m)}$. Two common methods to implement an MCMC sampler are single-move Gibbs sampling and the Metropolis–Hastings algorithm. Both methods are described in Subsection 3.4.1 and Subsection 3.4.2, respectively.

For a detailed account we refer to the relevant literature on Markov chain Monte Carlo methods, in particular Gamerman (1997), Liu (2001), and Robert and Casella (1999).

### 3.4.1 Single-Move Gibbs Sampling

In this subsection we briefly introduce Gibbs sampling in the context of classification without parameter estimation, following Chen and Liu (1996) who used single-move Gibbs sampling to sample allocations $\mathbf{S}$ from the posterior distribution $p(\mathbf{S}|\mathbf{y})$ given in Subsection 3.3.3.

The single-move Gibbs sampler starts from some preliminary classification $\mathbf{S}^{(0)}$. Within each sweep $m, m \geq 1$, of the Gibbs sampler, the old allocations $\mathbf{S} = \mathbf{S}^{(m-1)}$ are updated for each observation $\mathbf{y}_i$, for $i = 1, \ldots, N$. Starting with $i = 1$, a new classification $S_i^{new}$ is sampled, while holding the classifications $\mathbf{S}_{-i} = (S_1^{new}, \ldots, S_{i-1}^{new}, S_{i+1}, \ldots, S_N)$ of all other observations fixed. As not only $\mathbf{y}$, but also $\mathbf{S}_{-i}$ are assumed to be known, the appropriate posterior distribution for sampling $S_i^{new}$ is the conditional posterior distribution $p(S_i^{new}|\mathbf{S}_{-i}, \mathbf{y})$. Well-known properties of conditional distributions yield:

$$p(S_i^{new}|\mathbf{S}_{-i}, \mathbf{y}) = \frac{p(S_i^{new}, \mathbf{S}_{-i}|\mathbf{y})}{p(\mathbf{S}_{-i}|\mathbf{y})} \propto p(\mathbf{y}|S_i^{new}, \mathbf{S}_{-i})p(S_i^{new}, \mathbf{S}_{-i})$$
$$\propto p(\mathbf{y}_i|S_i^{new}, \mathbf{S}_{-i})p(S_i^{new}|\mathbf{S}_{-i}),$$

where constants independent of $S_i^{new}$ were dropped. This is a univariate discrete density with $K$ categories, which is easily sampled. Once $S_i^{new}$ has been simulated, the Gibbs sampler proceeds with sampling the next indicator $S_i^{new}$ after increasing $i$ by 1, until $i = N$. Then the new allocations are stored as $\mathbf{S}^{(m)} = (S_1^{new}, \ldots, S_N^{new})$, $m$ is increased by 1, and the whole procedure is repeated.

This sampling algorithm generates a sequence $\mathbf{S}^{(m)}, m = 1, 2, \ldots$ of classifications, which are obviously a Markov chain, as the distribution of $\mathbf{S}^{(m)}$ depends on $\mathbf{S}^{(m-1)}$, only:

$$p(\mathbf{S}^{(m)}|\mathbf{S}^{(m-1)}, \mathbf{y}) = \prod_{i=1}^{N} p(S_i^{(m)}|\mathbf{S}_{1:i-1}^{(m)}, \mathbf{S}_{i+1:N}^{(m-1)}, \mathbf{y}),$$

where $\mathbf{S}_{i:j}$ denotes the whole sequence $S_i, S_{i+1}, \ldots, S_j$. Well-known results from Markov chain theory guarantee that in the long run, as $m \to \infty$, the distribution of $\mathbf{S}^{(m)}$ converges to a stationary distribution, which could be shown to be equal to the desired marginal posterior $p(\mathbf{S}|\mathbf{y})$. When starting from an arbitrary allocation, the Markov chain will not be in equilibrium at the beginning, but will reach the stationary distribution after a suitable burn-in phase. Thus the first $M_0$ simulations are discarded before the simulated allocations may be used for posterior inference.

*Algorithm 3.1: Single-Move Gibbs Sampling of the Allocations* Start with some classification $\mathbf{S}$ and repeat the following steps for $m = 1, \ldots, M_0, \ldots, M+M_0$.

(a) Choose a certain observation $\mathbf{y}_i$, $i \in \{1, \ldots, N\}$, hold the most recent allocation of all observations but $\mathbf{y}_i$ fixed, and let $\mathbf{S}_{-i}$ be the sequence containing these allocations.

(b) Find a new allocation $S_i^{new}$ for the observation $\mathbf{y}_i$ in the following way. Determine the univariate discrete distribution

$$p(S_i^{new}|\mathbf{S}_{-i}, \mathbf{y}) \propto p(\mathbf{y}_i|S_i^{new}, \mathbf{S}_{-i})p(S_i^{new}|\mathbf{S}_{-i}), \tag{3.28}$$

for all possible values $S_i^{new} = 1, \ldots, K$. Sample $S_i^{new}$ from this distribution, and substitute the old allocation $S_i$ by the new allocation $S_i^{new}$.

Repeat these steps until the allocations of all observations are updated. Store the actual values of all allocations as $\mathbf{S}^{(m)}$, increase $m$ by one, and return to step (a).

Assume the current allocation of $\mathbf{y}_i$ is equal to $k : S_i = k$. Before sampling $S_i^{new}$ from the posterior given in (3.28), the likelihood $p(\mathbf{y}|\mathbf{S}_{-i}, S_i^{new})$, given by (3.25), and the prior $p(S_i^{new}|\mathbf{S}_{-i})$, given by (3.24), have to be evaluated for all values $S_i^{new} = l$, $l = 1, \ldots, K$. This is straightforward for $S_i^{new} = S_i = k$. Whenever the allocation changes (i.e., $S_i^{new} = l$ with $l \neq k$), the number of observations attached to component $k$ and $l$ need to be updated before applying (3.24):

$$N_k(S_i^{new}, \mathbf{S}_{-i}) = N_k(\mathbf{S}) - 1, \qquad N_l(S_i^{new}, \mathbf{S}_{-i}) = N_l(\mathbf{S}) + 1.$$

In a similar way, the statistics of the complete-data likelihood have to be updated before evaluating the likelihood $p(\mathbf{y}|\mathbf{S}_{-i}, S_i^{new})$ from (3.25) for $S_i^{new} = l$, where $l \neq k$. For mixtures of Poisson distributions, for instance, this reads:

$$\begin{aligned} b_k(S_i^{new}, \mathbf{S}_{-i}) &= b_k(\mathbf{S}) - 1, \qquad & b_l(S_i^{new}, \mathbf{S}_{-i}) &= b_l(\mathbf{S}) + 1, \\ a_k(S_i^{new}, \mathbf{S}_{-i}) &= a_k(\mathbf{S}) - y_i, \qquad & a_l(S_i^{new}, \mathbf{S}_{-i}) &= a_l(\mathbf{S}) + y_i. \end{aligned}$$

Similar simple updates are available for many other standard finite mixture models. For various other more complex mixture models, such as mixtures of regression models, Chen and Liu (1996) developed an efficient algorithm to compute the likelihood $p(\mathbf{y}|\mathbf{S}_{-i}, S_i^{new})$ recursively from $p(\mathbf{y}|\mathbf{S}_{-i}, S_i)$.

### Why Single-Move Gibbs Sampling Works

It is instructive to verify that single-move Gibbs sampling works, by showing that sampling $\mathbf{S}^{(m)}$ from $p(\mathbf{S}^{(m)}|\mathbf{S}^{(m-1)})$ yields a sample from $p(\mathbf{S}|\mathbf{y})$, once the chain reaches equilibrium, and $\mathbf{S}^{(m-1)}$ is drawn from $p(\mathbf{S}|\mathbf{y})$. Let $f(\mathbf{S}^{(m)})$ denote the density of the distribution of $\mathbf{S}^{(m)}$, which is given by

$$f(\mathbf{S}^{(m)}) = \sum_{\mathbf{S}^{(m-1)} \in \mathcal{S}_K} p(\mathbf{S}^{(m)}|\mathbf{S}^{(m-1)}, \mathbf{y})p(\mathbf{S}^{(m-1)}|\mathbf{y}) =$$

$$= \sum_{S_N^{(m-1)}=1}^{K} \cdots \sum_{S_2^{(m-1)}=1}^{K} \prod_{i=1}^{N} p(S_i^{(m)}|\mathbf{S}_{1:i-1}^{(m)}, \mathbf{S}_{i+1:N}^{(m-1)}, \mathbf{y}) \prod_{i=2}^{N} p(S_i^{(m-1)}|\mathbf{S}_{i+1:N}^{(m-1)}, \mathbf{y})$$

$$\cdot \left( \sum_{S_1^{(m-1)}=1}^{K} p(S_1^{(m-1)}|\mathbf{S}_{2:N}^{(m-1)}, \mathbf{y}) \right),$$

where the innermost term is obviously equal to 1. Therefore

$$f(\mathbf{S}^{(m)}) = \sum_{S_N^{(m-1)}=1}^{K} \cdots \sum_{S_3^{(m-1)}=1}^{K} \prod_{i=2}^{N} p(S_i^{(m)}|\mathbf{S}_{1:i-1}^{(m)}, \mathbf{S}_{i+1:N}^{(m-1)}, \mathbf{y})$$

$$\cdot \prod_{i=3}^{N} p(S_i^{(m-1)}|\mathbf{S}_{i+1:N}^{(m-1)}, \mathbf{y}) \left( \sum_{S_2^{(m-1)}=1}^{K} p(S_2^{(m-1)}|\mathbf{S}_{3:N}^{(m-1)}, \mathbf{y}) p(S_1^{(m)}|\mathbf{S}_{2:N}^{(m-1)}, \mathbf{y}) \right).$$

The innermost term is equal to $p(S_1^{(m)}|\mathbf{S}_{3:N}^{(m-1)}, \mathbf{y})$, therefore

$$f(\mathbf{S}^{(m)}) = \sum_{S_N^{(m-1)}=1}^{K} \cdots \sum_{S_4^{(m-1)}=1}^{K} \prod_{i=3}^{N} p(S_i^{(m)}|\mathbf{S}_{1:i-1}^{(m)}, \mathbf{S}_{i+1:N}^{(m-1)}, \mathbf{y})$$

$$\cdot \prod_{i=4}^{N} p(S_i^{(m-1)}|\mathbf{S}_{i+1:N}^{(m-1)}, \mathbf{y}) \left( \sum_{S_3^{(m-1)}=1}^{K} p(S_3^{(m-1)}|\mathbf{S}_{4:N}^{(m-1)}, \mathbf{y}) p(\mathbf{S}_{1:2}^{(m)}|\mathbf{S}_{3:N}^{(m-1)}, \mathbf{y}) \right).$$

The innermost term is equal to $p(\mathbf{S}_{1:2}^{(m)}|\mathbf{S}_{4:N}^{(m-1)}, \mathbf{y})$, therefore:

$$f(\mathbf{S}^{(m)}) = \sum_{S_N^{(m-1)}=1}^{K} \cdots \sum_{S_5^{(m-1)}=1}^{K} \prod_{i=4}^{N} p(S_i^{(m)}|\mathbf{S}_{1:i-1}^{(m)}, \mathbf{S}_{i+1:N}^{(m-1)}, \mathbf{y})$$

$$\cdot \prod_{i=5}^{N} p(S_i^{(m-1)}|\mathbf{S}_{i+1:N}^{(m-1)}, \mathbf{y}) \cdot \left( \sum_{S_4^{(m-1)}=1}^{K} p(S_4^{(m-1)}|\mathbf{S}_{5:N}^{(m-1)}, \mathbf{y}) p(\mathbf{S}_{1:3}^{(m)}|\mathbf{S}_{4:N}^{(m-1)}, \mathbf{y}) \right),$$

where the innermost term is equal to $p(\mathbf{S}_{1:3}^{(m)}|\mathbf{S}_{5:N}^{(m-1)}, \mathbf{y})$. This is repeated until we obtain:

$$f(\mathbf{S}^{(m)}) = \sum_{S_N^{(m-1)}=1}^{K} p(S_{N-1}^{(m)}|\mathbf{S}_{1:N-2}^{(m)}, S_N^{(m-1)}, \mathbf{y}) p(S_N^{(m)}|\mathbf{S}_{1:N-1}^{(m)}, \mathbf{y}) p(S_N^{(m-1)}|\mathbf{y})$$

$$\cdot \left( \sum_{S_{N-1}^{(m-1)}=1}^{K} p(S_{N-1}^{(m-1)}|S_N^{(m-1)}, \mathbf{y}) p(\mathbf{S}_{1:N-2}^{(m)}|\mathbf{S}_{N-1:N}^{(m-1)}, \mathbf{y}) \right),$$

which yields the desired result:

$$\begin{aligned} f(\mathbf{S}^{(m)}) &= p(S_N^{(m)}|\mathbf{S}_{1:N-1}^{(m)}, \mathbf{y}) \sum_{S_N^{(m-1)}=1}^{K} p(\mathbf{S}_{1:N-1}^{(m)}|S_N^{(m-1)}, \mathbf{y}) p(S_N^{(m-1)}|\mathbf{y}) \\ &= p(\mathbf{S}^{(m)}|\mathbf{y}). \end{aligned}$$

### 3.4.2 The Metropolis–Hastings Algorithm

Alternatively to the Gibbs sampler described in *Algorithm 3.1*, the Metropolis–Hastings algorithm may be applied to draw from the density $p(\mathbf{S}|\mathbf{y})$. Running a Gibbs sampler may be impractical if $K$ is large, as in (3.28) the probability $p(S_i^{new}|\mathbf{S}_{-i}, \mathbf{y})$ needs to be evaluated for all $S_i^{new} = 1, \ldots, K$.

Whereas the Gibbs sampler used the density $p(S_i^{new}|\mathbf{S}_{-i}, \mathbf{y})$ for proposing $S_i^{new}$, the Metropolis–Hastings algorithm uses an arbitrary discrete density $q(S_i^{new}|S_i)$, where $S_i$ is the current allocation, to propose $S_i^{new}$. Without modifications, the resulting Markov chain $\mathbf{S}^{(m)}$ would not draw from the desired posterior distribution $p(\mathbf{S}|\mathbf{y})$. To obtain draws from the desired distribution, the proposed allocation $S_i^{new}$ is not accepted in any case, but only with a certain probability $\alpha(S_i^{new}|S_i)$. If the new value is accepted, then $S_i^{(m)} = S_i^{new}$, otherwise $S_i^{new}$ is rejected and the chain does not move: $S_i^{(m)} = S_i$.

As pointed out by Chib and Greenberg (1995), the accept–reject step is necessary as $q(S_i^{new}|S_i)$ is not likely to fulfill the detailed balance condition. For instance, it may happen that

$$p(S_i|\mathbf{S}_{-i}, \mathbf{y})q(S_i^{new}|S_i) > p(S_i^{new}|\mathbf{S}_{-i}, \mathbf{y})q(S_i|S_i^{new}), \tag{3.29}$$

meaning that too many moves from $S_i$ to $S_i^{new}$, and too few moves from $S_i^{new}$ to $S_i$ are made. The probability $\alpha(S_i^{new}|S_i)$ of accepting a move from $S_i$ to $S_i^{new}$ is introduced, in order to ensure detailed balance. The acceptance probability $\alpha(S_i^{new}|S_i)$ is chosen precisely to ensure that the Markov chain $S_i^{(m)}$ is reversible with respect to $p(S_i|\mathbf{S}_{-i}, \mathbf{y})$. Following Chib and Greenberg (1995), $\alpha(S_i|S_i^{new})$ should be set to 1, if (3.29) holds, as moves from $S_i^{new}$ to $S_i$ are too rare. The reverse probability $\alpha(S_i^{new}|S_i)$ is then determined by forcing a detailed balance in (3.29),

$$p(S_i|\mathbf{S}_{-i}, \mathbf{y})q(S_i^{new}|S_i)\alpha(S_i^{new}|S_i) = p(S_i^{new}|\mathbf{S}_{-i}, \mathbf{y})q(S_i|S_i^{new}). \tag{3.30}$$

Thus $\alpha(S_i^{new}|S_i)$ which could not be larger than 1, is given by

$$\alpha(S_i^{new}|S_i) = \min\left(1, \frac{p(S_i^{new}|\mathbf{S}_{-i}, \mathbf{y})q(S_i|S_i^{new})}{p(S_i|\mathbf{S}_{-i}, \mathbf{y})q(S_i^{new}|S_i)}\right), \tag{3.31}$$

if $p(S_i|\mathbf{S}_{-i}, \mathbf{y})q(S_i^{new}|S_i) > 0$. Interestingly, other acceptance rules are possible (see Liu, 2001, Section 5), however, Peskun (1973) proves superiority of (3.31) in terms of statistical efficiency.

*Algorithm 3.2: Sampling the Allocations Through a Metropolis–Hastings Algorithm* Start with some classification $\mathbf{S}$ and repeat the following steps for $m = 1, \ldots, M_0, \ldots, M + M_0$.

(a) Choose a certain observation $\mathbf{y}_i$, $i \in \{1, \ldots, N\}$, hold the most recent allocations of all observations but $\mathbf{y}_i$ fixed, and let $\mathbf{S}_{-i}$ be the sequence containing these allocations.

(b) Find a new allocation $S_i^{new}$ for the observation $\mathbf{y}_i$ in the following way. Sample $S_i^{new}$ from a proposal density $q(S_i^{new}|S_i)$ and substitute the old allocation $S_i$ by the new allocation $S_i^{new}$ with probability $\min(1, r_i)$, where

$$r_i = \frac{p(\mathbf{y}|\mathbf{S}_{-i}, S_i^{new})p(S_i^{new}|\mathbf{S}_{-i})q(S_i|S_i^{new})}{p(\mathbf{y}|\mathbf{S}_{-i}, S_i)p(S_i|\mathbf{S}_{-i})q(S_i^{new}|S_i)}. \tag{3.32}$$

If $U_i < \min(1, r_i)$, where $U_i$ is random number from the $\mathcal{U}[0,1]$-distribution, then $S_i$ is substituted by $S_i^{new}$, otherwise leave $S_i$ unchanged.

Repeat these steps until the allocations of all observations are updated. Store the actual values of all allocations as $\mathbf{S}^{(m)}$, increase $m$ by one, and return to step (a).

If $q(S_i^{new}|S_i) = p(S_i^{new}|\mathbf{S}_{-i}, \mathbf{y})$, then $r_i = 1$, and the Metropolis–Hastings algorithm reduces to the Gibbs sampler described in *Algorithm 3.1.* To avoid the functional evaluations that are necessary to sample from this specific proposal density, much simpler proposal densities are used for the Metropolis–Hastings algorithm.

Some simplifications are possible when evaluating $r_i$. If $S_i^{new} = S_i$, the likelihood and the prior cancel, and $r_i$ is equal to the proposal ratio. If $S_i^{new} = l$ while $S_i = k$ with $k \neq l$, the acceptance ratio $r_i$ simplifies to

$$r_i = \frac{p(\mathbf{y}|\mathbf{S}_{-i}, S_i^{new})(N_l(\mathbf{S}) + 1 + e_{0,l})q(S_i|S_i^{new})}{p(\mathbf{y}|\mathbf{S}_{-i}, S_i)(N_k(\mathbf{S}) + e_{0,k})q(S_i^{new}|S_i)},$$

where $N_k(\mathbf{S})$ and $N_l(\mathbf{S})$ are the current numbers of allocations. For mixtures of Poisson distributions the likelihood ratio reduces to:

$$\frac{p(\mathbf{y}|\mathbf{S}_{-i}, S_i^{new})}{p(\mathbf{y}|\mathbf{S}_{-i}, S_i)} = \frac{\Gamma(a_k(\mathbf{S}) - y_i)\Gamma(a_l(\mathbf{S}) + y_i)b_k(\mathbf{S})^{a_k(\mathbf{S})}b_l(\mathbf{S})^{a_l(\mathbf{S})}}{\Gamma(a_k(\mathbf{S}))\Gamma(a_l(\mathbf{S}))(b_k(\mathbf{S}) - 1)^{a_k(\mathbf{S})-y_i}(b_l(\mathbf{S}) + 1)^{a_l(\mathbf{S})+y_i}}.$$

## 3.5 Parameter Estimation Through Data Augmentation and MCMC

Markov chain Monte Carlo sampling is not only useful for the purpose of sampling allocations, but also for parameter estimation.

### 3.5.1 Treating Mixture Models as a Missing Data Problem

As already discussed in Subsection 2.3.3, for mixture models from exponential families such as mixtures of Poisson distributions or mixtures of normal distributions, a conjugate analysis is feasible for the complete-data likelihood function (2.8) when the allocations $\mathbf{S} = (S_1, \ldots, S_N)$ are observed. For unknown allocations, however, this is not the case.

Following the seminal paper by Dempster et al. (1977), a mixture model may be seen as an incomplete data problem by introducing the allocations $\mathbf{S}$ as missing data. The benefit of this data augmentation (Tanner and Wong, 1987) is that conditional on $\mathbf{S}$ we are back in the conjugate setting of complete-data Bayesian estimation considered in Subsection 2.3.3. On the other hand, conditional on knowing the parameter $\boldsymbol{\vartheta}$, we are back to the classification problem studied in Section 2.2, where the posterior distribution of the allocations takes a very simple form. It is then rather straightforward to sample from the posterior (3.2) using Markov chain Monte Carlo methods, in particular Gibbs sampling. Early papers realizing the importance of Gibbs sampling for Bayesian estimation of mixture models are Evans et al. (1992), West (1992), Smith and Roberts (1993), Diebolt and Robert (1994), Escobar and West (1995), Mengersen and Robert (1996), and Raftery (1996b). We first give specific results for a mixture of Poisson distributions in Subsection 3.5.2 and then proceed with a discussion for more general finite mixture models in Subsection 3.5.3.

### 3.5.2 Data Augmentation and MCMC for a Mixture of Poisson Distributions

For $N$ observations $\mathbf{y} = (y_1, \dots, y_N)$, assumed to arise from a finite mixture of $K$ Poisson distributions, the mixture likelihood function $p(\mathbf{y}|\boldsymbol{\vartheta})$ is given by

$$p(\mathbf{y}|\boldsymbol{\vartheta}) = \prod_{i=1}^{N} p(y_i|\boldsymbol{\vartheta}) = \prod_{i=1}^{N} \left( \sum_{k=1}^{K} \eta_k f_P(y_i; \mu_k) \right), \tag{3.33}$$

where $f_P(y_i; \mu_k)$ is the density of a Poisson distribution with mean $\mu_k$. Although direct sampling from (3.33) is not easy, a straightforward method of sampling from (3.33) based on data augmentation is possible.

For each observation $y_i$, $i = 1, \dots, N$, the group indicator $S_i$ taking a value in $\{1, \dots, K\}$ is introduced as a missing observation. Conditional on knowing the group indicator $S_i$, the observation model for observation $y_i$ is a Poisson distribution with mean $\mu_{S_i}$:

$$y_i|\mu_1, \dots, \mu_K, S_i \sim \mathcal{P}(\mu_{S_i}). \tag{3.34}$$

All observations with the same group indicator $S_i$ equal to $k$, say, arise from the same $\mathcal{P}(\mu_k)$-distribution. Therefore the complete-data likelihood $p(\mathbf{y}, \mathbf{S}|\boldsymbol{\vartheta})$, which has been defined in (2.8), reads:

$$p(\mathbf{y}, \mathbf{S}|\boldsymbol{\vartheta}) = \prod_{k=1}^{K} \left( \prod_{i:S_i=k} f_P(y_i; \mu_k) \right) \left( \prod_{k=1}^{K} \eta_k^{N_k(\mathbf{S})} \right),$$

where $N_k(\mathbf{S}) = \#\{S_i = k\}$. $p(\mathbf{y}, \mathbf{S}|\boldsymbol{\vartheta})$, considered as a function of $\boldsymbol{\vartheta}$, is the product of $K + 1$ independent factors. Each of the first $K$ factors depends only on $\mu_k$, whereas the last factor depends on $\boldsymbol{\eta}$. Assuming independence a priori, the parameters $\mu_1, \ldots, \mu_K, \boldsymbol{\eta}$ are independent a posteriori given the complete data $(\mathbf{y}, \mathbf{S})$:

$$p(\mu_1, \ldots, \mu_K, \boldsymbol{\eta}|\mathbf{S}, \mathbf{y}) = \prod_{k=1}^{K} p(\mu_k|\mathbf{S}, \mathbf{y}) p(\boldsymbol{\eta}|\mathbf{S}).$$

Each of the conditional posteriors can be handled within the conjugate setting discussed in Subsection 2.3.3. We express prior knowledge about $\mu_k$ as a $\mathcal{G}(a_0, b_0)$-distribution. Then from Bayes' theorem:

$$p(\mu_k|\mathbf{S}, \mathbf{y}) \propto \left( \prod_{i:S_i=k} f_P(y_i; \mu_k) \right) p(\mu_k). \tag{3.35}$$

The posterior distribution $p(\mu_k|\mathbf{S}, \mathbf{y})$ is a $\mathcal{G}(a_k(\mathbf{S}), b_k(\mathbf{S}))$-distribution, where

$$a_k(\mathbf{S}) = a_0 + N_k(\mathbf{S})\overline{y}_k(\mathbf{S}), \qquad b_k(\mathbf{S}) = b_0 + N_k(\mathbf{S}), \tag{3.36}$$

and $N_k(\mathbf{S}) = \#\{S_i = k\}$ and $\overline{y}_k(\mathbf{S})$ are the number of observations and the mean in group $k$.

Based on assuming a Dirichlet $\mathcal{D}(e_0, \ldots, e_0)$-distribution for $\boldsymbol{\eta}$, the posterior distribution of the weight distribution $\boldsymbol{\eta}$ given $\mathbf{S}$ is a $\mathcal{D}(e_1(\mathbf{S}), \ldots, e_K(\mathbf{S}))$-distribution, where

$$e_k(\mathbf{S}) = e_0 + N_k(\mathbf{S}), \qquad k = 1, \ldots, K. \tag{3.37}$$

### MCMC Estimation Using Gibbs Sampling

MCMC estimation of a mixture of Poisson distributions under fixed hyperparameters $a_0$ and $b_0$ consists of the following steps.

*Algorithm 3.3: Gibbs Sampling for a Poisson Mixture* Start with some classification $\mathbf{S}^{(0)}$ and repeat the following steps for $m = 1, \ldots, M_0, \ldots, M + M_0$.

(a) Parameter simulation conditional on the classification $\mathbf{S}^{(m-1)}$:
  - (a1) Sample $\eta_1, \ldots, \eta_K$ from a $\mathcal{D}\left(e_1(\mathbf{S}^{(m-1)}), \ldots, e_K(\mathbf{S}^{(m-1)})\right)$-distribution, where $e_k(\mathbf{S}^{(m-1)})$ is given by (3.37).
  - (a2) For each $k = 1, \ldots, K$, sample $\mu_k$ from a $\mathcal{G}\left(a_k(\mathbf{S}^{(m-1)}), b_k(\mathbf{S}^{(m-1)})\right)$-distribution, where $a_k(\mathbf{S}^{(m-1)})$ and $b_k(\mathbf{S}^{(m-1)})$ are given by (3.36).

  Store the actual values of all parameters as $\boldsymbol{\vartheta}^{(m)} = (\mu_1^{(m)}, \ldots, \mu_K^{(m)}, \boldsymbol{\eta}^{(m)})$.

(b) Classification of each observation $y_i$ conditional on knowing $\boldsymbol{\vartheta}^{(m)}$: sample $S_i$ independently for each $i = 1, \ldots, N$ from the conditional posterior distribution $p(S_i|\boldsymbol{\vartheta}^{(m)}, y_i)$, which by the results of Subsection 2.2.1 is given by

$$p(S_i = k|\boldsymbol{\vartheta}^{(m)}, y_i) \propto (\mu_k^{(m)})^{y_i} e^{-\mu_k^{(m)}} \eta_k^{(m)}.$$

Store the actual values of all allocations as $\mathbf{S}^{(m)}$, increase $m$ by one, and return to step (a). Finally, the first $M_0$ draws are discarded.

**Hierarchical Priors**

Under the hierarchical prior (3.12), an additional block has to be added in *Algorithm 3.3*, where $b_0$ is sampled from the conditional posterior distribution $p(b_0|\mu_1, \ldots, \mu_K, \mathbf{S}, \mathbf{y})$, given by Bayes' theorem:

$$p(b_0|\mu_1, \ldots, \mu_K, \mathbf{S}, \mathbf{y}) \propto \prod_{k=1}^{K} p(\mu_k|b_0)p(b_0) \tag{3.38}$$

$$\propto \prod_{k=1}^{K} b_0^{a_0} \exp(-\mu_k b_0)\, p(b_0) \propto b_0^{g_0+Ka_0-1} \exp\left( -(G_0 + \sum_{k=1}^{K} \mu_k) b_0 \right).$$

Under a conjugate $\mathcal{G}(g_0, G_0)$-prior for $b_0$, this posterior is a Gamma distribution, depending on the data only indirectly through the component means.

Gibbs sampling requires the following modification of *Algorithm 3.3*. Select a starting value ${b_0}^{(0)}$ and run step (a2) conditional on ${b_0}^{(m-1)}$. A third step is added to sample the hyperparameter ${b_0}^{(m)}$:

(c) Sample $b_0^{(m)}$ from $p(b_0|\mu_1^{(m)}, \ldots, \mu_K^{(m)})$ given by (3.38):

$$b_0|\mu_1^{(m)}, \ldots, \mu_K^{(m)} \sim \mathcal{G}\left( g_0 + Ka_0, G_0 + \sum_{k=1}^{K} \mu_k^{(m)} \right). \tag{3.39}$$

### 3.5.3 Data Augmentation and MCMC for General Mixtures

As for the Poisson mixture, Bayesian estimation of a general mixture model through data augmentation estimates the augmented parameter $(\mathbf{S}, \boldsymbol{\vartheta})$ by sampling from the complete-data posterior distribution $p(\mathbf{S}, \boldsymbol{\vartheta}|\mathbf{y})$. This posterior is given by Bayes' theorem,

$$p(\mathbf{S}, \boldsymbol{\vartheta}|\mathbf{y}) \propto p(\mathbf{y}|\mathbf{S}, \boldsymbol{\vartheta})p(\mathbf{S}|\boldsymbol{\vartheta})p(\boldsymbol{\vartheta}), \tag{3.40}$$

thus the complete-data posterior is proportional to the complete-data likelihood likelihood $p(\mathbf{y}, \mathbf{S}|\boldsymbol{\vartheta})$ defined in (2.8) times the prior $p(\boldsymbol{\vartheta})$ on $\boldsymbol{\vartheta}$; see again Subsection 2.3.3 for more details. Sampling from the posterior (3.40) is most commonly carried out by the following MCMC sampling scheme, where $\boldsymbol{\vartheta}$ is sampled conditional on knowing $\mathbf{S}$, and $\mathbf{S}$ is sampled conditional on knowing $\boldsymbol{\vartheta}$. This scheme is formulated for the general case, where the observations $\mathbf{y}_i$ may be multivariate.

*Algorithm 3.4: Unconstrained MCMC for a Mixture Model* Start with some classification $\mathbf{S}^{(0)}$ and repeat the following steps for $m = 1, \ldots, M_0, \ldots, M + M_0$.

(a) Parameter simulation conditional on the classification $\mathbf{S}^{(m-1)}$:

(a1) Sample $\boldsymbol{\eta}$ from the $\mathcal{D}\left(e_1(\mathbf{S}^{(m-1)}), \ldots, e_K(\mathbf{S}^{(m-1)})\right)$-distribution, where $e_k(\mathbf{S}^{(m-1)})$ is given by (3.37).

(a2) Sample the component parameters $\boldsymbol{\theta}_1, \ldots, \boldsymbol{\theta}_K$ from the complete-data posterior $p(\boldsymbol{\theta}_1, \ldots, \boldsymbol{\theta}_K | \mathbf{S}^{(m-1)}, \mathbf{y})$.

Store the actual values of all parameters as $\boldsymbol{\vartheta}^{(m)} = (\boldsymbol{\theta}_1^{(m)}, \ldots, \boldsymbol{\theta}_K^{(m)}, \boldsymbol{\eta}^{(m)})$.

(b) Classification of each observation $\mathbf{y}_i$ conditional on knowing $\boldsymbol{\vartheta}^{(m)}$: sample $S_i$ independently for each $i = 1, \ldots, N$ from the conditional posterior distribution $p(S_i | \boldsymbol{\vartheta}^{(m)}, \mathbf{y}_i)$, which by the results of Subsection 2.2.1 is given by

$$p(S_i = k | \boldsymbol{\vartheta}^{(m)}, \mathbf{y}_i) \propto p(\mathbf{y}_i | \boldsymbol{\theta}_k^{(m)}) \eta_k^{(m)}. \tag{3.41}$$

Store the actual values of all allocations as $\mathbf{S}^{(m)}$, increase $m$ by one, and return to step (a). Finally, the first $M_0$ draws are discarded.

The structure of the posterior $p(\boldsymbol{\theta}_1, \ldots, \boldsymbol{\theta}_K | \mathbf{S}, \mathbf{y})$ depends on the specific distribution families appearing in the components of the mixture model and on the chosen priors. If the components come from an exponential family, the results of Subsection 3.2.3 will be helpful. Under the conditionally conjugate prior (3.9), the component parameters $\boldsymbol{\theta}_1, \ldots, \boldsymbol{\theta}_K$ are independent given $\mathbf{S}$ and may be sampled from the conditional posterior $p(\boldsymbol{\theta}_k | \mathbf{S}, \mathbf{y})$ given by (3.10) for each $k = 1, \ldots, K$.

The MCMC sampler described in *Algorithm 3.4* starts with sampling the parameter $\boldsymbol{\vartheta}$ based on allocations $\mathbf{S}^{(0)}$ defined by the investigator. Theoretically, it does not make any difference if the sampling steps (a) and (b) are interchanged, in which case the algorithm starts with sampling the allocations $\mathbf{S}$ based on a parameter $\boldsymbol{\vartheta}^{(0)}$. Practical MCMC convergence diagnostics for finite mixture models is considered by Robert et al. (1999).

### Hierarchical Priors

Under the hierarchical prior discussed in Subsection 3.2.4, an additional block has to be added in *Algorithm 3.4* to sample the hyperparameter $\boldsymbol{\delta}$ conditional on knowing $\boldsymbol{\theta}_1, \ldots, \boldsymbol{\theta}_K$ from

$$p(\boldsymbol{\delta} | \boldsymbol{\theta}_1, \ldots, \boldsymbol{\theta}_K) \propto \prod_{k=1}^{K} p(\boldsymbol{\theta}_k | \boldsymbol{\delta}) p(\boldsymbol{\delta}). \tag{3.42}$$

In many cases, this density will be of closed form. This leads to the following modification of *Algorithm 3.4*. Select a starting value $\boldsymbol{\delta}^{(0)}$ and run step (a2) conditional on $\boldsymbol{\delta}^{(m-1)}$. A third step is added to sample the hyperparameter $\boldsymbol{\delta}^{(m)}$ from $p(\boldsymbol{\delta} | \boldsymbol{\theta}_1^{(m)}, \ldots, \boldsymbol{\theta}_K^{(m)})$, given by (3.42).

### 3.5.4 MCMC Sampling Under Improper Priors

MCMC sampling under improper priors is possible as long as the conditional posterior $p(\boldsymbol{\vartheta}|\mathbf{S},\mathbf{y})$ is proper for all possible allocations. What happens, if unintentionally MCMC sampling is carried out under a prior like the improper product prior (3.4), where the posterior is improper (Natarajan and McCulloch, 1998)?

For a mixture of Poisson distributions, for instance, an improper product prior based on $\mu_k \sim \mathcal{G}(0,0)$ or $\mu_k \sim \mathcal{G}(0.5,0)$ leads to an improper posterior distribution by the results of Subsection 3.2.2. If $N_k(\mathbf{S}^{(m-1)}) = 0$ for a certain draw, then the conditional posterior $p(\mu_k|\mathbf{S}^{(m-1)},\mathbf{y})$ given by (3.36) is equal to the improper prior and the MCMC sampler breaks down when drawing $\mu_k^{(m)}$, warning us that something is not in order.

In other cases, it is possible to obtain sensible looking results when running data augmentation and MCMC under the product prior (3.4). Consider, for instance, a synthetic data set of size $N = 500$, simulated from a mixture of two Poisson distributions, where $\mu_1 = 1$, $\mu_2 = 5$, and $\eta_1 = 0.4$. We estimated $(\mu_1, \mu_2, \eta_1, \eta_2)$ under the uniform $\mathcal{D}(1,1)$ prior on $(\eta_1, \eta_2)$, with an improper $\mathcal{G}(0.5,0)$ as well as a proper $\mathcal{G}(0.01,0.01)$ prior on $\mu_1$ and $\mu_2$, running MCMC for 1 million iterations without problems. Furthermore, for both priors the resulting density estimates were indistinguishable. To understand this, consider the following representation of the posterior $p(\boldsymbol{\vartheta}|\mathbf{y})$,

$$p(\boldsymbol{\vartheta}|\mathbf{y}) = \sum_{\mathbf{S}\in\mathcal{S}_K} p(\boldsymbol{\vartheta}|\mathbf{S},\mathbf{y})p(\mathbf{S}|\mathbf{y}),$$

where the complete-data posterior $p(\boldsymbol{\vartheta}|\mathbf{S},\mathbf{y})$ is weighted by the posterior probability of the corresponding partition $\mathbf{S}$. If partitions $\mathbf{S}$, where the corresponding complete-data posterior is improper, have very low posterior probability, then it is very unlikely (though possible) that such a classification is selected during MCMC sampling. Therefore the *estimated* posterior

$$\hat{p}(\boldsymbol{\vartheta}|\mathbf{y}) = \frac{1}{M}\sum_{m=1}^{M} p(\boldsymbol{\vartheta}|\mathbf{S}^{(m)},\mathbf{y})$$

will be proper. Nevertheless, is not recommended to sample from improper posterior distributions in this way, as statistical inference drawn from such a posterior distribution lacks any theoretical justification.

### 3.5.5 Label Switching

The term *label switching* has been introduced into the literature on mixture models by Redner and Walker (1984) to describe the invariance of the mixture likelihood function under relabeling the components of a mixture model described in Subsection 2.4.2. Label switching is of no concern for maximum likelihood estimation, where the goal is to find one of the equivalent modes of the likelihood function. In the context of Bayesian estimation, however, label switching has to be addressed explicitly because in the course of sampling from the mixture posterior distribution, the labeling of the unobserved categories changes. Interestingly, the label switching problem was totally neglected in the early papers on MCMC estimation of finite mixture models and was addressed only later on by Celeux (1998), Celeux et al. (2000), Stephens (2000a, 2000b), Casella et al. (2000), and Frühwirth-Schnatter (2001b).

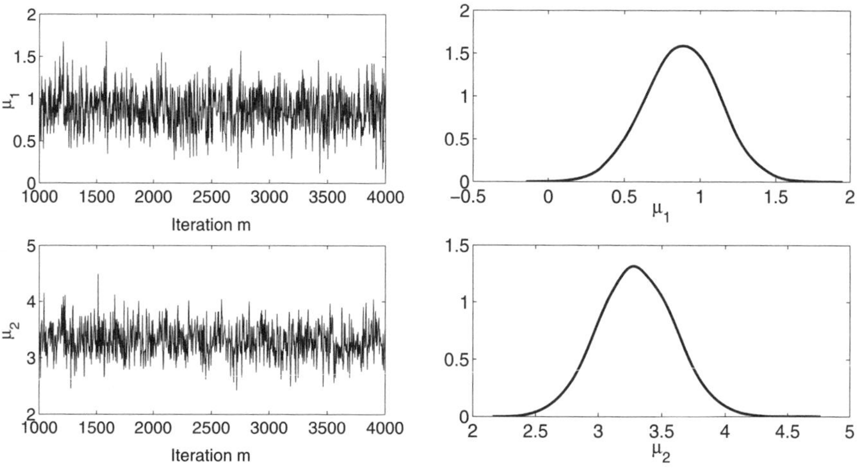


**Fig. 3.1.** HIDDEN AGE GROUPS — Synthetic Data Set 1; MCMC draws of $\mu_1$ and $\mu_2$ (left-hand side) and estimated marginal posterior densities of $\mu_1$ and $\mu_2$ (right-hand side)

### Some Illustration

For illustration, we reconsider the example of Subsection 2.4.3, where we simulated artificial data sets of length $N = 100$ from the following mixture of normals,

$$p(y) = 0.5 f_N(y; \mu_y, 1) + 0.5 f_N(y; \mu_o, 1),$$

where $\mu_y$ and $\mu_o$ are the mean of a random variable $Y$ in a younger and in an older subgroup in the population. For MCMC estimation of $\mu_1$ and $\mu_2$, we apply data augmentation as in Subsection 3.5.3 under the prior $p(\mu_k) \sim \mathcal{N}(0, 100)$. The details of step (a2) for the specific example of a mixture of normal distributions appear later in Subsection 6.2.4. The MCMC draws of $\mu_k$ as well as the estimated marginal densities $p(\mu_k|\mathbf{y})$ are plotted in Figure 3.1

and Figure 3.2 for two artificial data sets, where $\mu_y = 1$ and $\mu_o = 3$ for the first, and $\mu_y = 1$ and $\mu_o = 1.5$ for the second data set.

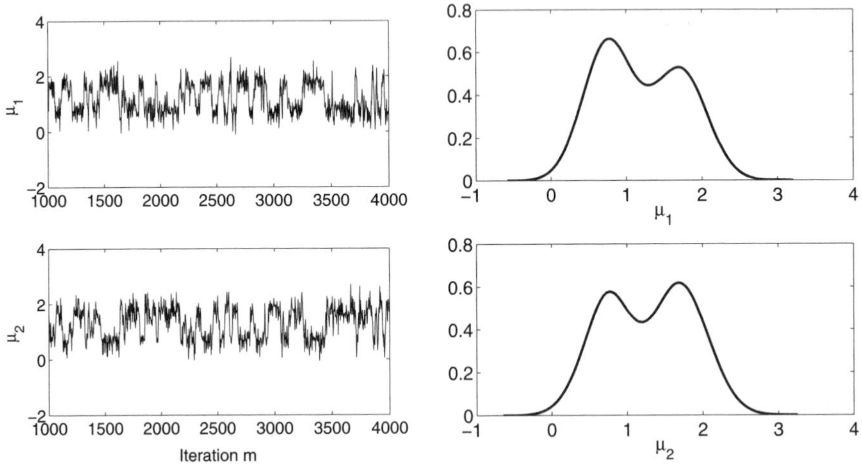


**Fig. 3.2.** HIDDEN AGE GROUPS — Synthetic Data Set 2; MCMC draws of $\mu_1$ and $\mu_2$ (left-hand side) and estimated marginal posterior densities of $\mu_1$ and $\mu_2$ (right-hand side)

For each data set we started at $\mu_1 = \mu_y$ and $\mu_2 = \mu_o$, which corresponds to labeling 1. For synthetic data set 1 the sampler stays within the modal region corresponding to this labeling, as this region is well separated from the region where the other labeling is valid; see again Figure 2.5. Note that the estimated marginal posterior densities in Figure 3.1 are unimodal and that Gibbs sampling leads implicitly to a unique labeling.

For synthetic data set 2, however, the marginal posterior densities are bimodal and the MCMC draws suffer from label switching. For this data set parameters around the nonidentifiability set $\mathcal{U}^E(\hat{\mu})$, where $\hat{\mu} = \overline{y}$, have considerable likelihood under both labelings; consider again Figure 2.6. Even if we start in the modal region corresponding to labeling 1, where $\mu_1 < \mu_2$, the sampler is likely to move into the area where $\mu_1 > \mu_2$. In this area, however, the parameter $(\mu_1, \mu_2)$ has higher likelihood if $\mu_1$ is associated with the older subgroup, rather than with the younger one. Therefore, when sampling the group indicators $\mathbf{S}$, there is a certain risk that the labeling changes and now $\mu_1$ is associated with the older subgroup. After such a label switching takes place, the sampler remains in the second modal region for a while until it returns to the area where $\mu_1 < \mu_2$. Then there exists considerable likelihood that the sampler switches back to labeling 1. This occasional change of labeling is obvious from the MCMC draws in Figure 3.2.

### 3.5.6 Permutation MCMC Sampling

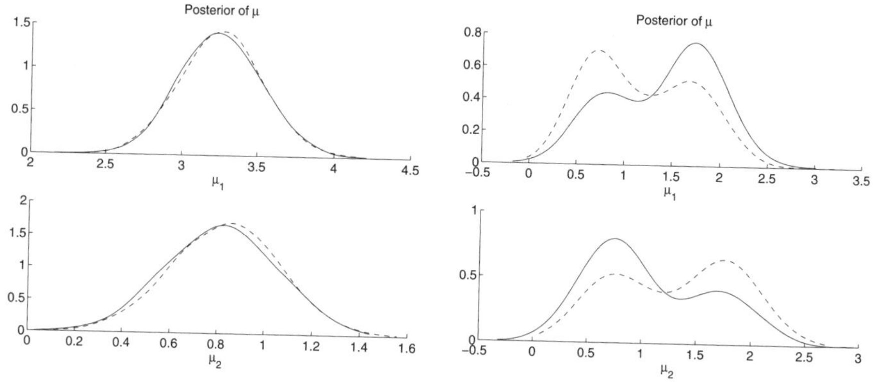


**Fig. 3.3.** HIDDEN AGE GROUPS — Synthetic Data Set 1 and 2; marginal posterior densities of $\mu_1$ and $\mu_2$ estimated from two different runs of Gibbs sampling under the same prior for Data Set 1 (left-hand side) and Data Set 2 (right-hand side)

The examples of the previous subsection demonstrated that the behavior of the Gibbs sampler described in *Algorithm 3.4* is somewhat unpredictable. For synthetic data set 1 it is trapped at one modal region, whereas it jumps from time to time to the other modal region for data set 2. In both cases the sampler did not explore the full mixture posterior distribution.

This matters especially when estimating marginal densities. Assume that we want to assess the influence of the prior $p(\boldsymbol{\vartheta})$ on the posterior distribution $p(\boldsymbol{\vartheta}|\mathbf{y})$. To do so, we usually compare the marginal posterior densities $p(\boldsymbol{\theta}_k|\mathbf{y})$ obtained under different prior distributions $p(\boldsymbol{\theta}_k)$. There, the marginal density is estimated from the MCMC draws by some kernel smoothing method.

For a mixture model it turns out that estimating the marginal density from the MCMC draws may lead to a poor estimate when *unbalanced* label switching takes place. It may even happen that although we assume the *same* prior distribution $p(\boldsymbol{\theta}_k)$, the marginal posterior densities $p(\boldsymbol{\theta}_k|\mathbf{y})$ estimated from different runs of the MCMC sampler, are very different. For illustration, Figure 3.3 compares estimates of the marginal density obtained from two different runs of full conditional Gibbs sampling for $M = 2000$ under the same prior for the two synthetic data sets considered earlier. The estimated marginal densities are nearly identical for data set 1, where no label switching took place. We observe a substantial difference in these densities for data set 2, the reason being that sampler did not explore the whole mixture posterior distribution as label switching took place only from time to time in an unbalanced manner.

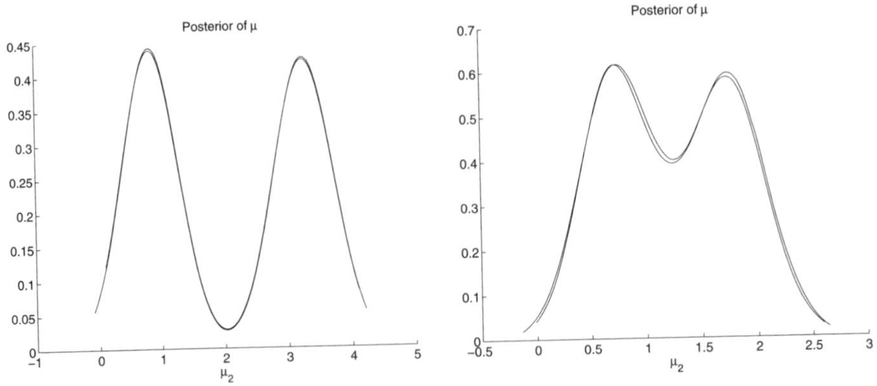


**Fig. 3.4.** HIDDEN AGE GROUPS — Synthetic Data Set 1 and 2; marginal posterior densities of $\mu_1$ and $\mu_2$ estimated from random permutation Gibbs sampling under the same prior for Data Set 1 (left-hand side) and Data Set 2 (right-hand side)

A simple, but efficient solution to obtain a sampler that explores the full mixture posterior distribution is to force balanced label switching by concluding each MCMC draw by a randomly selected permutation of the labeling. This method is called random permutation MCMC sampling (Frühwirth-Schnatter, 2001b).

*Algorithm 3.5: Random Permutation MCMC Sampling for a Finite Mixture Model* Start as described in *Algorithm 3.4.*

(a) and (b) are the same steps as in *Algorithm 3.4.*
(c) Conclude each draw by selecting randomly one of the $K!$ possible permutations $\rho_s(1),\dots,\rho_s(K)$ of the current labeling. This permutation is applied to $\boldsymbol{\eta}^{(m)}$, the component parameters $\boldsymbol{\theta}_1^{(m)},\dots,\boldsymbol{\theta}_K^{(m)}$, and the allocations $\mathbf{S}^{(m)}$:
  (c1) The group weights $\eta_1^{(m)},\dots,\eta_K^{(m)}$ are substituted by $\eta_{\rho_s(1)}^{(m)},\dots,\eta_{\rho_s(K)}^{(m)}$.
  (c2) The component parameters $\boldsymbol{\theta}_1^{(m)},\dots,\boldsymbol{\theta}_K^{(m)}$ are substituted by $\boldsymbol{\theta}_{\rho_s(1)}^{(m)}$, $\dots,\boldsymbol{\theta}_{\rho_s(K)}^{(m)}$.
  (c3) The allocations $S_i^{(m)}, i=1,\dots,N$, are substituted by $\rho_s(S_i^{(m)}), i=1,\dots,N$.

For illustration we consider once more the synthetic data sets 1 and 2. Let $(\mu_1^{(m)},\mu_2^{(m)},\mathbf{S}^{(m)})$ denote a draw obtained from Gibbs sampling as in *Algorithm 3.4*. To implement the random permutation Gibbs sampler, we perform a random permutation of the labels after each draw. For $K=2$, there are only two permutations, namely the identity $\rho_1(1)=1,\rho_1(2)=2$, and interchanging the labels: $\rho_2(1)=2,\rho_2(2)=1$. Thus with a probability of 0.5 the draws remain unchanged, whereas with probability 0.5 the labels are interchanged by substituting $(\mu_1, \mu_2)$ by $(\mu_2, \mu_1)$, and switching the allocations, which take the value 1, if they are 2, and take the value 2, if they are 1. Figure 3.4 shows the marginal posterior densities $p(\mu_k|\mathbf{y})$ estimated from random permutation MCMC sampling for both synthetic data sets. As expected from the theoretical considerations in Subsection 3.3.2, these densities are identical.

## 3.6 Other Monte Carlo Methods Useful for Mixture Models

In the previous section we focused on data augmentation and MCMC methods, but other Monte Carlo methods have been found to be useful for finite mixture models.

### 3.6.1 A Metropolis–Hastings Algorithm for the Parameters

Several authors (Celeux et al., 2000; Brooks, 2001; Viallefont et al., 2002) use a Metropolis–Hastings algorithm to generate a sample from the mixture posterior distribution $p(\boldsymbol{\vartheta}|\mathbf{y})$. This is feasible, because the Metropolis–Hastings algorithm requires knowledge of the mixture posterior density $p(\boldsymbol{\vartheta}|\mathbf{y})$ only up to a normalizing constant. The Metropolis–Hastings algorithm, introduced in Subsection 3.4.2 in the context of sampling allocations $\mathbf{S}$ from the posterior $p(\mathbf{S}|\mathbf{y})$, is implemented in the following manner to simulate $\boldsymbol{\vartheta}$ from the mixture posterior $p(\boldsymbol{\vartheta}|\mathbf{y})$.

*Algorithm 3.6: Sampling the Parameters of a Finite Mixture Through a Metropolis–Hastings Algorithm* Start with some parameter $\boldsymbol{\vartheta}^{(0)}$ and repeat the following steps for $m = 1, \ldots, M_0, \ldots, M + M_0$.

(a) Propose a new parameter $\boldsymbol{\vartheta}^{new}$ by sampling from a proposal density $q(\boldsymbol{\vartheta}|\boldsymbol{\vartheta}^{(m-1)})$.
(b) Move the sampler to $\boldsymbol{\vartheta}^{new}$ with probability $\min(1, A)$, where

$$A = \frac{p(\mathbf{y}|\boldsymbol{\vartheta}^{new})p(\boldsymbol{\vartheta}^{new})q(\boldsymbol{\vartheta}^{(m-1)}|\boldsymbol{\vartheta}^{new})}{p(\mathbf{y}|\boldsymbol{\vartheta}^{(m-1)})p(\boldsymbol{\vartheta}^{(m-1)})q(\boldsymbol{\vartheta}^{new}|\boldsymbol{\vartheta}^{(m-1)})}. \tag{3.43}$$

If $U < \min(1, A)$, where $U$ is a random number from the $\mathcal{U}[0,1]$-distribution, then accept $\boldsymbol{\vartheta}^{new}$ and set $\boldsymbol{\vartheta}^{(m)} = \boldsymbol{\vartheta}^{new}$, otherwise reject $\boldsymbol{\vartheta}^{new}$ and set $\boldsymbol{\vartheta}^{(m)} = \boldsymbol{\vartheta}^{(m-1)}$.

Increase $m$ by one, and return to step (a).

Hurn et al. (2003) use the following multivariate random walk proposal on a suitably transformed parameter $\phi(\boldsymbol{\vartheta})$, which is obtained from a log-transform on variance parameters and a logit transform on the weights,

$$\phi(\boldsymbol{\vartheta}^{new}|\boldsymbol{\vartheta}^{(m-1)}) = \phi(\boldsymbol{\vartheta}^{(m-1)}) + C\epsilon,$$

where $\epsilon$ follows a multivariate Cauchy distribution. $C$ is calibrated during a pilot-run to lead to an acceptance rate of about 40%.

An advantage of this method compared to data augmentation and MCMC is that sampling of the indicators is avoided. A disadvantage is that tuning the proposal density may require several pilot runs.

### 3.6.2 Importance Sampling for the Allocations

An alternative attempt at sampling from $p(\mathbf{S}|\mathbf{y})$ has been investigated in Casella et al. (2000). Rather than drawing from $p(\mathbf{S}|\mathbf{y})$, they draw a sequence $\mathbf{S}^{(1)},\dots,\mathbf{S}^{(L)}$ from an importance density $q(\mathbf{S})$. One way to construct the importance density is to ignore posterior correlation among the indicators, which is actually only introduced through the prior $p(\mathbf{S})$, and to use a density with independent components:

$$q(\mathbf{S}) = \prod_{i=1}^{N} q(S_i|\mathbf{y}_i), \qquad q(S_i|\mathbf{y}_i) \propto p(\mathbf{y}_i|S_i)p(S_i). \tag{3.44}$$

Under the conjugate Dirichlet prior $\boldsymbol{\eta} \sim \mathcal{D}(e_0,\dots,e_0)$, we obtain the following marginal prior for a single indicator $S_i$,

$$p(S_i) \propto \Gamma(1+e_0)\Gamma(e_0)^{K-1}, \tag{3.45}$$

and the marginal likelihood of $\mathbf{y}_i$ given $S_i = k$ results from (3.25),

$$p(\mathbf{y}_i|S_i = k) = \frac{p(\mathbf{y}_i|\boldsymbol{\theta}_k)p(\boldsymbol{\theta}_k)}{p(\boldsymbol{\theta}_k|\mathbf{y}_i)}, \tag{3.46}$$

where $p(\boldsymbol{\theta}_k|\mathbf{y}_i)$ is the posterior density from the single observation $\mathbf{y}_i$. The right-hand side of (3.46) may be evaluated for arbitrary $\boldsymbol{\theta}_k$, in particular for the posterior mode of $p(\boldsymbol{\theta}_k|\mathbf{y}_i)$. (3.46) is likely to be unstable for high-dimensional parameter $\boldsymbol{\theta}_k$, where the posterior $p(\boldsymbol{\theta}_k|\mathbf{y}_i)$ is not well defined from a single observation.

To improve the efficiency of importance sampling, Casella et al. (2000) use stratified importance sampling by decomposing the space of all possible allocations into all partition sets with identical allocation size $N_k(\mathbf{S})$. Casella et al. (2000) argue that among these partition sets only a few carry most of the weights.

Casella et al. (2000) use draws from the importance density to approximate the posterior expectation of any function $h(\boldsymbol{\vartheta})$ as explained, for instance, in Geweke (1989):

$$\mathrm{E}(h(\boldsymbol{\vartheta})|\mathbf{y}) \approx \frac{1}{L}\sum_{l=1}^{L} \mathrm{E}(h(\boldsymbol{\vartheta})|\mathbf{y},\mathbf{S}^{(l)})\frac{p(\mathbf{S}^{(l)}|\mathbf{y})}{q(\mathbf{S}^{(l)})}. \tag{3.47}$$

A certain objection to this approach is that the ergodic average (3.47) may be biased due to undetected label switching.

### 3.6.3 Perfect Sampling

Like MCMC, perfect sampling is based on the idea of constructing a Markov chain where the stationary distribution is equal to an untractable posterior distribution. Whereas MCMC exploits the fact that for an ergodic Markov chain the stationary distribution is also the limiting distribution, perfect sampling is an algorithm for generating independent draws from precisely the exact stationary distribution; see Casella et al. (2001) for an introduction.

The construction of a perfect sampler for mixture models is a delicate issue as the first attempt of Hobert et al. (1999) demonstrates where they applied perfect sampling to two- and three-component mixtures where the component parameters are known. Casella et al. (2002) extend these results to finite mixtures with an arbitrary number of components and unknown component parameters where the marginal posterior $p(\mathbf{S}|\mathbf{y})$ of the allocations is available explicitly up to a constant; see also Subsection 3.3.3.

## 3.7 Bayesian Inference for Finite Mixture Models Using Posterior Draws

From a Bayesian perspective, the posterior density $p(\boldsymbol{\vartheta}|\mathbf{y})$ contains all information provided by the data, and is the basis for drawing inference on any quantity of interest. If a sampling-based approach as described in Sections 3.5 and 3.6 is pursued for practical estimation, a sequence of draws $\{\boldsymbol{\vartheta}^{(m)}, m = 1, \ldots, M\}$ from the posterior distribution $p(\boldsymbol{\vartheta}|\mathbf{y})$ is available, which could be used to approximate all quantities of interest. In what follows, it is assumed that an appropriate amount of initial draws $M_0$ has been removed, if the draws were produced by an MCMC sampler.

### 3.7.1 Sampling Representations of the Mixture Posterior Density

It is sometimes helpful to visualize the mixture posterior density $p(\boldsymbol{\vartheta}|\mathbf{y})$, but producing a simple density plot is feasible only for very simple problems, where the unknown parameter $\boldsymbol{\vartheta}$ is at most bivariate. If the dimension of $\boldsymbol{\vartheta}$ exceeds two, other tools have been developed for visualizing the mixture posterior density $p(\boldsymbol{\vartheta}|\mathbf{y})$. Draws from the posterior density $p(\boldsymbol{\vartheta}|\mathbf{y})$ have been used as a sampling representation of the mixture posterior distribution, which is then visualized in an appropriate manner (Celeux et al., 2000; Frühwirth-Schnatter, 2001b; Hurn et al., 2003).

To illustrate the equivalence of a density plot and the sampling representation, Figure 3.5 compares the contours of the mixture posterior density $p(\mu_1, \mu_2|\mathbf{y})$ with MCMC draws $\mu_1^{(m)}$ and $\mu_2^{(m)}$ from $p(\mu_1, \mu_2|\mathbf{y})$ obtained from random permutation Gibbs sampling using *Algorithm 3.5* for the synthetic

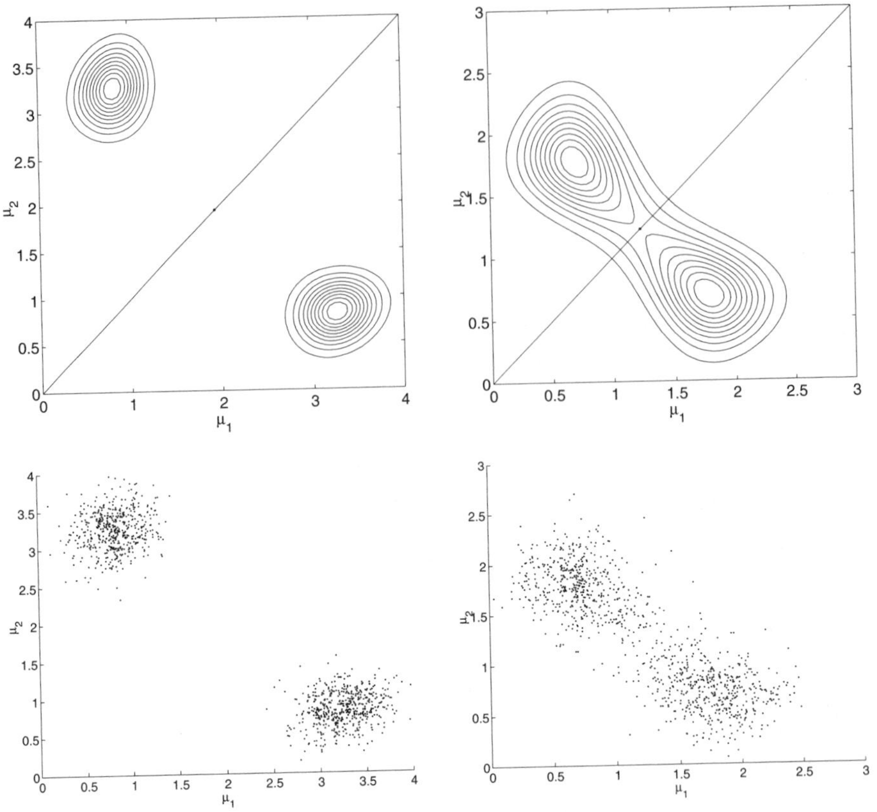


**Fig. 3.5.** HIDDEN AGE GROUPS — Synthetic Data Set 1 (left-hand side) and Data Set 2 (right-hand side); top: contours of the mixture posterior density $p(\mu_1, \mu_2|\mathbf{y})$, bottom: MCMC draws from the mixture posterior density $p(\mu_1, \mu_2|\mathbf{y})$ obtained from random permutation sampling)

data sets 1 and 2 discussed earlier in Subsection 3.5.6. By using the random permutation Gibbs sampler, rather than standard Gibbs sampling, the exploration of both modes of the posterior distribution is forced.

In particular, for higher-dimensional problems sampling representations are a very useful tool for visualizing the mixture posterior distribution. One interesting view is the bivariate marginal density $p(\theta_{k,j}, \theta_{k',j}|\mathbf{y})$, where $k \neq k'$, visualized for each $j = 1, \dots, d$, through scatter plots of the MCMC draws $(\theta_{k,j}^{(m)}, \theta_{k',j}^{(m)})$. By the results of Subsection 3.3.2, this density is the same for all pairs of $(k, k')$, thus $k = 1$ and $k' = 2$, or any other pair, may be selected, provided that the random permutation Gibbs sampler has been used. These figures allow us to study how much the $j$th element $\theta_{k,j}$ of the component parameter $\boldsymbol{\theta}_k$ differs among the various components. If this element is significantly different among all components, then this plot shows $K^2 - K = K(K-1)$

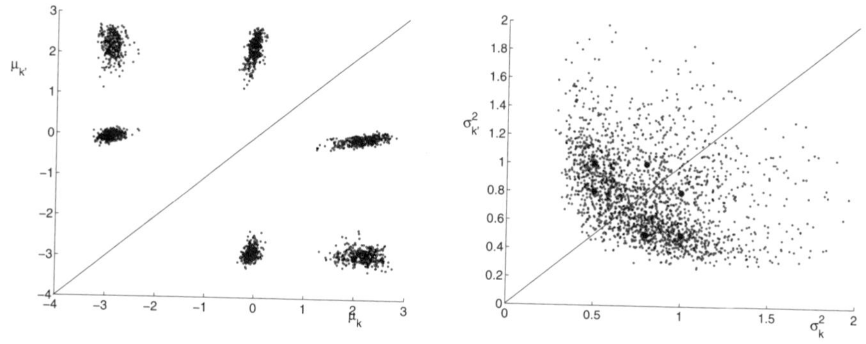

**Fig. 3.6.** Synthetic data of size $N = 500$ simulated from a mixture of three univariate normal distributions with $\eta_1 = 0.3$, $\eta_2 = 0.5$, $\mu_1 = -3$, $\mu_2 = 0$, $\mu_3 = 2$, $\sigma_1^2 = 1$, $\sigma_2^2 = 0.5$, $\sigma_3^2 = 0.8$; sampling representation of $p(\mu_k, \mu_{k'}|\mathbf{y})$ (left-hand side) and $p(\sigma_k^2, \sigma_{k'}^2|\mathbf{y})$ (right-hand side) based on random permutation Gibbs sampling

simulation clusters. If this element is nearly the same in all components, then this plot shows a single simulation cluster; see Figure 3.6 for illustration.

Another useful view is the bivariate marginal density $p(\theta_{k,j}, \theta_{k,j'}|\mathbf{y})$, which is visualized separately for each pair $j, j' = 1, \dots, d, j \neq j'$ through scatter plots of the MCMC draws $(\theta_{k,j}^{(m)}, \theta_{k,j'}^{(m)})$. By the results of Subsection 3.3.2, this density is the same for all $k = 1, \dots, K$, thus $k = 1$ or any other value may be selected. If the dimension of $\boldsymbol{\theta}_k$ is equal to two, this scatter plot is closely related to the point process representation of the underlying mixture distribution, discussed in Subsection 1.2.2. The MCMC draws will scatter around the points corresponding to the true point process representation, with the spread of the clouds representing the uncertainty of estimating the points; see Figure 3.7 for illustration. This is also true for multivariate component parameters, where the plots correspond to projections of the point process representation onto bivariate subspaces.

These figures allow us to study the component parameters in relation to each other without having to worry about label switching. In Figure 3.7, for instance, it becomes evident that the components differ mainly in the mean, that two components have nearly the same variance, whereas the third component has a variance which is slightly smaller.

For a mixture with a univariate component parameter $\theta_k$ a bivariate plot is not available. In this case $\theta_k^{(m)}$ may be plotted against $\eta_k^{(m)}$ or an auxiliary parameter $\psi^{(m)}$ which is drawn from a standard normal distribution.

### 3.7.2 Using Posterior Draws for Bayesian Inference

On the basis of the posterior density $p(\boldsymbol{\vartheta}|\mathbf{y})$, inference is drawn on quantities of interest such as the posterior mean $\mathrm{E}(\boldsymbol{\vartheta}|\mathbf{y})$, which commonly is used as a

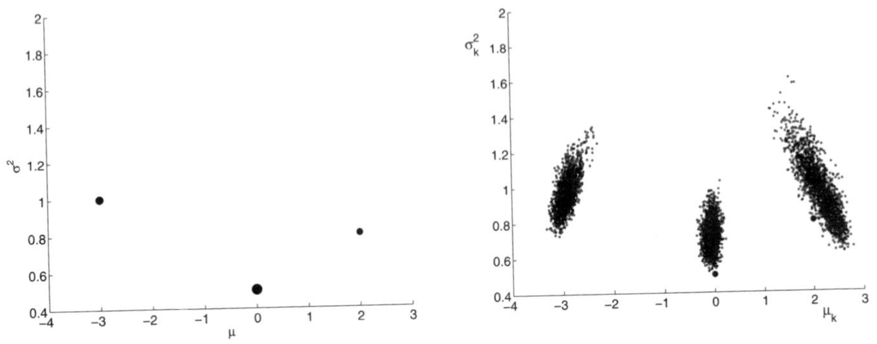


**Fig. 3.7.** Synthetic data of size $N = 500$ simulated from a mixture of three univariate normal distributions with $\eta_1 = 0.3$, $\eta_2 = 0.5$, $\mu_1 = -3$, $\mu_2 = 0$, $\mu_3 = 2$, $\sigma_1^2 = 1$, $\sigma_2^2 = 0.5$, $\sigma_3^2 = 0.8$; point process representation of the finite mixture distribution (left hand side) and point process representation of draws from $p(\mu_k, \sigma_k^2|\mathbf{y})$ based on random permutation Gibbs sampling (right-hand side)

point estimator of $\boldsymbol{\vartheta}$, or the predictive density $p(\mathbf{y}_f|\mathbf{y})$, which is a pointwise estimator of the density of the marginal distribution of the observed random variable $\mathbf{Y}$.

For finite mixture models, as for many other interesting and complex statistical models, no explicit expression is available for most quantities of interest, and draws $\{\boldsymbol{\vartheta}^{(m)}, m = 1, \ldots, M\}$ from the posterior density are used to approximate all quantities of interest. Consider, as an example, the posterior expectation

$$\mathrm{E}(h(\boldsymbol{\vartheta})|\mathbf{y}) = \int h(\boldsymbol{\vartheta}) p(\boldsymbol{\vartheta}|\mathbf{y}) d\boldsymbol{\vartheta}$$

of a function $h(\boldsymbol{\vartheta})$, which is approximated by averaging over the draws from the posterior distribution in the following way,

$$\overline{h}_M = \frac{1}{M} \sum_{m=1}^{M} h(\boldsymbol{\vartheta}^{(m)}). \tag{3.48}$$

Under mild conditions, $\overline{h}_M$ converges to $\mathrm{E}(h(\boldsymbol{\vartheta})|\mathbf{y})$ by the law of large numbers, even if the draws were generated by a Markov chain Monte Carlo method (Tierney, 1994). There are several questions associated with Bayesian inference based on posterior draws, in particular convergence diagnostics and choosing appropriate values of $M$, which are beyond the scope of this book, and are addressed, for example, in the excellent books by Robert and Casella (1999) and Liu (2001).

For finite mixture models, a specific issue arises that is related to the invariance of the posterior distribution discussed in Subsection 3.3.2 and the label switching problem discussed in Subsection 3.5.5. Bayesian inference for finite mixture models using posterior draws may be, but need not, be sensitive to label switching.

Label switching does not matter whenever the function $h(\boldsymbol{\vartheta})$ is invariant to relabeling the components of the mixture:

$$h(\boldsymbol{\vartheta}) = h(\tilde{\boldsymbol{\vartheta}}_s), \tag{3.49}$$

where $\tilde{\boldsymbol{\vartheta}}_s$ is the permuted parameter defined in (3.14). In such a case, averaging over the draws $h(\boldsymbol{\vartheta}^{(m)})$ as in (3.48) is evidently insensitive to label switching, and any of the methods discussed in Section 3.5 such as data augmentation and Gibbs sampling (*Algorithm 3.4*) or data augmentation and random permutation Gibbs sampling (*Algorithm 3.5*) may be used.

It is not always easy to identify functionals that are invariant to relabeling, in particular, if inference concerns the component parameters $(\boldsymbol{\theta}_k, \eta_k)$. Obvious estimators turn out to be sensitive to label switching, in which case it is necessary to identify the model before making an inference, as explained in detail in Subsection 3.7.7. Clustering of a single object $\mathbf{y}_i$, based on the posterior probability distribution $\Pr(S_i = k|\mathbf{y})$, into one of the $K$ hidden groups, is a further example of an inference problem where any kind of label switching matters; see Subsection 7.1.7 for more detail.

### 3.7.3 Predictive Density Estimation

A quantity that often is of interest when fitting a finite mixture model, is the posterior predictive density $p(\mathbf{y}_f|\mathbf{y})$ of a future realization $\mathbf{y}_f$, given the data $\mathbf{y}$, which is given by

$$p(\mathbf{y}_f|\mathbf{y}) = \int p(\mathbf{y}_f|\boldsymbol{\vartheta}) p(\boldsymbol{\vartheta}|\mathbf{y}) d\boldsymbol{\vartheta}.$$

This density is the posterior expectation of following function $h(\boldsymbol{\vartheta}) = p(\mathbf{y}_f|\boldsymbol{\vartheta})$,

$$p(\mathbf{y}_f|\boldsymbol{\vartheta}) = \sum_{k=1}^{K} \eta_k p(\mathbf{y}_f|\boldsymbol{\theta}_k), \tag{3.50}$$

which is invariant to relabeling the components of the mixture. Therefore, the density estimated from the MCMC draws,

$$\hat{p}(\mathbf{y}_f|\mathbf{y}) = \frac{1}{M} \sum_{m=1}^{M} \left( \sum_{k=1}^{K} \eta_k^{(m)} p(\mathbf{y}_f|\boldsymbol{\theta}_k^{(m)}) \right), \tag{3.51}$$

is robust against label switching. For illustration, consider Figure 3.8 which compares a histogram of the synthetic data sets 1 and 2 discussed earlier in Subsection 3.5.5 with the predictive density estimate $\hat{p}(y_f|\mathbf{y})$.

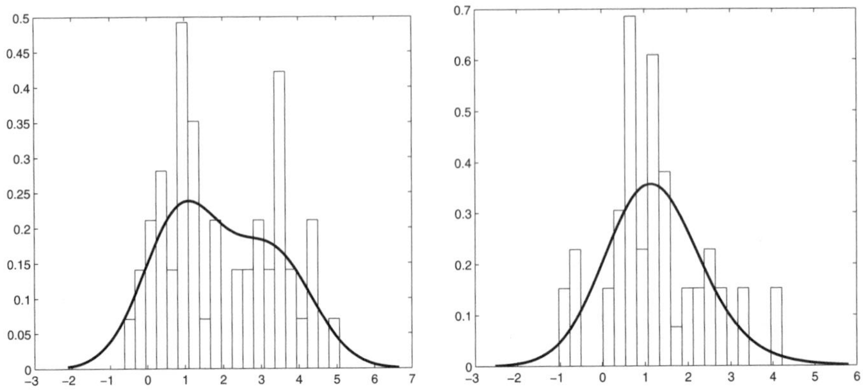

**Fig. 3.8.** HIDDEN AGE GROUPS — Synthetic Data Set 1 (left-hand side) and Synthetic Data Set 2 (right-hand side); predictive density estimate obtained from fitting a two-component normal mixture with $\mu_1, \mu_2$, and $\eta_1$ unknown (variance $\sigma_1^2 = \sigma_2^2 = 1$ fixed) in comparison to a histogram of the data

For univariate mixtures of normals, Richardson and Green (1997) studied MCMC estimation under various constraints, and observed that the predictive density estimator $\hat{p}(\mathbf{y}_f|\mathbf{y})$ differed significantly across the constraints, which is not surprising as a poor constraint introduces a bias; see again the discussion in Subsection 3.5.6. For this reason it is recommended to use draws from the unconstrained posterior when the mixture model is used for practical density estimation or as a smoothing device. Due to the invariance to relabeling, the estimator $\hat{p}(\mathbf{y}_f|\mathbf{y})$ could be based on Gibbs sampling (*Algorithm 3.4*) or random permutation Gibbs sampling (*Algorithm 3.5*).

### The Posterior Predictive Distribution of a Sequence

It is possible to predict a whole sequence $\mathbf{y}_f = (\mathbf{y}_{f,1}, \ldots, \mathbf{y}_{f,H})$ of length $H \geq 1$, given the data $\mathbf{y}$. The posterior predictive density $p(\mathbf{y}_f|\mathbf{y})$ of $\mathbf{y}_f$, conditional on the observations $\mathbf{y}$ is given by

$$p(\mathbf{y}_f|\mathbf{y}) = \int \prod_{h=1}^{H} p(\mathbf{y}_{f,h}|\boldsymbol{\vartheta})p(\boldsymbol{\vartheta}|\mathbf{y})d\boldsymbol{\vartheta}. \tag{3.52}$$

Analytical integration is not possible, but one could easily draw a sample from (3.52) if a sequence of draws from the posterior density $p(\boldsymbol{\vartheta}|\mathbf{y})$ is available, using the following algorithm.

*Algorithm 3.7: Sampling from the Posterior Predictive Distribution* Assume that a sequence of draws $\boldsymbol{\vartheta}^{(1)}, \ldots, \boldsymbol{\vartheta}^{(M)}$ from the posterior density $p(\boldsymbol{\vartheta}|\mathbf{y})$ is available. Perform the following two steps for $m = 1, \ldots, M$.

(a) Draw $H$ component indicators $S_1, \ldots, S_H$ independently from the discrete distribution $(\eta_1^{(m)}, \ldots, \eta_K^{(m)})$.
(b) For each $h = 1, \ldots, H$, sample $\mathbf{y}_{f,h}^{(m)}$ from the component density $p(\mathbf{y}|\boldsymbol{\theta}_h)$, where $\boldsymbol{\theta}_h = \boldsymbol{\theta}_{S_h}^{(m)}$. Define $\mathbf{y}_f^{(m)} = (\mathbf{y}_{f,1}^{(m)}, \ldots, \mathbf{y}_{f,H}^{(m)})$.

The sample $\mathbf{y}_f^{(1)}, \ldots, \mathbf{y}_f^{(M)}$ produced by this algorithm is a sample from the posterior predictive distribution $p(\mathbf{y}_f|\mathbf{y})$.

### 3.7.4 Individual Parameter Inference

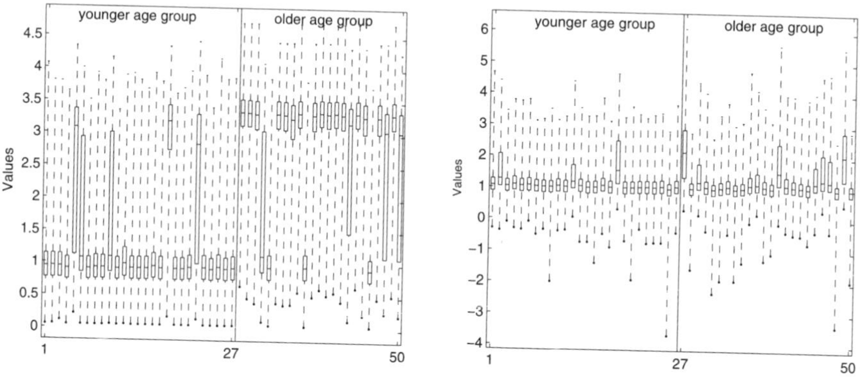


**Fig. 3.9.** HIDDEN AGE GROUPS — Synthetic Data Set 1 (left-hand side) and Synthetic Data Set 2 (right-hand side); reconstruction of the individual means $\mu_i^s$ for the two age groups obtained from fitting a two-component normal mixture with $\mu_1, \mu_2$, and $\eta_1$ unknown (variance $\sigma_1^2 = \sigma_2^2 = 1$ fixed); box plots of the posterior draws of $\mu_i^s$

Often it is of interest to make an inference about the individual parameters $\boldsymbol{\theta}_i^s$, which are defined for each subject $i$, $i = 1, \ldots, N$, by

$$\boldsymbol{\theta}_i^s = \sum_{k=1}^{K} \boldsymbol{\theta}_k I_{\{S_i=k\}}. \tag{3.53}$$

Obviously, $\boldsymbol{\theta}_i^s$ is invariant to relabeling the components of the mixture. Consequently, the sequence $\{(\boldsymbol{\theta}_1^{s,(m)}, \ldots, \boldsymbol{\theta}_N^{s,(m)}), m = 1, \ldots, M\}$, which is determined from the posterior draws $\{\boldsymbol{\vartheta}^{(m)}\}, m = 1, \ldots, M$ through the transformation (3.53),

$$\boldsymbol{\theta}_i^{s,(m)} = \boldsymbol{\theta}_{k_m}^{(m)}, \qquad k_m = S_i^{(m)},$$

contains $M$ draws from the joint posterior distribution $p(\boldsymbol{\theta}_1^s, \ldots, \boldsymbol{\theta}_N^s|\mathbf{y})$, which are insensitive to label switching. It is possible to visualize the individual parameters $\boldsymbol{\theta}_i^s$ through box-plots of $\{\boldsymbol{\theta}_i^{s,(m)}, m = 1, \ldots, M\}$ for each $i = 1, \ldots, N$, which estimate the marginal distribution $p(\boldsymbol{\theta}_i^s|\mathbf{y})$. To obtain a point estimator of $\boldsymbol{\theta}_i^s$, the expected value $\mathrm{E}(\boldsymbol{\theta}_i^s|\mathbf{y})$ is estimated from the posterior draws in an obvious way:

$$\hat{\boldsymbol{\theta}}_i^s = \frac{1}{M} \sum_{m=1}^{M} \boldsymbol{\theta}_i^{s,(m)}.$$

### An Illustrative Example

For illustration we consider the synthetic data sets 1 and 2, discussed earlier in Subsection 3.5.5. The true value of $\mu_i^s$ is equal to $\mu_y$ for the younger age group and equal to $\mu_o$ for the older age group. In Figure 3.9, box plots of $\mu_i^{s,(m)}$ are shown for both data sets, based on data augmentation and random permutation Gibbs sampling (*Algorithm 3.5*). Reconstruction of $\mu_i^s$ is rather precise for data set 1, whereas the lack of separation between the two groups leads to rather imprecise reconstructions for data set 2. This is, however, not due to any deficiencies of the sampling method, but due to a lack of information in the data.

## 3.7.5 Inference on the Hyperparameter of a Hierarchical Prior

Note that the hyperparameter $\boldsymbol{\delta}$ is invariant by definition, and may be easily estimated from the MCMC output by taking ergodic averages over the posterior draws $\boldsymbol{\delta}^{(1)}, \ldots, \boldsymbol{\delta}^{(M)}$.

## 3.7.6 Inference on Component Parameters

When making an inference about the component parameters $\boldsymbol{\theta}_1, \ldots, \boldsymbol{\theta}_K$, one is actually interested in an inference on the corresponding hidden groups in the population. Only under a unique labeling, does a fixed link exist between a hidden group with group-specific parameter $\boldsymbol{\theta}_G$ and a certain component in the mixture with component parameter $\boldsymbol{\theta}_k$. If this labeling remains the same throughout MCMC sampling, then the draws $\{\boldsymbol{\theta}_k^{(m)}, m = 1, \ldots, M\}$ may be regarded as posterior draws for the parameter $\boldsymbol{\theta}_G$, and it is possible to average over these draws to obtain a point estimator of the group-specific parameter $\boldsymbol{\theta}_G$:

$$\hat{\boldsymbol{\theta}}_G = \frac{1}{M} \sum_{m=1}^{M} \boldsymbol{\theta}_k^{(m)}. \qquad (3.54)$$

However, if label switching took place during sampling, then the hidden group parameter $\boldsymbol{\theta}_G$ no longer has to be associated with $\boldsymbol{\theta}_k$, but with another component parameter $\boldsymbol{\theta}_{k'}$. When averaging over the draws of $\boldsymbol{\theta}_k$ as in (3.54), a biased point estimator of the group-specific parameter $\boldsymbol{\theta}_G$ results, which is pulled toward the overall mean $\mathrm{E}(\boldsymbol{\theta}_k|\mathbf{y})$ of the unconstrained posterior.

To draw an inference about hidden groups by averaging over posterior draws, it is essential that these draws arise from a single labeling subspace $\mathcal{L}$. We denote such draws as $\boldsymbol{\vartheta}^{\mathcal{L},(m)} = (\boldsymbol{\theta}_1^{\mathcal{L},(m)}, \ldots, \boldsymbol{\theta}_K^{\mathcal{L},(m)}, \eta_1^{\mathcal{L},(m)}, \ldots, \eta_K^{\mathcal{L},(m)})$. These draws could be used to estimate the parameters in the hidden groups by

$$\hat{\boldsymbol{\theta}}_k = \frac{1}{M} \sum_{m=1}^{M} \boldsymbol{\theta}_k^{\mathcal{L},(m)}, \tag{3.55}$$

as well as the group sizes by

$$\hat{\eta}_k = \frac{1}{M} \sum_{m=1}^{M} \eta_k^{\mathcal{L},(m)}. \tag{3.56}$$

It is discussed in detail in Subsection 3.7.7 how to obtain posterior draws from a unique labeling subspace.

### Choosing Invariant Loss Functions

It should be noted that not all point estimators of $\boldsymbol{\vartheta}$ are sensitive to label switching. Whether this is the case depends on the underlying loss function. Within a decision-theoretic framework any point estimator $\boldsymbol{\vartheta}^\star$ is derived as that value which minimizes the expected posterior loss under a certain loss function $R(\hat{\boldsymbol{\vartheta}}, \boldsymbol{\vartheta})$:

$$\boldsymbol{\vartheta}^\star = \arg\min_{\hat{\boldsymbol{\vartheta}}} \mathrm{E}(R(\hat{\boldsymbol{\vartheta}}, \boldsymbol{\vartheta})|\mathbf{y}) = \int_{\Theta} R(\hat{\boldsymbol{\vartheta}}, \boldsymbol{\vartheta}) p(\boldsymbol{\vartheta}|\mathbf{y}) d\boldsymbol{\vartheta};$$

see Berger (1985) for a full account. If this framework is applied to finite mixture models, sensible estimators are obtained only if the loss function $R(\hat{\boldsymbol{\vartheta}}, \boldsymbol{\vartheta})$, which corresponds to $h(\boldsymbol{\vartheta})$ in (3.48), is invariant to relabeling the components of the mixture.

This leads immediately to problems with the quadratic loss-function $R(\hat{\boldsymbol{\vartheta}}, \boldsymbol{\vartheta}) = (\hat{\boldsymbol{\vartheta}} - \boldsymbol{\vartheta})'(\hat{\boldsymbol{\vartheta}} - \boldsymbol{\vartheta})$, which yields the posterior mean $\mathrm{E}(\boldsymbol{\vartheta}|\mathbf{y})$ as optimal estimator, and is for many other statistical models the most commonly used loss function. Evidently, the functional value of $R(\hat{\boldsymbol{\vartheta}}, \boldsymbol{\vartheta})$ changes when the components of the mixture are relabeled, leading to an ambiguous definition of the expected risk. Interestingly enough, it was realized earlier that the posterior mean $\mathrm{E}(\boldsymbol{\vartheta}|\mathbf{y})$ is not a sensible point estimator as it does not contain any component-specific information; see again (3.21).

The 0/1 loss function, for which the posterior mode turns out to be the optimal estimator (see, for instance, Zellner, 1971), is easily adapted to finite mixture models by defining that $R(\hat{\boldsymbol{\vartheta}}, \boldsymbol{\vartheta}) = 0$ iff $\hat{\boldsymbol{\vartheta}}$ and $\boldsymbol{\vartheta}$ are identical up to permutations, otherwise $R(\hat{\boldsymbol{\vartheta}}, \boldsymbol{\vartheta})$ is equal to 1. This loss function is invariant to relabeling, and the mode of the mixture posterior may be used for estimation. The posterior mode may be approximated from the posterior draws $\{\boldsymbol{\vartheta}^{(m)}, m = 1, \ldots, M\}$ through that value which maximizes the nonnormalized mixture posterior density $p^\star(\boldsymbol{\vartheta}|\mathbf{y}) = p(\mathbf{y}|\boldsymbol{\vartheta})p(\boldsymbol{\vartheta})$.

Various alternative loss functions have been considered for parameter estimation in mixture models. Celeux et al. (2000) consider loss functions that are based on the predictive density $p(\mathbf{y}_f|\boldsymbol{\vartheta})$ which is invariant to relabeling the components; see again (3.50). Examples include the integrated squared difference

$$R(\hat{\boldsymbol{\vartheta}}, \boldsymbol{\vartheta}) = \int_{\mathcal{Y}} (p(\mathbf{y}_f|\boldsymbol{\vartheta}) - p(\mathbf{y}_f|\hat{\boldsymbol{\vartheta}}))^2 d\mathbf{y}_f,$$

and the symmetrized Kullback–Leibler distance

$$R(\hat{\boldsymbol{\vartheta}}, \boldsymbol{\vartheta}) = \int_{\mathcal{Y}} \left( p(\mathbf{y}_f|\boldsymbol{\vartheta}) \log \frac{p(\mathbf{y}_f|\boldsymbol{\vartheta})}{p(\mathbf{y}_f|\hat{\boldsymbol{\vartheta}})} + p(\mathbf{y}_f|\hat{\boldsymbol{\vartheta}}) \log \frac{p(\mathbf{y}_f|\hat{\boldsymbol{\vartheta}})}{p(\mathbf{y}_f|\boldsymbol{\vartheta})} \right) d\mathbf{y}_f,$$

where in both cases integration reduces to summation for a discrete sample space $\mathcal{Y}$. In both cases the expected loss is given by an expression that contains expectations of terms such as $p(\mathbf{y}_f|\boldsymbol{\vartheta})$, $p(\mathbf{y}_f|\boldsymbol{\vartheta})^2$, or $\log p(\mathbf{y}_f|\boldsymbol{\vartheta})$, with respect to the posterior density $p(\boldsymbol{\vartheta}|\mathbf{y})$. The practical evaluation of these estimators is rather involved and Celeux et al. (2000) follow the two-step procedure of Rue (1995). In a first step, expectations with respect to the posterior density are evaluated using posterior draws and integration with respect to $\mathbf{y}_f$ is carried out using some numerical technique. In a second step, the minimization problem for the estimator $\boldsymbol{\vartheta}^\star$ is solved using simulated annealing. We refer to Celeux et al. (2000) for further computational details.

Dias and Wedel (2004) provide an empirical comparison of EM and MCMC performance, which includes different prior specifications and various procedures to deal with the label switching problem.

### 3.7.7 Model Identification

The parameter estimation problem discussed in Subsection 3.7.6 illustrated that care must be exercised when using draws from the mixture posterior density $p(\boldsymbol{\vartheta}|\mathbf{y})$ to estimate functionals of $\boldsymbol{\vartheta}$, which are not invariant to relabeling the components of the finite mixture. Inference on such functionals is sensible only if the posterior draws come from a unique labeling subspace of the unconstrained parameter space. The discussion of this subsection is devoted to the difficult task of identifying such draws.

Gibbs sampling as described in *Algorithm 3.4* may lead to implicit model identification if the $K!$ modal parts of the mixture posterior density are very well separated, and the sampler is trapped in one of modal regions; see again the discussion in Subsection 3.5.6. In this case the posterior draws obtained by *Algorithm 3.4* may be treated as coming from a unique labeling subspace, $\boldsymbol{\vartheta}^{\mathcal{L},(m)} = \boldsymbol{\vartheta}^{(m)}, m = 1, \ldots, M$, as was done for instance in Chib (1996). It is, however, not recommended to rely blindly on this implicit model identification, as the behavior of Gibbs sampling is unpredictable in this respect.

One strategy is to relabel the posterior draws $\{\boldsymbol{\vartheta}^{(m)}, m = 1, \ldots, M\}$ in such a way that draws $\{\boldsymbol{\vartheta}^{\mathcal{L},(m)}, m = 1, \ldots, M\}$ from a unique labeling subspace result. This may be achieved by isolating a sensible identifiability constraint through exploring the posterior draws (Frühwirth-Schnatter, 2001b) or by unsupervised clustering of the posterior draws (Celeux, 1998).

## Model Identification Through Identifiability Constraint

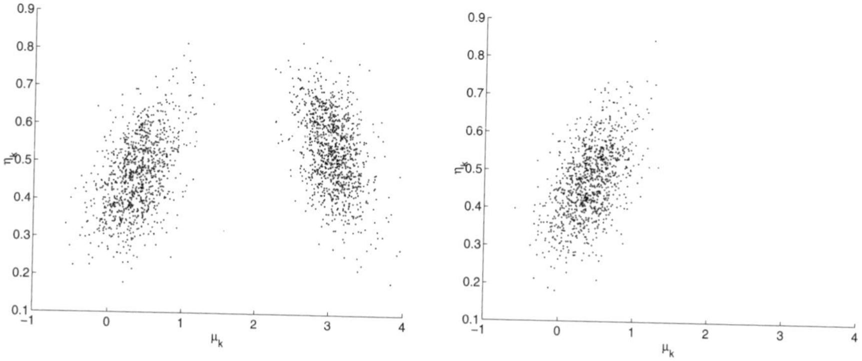


**Fig. 3.10.** HIDDEN AGE GROUPS — Synthetic Data Set 3; draws from the bivariate marginal distributions $p(\mu_k, \eta_k|\mathbf{y})$ (left-hand side); posterior draws of $(\mu_1, \eta_1)$ under the constraint $\mu_1 < \mu_2$ (right-hand side)

A common reaction to the label switching problem is to impose some formal identifiability constraint as in Subsection 1.3.3 within sampling-based Bayesian estimation (Albert and Chib, 1993; Richardson and Green, 1997). It has been realized only rather recently that an arbitrary formal identifiability constraint does not necessarily generate a unique labeling and that a poorly chosen constraint introduces a bias (Celeux, 1998; Celeux et al., 2000; Stephens, 2000b; Frühwirth-Schnatter, 2001b); recall also the discussion at the end of Subsection 2.4.3.

To identify sensible identifiability constraints, Frühwirth-Schnatter (2001b) explored the point process representation of the MCMC draws, introduced earlier in Subsection 3.7.1. Note that the constraint is only an indirect device to describe the differences between the components, and therefore is not necessarily unique. Various case studies where it is useful to explore the point process representation of the MCMC draws in this way may be found throughout the book; see also Frühwirth-Schnatter (2001a, 2001b), Kaufmann and Frühwirth-Schnatter (2002), and Frühwirth-Schnatter et al. (2004).

A straightforward method to impose a constraint on the posterior draws is to postprocess the MCMC draws that were generated from the mixture posterior. Whenever a draw does not satisfy the constraint, one permutes the labeling of the components such that the constraint is fulfilled (Richardson and Green, 1997; Stephens, 1997b; Frühwirth-Schnatter, 2001b). Frühwirth-Schnatter (2001b) also provides a formal proof that this method actually delivers a sample from the constrained posterior.

For illustration, we return to the synthetic data set 3 introduced at the end of Subsection 2.4.3. Figure 3.10 shows a sampling representation of the bivariate marginal distribution $p(\mu_k, \eta_k|\mathbf{y})$ for these data. From this scatter plot it is obvious that the component parameters differ mainly in the mean, whereas the weights are rather equal. The constraint $\mu_1 < \mu_2$ is actually able to impose a unique labeling.

#### Model Identification Through Unsupervised Clustering of the Posterior Draws

For higher-dimensional problems, in particular for multivariate mixtures, it is possible, but somewhat time-consuming to search for identifiability constraints in the MCMC output (Frühwirth-Schnatter et al., 2004). As a more automatic procedure, Celeux (1998) suggested permuting the MCMC draws obtained from unconstrained sampling by using a clustering procedure. His algorithm is an on-line $k$-means type algorithm with $K!$ clusters, which is initialized from the first 100 draws after reaching burn-in, by defining $K!$ reference centers from these draws. For each MCMC draw $\boldsymbol{\vartheta}^{(m)}$ the distance to each of these $K!$ centers is computed, which is then used to permute the labels.

#### Model Identification Through Clustering in the Point Process Representation

A related method is to search for clusters in the point process representation of the MCMC draws, introduced earlier in Subsection 3.7.1, which additionally provides some control over the important question of whether the model is overfitting the number of components.

Each of the MCMC simulations $\boldsymbol{\theta}_1^{(m)}, \dots, \boldsymbol{\theta}_K^{(m)}$ corresponds to a certain point process representation that will cluster around the point process representation of the underlying true finite mixture distribution. If the heterogeneity between the underlying points is large enough, $K$ simulation clusters will be present in the point process representation of the MCMC draws.

Permuting the labels of $\boldsymbol{\theta}_1^{(m)}, \ldots, \boldsymbol{\theta}_K^{(m)}$ does not change the point representation; it only changes the one-to-one correspondence between the component-specific draws and the simulation clusters. A unique labeling is achieved if all draws $\boldsymbol{\theta}_k^{(m)}$ are associated with the same simulation cluster for all $m = 1, \ldots, M$. This is achieved by applying a standard $k$-means clustering algorithm with $K$ clusters to a sample of size $MK$, formed from the MCMC draws $\{(\boldsymbol{\theta}_1^{(m)}, \ldots, \boldsymbol{\theta}_K^{(m)}), m = 1, \ldots, M\}$, with the posterior mode estimator $(\boldsymbol{\theta}_1^\star, \ldots, \boldsymbol{\theta}_K^\star)$ serving as a starting value for the cluster means. The clustering algorithm delivers a classification sequence $\{(\rho_m(1), \ldots, \rho_m(K)), m = 1, \ldots, M\}$, where $\rho_m(k)$ determines to which simulation cluster the MCMC draw $\boldsymbol{\theta}_k^{(m)}$ belongs.

If the simulation clusters are well separated, then the classification sequence $\{\rho_m(1), \ldots, \rho_m(K)\}$ is a permutation of $\{1, \ldots, K\}$; that is,

$$\sum_{k=1}^{K} \rho_m(k) = \frac{K(K+1)}{2}. \tag{3.57}$$

In this case it is possible to relabel the MCMC draw $\boldsymbol{\vartheta}^{(m)}$ through the permutation $\{\rho_m(1), \ldots, \rho_m(K)\}$; that is

$$\boldsymbol{\vartheta}^{\mathcal{L},(m)} = (\boldsymbol{\theta}_{\rho_m(1)}^{\mathcal{L},(m)}, \ldots, \boldsymbol{\theta}_{\rho_m(K)}^{\mathcal{L},(m)}, \eta_{\rho_m(1)}^{\mathcal{L},(m)}, \ldots, \eta_{\rho_m(K)}^{\mathcal{L},(m)}) \tag{3.58}$$

defined by:

$$\boldsymbol{\theta}_{\rho_m(k)}^{\mathcal{L},(m)} = \boldsymbol{\theta}_k^{(m)}, \qquad \eta_{\rho_m(k)}^{\mathcal{L},(m)} = \eta_k^{(m)}, \qquad k = 1, \ldots, K.$$

As all component-specific draws are associated with the same simulation cluster, the draws defined in (3.58) may be regarded as coming from an identified mixture model.

If in addition to $\boldsymbol{\vartheta}^{(m)}$, allocation variables $\mathbf{S}^{(m)} = (S_1^{(m)}, \ldots, S_N^{(m)})$ have been stored, then the same permutation could be used on them to define allocation under a unique labeling for each $i = 1, \ldots, N$:

$$S_i^{\mathcal{L},(m)} = \rho_m(S_i^{(m)}). \tag{3.59}$$

If $\{\rho_m(1), \ldots, \rho_m(K)\}$ is not a permutation of $\{1, \ldots, K\}$ (i.e., if (3.57) is violated for a considerable fraction of the MCMC draws), this is an indication that the mixture is overfitting the number of components, a problem that is discussed in Subsection 4.2.2.

## Further Approaches Toward Relabeling the MCMC Draws

Various authors found other ways of relabeling the MCMC draws useful. Stephens (1997b) suggested relabeling the MCMC output so that the estimated marginal posterior distributions of the parameters of interest are as close to unimodality as possible. Stephens (2000b) tackles the whole relabeling problem from a decision-theoretic viewpoint and shows that the relabeling strategies studied in Stephens (1997b) and Celeux (1998) may be viewed as an attempt to minimize the posterior expectation of a certain loss function.

# 4

# Statistical Inference for Finite Mixture Models Under Model Specification Uncertainty

## 4.1 Introduction

The decision to fit a finite mixture model to data will often result from careful consideration; sometimes, however, alternative models will be available, which then should be compared with models based on the selected finite mixture distribution.

Even if we stay within a certain family of mixture distributions, we may face model specification problems, the most important being the choice of $K$, the number of components. If it is impossible to assign a value to $K$ a priori with complete certainty, we are faced with the problem of estimating $K$ from the data. In many applications it is of substantial interest to test hypotheses about $K$, most important, to test heterogeneity $(K > 1)$ against homogeneity against $(K = 1)$. Testing for the number of components in a mixture model is known to be a difficult problem, because it involves inference for an overfitting mixture model where the true number of components is less than the number of components in the fitted mixture model. Parameter estimation in this case represent a nonregular problem, with the true parameter lying in a nondentifiable subset of the larger parameter space, as discussed in Section 4.2.

Many approaches have been put forward to deal with model specification uncertainty. Several informal methods for diagnosing mixtures such as diagnosing goodness-of-fit through implied moments or the predictive performance are reviewed in Section 4.3. Likelihood-based methods, in particular the likelihood ratio statistics and AIB and BIC are discussed in Section 4.4. Finally, Section 4.5 provides an introduction to Bayesian inference under model uncertainty. Its application to finite mixture models is, however, discussed in full detail in Chapter 5.

## 4.2 Parameter Estimation Under Model Specification Uncertainty

When a finite mixture distribution with $K$ components is used as part of a statistical model for real data, it may happen that the mixture distribution is misspecified.

### 4.2.1 Maximum Likelihood Estimation Under Model Specification Uncertainty

Consider two models $\mathcal{M}_K$ and $\mathcal{M}_\mathcal{R}$ where $\mathcal{M}_\mathcal{R}$ is a constrained version of $\mathcal{M}_K$, meaning that the parameter space $\Theta_\mathcal{R} \subset \Theta_K$. For finite mixture models, for instance, a constrained version of a univariate mixture of normal distributions could be a mixture where $\mu_k = 0$ in all groups, that is,

$$\Theta_\mathcal{R} = \{(\mu_1, \sigma_1^2, \ldots, \mu_K, \sigma_K^2, \boldsymbol{\eta}) \in \Theta_K : \mu_1 = \cdots = \mu_K = 0\}, \tag{4.1}$$

or a homoscedastic mixture with $\sigma_1^2 = \cdots = \sigma_K^2$, that is,

$$\Theta_\mathcal{R} = \{(\mu_1, \sigma_1^2, \ldots, \mu_K, \sigma_K^2, \boldsymbol{\eta}) \in \Theta_K : \sigma_1^2 = \cdots = \sigma_K^2\}. \tag{4.2}$$

Under uncertainty whether $\mathcal{M}_K$ or $\mathcal{M}_\mathcal{R}$ holds, one could fit the larger model, and provided that certain regularity conditions hold, standard ML theory still applies, meaning the ML estimator of model $\mathcal{M}_K$ is asymptotically normal around the true parameter $\boldsymbol{\vartheta}^{\text{true}}$, regardless of whether $\mathcal{M}_K$ or $\mathcal{M}_\mathcal{R}$ is true. These regularity conditions state that all points in $\mathcal{M}_\mathcal{R}$ lie in the interior of the parameter space $\Theta_K$, and that the true value $\boldsymbol{\vartheta}^{\text{true}}$ does not lie in a nonidentifiability set of $\Theta_K$.

For finite mixture models, these regularity conditions are usually fulfilled when both $\mathcal{M}_K$ and $\mathcal{M}_\mathcal{R}$ assume the correct number of components, and $\mathcal{M}_\mathcal{R}$ is obtained by putting constraints on $\boldsymbol{\theta}_1, \ldots, \boldsymbol{\theta}_K$ as in (4.1) or (4.2). In such a case all points in $\Theta_\mathcal{R}$ are interior points of the larger space $\Theta_K$ and the mixture is generically identifiable for all points in $\Theta_\mathcal{R}$ if it is generically identifiable for all points in $\Theta_K$. Consequently, as in Subsection 2.4.4, the mixture likelihood function is asymptotically normal around the true parameter $\boldsymbol{\vartheta}^{\text{true}} = (\boldsymbol{\theta}_1^{\text{true}}, \ldots, \boldsymbol{\theta}_K^{\text{true}}, \boldsymbol{\eta}^{\text{true}})$ (and all $K!$ permutations).

The standard regularity conditions, however, are violated when model uncertainty concerns the number of components. A finite mixture is said to be overfitting the number of components when the true data-generating mixture distribution contains only $K^{\text{true}} < K$ distinct, nonempty components. The difference $K - K^{\text{true}}$ between the assumed and the actual number of components defines the degree of overfitting.

Consider an overfitting mixture model where the data $\mathbf{y} = (\mathbf{y}_1, \ldots, \mathbf{y}_N)$ are generated by a mixture with $K^{\text{true}} = K-1$ distinct, nonempty components defined by parameter $\boldsymbol{\vartheta}_{K-1}^{\text{true}}$, and a mixture with $K$ components with parameter $\boldsymbol{\vartheta}_K$ has been fitted. It follows immediately from Subsection 1.3.2 that the true value of $\boldsymbol{\vartheta}_K$ lies in the nonidentifiability set $\mathcal{U}^Z(\boldsymbol{\vartheta}_{K-1}^{\text{true}}) \cup \mathcal{U}^E(\boldsymbol{\vartheta}_{K-1}^{\text{true}})$, defined in (1.28) and (1.30), which violates one of the regularity conditions mentioned above. Furthermore, one of these sets, namely $\mathcal{U}^E(\boldsymbol{\vartheta}_{K-1}^{\text{true}})$, contains points that lie on the boundary of the parameter space, namely all points where $\eta_k = 0$, which violates the other regularity condition. Consequently, standard ML theory no longer applies, and the surface of the mixture likelihood function of a model that is overfitting the number of components does not converge to an approximate normal distribution with full rank asymptotic covariance matrix, when considered as a function of $\boldsymbol{\vartheta}_K = (\boldsymbol{\theta}_1, \dots, \boldsymbol{\theta}_K, \eta_1, \dots, \eta_{K-1})$ (Li and Sedransk, 1988). This may lead to mixture likelihood functions that behave rather irregularly.

Nevertheless Feng and McCulloch (1996) proved convergence of the ML estimator $\hat{\boldsymbol{\vartheta}}_K$ to the true value in the following sense. There exists a point $\boldsymbol{\vartheta}_0(\hat{\boldsymbol{\vartheta}}_K) \in \mathcal{U}^Z(\boldsymbol{\vartheta}_{K-1}^{\text{true}}) \cup \mathcal{U}^E(\boldsymbol{\vartheta}_{K-1}^{\text{true}})$, such that the difference $\hat{\boldsymbol{\vartheta}}_K - \boldsymbol{\vartheta}_0(\hat{\boldsymbol{\vartheta}}_k)$ converges to 0. In the limit, as $N \to \infty$, the ML estimator is not unique, but equal to any point in $\mathcal{U}^Z(\boldsymbol{\vartheta}_{K-1}^{\text{true}}) \cup \mathcal{U}^E(\boldsymbol{\vartheta}_{K-1}^{\text{true}})$. Thus in the limit, the estimated mixture either has an empty component or two components are identical.

As mentioned earlier, all parameters in subset $\mathcal{U}^Z(\boldsymbol{\vartheta}_{K-1}^{\text{true}}) \cup \mathcal{U}^E(\boldsymbol{\vartheta}_{K-1}^{\text{true}})$ are related by $(K-1)!$ different ways of relabeling. As the likelihood function of an overfitting model of degree 1 converges to this set, the likelihood function will exhibit $K^{\text{true}}! = (K-1)!$ modes with increasing number of observations. Similar results hold for mixtures with a higher degree of overfitting. In the limit, as $N \to \infty$, the ML estimator is not unique, but equal to any point in $\mathcal{U}^Z(\boldsymbol{\vartheta}_{K-L}^{\text{true}}) \cup \mathcal{U}^E(\boldsymbol{\vartheta}_{K-L}^{\text{true}})$. Therefore the likelihood function will exhibit $K^{\text{true}}! = (K-L)!$ modes with increasing number of observations.

### A Helicopter Tour of the Mixture Likelihood Surface for an Overfitting Mixture

For illustration, consider a synthetic data set of $N = 100$ observations simulated from a single normal distribution with $\mu = 1$. Assume that the two-component normal mixture model

$$p(y) = \eta_1 f_N(y; \mu_1, 1) + \eta_2 f_N(y; \mu_2, 1)$$

is fitted. First, consider the mixture likelihood function of a two-component mixture model with $\mu_1$ and $\mu_2$ unknown, but equal weights ($\eta_1 = \eta_2 = 0.5$ fixed). The mixture likelihood function in Figure 4.1 is not roughly normal around two separated modes, which is typical for a mixture model that is overfitting. Next assume that $\eta_1$ and $\eta_2$ are unknown as well. Figure 4.2 shows the surface and the contours of the integrated likelihood $p(\mathbf{y}|\mu_1, \mu_2)$ where $\eta_2$ is integrated out:

$$p(\mathbf{y}|\mu_1, \mu_2) = \int_0^1 p(\mathbf{y}|\mu_1, \mu_2, \eta_2) d\eta_2.$$

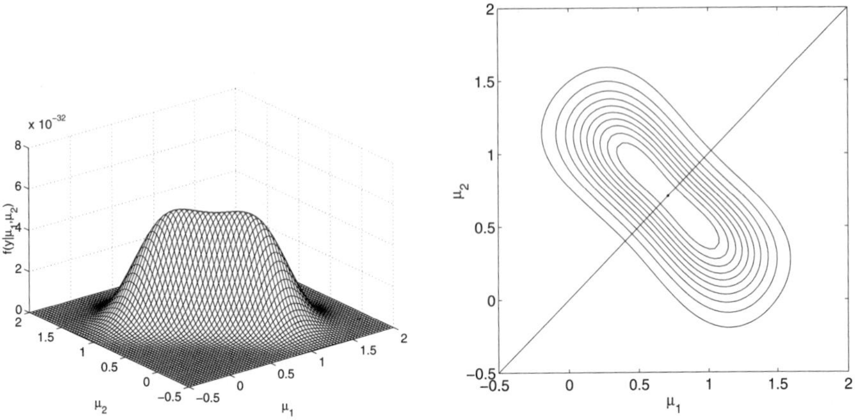


**Fig. 4.1.** Surface and contours of the mixture likelihood when fitting a mixture of two normal distributions with $\sigma_1^2 = \sigma_2^2 = 1$ and equal weights to 100 observations from the $\mathcal{N}(1, 1)$ distribution

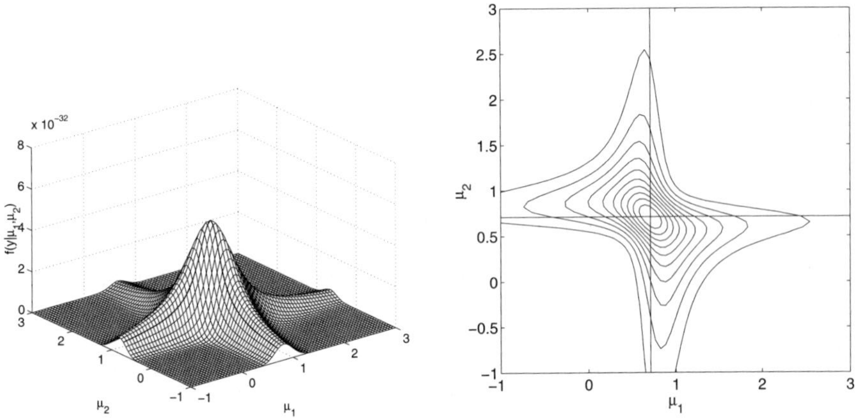


**Fig. 4.2.** Surface and contours of the integrated mixture likelihood when fitting a mixture of two normal distributions with $\sigma_1^2 = \sigma_2^2 = 1$ and unknown weights to 100 observations from the $\mathcal{N}(1, 1)$ distribution

In a Bayesian context, $p(\mathbf{y}|\mu_1, \mu_2)$ is the marginal likelihood of $(\mu_1, \mu_2)$ under a uniform prior on $\eta_2$. We find that the assumption of unknown weights exercises a considerable influence on the integrated mixture likelihood in comparison to the mixture likelihood with known weights appearing in Figure 4.1. The inclusion of the unknown weights helps the likelihood function to concentrate over the reduced model, which is the correct one for this data set. The integrated likelihood function has considerable mass along the axis $\mu_1 = \overline{y}$, $\mu_2$

arbitrary, and $\mu_1$ arbitrary, $\mu_2 = \overline{y}$, which correspond to the nonidentifiability set $\mathcal{U}^Z(\hat{\mu})$, where $\hat{\mu} = \overline{y}$ is the ML estimator, when fitting a single normal distribution.

### 4.2.2 Practical Bayesian Parameter Estimation for Overfitting Finite Mixture Models

Assume that practical Bayesian parameter estimation as discussed in Chapter 3 is applied to a finite model $\mathcal{M}_K$ with $\boldsymbol{\vartheta}_K = (\boldsymbol{\theta}_1, \dots, \boldsymbol{\theta}_K, \boldsymbol{\eta})$, which is overfitting the number of components. What happens if Bayesian estimation through Gibbs sampling and MCMC as described in *Algorithm 3.4* or *Algorithm 3.5* is applied? Most important, the resulting draws still form a Markov chain, with the posterior distribution $p(\boldsymbol{\vartheta}_K|\mathbf{y}, \mathcal{M}_K)$ of the overfitting model serving as stationary distribution. Like the mixture likelihood function discussed in Subsection 4.2.1, the mixture posterior distribution will center around the nonidentifiability sets $\mathcal{U}^E(\boldsymbol{\vartheta}_{K-1}^{\text{true}}) \cup \mathcal{U}^Z(\boldsymbol{\vartheta}_{K-1}^{\text{true}})$ defined in Subsection 1.3.2 unless a prior is applied that is informative enough to bound the posterior away from these sets. Therefore by exploring these draws, hypotheses about the true model structure may be deduced; see Subsection 4.3.1 for a more detailed discussion.

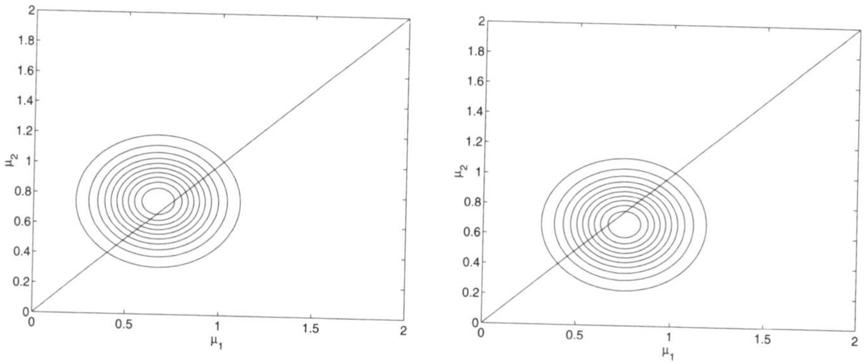

**Fig. 4.3.** Contours of the complete-data likelihood function for 100 observations simulated from a mixture of two normal distributions with $\sigma_1^2 = \sigma_2^2 = 1$, equal weights and $\mu_1 = \mu_2 = 1$; under labeling 1 (left-hand side) and under labeling 2 (right-hand side)

Some care must be exercised when the posterior draws are used for Bayesian inference as described in Subsection 3.7, because label switching may be present in the posterior draws. Unfortunately, label switching is unavoidable for a model that is overfitting the number of components. For illustration, consider a synthetic data set of $N = 100$ observations, simulated from the two-component normal mixture model

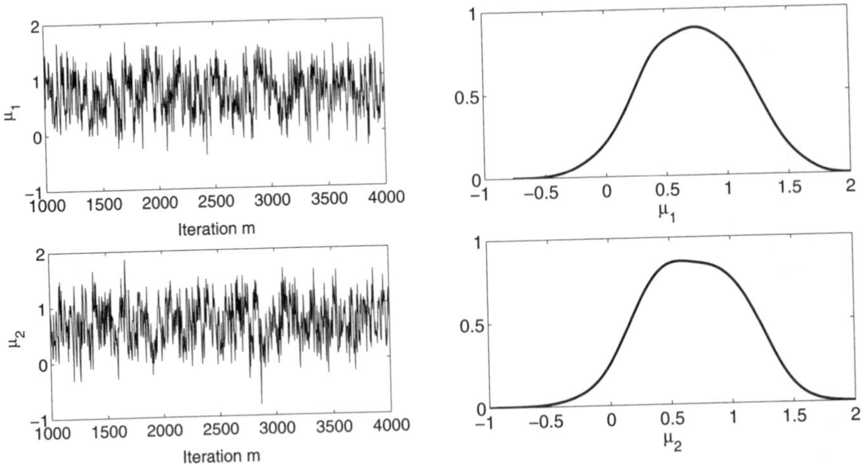


**Fig. 4.4.** Fitting a mixture of two normal distributions with $\sigma_1^2 = \sigma_2^2 = 1$ and equal weights to 100 observations simulated from a $\mathcal{N}(1,1)$-distribution; MCMC draws obtained by full conditional Gibbs sampling (left-hand side) and estimated marginal posterior densities (right-hand side) of $\mu_1$ (top) and $\mu_2$ (bottom)

$$p(y) = 0.5 f_N(y; \mu_1, 1) + 0.5 f_N(y; \mu_2, 1),$$

where $\mu_1 = \mu_2 = 1$. Figure 4.3 shows the contours of the complete-data likelihood function of these data under the two possible ways of labeling. As the likelihood functions are nearly identical for both ways of labeling, label switching is likely to occur frequently, even when running straightforward Gibbs sampling as is evident from Figure 4.4. Hence, label switching is natural for an overfitting model and could be regarded as an indicator that the assumed model is overfitting.

For a mixture model that is overfitting the number of components, formal identification as in Subsection 1.3.3 no longer is possible, as either one of the weights is zero, violating identifiability constraint (1.31), or as two of the component parameters are equal, violating identifiability constraint (1.33). Consequently, it is not sensible to try model identification for an overfitting mixture with $K$ components from the MCMC draws as in Subsection 3.7.7.

### Choosing Priors That Bound the Posterior Away from the Unidentifiability Sets

It may be desirable to bound the mixture posterior density $p(\boldsymbol{\vartheta}_K|\mathbf{y}, \mathcal{M}_K)$ of a finite mixture model with $K$ components away from the nonidentifiability sets $\mathcal{U}^Z(\boldsymbol{\vartheta}_{K-1})$ and $\mathcal{U}^E(\boldsymbol{\vartheta}_{K-1})$ defined in Subsection 1.3.2.

Under model specification uncertainty with respect to $K$, it is in general sensible to bound the mixture posterior away from the nonidentifiability set

$\mathcal{U}^Z(\boldsymbol{\vartheta}_{K-1})$ corresponding to a $K$-component mixture with one empty component by choosing a $\mathcal{D}(e_0, \ldots, e_0)$-prior for the weight distribution $\boldsymbol{\eta}$ with $e_0 > 1$ (e.g., $e_0 = 4$); see also Subsection 2.3.4. If the mixture model is overfitting, this will avoid sampling mixtures with empty components. Consequently, two of the component parameters will be pulled together to capture overfitting and observations are then allocated more or less randomly between these components. Therefore sampling of component parameters from the prior is avoided, which will increase stability of the sampler.

In some cases model selection may point to a finite mixture model, where the posterior density is not sufficiently bounded away from the nonidentifiability set $\mathcal{U}^E(\boldsymbol{\vartheta}_{K-1})$ corresponding to a $K$-component mixture with two identical components to achieve satisfactory model identification. Then it often helps to modify the prior on the component parameters by increasing prior shrinkage for elements of the component parameter $\boldsymbol{\theta}_k$ that are not extremely distinctive between the components. These elements could be identified from the point process representation of the MCMC draws, discussed earlier in Subsection 3.7.1. Simultaneously, too strong shrinkage for elements of the component parameter $\boldsymbol{\theta}_k$ that differ between the components should be avoided, because the invariant prior is centered at the nonidentifiability set $\mathcal{U}^E(\boldsymbol{\vartheta}_{K-1})$, and one may unintentionally force label switching by choosing a prior that is too informative for these elements of $\boldsymbol{\theta}_k$.

It is necessary to end these remarks with the warning that there is a price to be paid for avoiding nonidentifiability through the prior. If the finite mixture model is actually overfitting, the possibility of reducing this mixture to the true model is lost. In particular, for a mixture model that is overfitting the number of components, informative priors tend to force too many distinct components. Therefore these priors are mainly applied in a model specification context, where more than one mixture model is fitted to the data, and these priors should help to increase the discrimination between these models.

### 4.2.3 Potential Overfitting

It has been discussed in detail in Subsection 1.3.2 that a certain region of the parameter space $\Theta_K$ of a mixture distribution with $K$ components, which is centered around the nonidentifiability set $\mathcal{U}^Z(\hat{\boldsymbol{\vartheta}}_{K-1}) \cup \mathcal{U}^E(\hat{\boldsymbol{\vartheta}}_{K-1})$, is reserved for coping with data for which this mixture model is overfitting the number of components. We call this the region of *potential overfitting*. If the mixture is actually overfitting, then as discussed in Subsection 4.2.1, the mixture likelihood will become more and more concentrated over this region.

If we fit a mixture model with the correct number of components, then one would hope that the region of potential overfitting is bounded away from the main modal regions of the mixture likelihood function. This is true in the limit, as $N \rightarrow \infty$, because the ML estimator will converge to the true parameter, but for finite observations this need not be the case, even when the true number of components has been selected. If one of the true weights is rather small or two components are rather close to each other, then a more parsimonious mixture with less than the true number of components may explain the data equally well, in particular for small data sets. In this case the mixture likelihood function will exhibit additional modes over the regions of potential overfitting.

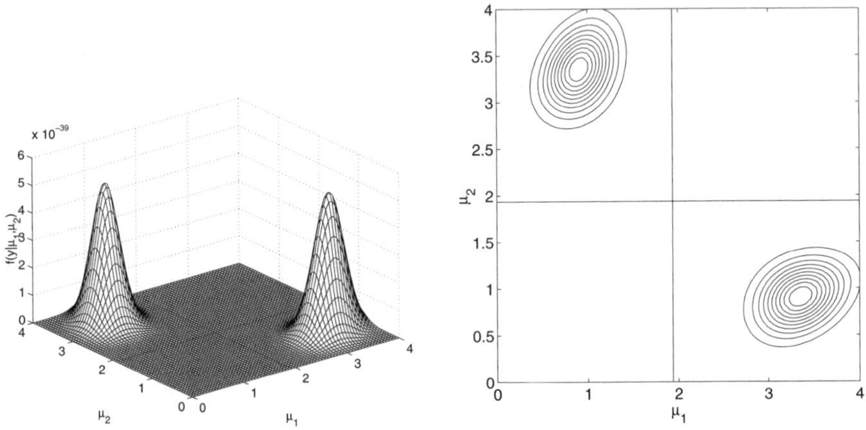


**Fig. 4.5.** HIDDEN AGE GROUPS — Synthetic Data Set 1; surface and contours of the integrated mixture likelihood $p(\mathbf{y}|\mu_1, \mu_2)$ for a two-component mixture model with unknown weights

For illustration, we reconsider synthetic data 1 and 2, simulated from (2.29). This time we fit a mixture of two normal distributions with $\mu_1$, $\mu_2$, and $\eta_2$ unknown. Adding the weights as unknown parameters may cause additional modes, especially if the data have a high likelihood also under the reduced model. These modes lie around parameters in the nonidentifiability set $\mathcal{U}^Z(\hat{\mu}, \hat{\sigma}^2) \cup \mathcal{U}^E(\hat{\mu}, \hat{\sigma}^2)$, where $(\hat{\mu}, \hat{\sigma}^2)$ is the ML estimator obtained when fitting a single normal distribution.

If the data are rather likely under the reduced model, then additional modal regions with considerable likelihood are present in the parameter space. For illustration, Figure 4.5 and Figure 4.6 show the surface and the contours of the mixture likelihood $p(\mathbf{y}|\mu_1, \mu_2)$ where $\eta_2$ is integrated out. For data set 1, the integrated mixture likelihood is not much different from the mixture likelihood appearing in Figure 2.5 where $\eta_2$ was assumed to be known, as parameters in the nonidentifiability set $\mathcal{U}^Z(\hat{\mu}, \hat{\sigma}^2) \cup \mathcal{U}^E(\hat{\mu}, \hat{\sigma}^2)$ are extremely unlikely. For data set 2, however, we find, in comparison to the mixture likelihood in Figure 2.6, additional local modes, making inference with regard to the component parameters rather difficult. Synthetic data set 2 demonstrates that the reduced model may be likely even if the data were generated from the more complex model.

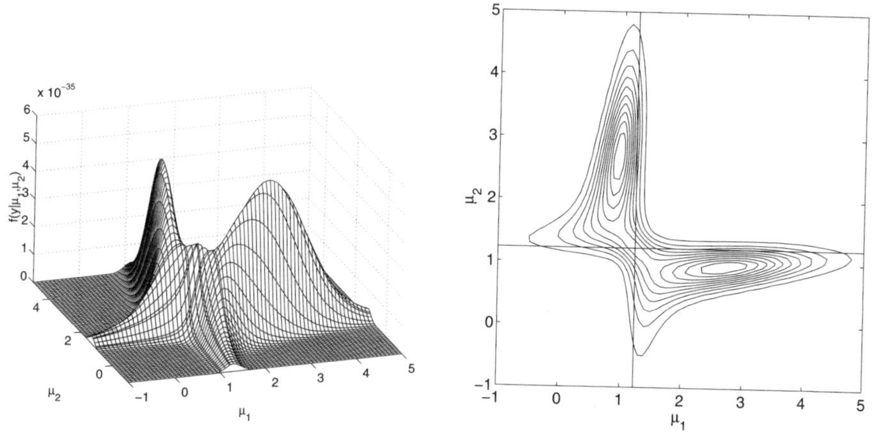


**Fig. 4.6.** HIDDEN AGE GROUPS — Synthetic Data Set 2; surface and contours of the integrated mixture likelihood $p(\mathbf{y}|\mu_1, \mu_2)$ for a two-component mixture model with unknown weights

### Bayesian Estimation

Under potential overfitting, MCMC draws from the mixture posterior density will come from different modal regions, part of them corresponding to the restricted model and part of them corresponding to the unrestricted model. This may cause label switching, even if the mixture is actually not overfitting ($K = K^{\text{true}}$). For synthetic data set 2, for instance, the two means are rather similar compared to the variance, and the two modes of the mixture likelihood functions are not well separated, causing label switching. Here label switching is a sign that the simpler model might also explain the data. If we are sure that we want to fit a model with two components, the only solution to avoid label switching is to choose a prior that helps to bound the mixture posterior away from the nonidentifiability set $\mathcal{U}^E(\mu)$ as has been explained in Subsection 4.2.2.

## 4.3 Informal Methods for Identifying the Number of Components

In this section various informal approaches for detecting the presence of mixtures and for identifying the number of components in a finite mixture are discussed. Most of these approaches are data oriented such as mode hunting in the sample histogram (Subsection 4.3.2) or comparing empirical properties of the data such as skewness or kurtosis with the corresponding quantities implied by the mixture (Subsection 4.3.3 and 4.3.4) whereas the approach discussed in Subsection 4.3.1 is parameter oriented and tries to learn about the presence of mixtures and the number of mixture components from the mixture posterior density.

### 4.3.1 Mode Hunting in the Mixture Posterior

One informal method for diagnosing mixtures is mode hunting in the mixture posterior density (Frühwirth-Schnatter, 2001b). It is based on the observation that with an increasing number of observations, the mixture likelihood function has $K!$ dominant modes if the data actually arise from a finite mixture distribution with $K$ components, and that less than $K!$ dominant modes are present if the finite mixture model is overfitting the number of components; see also Subsection 4.2.1.

Practically, mode hunting is carried out by exploring the sampling representations of the mixture posterior density, discussed earlier in Subsection 3.7.1. It is based on drawing from the mixture posterior using the random permutation Gibbs sampler described in *Algorithm 3.5*, rather than standard Gibbs sampling, to force the exploration of all modes of the mixture posterior density.

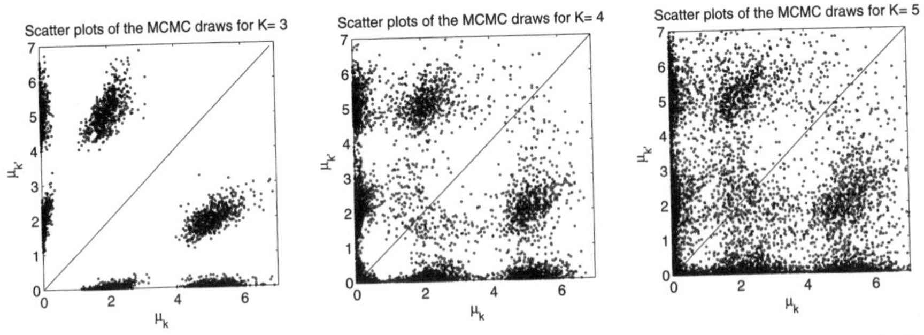


**Fig. 4.7.** Sampling representation of the posterior density $p(\mu_k, \mu_{k'}|\mathbf{y}, \mathcal{M}_K)$ for $K = 3$ (left-hand side), $K = 4$ (middle), and $K = 5$ (right-hand side)

For illustration, a data set of size $N = 500$ is simulated from a mixture of three Poisson distributions with $\mu_1 = 0.1$, $\mu_2 = 2$, $\mu_3 = 5$, $\eta_1 = 0.3$, $\eta_2 = 0.4$, and $\eta_3 = 0.3$. Poisson mixtures with $K = 3$ , $K = 4$, and $K = 5$ components are fitted and Bayesian estimation running random permutation Gibbs sampling for 2500 iterations is carried out under the priors $\boldsymbol{\eta} \sim \mathcal{D}(4, \dots, 4)$ and $\mu_k \sim \mathcal{G}(0.5, 0.5/\overline{y})$, where $\overline{y}$ is the mean of the observations.

Figure 4.7 shows the sampling representation of the bivariate marginal density $p(\mu_k, \mu_{k'}|\mathbf{y}, \mathcal{M}_K)$ for $K = 3$, $K = 4$, and $K = 5$. For $K = 3$ the finite mixture is not overfitting and the corresponding plot shows $K(K - 1) = 6$ simulation clusters as expected. For $K = 4$ the finite mixture is overfitting and less than $K(K - 1) = 12$ simulation clusters are visible. Under the prior

$\mathcal{D}(4,\ldots,4)$ on the weight distribution, empty components are rather unlikely, and the draws of the component parameters are pulled toward the nonidentifiability set $\mathcal{U}^Z(\boldsymbol{\vartheta}^{\text{true}})$ where two component means lie close to the diagonal $\mu_k = \mu_{k'}$, whereas the other component means are close to the modes of the three-component mixture posterior. The same is true for $K = 5$ although the figure is more fuzzy.

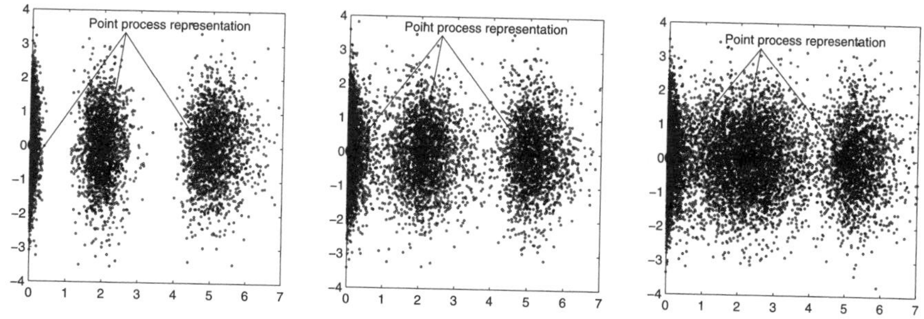


**Fig. 4.8.** Point process representation of the posterior density $p(\mu_k|\mathbf{y}, \mathcal{M}_K)$ for $K = 3$ (left-hand side), $K = 4$ (middle), and $K = 5$ (right-hand side)

Another useful plot for mode hunting in the mixture posterior distribution is the point process representation of the MCMC draws, explained earlier in Subsection 3.7.1. Note the point process representation of the overfitting and the true mixture distribution are identical. Therefore the point process representation of the MCMC draws will cluster around the point process representation of the true model even if the mixture is overfitting, as is evident from Figure 4.8. The number of simulation clusters in these figures clearly indicates that mixtures with $K = 4$ and $K = 5$ are overfitting, although the spread of these simulation clusters increases for $K = 4$ and $K = 5$. Therefore the point process representation is particularly useful for mode hunting.

### 4.3.2 Mode Hunting in the Sample Histogram

Historically seen, informal techniques for the detection of mixtures focused on hunting for modes in the sample histogram; see Titterington et al. (1985, Section 5.6) for some review. This might be appropriate for certain data sets, such as the FISHERY DATA that have been discussed in Subsection 1.1, where evidently several modes are present in the histogram. However, as stated by Everitt and Hand (1981, p.208), "Examination of the sample histogram is unlikely to be of much help in detecting the presence of a mixture — indeed, it might prove positively misleading."

Most important, as discussed in Subsection 1.2.2, many finite mixture distributions are unimodal, and the presence of a mixture distribution will remain

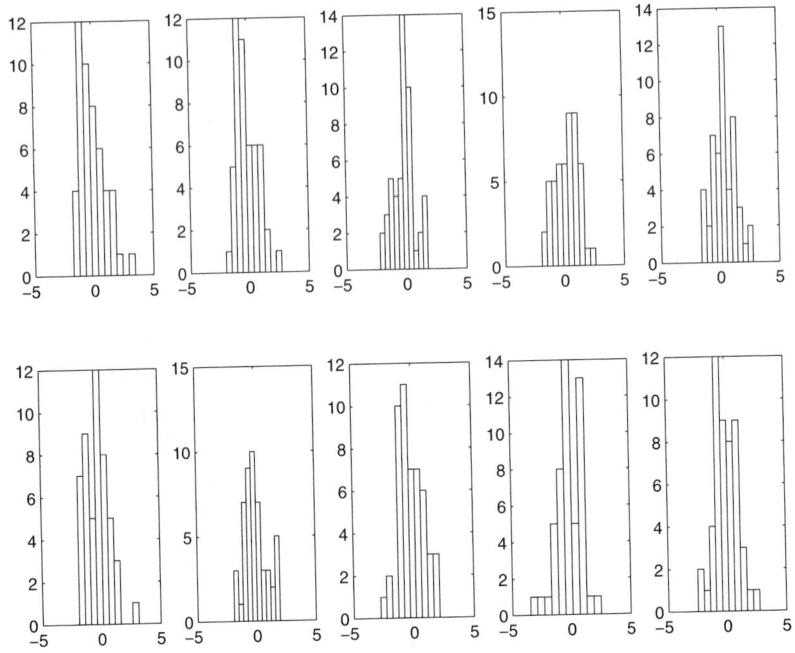

**Fig. 4.9.** Empirical componentwise marginal distribution of 50 observations generated from a standard multivariate normal distribution of dimension 10

undetected from the empirical histogram. In other cases mode hunting may be misleading as it tends to detect spurious clusters in the sample histogram (McLachlan and Peel, 2000, Section 1.8). Consider, as an example, 50 observations $\mathbf{y}_i$ generated from the ten-dimensional standard multivariate normal distribution $\mathcal{N}_{10}(\mathbf{0}, \mathbf{I}_{10})$ where the marginal distribution of each component of $\mathbf{y}_i$ displays a considerable amount of spurious multimodality in Figure 4.9.

### 4.3.3 Diagnosing Mixtures Through the Method of Moments

The method of moments is not only used for estimating the unknown parameter in a finite mixture model, as discussed in Subsection 2.4.1, but it has also been applied for diagnosing finite mixture models with respect to the number $K$ of components (Heckman et al., 1990; Dacunha-Castelle and Gassiat, 1997).

As in Subsection 2.4.1, several theoretical moments $\mathrm{E}(H_j(\mathbf{Y})|\boldsymbol{\vartheta}_K)$, $j = 1, 2, \ldots$, implied by a mixture model $\mathcal{M}_K$ with $K$ components, are compared with the corresponding sample moments $\overline{H}_j$. Mixtures with different numbers of components are evaluated by the impact increasing $K$ has on reducing the discrepancy between the theoretical moment $\mathrm{E}(H_j(\mathbf{Y})|\boldsymbol{\vartheta}_K)$ and the sample moment $\overline{H}_j$. To evaluate this discrepancy, $\boldsymbol{\vartheta}_K$ is substituted by an estimator, before the effect of adding another component is studied. The difficulty with diagnosing a mixture in this way is to decide whether the effect of adding another component is significant.

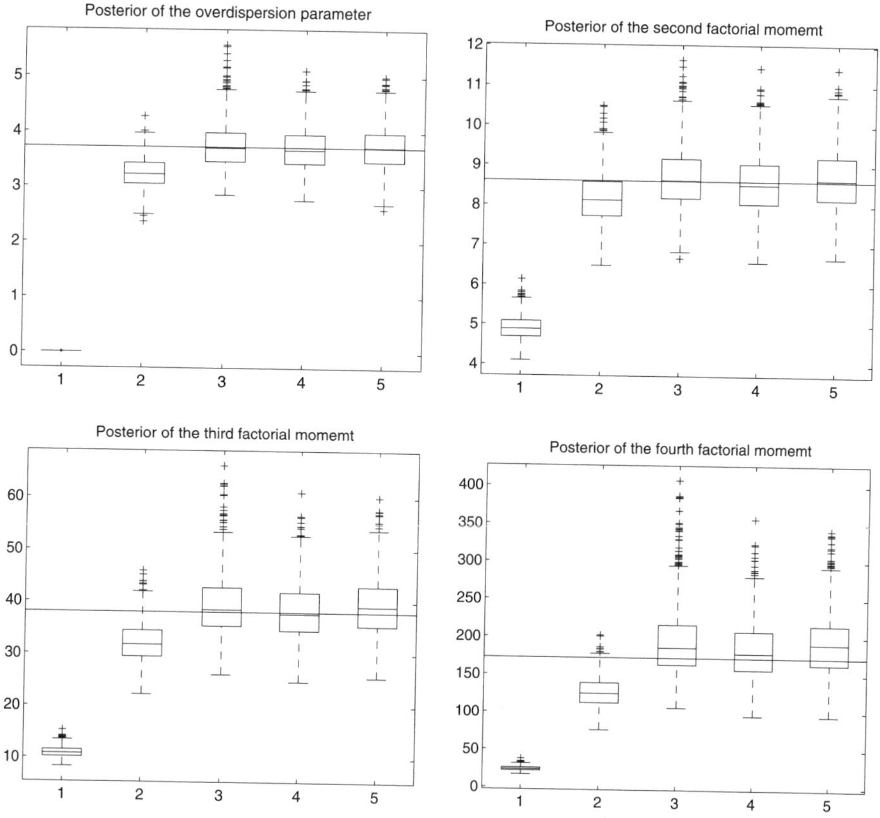


**Fig. 4.10.** Diagnosing mixtures of varying numbers of components for 500 observations simulated from a mixture of three Poisson distributions with $\mu_1 = 0.1$, $\mu_2 = 2$, $\mu_3 = 5$, $\eta_1 = 0.3$, $\eta_2 = 0.4$, and $\eta_3 = 0.3$; the posterior distribution of the overdispersion (top left), the second (top right), the third (bottom left), and the fourth (bottom right) factorial moment in comparison to the observed value (black horizontal line) for $K = 1, \dots, 5$

The rest of this subsection seeks to pursue a Bayesian variant of this approach, which provides an informal way of assessing this significance. As $h_j(\boldsymbol{\vartheta}_K) = \mathrm{E}(H_j(\mathbf{Y})|\boldsymbol{\vartheta}_K)$ is a function of $\boldsymbol{\vartheta}_K$, we consider the posterior density $p(h_j(\boldsymbol{\vartheta}_K)|\mathbf{y}, \mathcal{M}_K)$ of $h_j(\boldsymbol{\vartheta}_K)$ under model $\mathcal{M}_K$, rather than simply a point estimator $h_j(\hat{\boldsymbol{\vartheta}}_K)$ as in Heckman et al. (1990) and Dacunha-Castelle and Gassiat (1997). We compare this density with the sample moment $\overline{H}_j$ for each $j$

and for each $K$. If the observed value $\overline{H}_j$ is very unlikely under the posterior density $p(h_j(\boldsymbol{\vartheta}_K)|\mathbf{y}, \mathcal{M}_K)$, then the number of components is considered to be too small. If the effect of increasing $K$ on the corresponding posterior density $p(h_j(\boldsymbol{\vartheta}_K)|\mathbf{y}, \mathcal{M}_K)$ is negligible, then $K$ provides a sufficient number of components to fit the moment $\mathrm{E}(H_j(\mathbf{Y})|\boldsymbol{\vartheta}_K)$. Higher-order moments will typically require more components than lower-order moments. We may also consider the joint posterior distribution of two moments $(\mathrm{E}(H_j(\mathbf{Y})|\boldsymbol{\vartheta}_K), \mathrm{E}(H_m(\mathbf{Y})|\boldsymbol{\vartheta}_K))$, for instance, a skewness coefficient and excess kurtosis for univariate metric data, or the posterior distribution of some cumulative measure of discrepancy such as $\sum_j (\mathrm{E}(H_j(\mathbf{Y})|\boldsymbol{\vartheta}_K) - \overline{H}_j)^2$.

Any posterior density of interest is easily available when fitting a mixture model using Markov chain Monte Carlo or any other simulation-based method that yields a sequence of draws $\boldsymbol{\vartheta}_K^{(1)}, \ldots, \boldsymbol{\vartheta}_K^{(M)}$ from the posterior density $p(\boldsymbol{\vartheta}_K|\mathbf{y}, \mathcal{M}_K)$. The posterior distribution of $h_j(\boldsymbol{\vartheta}_K) = \mathrm{E}(H_j(\mathbf{Y})|\boldsymbol{\vartheta}_K)$ is approximated by the empirical distribution of the transformed draws $h_j(\boldsymbol{\vartheta}_K^{(1)}), \ldots, h_j(\boldsymbol{\vartheta}_K^{(M)})$. There is no need to identify the mixture model, as $\mathrm{E}(H_j(\mathbf{Y})|\boldsymbol{\vartheta}_K)$ is invariant to relabeling by definition. To study the effect of $K$ it is useful to consider Box plots of $h_j(\boldsymbol{\vartheta}_K^{(1)}), \ldots, h_j(\boldsymbol{\vartheta}_K^{(M)})$ for each $j$, whereas bivariate distributions are explored through scatter plots.

For illustration, we return to the illustrative example of Subsection 4.3.1. Figure 4.10 shows the posterior distribution of the overdispersion $B(\boldsymbol{\vartheta}) = \sum_{k=1}^{K} (\mu_k - \mu(\boldsymbol{\vartheta}))^2 \eta_k$, where $\mu(\boldsymbol{\vartheta}) = \mathrm{E}(Y|\boldsymbol{\vartheta})$, which is discussed in more detail in Subsection 9.2.2, and the second to the fourth factorial moment of the Poisson mixture in comparison to the observed values for an increasing number of components ($K = 1, \ldots, 5$). This figure indicates that three components are sufficient to capture the moments under investigation, because adding a fourth or a fifth component hardly changes the posterior distribution of these moments.

We end by noting that the same method could be applied to other data transformations such as the moment-generating function or the probability-generating function. For univariate continuous data, one could study the posterior distribution of $h_t(\boldsymbol{\vartheta}_K) = \mathrm{E}(\exp(tY)|\boldsymbol{\vartheta}_K) - \overline{H}_t$, where

$$\overline{H}_t = \frac{1}{N} \sum_{i=1}^{N} \exp(ty_i),$$

as a function of $t$ and $K$. Finally the Kullback–Leibler distance, or any other distance between the empirical distribution function of the data and a mixture model with $K$ components could be analyzed in a similar way, by evaluating the effect of increasing $K$ on the posterior distribution of these statistics.

### 4.3.4 Diagnosing Mixtures Through Predictive Methods

A further method for diagnosing mixtures is Bayesian posterior predictive model checking (Gelfand et al., 1992; Gelman et al., 1996) which is related to

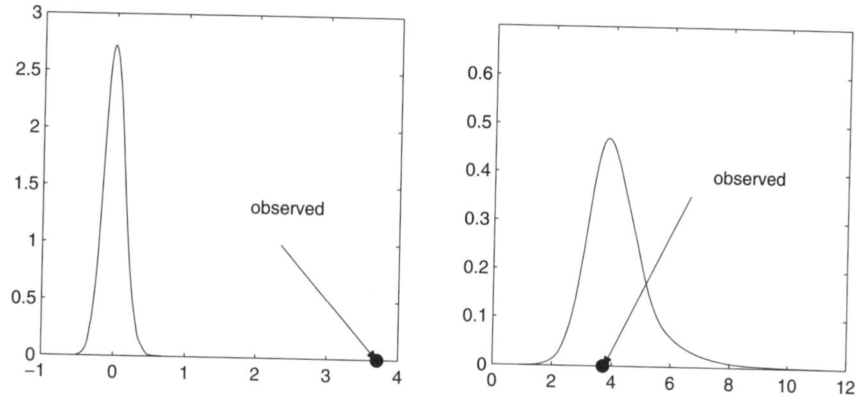


**Fig. 4.11.** Model checking for 500 observations simulated from a mixture of three Poisson distributions with $\mu_1 = 0.1$, $\mu_2 = 2$, $\mu_3 = 5$, $\eta_1 = 0.3$, $\eta_2 = 0.4$, and $\eta_3 = 0.3$ using the posterior distribution of the overdispersion statistic $T(\mathbf{y}) = s_y^2 - \overline{y}$ under a single Poisson distribution (left-hand side) and under a mixture of three Poisson distributions (right-hand side)

standard procedures of model checking, by considering one or several diagnostic statistics $T(\mathbf{y})$ of the observations $\mathbf{y} = (\mathbf{y}_1, \dots, \mathbf{y}_N)$, such as the skewness, that are likely to be extreme under certain misspecifications of the sampling distribution. The observed value $T(\mathbf{y})$ is compared with the posterior predictive distribution $p(T(\mathbf{y}_f)|\mathbf{y}, \mathcal{M}_K)$, where the statistic $T(\cdot)$ is evaluated at a sample $\mathbf{y}_f = (\mathbf{y}_{f,1}, \dots, \mathbf{y}_{f,N})$ of size $N$, predicted from the model under consideration. If the observed value $T(\mathbf{y})$ lies far out in the tails of the posterior predictive distribution $p(T(\mathbf{y}_f)|\mathbf{y}, \mathcal{M}_K)$, the data are diagnosed as being inconsistent with the model.

The posterior predictive distribution $p(T(\mathbf{y}_f)|\mathbf{y}, \mathcal{M}_K)$ is not available in closed form. It is approximated by a histogram of $T(\mathbf{y}_f^{(1)}), \dots, T(\mathbf{y}_f^{(M)})$, with $\mathbf{y}_f^{(1)}, \dots, \mathbf{y}_f^{(M)}$ being a sample from the posterior predictive distribution $p(\mathbf{y}_f|\mathbf{y}, \mathcal{M}_K)$, obtained by *Algorithm 3.7* with $H = N$.

For illustration, we return once more to the illustrative example of Subsection 4.3.1 and 4.3.3, and show in Figure 4.11 the posterior predictive distribution of the overdispersion statistic $T(\mathbf{y}) = s_y^2 - \overline{y}$ in comparison to the observed value. Model checking clearly reveals that a single Poisson distribution is misspecified, whereas a Poisson mixture with three components does not lead to any discrepancy.

A more formal method is based on the so-called Bayes $p$-value (Rubin, 1981, 1984; Meng, 1994), defined for a one-sided statistic $T(\mathbf{y})$ as the tail-area probability $\Pr(T(\mathbf{y}_f) > T(\mathbf{y})|\mathbf{y}, \mathcal{M}_K)$ which is estimated from posterior draws as

$$\Pr(T(\mathbf{y}_f) > T(\mathbf{y})|\mathbf{y}, \mathcal{M}_K) = \mathrm{E}(I_{\{T(\mathbf{y}_f)>T(\mathbf{y})\}}|\mathbf{y}, \mathcal{M}_K)$$
$$\approx \frac{1}{M} \sum_{m=1}^{M} I_{\{T(\mathbf{y}_f^{(m)})>T(\mathbf{y})\}}.$$

A Bayesian predictive approach that is slightly different from this approach is considered by Dey et al. (1995).

### 4.3.5 Further Approaches

Graphical tools for diagnosing mixtures such as quantile–quantile plots and normal probability plots are discussed in Fowlkes (1979). More recently, Lindsay and Roeder (1992) considered residual diagnostics for mixtures.

## 4.4 Likelihood-Based Methods

This section provides a short review of various likelihood-based methods that have been used to deal with model uncertainty in finite mixture models.

### 4.4.1 The Likelihood Ratio Statistic

Likelihood-based methods play a central role in testing parametric models and among these likelihood ratio tests is usually the preferred one, however, its application to finite mixture models creates some difficulty. Consider two nested finite mixture models $\mathcal{M}_K$ and $\mathcal{M}_{K+1}$, with $\mathcal{M}_K$ being the simpler model. A standard approach for testing between nested models is to apply a likelihood ratio test. First, the ML estimators $\hat{\boldsymbol{\vartheta}}_K$ and $\hat{\boldsymbol{\vartheta}}_{K+1}$, as well as the corresponding likelihood functions $p(\mathbf{y}|\hat{\boldsymbol{\vartheta}}_K, \mathcal{M}_K)$ and $p(\mathbf{y}|\hat{\boldsymbol{\vartheta}}_{K+1}, \mathcal{M}_{K+1})$ are determined for both models. Then the likelihood ratio statistic is defined as

$$\mathrm{LR} = -2 \left( \log p(\mathbf{y}|\hat{\boldsymbol{\vartheta}}_{K+1}, \mathcal{M}_{K+1}) - \log p(\mathbf{y}|\hat{\boldsymbol{\vartheta}}_K, \mathcal{M}_K) \right).$$

Under regularity conditions established by Wilks (1938), the likelihood ratio statistic asymptotically follows a $\chi^2_r$-distribution, if model $\mathcal{M}_K$ is true, with $r$ being equal to the number of constraints imposed on $\mathcal{M}_{K+1}$ to obtain $\mathcal{M}_K$. If the two finite mixture models differ only in the parameter structure, but both assume the correct number of components, these regularity conditions typically hold, and the LR statistic may be applied in a straightforward manner.

However, if $\mathcal{M}_K$ and $\mathcal{M}_{K+1}$ are finite mixture models with $K$ and $K+1$ components, respectively, testing the number of components of a mixture model through the likelihood ratio statistic (Wolfe, 1970; Hartigan, 1985) immediately leads to ambiguity because model $\mathcal{M}_K$ is obtained from model $\mathcal{M}_{K+1}$ in more than one way. The number of constraints is equal to 1, when imposing the constraint $\eta_{K+1}^{K+1} = 0$ on $\mathcal{M}_{K+1}$, and equal to $\dim(\boldsymbol{\theta}_k)$, when imposing the constraint $\boldsymbol{\theta}_K^{K+1} = \boldsymbol{\theta}_{K+1}^{K+1}$.

The main reason behind the difficulties with applying the likelihood ratio statistic to testing the number of components is the failure of the standard regularity conditions, on which the asymptotic $\chi^2$-distribution relies, if model $\mathcal{M}_K$ is actually true (Gosh and Sen, 1985; Hartigan, 1985). A lot of work has been done to establish the exact asymptotic distribution of LR under the null hypothesis, when testing homogeneity ($K = 1$) against heterogeneity ($K > 1$); see Titterington et al. (1985, Section 5.4) and McLachlan and Peel (2000, Section 6.4 and 6.5) for more details and a review of the relevant literature.

Various suggestions have been made to modify the likelihood ratio statistic in an appropriate manner. Aitkin and Rubin (1985) place a prior distribution on $\boldsymbol{\eta}$, and apply the LR test to the integrated likelihood $p(\mathbf{y}|\boldsymbol{\theta}_1, \ldots, \boldsymbol{\theta}_K)$, however, the asymptotic distribution of this test statistic under the null hypothesis is not necessarily a $\chi^2$-distribution. Chen et al. (2001) consider for mixtures with a univariate component parameter a modified likelihood ratio for testing homogeneity ($K = 1$) against heterogeneity ($K > 1$) which is based on using for $K > 1$ the following penalized likelihood function,

$$\log p^\star(\mathbf{y}|\boldsymbol{\vartheta}_K, \mathcal{M}_K) = \log p(\mathbf{y}|\boldsymbol{\vartheta}_K, \mathcal{M}_K) + C \sum_{k=1}^{K} \log(2\eta_k). \tag{4.3}$$

From a Bayesian viewpoint, the penalty term in (4.3) is nothing but the log of a Dirichlet prior $\mathcal{D}(C + 1, \ldots, C + 1)$, which is known from Subsection 2.3.4 to bound the weights away from zero for appropriate choices of $C$, and the penalized likelihood is, up to a constant, in fact equal to the log of the mixture posterior $\log p(\boldsymbol{\vartheta}_K|\mathbf{y}, \mathcal{M}_K)$. Chen et al. (2001) define the modified LR statistic by

$$\mathrm{LR}^\star = -2\Big(\log p^\star(\mathbf{y}|\hat{\boldsymbol{\vartheta}}_K, \mathcal{M}_K) - \log p(\mathbf{y}|\hat{\boldsymbol{\vartheta}}_1, \mathcal{M}_1)\Big),$$

where $\hat{\boldsymbol{\vartheta}}_K$ is the penalized ML estimator maximizing (4.3), and show that the asymptotic distribution of $\mathrm{LR}^\star$ under the null hypothesis of homogeneity is equal to a mixture of a $\chi^2_1$-distribution and a degenerate $\chi^2_0$-distribution with all its mass at 0:

$$\mathrm{LR}^\star \rightarrow_d \frac{1}{2}\chi^2_0 + \frac{1}{2}\chi^2_1.$$

The same asymptotic distribution results for the original LR statistic for mixtures when the component parameters are considered to be known (Titterington et al., 1985; Hartigan, 1985).

Chen et al. (2004) extend LR testing based on the penalized likelihood function to testing the null hypothesis of $K = 2$ against $K > 2$ and show that the asymptotic distribution of this modified LR statistic is equal to a mixture of $\chi^2_0$, $\chi^2_1$, and $\chi^2_2$ distributions.

### 4.4.2 AIC, BIC, and the Schwarz Criterion

Using heuristic arguments on how to account for model complexity, Akaike (1974) proposed a general criterion for model choice which is equivalent to choosing the model that maximizes

$$\log p(\mathbf{y}|\hat{\boldsymbol{\vartheta}}_K, \mathcal{M}_K) - d_K, \tag{4.4}$$

where $\hat{\boldsymbol{\vartheta}}_K$ is equal to the ML estimator of model $\mathcal{M}_K$. $d_K = \dim(\boldsymbol{\vartheta}_K)$ is equal to the dimension of the model and acts as a correction term without which one would choose the model that maximizes the likelihood function. In (4.4), $d_K$ introduces a severe penalty for high-dimensional models that provide little additional fit in terms of increasing the likelihood function in comparison to simpler models. Akaike's model choice procedure is commonly implemented by minimizing AIC defined by

$$\mathrm{AIC}_K = -2 \log p(\mathbf{y}|\hat{\boldsymbol{\vartheta}}_K, \mathcal{M}_K) + 2d_K. \tag{4.5}$$

Using asymptotic expansions rather than heuristics as did Akaike, Schwarz (1978) arrived at the conclusion to choose that model for which

$$\mathrm{SC}_K = \log p(\mathbf{y}|\hat{\boldsymbol{\vartheta}}_K, \mathcal{M}_K) - \frac{\log N}{2} d_K \tag{4.6}$$

is largest, where again $d_K = \dim(\boldsymbol{\vartheta}_K)$. Like Akaike's criterion, Schwarz's criterion (4.6) is often formulated in terms of minus twice the log likelihood function, also called BIC, which is commonly defined as

$$\mathrm{BIC}_K = -2 \log p(\mathbf{y}|\hat{\boldsymbol{\vartheta}}_K, \mathcal{M}_K) + (\log N) d_K, \tag{4.7}$$

although other definitions are in use. Evidently $\mathrm{SC}_K = -1/2\mathrm{BIC}_K$.

As the first term in $\mathrm{AIC}_K$ and $\mathrm{BIC}_K$ measures the goodness-of-fit, whereas the second term penalizes model complexity, one selects the model that minimizes either $\mathrm{AIC}_K$ or $\mathrm{BIC}_K$. Quantitatively, $\mathrm{AIC}_K$ and $\mathrm{BIC}_K$ differ only by the factor by which $d_K$ is multiplied. Qualitatively, both criteria provide a mathematical formulation of the principle of parsimony in model building, although for large data sets their behavior is rather different.

For $\log N > 2$, Akaike's criterion favors more complex models than Schwarz's criterion and has been shown to be inconsistent, choosing too-complex models even asymptotically (Bozdogan, 1987). Nevertheless, several authors (Bozdogan and Sclove, 1984; Solka et al., 1998; Liang et al., 1992) used AIC for finite mixture models to select the number $K$ of components and to choose among various constrained and unconstrained parameter structures.

Schwarz's criterion and BIC are consistent under certain regularity conditions, in particular for i.i.d. observations from exponential families; see the excellent review of Kass and Raftery (1995). In the context of choosing the number of components, it is not evident that the regularity conditions for deriving Schwarz's criterion through asymptotic expansions (Schwarz, 1978) actually hold. As discussed earlier in Subsection 4.2.1, asymptotic normality does not hold for a finite mixture model which is overfitting the number of components. Nevertheless, BIC has been applied by many authors to finite mixture models, among them Roeder and Wasserman (1997b), Fraley and Raftery (1998), and Dasgupta and Raftery (1998).

Leroux (1992a) proved that, asymptotically, choosing the number of components $K$ as to minimize $\mathrm{AIC}_K$ and $\mathrm{BIC}_K$ will not underestimate the true number of components. Simulation studies comparing various model choice criteria including AIC and BIC are found in Windham and Cutler (1992), Biernacki et al. (2000), McLachlan and Peel (2000, Section 6.11), Hawkins et al. (2001), and Hettmansperger and Thomas (2000). In general, BIC was found to be consistent under correct specification of the family of the component densities (Keribin, 2000), whereas BIC selected too many components when one of the true, normally distributed components was substituted by a different distribution, such as the uniform distribution. AIC tends to select too many components even for a correctly specified mixture.

### 4.4.3 Further Approaches

Several classification-based information criteria for choosing the number of components, such as the integrated classification likelihood (ICL) criterion (Biernacki et al., 2000), have been developed in the context of applying finite mixture models to clustering data, and are reviewed in Subsection 7.1.4.

Other methods are based on exploring the smallest eigenvalue of the information ratio matrix $\mathcal{I}_C^{-1}(\boldsymbol{\vartheta})\mathcal{I}(\boldsymbol{\vartheta})$, which is zero for an overfitting mixture model (Windham and Cutler, 1992) or minimizing the Kullback–Leibler or the $L_2$-distance between the true density and the fitted model (Miloslavsky and van der Laan, 2003).

## 4.5 Bayesian Inference Under Model Uncertainty

The rest of this section as well as Chapter 5 discusses how Bayesian inference is carried out under model uncertainty. This section is meant as a short introduction to Bayesian model selection; the issue of how to apply this approach to model uncertainty in finite mixture models is outlined in great detail in Chapter 5. Several recent reviews (Godsill, 2001; Green, 2003; Kadane and Lazar, 2004) deal with various aspects of the Bayesian approach to model selection; classical references are Zellner (1971), Jeffreys (1948), and Bernardo and Smith (1994).

### 4.5.1 Trans-Dimensional Bayesian Inference

Assume that $K_{\max}$ alternative models $\mathcal{M}_1, \ldots, \mathcal{M}_{K_{\max}}$ should be compared, given data $\mathbf{y} = (\mathbf{y}_1, \ldots, \mathbf{y}_N)$. In the Bayesian approach it is assumed that each model $\mathcal{M}_K$ is specified in terms of a sampling distribution $p(\mathbf{y}|\boldsymbol{\vartheta}_K, \mathcal{M}_K)$ and a prior distribution $p(\boldsymbol{\vartheta}_K|\mathcal{M}_K)$, where $\boldsymbol{\vartheta}_K \in \Theta_K$ denotes the collection of all unknown model parameters appearing in the specification of model $\mathcal{M}_K$. Because the dimension of $\boldsymbol{\vartheta}_K$ changes across the models, consider, for instance, finite mixture models that differ in the number of components, such problems have been termed trans-dimensional problem, where "The number of things you do not know is one of the things you do not know" (Roeder and Wasserman, 1997a).

Following Jeffreys (1948), a categorical model indicator may be introduced, which lives on the model space $\Omega = \{\mathcal{M}_1, \ldots, \mathcal{M}_{K_{\max}}\}$. The Bayesian approach requires the choice of a prior distribution over $\Omega$ in terms of prior model probabilities $p(\mathcal{M}_K)$ for all $\mathcal{M}_K \in \Omega$. Bayesian inference then is considered as a matter of joint inference about the model indicator and all model parameters $\boldsymbol{\vartheta}_1, \ldots, \boldsymbol{\vartheta}_{K_{\max}}$, given the data $\mathbf{y}$.

This is achieved by deriving the posterior density $p(\mathcal{M}_K, \boldsymbol{\vartheta}_1, \ldots, \boldsymbol{\vartheta}_{K_{\max}}|\mathbf{y})$ using Bayes' theorem:

$$\begin{aligned} &p(\mathcal{M}_K, \boldsymbol{\vartheta}_1, \ldots, \boldsymbol{\vartheta}_{K_{\max}}|\mathbf{y}) \propto \\ &p(\mathbf{y}|\mathcal{M}_K, \boldsymbol{\vartheta}_1, \ldots, \boldsymbol{\vartheta}_{K_{\max}})p(\boldsymbol{\vartheta}_1, \ldots, \boldsymbol{\vartheta}_{K_{\max}}|\mathcal{M}_K)p(\mathcal{M}_K), \end{aligned} \tag{4.8}$$

where, as usual, the normalizing constant has been dropped. This posterior distribution is rather complex, and there is little hope to obtain any analytical results for Bayesian inference.

A very useful approach for dealing with this posterior are trans-dimensional Markov chain Monte Carlo methods, which use a Markov chain to obtain draws from the joint posterior density $p(\mathcal{M}_K, \boldsymbol{\vartheta}_1, \ldots, \boldsymbol{\vartheta}_{K_{\max}}|\mathbf{y})$. These methods move through the model space $\Omega = \{\mathcal{M}_1, \ldots, \mathcal{M}_{K_{\max}}\}$ during sampling and, marginally, provide draws from the discrete posterior distribution $p(\mathcal{M}_K|\mathbf{y})$. Many different trans-dimensional MCMC methods have been proposed in the literature and their application to finite mixture models is discussed in much detail in Section 5.2.

### 4.5.2 Marginal Likelihoods

One should keep in mind that it is in general not really the joint density $p(\mathcal{M}_K, \boldsymbol{\vartheta}_1, \ldots, \boldsymbol{\vartheta}_{K_{\max}}|\mathbf{y})$, defined earlier in (4.8), which is the primary object of interest under model uncertainty, but more often this is the marginal posterior density $p(\mathcal{M}_K|\mathbf{y})$ that provides the posterior probability of the various models given the data. These probabilities are obtained from Bayes' rule (Bernardo and Smith, 1994) as

$$p(\mathcal{M}_K|\mathbf{y}) \propto p(\mathbf{y}|\mathcal{M}_K)p(\mathcal{M}_K), \tag{4.9}$$

where the so-called marginal likelihood $p(\mathbf{y}|\mathcal{M}_K)$ is given by

$$p(\mathbf{y}|\mathcal{M}_K) = \int_{\Theta_K} p(\mathbf{y}|\mathcal{M}_K, \boldsymbol{\vartheta}_K)p(\boldsymbol{\vartheta}_K|\mathcal{M}_K)d\boldsymbol{\vartheta}_K.$$

Evidently, Bayes' rule adapts for each model the prior probability $p(\mathcal{M}_K)$ in light of the data $\mathbf{y}$. The posterior distribution $p(\mathcal{M}_K|\mathbf{y})$ is available from Bayes' rule (4.9) only up to proportionality, and needs to be normalized over all models under consideration. To do so, it is necessary to derive the marginal likelihood $p(\mathbf{y}|\mathcal{M}_K)$ separately for each model $\mathcal{M}_K$ for $K = 1, \ldots, K_{\max}$.

For finite mixture models with more than one component the marginal likelihood $p(\mathbf{y}|\mathcal{M}_K)$ is not available in closed form, and obtaining a good numerical approximation to $p(\mathbf{y}|\mathcal{M}_K)$ is quite a challenging computational problem. Unfortunately, MCMC estimation as in Section 3.5 yields no direct information about the model indicator, and the marginal likelihood $p(\mathbf{y}|\mathcal{M}_K)$ has to be determined in a postprocessing manner. Various numerical approximation methods have been proposed in the literature, which are discussed in much detail in Sections 5.4 and 5.5.

### 4.5.3 Bayes Factors for Model Comparison

A useful way to compare two models $\mathcal{M}_1$ and $\mathcal{M}_2$ is the Bayes factor; see Kass and Raftery (1995) for an authoritative and comprehensive review. The odds ratio in favor of one of the models, say model $\mathcal{M}_1$, reads

$$\frac{p(\mathcal{M}_1|\mathbf{y})}{p(\mathcal{M}_2|\mathbf{y})} = \frac{p(\mathbf{y}|\mathcal{M}_1)\, p(\mathcal{M}_1)}{p(\mathbf{y}|\mathcal{M}_2)\, p(\mathcal{M}_2)},$$

and turns out to be the product of the prior odds and the ratio of the marginal likelihood, $B_{12} = p(\mathbf{y}|\mathcal{M}_1)/p(\mathbf{y}|\mathcal{M}_2)$, which is also called the Bayes factor.

**Table 4.1.** Relations among the log Bayes factor and the posterior probability for two models under comparison (assuming equal prior probability)

| $\log B_{12}$ | $\Pr(\mathcal{M}_1\|\mathbf{y})$ | $\Pr(\mathcal{M}_2\|\mathbf{y})$ | $\log B_{12}$ | $\Pr(\mathcal{M}_1\|\mathbf{y})$ | $\Pr(\mathcal{M}_2\|\mathbf{y})$ |
|---|---|---|---|---|---|
| –7 | 0.001 | 0.999 | 7 | 0.999 | 0.001 |
| –6 | 0.002 | 0.998 | 6 | 0.998 | 0.002 |
| –5 | 0.007 | 0.993 | 5 | 0.993 | 0.007 |
| –4 | 0.018 | 0.982 | 4 | 0.982 | 0.018 |
| –3 | 0.047 | 0.953 | 3 | 0.953 | 0.047 |
| –2 | 0.119 | 0.881 | 2 | 0.881 | 0.119 |
| –1 | 0.269 | 0.731 | 1 | 0.731 | 0.269 |
| 0 | 0.5 | 0.5 | 0 | 0.5 | 0.5 |

The Bayes factor provides a measure of whether the data $\mathbf{y}$ increased or decreased the odds on $\mathcal{M}_1$, relative to $\mathcal{M}_2$. Hence, $B_{12} > 1$ means that $\mathcal{M}_1$ is relatively more plausible than $\mathcal{M}_2$ in the light of the observed data, whereas the opposite is true if $B_{12} < 1$. In practice, it is usual to consider $\log B_{12}$, being the difference between the log of the marginal likelihoods.

Table 4.1 translates $\log B_{12}$ into the posterior probabilities of the two models under equal prior probability. If $\log B_{12}$ lies between $-1$ and 1, either of the two models has considerable posterior probability. $\log B_{12}$ around 3 gives a probability of around 95% for model $\mathcal{M}_1$, whereas $\log B_{12}$ lying around $-3$ gives the same probability for model $\mathcal{M}_2$. If $\log B_{12}$ goes to *minus* infinity, the posterior probability of hypothesis $\mathcal{M}_2$ goes to 1 and we may "reject" hypothesis $\mathcal{M}_1$ in favor of "accepting" hypothesis $\mathcal{M}_2$. If $\log B_{12}$ goes to *plus* infinity, the posterior probability of hypothesis $\mathcal{M}_1$ goes to 1, and we may "reject" hypothesis $\mathcal{M}_2$ in favor of "accepting" hypothesis $\mathcal{M}_1$.

This symmetry between the two hypotheses, which is also clearly visible in Table 4.1, is in sharp contrast to classical hypothesis testing such as the likelihood ratio test considered in Subsection 4.4.1, where the simpler model is considered to be the "null" hypothesis, which may be rejected, but never accepted. For a fundamental discussion of this remarkable difference between the Bayesian and the classical approach to hypothesis testing, see, for instance, Berger and Sellke (1987) and Berger and Delampady (1987), and the references therein.

The Bayes factor is in general a consistent tool of model selection. Assume that the Bayes factor is applied to nested models $\mathcal{M}_1$ and $\mathcal{M}_2$, with $\mathcal{M}_2$ being the more complex one. The following result on the asymptotic behavior of the Bayes factor is due to Kass and Vaidyanathan (1992) and holds under *Laplace regularity* in the sense of Kass et al. (1990). Crawford (1994, Theorem 3.1) verified that *Laplace regularity* holds for identifiable mixtures from the exponential family.

If model $\mathcal{M}_1$ is true, then with increasing $N$ the Bayes factor approaches infinity at a rate that depends on the difference in model complexity,

$$B_{12} = \mathcal{O}(N^{(d_2-d_1)/2}), \tag{4.10}$$

and the posterior probability $\Pr(\mathcal{M}_1|\mathbf{y})$ converges to 1. If model $\mathcal{M}_2$ is true, then for any fixed model parameter $\boldsymbol{\vartheta}_2$ the Bayes factor approaches minus infinity exponentially fast,

$$B_{12} = \exp\{-\mathcal{O}(N)\},$$

and the posterior probability $\Pr(\mathcal{M}_2|\mathbf{y})$ converges to 1.

It turns out that the Bayes factor implicitly penalizes model complexity if two nested models $\mathcal{M}_1$ and $\mathcal{M}_2$ provide a comparable fit for the data. More precisely, if the distance between the two models approaches 0 at the rate $1/\sqrt{N}$, then (4.10) holds even if model $\mathcal{M}_2$ is true (Kass and Vaidyanathan, 1992) and the Bayes factor prefers the simpler model $\mathcal{M}_1$ over the more complex model if $\mathcal{M}_2$ is close enough to $\mathcal{M}_1$. In this respect, Bayesian model selection is related to model choice criteria such as AIC, BIC, and the Schwarz criterion, which introduce an explicit penalty term for model complexity; see again Subsection 4.4.2. Under *Laplace regularity*, Schwarz's criterion may be derived as an asymptotic approximation to the marginal likelihood:

$$\log p(\mathbf{y}|\mathcal{M}_K) = \mathrm{SC}_K + \mathcal{O}(1); \tag{4.11}$$

see Gelfand and Dey (1994) for more details.

### 4.5.4 Formal Bayesian Model Selection

Model selection through Bayes factors is easily extended to deal with more than two models. If the number $K_{\max}$ of alternative models $\mathcal{M}_1, \ldots, \mathcal{M}_{K_{\max}}$ is not too large, the standard approach is to compare all possible models under consideration by computing the posterior probabilities $p(\mathcal{M}_K|\mathbf{y})$ for each model. Based on these posterior probabilities, the framework of Bayesian decision theory is used for model selection which is reviewed, for instance, in Bernardo and Smith (1994).

A basic requirement for applying a decision-theoretic approach to Bayesian model selection is the feasibility to quantify the loss made by a wrong decision. Let $R(\mathcal{M}_{\hat{K}}, \mathcal{M}_K)$ denote the loss made by selecting model $\mathcal{M}_{\hat{K}}$, if model $\mathcal{M}_K$ is true. The loss of selecting model $\mathcal{M}_{\hat{K}}$ is a random quantity, which depends on the unknown true model $\mathcal{M}_K$. By quantifying posterior evidence in favor of each model by the posterior probability $p(\mathcal{M}_K|\mathbf{y})$, one could determine the expected loss associated with selecting model $\mathcal{M}_{\hat{K}}$, for each $K = 1, \ldots, K_{\max}$:

$$\mathrm{E}(\mathcal{M}_{\hat{K}}) = \sum_{K=1}^{K_{\max}} R(\mathcal{M}_{\hat{K}}, \mathcal{M}_K) p(\mathcal{M}_K|\mathbf{y}).$$

The best strategy is then to select the model with the lowest expected loss. If one model, say $\mathcal{M}_{\hat{K}}$, dominates the others in the sense that $p(\mathcal{M}_{\hat{K}}|\mathbf{y}) = 1$, choosing this model will be optimal, regardless of the loss function. In all other cases, the decision of course depends on the loss function. In the absence of specific information about the actual loss, it is usual to consider the 0/1 loss function:

$$R(\mathcal{M}_{\hat{K}}, \mathcal{M}_K) = \begin{cases} 0, & \hat{K} = K, \\ 1, & \hat{K} \neq K. \end{cases},$$

which implies the following expected risk,

$$\mathrm{E}(\mathcal{M}_{\hat{K}}) = \sum_{K=1, K\neq\hat{K}}^{K_{\max}} p(\mathcal{M}_K|\mathbf{y}) = 1 - p(\mathcal{M}_{\hat{K}}|\mathbf{y}).$$

Thus choosing the model $\mathcal{M}_{\hat{K}}$ with the highest posterior probability $p(\mathcal{M}_{\hat{K}}|\mathbf{y})$ will minimize the risk under the 0/1 loss function. If all models have the same prior probability $p(\mathcal{M}_K)$, this strategy results in selecting the model with the highest marginal likelihood $p(\mathbf{y}|\mathcal{M}_K)$. The risk of this decision is equal to $1 - p(\mathcal{M}_{\hat{K}}|\mathbf{y})$, which is small only if $p(\mathcal{M}_{\hat{K}}|\mathbf{y})$ is close to 1.

### 4.5.5 Choosing Priors for Model Selection

Model selection is far more sensitive to choosing appropriate priors than is parameter estimation. This concerns in particular parameters that are unknown only in certain models. Consider, as above, two nested models $\mathcal{M}_1$ and $\mathcal{M}_2$, with $\mathcal{M}_2$ being the more complex one. Assume that the models are parameterized such that $\boldsymbol{\vartheta}_2$ splits as $\boldsymbol{\vartheta}_2 = (\boldsymbol{\vartheta}_1, \boldsymbol{\psi})$, with $\boldsymbol{\vartheta}_1$ being a parameter common to both models, whereas $\boldsymbol{\psi}$ is unconstrained under model $\mathcal{M}_2$ and constrained to $\boldsymbol{\psi} = \boldsymbol{\psi}_0$ under model $\mathcal{M}_1$. Verdinelli and Wasserman (1995) showed that in this case the Bayes factor in favor of model $\mathcal{M}_1$ is equal to:

$$B_{12} = \frac{p(\boldsymbol{\psi}_0|\mathbf{y}, \mathcal{M}_2)}{p(\boldsymbol{\psi}_0|\mathcal{M}_2)} \mathrm{E}\left(\frac{p(\boldsymbol{\vartheta}_1|\mathcal{M}_1)}{p(\boldsymbol{\vartheta}_1|\boldsymbol{\psi}_0, \mathcal{M}_2)}\right), \tag{4.12}$$

where the expectation is with respect to the posterior $p(\boldsymbol{\vartheta}_1|\boldsymbol{\psi}_0, \mathbf{y}, \mathcal{M}_2)$. The first ratio is commonly called the Savage–Dickey density ratio (Dickey, 1971).

Formula (4.12) allows us to study the effect of changing the spread of the prior $p(\boldsymbol{\psi}|\mathcal{M}_2)$ which is assumed to be centered at $\boldsymbol{\psi}_0$. Increasing the spread will hardly affect the posterior ordinate $p(\boldsymbol{\psi}_0|\mathcal{M}_2, \mathbf{y})$, whereas the prior ordinate $p(\boldsymbol{\psi}_0|\mathcal{M}_2)$ decreases, which in turn will increase evidence in favor of model $\mathcal{M}_1$. For the extreme case, where the prior $p(\boldsymbol{\psi}|\mathcal{M}_2)$ is becoming more and more vague, $\log B_{12}$ goes to infinity, and the simpler model will be chosen with probability 1, regardless of the data $\mathbf{y}$, the sample size $N$, and the true value of $\boldsymbol{\psi}$. This problem became known as Lindley's paradox (Lindley, 1957).

Such spurious evidence in favor of too-simple models is reported in Atkinson (1978) and Smith and Spiegelhalter (1980) in the context of comparing regression models. A simple example illustrates that comparable difficulties exist in the context of mixture models. Suppose that under model $\mathcal{M}_1$, $Y \sim \mathcal{N}(0, 1)$, whereas under $\mathcal{M}_2$, $Y \sim 0.5\mathcal{N}(-\mu, 1) + 0.5\mathcal{N}(\mu, 1)$, with prior $p(\mu|\mathcal{M}_2) \sim \mathcal{N}(0, \delta)$. Then given $N$ observations $y_1, \dots, y_N$ the Bayes factor of model $\mathcal{M}_1$ versus model $\mathcal{M}_2$ reads:

$$B_{12} = \sqrt{2\pi\delta} / \int \mathrm{e}^{-(N+1/\delta)\mu^2/2} \prod_{i=1}^{N} \frac{1}{2}(\mathrm{e}^{y_i \mu} + \mathrm{e}^{-y_i \mu}) d\mu.$$

By increasing the variance $\delta$ of the prior $p(\mu|\mathcal{M}_2)$, $1/\delta$ goes to zero and the denominator becomes independent of $\delta$, whereas the numerator increases with $\delta$. Evidence in favor of model $\mathcal{M}_1$ increases, regardless of the true value of $\mu$ and the number $N$ of observations, and consequently $\Pr(\mathcal{M}_1|\mathbf{y}) \rightarrow 1$, as $\delta \rightarrow \infty$, provided that the prior probability $\Pr(\mathcal{M}_1)$ remains fixed.

It is possible to detect spurious evidence in favor of the restricted model by exploring the marginal posterior $p(\boldsymbol{\psi}|\mathcal{M}_2, \mathbf{y})$ of $\boldsymbol{\psi}$ under model $\mathcal{M}_2$. If this posterior is clearly bounded away from $\boldsymbol{\psi}_0$ and the Bayes factor strongly favors model $\mathcal{M}_1$, this points at Lindley's paradox. A pragmatic way of reacting to

Lindley's paradox is to decrease the spread of the prior, however, if the spread is too small, a bias toward the restricted model might be introduced once more.

This suggests the existence of an optimal prior, which gives maximal evidence in favor of the more complex model, which was actually proven by Edwards et al. (1963) for the case where no unknown parameters appear in the simple model. For the more general case where unknown parameters are present and $\boldsymbol{\psi}$ is multivariate, Kass and Vaidyanathan (1992) gave an asymptotic approximation to this prior for null-orthogonal models.

For finite mixture models the issue of choosing sensible priors under model uncertainty is by far less understood, and we try to shed some light on this in Subsection 5.3.2.

### 4.5.6 Further Approaches

A Bayesian testing method for the number of components based on the Kullback–Leibler distance was proposed by Mengersen and Robert (1996); an extension was considered by Sahu and Cheng (2003). In Sahu and Cheng (2003), a stepwise procedure is proposed that starts with a mixture distribution with enough components to be overfitting with high probability. A collapsing approach based on the Kullback–Leibler distance between a mixture with $K$ and $K-1$ components is proposed that does not require refitting of the model at each time.

# 5

# Computational Tools for Bayesian Inference for Finite Mixtures Models Under Model Specification Uncertainty

## 5.1 Introduction

In this chapter it is assumed that $K_{\max}$ finite mixture models $\mathcal{M}_1, \ldots, \mathcal{M}_{K_{\max}}$ should be compared given data $\mathbf{y} = (\mathbf{y}_1, \ldots, \mathbf{y}_N)$. Typically $\mathcal{M}_K$ is a model based on a finite mixture distribution from a certain family with $K$ components, however, more general model choice problems may arise such as choosing the parameter structure.

There are basically two Bayesian approaches to deal with model specification uncertainty for a finite mixture model. One approach, which is reviewed in Section 5.2, is to apply trans-dimensional Markov chain Monte Carlo to obtain draws from the joint posterior density $p(\mathcal{M}_K, \boldsymbol{\vartheta}_1, \ldots, \boldsymbol{\vartheta}_{K_{\max}}|\mathbf{y})$. The second approach is to compute for all possible models the marginal likelihood $p(\mathbf{y}|\mathcal{M}_K)$, which is defined for a finite mixture model in Section 5.3, and to apply Bayes' rule to quantify posterior evidence in favor of each model. The computation of the marginal likelihood for a finite mixture model is quite a challenge. Section 5.4 discusses several simulation-based approaches, whereas Section 5.5 deals with approximations based on density ratios. Finally, it is discussed in Section 5.6, that from a theoretical point of view, both approaches are equivalent, and they are nothing but different computational tools designed to estimate exactly the same quantity, namely the marginal posterior distribution over the model space $\Omega = \{\mathcal{M}_1, \ldots, \mathcal{M}_{K_{\max}}\}$.

## 5.2 Trans-Dimensional Markov Chain Monte Carlo Methods

In this section it is assumed that $K_{\max}$ models $\mathcal{M}_1, \ldots, \mathcal{M}_{K_{\max}}$ should be compared given the data $\mathbf{y} = (\mathbf{y}_1, \ldots, \mathbf{y}_N)$, where $\mathcal{M}_K$ is a model based on a finite mixture distribution from a certain family with $K$ components, and the data are possibly vector-valued. During the past decade, trans-dimensional

MCMC methods emerged as a popular method of dealing with mixtures with an unknown number of components. Many such methods were suggested; see Green (2003) for an excellent review of these methods, and Godsill (2001) for a discussion of the relationship between MCMC methods for model uncertainty.

One early development appears in Carlin and Chib (1995), who applied product-space MCMC methods to a normal mixture model with an unknown number of components; see Subsection 5.2.1 for more details. The most influential work on trans-dimensional MCMC methods for finite mixture models has been the seminal paper by Richardson and Green (1997), who suggest applying the reversible jump Metropolis–Hastings algorithm, introduced by Green (1995), to the problem of choosing the number of components. Reversible jump MCMC is reviewed in Subsection 5.2.2. Phillips and Smith (1996) discuss a jump diffusion approach for selecting the number of components in a mixture model which is then applied to univariate mixtures of normal distributions. Stephens (2000a) applied birth and death MCMC methods to select the number of components in a mixture model; see Subsection 5.2.3 for more detail.

### 5.2.1 Product-Space MCMC

Product-space Markov chain Monte Carlo methods are based on Markov chains that sample the joint vector $(\mathcal{M}, \boldsymbol{\vartheta}_1, \ldots, \boldsymbol{\vartheta}_{K_{\max}})$ from the posterior $p(\mathcal{M}, \boldsymbol{\vartheta}_1, \ldots, \boldsymbol{\vartheta}_{K_{\max}}|\mathbf{y})$, which is derived by Bayes' theorem as

$$\begin{aligned} & p(\mathcal{M}, \boldsymbol{\vartheta}_1, \ldots, \boldsymbol{\vartheta}_{K_{\max}}|\mathbf{y}) \\ & \propto p(\mathbf{y}|\mathcal{M}, \boldsymbol{\vartheta}_1, \ldots, \boldsymbol{\vartheta}_{K_{\max}}) p(\boldsymbol{\vartheta}_1, \ldots, \boldsymbol{\vartheta}_{K_{\max}}|\mathcal{M}) p(\mathcal{M}). \end{aligned} \tag{5.1}$$

This is the density of a distribution over the product space:

$$\Omega \times \bigotimes_{K=1}^{K_{\max}} \Theta_K. \tag{5.2}$$

Full product-space MCMC methods live on this complex state space, and use Gibbsian-type transition kernels by iteratively sampling the model indicator $\mathcal{M}$ and the model parameters $(\boldsymbol{\vartheta}_1, \ldots, \boldsymbol{\vartheta}_{K_{\max}})$ from the appropriate conditional densities. This method was first discussed in full generality by Carlin and Chib (1995); various modifications were suggested later, for instance a reduced Gibbs sampler that resamples only the model parameter corresponding to the most recent model indicator (Green and O'Hagan, 2000) and product space MCMC based on a Metropolis–Hastings kernel (Dellaportas et al., 2002).

Carlin and Chib (1995) realized the need of specifying a joint prior $p(\boldsymbol{\vartheta}_1, \ldots, \boldsymbol{\vartheta}_{K_{\max}}|\mathcal{M})$ of *all model parameters* for all possible models $\mathcal{M}$, in order to obtain the *joint* posterior $p(\mathcal{M}, \boldsymbol{\vartheta}_1, \ldots, \boldsymbol{\vartheta}_{K_{\max}}|\mathbf{y})$ in (5.1). This is in sharp contrast to methods that directly determine the marginal posterior probabilities $\Pr(\mathcal{M}_K|\mathbf{y})$ based on marginal likelihoods, and require only the specification of the marginal prior density $p(\boldsymbol{\vartheta}_K|\mathcal{M}_K)$. To simplify the specification of this prior, they assume conditional independence among the model-specific parameters $(\boldsymbol{\vartheta}_1, \dots, \boldsymbol{\vartheta}_{K_{\max}})$. In this case the joint prior may be rewritten for a certain realization of the model indicator, for instance, $\mathcal{M} = \mathcal{M}_K$, as

$$p(\boldsymbol{\vartheta}_1, \dots, \boldsymbol{\vartheta}_{K_{\max}}|\mathcal{M}_K) = p(\boldsymbol{\vartheta}_K|\mathcal{M}_K) \prod_{j=1,j\neq K}^{K_{\max}} p(\boldsymbol{\vartheta}_j|\mathcal{M}_K). \tag{5.3}$$

Even under prior independence we need a prior for $\boldsymbol{\vartheta}_K$ not only under the assumption that the corresponding model $\mathcal{M}_K$ is true, but also for any other model parameter $\boldsymbol{\vartheta}_j$, $j \neq K$, for which $\mathcal{M}_K$ is not true, in order to obtain the *joint* posterior $p(\mathcal{M}_K, \boldsymbol{\vartheta}_1, \dots, \boldsymbol{\vartheta}_{K_{\max}}|\mathbf{y})$. Under this prior the joint posterior density reads:

$$\begin{aligned} p(\mathcal{M}_K, \boldsymbol{\vartheta}_1, \dots, \boldsymbol{\vartheta}_{K_{\max}}|\mathbf{y}) \propto\; & p(\mathbf{y}|\mathcal{M}_K, \boldsymbol{\vartheta}_K)p(\boldsymbol{\vartheta}_K|\mathcal{M}_K)p(\mathcal{M}_K) \\ & \cdot \prod_{j=1,j\neq K}^{K_{\max}} p(\boldsymbol{\vartheta}_j|\mathcal{M}_K). \end{aligned} \tag{5.4}$$

Carlin and Chib (1995) call the densities $p(\boldsymbol{\vartheta}_K|\mathcal{M} \neq \mathcal{M}_K)$ pseudo-priors, as they actually have no influence on the marginal posterior probabilities $\Pr(\mathcal{M}_K|\mathbf{y})$.

To sample from (5.4) Carlin and Chib (1995) implement a two-block Gibbs sampler by iteratively sampling the model indicator $\mathcal{M}$ from the conditional densities $p(\mathcal{M}|\boldsymbol{\vartheta}_1, \dots, \boldsymbol{\vartheta}_{K_{\max}}, \mathbf{y})$ and sampling all model parameters $\boldsymbol{\vartheta}_1, \dots, \boldsymbol{\vartheta}_{K_{\max}}$ from $p(\boldsymbol{\vartheta}_1, \dots, \boldsymbol{\vartheta}_{K_{\max}}|\mathcal{M}, \mathbf{y})$. The relevant conditional densities are obtained immediately from the joint distribution given in (5.4). The probability $p(\mathcal{M}_K|\boldsymbol{\vartheta}_1, \dots, \boldsymbol{\vartheta}_{K_{\max}}, \mathbf{y})$, for instance, reads:

$$\begin{aligned} & p(\mathcal{M}_K|\boldsymbol{\vartheta}_1, \dots, \boldsymbol{\vartheta}_{K_{\max}}, \mathbf{y}) \propto \\ & p(\mathbf{y}|\boldsymbol{\vartheta}_K, \mathcal{M}_K)p(\boldsymbol{\vartheta}_K|\mathcal{M}_K) \prod_{j=1,j\neq K}^{K_{\max}} p(\boldsymbol{\vartheta}_j|\mathcal{M}_K)p(\mathcal{M}_K). \end{aligned} \tag{5.5}$$

It is, however, not necessary to update *all* model-specific parameters at each sweep. Green and O'Hagan (2000) prove that it is a valid strategy to run a reduced sampler which updates the model indicator $\mathcal{M}$ and only the corresponding model parameter $\boldsymbol{\vartheta}_{\mathcal{M}}$, whereas the remaining model parameters $\boldsymbol{\vartheta}_j$ for $j \neq \mathcal{M}$ are held fixed. This sampler is described in the following *Algorithm 5.1.*

*Algorithm 5.1: Product-Space MCMC* Specify starting values $\boldsymbol{\vartheta}_1^{(0)}, \dots, \boldsymbol{\vartheta}_{K_{\max}}^{(0)}$ and select pseudo-priors $p(\boldsymbol{\vartheta}_K|\mathcal{M} \neq \mathcal{M}_K)$ for each $K = 1, \dots, K_{\max}$. For $m = 1, \dots, M$ iterate the following steps.

(a) Sample $\mathcal{M}^{(m)}$ from the discrete posterior $p(\mathcal{M}|\boldsymbol{\vartheta}_1^{(m-1)}, \ldots, \boldsymbol{\vartheta}_{K_{\max}}^{(m-1)}, \mathbf{y})$ given in (5.5).
(b) If $\mathcal{M}^{(m)} = K$, sample $\boldsymbol{\vartheta}_K^{(m)}$ from $p(\boldsymbol{\vartheta}_K|\mathcal{M}_K, \mathbf{y})$ given by

$$p(\boldsymbol{\vartheta}_K|\mathcal{M}_K, \mathbf{y}) \propto p(\mathbf{y}|\boldsymbol{\vartheta}_K, \mathcal{M}_K)p(\boldsymbol{\vartheta}_K|\mathcal{M}_K), \tag{5.6}$$

whereas all other model-specific parameters remain unchanged, $\boldsymbol{\vartheta}_j^{(m)} = \boldsymbol{\vartheta}_j^{(m-1)}$, for $j \neq K$.

## Choosing the Pseudo-Priors

The most difficult point in applying product-space MCMC methods are sensible choices of the pseudo-priors $p(\boldsymbol{\vartheta}_K|\mathcal{M} \neq \mathcal{M}_K)$. The pseudo-priors do not appear when updating the model parameters in *Algorithm 5.1*, but are relevant for sampling the model indicators from (5.5). The pseudo-priors may be chosen in any convenient way to improve the mixing of the resulting MCMC sampler.

Chib (2001) recommends choosing pseudo-priors $p(\boldsymbol{\vartheta}_K|\mathcal{M} \neq \mathcal{M}_K)$ that closely mimic the model-specific posterior $p(\boldsymbol{\vartheta}_K|\mathcal{M}_K, \mathbf{y})$. Although this contradicts intuition, it is the optimal choice with respect to the mixing properties of the sampler, as shown by Chib (2001, p.3634). If these two densities were identical, then the right-hand side of (5.5) would be equal to

$$p(\mathbf{y}|\boldsymbol{\vartheta}_K, \mathcal{M}_K)p(\boldsymbol{\vartheta}_K|\mathcal{M}_K)p(\mathcal{M}_K) \prod_{j=1, j\neq K}^{K_{\max}} p(\boldsymbol{\vartheta}_j|\mathcal{M}_j, \mathbf{y})$$

$$= p(\mathbf{y}|\mathcal{M}_K)p(\mathcal{M}_K) \prod_{j=1}^{K_{\max}} p(\boldsymbol{\vartheta}_j|\mathcal{M}_j, \mathbf{y}) \propto p(\mathbf{y}|\mathcal{M}_K)p(\mathcal{M}_K).$$

Under this specific choice of priors, one would actually sample the model indicators from the true marginal posterior $p(\mathcal{M}_K|\mathbf{y})$, and the Gibbs sampler reduces to i.i.d. sampling.

To avoid the specification of any pseudo-prior, Green and O'Hagan (2000) suggest choosing the same prior over all models, $p(\boldsymbol{\vartheta}_K|\mathcal{M}_j) = p(\boldsymbol{\vartheta}_K|\mathcal{M}_K)$ for all $j \neq K$ . In this case the pseudo-prior cancels from (5.5), and we are left with a sampler that works without having to specify any pseudo-priors. Green and O'Hagan (2000) report that the practical performance of this sampler is not very encouraging, which is not surprising in the light of the convincing recommendation of Chib (2001, p.3634) to select the model-specific posterior $p(\boldsymbol{\vartheta}_K|\mathcal{M}_K, \mathbf{y})$ rather than the model-specific prior $p(\boldsymbol{\vartheta}_K|\mathcal{M}_K)$ as pseudo-prior.

**Product-Space MCMC for Mixture Models**

To implement *Algorithm 5.1* for mixture models it necessary to sample from the posterior $p(\boldsymbol{\vartheta}_K|\mathcal{M}_K,\mathbf{y})$ given in (5.6), which does not belong to a well-known distribution family. Following Dellaportas et al. (2002), the sampler may be modified by first sampling $\mathcal{M}^{new}$ from a proposal $m(\mathcal{M}^{new}|\mathcal{M}^{old})$ that is independent of any model-specific parameters, with $\mathcal{M}^{old}$ being the current model indicator.

Conditional on $\mathcal{M}^{old} = \mathcal{M}_K$ and $\mathcal{M}^{new} = \mathcal{M}_j$, one samples $\boldsymbol{\vartheta}_j$ from a nondegenerate proposal density $q(\boldsymbol{\vartheta}_j|\mathcal{M}_K,\boldsymbol{\vartheta}_K,\mathcal{M}_j)$ which does not depend on $\boldsymbol{\vartheta}_j^{old}$. As shown by Dellaportas et al. (2002), this leads to a valid Metropolis–Hastings algorithm with the acceptance probability given by $\min(1,A)$, where

$$A = \frac{p(\mathbf{y}|\boldsymbol{\vartheta}_j,\mathcal{M}_j)p(\boldsymbol{\vartheta}_j|\mathcal{M}_j)p(\mathcal{M}_j)q(\boldsymbol{\vartheta}_K|\mathcal{M}_j,\boldsymbol{\vartheta}_j,\mathcal{M}_K)m(\mathcal{M}_K|\mathcal{M}_j)}{p(\mathbf{y}|\boldsymbol{\vartheta}_K,\mathcal{M}_K)p(\boldsymbol{\vartheta}_K|\mathcal{M}_K)p(\mathcal{M}_K)q(\boldsymbol{\vartheta}_j|\mathcal{M}_K,\boldsymbol{\vartheta}_K,\mathcal{M}_j)m(\mathcal{M}_j|\mathcal{M}_K)}.$$

## 5.2.2 Reversible Jump MCMC

Green (1995) introduced the reversible jump Metropolis–Hastings algorithm to sample from the posterior distribution $p(\mathcal{M},\boldsymbol{\vartheta}_1,\ldots,\boldsymbol{\vartheta}_{K_{\max}}|\mathbf{y})$ given by (5.1) for rather general model specification problems. A pedagogical description of the reversible jump formalism is found in Waagepetersen and Sorensen (2001). Reversible jump MCMC was applied in Richardson and Green (1997) to select the number of components for univariate mixtures of normal distributions, and since then found application to many other mixture distributions, such as mixtures of exponential distributions (Gruet et al., 1999), mixtures of Poisson distributions (Viallefont et al., 2002; Dellaportas et al., 2002), mixtures of spherical multivariate normal distributions (Marrs, 1998), and mixtures of arbitrary multivariate normal distributions (Dellaportas and Papageorgiou, 2006). Further recent applications are by Nobile and Green (2000), Green and Richardson (2001), Fernández and Green (2002), and Bottolo et al. (2003). Robert et al. (2000) consider selecting the number of states in a hidden Markov model.

Reversible jump MCMC is based on creating a Markov chain that lives on the state space

$$\mathcal{X} = \bigcup_{K=1}^{K_{\max}} (\mathcal{M}_K \times \Theta_k), \tag{5.7}$$

which is of smaller dimension than the product space defined in (5.2). The sampler "jumps" between the different models, by making moves from a current model, say $(\mathcal{M}_K,\boldsymbol{\vartheta}_K)$ to a new model $(\mathcal{M}_j,\boldsymbol{\vartheta}_j)$, while retaining detailed balance that ensures the correct limiting distribution, provided that the chain is irreducible and aperiodic. Although this resembles the Metropolis–Hastings algorithm of Dellaportas et al. (2002), which was briefly described at the end of Subsection 5.2.1, the transition kernels used in reversible jump MCMC are far more general, allowing in particular moves with carefully selected degenerate proposal densities $q(\boldsymbol{\vartheta}_j|\mathcal{M}_K, \boldsymbol{\vartheta}_K, \mathcal{M}_j)$.

Degenerate proposals arise in a natural way when jumping between nested models, as it is likely that the model parameters have some common features, and certain parameters of the new model may even be a deterministic function of some parameters of the old one. To capture potential relations between $\boldsymbol{\vartheta}_K$ and $\boldsymbol{\vartheta}_j$, the Metropolis–Hastings algorithm proposes values for $\boldsymbol{\vartheta}_j$ via a mapping $\boldsymbol{\vartheta}_j = g(\boldsymbol{\vartheta}_K, \mathbf{u})$ that depends on $\boldsymbol{\vartheta}_K$ and some random variable $\mathbf{u}$. In the standard Metropolis–Hastings algorithm all components of $\boldsymbol{\vartheta}_j$ have to be stochastic, given $\boldsymbol{\vartheta}_K$. The reversible jump MCMC Metropolis–Hastings algorithm is more flexible in this respect, as $\boldsymbol{\vartheta}_j$ may be completely or partially deterministic, given $\boldsymbol{\vartheta}_K$. We are, however, not completely free in our choice of $g(\boldsymbol{\vartheta}_K, \mathbf{u})$, as the following move from a mixture with $K$ components with model parameter $\boldsymbol{\vartheta}_K = (\boldsymbol{\theta}_1, \dots, \boldsymbol{\theta}_K, \eta_1, \dots, \eta_K)$ to a mixture with $j = K + 1$ components illustrates.

To obtain $\boldsymbol{\vartheta}_{K+1} = (\boldsymbol{\theta}_1, \dots, \boldsymbol{\theta}_{K+1}, \eta_1, \dots, \eta_{K+1})$, leave all components but component $k^\star$ in the original mixture unchanged. Component $k^\star$ is duplicated (index these new components by $k_1$ and $k_2$) by adding a component with the same component parameter $\boldsymbol{\theta}_{k^\star}$,

$$\boldsymbol{\theta}_{k_1} = \boldsymbol{\theta}_{k^\star}, \qquad \boldsymbol{\theta}_{k_2} = \boldsymbol{\theta}_{k^\star}, \tag{5.8}$$

whereas the weight $\eta_{k^\star}$ is split between components $k_1$ and $k_2$:

$$\eta_{k_1} = u_1 \eta_{k^\star}, \qquad \eta_{k_2} = (1 - u_1)\eta_{k^\star}. \tag{5.9}$$

$u_1$ is a random variable with a nondegenerate proposal density $q(u_1)$ defined on [0,1]. As it turns out, this is not a valid move. When we consider the reverse move from an arbitrary mixture with $K + 1$ components defined by $\boldsymbol{\vartheta}_{K+1}$ to a mixture with $K$ components, it is possible to merge the component weights $\eta_{k_1}$ and $\eta_{k_2}$ as in (5.9) to obtain $\eta_{k^\star}$. However, there is no way to obtain $\boldsymbol{\theta}_{k^\star}$ from $\boldsymbol{\theta}_{k_1}$ and $\boldsymbol{\theta}_{k_2}$ in such a way that the deterministic relation (5.8) is fulfilled, as the probability that $\boldsymbol{\theta}_{k_1}$ and $\boldsymbol{\theta}_{k_2}$ are identical is zero.

Most ingeniously, Green (1995) realized that a solution to the problem is simply adding noise to match dimensions. Consider, for instance, the following way of splitting component $k^\star$,

$$\boldsymbol{\theta}_{k_1} = \boldsymbol{\theta}_{k^\star} - \mathbf{u}_2, \qquad \boldsymbol{\theta}_{k_2} = \boldsymbol{\theta}_{k^\star} + \mathbf{u}_2,$$

where $\mathbf{u}_2$ is a random vector of the same dimension as $\boldsymbol{\theta}_{k^\star}$ with a nondegenerate proposal density $q(\mathbf{u}_2)$, centered at 0. Then in the reverse move, $\boldsymbol{\theta}_{k^\star}$ is defined as the mean of $\boldsymbol{\theta}_{k_1}$ and $\boldsymbol{\theta}_{k_2}$.

As a general rule, when moving to the higher model space, as many random variables $\mathbf{u}$ need to be drawn from a nondegenerate proposal $q(\mathbf{u})$ as correspond to the number of additional parameters:

$$\dim(\boldsymbol{\vartheta}_{K+1}) = \dim(\boldsymbol{\vartheta}_K) + \dim(\mathbf{u}).$$

These random variables are then used to construct $\boldsymbol{\vartheta}_{K+1}$ from $\boldsymbol{\vartheta}_K$ through a function $g_{K,K+1}(\boldsymbol{\vartheta}_K, \mathbf{u})$, which has to be invertible in order to guarantee reversibility.

### Reversible Jump MCMC for Finite Mixture Models

For finite mixture models reversible jump MCMC typically operates on the augmented parameter space, where the allocation variables $\mathbf{S}$ are included as unknown model parameters.

To implement reversible jump MCMC a first step is to design a strategy for moving between mixture models with different numbers of components. If the current model is a mixture with $K > 1$ components, then it is usual to reduce the searching strategy to moves that either preserve the number of components, or lead to a mixture with $K-1$ or $K+1$ components, respectively. It is possible to design more than one type of move to jump from $K$ to $K+1$ or $K-1$ components. Jumps are achieved by adding new components, deleting existing components, and splitting or merging existing components. The various moves could be scanned systematically or could be selected randomly. Reversible jump MCMC is implemented through the following algorithm which extends the algorithm suggested by Richardson and Green (1997) to general finite mixture models.

*Algorithm 5.2: Reversible Jump MCMC for a Finite Mixture Model* Start with a certain mixture model with $K$ components and select classifications $\mathbf{S}$ and a hyperparameter $\boldsymbol{\delta}$ for the prior $p(\boldsymbol{\vartheta}_K|\boldsymbol{\delta})$. Repeat the following steps for $m = 1, \ldots, M$.

(a) Perform the following dimension-preserving move.
  (a1) Update the model-specific parameter $\boldsymbol{\vartheta}_K = (\boldsymbol{\theta}_1^K, \ldots, \boldsymbol{\theta}_K^K, \boldsymbol{\eta}^K)$ as described in step (a) of *Algorithm 3.4*.
  (a2) Update the current allocation $\mathbf{S}$ as described in step (b) of *Algorithm 3.4*.
  (a3) Update the hyperparameter $\boldsymbol{\delta}$ of the prior $p(\boldsymbol{\vartheta}_K|\boldsymbol{\delta})$ if it is random.
(b) Perform the following dimension-changing move.
  (b1) Split one mixture component into two components or merge two components into one.
  (b2) Birth or death of an empty component.

In step (b1), the choice between splitting and merging is random in Richardson and Green (1997), as is the choice between birth and death in step (b2). Alternatively, Dellaportas and Papageorgiou (2006), in the context of multivariate mixtures, sweep through all steps randomly. In the latter case, no rejection step is needed as long as the probability of selecting a dimension-preserving move is independent of the current state of the Markov chain.

Ideally, the dimension-changing moves are designed to have a high probability of acceptance, so that the sampler explores the different models adequately. Although invariance will hold for arbitrary moves, the efficiency of the algorithm may depend crucially on the particular choice. As emphasized by Green (1995, p.715), *intuition can be used to choose moves that possibly induce good mixing behavior.* An important contribution to reversible jump MCMC is the work by Brooks et al. (2003) on the efficient construction of moves.

For practical implementation in the context of finite mixture models, it is sufficient to choose a matching function $\boldsymbol{\vartheta}_{K+1} = g_{K,K+1}(\boldsymbol{\vartheta}_K, \mathbf{u})$ together with a proposal density $q_{K,K+1}(\mathbf{u})$ to perform moves from $\mathcal{M}_K$ to $\mathcal{M}_{K+1}$, and the reverse move

$$(\boldsymbol{\vartheta}_K, \mathbf{u}) = g^{-1}_{K,K+1}(\boldsymbol{\vartheta}_{K+1}),$$

could be used to move from $\mathcal{M}_{K+1}$ to $\mathcal{M}_K$. Such moves form a reversible pair with acceptance probability $\min(1, A)$ and $\min(1, 1/A)$, respectively.

*Algorithm 5.3: Moving to a Mixture with $K+1$ Components in Reversible Jump MCMC* Assume that the current model $\mathcal{M}_K$ is a mixture with $K$ components, the model parameter being equal to $\boldsymbol{\vartheta}_K$. Choose a move of type $h$ with probability $m_h(\mathcal{M}_K, \boldsymbol{\vartheta}_K)$ which is allowed to depend on the current number $K$ of components and on $\boldsymbol{\vartheta}_K$. If this move suggests jumping to a mixture model $\mathcal{M}_{K+1}$ with $K+1$ components, proceed in the following way.

(a) Match the dimensions between the models: propose $\mathbf{u}$ from a proposal density $q_{K,K+1}(\mathbf{u})$, and determine $\boldsymbol{\vartheta}_{K+1}$ from

$$\boldsymbol{\vartheta}_{K+1} = g_{K,K+1}(\boldsymbol{\vartheta}_K, \mathbf{u}), \tag{5.10}$$

where $g_{K,K+1}(\cdot)$ may depend on the move type $h$.

(b) Reallocate the observations according to a proposal $q(\mathbf{S}^{new}|\mathbf{S}, \boldsymbol{\vartheta}_{K+1})$.

(c) Move to the finite mixture model $\mathcal{M}_{K+1}$ with component parameter $\boldsymbol{\vartheta}_{K+1}$ and allocations $\mathbf{S}^{new}$ with probability $\min(1, A)$, where $A$ depends on $\boldsymbol{\vartheta}_K$, $\boldsymbol{\vartheta}_{K+1}$, $\mathbf{S}$ and $\mathbf{S}^{new}$, and is equal to

$$A = (\text{likelihood ratio}) \times (\text{prior ratio}) \times (\text{proposal ratio}) \times |\text{Jacobian}|,$$

with

$$\text{likelihood ratio} = \prod_{i:S_i^{new} \neq S_i} \frac{p(\mathbf{y}_i|\boldsymbol{\theta}_{S_i^{new}})}{p(\mathbf{y}_i|\boldsymbol{\theta}_{S_i})}$$

$$\text{prior ratio} = \frac{p(\mathbf{S}^{new}|\boldsymbol{\vartheta}_{K+1}, \mathcal{M}_{K+1})p(\boldsymbol{\vartheta}_{K+1}|\mathcal{M}_{K+1})\Pr(\mathcal{M}_{K+1})}{p(\mathbf{S}|\boldsymbol{\vartheta}_K, \mathcal{M}_K)p(\boldsymbol{\vartheta}_K|\mathcal{M}_K)\Pr(\mathcal{M}_K)}$$

$$\text{proposal ratio} = \frac{m_h(\boldsymbol{\vartheta}_{K+1}, \mathcal{M}_{K+1})}{q(\mathbf{S}^{new}|\mathbf{S}, \boldsymbol{\vartheta}_{K+1})q_{K,K+1}(\mathbf{u})m_h(\boldsymbol{\vartheta}_K, \mathcal{M}_K)}$$

$$|\text{Jacobian}| = \left|\frac{\partial g_{K,K+1}(\boldsymbol{\vartheta}_K, \mathbf{u})}{\partial(\boldsymbol{\vartheta}_K, \mathbf{u})}\right|.$$

### Designing Split and Merge Moves

Split and merge moves typically are reversible pairs, that usually are constructed by first formulating how components are merged. Then random noise is added to achieve dimension matching. This random noise is used to define a split move that reduces to the original mixture when the split move is reversed.

Merging starts by choosing a pair $k_1$ and $k_2$ of components that are combined to form a component labeled $k^\star$, reducing the current number of components by one. In Richardson and Green (1997), this is a deterministic move, once $k_1$ and $k_2$ have been chosen. Whereas it appears natural to add the weights $\eta_{k_1}$ and $\eta_{k_2}$ to define $\eta_{k^\star}$,

$$\eta_{k^\star} = \eta_{k_1} + \eta_{k_2}, \tag{5.11}$$

there exists much more freedom in constructing the component parameter $\boldsymbol{\theta}_{k^\star}$ from $\boldsymbol{\theta}_{k_1}$ and $\boldsymbol{\theta}_{k_2}$. Richardson and Green (1997) show how moment matching could be used to construct $\boldsymbol{\theta}_{k^\star}$. The idea is that any move that leaves certain aspects of the implied marginal distribution unchanged is likely to be accepted. For a mixture of univariate normal distributions, for instance, Richardson and Green (1997) suggest a move that preserves the first two moments of the marginal density; see Subsection 6.4.2 for more detail.

In general, $\boldsymbol{\theta}_{k^\star}$ is defined in such a way that for a set of suitable functions $h_j : \mathcal{Y} \rightarrow \Re$, the expectation of $h_j(\mathbf{Y})$ with respect to the marginal distribution $p(\mathbf{y}|\boldsymbol{\vartheta}_K)$, as given by (1.19), is unchanged, when moving to $\boldsymbol{\vartheta}_{K-1}$:

$$\mathrm{E}(h_j(\mathbf{Y})|\boldsymbol{\vartheta}_K) = \mathrm{E}(h_j(\mathbf{Y})|\boldsymbol{\vartheta}_{K-1}).$$

When merging two components $k_1$ and $k_2$, while leaving all other components unchanged, this leads to the following set of equations,

$$\mathrm{E}(h_j(\mathbf{Y})|\boldsymbol{\theta}_{k^\star})\eta_{k^\star} = \mathrm{E}(h_j(\mathbf{Y})|\boldsymbol{\theta}_{k_1})\eta_{k_1} + \mathrm{E}(h_j(\mathbf{Y})|\boldsymbol{\theta}_{k_2})\eta_{k_2}. \tag{5.12}$$

The functions $h_j(\mathbf{Y})$ could be selected in such a way that important features of the component densities are not allowed to change arbitrarily. Typically, $\dim(\boldsymbol{\theta}_k)$ linearly independent functions will be needed to achieve reversibility. For univariate mixtures with a one-dimensional component parameter $\theta_k$, only a single function $h(Y)$, in most cases $h(Y) = Y$, is needed:

$$\eta_{k^\star}\mathrm{E}(Y|\theta_{k^\star}) = \eta_{k_1}\mathrm{E}(Y|\theta_{k_1}) + \eta_{k_2}\mathrm{E}(Y|\theta_{k_2}).$$

Next consider a split move, where a component $k^\star$ with parameters $(\boldsymbol{\theta}_{k^\star}, \eta_{k^\star})$ is chosen at random and split into two components with parameters $(\boldsymbol{\theta}_{k_1}, \eta_{k_1})$ and $(\boldsymbol{\theta}_{k_2}, \eta_{k_2})$, respectively. When reversing the split move, the conditions (5.11) and (5.12) need to be fulfilled, thus there are $\dim(\boldsymbol{\theta}_k) + 1$ degrees of freedom to match the dimension. One degree of freedom is used to split the weights as in Richardson and Green (1997):

$$\eta_{k_1} = \eta_{k^\star} u_1, \qquad \eta_{k_2} = \eta_{k^\star}(1 - u_1), \tag{5.13}$$

where $u_1$ is a random variable with density $q_1(u_1)$ defined on [0,1], for instance, $u_1 \sim \mathcal{B}(2,2)$.

To satisfy the remaining degrees of freedom, a random vector $\mathbf{u}_2$ of dimension $\dim(\boldsymbol{\theta}_k)$ with nondegenerate density $q_2(\mathbf{u}_2)$ is chosen. $\mathbf{u}_2$ is then used to construct $\boldsymbol{\theta}_{k_1}$ and $\boldsymbol{\theta}_{k_2}$ from $\boldsymbol{\theta}_{k^\star}$ in such a way that (5.12) is fulfilled:

$$\begin{aligned}\boldsymbol{\theta}_{k_1} &= g_1(\eta_{k^\star}, u_1, \boldsymbol{\theta}_{k^\star}, \mathbf{u}_2),\\ \boldsymbol{\theta}_{k_2} &= g_2(\eta_{k^\star}, u_1, \boldsymbol{\theta}_{k^\star}, \mathbf{u}_2).\end{aligned} \tag{5.14}$$

This is in general not as easy as it appears. The choice of the distribution of the random vector $\mathbf{u}_2$ is not totally free, in particular if some elements of $\boldsymbol{\theta}_k$ are subject to nonnegativity constraints, or even more complex constraints such as positive definiteness for covariance matrices $\boldsymbol{\Sigma}_k$ in a multivariate mixture of normals have to be respected; see Subsection 6.4.2 for more details. In this case the distribution of the random vector $\mathbf{u}_2$ must be chosen in such a way that $\boldsymbol{\theta}_{k_1}$ and $\boldsymbol{\theta}_{k_2}$ do not violate these constraints.

The split move is completed by splitting the allocations. For all $i \neq k^\star$, the allocation remains unchanged, thus $S_i^{new} = S_i$. Allocation of all observations in component $k^\star$ is done according to the standard classification rule:

$$\Pr(S_i^{new} = k_1 | S_i = k^\star, \boldsymbol{\vartheta}_{K+1}) = \frac{\eta_{k_1} p(\mathbf{y}_i|\boldsymbol{\theta}_{k_1})}{\eta_{k_1} p(\mathbf{y}_i|\boldsymbol{\theta}_{k_1}) + \eta_{k_2} p(\mathbf{y}_i|\boldsymbol{\theta}_{k_2})}.$$

For a mixture with $K$ components, a split move is selected with probability $b_K$, whereas a merge move is selected with probability $d_K = 1 - b_K$. Usually, an upper bound $K_{\max}$ has to be specified. Obviously, $d_1 = 0$ and $b_{K_{\max}} = 0$, and typically $d_K = b_K = 0.5$ for $1 < K < K_{\max}$. There is a random choice between splitting and combining, for $2 \leq K \leq K_{\max} - 1$, and of course no combining for $K = 1$, and no splitting for $K = K_{\max}$. It should be noted that this split move itself is independent of the actual number $K$ of components.

It is possible to simplify the acceptance probability, defined in (5.11), for a split move. The functions defined in (5.13) and (5.14) are sufficient to determine the determinant of the Jacobian appearing in (5.11), which reduces to

$$|\text{Jacobian}| = \left| \frac{\partial(\eta_{k_1}, \eta_{k_2}, \boldsymbol{\theta}_{k_1}, \boldsymbol{\theta}_{k_2})}{\partial(\eta_{k^\star}, u_1, \boldsymbol{\theta}_{k^\star}, \mathbf{u}_2)} \right|.$$

A bit of algebra is necessary to work out that the acceptance probability for a split move reduces to the following quantity,

$$\begin{aligned}A = &\prod_{i:S_i=k^\star} \frac{\eta_{k_1}/\eta_{k^\star} p(\mathbf{y}_i|\boldsymbol{\theta}_{k_1}) + \eta_{k_2}/\eta_{k^\star} p(\mathbf{y}_i|\boldsymbol{\theta}_{k_2})}{p(\mathbf{y}_i|\boldsymbol{\theta}_{k^\star})} \\ &\times \left(\frac{\eta_{k_1}\eta_{k_2}}{\eta_{k^\star}}\right)^{e_0-1} \frac{p(\boldsymbol{\theta}_{k_1}|\boldsymbol{\delta}) p(\boldsymbol{\theta}_{k_2}|\boldsymbol{\delta}) \Pr(\mathcal{M}_{K+1})}{B(e_0, Ke_0) p(\boldsymbol{\theta}_{k^\star}|\boldsymbol{\delta}) \Pr(\mathcal{M}_K)} \\ &\times \frac{d_{K+1}}{b_K q_1(u_1) q_2(\mathbf{u}_2)} \left| \frac{\partial(\eta_{k_1}, \eta_{k_2}, \boldsymbol{\theta}_{k_1}, \boldsymbol{\theta}_{k_2})}{\partial(\eta_{k^\star}, u_1, \boldsymbol{\theta}_{k^\star}, \mathbf{u}_2)} \right|,\end{aligned} \tag{5.15}$$

where $e_0$ is the parameter appearing in the $\mathcal{D}(e_0, \dots, e_0)$ prior on the weight distribution $\boldsymbol{\eta}$.

The merge move is easily obtained from the split move. Whereas the weights are added as in (5.11), the moment matching condition in (5.12) is used to define $\boldsymbol{\theta}_{k^\star}$. To compute the acceptance rate, $u_1$ and $\mathbf{u}_2$ have to be reconstructed. From (5.13) we obtain:

$$u_1 = \frac{\eta_{k_1}}{\eta_{k^\star}}.$$

$\mathbf{u}_2$ is obtained by inverting the functions $g_1$ and $g_2$ defined in (5.14). The acceptance probability for a combine move reads: $\min(1, A^{-1})$, where $A$ is the acceptance probability for a split move.

### Split and Merge Moves for Poisson Mixtures

For a mixture of Poisson distributions only a single function $h(Y)$ is needed to define the combine move, because $\theta_k = \mu_k$ is one-dimensional. Matching $h(Y) = Y$ as in (5.12) yields:

$$\eta_{k^\star}\mu_{k^\star} = \eta_{k_1}\mu_{k_1} + \eta_{k_2}\mu_{k_2}. \tag{5.16}$$

Only a univariate random variable $u_2$ is needed to match dimensions. When splitting a component with mean $\mu_{k^\star}$, the means $\mu_{k_1}$ and $\mu_{k_2}$ must satisfy (5.16), and both of them need to be positive. This implies the following constraint on $\mu_{k_1}$,

$$0 \leq \mu_{k_1} \leq \frac{\eta_{k^\star}}{\eta_{k_1}}\mu_{k^\star}.$$

To construct $\mu_{k_1}$, Viallefont et al. (2002) consider three different types of moves; one of them makes use of a univariate random variable $u_2$ with non-degenerate density defined on [0,1]; for example, $u_2 \sim \mathcal{B}(2, 2)$,

$$\mu_{k_1} = u_2 \frac{\eta_{k^\star}}{\eta_{k_1}}\mu_{k^\star} = \frac{u_2}{u_1}\mu_{k^\star}. \tag{5.17}$$

Substitute in (5.16) to obtain $\mu_{k_2}$:

$$\mu_{k_2} = (1 - u_2)\frac{\eta_{k^\star}}{\eta_{k_2}}\mu_{k^\star} = \frac{1 - u_2}{1 - u_1}\mu_{k^\star}. \tag{5.18}$$

The acceptance probability is computed as in (5.15). The three equations (5.13), (5.17), and (5.18) define the matching function needed to evaluate the Jacobian:

$$|\text{Jacobian}| = \frac{\eta_{k^\star}\mu_{k^\star}}{u_1(1 - u_1)}.$$

Solving (5.17) and (5.18) yields the following reverse move,

$$\mu_{k^\star} = \frac{\eta_{k_1}\mu_{k_1} + \eta_{k_2}\mu_{k_2}}{\eta_{k^\star}}, \qquad u_2 = \frac{\mu_{k_1}}{\mu_{k^\star}}\frac{\eta_{k_1}}{\eta_{k^\star}}.$$

We refer to Viallefont et al. (2002) for alternative ways of defining the split and merge move.

## Birth and Death Moves

Richardson and Green (1997) design birth and death moves for univariate mixtures of normals, which are easily extended to more general finite mixture models.

For a birth move, a nearly empty component is added. From Section 1.3.2 it is known that an empty component with an arbitrary component parameter may be added without changing the likelihood. Adding a component $k^\star$ with $\eta_{k^\star} = 0$ and sampling $\boldsymbol{\theta}_{k^\star}$ from some proposal density, however, is not a valid move. Dimension matching within a reversible move requires $\dim(\boldsymbol{\vartheta}_{K+1}) = \dim(\boldsymbol{\vartheta}_K) + \dim(\mathbf{u})$ for some random variable $\mathbf{u}$ of dimension $\dim(\boldsymbol{\theta}_k)+1$. Evidently $\boldsymbol{\theta}_{k^\star}$ plays in principle the role of $\mathbf{u}$, but one additional random variable is needed to match dimension. Richardson and Green (1997) suggested using this random variable to design component $k^\star$ as a component with a small weight, by sampling $\eta_{k^\star}$ from a proposal density $q(\eta_{k^\star})$ giving small weights, for instance, $\eta_{k^\star} \sim \mathcal{B}(1, K)$. To obtain a place for the new component, the existing weights are rescaled,

$$\eta_k^{new} = \begin{cases} \eta_k(1-\eta_{k^\star}), & k < k^\star, \\ \eta_{k^\star}, & k = k^\star, \\ \eta_{k-1}(1-\eta_{k^\star}), & k > k^\star, \end{cases}$$

but no other changes are made; in particular no observations are allocated to this component.

If $\boldsymbol{\theta}_{k^\star}$ is sampled from the prior $p(\boldsymbol{\theta}_{k^\star}|\mathcal{M}_{K+1})$, the acceptance rate for a birth move is $\min(1, A)$, where

$$A = \frac{\Pr(\mathcal{M}_{K+1})}{\Pr(\mathcal{M}_K)} \frac{\eta_{k^\star}^{e_0-1}(1-\eta_{k^\star})^{N+Ke_0-K}}{q(\eta_{k^\star}|\mathcal{M}_K)B(e_0, Ke_0)} \frac{(K+1)d_{K+1}}{(K_0+1)b_K}(1-\eta_{k^\star})^K, \quad (5.19)$$

because the likelihood ratio is 1, and the prior and the proposal for $\boldsymbol{\theta}_{k^\star}$ cancel. $d_{K+1}$ is the probability of choosing a death move if the current number of components is equal to $K+1$, $b_K$ is the probability of choosing a birth move if the current number of components is equal to $K$, and $K_0$ is the number of empty components before birth.

For a death move, a random choice is made between any empty component. A component $k$ is empty if no observations are allocated to that component within data augmentation ($N_k(\mathbf{S}) = \#\{S_i = k\} = 0$). No change occurs if none of the components is empty. The selected component $k^\star$ is deleted from the mixture and the remaining weights are readjusted to account for the deleted weight $\eta_{k^\star}$:

$$\eta_k^{new} = \frac{\eta_k}{1-\eta_{k^\star}}, \qquad k \neq k^\star.$$

As birth and death form a reversible pair, the acceptance rate for a death move is $\min(1, A^{-1})$, with $A$ being the acceptance probability for a birth given in (5.19).

### 5.2.3 Birth and Death MCMC Methods

Birth and death MCMC is a rather general simulation method, and was applied to univariate and multivariate mixtures of normal and $t$-distributions by Stephens (1997a) and Stephens (2000a). Hurn et al. (2003) deal with mixtures of regressions. Cappé et al. (2003) consider selecting the number of states in a hidden Markov model.

As mentioned already in Subsection 1.2.3, Stephens (2000a) noted that a finite mixture model may be viewed, in an abstract sense, as a marked point process in a general space. To sample from the posterior distribution of a finite mixture model with an unknown number of components, Stephens (2000a) modified simulation methods that were developed to simulate realizations of marked point processes. These methods regard a spatial point process as the invariant distribution of a continuous time spatial birth and death Markov process and are discussed, for instance, in Ripley (1977) and Geyer and Møller (1994).

Let $K$ be the current number of components of the finite mixture model and let $\boldsymbol{\vartheta}_K = (\boldsymbol{\theta}_1, \ldots, \boldsymbol{\theta}_K, \eta_1, \ldots, \eta_K)$ be the corresponding parameters. Births and deaths occur in continuous time. A birth corresponds to increasing the number of components by one, and a death means decreasing the number of components by one. A birth occurs at a constant rate $\lambda_b$. If a birth occurs, a point $(\eta_{K+1}, \boldsymbol{\theta}_{K+1})$ is added, and the weights $\eta_1, \ldots, \eta_K$ are adjusted. For each point $\boldsymbol{\theta}_k$, $k = 1, \ldots, K$, a death occurs at a rate $d(\boldsymbol{\theta}_k)$, which is low for components that are important for explaining the data, but high for components that do not help to explain the data. This relevance is measured in terms of the mixture likelihood $p(\mathbf{y}|\boldsymbol{\vartheta}_K)$ of the current mixture model in relation to the mixture likelihood of a mixture model without the component corresponding to $(\boldsymbol{\theta}_k, \eta_k)$. If a death occurs at $\boldsymbol{\theta}_k$, then the corresponding component is deleted, and the weights of the remaining components are adjusted.

*Algorithm 5.4: Birth and Death MCMC for a Mixture Model* Repeat the following steps.

(a) Simulate $(K, \boldsymbol{\eta}, \boldsymbol{\theta}_1, \ldots, \boldsymbol{\theta}_K)$ by running a birth and death process for fixed time $t_0$. Set $t = 0$.

(a1) Let the birth rate be equal to $b(t) \equiv \lambda_b$. Determine the actual death rate for each component:

$$d(\boldsymbol{\theta}_k) = \frac{p(\mathbf{y}|\mathcal{M}_{K-1}, \boldsymbol{\vartheta}_{-k})}{p(\mathbf{y}|\mathcal{M}_K, \boldsymbol{\vartheta}_K)} \frac{\lambda_b}{K} \frac{p(\mathcal{M}_{K-1})}{p(\mathcal{M}_K)}, \tag{5.20}$$

and determine the overall death rate $d(t) = \sum_{k=1}^K d(\boldsymbol{\theta}_k)$.

(a2) Simulate the arrival time to the next jump,

$$t^{new} = t + \mathcal{E}(1)/(b(t) + d(t)),$$

and proceed with step (b), if $t^{new} > t_0$. Otherwise simulate the type of jump with the appropriate probabilities,

$$\Pr(\text{birth}) = \frac{b(t)}{b(t) + d(t)}, \qquad \Pr(\text{death of } \boldsymbol{\theta}_k) = \frac{d(\boldsymbol{\theta}_k)}{b(t) + d(t)}.$$

(a3) Adjust the mixture model to reflect birth or death. For a death of component $k$ jump to $\boldsymbol{\vartheta}_{K-1}$ given by

$$\boldsymbol{\vartheta}_{K-1} = (\boldsymbol{\theta}_1, \ldots, \boldsymbol{\theta}_{k-1}, \boldsymbol{\theta}_{k+1}, \ldots, \boldsymbol{\theta}_K, \eta_1^{new}, \ldots, \eta_{K-1}^{new}),$$

where

$$\eta_j^{new} = \begin{cases} \dfrac{\eta_j}{(1-\eta_k)}, & j < k, \\ \dfrac{\eta_{j+1}}{(1-\eta_k)}, & j = k, \ldots, K-1. \end{cases} \tag{5.21}$$

For a birth of a new component jump to $\boldsymbol{\vartheta}_{K+1}$ given by

$$\boldsymbol{\vartheta}_{K+1} = (\boldsymbol{\theta}_1, \ldots, \boldsymbol{\theta}_K, \boldsymbol{\theta}_{K+1}, \eta_1^{new}, \ldots, \eta_{K+1}^{new}),$$

where

$$\eta_j^{new} = \begin{cases} \eta_j(1-\eta_{K+1}), & j \leq K, \\ \eta_{K+1}, & j = K+1. \end{cases} \tag{5.22}$$

The position of the new point $\boldsymbol{\theta}_{K+1}$ is proposed according to the prior $p(\boldsymbol{\theta}|\boldsymbol{\delta})$, whereas the mark is simulated from a beta distribution with prior mean equal to $1/K$, $\eta_{K+1} \sim \mathcal{B}(\gamma, K\gamma)$. Set $t = t^{new}$ and return to step (a1).

(b) Run several steps of full conditional Gibbs sampling for the current number of components.

(b1) Update the allocations $\mathbf{S}$, and, if necessary, parameters that are not component specific and the hyperparameters $\boldsymbol{\delta}$ of the component-specific priors.

(b2) Update $\boldsymbol{\eta}$ and $\boldsymbol{\theta}_1, \ldots, \boldsymbol{\theta}_K$.

Doubling the birth rate $\lambda_b$ is equivalent to doubling $t_0$, thus one is free to choose $t_0 = 1$. Larger values of $\lambda_b$ will result in better mixing, but will need more computation time. For a Poisson prior $K \sim \mathcal{P}(\lambda_n)$ with prior mean $\lambda_n$, Stephens (2000a) chooses $\lambda_b = \lambda_n$, in which case the ratio (5.20) reduces to the likelihood ratio. If necessary, sampling in step (b1) may be partitioned into several blocks. Step (b2) is not necessary, but recommended by Stephens (2000a) in order improve mixing. Hurn et al. (2003) substitute in step (b) full conditional Gibbs sampling by a marginal Metropolis–Hastings step which avoids sampling the allocations. The adjustment of the weights in (5.21) has been suggested by Stephens (1997a), but alternative strategies to adjust the weights, in order to satisfy the constraint, are sensible, for instance, adjusting the weights proportional to the distance.

To a certain degree, birth and death methods appear to be more natural and elegant than reversible jump methods, because they avoid calculating the Jacobian and sampling the allocation variables $\mathbf{S}$. To apply this simulation method to mixture models, however, a key feature is that the mixture likelihood $p(\mathbf{y}|\boldsymbol{\vartheta}_K, \mathcal{M}_K)$ is available in closed form.

## 5.3 Marginal Likelihoods for Finite Mixture Models

Using marginal likelihoods to select the number of components in a finite mixture model was considered by several authors; see, for instance, Bensmail et al. (1997) and Frühwirth-Schnatter (2004) for univariate and multivariate normal mixture models, Otter et al. (2002) for mixtures of multivariate regression models, and Lenk and DeSarbo (2000) and Frühwirth-Schnatter et al. (2004) for mixtures of random effects models.

### 5.3.1 Defining the Marginal Likelihood

As explained earlier in Subsection 4.5.2, the marginal posterior probability $p(\mathcal{M}_K|\mathbf{y})$ of a finite mixture model may be derived from the joint posterior density $p(\mathcal{M}_K, \boldsymbol{\vartheta}_1, \ldots, \boldsymbol{\vartheta}_{K_{\max}}|\mathbf{y})$ by marginalization with respect to all unknown model parameters:

$$p(\mathcal{M}_K|\mathbf{y}) \propto p(\mathbf{y}|\mathcal{M}_K)p(\mathcal{M}_K), \tag{5.23}$$

with $p(\mathbf{y}|\mathcal{M}_K)$ being the so-called marginal likelihood. If the joint prior $p(\boldsymbol{\vartheta}_1, \ldots, \boldsymbol{\vartheta}_{K_{\max}}|\mathcal{M}_K)$ is identical with prior (5.3), applied in Subsection 5.2.1 in the context of product space MCMC methods, then the marginal likelihood $p(\mathbf{y}|\mathcal{M}_K)$ is immediately available from (5.4):

$$p(\mathbf{y}|\mathcal{M}_K) = \int_{\Theta_K} p(\mathbf{y}|\mathcal{M}_K, \boldsymbol{\vartheta}_K)p(\boldsymbol{\vartheta}_K|\mathcal{M}_K)d\boldsymbol{\vartheta}_K. \tag{5.24}$$

This definition of the marginal likelihood, which involves the mixture likelihood function $p(\mathbf{y}|\boldsymbol{\vartheta}_K)$ and operates on a level where the unknown allocations $\mathbf{S}$ are integrated out, is not the only one available for a finite mixture model.

The same expression may be derived from the complete-data likelihood $p(\mathbf{y}|\mathbf{S}, \boldsymbol{\vartheta}_K, \mathcal{M}_K)p(\mathbf{S}|\boldsymbol{\eta}, \mathcal{M}_K)$ under a proper prior $p(\boldsymbol{\vartheta}_K)$, without making specific assumptions about the joint prior $p(\boldsymbol{\vartheta}_1, \ldots, \boldsymbol{\vartheta}_{K_{\max}}|\mathcal{M}_K)$:

$$p(\mathbf{y}|\mathcal{M}_K) = \int_{\mathcal{S}_K \times \Theta_K} p(\mathbf{y}|\mathbf{S}, \boldsymbol{\vartheta}_K)p(\mathbf{S}|\boldsymbol{\eta})p(\boldsymbol{\vartheta}_K)d(\mathbf{S}, \boldsymbol{\vartheta}_K). \tag{5.25}$$

For many finite mixture models, the dimensionality of (5.25) can be reduced by solving the integration with respect to the indicators $\mathbf{S}$ analytically which leads exactly to the definition given in (5.24). Marginalizing over $\mathbf{S}$ is possible for many finite mixture models, such as univariate and multivariate normal mixture models or mixtures of random effects models; see Frühwirth-Schnatter (2004). The drastic reduction of dimensionality in (5.24) compared to (5.25) will facilitate estimation of the marginal likelihood $p(\mathbf{y}|\mathcal{M}_K)$ using numerical methods.

For conditionally conjugate models, where

$$\int_{\Theta^K} p(\mathbf{y}|\mathbf{S}, \boldsymbol{\theta}_1, \ldots, \boldsymbol{\theta}_K)p(\boldsymbol{\theta}_1, \ldots, \boldsymbol{\theta}_K)d(\boldsymbol{\theta}_1, \ldots, \boldsymbol{\theta}_K)$$

may be solved analytically, one could alternatively marginalize over the parameters $\boldsymbol{\vartheta}_K$ as suggested by Nobile (2004). This yields the following definition of the marginal likelihood as the normalizing constant of the nonnormalized joint marginal posterior distribution of the indicators, studied earlier in Subsection 3.3.3;

$$p(\mathbf{y}|\mathcal{M}_K) = \sum_{\mathbf{S} \in \mathcal{S}_K} p(\mathbf{y}|\mathbf{S}, \mathcal{M}_K)p(\mathbf{S}|\mathcal{M}_K), \tag{5.26}$$

where $p(\mathbf{S}|\mathcal{M}_K)$ and $p(\mathbf{y}|\mathbf{S}, \mathcal{M}_K)$ are the prior and the likelihood defined in (3.24) and (3.25). Note that (5.26) is a summation over all partitions, and could be evaluated exactly for very small data sets with very few numbers of components.

Representation (5.26) has been exploited by Nobile (2004) to show that the marginal likelihood of two finite mixture models from the same parametric family, differing only in the number of components may be expressed in the following way,

$$p(\mathbf{y}|\mathcal{M}_K) = \frac{\Gamma(Ke_0)\Gamma((K-1)e_0 + N)}{\Gamma((K-1)e_0)\Gamma(Ke_0 + N)} p(\mathbf{y}|\mathcal{M}_{K-1}) + p(\mathbf{y}|\mathcal{M}_K^\star), \tag{5.27}$$

where $e_0$ is the parameter appearing in the $\mathcal{D}(e_0, \ldots, e_0)$-prior of the weight distribution $\boldsymbol{\eta}$. $p(\mathbf{y}|\mathcal{M}_K^\star)$ is that fraction of the marginal likelihood which accounts for partitions, where no component is left empty:

$$p(\mathbf{y}|\mathcal{M}_K^\star) = \sum_{\mathbf{S}\in\mathcal{S}_K^\star} p(\mathbf{y}|\mathbf{S},\mathcal{M}_K)p(\mathbf{S}|\mathcal{M}_K). \tag{5.28}$$

$\mathcal{S}_K^\star$ is that subspace of $\mathcal{S}_K$ which allocates at least one observation to each component:

$$\mathcal{S}_K^\star = \{\mathbf{S} \in \mathcal{S}_K : N_k(\mathbf{S}) > 0, k = 1, \dots, K\}.$$

(5.27) is a very important result, as it shows that the marginal likelihoods of finite mixture models with $K-1$ and $K$ components may be strongly linked together. To a certain extent, $p(\mathbf{y}|\mathcal{M}_K)$ reflects the probability mass associated with membership vectors $\mathbf{S}$ that allocate observations to fewer than $K$ components. In particular, if $\mathcal{M}_K$ is overfitting of degree 1, much of the marginal likelihood $p(\mathbf{y}|\mathcal{M}_{K-1})$ of model $\mathcal{M}_{K-1}$ will carry over to $p(\mathbf{y}|\mathcal{M}_K)$.

Consequently, Nobile (2004) doubted the usefulness of the marginal likelihood and the corresponding posterior distribution $p(K|\mathbf{y}) \propto p(\mathbf{y}|\mathcal{M}_K)p(K)$ for deciding how many components in a finite mixture model are needed to explain the data well. As a remedy, he suggests deriving the posterior distribution of the number of nonempty components; see Nobile (2004) for more details. Alternatively, it seems possible to control this link through selecting sensible prior distributions; see Subsection 5.3.2.

### 5.3.2 Choosing Priors for Selecting the Number of Components

**Table 5.1.** Artificial data of size $N$ generated from a mixture of two Poisson distributions with $\eta_1 = 0.3$, $\mu_1 = 1$, and $\mu_2 = 5$; average posterior probabilities for the true model, for too many and too few components under various $\mathcal{D}(e_0, \dots, e_0)$ priors on the weight distribution $\boldsymbol{\eta}$, and various priors on $K = 1, \dots, 5$: uniform prior (prior 1), truncated $\mathcal{P}(1)$ (prior 2), truncated $\mathcal{P}(2)$ (prior 3)

| | E(Pr($\mathcal{M}_2$\|$\mathbf{y}$)) | | | E(Pr($\mathcal{M}_3,\mathcal{M}_4,\mathcal{M}_5$\|$\mathbf{y}$)) | | | E(Pr($\mathcal{M}_1$\|$\mathbf{y}$)) | | |
|---|---|---|---|---|---|---|---|---|---|
| $N=20$ | prior 1 | prior 2 | prior 3 | prior 1 | prior 2 | prior 3 | prior 1 | prior 2 | prior 3 |
| $e_0=1$ | 0.090 | 0.416 | 0.237 | 0.905 | 0.484 | 0.740 | **0.004** | 0.057 | 0.014 |
| $e_0=4$ | 0.394 | **0.665** | 0.543 | 0.498 | **0.158** | 0.324 | 0.037 | 0.134 | 0.056 |
| $N=200$ | prior 1 | prior 2 | prior 3 | prior 1 | prior 2 | prior 3 | prior 1 | prior 2 | prior 3 |
| $e_0=1$ | 0.280 | 0.622 | 0.420 | 0.720 | 0.378 | 0.580 | 0 | 0 | 0 |
| $e_0=4$ | 0.708 | **0.893** | 0.800 | 0.292 | **0.107** | 0.200 | 0 | 0 | 0 |
| $N=2000$ | prior 1 | prior 2 | prior 3 | prior 1 | prior 2 | prior 3 | prior 1 | prior 2 | prior 3 |
| $e_0=4$ | 0.929 | **0.976** | 0.952 | 0.0708 | **0.024** | 0.048 | 0 | 0 | 0 |

Nobile (2004) showed that the Dirichlet prior on the weight distribution $\boldsymbol{\eta}$ will exercise considerable influence on the posterior distribution $p(K|\mathbf{y})$. Because the Dirichlet prior determines how likely empty components are (recall also the investigations in Subsection 2.3.4), this prior will exercise considerable influence on the link between $p(\mathbf{y}|\mathcal{M}_K)$ and $p(\mathbf{y}|\mathcal{M}_{K-1})$, given by (5.27), with the link decreasing when increasing $e_0$.

For illustration, Table 5.1 evaluates a simulation experiment where data sets of various sizes $N$ are simulated from a mixture of two Poisson distributions with $\eta_1 = 0.3$, $\mu_1 = 1$, and $\mu_2 = 5$. It turns out that the posterior probability of the true model is much smaller under $e_0 = 1$ than under $e_0 = 4$. Furthermore, the risk of choosing too many components is much higher for $e_0 = 1$.

This result holds for different prior distributions $p(K)$ on $K = 1, \ldots, 5$, which themselves have considerable influence on the posterior distribution $p(K|\mathbf{y})$. This is not surprising, because Nobile (2004) proved that a proper prior $p(K)$ over the number of components has to be chosen to obtain a proper posterior distribution $p(K|\mathbf{y})$, otherwise the posterior is improper.

In the context of choosing the number of components of a mixture model, it is common to choose the prior $K \sim \mathcal{P}(\lambda_n)$, truncated to $\{1, \ldots, K_{\max}\}$ (Richardson and Green, 1997; Stephens, 2000a), which penalizes more components. A special case of this prior results if $\lambda_n = 1$, in which case the prior is simply proportional to

$$p(K) \propto \frac{1}{K!}. \tag{5.29}$$

Nobile (2004) restricts the number of possible components to $\{1, \ldots, K_{\max}\}$, and assumes that all values within this set have equal prior probability. Such an assumption also implicitly underlies any approach that directly uses the marginal likelihood for selecting $K$, such as Bensmail et al. (1997), Lenk and DeSarbo (2000), and Frühwirth-Schnatter et al. (2004). In more standard model selection problems such a choice is driven by the hope of assuming prior information that is "objective" between competing models; see, for example, Berger (1985).

Table 5.1 demonstrates that combining the uniform $\mathcal{D}(1, \ldots, 1)$-prior on the weight distribution with a uniform prior on the number of components leads to an extremely high risk of choosing too many components, even for quite large data sets. The $\mathcal{P}(1)$ prior, which is proportional to (5.29), in combination with the $\mathcal{D}(4, \ldots, 4)$ outperforms for this simulation study all other priors if the major objective is to avoid overfitting. For the smallest data set, however, this prior has some risk of underfitting, which is the smallest for the uniform/uniform prior. Thus to a certain degree the optimal prior depends on the loss function associated with a wrong decision.

The simulation study was carried out under the hierarchical prior (3.12) with $g_0 = 0.5$ and $G_0 = 0.5$, to reduce sensitivity of the marginal likelihood to the prior distribution $p(\boldsymbol{\theta}_k|\boldsymbol{\delta})$ of the component parameters. Prior sensitivity of Bayes factors to $p(\boldsymbol{\theta}_k|\boldsymbol{\delta})$ was noted by several authors, for instance, Richardson and Green (1997), Stephens (2000a, Section 5), and Berkhof et al.

(2003), who demonstrated improvement by using a hierarchical prior, where $\boldsymbol{\delta}$ is treated as an unknown hyperparameter with a prior $p(\boldsymbol{\delta})$.

Stephens (2000a, Section 5), for instance, studied the influence of the prior distribution $p(\boldsymbol{\theta}_k|\boldsymbol{\delta})$ on posterior model probabilities $\Pr(\mathcal{M}_K|\mathbf{y})$ for mixtures of normal distributions, and found that "priors that appear only weakly informative for the components of the mixture may be highly informative for the number of components of the mixture." This influence is evident from (4.12) which shows that for finite mixture models with different numbers of components the prior $p(\boldsymbol{\theta}_k|\boldsymbol{\delta})$ put on the component parameters will strongly influence the ratio of any two marginal likelihoods.

### 5.3.3 Computation of the Marginal Likelihood for Mixture Models

In general, the computation of the marginal likelihood for complex statistical models is a nontrivial integration problem. Marginal likelihoods have been estimated using methods such as Chib's estimator (Chib, 1995; Chib and Jeliazkov, 2001), importance sampling-based on mixture approximations (Frühwirth-Schnatter, 1995, 2004), combining MCMC simulations and asymptotic approximation (DiCiccio et al., 1997), and bridge sampling (Meng and Wong, 1996; Meng and Schilling, 2002). Han and Carlin (2001) provide a comparative review of MCMC methods for computing Bayes factors and marginal likelihoods for model selection. Although these methods proved to be useful for a wide range of statistical models, some of them are apt to fail for models involving a finite mixture distribution (Neal, 1998; Frühwirth-Schnatter, 2004).

## 5.4 Simulation-Based Approximations of the Marginal Likelihood

In this section we study various simulation-based approximations to the integral defining the marginal likelihood of a mixture model:

$$p(\mathbf{y}|\mathcal{M}_K) = \int p(\mathbf{y}|\boldsymbol{\vartheta}_K)p(\boldsymbol{\vartheta}_K)d\boldsymbol{\vartheta}_K. \tag{5.30}$$

Simulation-based approximations to marginal likelihoods were considered by many authors, an early reference being Gelfand et al. (1992).

### 5.4.1 Some Background on Monte Carlo Integration

This subsection presents as much background on Monte Carlo integration as is needed later on. For a thorough discussion we refer to Geweke (1999) or Robert and Casella (1999).

Any integral that may be regarded as an expectation of a function $h(\boldsymbol{\vartheta}_K)$ with respect to a probability density $f(\boldsymbol{\vartheta}_K)$,

$$\mathrm{E}_f(h) = \int h(\boldsymbol{\vartheta}_K) f(\boldsymbol{\vartheta}_K) d\boldsymbol{\vartheta}_K, \tag{5.31}$$

could be approximated by the sample average of $h(\boldsymbol{\vartheta}_K^{(l)})$ over random draws $\boldsymbol{\vartheta}_K^{(1)}, \ldots, \boldsymbol{\vartheta}_K^{(L)}$ from $f(\boldsymbol{\vartheta}_K)$:

$$\overline{h}_f = \frac{1}{L} \sum_{l=1}^{L} h(\boldsymbol{\vartheta}_K^{(l)}). \tag{5.32}$$

Under very weak conditions on $h(\boldsymbol{\vartheta}_K)$, $\overline{h}_f$ will converge to $\mathrm{E}_f(h)$ by the law of large numbers.

The variance $\mathrm{Var}(\overline{h}_f)$ of the estimator $\overline{h}_f$ depends on three influence factors: First on the number $L$ of draws, second on the dependence among the draws, and finally on $\mathrm{Var}_f(h)$, the variance of $h(\boldsymbol{\vartheta}_K)$ with respect to the density $f(\boldsymbol{\vartheta}_K)$, which is defined in the usual way:

$$\mathrm{Var}_f(h) = \int (h(\boldsymbol{\vartheta}_K) - \mathrm{E}_f(h))^2 f(\boldsymbol{\vartheta}_K) d\boldsymbol{\vartheta}_K. \tag{5.33}$$

The precise dependence of $\mathrm{Var}(\overline{h}_f)$ on these factors reads:

$$\mathrm{Var}(\overline{h}_f) = \frac{\rho_h(0)}{L} \mathrm{Var}_f(h). \tag{5.34}$$

$\rho_h(0)$ is the normalized spectral density of the process $\{h(\boldsymbol{\vartheta}_K^{(l)}), l = 1, \ldots, L\}$ at the frequency 0; see, for instance, Geweke (1992). $\rho_h(0)$ takes the value 1 for i.i.d. draws from $f(\boldsymbol{\vartheta}_K)$, and may be considerably larger than 1, if the draws were simulated by a Markov chain. If the autocorrelations in this chain turn out to be very high, $\rho_h(0)$ may be influenced by the investigator only by choosing a different Markov chain for simulation.

$\mathrm{Var}_f(h)$ crucially depends on the variation of $h(\boldsymbol{\vartheta}_K)$ over regions with relatively high density $f(\boldsymbol{\vartheta}_K)$. If $\mathrm{Var}_f(h)$ is too large, it may be reduced by choosing a different set of functions $h(\boldsymbol{\vartheta}_K)$ and $f(\boldsymbol{\vartheta}_K)$, as is done for importance sampling; see Subsection 5.4.3. Finally, $L$ is the number of draws, and is under control of the investigator. If $\mathrm{Var}_f(h)$ is finite, then increasing $L$ will decrease the estimation error associated with estimating $\mathrm{E}_f(h)$ through $\overline{h}_f$.

### 5.4.2 Sampling-Based Approximations for Mixture Models

Frühwirth-Schnatter (2004) argued that sampling-based techniques are particularly useful for estimating the marginal likelihood of finite mixture models.

If a sampling-based estimator relies on MCMC draws, it is essential to use an MCMC technique that explores all modes of the unconstrained posterior, such as random permutation Gibbs sampling introduced in *Algorithm 3.5*. One should avoid samplers that are not well mixing over all possible modes of the posterior distribution, as most estimators of the marginal likelihood are prone to be biased when based on draws of a poorly mixing sampler.

For various sampling-based techniques discussed in this subsection, one has to select an importance density $q(\boldsymbol{\vartheta}_K) = q(\boldsymbol{\theta}_1, \dots, \boldsymbol{\theta}_K, \boldsymbol{\eta})$, from which it is easy to sample and that provides a rough approximation to the mixture posterior density $p(\boldsymbol{\vartheta}_K|\mathbf{y}, \mathcal{M}_K)$. As manual tuning of the importance density for each model under consideration is rather tedious, a method for choosing sensible importance densities in an unsupervised manner is needed.

Various suggestions for constructing an unsupervised importance density from the MCMC output have been put forward in the literature. DiCiccio et al. (1997), for instance, suggest constructing a normal importance density from the MCMC output, however, this will work only for well-behaved, regular problems with unimodal posterior densities. The multimodality of the mixture posterior density evidently rules out this method. Lenk and DeSarbo (2000) fit a nonnormal importance density to the MCMC draw of a constrained sampler, however, as shown by Frühwirth-Schnatter (2004) this method is sensitive to a poorly chosen constraint.

Frühwirth-Schnatter (2004) extends the unsupervised importance density suggested in Frühwirth-Schnatter (1995) for Gaussian state space models to finite mixture and Markov switching models. For finite mixture models, where the posterior $p(\boldsymbol{\vartheta}_K|\mathbf{y}, \mathcal{M}_K)$ is multimodal, a multimodal importance density arises in quite a natural way within the data augmentation algorithm considered in Subsection 3.5.3. Evidently, the posterior density $p(\boldsymbol{\vartheta}_K|\mathbf{y}, \mathcal{M}_K)$ may be expressed in the following way,

$$p(\boldsymbol{\vartheta}_K|\mathbf{y}, \mathcal{M}_K) = \sum_{\mathbf{S}\in\mathcal{S}_K} \prod_{k=1}^{K} p(\boldsymbol{\theta}_k|\mathbf{y}, \mathbf{S}) p(\boldsymbol{\eta}|\mathbf{S}) p(\mathbf{S}|\mathcal{M}_K, \mathbf{y}), \tag{5.35}$$

where $p(\boldsymbol{\theta}_k|\mathbf{y}, \mathbf{S})$ and $p(\boldsymbol{\eta}|\mathbf{S})$ are the complete-data posterior densities. Thus, if the complete-data posterior $p(\boldsymbol{\theta}_k|\mathbf{y}, \mathbf{S})$ is available in closed form, a random subsequence $\mathbf{S}^{(s)}, s = 1, \dots, S$ of the MCMC draws $\mathbf{S}^{(m)}, m = 1, \dots, M$, of the allocation vector $\mathbf{S}$ could be used to construct an importance density:

$$q(\boldsymbol{\vartheta}_K) = \frac{1}{S} \sum_{s=1}^{S} p(\boldsymbol{\eta}|\mathbf{S}^{(s)}) \prod_{k=1}^{K} p(\boldsymbol{\theta}_k|\mathbf{S}^{(s)}, \mathbf{y}). \tag{5.36}$$

Typically $S << M$, as (5.36) needs to be only a rough approximation to (5.35). Construction of the importance density (5.36) is fully automatic and may be easily incorporated into MCMC sampling; see Frühwirth-Schnatter (2001a, p.39) for further details.

As the construction is based on averaging over the conditional densities, where the allocations $\mathbf{S}$ are sampled from the unconstrained posterior with balanced label switching, the mixture importance density (5.36) will be multimodal. In order to reproduce all modes of the posterior it is essential to base the construction of the importance density on an MCMC method with balanced label switching.

A simplified version of (5.36) is available if the random sequence $\mathbf{S}^{(s)}, s = 1, \ldots, S$ is substituted by a deterministic sequence, which is derived from some optimal classification $\mathbf{S}^\star$ by considering all permutations of $\mathbf{S}^\star$, thus $S = K!$. Sampling from (5.36) reduces then to sampling from the complete-data density $\prod_{k=1}^{K} p(\boldsymbol{\theta}_k|\mathbf{S}^\star, \mathbf{y}, \mathcal{M}_K)p(\boldsymbol{\eta}|\mathbf{S}^\star)$, however the functional evaluation of $q(\boldsymbol{\vartheta}_K)$ still requires the evaluation of $K!$ complete-data densities for all permutations of the labels of $\boldsymbol{\vartheta}_K$.

### 5.4.3 Importance Sampling

Approximations of the marginal likelihood based on importance sampling have been considered by several authors, for instance, Geweke (1989), Frühwirth-Schnatter (1995), and Geweke (1999). Applications to finite mixture models appear in Frühwirth-Schnatter (2004).

A simple Monte Carlo approximation of the marginal likelihood (5.30) is obtained by:

$$\hat{p}_{MC}(\mathbf{y}|\mathcal{M}_K) = \frac{1}{L}\sum_{l=1}^{L} p(\mathbf{y}|\boldsymbol{\vartheta}_K^{(l)}), \tag{5.37}$$

where $\boldsymbol{\vartheta}_K^{(1)}, \ldots, \boldsymbol{\vartheta}_K^{(L)}$ is a sample from the prior $p(\boldsymbol{\vartheta}_K)$. The resulting estimator is rather inefficient if the likelihood is informative compared to the prior; see McCulloch and Rossi (1992). Importance sampling (Geweke, 1989) may be used to obtain a better approximation to the marginal likelihood by rewriting (5.30) as

$$p(\mathbf{y}|\mathcal{M}_K) = \int \frac{p(\mathbf{y}|\boldsymbol{\vartheta}_K)p(\boldsymbol{\vartheta}_K)}{q(\boldsymbol{\vartheta}_K)} q(\boldsymbol{\vartheta}_K) d\boldsymbol{\vartheta}_K, \tag{5.38}$$

where $q(\boldsymbol{\vartheta}_K)$ is a suitably chosen importance density; see Subsection 5.4.2. If a sample from $q(\boldsymbol{\vartheta}_K)$ is available,

$$\boldsymbol{\vartheta}_K^{(l)} \sim q(\boldsymbol{\vartheta}_K), \qquad l = 1, \ldots, L,$$

then the marginal likelihood is estimated by

$$\hat{p}_{IS}(\mathbf{y}|\mathcal{M}_K) = \frac{1}{L}\sum_{l=1}^{L} \frac{p(\mathbf{y}|\boldsymbol{\vartheta}_K^{(l)})p(\boldsymbol{\vartheta}_K^{(l)})}{q(\boldsymbol{\vartheta}_K^{(l)})}. \tag{5.39}$$

The variance of this estimator is given by (5.34):

$$\mathrm{Var}(\hat{p}_{IS}(\mathbf{y}|\mathcal{M}_K)) = \frac{p(\mathbf{y}|\mathcal{M}_K)^2}{L}\mathrm{Var}_q\left(\frac{p}{q}\right), \tag{5.40}$$

as the draws are independent. Because

$$\mathrm{Var}_q\left(\frac{p}{q}\right) = \mathrm{E}_q\left(\left(\frac{p}{q}\right)^2\right) - 1 = \mathrm{E}_p\left(\frac{p}{q}\right) - 1,$$

the estimator $\hat{p}_{IS}(\mathbf{y}|\mathcal{M}_K)$ has finite variance only, if

$$\mathrm{E}_p\left(\frac{p}{q}\right) = \int \frac{p(\boldsymbol{\vartheta}_K|\mathbf{y})^2}{q(\boldsymbol{\vartheta}_K)} d\boldsymbol{\vartheta}_K < \infty. \tag{5.41}$$

A sufficient, but not necessary, condition for this is that the ratio

$$\frac{p(\mathbf{y}|\boldsymbol{\vartheta}_K)p(\boldsymbol{\vartheta}_K)}{q(\boldsymbol{\vartheta}_K)}$$

is bounded, which implies that the tails of $q(\boldsymbol{\vartheta}_K)$ should be fat in comparison to the posterior density $p(\boldsymbol{\vartheta}_K|\mathbf{y})$. Otherwise the ratio of the nonnormalized posterior density over the importance density may be unbounded for a poorly chosen importance density, which highly increases the variance of the resulting estimator.

### 5.4.4 Reciprocal Importance Sampling

The marginal likelihood $p(\mathbf{y}|\mathcal{M}_K)$, defined in (5.30), is not directly available as an expectation with respect to the posterior density, thus straightforward approximations to the marginal likelihood from the MCMC output are not available.

A tricky method that expresses the marginal likelihood as the expectation with respect to the posterior has been introduced by Gelfand and Dey (1994). Rewrite Bayes' theorem, which is given by

$$p(\boldsymbol{\vartheta}_K|\mathbf{y},\mathcal{M}_K) = \frac{p(\mathbf{y}|\boldsymbol{\vartheta}_K)p(\boldsymbol{\vartheta}_K)}{p(\mathbf{y}|\mathcal{M}_K)}, \tag{5.42}$$

as

$$\frac{p(\boldsymbol{\vartheta}_K|\mathbf{y},\mathcal{M}_K)}{p(\mathbf{y}|\boldsymbol{\vartheta}_K)p(\boldsymbol{\vartheta}_K)} = \frac{1}{p(\mathbf{y}|\mathcal{M}_K)}.$$

Multiply both sides with an arbitrary density $q(\boldsymbol{\vartheta}_K)$, and integrate with respect to $\boldsymbol{\vartheta}_K$; then one obtains the following identity.

$$\frac{1}{p(\mathbf{y}|\mathcal{M}_K)} = \int \frac{q(\boldsymbol{\vartheta}_K)}{p(\mathbf{y}|\boldsymbol{\vartheta}_K)p(\boldsymbol{\vartheta}_K)} p(\boldsymbol{\vartheta}_K|\mathbf{y},\mathcal{M}_K)d\boldsymbol{\vartheta}_K.$$

Therefore the inverse of the marginal likelihood is equal to the posterior expectation of the ratio of an arbitrary importance density $q(\boldsymbol{\vartheta}_K)$ and the nonnormalized posterior density. This yields the following estimator of the marginal likelihood,

$$\hat{p}_{RI}(\mathbf{y}|\mathcal{M}_K) = \left( \frac{1}{M} \sum_{m=1}^{M} \frac{q(\boldsymbol{\vartheta}_K^{(m)})}{p(\mathbf{y}|\boldsymbol{\vartheta}_K^{(m)})p(\boldsymbol{\vartheta}_K^{(m)})} \right)^{-1}, \tag{5.43}$$

where $\boldsymbol{\vartheta}_K^{(1)}, \ldots, \boldsymbol{\vartheta}_K^{(M)}$ are simulations from the posterior $p(\boldsymbol{\vartheta}_K|\mathbf{y}, \mathcal{M}_K)$. Note that the importance density $q(\boldsymbol{\vartheta}_K)$ is only evaluated at the MCMC draws, but no draws from the importance density are required.

Because the draws are dependent, the variance of this estimator is given by (5.34):

$$\mathrm{Var}(\hat{p}_{RI}(\mathbf{y}|\mathcal{M}_K)^{-1}) = \frac{\rho_h(0)}{L} \mathrm{Var}_p \left( \frac{q}{p^\star} \right) = \frac{\rho_h(0)}{L p(\mathbf{y}|\mathcal{M}_K)^2} \mathrm{Var}_p \left( \frac{q}{p} \right), \tag{5.44}$$

where in (5.34) $h(\boldsymbol{\vartheta}_K|\mathbf{y}) = q(\boldsymbol{\vartheta}_K)/p(\mathbf{y}|\boldsymbol{\vartheta}_K)p(\boldsymbol{\vartheta}_K)$. Because

$$\mathrm{Var}_p \left( \frac{q}{p} \right) = \mathrm{E}_p \left( \left( \frac{q}{p} \right)^2 \right) - 1 = \mathrm{E}_q \left( \frac{q}{p} \right) - 1,$$

the estimator $\hat{p}_{RI}(\mathbf{y}|\mathcal{M}_K)^{-1}$ has finite variance only, if

$$\mathrm{E}_q \left( \frac{q}{p} \right) = \int \frac{q(\boldsymbol{\vartheta}_K)^2}{p(\boldsymbol{\vartheta}_K|\mathbf{y})} d\boldsymbol{\vartheta}_K < \infty. \tag{5.45}$$

A sufficient but not necessary condition for this is that the ratio

$$\frac{q(\boldsymbol{\vartheta}_K)}{p(\mathbf{y}|\boldsymbol{\vartheta}_K)p(\boldsymbol{\vartheta}_K)}$$

is bounded, which implies that the tails of $q(\boldsymbol{\vartheta}_K)$ should be thin in comparison to the tails of the posterior density $p(\boldsymbol{\vartheta}_K|\mathbf{y}, \mathcal{M}_K)$.

Approximations of the marginal likelihood of mixture models based on reciprocal importance sampling were considered by Lenk and DeSarbo (2000) and Frühwirth-Schnatter (2004).

### 5.4.5 Harmonic Mean Estimator

Another estimator of the marginal likelihood (5.30) that uses only samples from the posterior density is the harmonic mean estimator, introduced by Newton and Raftery (1994). Bayes' theorem (5.42) may be rewritten as

$$\frac{p(\boldsymbol{\vartheta}_K|\mathbf{y}, \mathcal{M}_K)}{p(\mathbf{y}|\boldsymbol{\vartheta}_K)} = \frac{p(\boldsymbol{\vartheta}_K)}{p(\mathbf{y}|\mathcal{M}_K)}. \tag{5.46}$$

If both sides are integrated with respect to $\boldsymbol{\vartheta}$, then following identity results,

$$\int \frac{p(\boldsymbol{\vartheta}_K|\mathbf{y},\mathcal{M}_K)}{p(\mathbf{y}|\boldsymbol{\vartheta}_K)} d\boldsymbol{\vartheta}_K = \frac{1}{p(\mathbf{y}|\mathcal{M}_K)}, \tag{5.47}$$

which again expresses the marginal likelihood as expectation with respect to the mixture posterior density. Thus if a sample $\boldsymbol{\vartheta}_K^{(1)},\dots,\boldsymbol{\vartheta}_K^{(M)}$ from the mixture posterior $p(\boldsymbol{\vartheta}_K|\mathbf{y},\mathcal{M}_K)$ is available, the harmonic mean estimator is given by

$$\hat{p}_{HM}(\mathbf{y}|\mathcal{M}_K) = \left(\frac{1}{M}\sum_{m=1}^{M}\frac{1}{p(\mathbf{y}|\boldsymbol{\vartheta}_K^{(m)})}\right)^{-1}. \tag{5.48}$$

The harmonic mean estimator is a very convenient estimator, as it only involves evaluating the mixture likelihood function $p(\mathbf{y}|\boldsymbol{\vartheta}_K^{(m)})$ for each posterior draw, which results as a byproduct from sampling $\mathbf{S}^{(m)}$ conditional on knowing $\boldsymbol{\vartheta}_K^{(m)}$ in the data augmentation algorithm *Algorithm 3.4*.

Nevertheless it is generally known to be unstable, and the simulation study in Subsection 5.4.7 shows that the harmonic mean estimator performs particularly poorly for finite mixture models. The harmonic mean estimator may be viewed as that special case of reciprocal importance sampling where the importance function in (5.43) is equal to the prior, $q(\boldsymbol{\vartheta}_K) = p(\boldsymbol{\vartheta}_K)$. From (5.45) we obtain that the variance of $\hat{p}_{HM}(\mathbf{y}|\mathcal{M}_K)^{-1}$ is finite, if and only if

$$\int \frac{p(\boldsymbol{\vartheta}_K)}{p(\mathbf{y}|\boldsymbol{\vartheta}_K)} d\boldsymbol{\vartheta}_K < \infty. \tag{5.49}$$

Condition (5.49) is often violated, as the ratio of the prior over the likelihood function often is unbounded; see also the very interesting work of Satagopan et al. (2000).

### A Harmonic Mean Estimator Based on the Integrated Classification Likelihood

If the integrated classification likelihood $p(\mathbf{y}|\mathbf{S})$ is available in closed form as discussed in Subsection 3.3.3, then the harmonic mean estimator could be applied to the posterior distribution $p(\mathbf{S}|\mathbf{y},\mathcal{M}_K)$. Bayes' theorem may be rewritten as

$$\frac{p(\mathbf{S}|\mathbf{y},\mathcal{M}_K)}{p(\mathbf{y}|\mathbf{S})} = \frac{p(\mathbf{S})}{p(\mathbf{y}|\mathcal{M}_K)}.$$

If both sides are integrated with respect to $\mathbf{S}$, then one obtains the identity

$$\sum_{\mathbf{S}\in\mathcal{S}_K} \frac{p(\mathbf{S}|\mathbf{y},\mathcal{M}_K)}{p(\mathbf{y}|\mathbf{S})} = \frac{1}{p(\mathbf{y}|\mathcal{M}_K)},$$

which expresses the marginal likelihood as expectation with respect to the posterior $p(\mathbf{S}|\mathbf{y}, \mathcal{M}_K)$. This yields the following modified harmonic mean estimator,

$$\hat{p}_{HM,2}(\mathbf{y}|\mathcal{M}_K) = \left( \frac{1}{M} \sum_{m=1}^{M} \frac{1}{p(\mathbf{y}|\mathbf{S}^{(m)})} \right)^{-1}, \tag{5.50}$$

where $\{\mathbf{S}^{(m)}, m = 1, \ldots, M\}$ is a sequence of posterior draws for the allocations $\mathbf{S}$, which are obtained either by sampling from the marginal posterior $p(\mathbf{S}|\mathbf{y}, \mathcal{M}_K)$ as in *Algorithm 3.1* or *Algorithm 3.2*, or by data augmentation as in *Algorithm 3.4*.

### 5.4.6 Bridge Sampling Technique

Bridge sampling was introduced into statistics by Meng and Wong (1996) as a simulation-based technique for computing ratios of normalizing constants. DiCiccio et al. (1997) suggested, for rather simple model selection problems, applying bridge sampling to the problem of computing a marginal likelihood. Frühwirth-Schnatter (2004) investigated the application of bridge sampling to compute the marginal likelihood of a finite mixture model in full detail.

Bridge sampling generalizes the method of importance sampling discussed in Subsection 5.4.3. Like importance sampling, bridge sampling is based on an i.i.d. sample from an importance density, however, this sample is combined with the MCMC draws from the posterior density in an appropriate way. One might wonder why this extension is sensible. Importance sampling may be unstable if the ratio of the nonnormalized posterior density over the importance density is unbounded; see again Subsection 5.4.3. An important advantage of bridge sampling is that the variance of the resulting estimator depends on a ratio that is bounded regardless of the tail behavior of the underlying importance density. This allows far more flexibility in the construction of the importance density.

#### The Method

Let $q(\boldsymbol{\vartheta}_K)$ be a probability density function with known normalizing constant, which has been called the importance density as for the simulation-based approximations discussed earlier. Let $\alpha(\boldsymbol{\vartheta}_K)$ be an arbitrary function such that

$$C_\alpha = \int \alpha(\boldsymbol{\vartheta}_K) p(\boldsymbol{\vartheta}_K|\mathbf{y}, \mathcal{M}_K) q(\boldsymbol{\vartheta}_K) d\boldsymbol{\vartheta}_K > 0. \tag{5.51}$$

Bridge sampling is based on the following result,

$$1 = \frac{\int \alpha(\boldsymbol{\vartheta}_K) p(\boldsymbol{\vartheta}_K|\mathbf{y}, \mathcal{M}_K) q(\boldsymbol{\vartheta}_K) d\boldsymbol{\vartheta}_K}{\int \alpha(\boldsymbol{\vartheta}_K) q(\boldsymbol{\vartheta}_K) p(\boldsymbol{\vartheta}_K|\mathbf{y}, \mathcal{M}_K) d\boldsymbol{\vartheta}_K} = \frac{\mathrm{E}_q(\alpha(\boldsymbol{\vartheta}_K) p(\boldsymbol{\vartheta}_K|\mathbf{y}, \mathcal{M}_K))}{\mathrm{E}_p(\alpha(\boldsymbol{\vartheta}_K) q(\boldsymbol{\vartheta}_K))},$$

where $\mathrm{E}_f(h(\boldsymbol{\vartheta}_K))$ is the expectation of $h(\boldsymbol{\vartheta}_K)$ with respect to the density $f(\boldsymbol{\vartheta}_K)$. Substituting $p(\boldsymbol{\vartheta}_K|\mathbf{y}, \mathcal{M}_K) = p^\star(\boldsymbol{\vartheta}_K|\mathbf{y}, \mathcal{M}_K)/p(\mathbf{y}|\mathcal{M}_K)$, where $p^\star(\boldsymbol{\vartheta}_K|\mathbf{y}, \mathcal{M}_K) = p(\mathbf{y}|\boldsymbol{\vartheta}_K)p(\boldsymbol{\vartheta}_K)$ is the nonnormalized posterior, yields the key identity for bridge sampling:

$$p(\mathbf{y}|\mathcal{M}_K) = \frac{\mathrm{E}_q(\alpha(\boldsymbol{\vartheta}_K)p^\star(\boldsymbol{\vartheta}_K|\mathbf{y}, \mathcal{M}_K))}{\mathrm{E}_p(\alpha(\boldsymbol{\vartheta}_K)q(\boldsymbol{\vartheta}_K))}. \tag{5.52}$$

To estimate the marginal likelihood for a given function $\alpha(\boldsymbol{\vartheta}_K)$, the expectations on the right-hand side of (5.52) are substituted by sample averages. The denominator, which is an expectation with respect to the importance density $q(\boldsymbol{\vartheta}_K)$, is approximated using i.i.d. draws $\{\tilde{\boldsymbol{\vartheta}}_K^{(l)}, l = 1, \ldots, L\}$ from $q(\boldsymbol{\vartheta}_K)$, whereas the numerator, which is an expectation with respect to the mixture posterior density $p(\boldsymbol{\vartheta}_K|\mathbf{y}, \mathcal{M}_K)$ is approximated using Markov chain Monte Carlo draws $\{\boldsymbol{\vartheta}_K^{(m)}, m = 1, \ldots, M\}$ from $p(\boldsymbol{\vartheta}_K|\mathbf{y}, \mathcal{M}_K)$. The resulting estimator $\hat{p}(\mathbf{y}|\mathcal{M}_K)$ is called the general bridge sampling estimator:

$$\hat{p}(\mathbf{y}|\mathcal{M}_K) = \frac{L^{-1}\sum_{l=1}^{L} \alpha(\tilde{\boldsymbol{\vartheta}}_K^{(l)})p^\star(\tilde{\boldsymbol{\vartheta}}_K^{(l)}|\mathbf{y}, \mathcal{M}_K)}{M^{-1}\sum_{m=1}^{M} \alpha(\boldsymbol{\vartheta}_K^{(m)})q(\boldsymbol{\vartheta}_K^{(m)})}. \tag{5.53}$$

Various simulation-based methods discussed earlier result as special cases by appropriate choices of $\alpha(\boldsymbol{\vartheta}_K)$, namely the importance sampling estimator (5.39) with $\alpha(\boldsymbol{\vartheta}_K) = 1/q(\boldsymbol{\vartheta}_K)$ and the reciprocal importance sampling estimator (5.43) with $\alpha(\boldsymbol{\vartheta}_K) = 1/p^\star(\boldsymbol{\vartheta}_K|\mathbf{y}, \mathcal{M}_K)$. The general bridge sampling estimator is more general than these methods insofar as it makes use of samples from the importance density $q(\boldsymbol{\vartheta}_K)$ and MCMC draws from the posterior density $p(\boldsymbol{\vartheta}_K|\mathbf{y}, \mathcal{M}_K)$.

Meng and Wong (1996) discuss an asymptotically optimal choice of $\alpha(\boldsymbol{\vartheta}_K)$, which minimizes the expected relative error of the estimator $\hat{p}(\mathbf{y}|\mathcal{M}_K)$ for i.i.d. draws from $p(\boldsymbol{\vartheta}_K|\mathbf{y}, \mathcal{M}_K)$ and $q(\boldsymbol{\vartheta}_K)$:

$$\alpha(\boldsymbol{\vartheta}_K) \propto \frac{1}{Lq(\boldsymbol{\vartheta}_K) + Mp(\boldsymbol{\vartheta}_K|\mathbf{y}, \mathcal{M}_K)}. \tag{5.54}$$

We refer to the corresponding estimator as the bridge sampling estimator $\hat{p}_{BS}(\mathbf{y}|\mathcal{M}_K)$.

### Practical Computation

It turns out that the optimal choice of $\alpha(\boldsymbol{\vartheta}_K)$, given by (5.54), depends on the *normalized* mixture posterior $p(\boldsymbol{\vartheta}_K|\mathcal{M}_K, \mathbf{y})$, thus to estimate the normalizing constant, we need to know the normalizing constant. To solve this problem, Meng and Wong (1996) apply an iterative procedure to obtain $\hat{p}_{BS}(\mathbf{y}|\mathcal{M}_K)$ as the limit of a sequence $\hat{p}_{BS,t}$ as $t \to \infty$. Based on the most recent estimate $\hat{p}_{BS,t-1}$ of the normalizing constant, the posterior is normalized, and a new estimate $\hat{p}_{BS,t}$ is computed from (5.53). This leads to the recursion given below in *Algorithm 5.5.*

*Algorithm 5.5: Bridge Sampling Estimator of the Marginal Likelihood of a Mixture Model*

(a) *Simulation step.* Run an MCMC sampler to obtain draws $\{\boldsymbol{\vartheta}_K^{(m)}, m = 1, \ldots, M\}$ from the mixture posterior $p(\boldsymbol{\vartheta}_K|\mathbf{y}, \mathcal{M}_K)$. Choose an importance density $q(\boldsymbol{\vartheta}_K)$, and draw $\tilde{\boldsymbol{\vartheta}}_K^{(l)}, l = 1, \ldots, L$ independently from the importance density $q(\boldsymbol{\vartheta}_K)$.
(b) *Functional Evaluation.* Evaluate both the nonnormalized mixture posterior $p^\star(\boldsymbol{\vartheta}_K|\mathbf{y}, \mathcal{M}_K)$ and the importance density $q(\boldsymbol{\vartheta}_K)$ at all draws from the posterior as well as at all draws from the importance density.
(c) *Iteration.* Use the functional values to determine a starting value for $\hat{p}_{BS,0}$. Run the following recursion until convergence.

$$\hat{p}_{BS,t} = \frac{L^{-1} \displaystyle\sum_{l=1}^{L} \frac{p^\star(\tilde{\boldsymbol{\vartheta}}_K^{(l)}|\mathbf{y}, \mathcal{M}_K)}{Lq(\tilde{\boldsymbol{\vartheta}}_K^{(l)}) + Mp^\star(\tilde{\boldsymbol{\vartheta}}_K^{(l)}|\mathbf{y}, \mathcal{M}_K)/\hat{p}_{BS,t-1}}}{M^{-1} \displaystyle\sum_{m=1}^{M} \frac{q(\boldsymbol{\vartheta}_K^{(m)})}{Lq(\boldsymbol{\vartheta}_K^{(m)}) + Mp^\star(\boldsymbol{\vartheta}_K^{(m)}|\mathbf{y}, \mathcal{M}_K)/\hat{p}_{BS,t-1}}}. \tag{5.55}$$

Note that the functional values appearing in (5.55) are fixed for all iterations, apart from rescaling the functional evaluations of the nonnormalized posterior $p^\star(\boldsymbol{\vartheta}_K|\mathbf{y}, \mathcal{M}_K)$ in the nominator with the most recent estimate $\hat{p}_{BS,t-1}$ of the normalizing constant. Iteration (5.55) is typically very fast in practice. Either the importance sampling estimator (5.39) or the reciprocal importance sampling estimator (5.43) may be used as starting values for $\hat{p}_{BS,0}$. As both estimators use the same functional values as the bridge sampling estimator, the computation of these two estimators is possible with practically no additional computational effort.

### Relative Mean-Squared Errors

The performance of the various estimators described in the previous section may be measured in terms of the expected relative mean-squared error ($RE^2$) as in Chen et al. (2000):

$$RE^2(\hat{p}(\mathbf{y}|\mathcal{M}_K)) = \frac{\mathrm{E}((\hat{p}(\mathbf{y}|\mathcal{M}_K) - p(\mathbf{y}|\mathcal{M}_K))^2)}{p(\mathbf{y}|\mathcal{M}_K)^2}. \tag{5.56}$$

In (5.56) the data $\mathbf{y}$ are considered to be fixed, whereas $\hat{p}(\mathbf{y}|\mathcal{M}_K)$ is a random variable, depending on the random sequences $\tilde{\boldsymbol{\vartheta}}_K^{(1)}, \ldots, \tilde{\boldsymbol{\vartheta}}_K^{(L)}$ and $\boldsymbol{\vartheta}_K^{(1)}, \ldots, \boldsymbol{\vartheta}_K^{(M)}$. The expectation in (5.56) is taken with respect to the joint distribution of these random sequences. Chen et al. (2000) derived approximate relative mean-squared errors by the $\delta$-method under the assumption that both random sequences are i.i.d. draws from the densities $q(\boldsymbol{\vartheta}_K)$ and $p(\boldsymbol{\vartheta}_K|\mathbf{y}, \mathcal{M}_K)$, respectively. In Frühwirth-Schnatter (2004), the results of Chen et al. (2000) are extended to account explicitly for the autocorrelation in the MCMC draws. For the general bridge sampling estimator $\hat{p}(\mathbf{y}|\mathcal{M}_K)$ defined in (5.53), the following holds,

$$\hat{RE}^2(\hat{p}(\mathbf{y}|\mathcal{M}_K)) = \frac{1}{L}\frac{\mathrm{Var}_q(\alpha p)}{C_\alpha^2} + \frac{\rho_h(0)}{M}\frac{\mathrm{Var}_p(\alpha q)}{C_\alpha^2}, \tag{5.57}$$

where $\rho_h(0)$ is the normalized spectral density at frequency 0 of the process $h_m = \alpha(\boldsymbol{\vartheta}_K^{(m)})q(\boldsymbol{\vartheta}_K^{(m)})$. For the special choices of $\alpha(\cdot)$ discussed above, (5.57) simplifies to:

$$\begin{aligned}
\hat{RE}^2(\hat{p}_{IS}(\mathbf{y}|\mathcal{M}_K)) &= \frac{1}{L}\mathrm{Var}_q\left(\frac{p}{q}\right) \\
\hat{RE}^2(\hat{p}_{RI}(\mathbf{y}|\mathcal{M}_K)) &= \frac{\rho_{h_1}(0)}{M}\mathrm{Var}_p\left(\frac{q}{p}\right) \\
\hat{RE}^2(\hat{p}_{BS}(\mathbf{y}|\mathcal{M}_K)) &= \frac{1}{L}\frac{\mathrm{Var}_q\left(\frac{p}{w_q q + w_p p}\right)}{E_q^2\left(\frac{p}{w_q q + w_p p}\right)} + \frac{\rho_{h_2}(0)}{M}\frac{\mathrm{Var}_p\left(\frac{q}{w_q q + w_p p}\right)}{E_p^2\left(\frac{q}{w_q q + w_p p}\right)},
\end{aligned} \tag{5.58}$$

where $w_q = L/(L+M)$, $w_p = M/(L+M)$, $h_1(\boldsymbol{\vartheta}_K) = q(\boldsymbol{\vartheta}_K)/p(\boldsymbol{\vartheta}_K|\mathbf{y}, \mathcal{M}_K)$, and $h_2(\boldsymbol{\vartheta}_K) = q(\boldsymbol{\vartheta}_K)/(w_q q(\boldsymbol{\vartheta}_K) + w_p p(\boldsymbol{\vartheta}_K|\mathbf{y}, \mathcal{M}_K))$.

The expected relative mean-squared error ($RE^2$) is a useful approximation to the expected absolute mean-squared error of $\log \hat{p}(\mathbf{y}|\mathcal{M}_K)$; see Chen et al. (2000):

$$\mathrm{E}\left(\log \hat{p}(\mathbf{y}|\mathcal{M}_K) - \log p(\mathbf{y}|\mathcal{M}_K)\right)^2 \approx RE^2(\hat{p}(\mathbf{y}|\mathcal{M}_K)). \tag{5.59}$$

In order to assess the accuracy of the various estimators of the marginal likelihood in practice, it is necessary to estimate unknown quantities appearing in (5.58) such as the spectral density at 0 or various variances. In our case studies we use the sample variance of $h_l = h(\tilde{\boldsymbol{\vartheta}}_K^{(l)}), l = 1, \dots, L$ to estimate the variance $\mathrm{Var}_q(h(\boldsymbol{\vartheta}_K))$ with respect to $q(\boldsymbol{\vartheta}_K)$, whereas we use the sample variance of $h_m = h(\boldsymbol{\vartheta}_K^{(m)}), m = 1, \dots, M$ to estimate the variance $\mathrm{Var}_p(h(\boldsymbol{\vartheta}_K))$ with respect to $p(\boldsymbol{\vartheta}_K|\mathbf{y}, \mathcal{M}_K)$. Note that $h_l, l = 1, \dots, L$ and $h_m, m = 1, \dots, M$ are exactly the functional evaluations available from computing the various estimators. To estimate $\rho_h(0)$ we use:

$$\begin{aligned}
\widehat{\rho_h(0)} &= 1 + 2 \cdot \sum_{s=1}^{S}\left(1 - \frac{s}{q+1}\right) r_s, \\
r_s &= M^{-1}\sum_{m=s+1}^{M}\frac{(h_m - \overline{h})(h_{m-s} - \overline{h})}{s_h^2},
\end{aligned}$$

where $\overline{h}$ and $s_h^2$ are the sample mean and the sample variance of $h_m$; see Chib (1995) for details.

### 5.4.7 Comparison of Different Simulation-Based Estimators

The results of Subsection 5.4.6 allow a deeper understanding of the circumstances under which the various estimators will be stable or unstable. If the tails of the importance function $q(\boldsymbol{\vartheta}_K)$ are thinner than the tails of the posterior $p(\boldsymbol{\vartheta}_K|\mathbf{y}, \mathcal{M}_K)$ in some direction, then in (5.58):

$$\frac{p}{q} \rightarrow \infty, \quad \frac{q}{p} \rightarrow 0, \quad \frac{p}{w_p p + w_q q} \rightarrow \frac{1}{w_p}, \quad \frac{q}{w_p p + w_q q} \rightarrow 0.$$

As the ratio $p/q$ is unbounded, the importance sampling estimator is expected to be unstable and to exhibit high standard errors compared to the two other estimators for which the relevant ratios are bounded.

If, on the other hand, the tails of the importance function $q(\boldsymbol{\vartheta}_K)$ are fatter than the tails of the posterior $p(\boldsymbol{\vartheta}_K|\mathbf{y}, \mathcal{M}_K)$ in some direction, then:

$$\frac{p}{q} \rightarrow 0, \quad \frac{q}{p} \rightarrow \infty, \quad \frac{p}{w_p p + w_q q} \rightarrow 0, \quad \frac{q}{w_p p + w_q q} \rightarrow \frac{1}{w_q}.$$

As this time the ratio $q/p$ is unbounded; the reciprocal importance sampling estimator is expected to be unstable and to exhibit high standard errors compared to the two other estimators for which the relevant ratios are bounded.

To sum up, both importance sampling as well as reciprocal importance sampling are sensitive to the tail behavior of the importance density $q(\boldsymbol{\vartheta}_K)$ relative to the posterior $p(\boldsymbol{\vartheta}_K|\mathbf{y}, \mathcal{M}_K)$. The bridge sampling estimator $\hat{p}_{BS}$, however, is much more robust in this respect. Even for importance functions having fat tails in one direction and thin tails in the other, the ratios appearing in (5.58) are bounded. This robustness of the "optimal" bridge sampling estimator against the tail behavior of the importance density is crucial especially for unsupervised choices of $q(\boldsymbol{\vartheta}_K)$.

Final comments on the harmonic mean estimator (5.48) which is commonly known to be unstable are added. The harmonic mean estimator is that special case of the reciprocal importance sampling estimator (5.43) where $q(\boldsymbol{\vartheta}_K)$ is equal to the prior $p(\boldsymbol{\vartheta}_K)$. Consequently, whenever the prior is likely to have fatter tails than the posterior, we may rule out the harmonic mean estimator (5.48) as a reliable device for estimating marginal likelihoods.

#### Performance in a Simulation Experiment

To compare the various estimators of the marginal likelihood, a simulation experiment is performed. One hundred data sets each consisting of 20 observations were simulated from the two-component Poisson mixture model

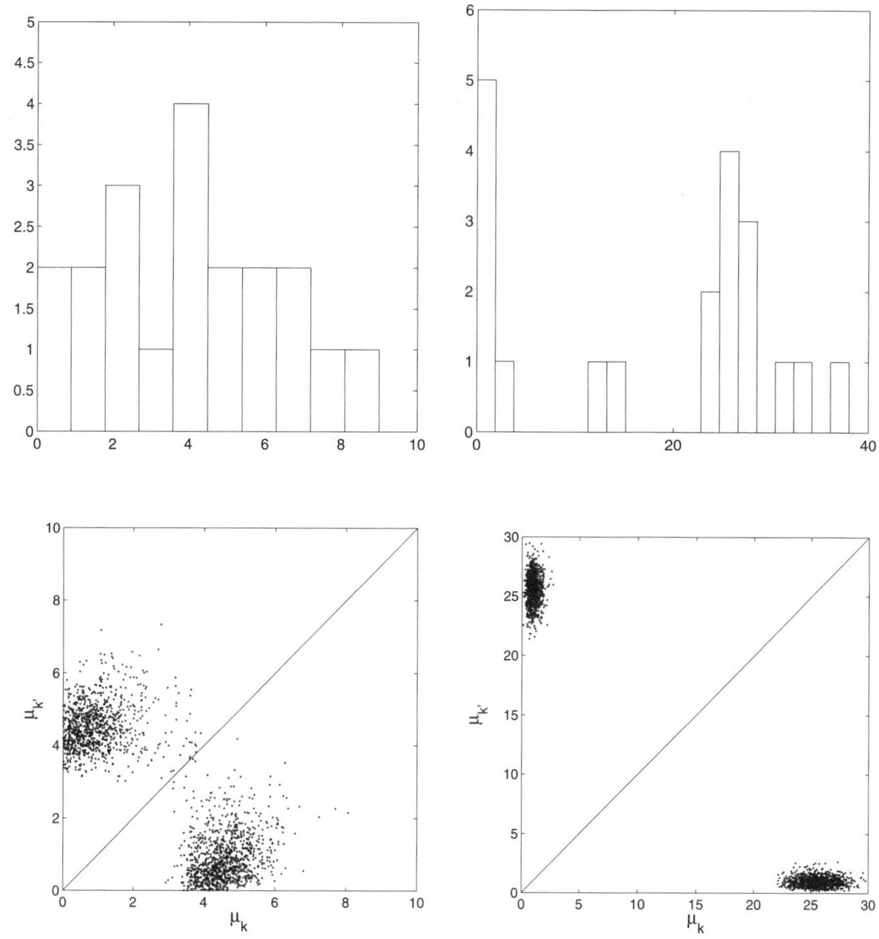


**Fig. 5.1.** Top: randomly selected data set used in the simulation experiment with $\mu_2 = 5$ (left) and $\mu_2 = 25$ (right); bottom: corresponding marginal mixture posterior $p(\mu_k, \mu_{k'}|\mathbf{y})$

$$Y \sim \eta_1 \mathcal{P}(\mu_1) + \eta_2 \mathcal{P}(\mu_2) . \tag{5.60}$$

Choosing the sample size as small as $N = 20$ allows *exact* computation of the marginal likelihood $p(\mathbf{y}|\mathcal{M}_2)$ through (5.26) for each simulated data set $\mathbf{y} = (y_1, \dots, y_{20})$. This provides a "gold standard" against which the various methods of *estimating* the marginal likelihood $p(\mathbf{y}|\mathcal{M}_2)$ could be compared.

For all simulation experiments, the same values $\eta_1 = 0.3$, $\eta_2 = 0.7$, and $\mu_1 = 1$ were used to simulate the data from (5.60), whereas $\mu_2$ took three different values. In the first simulation experiment $\mu_2 = 5$, leading to data that do not show well-separated clusters, whereas in the second simulation exper-

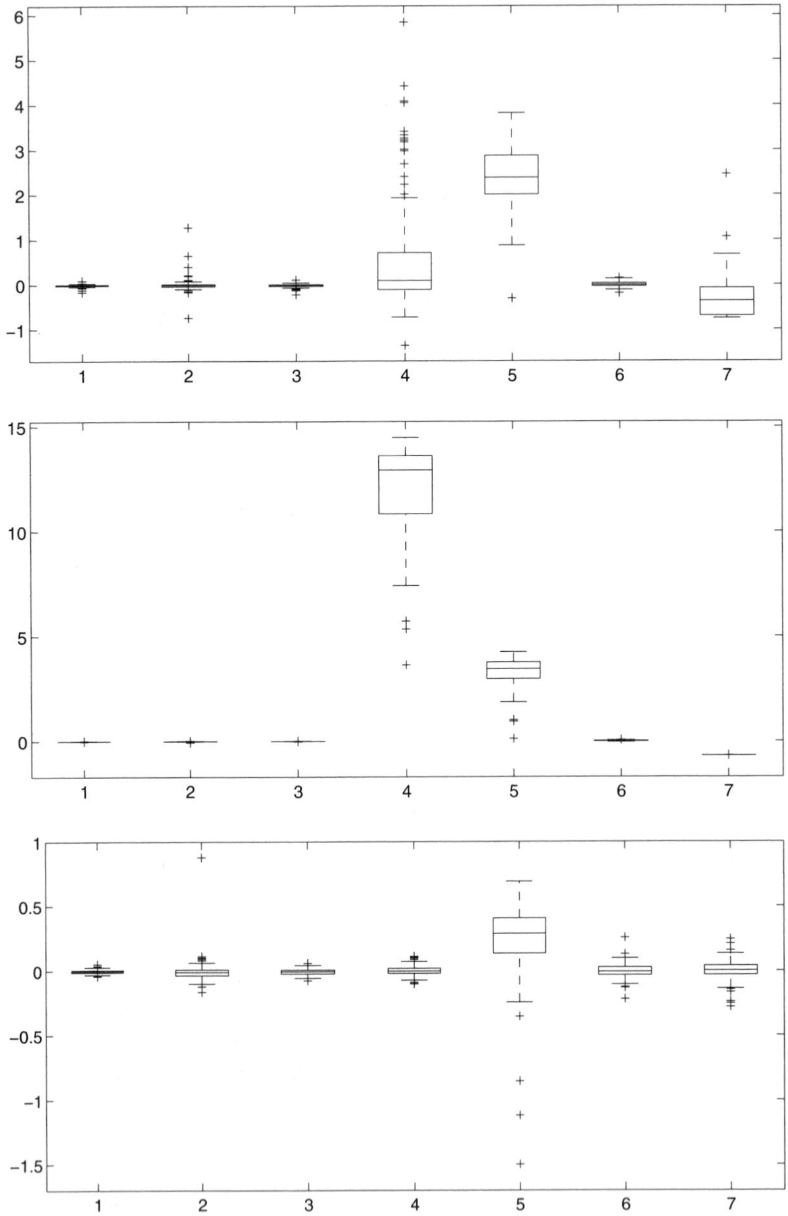


**Fig. 5.2.** Simulation experiment to evaluate different estimators of the marginal likelihood $p(\mathbf{y}|\mathcal{M}_2)$ for data simulated from a mixture of two Poisson distributions with $\eta_1 = 0.3$, $\mu_1 = 1$, and $\mu_2 = 5$ (top), $\mu_2 = 25$ (middle), and $\mu_2 = 1$ (bottom); the box plots show the distribution of the estimation error $\log \hat{p}(\mathbf{y}|\mathcal{M}_2) - \log p(\mathbf{y}|\mathcal{M}_2)$ over 100 simulated data sets ($1 \ldots \hat{p}_{BS}(\mathbf{y}|\mathcal{M}_2)$, $2 \ldots \hat{p}_{IS}(\mathbf{y}|\mathcal{M}_2)$, $3 \ldots \hat{p}_{RI}(\mathbf{y}|\mathcal{M}_2)$, $4 \ldots \hat{p}_{HM}(\mathbf{y}|\mathcal{M}_2)$, $5 \ldots \hat{p}_{HM,2}(\mathbf{y}|\mathcal{M}_2)$, $6 \ldots \hat{p}^{\star}_{CH}(\mathbf{y}|\mathcal{M}_2)$, $7 \ldots \hat{p}_{CH}(\mathbf{y}|\mathcal{M}_2)$)

iment $\mu_2 = 25$ generates well-separated clusters in the data; see Figure 5.1. For the third simulation experiment $\mu_2 = 1$, thus the mixture is overfitting the number of components, with the true number being equal to 1.

For all three simulation experiments, $\boldsymbol{\vartheta}_2 = (\mu_1, \mu_2, \eta_1, \eta_2)$ is estimated for each of the 100 data sets based on the priors $\mu_k \sim \mathcal{G}(1,1)$ and $\boldsymbol{\eta} \sim \mathcal{D}(4,4)$, using the random permutation Gibbs sampler described in *Algorithm 3.5*. For the first simulation experiment the two modes of the mixture posterior density $p(\mu_1, \mu_2|\mathbf{y})$ are not extremely well separated, which is, however, the case for the second simulation experiment; see again Figure 5.1.

**Table 5.2.** Performance of different estimators of the marginal likelihood $p(\mathbf{y}|\mathcal{M}_2)$ for data simulated from a mixture of two Poisson distributions with $\eta_1 = 0.3$, $\mu_1 = 1$, and $\mu_2 = 5$

| Estimator | Average Bias | Average MSE |
|---|---|---|
| $\log \hat{p}_{BS}(\mathbf{y}\|\mathcal{M}_2)$ | −0.0077148 | 0.00081305 |
| $\log \hat{p}_{IS}(\mathbf{y}\|\mathcal{M}_2)$ | +0.0068868 | 0.030403 |
| $\log \hat{p}_{RI}(\mathbf{y}\|\mathcal{M}_2)$ | −0.014403 | 0.0017834 |
| $\log \hat{p}_{HM,2}(\mathbf{y}\|\mathcal{M}_2)$ | +0.5845 | 2.0435 |
| $\log \hat{p}_{HM}(\mathbf{y}\|\mathcal{M}_2)$ | +2.3598 | 6.0528 |
| $\log \hat{p}_{CH}(\mathbf{y}\|\mathcal{M}_2)$ | −0.30264 | 0.31126 |
| $\log \hat{p}_{CH}(\mathbf{y}\|\mathcal{M}_2) + \log 2$ | +0.3905 | 0.5819 |
| $\log \hat{p}^{\star}_{CH}(\mathbf{y}\|\mathcal{M}_2)$ | −5.6487e-005 | 0.0035683 |

**Table 5.3.** Performance of different estimators of the marginal likelihood $p(\mathbf{y}|\mathcal{M}_2)$ for data simulated from a mixture of two Poisson distributions with $\eta_1 = 0.3$, $\mu_1 = 1$, and $\mu_2 = 25$

| Estimator | Average Bias | Average MSE |
|---|---|---|
| $\log \hat{p}_{BS}(\mathbf{y}\|\mathcal{M}_2)$ | −0.00068049 | 3.3692e-005 |
| $\log \hat{p}_{IS}(\mathbf{y}\|\mathcal{M}_2)$ | −0.0011205 | 0.00011117 |
| $\log \hat{p}_{RI}(\mathbf{y}\|\mathcal{M}_2)$ | −0.00058584 | 3.2195e-005 |
| $\log \hat{p}_{HM,2}(\mathbf{y}\|\mathcal{M}_2)$ | +12.184 | 153.36 |
| $\log \hat{p}_{HM}(\mathbf{y}\|\mathcal{M}_2)$ | +3.2745 | 11.268 |
| $\log \hat{p}_{CH}(\mathbf{y}\|\mathcal{M}_2)$ | −0.69304 | 0.4803 |
| $\log \hat{p}_{CH}(\mathbf{y}\|\mathcal{M}_2) + \log 2$ | +0.00010925 | 1.559e-006 |
| $\log \hat{p}^{\star}_{CH}(\mathbf{y}\|\mathcal{M}_2)$ | +0.00049045 | 0.00028114 |

First we computed the importance sampling estimator $\hat{p}_{IS}(\mathbf{y}|\mathcal{M}_2)$, defined in (5.39), and the reciprocal importance sampling estimator $\hat{p}_{RI}(\mathbf{y}|\mathcal{M}_2)$, defined in (5.43). Starting from the importance sampling estimator, we derived the bridge sampling estimator $\hat{p}_{BS}(\mathbf{y}|\mathcal{M}_2)$ as the limit of $\hat{p}_{BS,t}$, defined in

**Table 5.4.** Performance of different estimators of the marginal likelihood $p(\mathbf{y}|\mathcal{M}_2)$ for data simulated from a single Poisson distribution with $\mu_1 = 1$

| Estimator | Average Bias | Average MSE |
|---|---|---|
| $\log \hat{p}_{BS}(\mathbf{y}\|\mathcal{M}_2)$ | −0.0015064 | 0.00022875 |
| $\log \hat{p}_{IS}(\mathbf{y}\|\mathcal{M}_2)$ | −0.0025096 | 0.0098332 |
| $\log \hat{p}_{RI}(\mathbf{y}\|\mathcal{M}_2)$ | −0.0049795 | 0.00057365 |
| $\log \hat{p}_{HM,2}(\mathbf{y}\|\mathcal{M}_2)$ | −0.00027147 | 0.0015687 |
| $\log \hat{p}_{HM}(\mathbf{y}\|\mathcal{M}_2)$ | +0.22928 | 0.15127 |
| $\log \hat{p}_{CH}(\mathbf{y}\|\mathcal{M}_2)$ | −0.00051896 | 0.0066741 |
| $\log \hat{p}_{CH}(\mathbf{y}\|\mathcal{M}_2) + \log 2$ | +0.69263 | 0.48641 |
| $\log \hat{p}^{\star}_{CH}(\mathbf{y}\|\mathcal{M}_2)$ | −0.0038621 | 0.0032436 |

(5.55), as $t \to \infty$. Each of these estimators is computed with $M = L = 2000$, after a burn-in of 500, and is based on the importance density (5.36), constructed from $S = 10$ randomly selected draws of the random permutation Gibbs sampler. The tail behavior of the importance density in relation to the mixture posterior density will be quite different for the various simulation experiments. The two groups in the data are not well separated for the first simulation experiment, hence there will be quite a large uncertainty in posterior classification. The importance density (5.36) will provide a poor fit to the mixture posterior $p(\boldsymbol{\vartheta}_2|\mathbf{y}, \mathcal{M}_2)$ in particular in the tails of both densities. For the second simulation experiment the groups are very well separated, and the importance density will be a good approximation to the mixture posterior. Finally, we consider the two harmonic mean estimators $\hat{p}_{HM}(\mathbf{y}|\mathcal{M}_2)$ and $\hat{p}_{HM,2}(\mathbf{y}|\mathcal{M}_2)$, defined in (5.48) and (5.50), respectively, where $M$ is the same as before.

For each simulation experiment, the various estimators $\hat{p}(\mathbf{y}|\mathcal{M}_2)$ are compared in Figure 5.2 through the distribution of the estimation error $\log \hat{p}(\mathbf{y}|\mathcal{M}_2) - \log p(\mathbf{y}|\mathcal{M}_2)$ over the 100 replications, and in Tables 5.2 to 5.4 through two numerical measures of performance: first, the average relative bias measured by the average estimation error $\log \hat{p}(\mathbf{y}|\mathcal{M}_2) - \log p(\mathbf{y}|\mathcal{M}_2)$, and second, the relative mean-squared error measured by the average of the squared estimation error $(\log \hat{p}(\mathbf{y}|\mathcal{M}_2) - \log p(\mathbf{y}|\mathcal{M}_2))^2$.

The two harmonic mean estimators are the worst among the estimators considered. $\hat{p}_{HM,2}(\mathbf{y}|\mathcal{M}_2)$ is biased in favor of a Poisson mixture model with two components for all simulation experiments. $\hat{p}_{HM}(\mathbf{y}|\mathcal{M}_2)$ shows extreme bias in favor of a Poisson mixture model with two components for $\mu_2 = 25$, which disappears for the overfitting mixture. The box plot of the estimation error in Figure 5.2 shows that the other simulation-based estimators are much more reliable than the two harmonic mean estimators. Among these estimators, the bridge sampling estimator $\hat{p}_{BS}(\mathbf{y}|\mathcal{M}_2)$ is the most reliable one in terms of the MSE. The MSE of the importance sampling estimator $\hat{p}_{IS}(\mathbf{y}|\mathcal{M}_2)$ is considerably higher in particular for the first simulation experiment, reflecting the poor fit of the importance density in the tails. The reciprocal importance sampling estimator $\hat{p}_{RI}(\mathbf{y}|\mathcal{M}_2)$ is better than the importance sampling estimator, but in most cases worse than the bridge sampling estimator.[1]

### 5.4.8 Dealing with Hierarchical Priors

Simulation-based approximations of the marginal likelihood are easily extended to deal with hierarchical priors; see Frühwirth-Schnatter (2004). Because both the complete-data likelihood $p(\mathbf{y}|\mathbf{S}, \boldsymbol{\theta}_1, \ldots, \boldsymbol{\theta}_K)p(\mathbf{S}|\boldsymbol{\eta})$ as well as the mixture likelihood $p(\mathbf{y}|\boldsymbol{\vartheta}_K)$ are independent of $\boldsymbol{\delta}$ given $\boldsymbol{\vartheta}_K$, no modifications are necessary, if an analytical expression for the marginal prior

$$p(\boldsymbol{\theta}_1, \ldots, \boldsymbol{\theta}_K) = \int \prod_{k=1}^{K} p(\boldsymbol{\theta}_k|\boldsymbol{\delta})p(\boldsymbol{\delta})d\boldsymbol{\delta}$$

is available. Otherwise the hyperparameter $\boldsymbol{\delta}$ could be added to the set of unknown parameters before applying the bridge sampling technique to $\boldsymbol{\vartheta} = (\boldsymbol{\theta}_1, \ldots, \boldsymbol{\theta}_K, \boldsymbol{\eta}, \boldsymbol{\delta})$. As $\boldsymbol{\delta}$ is independent of $\mathbf{S}$ given $\boldsymbol{\theta}_1, \ldots, \boldsymbol{\theta}_K$, the importance density may be chosen as $q(\boldsymbol{\theta}_1, \ldots, \boldsymbol{\theta}_K, \boldsymbol{\eta}, \boldsymbol{\delta}) = q(\boldsymbol{\theta}_1, \ldots, \boldsymbol{\theta}_K, \boldsymbol{\eta})q(\boldsymbol{\delta})$, where $q(\boldsymbol{\theta}_1, \ldots, \boldsymbol{\theta}_K, \boldsymbol{\eta})$ is equal to the mixture importance density (5.36) and $q(\boldsymbol{\delta})$ is obtained by fitting a unimodal importance density to the MCMC draws $\boldsymbol{\delta}^{(m)}$.

## 5.5 Approximations to the Marginal Likelihood Based on Density Ratios

### 5.5.1 The Posterior Density Ratio

The posterior density ratio provides a strikingly simple way to compute the marginal likelihood $p(\mathbf{y}|\mathcal{M}_K)$ of a model $\mathcal{M}_K$, defined through a likelihood function $p(\mathbf{y}|\boldsymbol{\vartheta}_K)$ and a prior $p(\boldsymbol{\vartheta}_K)$. It exploits a formal equivalence between the marginal likelihood $p(\mathbf{y}|\mathcal{M}_K)$ and the normalizing constant appearing in the definition of the posterior distribution $p(\boldsymbol{\vartheta}_K|\mathbf{y}, \mathcal{M}_K)$ through Bayes' theorem:

$$p(\boldsymbol{\vartheta}_K|\mathbf{y}, \mathcal{M}_K) = \frac{p(\mathbf{y}|\boldsymbol{\vartheta}_K)p(\boldsymbol{\vartheta}_K)}{\int p(\mathbf{y}|\boldsymbol{\vartheta}_K)p(\boldsymbol{\vartheta}_K)d\boldsymbol{\vartheta}_K}.$$

Therefore the following identity, also known as the candidate's formula (Besag, 1989), holds for arbitrary $\boldsymbol{\vartheta}_K$,

$$p(\mathbf{y}|\mathcal{M}_K) = \frac{p(\mathbf{y}|\boldsymbol{\vartheta}_K)p(\boldsymbol{\vartheta}_K)}{p(\boldsymbol{\vartheta}_K|\mathbf{y}, \mathcal{M}_K)} = \frac{p^\star(\boldsymbol{\vartheta}_K|\mathbf{y}, \mathcal{M}_K)}{p(\boldsymbol{\vartheta}_K|\mathbf{y}, \mathcal{M}_K)}. \tag{5.61}$$

[1] The figure as well as the tables contains further estimators of the marginal likelihood, which is explained later on.

Note that the density ratio on the right-hand side may be computed for an arbitrary value of $\boldsymbol{\vartheta}_K$, each of which yields the same marginal likelihood. With the help of (5.61), computation of the marginal likelihood as the ratio of the nonnormalized posterior over the normalized posterior $p(\boldsymbol{\vartheta}_K|\mathbf{y}, \mathcal{M}_K)$, is straightforward in cases where the posterior density $p(\boldsymbol{\vartheta}_K|\mathbf{y}, \mathcal{M}_K)$ belongs to a well-known distribution family. For mixture models this is only the case for mixtures from the exponential family with one component. To give an example, consider $N$ observations from a single $\mathcal{P}(\mu)$-distribution (model $\mathcal{M}_1$), with prior $\mu \sim \mathcal{G}(a_0, b_0)$, where the posterior reads $\mu|\mathbf{y} \sim \mathcal{G}(a_N, b_N)$ with $a_N = a_0 + \sum_{i=1}^N y_i$ and $b_N = N + b_0$. The posterior density ratio yields:

$$p(\mathbf{y}|\mathcal{M}_1) = \frac{b_0^{a_0}\Gamma(a_N)}{b_N^{a_N}\Gamma(a_0)\prod_{i=1}^N \Gamma(y_i+1)}. \tag{5.62}$$

For finite mixture models with more than one component, the posterior ordinate $p(\boldsymbol{\vartheta}_K|\mathbf{y}, \mathcal{M}_K)$ is no longer available in closed form, and the posterior density ratio formula is not directly applicable. Various approximations, however, have been suggested, which are based on approximating the posterior ordinate. Due to the multimodality of the posterior distribution this is a rather challenging problem.

### 5.5.2 Chib's Estimator

Chib (1995) suggested an approximation to (5.61) which is based on substituting the unknown posterior ordinate $p(\boldsymbol{\vartheta}_K^\star|\mathbf{y}, \mathcal{M}_K)$ by a suitable estimate:

$$\hat{p}_{CH}(\mathbf{y}|\mathcal{M}_K) \approx \frac{p(\mathbf{y}|\boldsymbol{\vartheta}_K^\star)p(\boldsymbol{\vartheta}_K^\star)}{\hat{p}(\boldsymbol{\vartheta}_K^\star|\mathbf{y}, \mathcal{M}_K)}. \tag{5.63}$$

Various estimators are available, depending on whether the conditional densities appearing in the transition kernel of the Markov chain have a known normalizing constant (Chib, 1995) or not (Chib and Jeliazkov, 2001).

For mixture models, an approximation to $p(\boldsymbol{\vartheta}_K^\star|\mathbf{y}, \mathcal{M}_K)$ arises in quite a natural way within the data augmentation algorithm considered in Subsection 3.5.3, as the posterior density may be expressed as

$$p(\boldsymbol{\vartheta}_K^\star|\mathbf{y}, \mathcal{M}_K) = \sum_{\mathbf{S}\in\mathcal{S}_K} \prod_{k=1}^K p(\boldsymbol{\theta}_k^\star|\mathbf{y}, \mathbf{S})p(\boldsymbol{\eta}^\star|\mathbf{S})p(\mathbf{S}|\mathcal{M}_K, \mathbf{y}),$$

where $p(\boldsymbol{\theta}_k|\mathbf{y}, \mathbf{S})$ and $p(\boldsymbol{\eta}|\mathbf{S})$ are the complete-data posterior densities. If the complete-data posterior $p(\boldsymbol{\theta}_k|\mathbf{y}, \mathbf{S})$ is available in closed form, MCMC draws $\mathbf{S}^{(m)}, m = 1, \ldots, M$, of $\mathbf{S}$ could be used to estimate the posterior ordinate $p(\boldsymbol{\vartheta}_K^\star|\mathbf{y}, \mathcal{M}_K)$ by

$$\hat{p}(\boldsymbol{\vartheta}_K^\star|\mathbf{y}, \mathcal{M}_K) = \frac{1}{M}\sum_{m=1}^M \prod_{k=1}^K p(\boldsymbol{\theta}_k^\star|\mathbf{S}^{(m)}, \mathbf{y})p(\boldsymbol{\eta}^\star|\mathbf{S}^{(m)}). \tag{5.64}$$

This estimator is simulation-consistent, as the density $\hat{p}(\boldsymbol{\vartheta}_K^\star|\mathbf{y}, \mathcal{M}_K)$ converges to $p(\boldsymbol{\vartheta}_K^\star|\mathbf{y}, \mathcal{M}_K)$ by the strong law of large numbers as $M$ goes to infinity.

The estimator (5.64) could be applied immediately for practically all standard mixtures from the exponential family with conditionally conjugate priors with fixed hyperparameters. For random hyperparameter $\boldsymbol{\delta} \sim p(\boldsymbol{\delta})$, a minor modification is necessary. First the joint posterior of $(\boldsymbol{\vartheta}_K, \boldsymbol{\delta})$ is decomposed as $p(\boldsymbol{\vartheta}_K, \boldsymbol{\delta}|\mathbf{y}, \mathcal{M}_K) = p(\boldsymbol{\vartheta}_K|\mathbf{y}, \boldsymbol{\delta}, \mathcal{M}_K)p(\boldsymbol{\delta}|\boldsymbol{\vartheta}_K)$. If the posterior $p(\boldsymbol{\delta}|\boldsymbol{\vartheta}_K)$ is of closed form, then the posterior density ratio formula reads:

$$\hat{p}_{CH}(\mathbf{y}|\mathcal{M}_K) \approx \frac{p(\mathbf{y}|\boldsymbol{\vartheta}_K^\star)p(\boldsymbol{\vartheta}_K^\star, \boldsymbol{\delta}^\star)}{p(\boldsymbol{\delta}^\star|\boldsymbol{\vartheta}_K^\star)\hat{p}(\boldsymbol{\vartheta}_K^\star|\mathbf{y}, \boldsymbol{\delta}, \mathcal{M}_K)},$$

where $p(\boldsymbol{\vartheta}_K^\star, \boldsymbol{\delta}^\star)$ is equal to the hierarchical prior (3.11) and

$$\hat{p}(\boldsymbol{\vartheta}_K^\star|\mathbf{y}, \mathcal{M}_K) = \frac{1}{M}\sum_{m=1}^{M}\prod_{k=1}^{K} p(\boldsymbol{\theta}_k^\star|\mathbf{S}^{(m)}, \boldsymbol{\delta}^{(m)}, \mathbf{y})p(\boldsymbol{\eta}^\star|\mathbf{S}^{(m)}).$$

Chib's estimator has to be modified in cases where the elements of $\boldsymbol{\theta}_k$ are sampled in different blocks, as will be the case for mixtures with normal components under independence priors; see Chib (1995) for computational details. Finally, these estimators cannot be applied to models where the complete-data posterior $p(\boldsymbol{\theta}_k|\mathbf{S}, \mathbf{y})$ is nonstandard. This is rarely the case for standard mixture models, but happens, for instance, for mixtures of nonnormal regression models; see Section 9.4. In this case, the estimator of Chib and Jeliazkov (2001) may be implemented.

### Chib's Estimators for Poisson Mixtures

For Poisson mixtures with $K$ components, where $\boldsymbol{\vartheta}_K^\star = (\mu_1^\star, \dots, \mu_K^\star, \boldsymbol{\eta}^\star)$, the function $p(\mathbf{y}|\boldsymbol{\vartheta}_K^\star)$ appearing in the estimator (5.63) is the mixture likelihood of a Poisson mixture with $K$ components, defined earlier in (2.27), whereas the other quantities are given by

$$p(\boldsymbol{\vartheta}_K^\star) = f_D(\boldsymbol{\eta}^\star; e_0, \dots, e_0)\prod_{k=1}^{K} f_G(\mu_k^\star; a_0, b_0),$$

$$\hat{p}(\boldsymbol{\vartheta}_K^\star|\mathbf{y}, \mathcal{M}_K) = \frac{1}{M}\sum_{m=1}^{M} f_D(\boldsymbol{\eta}^\star; e_0 + N_1^{(m)}, \dots, e_0 + N_K^{(m)}) \times \prod_{k=1}^{K} f_G(\mu_k^\star; a_0 + N_k^{(m)}\overline{y}_k^{(m)}, b_0 + N_k^{(m)}).$$

Thus to evaluate the posterior density ratio we need to store only the group sizes $N_k^{(m)} = N_k(\mathbf{S}^{(m)})$ and the group means $\overline{y}_k^{(m)} = \overline{y}_k(\mathbf{S}^{(m)})$ for $k = 1, \dots, K$ for each classification $\mathbf{S}^{(m)}$ during MCMC sampling.

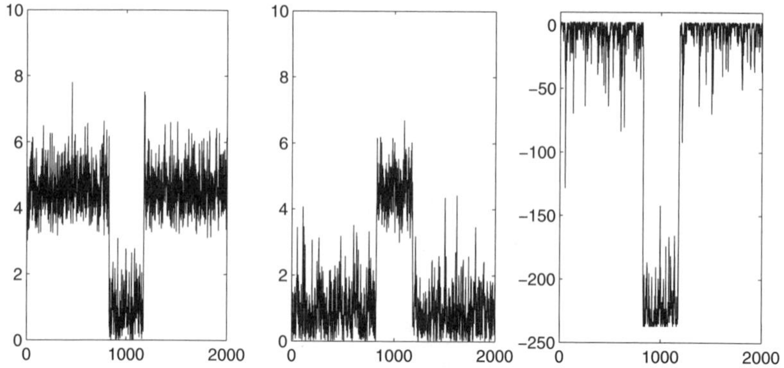


**Fig. 5.3.** Behavior of the Gibbs sampler for a single data set used in the simulation experiment with $\mu_2 = 5$; MCMC draws for $\mu_1^{(m)}$ (left), MCMC draws for $\mu_2^{(m)}$ (middle), and functional evaluation $p(\boldsymbol{\vartheta}_2^\star|\mathbf{S}^{(m)}, \mathbf{y})$ in Chib's estimator of the marginal likelihood (right)

## Problems with Chib's Estimator for Mixture Models

Chib's estimator of the marginal likelihood proved to be very useful for a wide range of statistical and econometric models, and consequently the method was applied in Chib (1995) and Chib (1996) to estimate the marginal likelihood also for finite mixture models, in particular for Markov switching models. In these early applications, it was not yet evident that Chib's estimator is apt to fail when being applied to finite mixture models, as the posterior density of such a model is highly irregular due to lack of identification for these models. This problem was first noted by Neal (1998), and explained later on in much detail in Frühwirth-Schnatter (2004).

What is the problem with Chib's estimator? The estimator (5.64) is based on estimating marginal densities from averaging conditional densities. Thus it is essential to make sure that $\mathbf{S}$ mixes well over the whole parameter space. For the estimator (5.64) to be correct, the Markov chain $(\mathbf{S}^{(m)}, \boldsymbol{\vartheta}_K^{(m)})$ used for simulation, has to explore all $K!$ modal regions that exist because of the nonidentifiability of the mixture components. The applications that appeared both in Chib (1995) and Chib (1996) were based on the Gibbs output, which in general *does not* visit all modes. This introduces a bias when estimating the marginal likelihood.

For illustration we apply Chib's estimator to the synthetic data discussed earlier in Subsection 5.4.7. Despite the simplicity of the model (the whole parameter $\boldsymbol{\vartheta}_2$ may be sampled within one block), the resulting estimator $\hat{p}_{CH}(\mathbf{y}|\mathcal{M}_2)$ shows considerable bias for the simulation experiments with $\mu_2 = 5$ and $\mu_2 = 25$ over 100 replications; see Tables 5.2 and 5.3.

From where does the bias come? Integration in the definition of the marginal likelihood is over the unconstrained parameter space $\Theta_K$ which is the union of $K!$ subspaces $\mathcal{L}_s$, $s = 1, \ldots, K!$, differing only in the labeling of the $K$ states of $S_i$. Therefore the marginal likelihood may be expressed as

$$p(\mathbf{y}|\mathcal{M}_K) = \sum_{s=1}^{K!} \int_{\mathcal{L}_s} p(\mathbf{y}|\boldsymbol{\vartheta}_K, \mathbf{S})p(\boldsymbol{\vartheta}_K, \mathbf{S})d(\boldsymbol{\vartheta}_K, \mathbf{S}) =$$

$$K! \int_{\mathcal{L}_{s_0}} p(\mathbf{y}|\boldsymbol{\vartheta}_K, \mathbf{S})p(\boldsymbol{\vartheta}_K, \mathbf{S})d(\boldsymbol{\vartheta}_K, \mathbf{S}),$$

where $s_0$ is an arbitrary value between 1 and $K!$. Thus to compute the marginal likelihood simulations from *all* labeling subspaces are required. Alternatively, if simulations from a unique labeling subspace are used, the marginal likelihood results only after multiplying with the factor $K!$.

If the modes of the mixture posterior density are extremely well separated and the Gibbs sampler sticks at one of the labeling subspaces $\mathcal{L}_{s_0}$ instead of exploring *all* labeling subspaces, $\hat{p}_{CH}(\mathbf{y}|\mathcal{M}_K)$ estimates

$$\int_{\mathcal{L}_{s_0}} p(\mathbf{y}|\boldsymbol{\vartheta}_K, \mathbf{S})p(\boldsymbol{\vartheta}_K, \mathbf{S})d(\boldsymbol{\vartheta}_K, \mathbf{S})$$

instead of $p(\mathbf{y}|\mathcal{M}_K)$, and consequently the expected bias is equal to:

$$E(\log \hat{p}_{CH}(\mathbf{y}|\mathcal{M}_K) - \log p(\mathbf{y}|\mathcal{M}_K)) = -\log K!. \qquad (5.65)$$

As expected from (5.65), the bias of Chib's estimator $\log \hat{p}_{CH}(\mathbf{y}|\mathcal{M}_2)$ for $\mu_2 = 25$, given in Table 5.3, turns out to be practically equal to $-\log K! = -\log 2 = -0.69315$.

For this simulation experiment there was hardly any misclassification risk, as the groups are very well separated. After convergence, the posterior partition of the data is extremely stable. In this case the mixture posterior density is close to the product of complete-data posteriors and the posterior density ratio is, apart from the bias, very precise. If we correct for this bias, we obtain the estimator $\log \hat{p}_{CH}(\mathbf{y}|\mathcal{M}_2) + \log 2$, which gives an extremely precise answer; see Table 5.3.

The problem with this bias-corrected estimator is that in cases where *occasional* label switching occurs, as for the simulation experiment with $\mu_2 = 5$, it is more difficult to quantify the bias and to correct for it. Label switching is evident for this simulation experiment from Figure 5.3, which shows the simulated paths of $\mu_1$ and $\mu_2$ for a randomly selected data set, as well as occasional switching in the functional values $p(\boldsymbol{\vartheta}_2^\star|\mathbf{S}^{(m)}, \mathbf{y})$, which causes an unpredictable bias of $\log \hat{p}_{CH}(\mathbf{y}|\mathcal{M}_2)$. From Table 5.2 we find that the bias is smaller than $-\log 2$, thus the bias-corrected estimator $\log \hat{p}_{CH}(\mathbf{y}|\mathcal{M}_2) + \log 2$ overrates the bias.

Finally, the bias of $\log \hat{p}_{CH}(\mathbf{y}|\mathcal{M}_2)$ disappears for the overfitting model with $\mu_2 = 1$ (see Table 5.4), and adding the bias correction $\log 2$ would be totally inappropriate.

### Variants of Chib's Estimator That Account for Nonidentifiability

Various authors suggested modifications of Chib's estimator that account for nonidentifiability, and lead to a simulation-consistent estimator of the posterior ordinate under the unconstrained model.

Given a sequence $(\mathbf{S}^{(m)}, \boldsymbol{\vartheta}_K^{(m)})$ of draws from a sampler, where label switching is not encouraged, a simulation-consistent estimator is given by considering all permutations of the posterior draws (Berkhof et al., 2003):

$$\hat{p}(\boldsymbol{\vartheta}_K^\star|\mathbf{y}, \mathcal{M}_K) = \frac{1}{M} \sum_{m=1}^{M} \frac{1}{K!} \sum_{s=1}^{K!} p(\boldsymbol{\vartheta}_K^\star|\rho_s(\mathbf{S}^{(m)}), \mathbf{y}).$$

A simplified version of this results, if for each $m$, not all, but only a single permutation $s(m)$ is selected:

$$\hat{p}(\boldsymbol{\vartheta}_K^\star|\mathbf{y}, \mathcal{M}_K) = \frac{1}{M} \sum_{m=1}^{M} p(\boldsymbol{\vartheta}_K^\star|\rho_{s(m)}(\mathbf{S}^{(m)}), \mathbf{y}). \tag{5.66}$$

One such estimator appears in Neal (1998), where the permutation $s(m)$ is determined in a deterministic manner by $s(m) = m \bmod K!$. Thus for $K = 2$, $s(m) = 1$, iff $m$ is odd, and $s(m) = 2$, iff $m$ is even.

A further modified Chib's estimator, denoted by $\hat{p}_{CH}^\star(\mathbf{y}|\mathcal{M}_K)$, results when the density estimate (5.66) is based on the output of the random permutation sampler introduced in *Algorithm 3.5*, where $s(m)$ is selected randomly from $1, \ldots, K!$. For this estimator the bias automatically disappears due to balanced label switching; see for illustration the evaluation of this estimator for the simulation experiments in Figure 5.2 and Tables 5.2 to 5.4. The inaccuracy of this estimator, however, measured in terms of the MSE, is considerably larger than for bridge sampling or importance sampling.

### 5.5.3 Laplace Approximation

Laplace approximation is a widely used method of approximating the marginal likelihood, applied in Kass et al. (1988), Tierney et al. (1989), and Kass and Vaidyanathan (1992); see also the reviews in Kass and Raftery (1995) and Raftery (1996a). It may be viewed as a numerical approximation to the marginal likelihood, obtained by substituting the mixture posterior density $p(\boldsymbol{\vartheta}_K|\mathbf{y}, \mathcal{M}_K)$ in the density ratio formula (5.61) by the local normal density

$$p(\boldsymbol{\vartheta}_K|\mathbf{y}, \mathcal{M}_K) \approx f_N(\boldsymbol{\vartheta}_K; \boldsymbol{\vartheta}_K^\star, \boldsymbol{\Sigma}). \tag{5.67}$$

In (5.67), $\boldsymbol{\vartheta}_K^\star$ is the posterior mode, whereas $\boldsymbol{\Sigma}^{-1}$ is minus the Hessian matrix of the log posterior, evaluated at $\boldsymbol{\vartheta}_K^\star$:

$$\boldsymbol{\Sigma}^{-1} = - \left. \frac{\partial^2 \log(p(\mathbf{y}|\boldsymbol{\vartheta}_K)p(\boldsymbol{\vartheta}_K))}{\partial \boldsymbol{\vartheta}_K^2} \right|_{\boldsymbol{\vartheta}_K = \boldsymbol{\vartheta}_K^\star}. \tag{5.68}$$

Thus the log of the marginal likelihood is approximated by the following expression,

$$\log \hat{p}_L(\mathbf{y}|\mathcal{M}_K) \approx \log p(\mathbf{y}|\boldsymbol{\vartheta}_K^\star) + \log p(\boldsymbol{\vartheta}_K^\star) + \frac{d_K}{2}\log(2\pi) + 0.5 \cdot \log|\boldsymbol{\Sigma}|, \tag{5.69}$$

where $d_K = \dim(\boldsymbol{\vartheta}_K^\star)$. The approximation error of (5.69) is not under control, as it depends on the accuracy of the local normal approximation, but an asymptotic justification is obtained under the same regularity conditions that guarantee asymptotic normality of the posterior density. For mixture models where the number of components is overfitting and asymptotic normality does not hold, there is a certain lack of justifying the use of (5.69) in this case.

A practical limitation is that the Hessian of the log of the mixture posterior, required in (5.68) to define $\boldsymbol{\Sigma}^{-1}$, is often not available. Raftery (1996b) and Lewis and Raftery (1997) use posterior simulations to estimate posterior mode and posterior curvature, by using robust estimators of location and scale. They call this procedure the Laplace–Metropolis estimator.

## 5.6 Reversible Jump MCMC Versus Marginal Likelihoods?

Is Bayes' rule in combination with a good method for computing marginal likelihoods or is trans-dimensional MCMC a better way of dealing with model specification problems such as selecting the number of components? From a theoretical point of view, both approaches are equivalent, as they are nothing but different computational tools for obtaining an estimator of the posterior model probability for each model under investigation. If both methods use the same model space $\Omega = \{\mathcal{M}_1, \ldots, \mathcal{M}_{K_{\max}}\}$, the same prior model probabilities $p(\mathcal{M}_1), \ldots, p(\mathcal{M}_{K_{\max}})$, and the same collection of priors $p(\boldsymbol{\vartheta}_1|\mathcal{M}_1), \ldots, p(\boldsymbol{\vartheta}_{K_{\max}}|\mathcal{M}_{K_{\max}})$, then both approaches are approximating the same posterior model probabilities $p(\mathcal{M}_1|\mathbf{y}), \ldots, p(\mathcal{M}_{K_{\max}}|\mathbf{y})$.

To this aim, trans-dimensional MCMC methods use draws from the discrete posterior distribution $p(\mathcal{M}|\mathbf{y})$ over $\mathcal{M} \in \Omega$, which are obtained by running a Markov chain on a much larger space than $\Omega$. Provided that this Markov chain mixes sufficiently well, the posterior probability $p(\mathcal{M}_K|\mathbf{y})$ of model $\mathcal{M}_K$ may be estimated from the Markov chain draws $\mathcal{M}^{(1)}, \ldots, \mathcal{M}^{(M)}$ as

$$\hat{p}_{RJ}(\mathcal{M}_K|\mathbf{y}) = \frac{1}{M}\sum_{m=1}^{M} I_{\{\mathcal{M}^{(m)}=\mathcal{M}_K\}}. \tag{5.70}$$

By the law of large numbers, $\hat{p}_{RJ}(\mathcal{M}_K|\mathbf{y})$ will converge to $p(\mathcal{M}_K|\mathbf{y})$, if $M$ increases to infinity. On the other hand, Bayes' rule (4.9) yields:

$$\hat{p}_{ML}(\mathcal{M}_K|\mathbf{y}) = \frac{p(\mathcal{M}_K)\hat{p}_{ML}(\mathbf{y}|\mathcal{M}_K)}{\sum_{j=1}^{K_{\max}} p(\mathcal{M}_j)\hat{p}_{ML}(\mathbf{y}|\mathcal{M}_j)}, \tag{5.71}$$

where $\hat{p}_{ML}(\mathbf{y}|\mathcal{M}_K)$ is one of the many estimators of the marginal likelihood $p(\mathbf{y}|\mathcal{M}_K)$ that have been discussed in this chapter. Most of these estimators are based on Monte Carlo simulation methods, and converge to the true value $p(\mathbf{y}|\mathcal{M}_K)$, as the number of draws increases. Therefore $\hat{p}_{ML}(\mathcal{M}_K|\mathbf{y})$ will converge to $p(\mathcal{M}_K|\mathbf{y})$ if the number of draws increases to infinity.

In practice, limited computing resources will limit the maximum number of possible draws for both approaches, and the estimated posterior model probabilities $\hat{p}_{RJ}(\mathcal{M}_K|\mathbf{y})$ and $\hat{p}_{ML}(\mathcal{M}_K|\mathbf{y})$ will be different. Computational accuracy of each computational method in terms of the simulation error $\hat{p}(\mathcal{M}_K|\mathbf{y}) - p(\mathcal{M}_K|\mathbf{y})$ will then become important.

If the number of alternative models is large, then the computation of the marginal likelihood for each model is prohibitive, and trans-dimensional MCMC methods are the only way to cope with model specification uncertainty. Think, for instance, of an $r$-variate mixture of $K$ normal distributions with component variance–covariance matrices $\mathbf{\Sigma}_k$, $k = 1\ldots, K$ that have $r(r+1)/2$ distinct elements if $\mathbf{\Sigma}_k$ is unconstrained. For reason of parsimony one might want to restrict some of elements of $\mathbf{\Sigma}_k$ (or $\mathbf{\Sigma}_k^{-1}$) to zero. For each $k = 1, \ldots, K$, there are $r(r-1)/2$ ways of choosing zero elements in $\mathbf{\Sigma}_k^{-1}$, without losing positive definiteness. If $r$ is large, it is not possible to compare all possible models through marginal likelihoods; for instance, for $r = 15$ $105K$ different marginal likelihoods would have to be computed.

### A Simulation Experiment

For illustration we return to the synthetic data discussed in Subsection 5.4.7, and assume $K_{\max} = 2$ and $\Pr(\mathcal{M}_1) = \Pr(\mathcal{M}_2)$. As both marginal likelihoods $p(\mathbf{y}|\mathcal{M}_1)$ and $p(\mathbf{y}|\mathcal{M}_2)$ are available in closed form, the exact posterior probabilities $p(\mathcal{M}_1|\mathbf{y})$ and $p(\mathcal{M}_2|\mathbf{y})$ may be computed for each synthetic data set, and may be compared to the reversible jump estimators $\hat{p}_{RJ}(\mathcal{M}_1|\mathbf{y})$ and $\hat{p}_{RJ}(\mathcal{M}_2|\mathbf{y})$ defined in (5.70) and the bridge sampling estimator $\hat{p}_{BS}(\mathcal{M}_2|\mathbf{y})$ defined in (5.71).

The reversible jump estimators are based on 4000 MCMC draws, after a burn-in of 2000, whereas the bridge sampling estimator is based on 2000 MCMC draws from the posterior of model $\mathcal{M}_2$, using a burn-in of 500 draws, and 2000 draws from the importance density.

In Figure 5.4, the estimators are compared through the distribution of the estimation error $\hat{p}(\mathcal{M}_{K^{\text{true}}}|\mathbf{y}) - p(\mathcal{M}_{K^{\text{true}}}|\mathbf{y})$ over 100 replications from a simulation experiment, where 20 observations are simulated from a mixture of two Poisson distributions with $\mu_1 = 1$ and $\mu_2 = 25$, and for a simulation experiment, where 20 observations are simulated from a single Poisson distribution with $\mu = 1$. Choosing the sample size $N$ as small as 20 allows the exact computation of the posterior probabilities; see Subsection 5.4.7. Both estimators

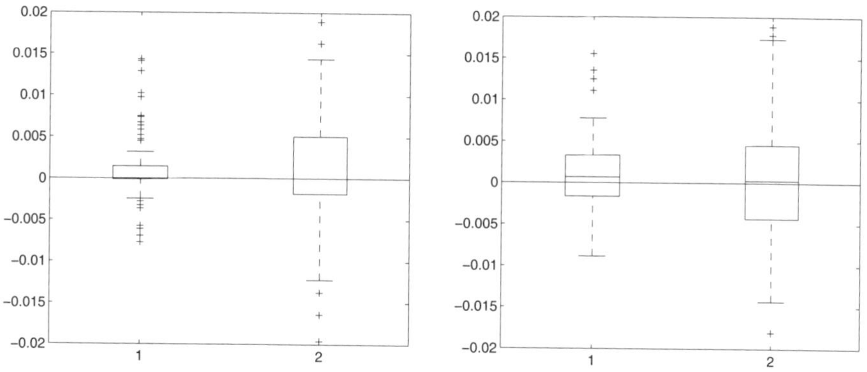


**Fig. 5.4.** Simulation experiment in computing the model posterior probabilities $p(\mathcal{M}_1|\mathbf{y})$ and $p(\mathcal{M}_2|\mathbf{y})$ for 20 observations simulated from a mixture of two Poisson distributions with $\eta_1 = 0.3$, $\mu_1 = 1$, and $\mu_2 = 25$ (left-hand side) and 20 observations simulated from a single Poisson distribution with $\mu = 1$ (right-hand side); the box plots show the distribution of the absolute estimation error $\hat{p}(\mathcal{M}_{K^{\text{true}}}|\mathbf{y}) - p(\mathcal{M}_{K^{\text{true}}}|\mathbf{y})$ over 100 simulated data sets (1 $\ldots \hat{p}_{BS}(\mathcal{M}_{K^{\text{true}}}|\mathbf{y})$, 2 $\ldots \hat{p}_{RJ}(\mathcal{M}_{K^{\text{true}}}|\mathbf{y})$)

are accurate with the estimation error in the probability being in most cases less than 0.01. In a relative comparison, however, it turns out that bridge sampling is considerably more efficient than reversible jump MCMC, with MSE being equal to 0.0014 and 0.0085, given an inefficiency factor of about 6 for reversible jump MCMC, when both methods are based on roughly the same number of draws.

# 6

# Finite Mixture Models with Normal Components

## 6.1 Finite Mixtures of Normal Distributions

Data for which two or more normal distributions are mixed occur frequently in many areas of applied statistics such as biology, economics, marketing, medicine, or physics.

### 6.1.1 Model Formulation

A frequently used finite mixture model for univariate continuous data $y_1, \dots, y_N$ is to assume that the observations are i.i.d. realizations from a random variable $Y$, following a mixture of $K$ univariate normal distributions. The density of this distribution is given by

$$p(y|\boldsymbol{\vartheta}) = \eta_1 f_N(y; \mu_1, \sigma_1^2) + \cdots + \eta_K f_N(y; \mu_K, \sigma_K^2), \tag{6.1}$$

with $f_N(y; \mu_k, \sigma_k^2)$ being the density of a univariate normal distribution. Finite mixtures of univariate normal distributions are generically identifiable (Teicher, 1963). In its most general form, the model is parameterized in terms of $3K - 1$ distinct model parameters $\boldsymbol{\vartheta} = (\boldsymbol{\theta}_1, \dots, \boldsymbol{\theta}_K, \eta_1, \dots, \eta_K)$, where $\boldsymbol{\theta}_k = (\mu_k, \sigma_k^2)$. More specific finite mixture models are obtained by putting constraints either on $\mu_1, \dots, \mu_K$ or $\sigma_1^2, \dots, \sigma_K^2$; see also Subsection 6.4.1.

Mixtures of normals are easily extended to deal with multivariate continuous observations $\mathbf{y}_1, \dots, \mathbf{y}_N$, where $\mathbf{y}_i$ is an $r$-dimensional vector. Typically the various elements $y_{i1}, \dots, y_{ir}$ of $\mathbf{y}_i$ measure $r$ features for a unit $i$ drawn from a population. A frequently used finite mixture model for multivariate data $\mathbf{y} = (\mathbf{y}_1, \dots, \mathbf{y}_N)$ is to assume that the observations are i.i.d. realizations from a multivariate random variable $\mathbf{Y}$ of dimension $r$, arising from a mixture of $K$ multivariate normal distributions. The density of the distribution of $\mathbf{Y}$ is given by

$$p(\mathbf{y}|\boldsymbol{\vartheta}) = \eta_1 f_N(\mathbf{y}; \boldsymbol{\mu}_1, \boldsymbol{\Sigma}_1) + \cdots + \eta_K f_N(\mathbf{y}; \boldsymbol{\mu}_K, \boldsymbol{\Sigma}_K), \tag{6.2}$$

with $f_N(\mathbf{y}; \boldsymbol{\mu}_k, \boldsymbol{\Sigma}_k)$ being the density of a multivariate normal distribution with mean $\boldsymbol{\mu}_k$ and variance–covariance matrix $\boldsymbol{\Sigma}_k$. Finite mixtures of multivariate normal distributions are generically identifiable (Yakowitz and Spragins, 1968).

A multivariate mixture of normal distributions with general variance–covariance matrices $\boldsymbol{\Sigma}_1, \ldots, \boldsymbol{\Sigma}_K$ is highly parameterized in terms of $K(r + r(r+1)/2 + 1) - 1$ distinct model parameters. When fitting a mixture of 3 multivariate normal distributions to a data set containing 10-dimensional observations $\mathbf{y}_i$, one has to estimate as many as 198 distinct parameters! Thus an unconstrained multivariate mixture may turn out to be too general in various situations. Other interesting multivariate finite mixture models are obtained by putting certain constraints on the variance–covariance matrices $\boldsymbol{\Sigma}_1, \ldots, \boldsymbol{\Sigma}_K$. Such finite mixture models are discussed in Subsection 6.4.1.

### Capturing Between- and Within-Group Heterogeneity

For a multivariate mixture, the expected value $\mathrm{E}(\mathbf{Y}|\boldsymbol{\vartheta}) = \sum_{k=1}^K \eta_k \boldsymbol{\mu}_k$ defines the overall center of the distribution of $\mathbf{Y}$, whereas the spread of the distribution is measured by the variance–covariance matrix of $\mathbf{Y}$, which may be written as

$$\begin{aligned} \mathrm{Var}(\mathbf{Y}|\boldsymbol{\vartheta}) &= \mathrm{E}(\mathrm{Var}(\mathbf{Y}|\mathbf{S}, \boldsymbol{\vartheta})) + \mathrm{Var}(\mathrm{E}(\mathbf{Y}|\mathbf{S}, \boldsymbol{\vartheta})) \\ &= \sum_{k=1}^K \eta_k \boldsymbol{\Sigma}_k + \sum_{k=1}^K \eta_k (\boldsymbol{\mu}_k - \mathrm{E}(\mathbf{Y}|\boldsymbol{\vartheta}))(\boldsymbol{\mu}_k - \mathrm{E}(\mathbf{Y}|\boldsymbol{\vartheta}))'. \end{aligned} \tag{6.3}$$

Thus the total variance $\mathrm{Var}(\mathbf{Y}|\boldsymbol{\vartheta})$ arises from two sources of variability, namely within-group heterogeneity and between-group heterogeneity. The measure of within-group heterogeneity is a weighted average of within-group variability $\boldsymbol{\Sigma}_k$. The measure of between-group heterogeneity is based on a weighted distance of the group mean $\boldsymbol{\mu}_k$ from the overall mean $\mathrm{E}(\mathbf{Y}|\boldsymbol{\vartheta})$. In both cases the weights are equal to group sizes.

For a mixture, where $\sum_{k=1}^K \eta_k \boldsymbol{\Sigma}_k$ is much smaller than $\mathrm{Var}(\mathbf{Y}|\boldsymbol{\vartheta})$ in a suitable matrix norm, most of the variability of $\mathbf{Y}$ results from unobserved between-group heterogeneity. Common measures of how much variability may be addressed to unobserved between-group heterogeneity, are the following two coefficients of determination, derived from (6.3),

$$R_t^2(\boldsymbol{\vartheta}) = 1 - \mathrm{tr}\left(\sum_{k=1}^K \eta_k \boldsymbol{\Sigma}_k\right) / \mathrm{tr}\left(\mathrm{Var}(\mathbf{Y}|\boldsymbol{\vartheta})\right), \tag{6.4}$$

$$R_d^2(\boldsymbol{\vartheta}) = 1 - \left|\sum_{k=1}^K \eta_k \boldsymbol{\Sigma}_k\right| / |\mathrm{Var}(\mathbf{Y}|\boldsymbol{\vartheta})|. \tag{6.5}$$

For the limiting case of equal means, $\boldsymbol{\mu}_1 = \cdots = \boldsymbol{\mu}_K$, these coefficients are equal to 0, whereas these coefficients are close to one for well-separated groups; see Figure 6.1 for an illustration.

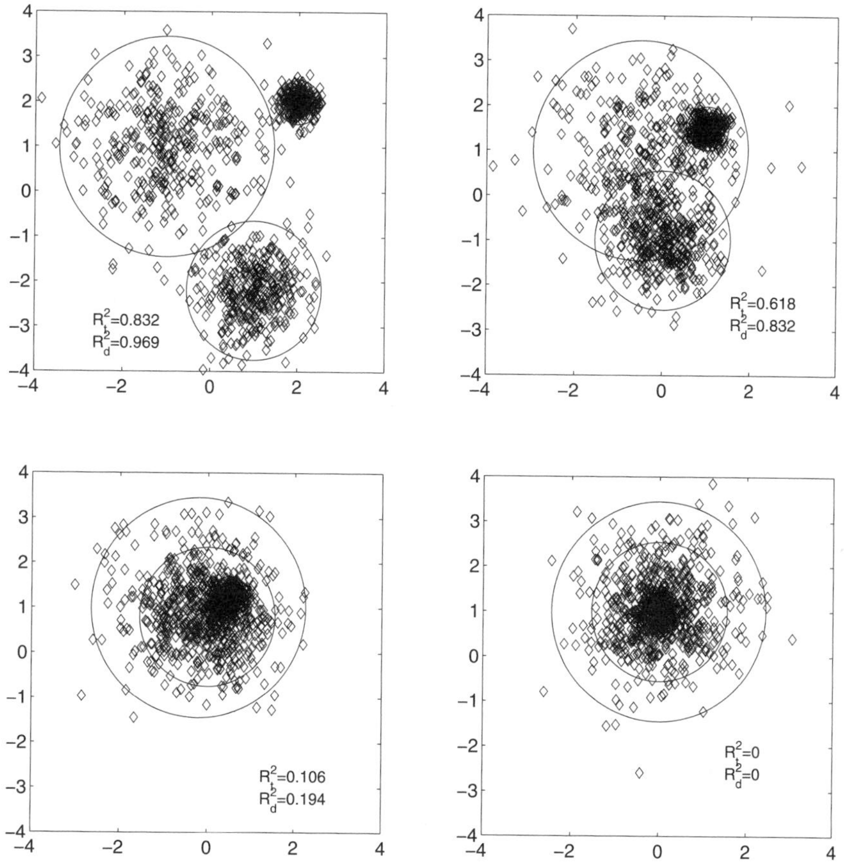


**Fig. 6.1.** 1000 observations simulated from a spherical mixture of three bivariate normal distributions with $\eta_1 = \eta_2 = \eta_3 = 1/3$, $(\sigma_1^2, \sigma_2^2, \sigma_3^2) = (1, 0.4, 0.04)$; with different locations for the means; $R_t^2$ and $R_d^2$ are coefficients of unobserved heterogeneity explained by the mixture, defined in (6.4) and (6.5)

### 6.1.2 Parameter Estimation for Mixtures of Normals

Suppose that a data set $\mathbf{y}$ is available, which consists of $N$ i.i.d. observations of a random variable distributed according to a mixture of normal distributions, thus for univariate data $\mathbf{y} = \{y_1, \dots, y_N\}$, whereas for multivariate data $\mathbf{y} = \{\mathbf{y}_1, \dots, \mathbf{y}_N\}$.

This subsection is concerned with the estimation of the component parameters and the weight distribution $\boldsymbol{\eta} = (\eta_1, \dots, \eta_K)$ of the underlying mixture distribution for fixed $K$, based on the data $\mathbf{y}$. For univariate mixtures the component means $\boldsymbol{\mu} = (\mu_1, \dots, \mu_K)$ and the component variances $\boldsymbol{\sigma^2} = (\sigma_1^2, \ldots, \sigma_K^2)$; for multivariate mixtures the component mean vectors $\boldsymbol{\mu} = (\boldsymbol{\mu}_1, \ldots, \boldsymbol{\mu}_K)$ and the component variance–covariance matrices $\boldsymbol{\Sigma} = (\boldsymbol{\Sigma}_1, \ldots, \boldsymbol{\Sigma}_K)$ have to be estimated.

Pioneering work on the estimation of mixtures of normals is based on the method of moments (Pearson, 1894; Charlier and Wicksell, 1924). The method of moments estimator suggested by Day (1969) turned out to be inefficient compared to maximum likelihood estimation, both for univariate as well as multivariate mixtures of normals. A more attractive method of moments estimator was suggested by Lindsay (1989) for univariate and by Lindsay and Basak (1993) for multivariate mixtures of normals.

Maximum likelihood (ML) estimation was used for a mixture of two univariate normal distributions with $\sigma_1^2 = \sigma_2^2$ as early as Rao (1948); further pioneering work was done by Hasselblad (1966). Wolfe (1970) suggested an iterative scheme for practical ML estimation of multivariate mixtures of normals that is essentially an early variant of the EM algorithm.

## EM Algorithm

Nowadays, the EM algorithm, which was introduced in Subsection 2.4.4, is the preferred method for practical ML estimation of univariate and multivariate mixtures. For univariate mixtures of normals the M-step reads:

$$\mu_k^{(m)} = \frac{1}{n_k} \sum_{i=1}^{N} \hat{D}_{ik}^{(m)} y_i,$$

$$(\sigma_k^2)^{(m)} = \frac{1}{n_k} \sum_{i=1}^{N} \hat{D}_{ik}^{(m)} \left(y_i - \mu_k^{(m)}\right)^2,$$

where $\hat{D}_{ik}{}^{(m)}$ and $n_k$ have been defined in (2.32) and (2.33), respectively. For multivariate mixtures the M-step reads:

$$\boldsymbol{\mu}_k^{(m)} = \frac{1}{n_k} \sum_{i=1}^{N} \hat{D}_{ik}^{(m)} \mathbf{y}_i,$$

$$\boldsymbol{\Sigma}_k^{(m)} = \frac{1}{n_k} \sum_{i=1}^{N} \hat{D}_{ik}^{(m)} \left(\mathbf{y}_i - \boldsymbol{\mu}_k^{(m)}\right) \left(\mathbf{y}_i - \boldsymbol{\mu}_k^{(m)}\right)'.$$

A certain difficulty with ML estimation is that the EM algorithm breaks down, whenever $(\sigma_k^2)^{(m)}$ is (numerically) zero or $\boldsymbol{\Sigma}_k{}^{(m)}$ is singular or nearly singular, which happens when $\hat{D}_{ik}{}^{(m)}$ is close to zero for too many observations. Then at the next iteration the computation of $\hat{D}_{ik}{}^{(m+1)}$ through (2.32) is no longer possible. Such difficulties arise in particular if the EM algorithm is applied to a finite mixture of normals overfitting the number of components.

### Unboundedness of the Mixture Likelihood Function

A further difficulty with ML estimation, first noted by Kiefer and Wolfowitz (1956) for univariate mixtures of normals, is that the mixture likelihood function

$$p(\mathbf{y}|\boldsymbol{\mu}, \boldsymbol{\sigma}^2, \boldsymbol{\eta}) = \prod_{i=1}^{N} \left( \sum_{k=1}^{K} \eta_k f_N(y_i; \mu_k, \sigma_k^2) \right) \tag{6.6}$$

is unbounded and has many local spurious modes; see Subsection 6.1.3 for an illustration. As first noted by Day (1969), the unboundedness of the mixture likelihood function is also relevant for the multivariate mixtures of normals, as each observation $\mathbf{y}_i$ gives rise to a singularity on the boundary of the parameter space.

Thus the ML estimator as global maximizer of the mixture likelihood function does not exist. Nevertheless, statistical theory outlined in Kiefer (1978) guarantees that a particular local maximizer of the mixture likelihood function is consistent, efficient, and asymptotically normal if the mixture is not overfitting. Several local maximizers may exist for a given sample, and a major difficulty with the ML approach is to identify if the correct one has been found. Titterington et al. (1985, pp.97), for instance, refer to various empirical studies that report convergence to singularities and spurious local modes using the EM algorithm, even if the true parameter was used as the starting value.

To avoid these problems, Hathaway (1985) considers constrained ML estimation of univariate mixtures of normals based on the inequality constraint

$$\min_{k,j} \frac{\sigma_k}{\sigma_j} \geq c > 0, \tag{6.7}$$

and proves strong consistency of the resulting estimator. For multivariate mixtures of normals, Hathaway (1985) suggests constraining all eigenvalues of $\boldsymbol{\Sigma}_k \boldsymbol{\Sigma}_j^{-1}$ to be greater than a positive constraint.

### Bayesian Parameter Estimation

Geisser and Cornfield (1963), Geisser (1964), and Binder (1978) give pioneering papers discussing a Bayesian approach to classification, clustering, and discrimination analysis based on multivariate mixtures, however, practical application was in general infeasible at that time. Lavine and West (1992) discuss practical Bayesian estimation of multivariate mixtures using data augmentation and Gibbs sampling. This approach is described in detail in Section 6.3.

The main difference between the ML approach and the Bayesian approach lies in the use of a proper prior distribution on the component parameter, in particular on the component variances, which usually takes the form of an inverted Gamma prior, $\sigma_k^2 \sim \mathcal{G}^{-1}(c_0, C_0)$, for univariate mixtures and an inverted Wishart prior, $\mathbf{\Sigma}_k^{-1} \sim \mathcal{W}_r(c_0, \mathbf{C}_0)$ with $c_0 > (r+1)/2$, for multivariate mixtures. This has two desirable effects in comparison to ML estimation.

First, within Gibbs sampling, which may be seen as a kind of Bayesian version of the EM algorithm, the conditional posterior distribution of $\sigma_k^2$ or $\mathbf{\Sigma}_k$ is always proper, and sampling yields a well-defined variance even if the corresponding group is empty or contains too few observations to obtain a well-defined sample variance or variance–covariance matrix.

Second, as mentioned above, it is complete ignorance about the variance ratio that causes the unboundedness of the mixture likelihood function, and again the Bayesian approach is helpful in this respect, as it allows us to include *some* prior information on this ratio, however vague this might be. In comparison to the mixture likelihood function the posterior density is much more regular. It is shown in Subsection 6.2.2 that for univariate mixtures of normals the ratio $\sigma_k^2\sigma_j^{-2}$ is bounded away from 0 under suitable inverted Gamma priors. For multivariate mixtures of normals the introduction of the inverted Wishart prior has a smoothing effect on the eigenvalues of $\mathbf{\Sigma}_k\mathbf{\Sigma}_j^{-1}$, which will be bounded away from 0. Hence the introduction of a proper prior on the component variances is related to the constraints introduced by Hathaway (1985).

### 6.1.3 The Kiefer–Wolfowitz Example

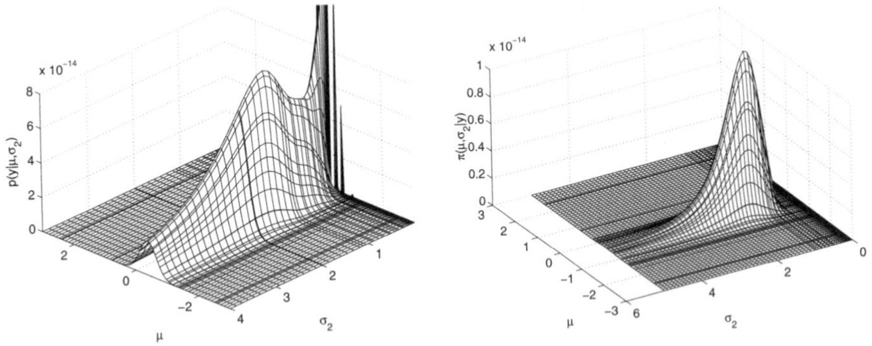


**Fig. 6.2.** KIEFER–WOLFOWITZ EXAMPLE — simulated data set ($N = 20$); surface plot of the mixture likelihood function $\log p(\mathbf{y}|\mu, \sigma_2)$ (left-hand side) in comparison to the posterior density $p(\mu, \sigma_2|\mathbf{y})$ under a $\mathcal{G}(1, 4)$-prior (right-hand side)

We reconsider the following mixture of two normal distributions,

$$Y \sim (1 - \eta_2)\mathcal{N}(\mu, 1) + \eta_2\mathcal{N}(\mu, \sigma_2^2), \tag{6.8}$$

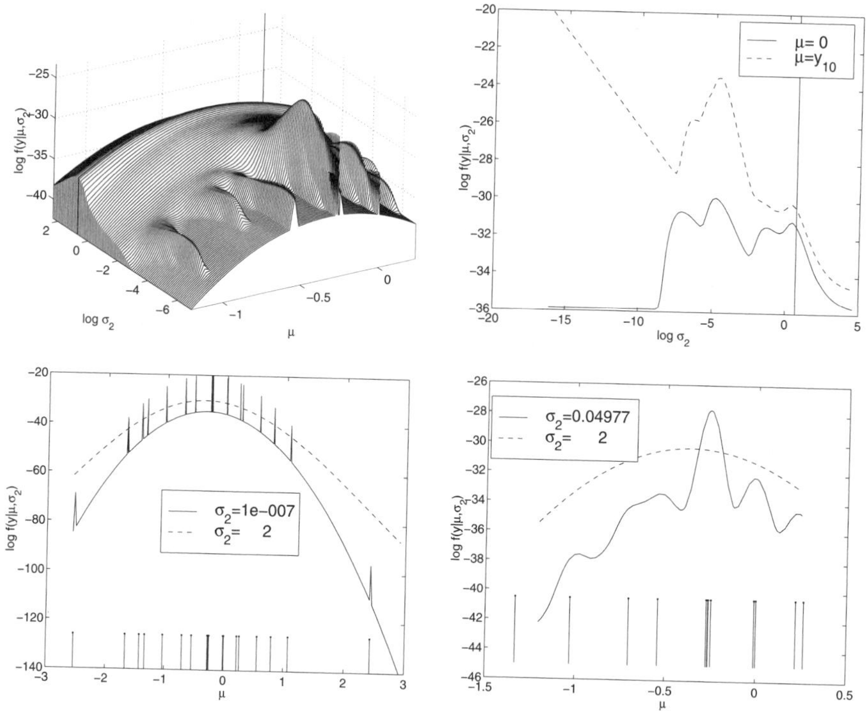


**Fig. 6.3.** KIEFER–WOLFOWITZ EXAMPLE — simulated data set ($N = 20$); various detailed views of the log mixture likelihood function $\log p(\mathbf{y}|\mu, \sigma_2)$; top left: zoom of surface plot for very small value of $\sigma_2$ (the vertical line indicates the position of the true parameters); top right: profile over $\sigma_2$ for $\mu$ fixed (full line: $\mu = \mu^{\text{true}} = 0$, dashed line: $\mu = y_{10} = -0.26627$; bottom: profile over $\mu$ for $\sigma_2$ fixed (dashed line: $\sigma_2 = \sigma_2^{\text{true}} = 2$, full line: $\sigma_2 = 10^{-7}$ (left-hand side) and $\sigma_2 = 0.05$ (right-hand side)); the bar diagram at the bottom of both figures shows the position of the data

where $\eta_2$ is fixed, whereas $\mu$ and $\sigma_2^2$ are unknown, which was used by Kiefer and Wolfowitz (1956) to show that each observation in an arbitrary data set $\mathbf{y} = (y_1, \dots, y_N)$, of arbitrary size $N$, generates a singularity in the mixture likelihood function (6.6). Whenever $\mu = y_i$, then as $\sigma_2^2 \to 0$, the mixture likelihood $p(\mathbf{y}|\mu = y_i, \sigma_2^2)$ is dominated by a term proportional to a constant times $1/\sigma_2^2$. Therefore:

$$\lim_{\sigma_2^2 \to 0} p(\mathbf{y}|\mu = y_i, \sigma_2^2) = \infty.$$

For illustration, we simulated $N = 20$ observation from model (6.8), with $\eta_2 = 0.2$, $\mu = 0$, and $\sigma_2^2 = 4$. The observations, ordered by size are:

| | | | | |
|---|---|---|---|---|
| −2.541 | −1.664 | −1.6636 | −1.4242 | −1.3335 |
| −1.029 | −0.70355 | −0.54121 | −0.27467 | −0.26627 |
| −0.26401 | −0.24944 | −0.011286 | −0.00081703 | 0.2189 |
| 0.26166 | 0.56176 | 0.79315 | 1.0727 | 2.4269 |

The left-hand side plot in Figure 6.2 shows the surface of the mixture likelihood function $p(\mathbf{y}|\mu, \sigma_2)$, regarded as a function of $\mu$ and $\sigma_2$ (rather than $\sigma_2^2$). There is a local mode around the true value $(\mu, \sigma_2) = (0, 2)$, however, the mixture likelihood is unbounded over a region corresponding to very small values of $\sigma_2$. Figure 6.3 zooms on this part of the parameter space, where we find many spurious local modes in the surface plot of $\log p(\mathbf{y}|\mu, \sigma_2)$. The profile over $\sigma_2$, for $\mu$ fixed at observation $y_{10} = -0.26627$, demonstrates that the log mixture likelihood is unbounded as $\sigma_2$ goes to 0, whereas for $\mu = \mu^{\text{true}} = 0$ (and any other value $\mu \neq y_i$), the log mixture likelihood is bounded as $\sigma_2$ goes to 0. The profile over $\mu$ for $\sigma_2$ fixed at the very small value $10^{-7}$ shows that the mixture likelihood surface has a spike whenever $\mu$ is close to one of the observations. For $\sigma_2$ fixed at the slightly larger value 0.05, which is still far from the true value, the profile over $\mu$ shows that any subset of observations which are sufficiently close to each other generate a local mode in the mixture likelihood. The local mode around $\mu$ equal to the average of the three similar observations $y_9, y_{10}, y_{11}$ has a likelihood value that is even bigger than the local mode around the true parameter value $(\mu, \sigma_2) = (0, 2)$.

This pathological part of the parameter corresponds to mixtures that fit one component to a small group of similar observations, whereas all other observations are assumed to belong the second component. The surface plot in Figure 6.2 shows that for this small data set the region corresponding to this pathological part of the likelihood is not well separated from the modal region around the true value. Thus the EM algorithm, or any other numerical method for maximizing this likelihood, certainly has a high risk of being trapped at one of the spurious local modes, or of diverging to 0, even if started at the true value.

Let us now assume that the mixture likelihood $p(\mathbf{y}|\mu, \sigma_2)$ is combined with the prior $p(\mu, \sigma_2^2) \propto p(\sigma_2^2)$, where $\sigma_2^2 \sim \mathcal{G}^{-1}(1, 4)$. We stay noninformative about $\mu$, which is a parameter common to both groups. The right-hand side of Figure 6.2 shows the posterior density $p(\mu, \sigma_2|\mathbf{y})$ under this prior for the simulated data set with $N = 20$. As shown in Subsection 6.2.2, the inverted Gamma prior introduces a constraint comparable to (6.7), and keeps the variance sufficiently bounded away from 0 to cut out all singularities and local modes, which were apparent for the comparable surface of the mixture likelihood $p(\mathbf{y}|\mu, \sigma_2)$ in Figure 6.2.

### 6.1.4 Applications of Mixture of Normal Distributions

Statistical modeling based on finite mixtures of normal distributions has a long history dating back to the application for outlier detection in Newcomb

(1886) and modeling unobserved discrete factors in a biological sample in Pearson (1894). Finite mixtures of normal distributions found widespread application in many areas of applied statistics such as biology, marketing, medicine, physics, and economics. More recently, mixture models are applied in bioinformatics and genetics, see, for instance, Delmar et al. (2005) and Tadesse et al. (2005).

Given such a tremendous amount of applied work, it seems impossible to provide an exhaustive list of important references. A comprehensive review of applications of univariate normal mixture models until the mid-eighties appears in Titterington et al. (1985, Chapter 2). More recent references are to be found in the comprehensive bibliography of McLachlan and Peel (2000). Additional useful references appear in Böhning (2000) and in a special issue of the journal *Computational Statistics and Data Analysis* (2003) publishing selected papers from a 2001 Workshop on finite mixture modeling in Hamburg.

Chapter 7 discusses several statistical applications of mixtures of normal distributions in such diverse areas as cluster analysis, outlier detection, discriminant analysis, and density estimation.

## 6.2 Bayesian Estimation of Univariate Mixtures of Normals

### 6.2.1 Bayesian Inference When the Allocations Are Known

This subsection derives the posterior distribution of $\mu_k, \sigma_k^2$ given the complete data $\mathbf{S}, \mathbf{y}$, by combining the information from all observations in group $k$. The derivation is based on a standard result in Bayesian inference outlined, for instance, in Antelman (1997) and Box and Tiao (1973).

Relevant group-specific quantities are the number $N_k(\mathbf{S})$ of observations in group $k$, the group mean $\overline{y}_k(\mathbf{S})$, and the within-group variance $s_{y,k}^2(\mathbf{S})$:

$$N_k(\mathbf{S}) = \#\{i : S_i = k\}, \tag{6.9}$$

$$\overline{y}_k(\mathbf{S}) = \frac{1}{N_k(\mathbf{S})} \sum_{i:S_i=k} y_i, \tag{6.10}$$

$$s_{y,k}^2(\mathbf{S}) = \frac{1}{N_k(\mathbf{S})} \sum_{i:S_i=k} (y_i - \overline{y}_k(\mathbf{S}))^2. \tag{6.11}$$

Note that any of these quantities depends on the classification $\mathbf{S}$.

Assume that for observation $y_i$ the allocation $S_i$ is equal to $k$. Then the observational model for observation $y_i$ is a normal distribution with mean $\mu_k$ and variance $\sigma_k^2$, and the contribution of $y_i$ to the complete-data likelihood function $p(\mathbf{y}|\boldsymbol{\mu}, \boldsymbol{\sigma}^2, \mathbf{S})$ is equal to:

$$\frac{1}{\sqrt{2\pi\sigma_k^2}} \exp\left(-\frac{1}{2\sigma_k^2}(y_i - \mu_k)^2\right).$$

After combining the information from all observations, we obtain a complete-data likelihood function with $K$ independent factors, each carrying all information about the parameters in a certain group:

$$p(\mathbf{y}|\boldsymbol{\mu},\boldsymbol{\sigma}^2,\mathbf{S}) = \prod_{k=1}^{K}\prod_{i:S_i=k}\left(\frac{1}{2\pi\sigma_k^2}\right)^{N_k(\mathbf{S})/2}\exp\left(-\frac{1}{2}\sum_{i:S_i=k}\frac{(y_i-\mu_k)^2}{\sigma_k^2}\right). \tag{6.12}$$

In a Bayesian analysis each of these factors is combined with a prior. When holding the variance $\sigma_k^2$ fixed, the complete-data likelihood function, regarded as a function of $\mu_k$, is the kernel of a univariate normal distribution. Under the conjugate prior $\mu_k \sim \mathcal{N}(b_0, B_0)$, the posterior density of $\mu_k$ given $\sigma_k^2$ and $N_k(\mathbf{S})$ observations assigned to this group, is again a density from the normal distribution, $\mu_k|\sigma_k^2, \mathbf{S}, \mathbf{y} \sim \mathcal{N}(b_k(\mathbf{S}), B_k(\mathbf{S}))$, where

$$B_k(\mathbf{S})^{-1} = B_0^{-1} + \sigma_k^{-2}N_k(\mathbf{S}), \tag{6.13}$$

$$b_k(\mathbf{S}) = B_k(\mathbf{S})(\sigma_k^{-2}N_k(\mathbf{S})\overline{y}_k(\mathbf{S}) + B_0^{-1}b_0), \tag{6.14}$$

with $N_k(\mathbf{S})\overline{y}_k(\mathbf{S})$ being defined as zero for an empty group with $N_k(\mathbf{S}) = 0$.

When holding the mean $\mu_k$ fixed, the complete-data likelihood function, regarded as a function in $\sigma_k^2$, is the kernel of an inverted Gamma density; see Appendix A.1.6 for more details on this distribution family. Under the conjugate inverted Gamma prior $\sigma_k^2 \sim \mathcal{G}^{-1}(c_0, C_0)$, the posterior density of $\sigma_k^2$ given $\mu_k$ and $N_k(\mathbf{S})$ observations assigned to this group is again a density from an inverted Gamma distribution, $\sigma_k^2|\mu_k, \mathbf{S}, \mathbf{y} \sim \mathcal{G}^{-1}(c_k(\mathbf{S}), C_k(\mathbf{S}))$, where

$$c_k(\mathbf{S}) = c_0 + \frac{1}{2}N_k(\mathbf{S}), \tag{6.15}$$

$$C_k(\mathbf{S}) = C_0 + \frac{1}{2}\sum_{i:S_i=k}(y_i-\mu_k)^2. \tag{6.16}$$

If both $\mu_k$ and $\sigma_k^2$ are unknown, a closed-form solution for the joint posterior $p(\mu_k, \sigma_k^2|\mathbf{S}, \mathbf{y})$ exists only if the prior variance of $\mu_k$ depends on $\sigma_k^2$ through $B_{0,k} = \sigma_k^2/N_0$. Then the joint posterior $p(\boldsymbol{\mu}, \boldsymbol{\sigma}^2|\mathbf{S}, \mathbf{y})$ factors as

$$p(\boldsymbol{\mu}, \boldsymbol{\sigma}^2|\mathbf{S}, \mathbf{y}) = \prod_{k=1}^{K} p(\mu_k|\sigma_k^2, \mathbf{y}, \mathbf{S})p(\sigma_k^2|\mathbf{y}, \mathbf{S}), \tag{6.17}$$

where the posterior density of $\mu_k$ given $\sigma_k^2$ arises from an $\mathcal{N}(b_k(\mathbf{S}), B_k(\mathbf{S}))$ distribution with

$$B_k(\mathbf{S}) = \frac{1}{N_k(\mathbf{S}) + N_0}\sigma_k^2, \tag{6.18}$$

$$b_k(\mathbf{S}) = \frac{N_0}{N_k(\mathbf{S}) + N_0}b_0 + \frac{N_k(\mathbf{S})}{N_k(\mathbf{S}) + N_0}\overline{y}_k(\mathbf{S}), \tag{6.19}$$

whereas the marginal posterior of $\sigma_k^2$ is the inverted Gamma distribution $\mathcal{G}^{-1}\left(c_k(\mathbf{S}), C_k(\mathbf{S})\right)$, where $c_k(\mathbf{S})$ is the same as in (6.15), however,

$$C_k(\mathbf{S}) = C_0 + \frac{1}{2}\left(N_k(\mathbf{S})s_{y,k}^2(\mathbf{S}) + \frac{N_k(\mathbf{S})N_0}{N_k(\mathbf{S}) + N_0}(\overline{y}_k(\mathbf{S}) - b_0)^2\right). \quad (6.20)$$

### 6.2.2 Standard Prior Distributions

Diebolt and Robert (1994), Raftery (1996b), and Bensmail et al. (1997), among many others, consider the *conditionally conjugate prior*

$$p(\mu_1, \dots, \mu_K, \sigma_1^2, \dots, \sigma_K^2) = \prod_{k=1}^{K} p(\mu_k|\sigma_k^2)p(\sigma_k^2), \quad (6.21)$$

where $\mu_k|\sigma_k^2 \sim \mathcal{N}\left(b_0, \sigma_k^2/N_0\right)$ and $\sigma_k^2 \sim \mathcal{G}^{-1}\left(c_0, C_0\right)$. This prior implies that a priori the component parameters $\boldsymbol{\theta}_k = (\mu_k, \sigma_k^2)$ are pairwise independent across the groups, whereas within each group $\mu_k$ and $\sigma_k^2$ are dependent. As explained in Subsection 6.2.1, this prior offers the advantage of being conditionally conjugate with respect to the complete-data likelihood function, leading to a closed-form posterior $p(\boldsymbol{\mu}, \boldsymbol{\sigma}^2|\mathbf{S}, \mathbf{y})$.

For practical Bayesian estimation the hyperparameters $\boldsymbol{\delta} = (b_0, N_0, c_0, C_0)$ have to be selected carefully, as they may exercise considerable influence on the posterior distribution. A proper prior results if $N_0 > 0$ and $c_0 > 0$. Raftery (1996b) uses the following data-dependent hyperparameters: $b_0 = \overline{y}$, $N_0 = 2.6/(y_{\max} - y_{\min})$, $c_0 = 1.28$, and $C_0 = 0.36s_y^2$, whereas Bensmail et al. (1997) use $b_0 = \overline{y}$, $N_0 = 1$, $c_0 = 2.5$, and $C_0 = 0.5s_y^2$.

Another commonly used prior, called the *independence prior*, assumes that $\mu_k$ and $\sigma_k^2$ are independent a priori:

$$p(\mu_1, \dots, \mu_K, \sigma_1^2, \dots, \sigma_K^2) = \prod_{k=1}^{K} p(\mu_k) \prod_{k=1}^{K} p(\sigma_k^2), \quad (6.22)$$

where $\mu_k \sim \mathcal{N}\left(b_0, B_0\right)$ and $\sigma_k^2 \sim \mathcal{G}^{-1}\left(c_0, C_0\right)$. This prior, where the posterior $p(\boldsymbol{\mu}, \boldsymbol{\sigma}^2|\mathbf{S}, \mathbf{y})$ is no longer of closed form, has been applied, for instance, in Escobar and West (1995) and Richardson and Green (1997). More general priors, where some prior dependence among the component parameters $\boldsymbol{\theta}_k = (\mu_k, \sigma_k^2)$ is introduced, are discussed in Subsection 6.2.6.

### 6.2.3 The Influence of the Prior on the Variance Ratio

It is a definite advantage of the Bayesian approach in the context of normal mixture models, that selecting a prior on $\sigma_1^2, \dots, \sigma_K^2$ corresponds to imposing a structure on the variances, which keeps them sufficiently away from 0 to overrule spurious local modes and the unboundedness of the mixture likelihood function. Under the inverted Gamma prior $\sigma_k^2 \sim \mathcal{G}^{-1}(c_0, C_0)$, the ratio of any two variances $\sigma_k^2$ and $\sigma_l^2$ with $k \neq l$ follows an $\mathrm{F}(2c_0, 2c_0)$-distribution:

$$\frac{\sigma_k^2}{\sigma_l^2} \sim \mathrm{F}(2c_0, 2c_0),$$

which is obtained using result (A.10) in Appendix A.1.5, after rewriting the inverted Gamma prior as $\sigma_k^2 \sim (2c_0)C_0/W_k$, $W_k \sim \chi^2_{2c_0}$. As the prior is invariant, the ratio $\sigma_k^2/\sigma_l^2$ has the same distribution for any combination of $k$ and $l$ with $k \neq l$.

The choice of $c_0$ has considerable influence on the ratio of any two variances and could be chosen under the aspect of keeping this ratio within certain limits. It is now relevant, how close this ratio is allowed to be to zero for two variances $\sigma_k^2$ and $\sigma_l^2$, where $\sigma_k^2 < \sigma_l^2$. From the properties of the $\mathrm{F}(2c_0, 2c_0)$-distribution discussed in Appendix A.1.5, this density behaves as

$$\left(\frac{\sigma_k^2}{\sigma_l^2}\right)^{c_0-1}$$

as $\sigma_k^2/\sigma_l^2 \to 0$. The density is unbounded at 0 for $c_0 < 1$ and the ratio may be arbitrarily close to 0. The density is bounded at 0 for $c_0 = 1$ and is equal to 0 for $c_0 > 1$. For $1 \leq c_0 < 2$, the first derivative of the density is infinite at 0, and the ratio may be rather close to 0. For $c_0 = 2$, the first derivative is bounded, whereas for $c_0 > 2$ the first derivative is equal to 0 and the ratio is bounded away from 0. Furthermore the ratio has a finite variance if $c_0 > 2$.

Thus the $\mathrm{F}(2c_0, 2c_0)$-distribution is a stochastic version of the deterministic constraint (6.7) introduced by Hathaway (1985). In Table 6.1 we derive a $c$ for which a constraint comparable to (6.7) holds with high probability, say 95% for $K = 2$. Note that $c$ is simply given by the 95%-percentile of the $\mathrm{F}(2c_0, 2c_0)$-distribution. For $c_0 = 2.5$, for instance, this ratio is roughly bounded by 0.1.

**Table 6.1.** Lower bound for the ratio of two variances depending on the hyperparameter $c_0$ of the inverted Gamma prior

| $c_0$ | 0.5 | 1 | 1.5 | 2 | 2.5 | 3 | 4 | 5 |
|---|---|---|---|---|---|---|---|---|
| $c$ | 0.00025 | 0.010 | 0.034 | 0.063 | 0.091 | 0.118 | 0.166 | 0.206 |

### 6.2.4 Bayesian Estimation Using MCMC

Many authors, in particular Diebolt and Robert (1994) and Raftery (1996b), have considered Gibbs sampling based on data augmentation, as described for general mixtures in Section 3.5, to estimate the parameters of a univariate mixture of normals. Full conditional Gibbs sampling proceeds along the lines indicated by *Algorithm 3.4*, where the results of Subsection 6.2.1 are used to sample the mean $\mu_k$ and variance $\sigma_k^2$ in each group, conditional upon a given allocation vector $\mathbf{S}$.

*Algorithm 6.1: Two-Block Gibbs Sampling for a Univariate Gaussian Mixture*

(a) Parameter simulation conditional on the classification $\mathbf{S} = (S_1, \dots, S_N)$:
  (a1) Sample $\boldsymbol{\eta}$ from the conditional Dirichlet posterior $p(\boldsymbol{\eta}|\mathbf{S})$ as in *Algorithm 3.4*.
  (a2) Sample $\sigma_k^2$ in each group $k$ from a $\mathcal{G}^{-1}\left(c_k(\mathbf{S}), C_k(\mathbf{S})\right)$-distribution.
  (a3) Sample $\mu_k$ in each group $k$ from an $\mathcal{N}\left(b_k(\mathbf{S}), B_k(\mathbf{S})\right)$-distribution.

(b) Classification of each observation $y_i$, for $i = 1, \dots, N$, conditional on knowing $\boldsymbol{\mu}, \boldsymbol{\sigma^2}$, and $\boldsymbol{\eta}$:

$$\Pr(S_i = k|\boldsymbol{\mu}, \boldsymbol{\sigma^2}, \boldsymbol{\eta}, y_i) \propto \frac{1}{\sqrt{2\pi\sigma_k^2}} \exp\left\{-\frac{(y_i - \mu_k)^2}{2\sigma_k^2}\right\} \eta_k. \quad (6.23)$$

The precise form of $b_k(\mathbf{S})$, $B_k(\mathbf{S})$, $c_k(\mathbf{S})$, and $C_k(\mathbf{S})$ in steps (a2) and (a3) depends upon the chosen prior distribution family. For the independence prior $b_k(\mathbf{S})$ and $B_k(\mathbf{S})$ are given by (6.13) and (6.14), whereas $c_k(\mathbf{S})$ and $C_k(\mathbf{S})$ are available from (6.15) and (6.16). Under the conditionally conjugate prior, $\mu_k$ and $\sigma_k^2$ are sampled jointly, as the marginal density $\sigma_k^2|\mathbf{S}, \mathbf{y}$ is available in closed form. $b_k(\mathbf{S})$ and $B_k(\mathbf{S})$ are given by (6.18) and (6.19), whereas $c_k(\mathbf{S})$ and $C_k(\mathbf{S})$ are available from (6.15) and (6.20).

*Algorithm 6.1* is started with parameter estimation based on some preliminary classification, which may be obtained from partitioning the ordered data into $K$ groups. The sampling order may be reversed, by starting with a classification based on the starting values $\eta_k = 1/K$, $\mu_k = Q_{k/K+1}$, where $Q_\alpha$ is the empirical $\alpha$-percentile of the data, and $\sigma_k^2 = R^2$, where $R$ is a robust estimator of the scale, such as the median absolute deviation, $R = \text{med}\, |y_i - \text{med}\, y|$, or the interquartile range, $R = Q_{0.75} - Q_{0.25}$. Alternatively, one could adopt random starting values as in McLachlan and Peel (2000, Section 2.12.2), where $\mu_k$ is sampled from $\mathcal{N}\left(\overline{y}, s_y^2\right)$ and $\sigma_k^2 = s_y^2$.

## Classification Without Parameter Estimation

As already discussed in Subsection 3.3.3, Chen and Liu (1996) showed that for many mixture models Bayesian clustering is possible without explicit parameter estimation, using directly the marginal posterior $p(\mathbf{S}|\mathbf{y})$ given by $p(\mathbf{S}|\mathbf{y}) \propto p(\mathbf{y}|\mathbf{S})p(\mathbf{S})$. For a mixture of univariate normal distributions, $p(\mathbf{y}|\mathbf{S})$ is given by

$$p(\mathbf{y}|\mathbf{S}) \propto \prod_{k=1}^{K} \frac{\Gamma(c_k(\mathbf{S}))}{(C_k(\mathbf{S}))^{c_k(\mathbf{S})}\sqrt{N_k(\mathbf{S}) + N_0}}, \quad (6.24)$$

where $c_k(\mathbf{S})$, and $C_k(\mathbf{S})$ are the posterior moments of the full conditional posterior density $p(\sigma_k^2|\mathbf{y}, \mathbf{S})$ under the conditionally conjugate prior.

### 6.2.5 MCMC Estimation Under Standard Improper Priors

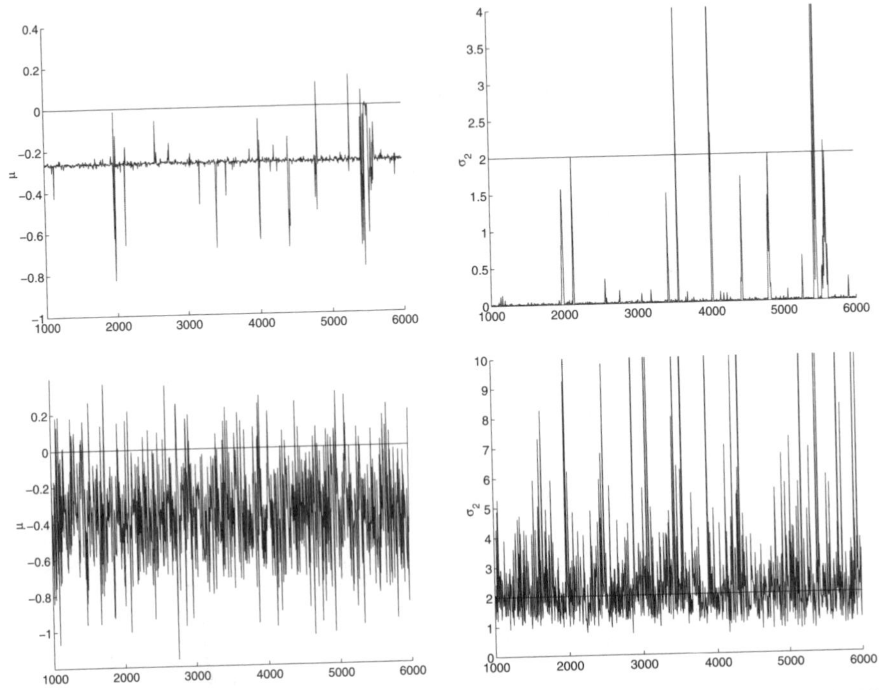


**Fig. 6.4.** KIEFER–WOLFOWITZ EXAMPLE — simulated data set ($N = 20$); MCMC draws of $\mu$ and $\sigma_2$ from the posterior $p(\mu, \sigma_2|\mathbf{y})$ based on different priors; top: improper prior $p(\mu, \sigma_2^2) \propto 1/\sigma_2^2$; bottom: $p(\mu) \propto$ constant, $\sigma_2^2 \sim \mathcal{G}^{-1}(1, 4)$; the horizontal line indicates the true values

One has to be very careful when running MCMC under an improper prior, because such a prior may cause an improper mixture posterior distribution, as discussed in Subsection 3.2.2. We also refer to Hobert and Casella (1998) for a general discussion of MCMC simulations from improper posterior distributions.

Nevertheless, Diebolt and Robert (1994) suggested using improper priors and to reject classifications $\mathbf{S}$ which lead to problems. Under the improper prior $p(\mu_k, \sigma_k^2) \propto 1/\sigma_k^2$, for instance, the posterior moments in steps (a2) and (a3) of *Algorithm 6.1* are given by $b_k(\mathbf{S}) = \overline{y}_k(\mathbf{S})$, $B_k(\mathbf{S}) = 1/N_k(\mathbf{S})$, $c_k(\mathbf{S}) = (N_k(\mathbf{S}) - 1)/2$, and $C_k(\mathbf{S}) = N_k(\mathbf{S})/2s_{y,k}^2(\mathbf{S})$. Thus classifications are rejected whenever one group is empty ($N_k(\mathbf{S}) = 0$), or whenever the within-group variance $s_{y,k}^2(\mathbf{S})$ is 0, either because the group contains only a single observation or two observations with identical value, as may happen for rounded data.

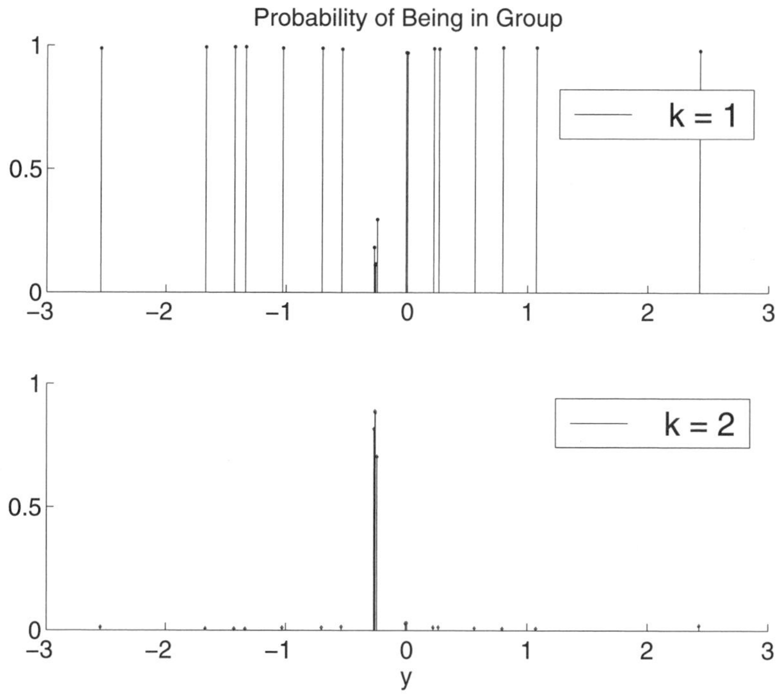


**Fig. 6.5.** KIEFER–WOLFOWITZ EXAMPLE — simulated data set ($N = 20$); posterior classification of the observations for the improper prior $p(\mu, \sigma_2^2) \propto 1/\sigma_2^2$

Whereas these classifications are easily detected, it is hard to identify classifications that assign a few similar observations into one group, and act as trapping states for the sampler around a local mode of the mixture likelihood function. As these spurious local modes disappear, when choosing a proper prior on the variances, the problem of trapping states of the MCMC sampler is easy to circumvent, and the use of an improper prior in combination with MCMC estimation of finite mixtures is generally not recommended.

## MCMC Estimation of the Kiefer–Wolfowitz Example

For further illustration, we discuss MCMC estimation for the data simulated for the Kiefer–Wolfowitz example, discussed earlier in Subsection 6.1.3, under the improper prior $p(\mu, \sigma_2^2) \propto 1/\sigma_2^2$. As recommended by Diebolt and Robert (1994), classifications $\mathbf{S}$ leading to empty and zero-variance groups are rejected, nevertheless the MCMC draws in Figure 6.4 clearly indicate that the sampler is trapped at one of the local modes of the mixture likelihood function found earlier in Figure 6.3. The posterior classifications in Figure 6.5 show that the sampler with high probability forms one group from three observations that are extremely close to each other. Including a $\mathcal{G}^{-1}(1, 4)$-prior on $\sigma_2^2$

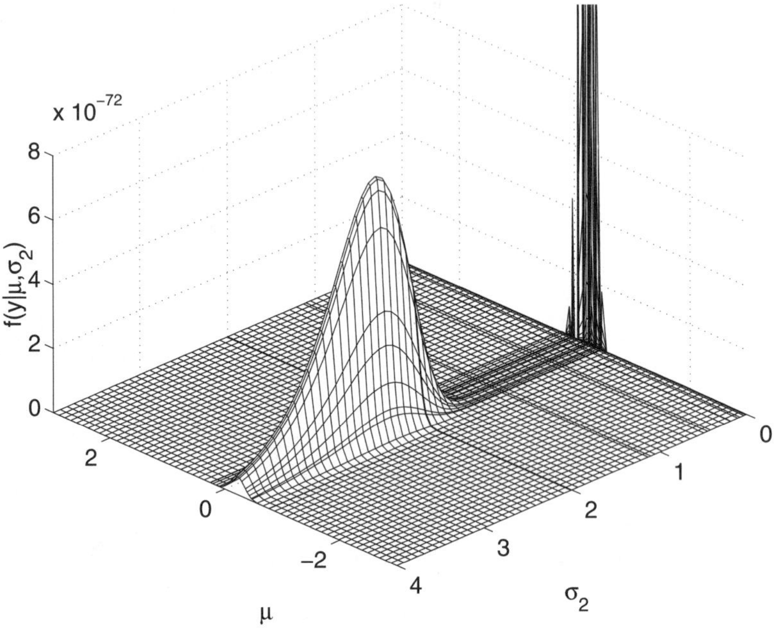


**Fig. 6.6.** KIEFER–WOLFOWITZ EXAMPLE — simulated data; surface of the nonnormalized posterior $p(\mu, \sigma_2|\mathbf{y})$ under the improper prior $p(\mu, \sigma_2^2) \propto 1/\sigma_2^2$ for $N = 100$

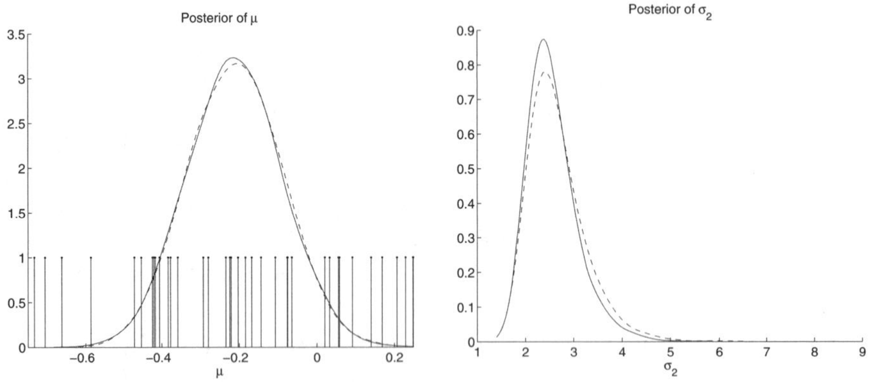


**Fig. 6.7.** KIEFER–WOLFOWITZ EXAMPLE — simulated data set ($N = 100$); posterior densities $p(\mu|\mathbf{y})$ and $p(\sigma_2|\mathbf{y})$ estimated from the MCMC draws for the improper prior (dashed line) and the $\mathcal{G}(1, 4)$-prior (full line)

improves the performance of the MCMC sampler considerably. The bottom of Figure 6.4 indicates that we now obtain draws from the modal region close to the true value.

For a larger data set with $N = 100$ rather than $N = 20$, it is possible to run an MCMC sampler without being trapped even for the improper prior. The risk of jumping into the critical area is very small for this data set; see Figure 6.6. When comparing the marginal posterior densities $p(\mu|\mathbf{y})$ and $p(\sigma_2|\mathbf{y})$ estimated from the MCMC draws in Figure 6.7, we find little difference between the improper posterior based on a noninformative prior and the proper posterior based on the $\mathcal{G}^{-1}(1,4)$-prior. As discussed earlier in Subsection 3.5.4, MCMC sampling under an improper prior may lead to a proper posterior density estimate if the sampler avoids the critical part of the parameter space, simply because it is well separated from the modal region around the true value. However, it is not recommended to rely on this, as an unaccountable risk remains that the sampler jumps to this critical part when the MCMC sampler is run long enough.

### 6.2.6 Introducing Prior Dependence Among the Components

Rather than assuming fixed hyperparameters $\boldsymbol{\delta}$, an alternative method is to treat $\boldsymbol{\delta}$ as an unknown hyperparameter with a prior $p(\boldsymbol{\delta})$ of its own. As mentioned in Subsection 3.2.4, this introduces prior dependence among the component parameters. Such priors have been applied to finite mixtures of univariate normal distributions in particular by Richardson and Green (1997) and Roeder and Wasserman (1997b).

In Richardson and Green (1997) the hyperparameter $C_0$ is treated as an unknown hyperparameter with a prior of its own, whereas the other hyperparameters remain fixed. The joint prior $p(\mu_1, \ldots, \mu_K, \sigma_1^2, \ldots, \sigma_K^2, C_0)$ takes the form of a *hierarchical independence prior*:

$$p(\mu_1, \ldots, \mu_K, \sigma_1^2, \ldots, \sigma_K^2, C_0) = \prod_{k=1}^{K} p(\mu_k) \prod_{k=1}^{K} p(\sigma_k^2|C_0)p(C_0), \quad (6.25)$$

where $\mu_k \sim \mathcal{N}(b_0, B_0)$, $\sigma_k^2 \sim \mathcal{G}^{-1}(c_0, C_0)$, and $C_0 \sim \mathcal{G}(g_0, G_0)$. For their case studies, Richardson and Green (1997) select $c_0 = 2$, $g_0 = 0.2$, $G_0 = 10/R^2$, $b_0 = m$, and $B_0 = R^2$, where $m$ and $R$ are the midpoint and the length of the observation interval.

Roeder and Wasserman (1997b) extend this concept further by using hierarchical priors with improper priors for the hyperparameters. They suggest a partly proper prior for the variances that is equal to the hierarchical prior (6.25), with $p(C_0) \propto C_0^{-1}$ being the standard improper prior for the scale parameter. Roeder and Wasserman (1997b) prove that the marginal prior distribution of each $\sigma_k^2$ has the usual improper reference prior, $p(\sigma_k^2) \propto \sigma_k^{-2}$, nevertheless the posterior distribution is proper. Roeder and Wasserman (1997b) suggest also a hierarchical prior for the means. They assume that the mean

$b_0$ appearing in the normal prior on the group mean $\mu_k$ is an unknown hyperparameter with improper prior: $p(b_0) \propto$ constant.

Concerning MCMC estimation, under any hierarchical prior an additional step has to be added to sample the random hyperparameters $\boldsymbol{\delta}$. For the hierarchical prior (6.25), for instance, the marginal prior $p(\sigma_1^2, \ldots, \sigma_K^2)$, where $C_0$ is integrated out, is not conjugate for the complete-data likelihood, and an additional step has to be added in *Algorithm 6.1*, to sample $C_0$ from $p(C_0|\mathbf{S}, \boldsymbol{\mu}, \boldsymbol{\sigma^2}, \mathbf{y})$, which is given by Bayes' theorem as

$$p(C_0|\mathbf{S}, \boldsymbol{\mu}, \boldsymbol{\sigma^2}, \mathbf{y}) \propto \prod_{k=1}^{K} p(\sigma_k^2|C_0)p(C_0) \propto$$

$$\prod_{k=1}^{K} \left( C_0^{c_0} \exp\left\{ -\frac{C_0}{\sigma_k^2} \right\} \right) C_0^{g_0-1} \exp\{-G_0 C_0\} .$$

Obviously this is the kernel of a $\mathcal{G}\left(g_N, G_N\right)$- density with:

$$g_N = g_0 + Kc_0, \qquad G_N = G_0 + \sum_{k=1}^{K} \frac{1}{\sigma_k^2}.$$

If $b_0$ is an unknown hyperparameter with improper prior $p(b_0) \propto$ constant, then $b_0$ is sampled in an additional step from $p(b_0|\mathbf{S}, \boldsymbol{\mu}, \boldsymbol{\sigma^2}, \mathbf{y})$, given by

$$b_0|\boldsymbol{\mu}, \boldsymbol{\sigma^2}, \mathbf{y}, \mathbf{S} \sim \mathcal{N}\left( \frac{1}{K} \sum_{k=1}^{K} \mu_k, B_0^{-1}/K \right).$$

Finally, under a hierarchical prior some starting value for the random hyperparameter $\boldsymbol{\delta}$ has to be provided.

### Prior Dependence Through Reparameterization

Mengersen and Robert (1996) suggest reparameterizing a mixture of two normal distributions in terms of a perturbation of a global normal distribution $\mathcal{N}\left(\mu, \tau^2\right)$:

$$Y \sim p\mathcal{N}\left(\mu, \tau^2\right) + (1-p)\mathcal{N}\left(\mu + \tau\delta, \tau^2\omega^2\right).$$

Robert and Mengersen (1999) extend this idea to the general K-component normal mixture. They introduce a sequential parameterization, where for $k > 1$ the $k$th component is expressed as a local perturbation of component $k-1$:

$$\begin{aligned} Y \sim p\mathcal{N}\left(\mu, \tau^2\right) + \sum_{k=1}^{K-2} (1-p)(1-q_1)\cdots(1-q_{k-1})q_k\mathcal{N}(\mu + \tau\delta_1 + \\ \cdots + \tau\cdots\omega_{k-1}\delta_k, \tau^2\omega_1\cdots\omega_k) + (1-p)(1-q_1)\cdots(1-q_{K-2})\mathcal{N}(\mu + \\ \tau\delta_1 + \cdots + \tau\cdots\omega_{K-2}\delta_{K-1}, \tau^2\omega_1\cdots\omega_{K-1}), \end{aligned} \tag{6.26}$$

with the convention $\omega_0 = 1$ and $q_0 = 0$. This parameterization allows us to choose partly proper priors. The prior on $\mu$ and $\tau$ is improper, $p(\mu, \tau) \propto 1/\tau$, the prior on $p$ and $q_1, \ldots, q_{K-2}$ is uniform, and the prior on $\delta_k$ is slightly informative assuming $\delta_k \sim \mathcal{N}\left(0, \delta^2\right)$ for $k = 1, \ldots, K-1$. The prior on $\omega_k$ is either uniform, $\omega_k \sim \mathcal{U}\left[0,1\right]$, to force the component variance to be decreasing in $k$, or equal to $\omega_k \sim 0.5\mathcal{U}\left[0,1\right] + 0.5Pa(1,2)$, where $Pa(1,2)$ is uniform on $1/\omega_k^2$. Note that $\delta^2$, the prior variance of $\delta_k$, is the only parameter that remains for tuning. Robert and Titterington (1998) express this prior in terms of a prior on $\mu_1, \ldots, \mu_K$ and $\sigma_1^2, \ldots, \sigma_K^2$ in the standard parameterization:

$$p(\mu_1, \ldots, \mu_K, \sigma_1^2, \ldots, \sigma_K^2) \propto \frac{\sigma_K^2}{\sigma_1^3} \prod_{k=2}^{K} \frac{1}{\sigma_k^2} \exp\{-1/2 \frac{(\mu_k - \mu_{k-1})^2}{2\delta^2 \sigma_{k-1}^2}\},$$

and show that this partly proper prior leads to a proper posterior distribution.

Concerning MCMC estimation, Mengersen and Robert (1996) show how full conditional Gibbs sampling may be applied for $K = 2$. For general $K$, Robert and Mengersen (1999) use a hybrid MCMC method combining Gibbs steps for $p$, $q_1, \ldots, q_{K-2}$, $\mu$, $\delta_1, \ldots, \delta_{K-1}$, and $\omega_{K-1}^2$ with Metropolis–Hastings steps based on tailor-made proposals for $\omega_1^2, \ldots, \omega_{K-2}^2$ and $\tau$. Robert and Titterington (1998) show for the more general setting of hidden Markov models, of which mixture models are a special case, that full conditional Gibbs sampling may be applied for all parameters in this model.

### A Markov Prior Based on Ordered Means

Roeder and Wasserman (1997b) suggest a Markov prior based on the ordered means. Under the assumption that $\mu_1 < \cdots < \mu_K$, they assume that $p(\mu_1) \propto$ constant, whereas each $\mu_k$ with $k > 1$ has a normal prior with mean $\mu_{k-1}$ that is truncated at $\mu_{k-1}$:

$$\mu_k | \mu_{k-1}, \sigma_k^2, \sigma_{k-1}^2 \sim \mathcal{N}\left(\mu_{k-1}, \phi^2(\sigma_{k-1}^{-2} + \sigma_k^{-2})/2\right) I_{\{\mu_k > \mu_{k-1}\}},$$

where $\phi = 5$ is a default value in their applications. MCMC sampling under this prior may be carried out by running a single-move Gibbs sampler. For more details on the corresponding conditional densities we refer to Roeder and Wasserman (1997b). As pointed out by Roeder and Wasserman (1997b), this prior makes it difficult to handle mixtures, where two components have the same mean.

### 6.2.7 Further Sampling-Based Approaches

MCMC methods are not practicable in a dynamic setting when a sequence of posterior densities $p(\boldsymbol{\vartheta}|\mathbf{y}^t)$, based on a set of observations $\mathbf{y}^t = (\mathbf{y}_1, \ldots, \mathbf{y}_t)$ which increases with $t$ is involved, because MCMC methods require rerunning a new chain for each $p(\boldsymbol{\vartheta}|\mathbf{y}^t)$ without using draws from $p(\boldsymbol{\vartheta}|\mathbf{y}^{t-1})$. In such a setting, sequential importance sampling schemes, usually referred to as a "particle filter method" are used which are reviewed, for instance, in Doucet et al. (2001). Recently, Chopin (2002) showed that sequential importance sampling can be useful also for estimation in a static scenario and demonstrates this through an application to a mixture of normal distributions.

Other flexible sampling-based approaches that were applied to parameter estimation for univariate mixtures of normals for the purpose of illustration include adaptive radial-based direction sampling (Bauwens et al., 2006) and hit-and-run sampling in combination with the ratio-of-uniform method (Karawatzki et al., 2005).

### 6.2.8 Application to the Fishery Data

For the FISHERY DATA, Titterington et al. (1985) assume that for a certain age group indexed by $k$, the length of a fish follows a normal distribution $\mathcal{N}\left(\mu_k, \sigma_k^2\right)$, however, age is unobserved. Assuming that a randomly selected fish belongs to age group $k$ with probability $\eta_k$, the marginal distribution of the length follows a finite mixture of normal distributions with different means and different variances:

$$Y_i \sim \eta_1 \mathcal{N}\left(\mu_1, \sigma_1^2\right) + \cdots + \eta_K \mathcal{N}\left(\mu_K, \sigma_K^2\right). \tag{6.27}$$

For illustration, we fit finite mixtures of $K$ normal distributions with increasing number $K$ of potential groups to the FISHERY DATA introduced in Subsection 1.1. We assume a $\mathcal{D}(4, \dots, 4)$-prior for $\boldsymbol{\eta}$, and apply the hierarchical independence prior introduced by Richardson and Green (1997); see Subsection 6.2.6.

**Table 6.2.** FISHERY DATA, normal mixtures with $K = 4$ (hierarchical independence prior)

| Parameter | Posterior Mean | SD | 95% Confidence Region | |
|---|---|---|---|---|
| | | | Lower | Upper |
| $\mu_1$ | 3.30 | 0.119 | 3.11 | 3.58 |
| $\mu_2$ | 5.23 | 0.081 | 5.06 | 5.39 |
| $\mu_3$ | 7.29 | 0.244 | 6.64 | 7.63 |
| $\mu_4$ | 8.78 | 0.771 | 7.36 | 10.21 |
| $\sigma_1^2$ | 0.159 | 0.096 | 0.063 | 0.416 |
| $\sigma_2^2$ | 0.312 | 0.081 | 0.181 | 0.498 |
| $\sigma_3^2$ | 0.488 | 0.327 | 0.122 | 1.406 |
| $\sigma_4^2$ | 2.82 | 1.128 | 1.022 | 5.229 |
| $\eta_1$ | 0.115 | 0.023 | 0.076 | 0.164 |
| $\eta_2$ | 0.469 | 0.059 | 0.336 | 0.569 |
| $\eta_3$ | 0.222 | 0.068 | 0.102 | 0.377 |
| $\eta_4$ | 0.195 | 0.074 | 0.084 | 0.368 |

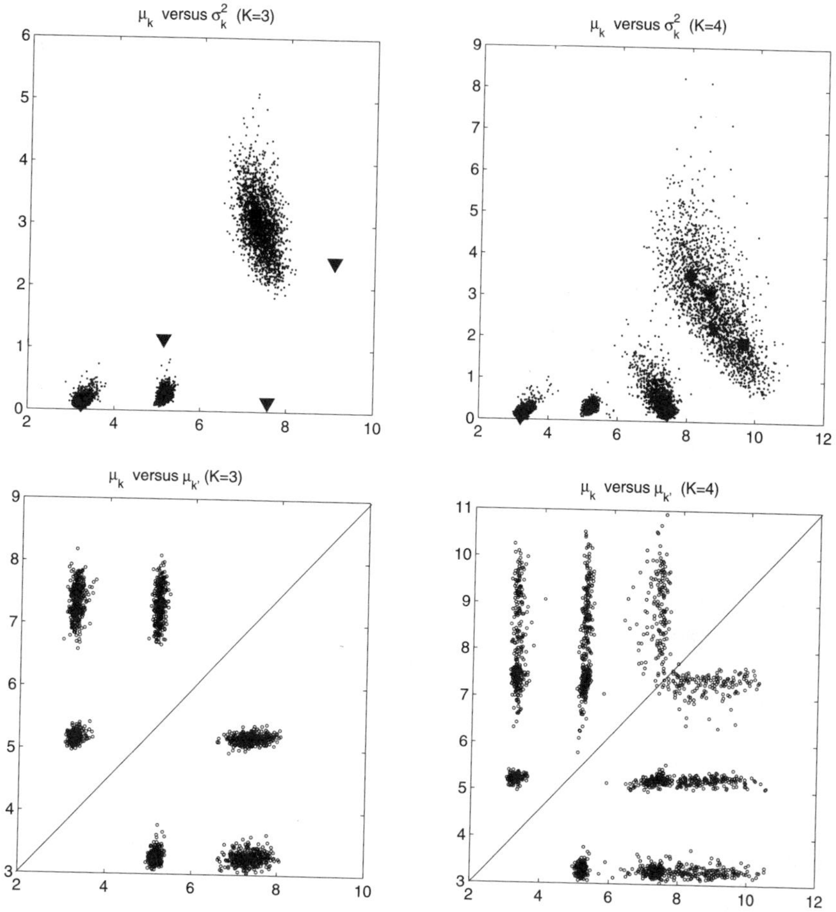


**Fig. 6.8.** FISHERY DATA, normal mixtures with $K = 3$ (left-hand side) and $K = 4$ (right-hand side) components (hierarchical independence prior); MCMC draws obtained from the random permutation sampler; top: $\mu_k$ plotted against $\sigma_k^2$; bottom: $\mu_k$ plotted against $\mu_{k'}$

For estimation, we use *Algorithm 6.1* and store 5000 MCMC draws after a burn-in phase of 2000 draws. We estimate unidentified mixtures with $K = 3$ and $K = 4$ components, respectively. MCMC estimation for an unidentified mixture is carried out with the help of the random permutation Gibbs sampler. The top of Figure 6.8 shows two-dimensional scatter plots of the MCMC draws $(\mu_k^{(m)}, \sigma_k^{(2,m)})$ for an arbitrary $k$ for $K = 3$ and $K = 4$. For $K = 3$ and $K = 4$ there are, respectively, three and four clearly separated clusters visible in the MCMC draws. As expected, the mean of the height is different between the groups. The variances are nearly identical for the groups with smaller fish, but rather large for the group with the largest fish. For $K = 4$, group 1 and group 2 remained roughly where they have been for a mixture with $K = 3$, whereas the third group has been split into two separate groups. Interestingly, the variance of the third group is now comparable with the variance of groups 1 and 2; only the variance of the fourth group is considerably larger.

The MCMC draws from the marginal bivariate density $p(\mu_k, \mu_{k'}|\mathbf{y})$, with $k$ and $k'$ arbitrary, but different, shown in the bottom of Figure 6.8, indicate that for $K = 3$ the $K(K-1) = 6$ modes are all bounded away from the line $\mu_k = \mu_{k'}$ which corresponds to a model where two components have equal means. Therefore the constraint $\mu_1 < \mu_2 < \mu_3$ induces a unique labeling. After reordering the MCMC draws according to this constraint by the permutation sampler, we obtain MCMC draws which could be used for category-specific inference. For $K = 4$, the MCMC draws from the marginal bivariate density $p(\mu_k, \mu_{k'}|\mathbf{y})$ in Figure 6.8, however, indicate that $K(K-1) = 12$ possible modes are not all bounded away from the line $\mu_k = \mu_{k'}$ which corresponds to a model where two components have equal means. Thus the constraint $\mu_1 < \cdots < \mu_4$ does not induce a unique labeling. Postprocessing the MCMC draws through unsupervised clustering as explained in Subsection 3.7.7, however, leads to an identified mixture model. These MCMC draws were used for category-specific inference, in particular for parameter estimation for $K = 4$ given in Table 6.2.

The prior exercises considerable influence for these data. For the conditionally conjugate prior introduced by Bensmail et al. (1997) (see Subsection 6.2.2), for instance, shrinkage is much stronger, whereas the prior of Raftery (1996b) completely failed.

## 6.3 Bayesian Estimation of Multivariate Mixtures of Normals

### 6.3.1 Bayesian Inference When the Allocations Are Known

In this subsection the conditional posterior distribution of $\boldsymbol{\mu}_k, \boldsymbol{\Sigma}_k|\mathbf{S}, \mathbf{y}$ is derived by combining the information from all observations in group $k$. Relevant group-specific quantities are the number $N_k(\mathbf{S})$ of observations in group $k$, the group mean $\overline{\mathbf{y}}_k(\mathbf{S})$, and the unexplained within-group variability $\boldsymbol{W}_k(\mathbf{S})$:

$$N_k(\mathbf{S}) = \#\{i : S_i = k\}, \tag{6.28}$$

$$\overline{\mathbf{y}}_k(\mathbf{S}) = \frac{1}{N_k(\mathbf{S})} \sum_{i:S_i=k} \mathbf{y}_i, \tag{6.29}$$

$$\boldsymbol{W}_k(\mathbf{S}) = \sum_{i:S_i=k} (\mathbf{y}_i - \overline{\mathbf{y}}_k(\mathbf{S}))(\mathbf{y}_i - \overline{\mathbf{y}}_k(\mathbf{S}))'. \tag{6.30}$$

Assume that for observation $\mathbf{y}_i$, the allocation is equal to $k$, $S_i = k$. Then the contribution of $\mathbf{y}_i$ to the complete-data likelihood function $p(\mathbf{y}|\boldsymbol{\mu}, \boldsymbol{\Sigma}, \mathbf{S})$ is equal to

$$p(\mathbf{y}_i|\boldsymbol{\mu}_k, \boldsymbol{\Sigma}_k) = c|\boldsymbol{\Sigma}_k|^{-1/2}\exp\left(-\frac{1}{2}(\mathbf{y}_i - \boldsymbol{\mu}_k)'\boldsymbol{\Sigma}_k^{-1}(\mathbf{y}_i - \boldsymbol{\mu}_k)\right),$$

where $c = (2\pi)^{-r/2}$, and the complete-data likelihood function has $K$ independent factors, each carrying all information about the parameters in a certain group:

$$p(\mathbf{y}|\boldsymbol{\mu}, \boldsymbol{\Sigma}, \mathbf{S}) = \tag{6.31}$$
$$c^N \prod_{k=1}^{K} |\boldsymbol{\Sigma}_k|^{-N_k(\mathbf{S})/2}\exp\left(-\frac{1}{2}\sum_{i:S_i=k} (\mathbf{y}_i - \boldsymbol{\mu}_k)'\boldsymbol{\Sigma}_k^{-1}(\mathbf{y}_i - \boldsymbol{\mu}_k)\right).$$

In a Bayesian analysis each of these factors is combined with a prior. When holding the variance–covariance matrix $\boldsymbol{\Sigma}_k$ fixed, then the complete-data likelihood function, regarded as a function of $\boldsymbol{\mu}_k$, is the kernel of a multivariate normal distribution. Under the conjugate prior $\boldsymbol{\mu}_k \sim \mathcal{N}_r(\mathbf{b}_0, \mathbf{B}_0)$, the posterior density of $\boldsymbol{\mu}_k$ given $\boldsymbol{\Sigma}_k$ and $N_k(\mathbf{S})$ observations assigned to this group, is again a density from the multivariate normal distribution, $\boldsymbol{\mu}_k|\boldsymbol{\Sigma}_k, \mathbf{S}, \mathbf{y} \sim \mathcal{N}_r(\mathbf{b}_k(\mathbf{S}), \mathbf{B}_k(\mathbf{S}))$, where

$$\mathbf{B}_k(\mathbf{S}) = (\mathbf{B}_0^{-1} + N_k(\mathbf{S})\boldsymbol{\Sigma}_k^{-1})^{-1}, \tag{6.32}$$
$$\mathbf{b}_k(\mathbf{S}) = \mathbf{B}_k(\mathbf{S})(\mathbf{B}_0^{-1}\mathbf{b}_0 + \boldsymbol{\Sigma}_k^{-1}N_k(\mathbf{S})\overline{\mathbf{y}}_k(\mathbf{S})). \tag{6.33}$$

When holding the mean $\boldsymbol{\mu}_k$ fixed, the complete-data likelihood function, regarded as a function in $\boldsymbol{\Sigma}_k^{-1}$, is the kernel of a Wishart density; see Appendix A.1.14 for more detail on this distribution family. Under the conjugate Wishart prior $\boldsymbol{\Sigma}_k^{-1} \sim \mathcal{W}_r(c_0, \mathbf{C}_0)$, the posterior density of $\boldsymbol{\Sigma}_k$ given $\boldsymbol{\mu}_k$ and $N_k(\mathbf{S})$ observations assigned to this group, is again a density from the Wishart distribution, $\boldsymbol{\Sigma}_k^{-1}|\boldsymbol{\mu}_k, \mathbf{S}, \mathbf{y} \sim \mathcal{W}_r(c_k(\mathbf{S}), \mathbf{C}_k(\mathbf{S}))$, where

$$c_k(\mathbf{S}) = c_0 + \frac{N_k(\mathbf{S})}{2}, \tag{6.34}$$
$$\mathbf{C}_k(\mathbf{S}) = \mathbf{C}_0 + \frac{1}{2}\sum_{i:S_i=k} (\mathbf{y}_i - \boldsymbol{\mu}_k)(\mathbf{y}_i - \boldsymbol{\mu}_k)'. \tag{6.35}$$

If both $\boldsymbol{\mu}_k$ and $\boldsymbol{\Sigma}_k$ are unknown, a closed-form solution for the joint posterior $p(\boldsymbol{\mu}_k, \boldsymbol{\Sigma}_k|\mathbf{S}, \mathbf{y})$ exists only if the prior of $\boldsymbol{\mu}_k$ is restricted by assuming that the prior covariance matrix depends on $\boldsymbol{\Sigma}_k$ through $\mathbf{B}_{0,k} = \boldsymbol{\Sigma}_k/N_0$. Then the joint posterior factors as $p(\boldsymbol{\mu}_k|\boldsymbol{\Sigma}_k, \mathbf{y}, \mathbf{S})p(\boldsymbol{\Sigma}_k|\mathbf{y}, \mathbf{S})$, where the density of $\boldsymbol{\mu}_k$ given $\boldsymbol{\Sigma}_k$ arises from a $\mathcal{N}_r(\mathbf{b}_k(\mathbf{S}), \mathbf{B}_k(\mathbf{S}))$ distribution with

$$\mathbf{B}_k(\mathbf{S}) = \frac{1}{N_k(\mathbf{S}) + N_0}\boldsymbol{\Sigma}_k, \tag{6.36}$$
$$\mathbf{b}_k(\mathbf{S}) = \frac{N_0}{N_k(\mathbf{S}) + N_0}\mathbf{b}_0 + \frac{N_k(\mathbf{S})}{N_k(\mathbf{S}) + N_0}\overline{\mathbf{y}}_k(\mathbf{S}), \tag{6.37}$$

whereas the marginal posterior of $\boldsymbol{\Sigma}_k^{-1}$ is a $\mathcal{W}_r(c_k(\mathbf{S}), \mathbf{C}_k(\mathbf{S}))$-distribution, where $c_k(\mathbf{S})$ is the same as in (6.34), however,

$$\mathbf{C}_k(\mathbf{S}) = \mathbf{C}_0 + \frac{1}{2}\left(\frac{N_k(\mathbf{S})N_0}{N_k(\mathbf{S}) + N_0}(\overline{\mathbf{y}}_k(\mathbf{S}) - \mathbf{b}_0)(\overline{\mathbf{y}}_k(\mathbf{S}) - \mathbf{b}_0)'\right) + \frac{1}{2}\boldsymbol{W}_k(\mathbf{S}). \tag{6.38}$$

### 6.3.2 Prior Distributions

#### Standard Prior Distributions

Binder (1978), Lavine and West (1992), Robert (1996), and Bensmail et al. (1997) consider the *conditionally conjugate prior*

$$p(\boldsymbol{\mu}_1, \ldots, \boldsymbol{\mu}_K, \boldsymbol{\Sigma}_1, \ldots, \boldsymbol{\Sigma}_K) = \prod_{k=1}^{K} p(\boldsymbol{\mu}_k|\boldsymbol{\Sigma}_k)p(\boldsymbol{\Sigma}_k), \tag{6.39}$$

where $\boldsymbol{\mu}_k|\boldsymbol{\Sigma}_k \sim \mathcal{N}_r\left(\mathbf{b}_0, \boldsymbol{\Sigma}_k/N_0\right)$, and $\boldsymbol{\Sigma}_k^{-1} \sim \mathcal{W}_r\left(c_0, \mathbf{C}_0\right)$. As explained in Subsection 6.3.1, this prior has the advantage that the joint posterior $p(\boldsymbol{\mu}, \boldsymbol{\Sigma}|\mathbf{S}, \mathbf{y})$ is available in closed form, conditional on knowing the allocations $\mathbf{S}$. This prior is proper if $N_0 > 1$ and $c_0 > (r-1)/2$. Binder (1978) and Lavine and West (1992) choose hyperparameters $N_0$ and $c_0$ that result in an improper prior, which is not recommended both for the theoretical and practical reasons discussed earlier. Robert (1996) chooses $\mathbf{b}_0 = \overline{\mathbf{y}}$, $N_0 = 1$, $c_0 = 3$, and $\mathbf{C}_0 = 0.75\mathbf{S}_\mathbf{y}$, where $\overline{\mathbf{y}}$ and $\mathbf{S}_\mathbf{y}$ are the sample mean and the sample variance–covariance matrix, respectively. Bensmail et al. (1997) choose the same prior on $\boldsymbol{\mu}_k$, but a slightly different prior on the covariance matrices, namely $c_0 = 2.5$ and $\mathbf{C}_0 = 0.5\mathbf{S}_\mathbf{y}$.

#### Understanding the Hyperparameters

$c_0$ will influence the prior distribution of the eigenvalues of $\boldsymbol{\Sigma}_k\boldsymbol{\Sigma}_j^{-1}$ which are given by $\lambda_{k,l}/\lambda_{j,m}$, where $\lambda_{k,l}$ and $\lambda_{j,m}$ are the eigenvalues of $\boldsymbol{\Sigma}_k$ and $\boldsymbol{\Sigma}_j$. These eigenvalues are bounded away from 0, if $c_0 > 2 + (r-1)/2$. This suggests choosing $c_0$ as a function of $r$.

If $\mathbf{C}_0 = \phi\mathbf{S}_\mathbf{y}$ as in Robert (1996) and Bensmail et al. (1997), then $\phi$ influences the prior expectation of the amount of heterogeneity $\mathrm{E}(R_t^2(\boldsymbol{\vartheta}))$, explained by differences in the group means:

$$\mathrm{E}(R_t^2(\boldsymbol{\vartheta})) = 1 - \frac{\phi}{c_0 - (r+1)/2}. \tag{6.40}$$

If for instance $c_0 = 2.5 + (r-1)/2$, then choosing $\phi = 0.75$ corresponds to a prior expectation of 50 percent of explained heterogeneity, whereas choosing $\phi = 0.5$ leads to prior expectation of 2/3 explained heterogeneity.

### Prior Dependence Between the Components

Stephens (1997a) considers an extension of the hierarchical independence prior introduced by Richardson and Green (1997) for univariate mixtures of normals, by generalizing (6.25) in the following way to multivariate mixtures:

$$p(\boldsymbol{\mu}_1, \ldots, \boldsymbol{\mu}_K, \boldsymbol{\Sigma}_1, \ldots, \boldsymbol{\Sigma}_K) = \prod_{k=1}^{K} p(\boldsymbol{\mu}_k) \prod_{k=1}^{K} p(\boldsymbol{\Sigma}_k | \mathbf{C}_0), \tag{6.41}$$

where $\boldsymbol{\mu}_k \sim \mathcal{N}_r(\mathbf{b}_0, \mathbf{B}_0)$, and $\boldsymbol{\Sigma}_k^{-1} \sim \mathcal{W}_r(c_0, \mathbf{C}_0)$. If $\mathbf{C}_0$ is a fixed hyperparameter, the *independence prior* results, where all component parameters $\boldsymbol{\mu}_1, \ldots, \boldsymbol{\mu}_K, \boldsymbol{\Sigma}_1, \ldots, \boldsymbol{\Sigma}_K$ are pairwise independent a priori. For the hierarchical independence prior, prior dependence between $\boldsymbol{\Sigma}_1, \ldots, \boldsymbol{\Sigma}_K$ is introduced by assuming that the scale matrix $\mathbf{C}_0$ of the inverted Wishart prior is a random hyperparameter with a prior of its own, $\mathbf{C}_0 \sim \mathcal{W}_r(g_0, \mathbf{G}_0)$. As emphasized by Stephens (1997a), the posterior distribution will be proper, even if the prior on the scale matrix $\mathbf{C}_0$ is improper.

Note that the marginal prior $p(\boldsymbol{\Sigma}_1, \ldots, \boldsymbol{\Sigma}_K)$, where $\mathbf{C}_0$ is integrated out, takes the form:

$$p(\boldsymbol{\Sigma}_1, \ldots, \boldsymbol{\Sigma}_K) = \tag{6.42}$$
$$\frac{|\mathbf{G}_0|^{g_0} \Gamma_r(g_K)}{\Gamma_r(g_0)\Gamma_r(c_0)^K} \left| \mathbf{G}_0 + \sum_{k=1}^{K} \boldsymbol{\Sigma}_k^{-1} \right|^{-g_K} \left( \prod_{k=1}^{K} |\boldsymbol{\Sigma}_k^{-1}| \right)^{c_0+(r+1)/2},$$

where $g_K = g_0 + Kc_0$ and $\Gamma_r(\alpha)$ is the generalized Gamma function defined in Appendix A.1, formula (A.26).

Concerning the hyperparameters, Stephens (1997a) considered robust estimators of location and scale and chose the midpoint and the length of the observation interval for the different components of $\mathbf{y}_i$ to define location and spread. For bivariate mixtures this prior reads $c_0 = 3$, $g_0 = 0.3$, and $\mathbf{b}_0$, $\mathbf{B}_0$ and $\mathbf{G}_0$ defined by:

$$\mathbf{b}_0 = \begin{pmatrix} m_1 \\ m_2 \end{pmatrix}, \quad \mathbf{B}_0 = \begin{pmatrix} R_1^2 & 0 \\ 0 & R_2^2 \end{pmatrix}, \quad \mathbf{G}_0 = \begin{pmatrix} \frac{100 g_0}{c_0 R_1^2} & 0 \\ 0 & \frac{100 g_0}{c_0 R_2^2} \end{pmatrix},$$

where $m_l$ and $R_l$ are the midpoint and the length of the observation interval of the $l$th component of $\mathbf{y}_i$. This prior is easily extended to higher-dimensional mixtures.

## 6.3.3 Bayesian Parameter Estimation Using MCMC

Several authors, in particular Lavine and West (1992), Bensmail et al. (1997), Stephens (1997a), and Frühwirth-Schnatter et al. (2004), have considered Gibbs sampling, based on data augmentation, as described for general mixtures in Section 3.5, to estimate the parameters of a multivariate mixture of normals. Full conditional Gibbs sampling proceeds along the lines indicated by *Algorithm 3.4*, where the results of Subsection 6.3.1 are used to sample the mean $\boldsymbol{\mu}_k$ and variance–covariance matrix $\boldsymbol{\Sigma}_k$ in each group, conditional upon a given allocation vector $\mathbf{S}$.

*Algorithm 6.2: Two-Block Gibbs Sampling for a Multivariate Gaussian Mixture*

(a) Parameter simulation conditional on the classification $\mathbf{S} = (S_1, \ldots, S_N)$:
  (a1) Sample $\boldsymbol{\eta}$ from the conditional Dirichlet posterior $p(\boldsymbol{\eta}|\mathbf{S})$ as in *Algorithm 3.4*.
  (a2) Sample $\boldsymbol{\Sigma}_k^{-1}$ in each group $k$ from a $\mathcal{W}_r\left(c_k(\mathbf{S}), \mathbf{C}_k(\mathbf{S})\right)$-distribution.
  (a3) Sample $\boldsymbol{\mu}_k$ in each group $k$ from an $\mathcal{N}_r\left(\mathbf{b}_k(\mathbf{S}), \mathbf{B}_k(\mathbf{S})\right)$-distribution.

(b) Classification of each observation $\mathbf{y}_i$, for $i = 1, \ldots, N$, conditional on knowing $\boldsymbol{\mu}, \boldsymbol{\Sigma}$, and $\boldsymbol{\eta}$:

$$\Pr(S_i = k|\boldsymbol{\mu}, \boldsymbol{\Sigma}, \boldsymbol{\eta}, \mathbf{y}_i) \propto f_N(\mathbf{y}_i; \boldsymbol{\mu}_k, \boldsymbol{\Sigma}_k)\eta_k. \tag{6.43}$$

The precise form of $\mathbf{b}_k(\mathbf{S})$, $\mathbf{B}_k(\mathbf{S})$, $c_k(\mathbf{S})$, and $\mathbf{C}_k(\mathbf{S})$ in steps (a2) and (a3) depends upon the chosen prior distribution family. For the independence prior and the hierarchical independence prior $\mathbf{b}_k(\mathbf{S})$ and $\mathbf{B}_k(\mathbf{S})$ are given by (6.32) and (6.33), whereas $c_k(\mathbf{S})$ and $\mathbf{C}_k(\mathbf{S})$ are available from (6.34) and (6.35). Under the conditionally conjugate prior and the hierarchical conditionally conjugate prior, $\boldsymbol{\mu}_k$ and $\boldsymbol{\Sigma}_k$ may be sampled jointly, as the marginal density $\boldsymbol{\Sigma}_k^{-1}|\mathbf{S}, \mathbf{y}$ is available in closed form. $\mathbf{b}_k(\mathbf{S})$ and $\mathbf{B}_k(\mathbf{S})$ are given by (6.36) and (6.37), whereas $c_k(\mathbf{S})$ and $\mathbf{C}_k(\mathbf{S})$ are available from (6.34) and (6.38).

Note that the sampling order in *Algorithm 6.2* could be reversed, by starting with classification based on starting values for $\boldsymbol{\mu}$, $\boldsymbol{\Sigma}$, and $\boldsymbol{\eta}$. Whereas choosing starting values is not a problem for univariate mixtures, for multivariate mixtures it is essential to use some carefully selected starting value, in particular when $r$ is large. As shown, for instance, in Justel and Peña (1996), Gibbs sampling may be trapped at a suboptimal solution when starting with poor parameter estimates and may never converge.

In general, it is easier to find a rough classification of the data than good starting values for rather high-dimensional variance–covariance matrices. One solution, dating back at least to Friedman and Rubin (1967) is to utilize repeated random partitioning of the observations into $K$ groups. Another useful solution, suggested by Everitt and Hand (1981) is to run a pilot analysis and to use some simple clustering technique to obtain a starting value for $\mathbf{S}$.

Under any hierarchical prior an additional step has to be added to *Algorithm 6.2*, to sample the scale matrix $\mathbf{C}_0$ of the hierarchical prior from the conditional posterior $p(\mathbf{C}_0|\boldsymbol{\Sigma}_1, \ldots, \boldsymbol{\Sigma}_K, \mathbf{y})$, which takes the form of the following Wishart distribution,

$$\mathbf{C}_0 \sim \mathcal{W}_r\left(g_0 + Kc_0, \mathbf{G}_0 + \sum_{k=1}^{K} \boldsymbol{\Sigma}_k^{-1}\right). \tag{6.44}$$

### Classification Without Parameter Estimation

As already discussed in Subsection 3.3.3, Chen and Liu (1996) showed that for many mixture models Bayesian clustering is possible without explicit parameter estimation, using directly the marginal posterior $p(\mathbf{S}|\mathbf{y})$ given by $p(\mathbf{S}|\mathbf{y}) \propto p(\mathbf{y}|\mathbf{S})p(\mathbf{S})$. For a mixture of multivariate normal distributions, $p(\mathbf{y}|\mathbf{S})$ is given by

$$p(\mathbf{y}|\mathbf{S}) \propto \prod_{k=1}^{K} \frac{\Gamma_d(c_k(\mathbf{S}))}{|\mathbf{C}_k(\mathbf{S})|^{c_k(\mathbf{S})}}, \qquad (6.45)$$

where $c_k(\mathbf{S})$, and $\mathbf{C}_k(\mathbf{S})$ are the posterior moments of the full conditional posterior density $p(\mathbf{\Sigma}_k|\mathbf{y}, \mathbf{S})$ under the conditionally conjugate prior, and $\Gamma_d(\cdot)$ is the generalized Gamma function, defined in (A.26).

### 6.3.4 Application to Fisher's Iris Data

This example involves a data set well known in multivariate analysis, namely Fisher's Iris Data. This data set consists of 150 four-dimensional observations of three species of iris (*iris setosa, iris versicolour, iris virginica*). The measurements taken for each plant are *sepal length, sepal width, petal length* and *petal width.*[1] ML estimation of these data has been considered by many authors, for instance, by Everitt and Hand (1981). Everitt and Hand (1981, p.43), when fitting a three-component mixture with unconstrained component variance-covariances matrices to Fisher's Iris Data, report convergence to the singularities of the likelihood surface for certain starting values.

We reanalyze these data using a Bayesian approach, based on the hierarchical independence prior, and fit a three-component multivariate normal mixture with unconstrained component variance–covariance matrices to these data. The posterior draws of $\mu_{k,1}$ versus $\mu_{k,2}$ and $\mu_{k,3}$ versus $\mu_{k,4}$ in Figure 6.9 show three clear simulation clusters. Thus identification could be achieved using unsupervised clustering of the MCMC draws of $(\mu_{k,1}, \mu_{k,2}, \mu_{k,3}, \mu_{k,4})$ as explained in Subsection 3.7.7. The corresponding estimates are shown in Table 6.3, whereas Figure 6.10 shows the position of the estimated mixture components relative to the data.

## 6.4 Further Issues

### 6.4.1 Parsimonious Finite Normal Mixtures

If the number of observations $N$ is small compared to the number $r$ of features, measured for each $\mathbf{y}_i$, efficient estimation of unconstrained variance–

[1] Data from ftp://ftp.ics.uci.edu/pub/machine-learning-databases/iris/iris.names. These data differ from the data presented in Fisher's article; errors in the 35th sample in the fourth feature and in the 38th sample in the second and third features were identified by Steve Chadwick (spchadwick@espeedaz.net).

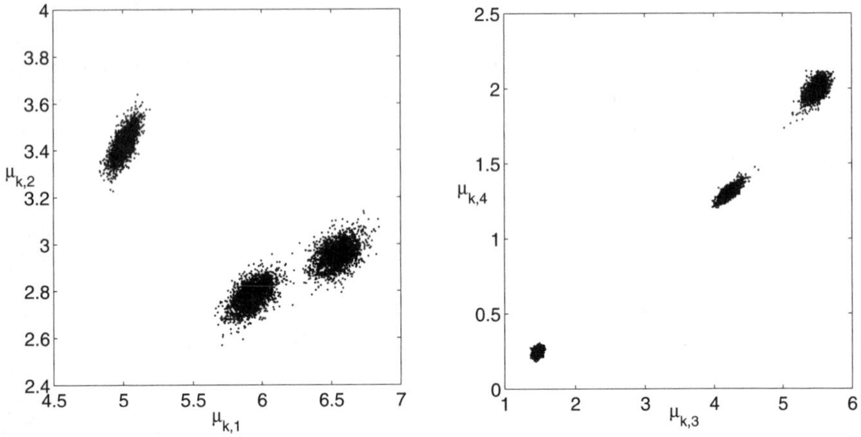


**Fig. 6.9.** FISHER'S IRIS DATA, multivariate normal mixture with $K = 3$ components (hierarchical independence prior); MCMC draws obtained from the random permutation sampler; left-hand side: $\mu_{k,1}$ plotted against $\mu_{k,2}$; right-hand side: $\mu_{k,3}$ plotted against $\mu_{k,4}$

**Table 6.3.** FISHER'S IRIS DATA, multivariate normal mixture with $K = 3$ components (hierarchical independence prior); posterior expectation of certain characteristics of the covariance matrices $\mathbf{\Sigma}_k$, identification achieved through unsupervised clustering

| Group $k$ | E(Eigenvalues of $\mathbf{\Sigma}_k\|\mathbf{y}$) | E(tr $(\mathbf{\Sigma}_k)$ $\|\mathbf{y}$) | E(det $\mathbf{\Sigma}_k\|\mathbf{y}$) |
|---|---|---|---|
| 1 | 0.179 0.0361 0.0218 0.00731 | 0.244 | 2.84e-006 |
| 2 | 0.381 0.0526 0.0325 0.00686 | 0.473 | 1.26e-005 |
| 3 | 0.512 0.0818 0.0436 0.0236 | 0.661 | 0.00012 |

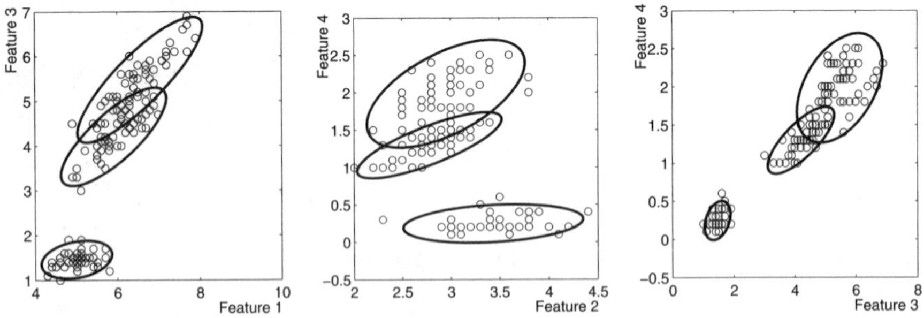


**Fig. 6.10.** FISHER'S IRIS DATA, multivariate normal mixture with $K = 3$ components (hierarchical independence prior); fitted mixture model in relation to the data, identification achieved through unsupervised clustering

covariance matrices $\mathbf{\Sigma}_k$ is not possible, and the performance of the corresponding classification rules will be disappointing (see, e.g., McLachlan and Basford, 1988). Gain in efficiency may be achieved as in Friedman (1989) by shrinking each $\mathbf{\Sigma}_k$ toward a common variance–covariance matrix $\mathbf{\Sigma}$, through a hierarchical prior with large degrees of freedom $c_0$.

Various authors studied models for component variance–covariance matrices that are more parsimonious than unconstrained variance–covariance matrices in terms of the number of parameters, and the most useful ones are briefly reviewed here.

### Constrained Mixtures

The easiest way to achieve parsimony is to put simple constraints on the variance–covariance matrices. For a homoscedastic mixture, the variance–covariance matrices are restricted to be the same in each component, $\mathbf{\Sigma}_k \equiv \mathbf{\Sigma}$. A spherical mixture is one where $\mathbf{\Sigma}_k \equiv \sigma_k^2 \mathbf{I}_r$. In the isotropic case all variance–covariance matrices are restricted to the same scaled identity matrix, $\mathbf{\Sigma}_k \equiv \sigma^2 \mathbf{I}_r$. For an illustration of these various types of multivariate normal mixtures see the simulated data sets in Figure 6.11.

For MCMC estimation of constrained mixtures, step (a2) in *Algorithm 6.2* has to be modified in an appropriate way. The following results hold for conditionally conjugate priors; modifications for independence priors are straightforward. For a homoscedastic mixture with prior $\mathbf{\Sigma}^{-1} \sim \mathcal{W}_r(c_0, \mathbf{C}_0)$, step (a2) reduces to sampling $\mathbf{\Sigma}^{-1}$ from a $\mathcal{W}_r(c_N, \mathbf{C}_N(\mathbf{S}))$, where $c_N = c_0 + N/2$, and

$$\mathbf{C}_N(\mathbf{S}) = \mathbf{C}_0 + \frac{1}{2}\sum_{k=1}^{K}\left(\boldsymbol{W}_k(\mathbf{S}) + \frac{N_0 N_k(\mathbf{S})}{N_0 + N_k(\mathbf{S})}(\overline{\mathbf{y}}_k(\mathbf{S}) - \mathbf{b}_0)(\overline{\mathbf{y}}_k(\mathbf{S}) - \mathbf{b}_0)'\right).$$

For $r = 1$ the Wishart distribution reduces to the Gamma distribution; see Subsection A.1.14. For a spherical mixture under prior $\sigma_k^2 \sim \mathcal{G}^{-1}(c_0, C_0)$ step (a2) reduces to sampling $\sigma_k^2$ in each group $k$ as $\sigma_k^2 \sim \mathcal{G}^{-1}(c_k(\mathbf{S}), C_k(\mathbf{S}))$, where $c_k(\mathbf{S}) = c_0 + N_k(\mathbf{S})r/2$ and

$$C_k(\mathbf{S}) = C_0 + \frac{1}{2}\left(\mathrm{tr}\left(\boldsymbol{W}_k(\mathbf{S})\right) + \frac{N_0 N_k(\mathbf{S})}{N_0 + N_k(\mathbf{S})}(\overline{\mathbf{y}}_k(\mathbf{S}) - \mathbf{b}_0)'(\overline{\mathbf{y}}_k(\mathbf{S}) - \mathbf{b}_0)\right).$$

For an isotropic mixture with prior $\sigma^2 \sim \mathcal{G}^{-1}(c_0, C_0)$, step (a2) reduces to sampling of $\sigma^2$ as $\sigma^2 \sim \mathcal{G}^{-1}(c_N, C_N(\mathbf{S}))$, where $c_N = c_0 + Nr/2$, and

$$C_N(\mathbf{S}) = C_0 + \frac{1}{2}\sum_{k=1}^{K}\left(\mathrm{tr}\left(\boldsymbol{W}_k(\mathbf{S})\right) + \frac{N_0 N_k(\mathbf{S})}{N_0 + N_k(\mathbf{S})}(\overline{\mathbf{y}}_k(\mathbf{S}) - \mathbf{b}_0)'(\overline{\mathbf{y}}_k(\mathbf{S}) - \mathbf{b}_0)\right).$$

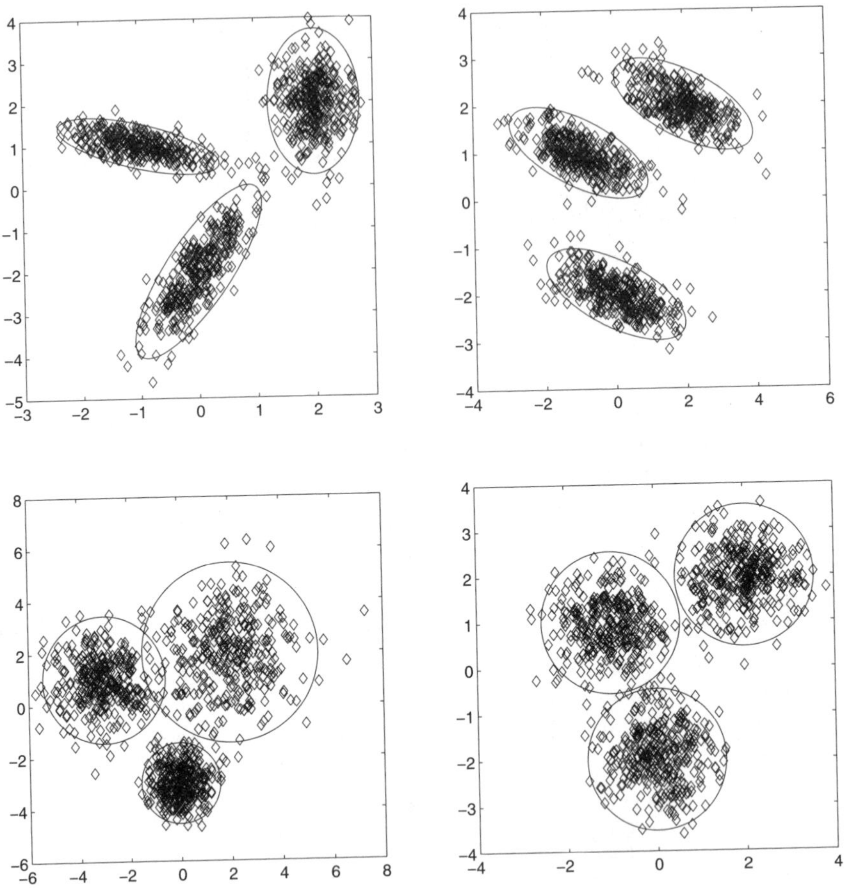

**Fig. 6.11.** 1000 observations simulated from a mixture of three bivariate normal distributions with $\eta_1 = \eta_2 = \eta_3 = 1/3$, means located at $(-1, 1)$, $(0, -2)$, and $(2, 2)$, and different covariance structures: unrestricted covariances (top, left-hand side), homoscedastic covariances (top, right-hand side), spherical mixture (bottom, left-hand side), isotropic mixture (bottom, right-hand side); full line: 95%-confidence region for each component

## Parsimony Based on the Eigenvalue Decomposition

Banfield and Raftery (1993) suggested parsimonious modeling of the variance–covariance matrices $\mathbf{\Sigma}_k$, based on the spectral decomposition of $\mathbf{\Sigma}_k$:

$$\mathbf{\Sigma}_k = \lambda_k \boldsymbol{D}_k \boldsymbol{A}_k \boldsymbol{D}_k^{'}, \tag{6.46}$$

where $\lambda_k$ is the largest eigenvalue of $\mathbf{\Sigma}_k$, $\boldsymbol{A}_k$ is a diagonal matrix containing the decreasing eigenvalues of $\mathbf{\Sigma}_k$, divided by the largest eigenvalue $\lambda_k$, and

$\boldsymbol{D}_k$ is an orthogonal matrix with the columns corresponding to the normalized eigenvectors of $\boldsymbol{\Sigma}_k$. A slightly different parameterization has been considered by Bensmail and Celeux (1996), where $\lambda_k = |\boldsymbol{\Sigma}_k|^{1/r}$ and $\boldsymbol{A}_k$ contains the normalized eigenvalues in decreasing order. In (6.46), $\lambda_k$ controls the volume occupied by the cluster in the sample space. The elements of $\boldsymbol{A}_k$ control the shape with $\boldsymbol{A}_k = \text{Diag}\,(1 \cdots 1)$ corresponding to spherical clusters, whereas a cluster with $\boldsymbol{A}_k = \text{Diag}\,(1\, A_{k2} \cdots A_{kr})$, where $A_{kj} << 1$, will be concentrated around a line in the sample. Finally, $\boldsymbol{D}_k$ determines the orientation with respect to the coordinate axes, with $\boldsymbol{D}_k = \mathbf{I}_r$ corresponding to the case where the principal components are parallel to the coordinate axes. Therefore when clusters share certain parameters $\lambda_k$, $\boldsymbol{A}_k$, and $\boldsymbol{D}_k$, they share certain geometric properties, and parsimony is achieved through imposing constraints on $\lambda_k$, $\boldsymbol{A}_k$, and $\boldsymbol{D}_k$. Bensmail and Celeux (1996) discuss in total 14 different models. ML estimation of these models, where all relevant matrices are estimated from the data is discussed in Celeux and Govaert (1995), whereas a fully Bayesian analysis using MCMC estimation is discussed in Bensmail et al. (1997).

These models were applied to clustering multivariate data (Banfield and Raftery, 1993; Celeux and Govaert, 1995; Bensmail et al., 1997) and discriminant analysis (Bensmail and Celeux, 1996); see also Chapter 7.

### 6.4.2 Model Selection Problems for Mixtures of Normals

The most common model selection problem is naturally selecting the number $K$ of components. But also choosing an appropriate structure of the component variance–covariance matrices is important.

#### Model Selection Based on Marginal Likelihoods

Bensmail et al. (1997) used the marginal likelihood $p(\mathbf{y}|\mathcal{M}_K)$, which has been defined in Section 5.3, to treat the problem of choosing the number of components and selecting between different variance–covariance models for $\boldsymbol{\Sigma}_k$ simultaneously. They illustrate for various simulated data sets that the marginal likelihood is able to find the true number of components, when no misspecifications are present. For numerical approximation, Bensmail et al. (1997) use the Laplace–Metropolis estimator (Lewis and Raftery, 1997); see also Subsection 5.5.3 for more detail on this estimator.

Frühwirth-Schnatter (2004) compared various estimators of the marginal likelihood for bivariate mixtures of normals, and found that the bridge sampling estimator, which has been described in Subsection 5.4.6, gave accurate estimates of the marginal likelihood, if $K$ is not too large. The bridge sampling estimator may be applied to univariate and multivariate mixtures of normals in a rather straightforward manner. We give more details for a multivariate mixture. The importance density, defined in (5.36) takes the form of a mixture of normal inverted Wishart densities:

$$q(\boldsymbol{\mu}, \boldsymbol{\Sigma}, \boldsymbol{\eta}) = \frac{1}{S} \sum_{s=1}^{S} p(\boldsymbol{\eta}|\mathbf{S}^{(s)}) \prod_{k=1}^{K} p(\boldsymbol{\mu}_k|\boldsymbol{\Sigma}_k, \mathbf{S}^{(s)}, \mathbf{y}) q(\boldsymbol{\Sigma}_k|\mathbf{S}^{(s)}, \mathbf{y}),$$

where $p(\boldsymbol{\mu}_k|\boldsymbol{\Sigma}_k, \mathbf{S}^{(s)}, \mathbf{y})$ is the full conditional multivariate normal distribution appearing in step (a3) in *Algorithm 6.2*. Under the conditionally conjugate prior it is possible to choose $q(\boldsymbol{\Sigma}_k|\mathbf{S}^{(s)}, \mathbf{y}) = p(\boldsymbol{\Sigma}_k|\mathbf{S}^{(s)}, \mathbf{y})$, with $p(\boldsymbol{\Sigma}_k|\mathbf{S}^{(s)}, \mathbf{y})$ being the marginal inverted Wishart density appearing in step (a2). For the independence prior, we choose $q(\boldsymbol{\Sigma}_k|\mathbf{S}^{(s)}, \mathbf{y}) = p(\boldsymbol{\Sigma}_k|\boldsymbol{\mu}^{(s-1)}, \mathbf{S}^{(s)}, \mathbf{y})$, with $p(\boldsymbol{\Sigma}_k|\boldsymbol{\mu}^{(s-1)}, \mathbf{S}^{(s)}, \mathbf{y})$ being the inverted Wishart density appearing in step (a2) in *Algorithm 6.2*. Under a hierarchical prior, the marginal prior $p(\boldsymbol{\Sigma}_1, \ldots, \boldsymbol{\Sigma}_K)$ given in (6.42) has to be used in evaluating the bridge sampling estimator.

### AIC and BIC

AIC and BIC, defined earlier in Subsection 4.4.2, are frequently used for model choice, especially in the context of multivariate normal mixtures; see in particular Fraley and Raftery (1998), Dasgupta and Raftery (1998), Yeung et al. (2001), and Fraley and Raftery (2002) for an application of BIC and Bozdogan and Sclove (1984) for an application of AIC. For an unconstrained mixture with $K$ components, one chooses the model $\mathcal{M}_K$ which minimizes

$$\begin{aligned} \mathrm{AIC}_K &= -2 \log p(\mathbf{y}|\hat{\boldsymbol{\vartheta}}_K, \mathcal{M}_K) + 2(K(r+1)(r/2+1)-1), \\ \mathrm{BIC}_K &= -2 \log p(\mathbf{y}|\hat{\boldsymbol{\vartheta}}_K, \mathcal{M}_K) + \log(N)(K(r+1)(r/2+1)-1). \end{aligned}$$

To obtain the ML estimator $\hat{\boldsymbol{\vartheta}}_K$ one could run the EM algorithm; see Subsection 6.1.2.

### Model Choice Based on Model Space MCMC

Model space MCMC methods are usually applied to selecting the number of components for unconstrained mixture of normals, whereas variance–covariance selection using model space MCMC methods is still an open issue. Stephens (1997a, 2000a) applied birth and death MCMC methods, reviewed in Subsection 5.2.3, to univariate and multivariate mixtures of normals. The implementation of *Algorithm 5.4* for mixtures of normals is in principle straightforward.

The most popular model space MCMC method for mixtures of normals, however, is reversible jump MCMC, reviewed in Subsection 5.2.2, and has been applied by Richardson and Green (1997), Marrs (1998), and Dellaportas and Papageorgiou (2006).

To apply *Algorithm 5.2* to univariate mixtures of normals, it is necessary to design a suitable split/merge move along the lines discussed in Subsection 5.2.2. Using the functions $h_1(Y) = Y$ and $h_2(Y) = Y^2$, when matching the moments in (5.12), leaves the mean and the variance of the marginal distribution of $Y$ unchanged. For a univariate mixture of normals, (5.12) reduces to

$$\eta_{k^\star}\mu_{k^\star} = \eta_{k_1}\mu_{k_1} + \eta_{k_2}\mu_{k_2}, \tag{6.47}$$
$$\eta_{k^\star}(\mu_{k^\star}^2 + \sigma_{k^\star}^2) = \eta_{k_1}(\mu_{k_1}^2 + \sigma_{k_1}^2) + \eta_{k_2}(\mu_{k_2}^2 + \sigma_{k_2}^2). \tag{6.48}$$

For the reverse split move, there are three degrees of freedom, and a random variable $\mathbf{u} = (u_1, u_2, u_3)$ of dimension 3 is introduced, to specify the new pairs. There are various ways in which a split move may be implemented. In Richardson and Green (1997), $u_1 \sim \mathcal{B}(2,2)$, $u_2 \sim \mathcal{B}(2,2)$, and $u_3 \sim \mathcal{B}(1,1)$, and splitting is made according to the following rule, which fulfills (6.47) and (6.48) and leads to positive variances,

$$\eta_{k_1} = \eta_{k^\star} u_1, \qquad \eta_{k_2} = \eta_{k^\star}(1 - u_1), \tag{6.49}$$
$$\mu_{k_1} = \mu_{k^\star} - u_2\sigma_{k^\star}\sqrt{\frac{\eta_{k_2}}{\eta_{k_1}}}, \qquad \mu_{k_2} = \mu_{k^\star} + u_2\sigma_{k^\star}\sqrt{\frac{\eta_{k_1}}{\eta_{k_2}}}, \tag{6.50}$$
$$\sigma_{k_1}^2 = u_3(1 - u_2^2)\sigma_{k^\star}^2\frac{\eta_{k^\star}}{\eta_{k_1}}, \qquad \sigma_{k_2}^2 = (1 - u_3)(1 - u_2^2)\sigma_{k^\star}^2\frac{\eta_{k^\star}}{\eta_{k_2}}. \tag{6.51}$$

The determinant of the Jacobian for this split move is given by

$$\frac{\eta_{k^\star}|\mu_{k_1} - \mu_{k_2}|\sigma_{k_1}^2\sigma_{k_2}^2}{u_3(1 - u_3)u_2(1 - u_2^2)\sigma_{k^\star}^2}.$$

Marrs (1998) extended the reversible jump MCMC algorithm to multivariate spherical Gaussian mixtures, where $\mathbf{\Sigma}_k = \sigma_k^2\mathbf{I}_r$, which reduced for $r = 1$ to a merge/split move for univariate mixtures of normal distributions, that is different from the one described above. When merging two components, Marrs (1998) uses the following rule,

$$\eta_{k^\star}\boldsymbol{\mu}_{k^\star} = \eta_{k_1}\boldsymbol{\mu}_{k_1} + \eta_{k_2}\boldsymbol{\mu}_{k_2}, \tag{6.52}$$

to merge the component means, which reduces to (6.47) for $r = 1$, but the rule

$$\eta_{k^\star}\sigma_{k^\star}^2 = \eta_{k_1}\sigma_{k_1}^2 + \eta_{k_2}\sigma_{k_2}^2,$$

rather than (6.48) to define the variance of the new component. The split move reads:

$$\boldsymbol{\mu}_{k_1} = \boldsymbol{\mu}_{k^\star} - \mathbf{u}_2\sigma_{k^\star}\sqrt{\frac{\eta_{k_2}}{\eta_{k_1}}}, \qquad \boldsymbol{\mu}_{k_2} = \boldsymbol{\mu}_{k^\star} + \mathbf{u}_2\sigma_{k^\star}\sqrt{\frac{\eta_{k_1}}{\eta_{k_2}}},$$
$$\sigma_{k_1}^2 = u_3\sigma_{k^\star}^2\frac{\eta_{k^\star}}{\eta_{k_1}}, \qquad \sigma_{k_2}^2 = (1 - u_3)\sigma_{k^\star}^2\frac{\eta_{k^\star}}{\eta_{k_2}},$$

where $u_1$, $\mathbf{u}_2 = (u_{21}, \dots, u_{2r})$, and $u_3$ all are drawn from a $\mathcal{B}(2,2)$ distribution, and the components of $\mathbf{u}_2$ have a probability of 0.5 of being negative. The determinant of the Jacobian for this split move is given by

$$\frac{\eta_{k^\star}\sigma_{k^\star}^{r+1}}{2(u_1(1-u_1))^{(r+1)/2}\sqrt{u_3(1-u_3)}}.$$

The acceptance rate of a combine move will be strongly influenced by combining similar components rather than different ones. For mixtures of normal distributions, Richardson and Green (1997) choose at random a pair $(k_1, k_2)$ among all components that are adjacent in terms of their means $\mu_{k_1}$ and $\mu_{k_2}$. This is possible, as Richardson and Green (1997) do not sample from the unconstrained posterior, but impose an order constraint on the means. In an unconstrained setting, the procedure of Marrs (1998) is useful, which first selects one component $k_1$ randomly among all components, and then computes a distance $r(\boldsymbol{\theta}_{k_1}; \boldsymbol{\theta}_k)$ between $k_1$ and any other component $k \neq k_1$. The second component $k_2$ is then chosen with probability proportional to $1/r(\boldsymbol{\theta}_{k_1}; \boldsymbol{\theta}_{k_2})$.

Dellaportas and Papageorgiou (2006) adopt the reversible jump MCMC to general multivariate mixtures and construct split and merge moves, in a manner similar to Richardson and Green (1997). As the dimension of $\mathbf{y}_i$ increases, the difference in the number of parameters between a mixture with $K$ and $K+1$ components, which is equal to $(r+1)(r/2+1)$, increases with order $r^2$. The main challenge is to split the covariance matrix in such a way that the overall dispersion remains relatively constant, whereas the new matrices are positive definite. To solve this problem, Dellaportas and Papageorgiou (2006) operate in the space of the $r$ eigenvectors and eigenvalues of the covariance matrix, and propose permuting the current eigenvectors through randomly chosen permutation matrices.

To avoid the complicated splitting of $\boldsymbol{\mu}_k$ and $\boldsymbol{\Sigma}_k$ in a split move, Tadesse et al. (2005) marginalize with respect to $\boldsymbol{\mu} = (\boldsymbol{\mu}_1, \ldots, \boldsymbol{\mu}_K)$ and $\boldsymbol{\Sigma} = (\boldsymbol{\Sigma}_1, \ldots, \boldsymbol{\Sigma}_K)$ which is possible under the conjugate prior (6.39) and operate only on the space of all allocations $\mathbf{S} = (S_1, \ldots, S_N)$, using the marginal likelihood $p(\mathbf{y}|\mathbf{S})$ given in (6.45). Split and merge moves are defined directly through the allocations $S_i$. The resulting model space MCMC sampler is closely related to the Markov chain Monte Carlo model comparison ($\mathrm{MC}^3$) algorithm of Madigan and York (1995). Because this algorithm involves rather simple acceptance rates, it is a promising alternative to the full reversible jump method described above.

# 7

# Data Analysis Based on Finite Mixtures

This chapter lays out the more practical side of mixtures of normal distributions and discusses several statistical topics from the viewpoint of finite mixture modeling, such as cluster analysis in Section 7.1, outlier detection in Section 7.2, robust modeling based on mixture of Student-$t$ distributions in Section 7.3, and discriminant analysis and density estimation in Section 7.4.

## 7.1 Model-Based Clustering

### 7.1.1 Some Background on Cluster Analysis

The main purpose of cluster analysis is to group previously unstructured data $\{\mathbf{y}_1, \ldots, \mathbf{y}_N\}$, where $\mathbf{y}_i$ is an $r$-dimensional vector, into groups containing data that are similar in some sense (Kaufman and Rousseeuw, 1990; Everitt et al., 2001). In applying cluster analysis one expects that meaningful groups exist among the data, however, there are no external criteria by which to define them. Rather cluster analysis relies on an internal criterion defined solely from the data that evaluates each partition $\mathbf{S}$ of the $N$ observations into $K$ groups through some numerical function $c(\mathbf{S})$, measuring adequacy of this particular partition. By searching for that partition which minimizes (or maximizes) $c(\mathbf{S})$, the data themselves suggest sensible groupings.

The choice of an appropriate clustering criterion, however, has been the subject of much controversy; see Everitt et al. (2001, Chapter 5) for an extensive review. Clustering criteria are often based on decomposing the total variance $\mathbf{T}$ of the data around the overall mean $\overline{\mathbf{y}}$,

$$\mathbf{T} = \sum_{i=1}^{N} (\mathbf{y}_i - \overline{\mathbf{y}})(\mathbf{y}_i - \overline{\mathbf{y}})',$$

into the total within-group variance $\boldsymbol{W}(\mathbf{S})$ and the total between-group variance $\boldsymbol{B}(\mathbf{S})$:

$$\mathbf{T} = \boldsymbol{W}(\mathbf{S}) + \boldsymbol{B}(\mathbf{S}), \tag{7.1}$$

where

$$\boldsymbol{W}(\mathbf{S}) = \sum_{k=1}^{K} \boldsymbol{W}_k(\mathbf{S}), \tag{7.2}$$

$$\boldsymbol{B}(\mathbf{S}) = \sum_{k=1}^{K} N_k(\mathbf{S})(\overline{\mathbf{y}}_k(\mathbf{S}) - \overline{\mathbf{y}})(\overline{\mathbf{y}}_k(\mathbf{S}) - \overline{\mathbf{y}})', \tag{7.3}$$

and $N_k(\mathbf{S})$, $\overline{\mathbf{y}}_k(\mathbf{S})$, and $\boldsymbol{W}_k(\mathbf{S})$ are the group sizes, the group means, and the within-group variance for partition $\mathbf{S}$, defined earlier in (6.28) to (6.30).

A natural criterion for the evaluation of a certain partition $\mathbf{S}$ is the amount of heterogeneity explained by this particular partition, measured in terms of an empirical coefficient of determination such as

$$1 - \frac{\mathrm{tr}\,(\boldsymbol{W}(\mathbf{S}))}{\mathrm{tr}\,(\mathbf{T})} \qquad \text{or} \qquad 1 - \frac{|\boldsymbol{W}(\mathbf{S})|}{|\mathbf{T}|}.$$

In the first case, one would search for the partition $\mathbf{S}$ that minimizes $\mathrm{tr}\,(\boldsymbol{W}(\mathbf{S}))$, which is perhaps the most commonly used criterion in cluster analysis. In the second case one would search for the partition $\mathbf{S}$ that minimizes $|\boldsymbol{W}(\mathbf{S})|$, a criterion suggested by Friedman and Rubin (1967).

Although these criteria seem to be heuristic, we note already at this point that the variance decomposition in (7.1) is closely related to the variance decomposition of a multivariate mixture of normal distributions, given earlier in (6.3), if the parameters $\eta_k$, $\boldsymbol{\mu}_k$, and $\boldsymbol{\Sigma}_k$ are substituted by the complete-data ML estimator $N_k(\mathbf{S})/N$, $\overline{\mathbf{y}}_k(\mathbf{S})$, and $\boldsymbol{W}_k(\mathbf{S})/N_k(\mathbf{S})$. This close relationship between heuristic clustering criteria and mixture modeling indicates that clustering may be directly based on finite mixture models.

### 7.1.2 Model-Based Clustering Using Finite Mixture Models

Many authors prefer probabilistic models for clustering multivariate data to cluster analysis based on ad hoc criteria; see McLachlan and Basford (1988), Bock (1996), and Fraley and Raftery (2002) for a comprehensive review. In model-based clustering it is assumed that a collection of multivariate observations $\mathbf{y}_1, \ldots, \mathbf{y}_N$, where $\mathbf{y}_i$ is an $r$-dimensional vector, arises from a population with $K$ subgroups (clusters) each of relative size $\eta_k$. Within each cluster the data are generated by a multivariate distribution $p(\mathbf{y}_i|\boldsymbol{\theta}_k)$ with group specific parameter $\boldsymbol{\theta}_k$. As group membership is unknown, $\mathbf{y}_i$ is regarded as the realization of a random variable $\mathbf{Y}$ from a multivariate mixture distribution:

$$p(\mathbf{y}_i|\boldsymbol{\vartheta}) = \sum_{k=1}^{K} \eta_k p(\mathbf{y}_i|\boldsymbol{\theta}_k).$$

In this case each of the $K$ components of the mixture corresponds to one the $K$ clusters. In a clustering context little information about the parameters of the component densities is available, and classification has to be carried out simultaneously with parameter estimation by jointly quantifying information about $(\boldsymbol{\theta}_1, \dots, \boldsymbol{\theta}_K)$ and the allocation variables $\mathbf{S}$, which provide a basis for posterior classification of the observations $\{\mathbf{y}_1, \dots, \mathbf{y}_N\}$ into the different components.

When observing a random variable with metric features commonly a multivariate mixture of normal distributions is applied for clustering (Wolfe, 1970; Scott and Symons, 1971; Binder, 1978; Symons, 1981; Banfield and Raftery, 1993). Scott and Symons (1971) showed that many heuristic data-based clustering criteria are equivalent to finding the optimal classification for a certain mixture of normal distributions; see Subsection 7.1.3.

Practical application of model-based clustering using Gaussian mixtures include character recognition (Murtagh and Raftery, 1984), tissue segmentation (Banfield and Raftery, 1993), minefield and seismic fault detection (Dasgupta and Raftery, 1998), clustering gene expression data (Yeung et al., 2001), and classification of astronomical data (Celeux and Govaert, 1995). The model-based clustering framework is, however, more general than that and has been applied to clustering under outliers and noise (see Subsection 7.2.4), to clustering discrete multivariate data (see Subsection 9.5.1), as well as to mixed-mode data (see Subsection 9.6.1).

Several decades ago, Everitt (1979) addressed some nagging problems in practical cluster analysis that were unresolved at that time. These problems are selecting a suitable clustering criterion, selecting the number of clusters, and computational issues such as identifying a sensible search strategy for finding the optimal partition of the data and choosing sensible starting values. It has to be admitted that partly these problems are still relevant today. In particular the problem of choosing the number of clusters is still unsolved despite tremendous work done in this area; see Subsection 7.1.4 and Gordon (1999) for a recent review.

Practical cluster analysis through mixtures of normals is often based on the EM algorithm (Wolfe, 1970). Variants of the EM algorithm that were found to be useful for clustering based on finite mixture models include the stochastic EM (SEM) algorithm (Celeux and Diebolt, 1985), where the indicators in the E-step are simulated rather than estimated, and the classification EM (CEM) algorithm (Celeux and Govaert, 1992), which performs classification based on the estimated indicators prior to the M-step.

Scott and Symons (1971), Binder (1978), and Symons (1981) studied clustering based on a multivariate mixture of normal distributions from a Bayesian point of view, however, the practicability was limited in these days and approximations had to be applied (Binder, 1981). With the availability of powerful MCMC methods, interest in the Bayesian approach to model-based clustering has been renewed (Chen and Liu, 1996; Bensmail et al., 1997; Stephens, 2000b; Frühwirth-Schnatter and Kaufmann, 2006b). There are various reasons why one might be interested in adopting a Bayesian approach to model-based clustering using finite mixtures. First, as discussed in Subsection 6.1.2, using the EM algorithm for ML maximization may lead to degenerate solutions, which are less likely under Bayesian estimation using proper priors. Second, there exists a more principled way of posterior classification of the objects into the different clusters, as explained in Subsection 7.1.7.

Finally, for practical Bayesian cluster analysis, powerful Markov chain Monte Carlo methods may be applied, such as sampling the allocation without sampling the parameters (see *Algorithm 3.1* and *Algorithm 3.2*), or data augmentation and Gibbs sampling, which deliver draws of both the unknown parameters and the allocations (see *Algorithm 3.4* and *Algorithm 3.5*). These are stochastic search algorithms, based on ergodic Markov chains, which converge theoretically to the desired posterior distribution, regardless of the starting partition. Consider, for instance, *Algorithm 3.2* with a symmetric proposal, where $q(S_i^{new}|S_i) = q(S_i|S_i^{new})$; the ratio is equivalent to the ratio of the two classification probabilities:

$$r_i = \frac{p(\mathbf{S}_{-i}, S_i^{new}|\mathbf{y})}{p(\mathbf{S}_{-i}, S_i|\mathbf{y})}.$$

In this case, any proposed classification that increases the joint marginal posterior probability will be accepted with probability one. If we proposed a classification with lower marginal posterior probability, then it will be accpeted with a probability that is equal to the posterior odds of the new versus the old classification. Thus the Metropolis–Hastings algorithm may be regarded as a stochastic version of hill climbing methods that are commonly applied in optimizing clustering criteria.

Sensible starting values for $\mathbf{S}^{(0)}$ may be obtained using common heuristic clustering methods such as $K$ means or hierarchical clustering. Nevertheless, dependence on the initial partition $\mathbf{S}^{(0)}$ is not an unlikely occurrence.

A certain challenge for both Bayesian and ML approaches is clustering large data sets because for each sweep of the MCMC sampler or for each iteration of the EM algorithm one needs to evaluate the likelihood of $\mathbf{y}_i$ for all $i = 1, \dots, N$ under $K$ different assumptions concerning $S_i$. This may be rather time consuming for large data sets. Banfield and Raftery (1993) suggest applying model-based clustering only to a small subset of the data and to use the resulting mixture approximation to classify the remaining observations. Problems arise here with small clusters because these may not be represented in the chosen subset. To reduce complexity while involving all observations, Roeder and Wasserman (1997b, p.897) suggest applying data augmentation and Gibbs sampling and updating only a small fraction of randomly selected allocations, for instance, five percent, rather than sampling all indicators $\mathbf{S}$ at each sweep of the sampler. Posse (2001) discusses hierarchical model-based clustering for large data sets.

A final challenge is clustering of high-dimensional data which is addressed in Subsection 7.4.1.

### 7.1.3 The Classification Likelihood and the Bayesian MAP Approach

In the classification likelihood approach (Scott and Symons, 1971; Symons, 1981), the likelihood function $p(\mathbf{y}, \mathbf{S}|\boldsymbol{\vartheta}) = p(\mathbf{y}|\mathbf{S}, \boldsymbol{\vartheta})p(\mathbf{S}|\boldsymbol{\vartheta})$ of the complete data, which has been defined as

$$p(\mathbf{y}|\mathbf{S}, \boldsymbol{\vartheta})p(\mathbf{S}|\boldsymbol{\vartheta}) = \prod_{i=1}^{N} p(\mathbf{y}_i|\boldsymbol{\theta}_{S_i}) \prod_{k=1}^{K} \eta_k^{N_k(\mathbf{S})}, \tag{7.4}$$

is regarded as a function of both the unknown allocations $\mathbf{S}$ and the unknown parameters $\boldsymbol{\vartheta}$, which are then estimated by maximizing $p(\mathbf{y}, \mathbf{S}|\boldsymbol{\vartheta})$ jointly with respect to $\mathbf{S}$ and $\boldsymbol{\vartheta}$. The classification likelihood approach is often applied under the assumption that the weight distribution is equal to the uniform distribution, in which case $\eta_k = 1/K$, $p(\mathbf{S}|\boldsymbol{\eta})$ reduces to

$$p(\mathbf{S}) \propto \prod_{k=1}^{K} \left(\frac{1}{K}\right)^{N_k(\mathbf{S})} = \left(\frac{1}{K}\right)^{N} \propto \text{constant},$$

and the classification likelihood function is equal to the sampling distribution $p(\mathbf{y}|\mathbf{S}, \boldsymbol{\vartheta})$.

The classification likelihood approach is generally known to yield inconsistent estimates of the component parameters $\boldsymbol{\theta}_1, \dots, \boldsymbol{\theta}_K$ and the weight distribution $\boldsymbol{\eta}$ (Bryant and Williamson, 1978) which is not the case when maximizing the mixture likelihood function $p(\mathbf{y}|\boldsymbol{\vartheta})$, where $\mathbf{S}$ is integrated out. Celeux and Govaert (1993) compared the two approaches and found that the classification likelihood approach tends to be better for small samples, whereas the mixture likelihood approach is clearly preferable for large samples.

Despite these practical shortcomings, the classification likelihood approach is very useful for a theoretical understanding of the mixture model structure underlying heuristic clustering criteria which are formulated without reference to a probabilistic model, as exemplified by Scott and Symons (1971), Symons (1981), and Celeux and Govaert (1991).

#### Relation to Common Clustering Criteria

Scott and Symons (1971) realized that maximizing the classification likelihood function of a multivariate mixture of normal distributions with uniform weight distribution under different assumptions about component variance–covariance matrices is related to common clustering criteria such as the $\text{tr}\,(\boldsymbol{W}(\mathbf{S}))$ or the $|\boldsymbol{W}(\mathbf{S})|$ criteria. For a mixture of normal distributions, the log of the sampling distribution reads:

$$\log p(\mathbf{y}|\mathbf{S}, \boldsymbol{\mu}, \boldsymbol{\Sigma}) =$$
$$-\frac{1}{2}\sum_{k=1}^{K}\left(N_k \log|\boldsymbol{\Sigma}_k| + \sum_{i:S_i=k} (\mathbf{y}_i - \boldsymbol{\mu}_k)' \boldsymbol{\Sigma}_k^{-1} (\mathbf{y}_i - \boldsymbol{\mu}_k)\right) + \text{constant}.$$

Maximizing $p(\mathbf{y}|\mathbf{S}, \boldsymbol{\mu}, \boldsymbol{\Sigma})$ reduces to maximizing $p(\mathbf{y}|\mathbf{S}, \hat{\boldsymbol{\mu}}(\mathbf{S}), \boldsymbol{\Sigma})$, where $\hat{\boldsymbol{\mu}}_k(\mathbf{S}) = \overline{\mathbf{y}}_k(\mathbf{S})$, or equivalently:

$$-\frac{1}{2}\sum_{k=1}^{K} \left(N_k(\mathbf{S})\log|\boldsymbol{\Sigma}_k| + \operatorname{tr}\left(\boldsymbol{W}_k(\mathbf{S})\boldsymbol{\Sigma}_k^{-1}\right)\right) + \text{constant}, \tag{7.5}$$

where $\boldsymbol{W}_k(\mathbf{S})$ is the measure of within-group variability defined earlier in (6.30).

Different optimal solutions are obtained for different assumptions about $\boldsymbol{\Sigma}_k$. For a homogeneous mixture of normals ($\boldsymbol{\Sigma}_k = \boldsymbol{\Sigma}$) with $\boldsymbol{\Sigma}$ unknown, (7.5) reduces to

$$-\frac{1}{2}\left(N\log|\boldsymbol{\Sigma}| + \operatorname{tr}\left(\boldsymbol{\Sigma}^{-1}\sum_{k=1}^{K}\boldsymbol{W}_k(\mathbf{S})\right)\right) + \text{constant},$$

which is maximized for a fixed allocation $\mathbf{S}$ for $\hat{\boldsymbol{\Sigma}}(\mathbf{S}) = \boldsymbol{W}(\mathbf{S})/N$, where $\boldsymbol{W}(\mathbf{S}) = \sum_{k=1}^{K}\boldsymbol{W}_k(\mathbf{S})$. Finding the optimal allocation reduces to maximizing

$$c(\mathbf{S}) = -\frac{N}{2}\log|\boldsymbol{W}(\mathbf{S})| + \text{constant}, \tag{7.6}$$

with respect to $\mathbf{S}$, or equivalently minimizing $|\boldsymbol{W}(\mathbf{S})|$, which is the clustering criterion suggested by Friedman and Rubin (1967). For an isotropic mixture of normals ($\boldsymbol{\Sigma}_k = \sigma^2\mathbf{I}$) with $\sigma^2$ unknown, (7.5) reduces to

$$-\frac{1}{2}\left(rN\log\sigma^2 + \operatorname{tr}\left(\boldsymbol{W}(\mathbf{S})\right)/\sigma^2\right) + \text{constant},$$

which is maximized for a fixed allocation $\mathbf{S}$ for $\hat{\sigma}^2(\mathbf{S}) = \operatorname{tr}\left(\boldsymbol{W}(\mathbf{S})\right)/N$. Finding the optimal allocation reduces to maximizing

$$c(\mathbf{S}) = -\frac{rN}{2}\log\operatorname{tr}\left(\boldsymbol{W}(\mathbf{S})\right) + \text{constant}, \tag{7.7}$$

with respect to $\mathbf{S}$, or equivalently minimizing $\operatorname{tr}\left(\boldsymbol{W}(\mathbf{S})\right)$, which is the most commonly used clustering criterion.

It has been noted in several empirical studies, that the $\operatorname{tr}\left(\boldsymbol{W}(\mathbf{S})\right)$ criterion imposes a spherical structure on the grouping even if the true groups are of different shape, whereas the $|\boldsymbol{W}(\mathbf{S})|$ criterion allows for elliptical clusters. This is not surprising in light of the results derived above.

For a heterogeneous mixture with an unconstrained variance–covariance matrix, (7.5) is maximized for a fixed allocation $\mathbf{S}$ for $\hat{\boldsymbol{\Sigma}}_k(\mathbf{S}) = \boldsymbol{W}_k(\mathbf{S})/N_k(\mathbf{S})$. Finding the optimal allocation reduces to maximizing

$$c(\mathbf{S}) = -\frac{1}{2}\sum_{k=1}^{K} N_k(\mathbf{S})\log\left|\frac{\boldsymbol{W}_k(\mathbf{S})}{N_k(\mathbf{S})}\right| + \text{constant}, \tag{7.8}$$

with respect to $\mathbf{S}$, or equivalently minimizing

$$\prod_{k=1}^{K} \left| \frac{\boldsymbol{W}_k(\mathbf{S})}{N_k(\mathbf{S})} \right|^{N_k(\mathbf{S})}, \tag{7.9}$$

which is a criterion related to the one suggested by Scott and Symons (1971).[1]

For spherical mixtures of normals ($\boldsymbol{\Sigma}_k = \sigma_k^2 \mathbf{I}$) with $\sigma_1^2, \dots, \sigma_K^2$ unknown, (7.5) reduces to

$$-\frac{1}{2} \sum_{k=1}^{K} \left( N_k(\mathbf{S}) \log \sigma_k^2 + \operatorname{tr}\left( \boldsymbol{W}_k(\mathbf{S}) / \sigma_k^2 \right) \right) + \text{constant},$$

which is maximized for a fixed allocation $\mathbf{S}$ for $\hat{\sigma}_k^2(\mathbf{S}) = \operatorname{tr}\left( \boldsymbol{W}_k(\mathbf{S}) \right) / N_k(\mathbf{S})$. Finding the optimal allocation reduces to maximizing

$$c(\mathbf{S}) = -\frac{1}{2} \sum_{k=1}^{K} N_k(\mathbf{S}) \log \operatorname{tr} \left( \frac{\boldsymbol{W}_k(\mathbf{S})}{N_k(\mathbf{S})} \right) + \text{constant}, \tag{7.10}$$

with respect to $\mathbf{S}$, or equivalently minimizing

$$\prod_{k=1}^{K} \operatorname{tr} \left( \frac{\boldsymbol{W}_k(\mathbf{S})}{N_k(\mathbf{S})} \right)^{N_k(\mathbf{S})}, \tag{7.11}$$

which is the clustering criterion suggested by Banfield and Raftery (1993). It has been noted by several authors (Binder, 1978; Scott and Symons, 1971) that both the $\operatorname{tr}(\boldsymbol{W}(\mathbf{S}))$ as well as the $|\boldsymbol{W}(\mathbf{S})|$ criterion tend to give clusters of equal size, which is not surprising because they are based on the classification likelihood of a mixture with uniform weight distribution.

In later work, Symons (1981) considered maximizing the classification likelihood (7.4) for a multivariate mixture of normals for an unconstrained weight distribution with respect to $\boldsymbol{\vartheta}$ and $\mathbf{S}$. Maximizing (7.4) for a fixed classification $\mathbf{S}$ leads to $\hat{\eta}_k(\mathbf{S}) = N_k(\mathbf{S})/N$, regardless of $\boldsymbol{\Sigma}_k$, whereas the estimators $\hat{\boldsymbol{\mu}}_k(\mathbf{S})$ and $\hat{\boldsymbol{\Sigma}}_k(\mathbf{S})$ are the same as for a uniform weight distribution. It is evident that the presence of $p(\mathbf{S}|\boldsymbol{\eta})$ changes the clustering criterion which now requires maximizing

$$c(\mathbf{S}) + \sum_{k=1}^{K} N_k(\mathbf{S}) \log N_k(\mathbf{S}), \tag{7.12}$$

where $c(\mathbf{S})$ is one of the clustering criteria defined in (7.6) to (7.10) and the second term results from maximizing $\log p(\mathbf{S}|\boldsymbol{\eta})$ with respect to $\boldsymbol{\eta}$ for a

[1] The criterion suggested by Scott and Symons (1971) reads $\prod_{k=1}^{K} |\boldsymbol{W}_k(\mathbf{S})|^{N_k(\mathbf{S})}$, however, as noted by Banfield and Raftery (1993) this is not obtained when maximizing (7.8).

given classification $\mathbf{S}$. For homogeneous mixtures, for instance, this leads to minimizing

$$N \log |\boldsymbol{W}(\mathbf{S})| - 2 \sum_{k=1}^{K} N_k(\mathbf{S}) \log N_k(\mathbf{S}),$$

which is a criterion suggested by Symons (1981). Similar criteria are obtained for the other model assumptions concerning the covariances.

### Bayesian Maximum A Posteriori (MAP) Classification

Bayesian maximum a posteriori (MAP) classification has been suggested by Symons (1981) and is based on maximizing the joint posterior

$$p(\boldsymbol{\vartheta}, \mathbf{S}|\mathbf{y}) \propto p(\mathbf{y}|\boldsymbol{\vartheta}, \mathbf{S}) p(\mathbf{S}|\boldsymbol{\vartheta}) p(\boldsymbol{\vartheta}), \tag{7.13}$$

simultaneously with respect to $\boldsymbol{\vartheta}$ and $\mathbf{S}$, where $p(\mathbf{y}|\boldsymbol{\vartheta}, \mathbf{S}) p(\mathbf{S}|\boldsymbol{\vartheta})$ is equal to the classification likelihood. Bayesian MAP classification is equivalent to the classification likelihood under the flat prior $p(\boldsymbol{\vartheta}) \propto$ constant and is related to, but different from the classification likelihood approaches for any other prior. If $p(\boldsymbol{\vartheta})$ is a proper prior, the resulting clustering criteria show a certain robustness under small and nearly empty groups.

Due to the close relationship between model-based clustering and mixture modeling, an approximation to the Bayesian MAP classifier is available when estimating the underlying mixture model using data augmentation and Gibbs sampling as described in *Algorithm 3.4* and *Algorithm 3.5*, because it is possible to evaluate the nonnormalized posterior $p(\boldsymbol{\vartheta}^{(m)}, \mathbf{S}^{(m)}|\mathbf{y})$ for each MCMC draw, and to keep track of the classification that gave the highest posterior density. Most conveniently, the Bayesian MAP classifier is invariant to label switching.

## 7.1.4 Choosing Clustering Criteria and the Number of Components

Two basic model selection problems arising in practical cluster analysis are the selection of an appropriate cluster criterion and the determination of the number of clusters. Symons (1981, p.41), when comparing various criteria for simulated and real data, comes to the conclusion that "There seems to be no simple recommendation to guide the use of these criteria [...]. The most appropriate criterion or approach depends upon the knowledge of the structure of the data set. [...] the only reasonable a priori suggestion is to compare results obtained from several approaches."

Model-based clustering is rather helpful in this respect, as common heuristic clustering criteria are substituted by assumptions about the component parameters, such as $\boldsymbol{\Sigma}_k$ in a multivariate Gaussian mixture, and about the distribution of the allocations. Hence comparing clustering criteria is cast into the problem of selecting among different statistical models.

It is generally recommended to select the clustering criterion and the number of clusters simultaneously (Fraley and Raftery, 1998). If the structure of the component parameters is misspecified, there will be a certain trade-off between the number of components and the clustering method. If model selection is carried out, for instance, for a multivariate mixture of normals where the covariance structures are too simple, more clusters will be needed to represent the distribution of the data.

### Diagnosing the Within- and Between-Cluster Variance

Many authors tried to diagnose the number of clusters using some method based on the within- and between-cluster sum of squares $W(K)$ and $B(K)$ resulting from fitting a $K$ component mixture. In various simulation studies (Milligan and Cooper, 1985; Tibshirani et al., 2001; Sugar and James, 2003) two simple methods were among those performing best, namely maximizing over $K$ an analysis of variance type statistic (Calinski and Harabasz, 1974):

$$\mathrm{CH}(K) = \frac{B(K)/(K-1)}{W(K)/(N-K)},$$

and maximizing over $K$ a statistic measuring the rate of change in distortion (Krzanowski and Lai, 1985):

$$\mathrm{KL}(K) = \left| \frac{\mathrm{DIFF}(K)}{\mathrm{DIFF}(K+1)} \right|,$$
$$\mathrm{DIFF}(K) = (K-1)^{2/r} W(K-1) - K^{2/r} W(K).$$

A recent approach along these lines is the gap statistic (Tibshirani et al., 2001) which is based on the idea of standardizing $\log(W(K))$ with its expected value under an appropriate reference distribution. $M$ data sets are simulated from a uniform distribution over the range of the original data and the gap statistic is defined as

$$\mathrm{Gap}(K) = \frac{1}{M} \sum_{m=1}^{M} \log(W_m^\star(K)) - \log(W(K)),$$

with $W_m^\star(K)$ being the within-cluster sum of squares for the $m$th simulated data set. To avoid unnecessary clusters, an estimate of the standard deviation $s_K$ of $\log(W_m^\star(K))$ is produced, and the smallest value such that $\mathrm{Gap}(K) > \mathrm{Gap}(K+1) - s_K$ is chosen as the number of clusters.

Sugar and James (2003) suggest a method based on the minimum achievable distortion associated with fitting $K$ centers $\boldsymbol{\mu}_1, \dots, \boldsymbol{\mu}_K$ to the data:

$$d_K = \frac{1}{r} \min_{\boldsymbol{\mu}_1,\dots,\boldsymbol{\mu}_K} \mathrm{E}\left( (\mathbf{Y} - \boldsymbol{\mu}_k)' \boldsymbol{\Sigma}^{-1} (\mathbf{Y} - \boldsymbol{\mu}_k) \right),$$

which is a measure of within-cluster dispersion based on the average distance, per dimension, between each observation and its closest cluster center. The minimum achievable distortion $d_K$ is monotone decreasing in $K$ with the decrease being relatively small when unnecessary clusters are added. A simple approach would plot $d_K$ against $K$ and look for that $K$ where an elbow in the curve occurs. Sugar and James (2003) show, both theoretically and empirically, that this distortion curve, when transformed to the power $2/r$, exhibits a sharp jump at the true number of clusters. In practice, $d_K$ is estimated using $\hat{d}_K$, the minimum distortion obtained by applying the $k$-means clustering algorithm (Hartigan, 1975). Therefore $K$ is selected such that

$$J_K = \hat{d}_K^{2/r} - \hat{d}_{K-1}^{2/r},$$

with $\hat{d}_0^{2/r} \equiv 0$ is largest.

The simulation studies in Tibshirani et al. (2001) and Sugar and James (2003) demonstrate that both the gap statistic as well as the jump statistic work very well provided that the clusters do not overlap too severely.

## Model Selection Based on Marginal Likelihoods

As each combination of a certain number $K$ of components and a certain clustering method corresponds to a different statistical model, the machinery of Bayesian model selection as described in Section 4.5 could be used to select jointly the clustering criterion and the number of components in the mixture (Bensmail et al., 1997).

The Bayesian approach requires the choice of a prior distribution $p(\boldsymbol{\vartheta}_K)$ for all unknown parameters $\boldsymbol{\vartheta}_K$ appearing in each model $\mathcal{M}_K$ and the choice of prior probabilities for each model $\mathcal{M}_K$ as well as a loss function quantifying the consequences of making a wrong decision. For the matter of convenience, often default values are chosen, in particular a uniform prior over all models and a 0/1 loss function, leading to choosing that model $\mathcal{M}_K$ that maximizes the marginal likelihood $p(\mathbf{y}|\mathcal{M}_K)$. As discussed in great length in Subsection 4.5.5, any of these assumptions may exercise considerable influence on the model selection procedure, even if the true data-generating mechanism is among the models under investigation.

A numerical challenge with applying the Bayesian approach is the evaluation of the marginal likelihood $p(\mathbf{y}|\mathcal{M}_K)$. Approximations were based on the Laplace–Metropolis estimator (Bensmail et al., 1997; see also Subsection 5.5.3) and bridge sampling (Frühwirth-Schnatter, 2004; Frühwirth-Schnatter et al., 2004; Frühwirth-Schnatter and Kaufmann, 2006b; see Subsection 5.4.6). To avoid these numerical issues, often a rough approximation to the marginal likelihood $p(\mathbf{y}|\mathcal{M}_K)$ based on $\mathrm{BIC}_K$ is used (Fraley and Raftery, 1998; Dasgupta and Raftery, 1998; see also Subsection 4.4.2).

### Classification-Based Information Criteria

AIC and BIC do not take into account that in a clustering context a finite mixture model is fitted with the hope of finding a good partition of the data. Several specific criteria that involve the quality of the resulting partition have been developed with the specific aim of selecting the number of components of a finite mixture model in model-based clustering.

Let $D_{ik}$ be the 0/1 coding of the allocation $S_i$, introduced in (2.31), let $\hat{D}_{ik}$ be the EM estimator of $D_{ik}$, derived in (2.32), and let $\hat{\mathbf{D}} = (\hat{D}_{ik})$ be the corresponding $(N \times K)$ fuzzy classification matrix. By (2.28) the log of the complete-data likelihood $p(\mathbf{y}, \hat{\mathbf{D}}|\boldsymbol{\vartheta}_K)$ where the unknown allocations $\mathbf{S}$, coded through $\mathbf{D}$, are substituted by $\hat{\mathbf{D}}$, is equal to the log of the mixture likelihood function $p(\mathbf{y}|\boldsymbol{\vartheta}_K)$ penalized by the entropy $\mathrm{EN}(\boldsymbol{\vartheta}_K)$, defined in (2.7):

$$\log p(\mathbf{y}, \hat{\mathbf{D}}|\boldsymbol{\vartheta}_K) = \log p(\mathbf{y}|\boldsymbol{\vartheta}_K) - \mathrm{EN}(\boldsymbol{\vartheta}_K). \tag{7.14}$$

Evaluating this relationship at the ML estimator $\hat{\boldsymbol{\vartheta}}_K$ yields the classification likelihood information criterion $\mathrm{CLC}_K$ (Biernacki and Govaert, 1997):

$$\mathrm{CLC}_K = -2 \log p(\mathbf{y}|\hat{\boldsymbol{\vartheta}}_K) + 2\mathrm{EN}(\hat{\boldsymbol{\vartheta}}_K), \tag{7.15}$$

where the entropy $\mathrm{EN}(\hat{\boldsymbol{\vartheta}}_K)$ defined by

$$\mathrm{EN}(\hat{\boldsymbol{\vartheta}}_K) = -\sum_{i=1}^{N}\sum_{k=1}^{K} \Pr(S_i = k|\mathbf{y}_i, \hat{\boldsymbol{\vartheta}}_K) \log \Pr(S_i = k|\mathbf{y}_i, \hat{\boldsymbol{\vartheta}}_K), \tag{7.16}$$

is used as a penalty rather than model complexity as in $\mathrm{AIC}_K$ or $\mathrm{BIC}_K$. $\mathrm{EN}(\hat{\boldsymbol{\vartheta}}_K)$ measures the inability of the fitted $K$-component mixture model to provide a good partition of the data. It is close to 0 if the resulting clusters are well separated and will have a large value if this is not the case. Celeux and Soromenho (1996) use $\mathrm{EN}(\hat{\boldsymbol{\vartheta}}_K)$ directly, normalized by $\log p(\mathbf{y}|\hat{\boldsymbol{\vartheta}}_K) - \log p(\mathbf{y}|\hat{\boldsymbol{\vartheta}}_1)$, however, this criterion fails to test $K = 1$ against $K > 1$, because $\mathrm{EN}(\hat{\boldsymbol{\vartheta}}_K) = 0$ for $K = 1$.

Biernacki and Govaert (1997) consider minimizing $\mathrm{CLC}_K$ as a criterion for selecting the number $K$ of clusters. They conclude that this works only for well-separated clusters with a fixed weight distribution. If the weight distribution is unknown, $\mathrm{CLC}_K$ tends to overestimate the correct number of clusters, because the complete-data likelihood function does not account for model complexity.

To overcome these shortcomings, Biernacki et al. (2000) introduced the integrated classification likelihood (ICL) criterion. They started with the log of the marginal likelihood under the assumption that the allocations $\mathbf{S}$, coded through the dummy variables $\mathbf{D}$, are known:

$$\log p(\mathbf{y}, \mathbf{D}|\mathcal{M}_K) = \log p(\mathbf{y}|\mathbf{D}, \mathcal{M}_K) + \log p(\mathbf{D}|\mathcal{M}_K), \tag{7.17}$$

which they call the integrated classification likelihood, because the component parameters and the weight distribution are integrated out. Both quantities on the right-hand side of (7.17) may be computed explicitly for conjugate priors on $p(\boldsymbol{\vartheta}_K)$; see Subsection 3.3.3. Under the conditionally conjugate prior $\boldsymbol{\eta} \sim \mathcal{D}(e_0, \dots, e_0)$, for instance, $p(\mathbf{D}|\mathcal{M}_K)$ is given by

$$p(\mathbf{D}|\mathcal{M}_K) = \frac{\Gamma(Ke_0) \prod_{k=1}^{K} \Gamma(N_k(\mathbf{D}) + e_0)}{\Gamma(N + Ke_0)\Gamma(e_0)^K}, \tag{7.18}$$

where $N_k(\mathbf{D}) = N_k(\mathbf{S})$ may be written in the following way,

$$N_k(\mathbf{D}) = \sum_{i=1}^{N} D_{ik}.$$

A similar closed form is available for $p(\mathbf{y}|\mathbf{D}, \mathcal{M}_K)$, which involves the hyperparameters of the prior $p(\boldsymbol{\theta}_k|\mathcal{M}_K)$ and group-specific data summaries such as $\overline{\mathbf{y}}_k(\mathbf{D}) = \overline{\mathbf{y}}_k(\mathbf{S})$ which may be written as

$$\overline{\mathbf{y}}_k(\mathbf{D}) = \sum_{i=1}^{N} D_{ik}\mathbf{y}_i.$$

The first basic assumption made by Biernacki et al. (2000) is that the integrated classification likelihood $\log p(\mathbf{y}, \mathbf{D}|\mathcal{M}_K)$ could be used for selecting the number of components, even if unknown allocations $\mathbf{S}$, coded through $\mathbf{D}$, are substituted by an estimate $\mathbf{D}^\star$ which need not necessarily be equal to the fuzzy classification matrix $\hat{\mathbf{D}} = (\hat{D}_{ik})$. To evaluate this criterion, they choose $e_0 = 0.5$ to compute $p(\mathbf{D}^\star|\mathcal{M}_K)$ exactly from (7.18), whereas $-2\log p(\mathbf{y}|\mathbf{D}^\star, \mathcal{M}_K)$, which is the marginal likelihood of a complete-data problem, where regularity conditions hold, is approximated by BIC:

$$-2\log p(\mathbf{y}|\mathbf{D}^\star, \mathcal{M}_K) \approx -2\log p(\mathbf{y}|\mathbf{D}^\star, \boldsymbol{\theta}_1^\star, \dots, \boldsymbol{\theta}_K^\star, \mathcal{M}_K) + d_K^C \log N,$$

with $\boldsymbol{\theta}_1^\star, \dots, \boldsymbol{\theta}_K^\star$ being the complete-data estimator based on $\mathbf{D}^\star$, and $d_K^C$ being the number of distinct elements in $\boldsymbol{\theta}_1, \dots, \boldsymbol{\theta}_K$. If $\mathbf{D}^\star = \hat{\mathbf{D}}$, then $\boldsymbol{\theta}_k^\star = \hat{\boldsymbol{\theta}}_k$ is equal to the ML estimator of $\boldsymbol{\theta}_k$ in model $\mathcal{M}_K$. By using the BIC approximation, Biernacki et al. (2000) avoid having to choose an explicit hyperparameter for the prior on $\boldsymbol{\theta}_k$. The BIC approximation will be accurate as long as the estimated number of allocations $N_k(\mathbf{D}^\star)$ is not too small. The resulting integrated classification likelihood (ICL) criterion reads:

$$\begin{aligned} \mathrm{ICL}_K = & -2\log p(\mathbf{y}|\mathbf{D}^\star, \boldsymbol{\theta}_1^\star, \dots, \boldsymbol{\theta}_K^\star, \mathcal{M}_K) + d_K^C \log N \\ & -2\log\left(\frac{\Gamma(K/2) \prod_{k=1}^{K} \Gamma(N_k(\mathbf{D}^\star) + 0.5)}{\Gamma(N + K/2)\Gamma(0.5)^K}\right). \end{aligned} \tag{7.19}$$

McLachlan and Peel (2000, p.216) analyze this criterion for the special case where $\mathbf{D}^\star = \hat{\mathbf{D}}$ and $\boldsymbol{\theta}_k^\star = \hat{\boldsymbol{\theta}}_k$. Based on evaluating (7.14) at these estimates:

$$\log p(\mathbf{y}, \hat{\mathbf{D}}|\hat{\boldsymbol{\vartheta}}_K) = \log p(\mathbf{y}|\hat{\mathbf{D}}, \hat{\boldsymbol{\vartheta}}_K) + \log p(\hat{\mathbf{D}}|\hat{\boldsymbol{\vartheta}}_K) = \log p(\mathbf{y}|\hat{\boldsymbol{\vartheta}}_K) - \mathrm{EN}(\hat{\boldsymbol{\vartheta}}_K),$$

they obtain

$$\begin{aligned}\mathrm{ICL}_K = & -2\log p(\mathbf{y}|\hat{\boldsymbol{\vartheta}}_K, \mathcal{M}_K) + 2\mathrm{EN}(\hat{\boldsymbol{\vartheta}}_K) + d_K^C \log N - 2\log p(\hat{\mathbf{D}}|\hat{\boldsymbol{\vartheta}}_K) \\ & -2\log\left(\frac{\Gamma(K/2)\prod_{k=1}^K \Gamma(N_k(\hat{\mathbf{D}}) + 0.5)}{\Gamma(N+K/2)\Gamma(0.5)^K}\right).\end{aligned}$$

Using Stirling's formula to evaluate $\Gamma(N_k(\hat{\mathbf{D}}) + 0.5)$ and $\Gamma(N + K/2)$ and ignoring terms that are $\mathcal{O}(1)$, McLachlan and Peel (2000, p.216) are able to show that for $N_k(\hat{\mathbf{D}})$ large enough, $\mathrm{ICL}_K$ is approximately equal to

$$\begin{aligned}\text{ICL-BIC}_K &= -2\log p(\mathbf{y}|\hat{\boldsymbol{\vartheta}}_K, \mathcal{M}_K) + d_K \log N + 2\mathrm{EN}(\hat{\boldsymbol{\vartheta}}_K) \\ &= \mathrm{BIC}_K + 2\mathrm{EN}(\hat{\boldsymbol{\vartheta}}_K).\end{aligned} \tag{7.20}$$

Therefore, the integrated classification likelihood and its asymptotic variant penalize not only model complexity, but also the failure of the mixture model to provide a classification in well-separated clusters.

### Misspecifying the Component Densities for Well-Separated Clusters

When fitting a finite mixture model to data forming $K$ well-separated clusters, due to the consistency mentioned earlier, the number of components in the mixture selected by the marginal likelihood will be equal to the number of clusters if the component densities are correctly specified. It is, however, important to distinguish between clusters in the data and the components in the mixture model, if the component densities are misspecified. If the distribution in the groups is not perfectly Gaussian, then two or even more components in the normal mixture will be needed to capture skewness or fat tails in this particular group. Consequently, the number of components in the fitted mixture need not be the number of hidden groups or segments.

Simulation studies reported in (Biernacki and Govaert, 1997), Biernacki et al. (2000), and McLachlan and Peel (2000, Section 6.11) show that $\mathrm{BIC}_K$ will overrate the number of clusters under misspecification of the component density, whereas the $\mathrm{ICL}_K$ criterion defined in (7.19), respectively, its large cluster approximation $\text{ICL-BIC}_K$ defined in (7.20), is able to identify the correct number of clusters even when the component densities are misspecified.

### Overlapping Clusters

One of the main problems in choosing the numbers of clusters seems to be that there is no clear definition of a "cluster". In most applications of cluster analysis the primary goal is to identify distinct clusters that split an otherwise heterogeneous sample into groups of "homogeneous" objects. Dissection of a homogeneous sample into two or more groups is not the primary goal of cluster analysis. Ideally, one would hope to learn from the data that they do not contain any distinct cluster.

Most methods for identifying the number of clusters tend to fail for overlapping clusters; see, for instance, Tibshirani et al. (2001). For a small simulation study, where 50 observations were simulated from each of two bivariate normal distributions with means $\boldsymbol{\mu}_1 = (0\,0)$, $\boldsymbol{\mu}_2 = (\delta\,0)$, and $\boldsymbol{\Sigma}_1 = \boldsymbol{\Sigma}_2 = \mathbf{I}_2$, Tibshirani et al. (2001) found that the probability of selecting one cluster is roughly equal to the overlap, defined by the expected proportion of data points that were closer to the wrong component means, which is given by $1 - \Phi(\delta/2)$.

Leisch (2004) tries to find well-separated clusters in the data by identifying data points that are grouped into one of a collection of similar components.

### 7.1.5 Model Choice for the Fishery Data

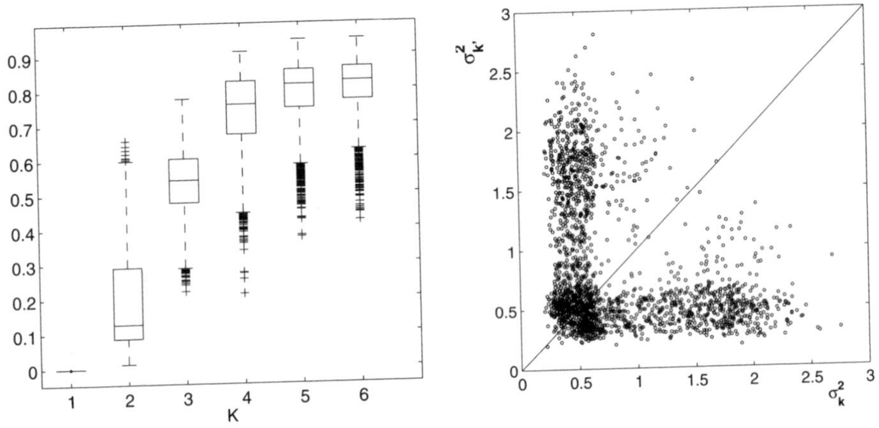


**Fig. 7.1.** FISHERY DATA modeled by a heterogeneous mixture of normal distribution; box-plots representing the posterior distribution of the within-group variance $\sum_{k=1}^{K} \eta_k \sigma_k^2$ for increasing number $K$ of components (left-hand side); scatter plot of the MCMC draws $\sigma_k^{(2,m)}$ versus $\sigma_{k'}^{(2,m)}$ for $K = 4$ (right-hand side)

We continue with the FISHERY DATA, already studied in Subsection 6.2.8, and use a Bayesian approach for selecting the number of components and for variance selection.

We start with exploratory Bayesian posterior analysis. First we study the influence of increasing $K$ on the posterior distribution $p(W(\boldsymbol{\vartheta}_K)|\mathbf{y})$ of the within-group variance $W(\boldsymbol{\vartheta}_K) = \sum_{k=1}^{K} \eta_k \sigma_k^2$, which is based on diagnosing mixtures through the posterior distribution of implied moments as discussed in Subsection 4.3.3. Figure 7.1 shows that increasing $K$ up to 4 reduces within-group variance considerably, which is not the case when adding a fifth or more components. This suggests choosing $K = 4$. To diagnose the assumption that the variances are heterogeneous, we explore the scatter plot of the MCMC draws $\sigma_k^{(2,m)}$ versus $\sigma_{k'}^{(2,m)}$ in Figure 7.1 for $K = 4$. If the variances were the same in all groups, this scatter plot would show just a single simulation cluster. As we find three simulation clusters we may conclude that the variances are different in at least two groups.

**Table 7.1.** FISHERY DATA, marginal likelihood $p(\mathbf{y}|K, \mathcal{V}_l)$ for $K = 1$ to $K = 5$ number of components for normal mixtures with unequal ($\mathcal{V}_1$) and equal variances ($\mathcal{V}_2$)

| $p(\mathbf{y}\|K, \mathcal{V}_1)$ | $K=1$ | $K=2$ | $K=3$ | $K=4$ | $K=5$ |
|---|---|---|---|---|---|
| Unconstrained variances | | | | | |
| $\log \hat{p}_{BS}(\mathbf{y}\|K, \mathcal{V}_1)$ | –535.11 | –525.66 | –521.44 | –518.74 | –521.06 |
| $\log \hat{p}_{IS}(\mathbf{y}\|K, \mathcal{V}_1)$ | –535.11 | –525.72 | –522.04 | –519.19 | –520.18 |
| $\log \hat{p}_{RI}(\mathbf{y}\|K, \mathcal{V}_1)$ | –535.11 | –525.70 | –521.46 | –519.99 | –527.60 |
| Constrained variances | | | | | |
| $\log \hat{p}_{BS}(\mathbf{y}\|K, \mathcal{V}_2)$ | –535.11 | –529.03 | –528.92 | –529.41 | –526.69 |
| $\log \hat{p}_{IS}(\mathbf{y}\|K, \mathcal{V}_2)$ | –535.11 | –529.03 | –530.24 | –530.15 | –526.61 |
| $\log \hat{p}_{RI}(\mathbf{y}\|K, \mathcal{V}_2)$ | –535.11 | –529.04 | –529.07 | –530.25 | –529.16 |

**Table 7.2.** FISHERY DATA, posterior probabilities $\Pr(K, \mathcal{V}_l|\mathbf{y})$ based on the truncated Poisson prior $p(K, \mathcal{V}_l) \propto f_P(K; 1)$, AIC and BIC for $K = 1$ to $K = 5$ number of components for normal mixtures with unequal ($\mathcal{V}_1$) and equal variances ($\mathcal{V}_2$)

| $\Pr(K, \mathcal{V}_l\|\mathbf{y})$ | $K=1$ | $K=2$ | $K=3$ | $K=4$ | $K=5$ |
|---|---|---|---|---|---|
| Unconstrained variances | 0 | 0.01 | 0.21 | **0.77** | 0.01 |
| Constrained variances | 0 | 0 | 0 | 0 | 0 |
| AIC | $K=1$ | $K=2$ | $K=3$ | $K=4$ | $K=5$ |
| Unconstrained variances | –529.19 | –518.65 | –515.31 | **–499.59** | –499.86 |
| Constrained variances | –529.19 | –519.64 | –518.01 | –513.72 | –503.49 |
| BIC | $K=1$ | $K=2$ | $K=3$ | $K=4$ | $K=5$ |
| Unconstrained variances | –532.74 | –527.51 | –529.49 | **–519.08** | –524.67 |
| Constrained variances | –532.74 | –526.73 | –528.65 | –527.90 | –521.22 |

This exploratory result is supported by formal model selection. Table 7.1 shows the log of the marginal likelihood $p(\mathbf{y}|K, \mathcal{V}_l)$ for various numbers of components, each in combination with unequal variances ($\mathcal{V}_1$) and equal variances ($\mathcal{V}_2$). Estimation is based on various simulation-based approximations of the marginal likelihood that were discussed in Section 5.4, namely bridge sampling, importance sampling, and reciprocal importance sampling. The importance density is constructed from the MCMC draws with $S = M_0 K!$, where $M_0 = 100$ for $K \leq 2$ and $M_0 = 5$ otherwise. The estimators are based on $M = 5000$ MCMC draws and $L = 5000$ draws from the importance density. Table 7.2, showing the posterior probabilities $\Pr(K, \mathcal{V}_l|\mathbf{y})$ based on the truncated Poisson prior $p(K, \mathcal{V}_l) \propto f_P(K; 1)$, suggests choosing a four-component mixture with unequal variances. The same table shows that AIC and BIC lead to the same conclusion.

### 7.1.6 Model Choice for Fisher's Iris Data

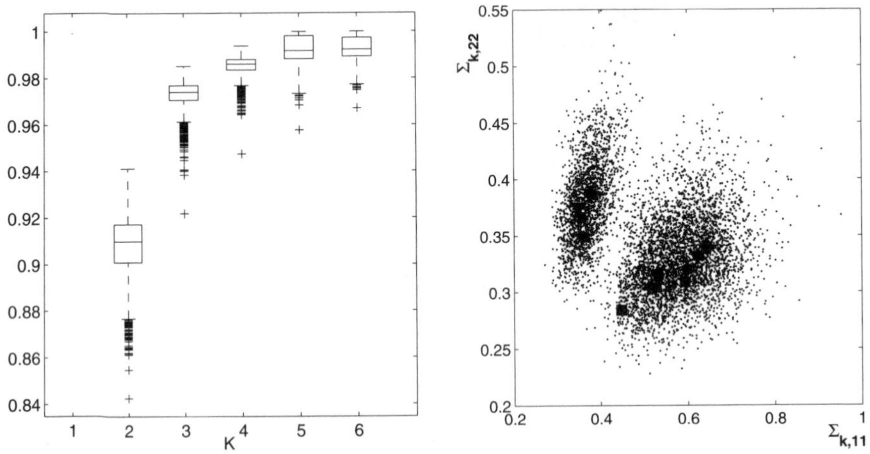


**Fig. 7.2.** FISHER'S IRIS DATA modeled by a heterogeneous mixture of multivariate normal distributions; box-plots representing the posterior distribution of the within-group variance–covariance matrix $W(\boldsymbol{\vartheta}_K) = |\sum_{k=1}^K \eta_k \boldsymbol{\Sigma}_k|$ under heterogeneous variance–covariance matrices for increasing number $K$ of components (left-hand side); scatter plot of the MCMC draws $\Sigma_{k,11}^{(m)}$ and $\Sigma_{k,22}^{(m)}$ obtained from a heterogeneous normal mixture with three components (right-hand side)

We continue with FISHER'S IRIS DATA, already studied in Subsection 6.3.4, and use a Bayesian approach for selecting the number of components as well as for covariance selection in a multivariate mixture of normal distributions with heterogeneous or homogeneous covariance matrices.

We start with exploratory Bayesian posterior analysis. First we study the influence of increasing $K$ on the posterior distribution of the within-group variance, which is based on diagnosing mixtures through the posterior distribution of implied moments as discussed in Subsection 4.3.3. Figure 7.1

shows that increasing $K$ up to 3 reduces within-group variance, but adding a further component does not reduce unexplained heterogeneity. This would suggest choosing $K = 3$. To explore the covariance matrices with respect to heterogeneity, we consider the scatter plot of $\Sigma_{k,11}^{(m)}$ versus $\Sigma_{k,22}^{(m)}$ in Figure 7.2 for a mixture with $K = 3$ components. Because we find two simulation clusters we may conclude that the covariance matrices are not identical in the various components.

**Table 7.3.** FISHER'S IRIS DATA, various estimators of the marginal likelihood $p(\mathbf{y}|K, \mathcal{V}_l)$ for $K = 1$ to $K = 5$ number of components and heteroscedastic ($\mathcal{V}_1$) and homoscedastic covariance matrices ($\mathcal{V}_2$)

| $p(\mathbf{y}\|K, \mathcal{V}_1)$ | $K=1$ | $K=2$ | $K=3$ | $K=4$ | $K=5$ |
|---|---|---|---|---|---|
| Heteroscedastic covariance matrices | | | | | |
| $\log \hat{p}_{BS}(\mathbf{y}\|K, \mathcal{V}_1)$ | –430.11 | –302.27 | **–294.46** | –297.87 | –307.65 |
| $\log \hat{p}_{IS}(\mathbf{y}\|K, \mathcal{V}_1)$ | –430.11 | –302.27 | **–294.63** | –299.22 | –306.53 |
| $\log \hat{p}_{RI}(\mathbf{y}\|K, \mathcal{V}_1)$ | –430.11 | –302.28 | **–294.56** | –308.32 | –320.52 |
| Homoscedastic covariance matrices | | | | | |
| $\log \hat{p}_{BS}(\mathbf{y}\|K, \mathcal{V}_2)$ | –430.11 | –360.30 | –335.37 | –319.65 | –323.92 |
| $\log \hat{p}_{IS}(\mathbf{y}\|K, \mathcal{V}_2)$ | –430.11 | –360.30 | –335.40 | –319.54 | –323.50 |
| $\log \hat{p}_{RI}(\mathbf{y}\|K, \mathcal{V}_2)$ | –430.11 | –360.31 | –335.35 | –320.00 | –329.15 |

**Table 7.4.** FISHER'S IRIS DATA, posterior probabilities $\Pr(K, \mathcal{V}_l|\mathbf{y})$ based on the prior $p(K, \mathcal{V}_l) \propto f_P(K; 1)$, AIC and BIC for $K = 1$ to $K = 5$ number of components, and heteroscedastic ($\mathcal{V}_1$) and homoscedastic covariance matrices ($\mathcal{V}_2$)

| $\Pr(K, \mathcal{V}_l\|\mathbf{y})$ | $K=1$ | $K=2$ | $K=3$ | $K=4$ | $K=5$ |
|---|---|---|---|---|---|
| Heteroscedastic covariance matrices | 0 | 0.001 | **0.991** | 0.008 | 0.01 |
| Homoscedastic covariance matrices | 0 | 0 | 0 | 0 | 0 |
| AIC | $K=1$ | $K=2$ | $K=3$ | $K=4$ | $K=5$ |
| Unconstrained variances | –393.91 | –243.35 | –224.19 | –221.17 | **–219.02** |
| Constrained variances | –393.91 | –315.45 | –280.35 | –279.33 | –251.28 |
| BIC | $K=1$ | $K=2$ | $K=3$ | $K=4$ | $K=5$ |
| Unconstrained variances | –414.99 | **–287.01** | –290.42 | –309.98 | –330.42 |
| Constrained variances | –414.99 | –344.05 | –316.48 | –322.99 | –302.46 |

Again this exploratory result is supported by formal model selection. Table 7.3 shows the log of the marginal likelihood for various numbers of components, each in combination with heterogeneous and homogeneous covariance matrices. Estimation is based on various simulation-based approximations of the marginal likelihood that were discussed in Section 5.4, namely bridge sampling, importance sampling, and reciprocal importance sampling. The importance density is constructed from the MCMC draws with $S = M_0K!$, where $M_0 = 100$ for $K \leq 2$ and $M_0 = 5$ otherwise. The estimators are based on $M = 5000$ MCMC draws and $L = 5000$ draws from the importance density.

Table 7.4, showing the posterior probabilities $\Pr(K, \mathcal{V}_l|\mathbf{y})$ based on the truncated Poisson prior $p(K, \mathcal{V}_l) \propto f_P(K; 1)$, suggests choosing a three-component heterogeneous mixture. The same table shows that this time AIC and BIC lead to different conclusions. Whereas AIC overrates the number of components by choosing $K = 5$, BIC underrates the number of components by choosing $K = 2$.

### 7.1.7 Bayesian Clustering Based on Loss Functions

As for any cluster analysis, the goal of Bayesian cluster analysis is to find an optimal partition of the data by estimating a sensible allocation $\hat{S}_i$ for each observation $\mathbf{y}_i$. In pioneering work, Binder (1978) cast this problem into a decision-theoretic framework, by formulating a loss function $R(\hat{\mathbf{S}}, \mathbf{S})$ which quantifies the loss made by choosing the allocations $\hat{\mathbf{S}} = (\hat{S}_1, \dots, \hat{S}_N)$ for observations with true allocation $\mathbf{S}$. The optimal allocation $\hat{\mathbf{S}}$ is chosen in such a way that the expected loss $\mathrm{E}(R(\hat{\mathbf{S}}, \mathbf{S})|\mathbf{y})$ is minimized. However, Binder (1978) realized that the choice of a sensible loss function in a clustering context is nontrivial due to the relabeling problem.

#### The 0/1 Loss Function

The simple 0/1-loss function considered by Binder (1978) assigns zero loss, iff the estimated classification is perfect up to relabeling the groups, otherwise this loss function is indifferent to any form of misclassification. More formally, $R(\hat{\mathbf{S}}, \mathbf{S}) = 0$, iff there exists a permutation $\rho_s(\cdot)$ of the labels of $\hat{S}_i$, such that $S_i = \rho_s(\hat{S}_i)$ for all $i = 1, \dots, N$, otherwise $R(\hat{\mathbf{S}}, \mathbf{S}) = 1$. The expected loss is given by

$$\mathrm{E}(R(\hat{\mathbf{S}}, \mathbf{S})|\mathbf{y}) = 1 - \Pr(S_1 = \rho_s(\hat{S}_1), \dots, S_N = \rho_s(\hat{S}_N)|\mathbf{y}),$$

which is minimized by the mode of the marginal posterior probability distribution $p(\mathbf{S}|\mathbf{y})$. As shown in Subsection 3.3.4, this posterior is invariant to relabeling, $p(S_1, \dots, S_N|\mathbf{y}) = p(\rho_s(\hat{S}_1), \dots, \rho_s(\hat{S}_N)|\mathbf{y})$, for all $s = 1, \dots, K!$ permutations, thus there exist at least $K!$ equivalent allocations, which are all optimal with respect to the 0/1-loss function.

The practical computation of the optimal allocation is possible for mixture models from the exponential family under conditionally conjugate priors, as for these finite mixture models the posterior ordinate $p(\mathbf{S}|\mathbf{y})$ is available (up to a normalizing constant) in closed form; see again Subsection 3.3.3. Several sampling strategies to draw from this posterior have been discussed in Subsections 3.4.1 and 3.4.2.

### Loss Functions Based on the Misclassification Rate

Binder (1978) considered a further loss function, based on the misclassification rate, which has become the most commonly applied in finite mixture modeling; see, for instance, McLachlan and Basford (1988) and Richardson and Green (1997).

Misclassification occurs whenever an observation is assigned to the $k$th group, although it arises from the $l$th group. Evidently the notion misclassification is sensitive to group labeling. An allocation $\hat{\mathbf{S}}$ with misclassification rate 0, meaning that $S_i = \hat{S}_i$ for all $i = 1, \ldots, N$, obtains a misclassification rate of 100 percent simply by relabeling $S_1, \ldots, S_N$ through some permutation $\rho_s(\cdot)$. Thus Bayesian classification based on the misclassification rate is sensible only if $S_i$ is an indicator with unique labeling, which is denoted by $S_i^{\mathcal{L}}$.

For any pair $(k, l)$ with $k \neq l$, let $c_{kl}$ be the loss occurring whenever an observation is assigned to the $k$th group, when it actually arises from the $l$th group, and let $n_{kl}(\hat{\mathbf{S}}, \mathbf{S})$ be the corresponding number of misclassifications:

$$n_{kl}(\hat{\mathbf{S}}, \mathbf{S}) = \sum_{i=1}^{N} I_{\{\hat{S}_i = k, S_i^{\mathcal{L}} = l\}}.$$

Then the corresponding loss function is given by

$$R(\hat{\mathbf{S}}, \mathbf{S}) = \sum_{k \neq l} c_{kl} \sum_{i=1}^{N} I_{\{\hat{S}_i = k, S_i^{\mathcal{L}} = l\}}.$$

If $c_{kl} = cI_{\{k \neq l\}}$, then this loss function is proportional to the total misclassification rate. The expected loss is given by

$$\mathrm{E}(R(\hat{\mathbf{S}}, \mathbf{S})|\mathbf{y}) = \sum_{i=1}^{N} \sum_{l=1, l \neq \hat{S}_i}^{K} c_{\hat{S}_i, l} \Pr(S_i^{\mathcal{L}} = l|\mathbf{y}),$$

and is minimized by choosing the allocations $\hat{S}_i, i = 1, \ldots, N$ independently for each observation such that

$$\sum_{l=1, l \neq k}^{K} c_{kl} \Pr(S_i^{\mathcal{L}} = l|\mathbf{y})$$

is minimized at $\hat{S}_i = k$. If $c_{kl} = cI_{\{k \neq l\}}$, then this is equivalent with choosing $\hat{S}_i = k$, where $k$ maximizes the individual posterior classification probabilities $\Pr(S_i^{\mathcal{L}} = k|\mathbf{y})$.

In this case the misclassification risk is equal to $1 - \max_k \Pr(S_i^{\mathcal{L}} = k|\mathbf{y})$. For a perfect classification of $\mathbf{y}_i$ into one of the $K$ groups, this value is zero, otherwise it measures the difficulty of assigning a certain subject $\mathbf{y}_i$ into a single group. As

$$\max_k \Pr(S_i^{\mathcal{L}} = k|\mathbf{y}) \geq \frac{1}{K},$$

the risk is limited by $(K-1)/K$.

It remains to discuss how to estimate the individual posterior classification probabilities $\Pr(S_i^{\mathcal{L}} = k|\mathbf{y}), k = 1, \dots, K$ for $i = 1, \dots, N$. When sampling from the mixture posterior distribution using data augmentation and MCMC or marginal sampling, it is quite tempting to estimate the classification probability $\Pr(S_i^{\mathcal{L}} = k|\mathbf{y})$ from the MCMC draws $S_i^{(m)}$ by the relative frequency of the event $\{S_i^{(m)} = k\}$:

$$\Pr(S_i^{\mathcal{L}} = k|\mathbf{y}) \approx \frac{1}{M} \sum_{m=1}^{M} I_{\{S_i^{(m)}=k\}}. \tag{7.21}$$

However, as the sampler draws from the unconstrained posterior, the right-hand side of (7.21) converges to the probability $\Pr(S_i = k|\mathbf{y})$, which by the results of Subsection 3.3.4 is equal to $1/K$ for all observations and all $k = 1, \dots, K$:

$$\lim_{M\to\infty} \frac{1}{M} \sum_{m=1}^{M} I_{\{S_i^{(m)}=k\}} = \Pr(S_i = k|\mathbf{y}) = \frac{1}{K}.$$

Thus for a well-mixing sampler, the estimator (7.21) yields useless results, as the observations are assigned randomly to the $K$ groups. Binder (1978) seems to be the first who realized that any assignment rule based on $\Pr(S_i = k|\mathbf{y})$ fails to classify the objects.

As mentioned in Subsection 3.7.7, Gibbs sampling as described in *Algorithm 3.4* may lead to an implicit labeling if trapped at a single modal region of the posterior density, in which case estimator (7.21) will be sensible. This is, however, not necessarily the case, and in general estimator (7.21) runs the risk of yielding useless results.

In general, it is necessary to identify the mixture model to obtain draws $S_i^{\mathcal{L},(m)}, m = 1, \dots, M$ with a unique labeling. As outlined in Subsection 3.7.7, it is possible to derive draws $S_i^{\mathcal{L},(m)}$ with a unique labeling by relabeling $S_i^{(m)}$ jointly with $\boldsymbol{\vartheta}^{(m)}$. For such draws it is possible to estimate $\Pr(S_i^{\mathcal{L}} = k|\mathbf{y})$ by the corresponding relative frequency:

$$\Pr(S_i^{\mathcal{L}} = k|\mathbf{y}) \approx \frac{1}{M} \sum_{m=1}^{M} I_{\{S_i^{\mathcal{L},(m)}=k\}}.$$

If for storage reasons only parameter draws $\boldsymbol{\vartheta}^{\mathcal{L},(m)}$ from a unique labeling subspace are available, an alternative estimator of the marginal classification probability $\Pr(S_i = k|\mathbf{y})$ is given by

$$\Pr(S_i^{\mathcal{L}} = k|\mathbf{y}) \approx \frac{1}{M} \sum_{m=1}^{M} \eta_k^{\mathcal{L},(m)} p(\mathbf{y}_i|\boldsymbol{\theta}_k^{\mathcal{L},(m)}).$$

### Loss Functions Based on the Posterior Similarity Matrix

Binder (1978) and Hurn et al. (2003) considered loss functions based on the posterior similarity matrix $\Pr(S_i = S_j|\mathbf{y})$. The posterior similarity matrix is an $(N \times N)$ matrix, defined from the joint posterior $p(\mathbf{S}|\mathbf{y})$ for all observations as the probability that any two observations belong to the same component. $\Pr(S_i = S_j|\mathbf{y})$ may be expressed as

$$\Pr(S_i = S_j|\mathbf{y}) = \sum_{k=1}^{K} \Pr(S_i = k, S_j = k|\mathbf{y}),$$

and has the advantage of being invariant to relabeling the groups. Therefore this probability may be estimated from the MCMC draws without bothering about identification as

$$\Pr(S_i = S_j|\mathbf{y}) \approx \frac{1}{M} \sum_{m=1}^{M} I_{\{S_i^{(m)} = S_j^{(m)}\}}. \tag{7.22}$$

For clustering, Binder (1978) considers the following loss function,

$$R(\hat{\mathbf{S}}, \mathbf{S}) = \sum_{i<j} l_1 I_{\{\hat{S}_i \neq \hat{S}_j\}} I_{\{S_i = S_j\}} + l_2 I_{\{\hat{S}_i = \hat{S}_j\}} I_{\{S_i \neq S_j\}},$$

with $l_1$ being the loss associated with assigning two observations to different groups, although they belong to the same group, and $l_2$ being the loss associated with assigning two observations to the same group, although they belong to different groups. As noted by Binder (1978), this loss function evaluates allocations $\hat{\mathbf{S}}$ with respect to their ability to achieve the two main goals of cluster analysis, stated in Cormack (1971). Whereas $l_1$ penalizes the lack of internal cohesion, $l_2$ penalizes lack of external isolation, and the ratio $l_1/l_2$ may be chosen with regard to the relative importance of these goals. The same loss function as in (7.23) with $l_1 = l_2 = 1$ is applied in Hurn et al. (2003) in the context of switching regression models.

The optimal allocation $\hat{\mathbf{S}}$ is chosen such that the expected loss,

$$\mathrm{E}(R(\hat{\mathbf{S}}, \mathbf{S})|\mathbf{y}) = \sum_{i<j} I_{\{\hat{S}_i = \hat{S}_j\}} \left(l_2 - \Pr(S_i = S_j|\mathbf{y})(l_1 + l_2)\right), \tag{7.23}$$

is minimized. There exists no explicit solution to this minimization problem, and neither Binder (1978) nor Hurn et al. (2003) mention how numerical optimization may be carried out in an efficient manner. Note that the posterior similarity matrix $\Pr(S_i = S_j|\mathbf{y})$ may be estimated from (7.22) and that (7.23) may be evaluated for all MCMC draws to identify (nearly) optimal allocations.

### 7.1.8 Clustering for Fisher's Iris Data

We continue with FISHER'S IRIS DATA, already studied in Subsections 6.3.4 and 7.1.6. As for these data the species of each observation is known, it is illuminating to study the misclassification risk of using a mixture of three normal distributions with an unrestricted variance–covariance matrix. The finite mixture model is fitted exactly as described in Subsection 6.3.4.

We compare various Bayesian clustering strategies, namely minimizing the misclassification rate, which is derived from the model identified in Subsection 6.3.4, Bayesian classification based on the posterior similarity matrix, and the classification obtained from Bayesian MAP classification under the hierarchical independence prior. The resulting estimators are denoted by $\hat{S}_{i,\mathcal{L}}$, $\hat{S}_{i,SM}$, and $\hat{S}_{i,MAP}$, respectively.

The first two methods give exactly the same result, whereas the third method leads to a different classification only for a single observation. Table 7.5 shows all observations that are misclassified for at least one of these classifications. Classification is perfect for the *iris setosa* and the *iris virginica* group. Four plants out of the *iris versicolour* group are misclassified into the *iris virginica* group for $\hat{S}_{i,\mathcal{L}}$ and $\hat{S}_{i,SM}$; an additional misclassification of the same nature occurs for $\hat{S}_{i,MAP}$.

**Table 7.5.** FISHER'S IRIS DATA, normal mixtures with $K = 3$ (hierarchical independence prior), observations misclassified through different Bayesian clustering strategies ($S_i^{\text{true}} = 2$ for all observations), all other observations are correctly classified

| $i$ | 69 | 71 | 73 | 78 | 84 |
|---|---|---|---|---|---|
| $\hat{S}_{i,\mathcal{L}}$ | **3** | **3** | **3** | 2 | **3** |
| $\hat{S}_{i,SM}$ | **3** | **3** | **3** | 2 | **3** |
| $\hat{S}_{i,MAP}$ | **3** | **3** | **3** | **3** | **3** |

## 7.2 Outlier Modeling

Numerous suggestions have been made how to deal with outliers. Peña and Guttman (1993) provide a review article of finite mixtures and other probabilistic methods of outlier detection.

### 7.2.1 Outlier Modeling Using Finite Mixtures

Newcomb (1886), which is one of the earliest statistical applications of mixture models, used a normal mixture to model aberrant observations in astronomical data of transits of Mercury. Another famous univariate data set containing outliers that has been analyzed by mixtures of normals is DARWIN'S DATA on the differences in heights between pairs of self-fertilized and cross-fertilized plants grown in the same conditions. The observations, ordered by size, are given by:

$$-67, -48, 6, 8, 14, 16, 23, 24, 28, 29, 41, 49, 56, 60, 75.$$

The general idea of the finite mixture approach to outlier modeling is nicely described by Box and Tiao (1968, p.119): "There is a small probability $\eta_2$ that any given observation was *not* generated by the central stochastic model as well as a complementary prior probability $(1-\eta_2)$ that it *was* so generated."[2]

One early example is the variance inflation model introduced by Tukey (1960):

$$Y \sim (1-\eta_2)\mathcal{N}\left(\mu, \sigma^2\right) + \eta_2 \mathcal{N}\left(\mu, k\sigma^2\right), \tag{7.24}$$

where it is assumed that the data come from a central $\mathcal{N}\left(\mu, \sigma^2\right)$-distribution with high probability $(1-\eta_2)$, and with low probability $\eta_2$ from a contaminated distribution, $\mathcal{N}\left(\mu, k\sigma^2\right)$, where $k >> 1$. Probability $\eta_2$ typically is a small proportion of outliers in the sample. Guttman (1973) introduced the location shift model, with additive, rather than multiplicative, contamination:

$$Y \sim (1-\eta_2)\mathcal{N}\left(\mu, \sigma^2\right) + \eta_2 \mathcal{N}\left(\mu + k, \sigma^2\right). \tag{7.25}$$

Both models are evidently special cases of the univariate mixtures of normals and may be extended to handle multiple outliers that tend to clump (Aitkin and Wilson, 1980) and to deal with outliers in a linear regression model (Box and Tiao, 1968; Abraham and Box, 1978); see also Subsection 8.2.4.

### 7.2.2 Bayesian Inference for Outlier Models Based on Finite Mixtures

Approximate Bayesian estimation of outlier models based on finite mixture distributions without using MCMC methods has been discussed in many papers (Box and Tiao, 1968; Guttman, 1973; Abraham and Box, 1978; Guttman et al., 1978; Peña and Tiao, 1992). In most of these papers the parameters of the contaminated component, $\eta_2$ and $k$, are assumed to be known because of the computational burden associated with making $\eta_2$ and $k$ unknown outside the framework of MCMC.

Verdinelli and Wasserman (1991) were the first who applied data augmentation and Gibbs sampling for univariate normal observations with outliers with $\eta_2$ being random, while holding $k$ fixed. Justel and Peña (1996) show that this works well for an isolated outlier in a univariate sample, however, under the existence of outliers that mask or swamp other observations the sampler tends to be trapped at some local mode and turns out to be sensitive

[2] Notation changed.

to choosing appropriate starting values. Evans et al. (1992) implement a fully Bayesian analysis of outlier models including both $\eta_2$ and $k$ as unknown parameters for univariate mixtures of normals as in described in *Algorithm 6.1.*

### Choosing Nonsymmetric Priors

**Table 7.6.** Choosing the prior $\boldsymbol{\eta} \sim \mathcal{D}(a, 1)$ in an outlier model such that the proportion of outliers (in percent) is less than or equal to $\alpha$ with a priori probability equal to 95%

| $a$ | $\alpha$ | $a$ | $\alpha$ | $a$ | $\alpha$ | $a$ | $\alpha$ | $a$ | $\alpha$ |
|---|---|---|---|---|---|---|---|---|---|
| 5 | 45.1 | 9 | 28.3 | 13 | 20.6 | 24 | 11.7 | 57 | 5.12 |
| 6 | 39.3 | 10 | 25.9 | 15 | 18.1 | 28 | 10.1 | 71 | 4.13 |
| 7 | 34.8 | 11 | 23.8 | 18 | 15.3 | 35 | 8.2 | 100 | 2.95 |
| 8 | 31.2 | 12 | 22.1 | 20 | 13.9 | 41 | 7.05 | 200 | 1.49 |

When finite mixture models are applied to dealing with outliers, the components are typically not symmetric, because $\eta_2 << \eta_1$, and it is sensible to use nonsymmetric priors rather than invariant priors (Verdinelli and Wasserman, 1991). The prior constraint $\eta_2 < \eta_1$ is forced by choosing a $\mathcal{D}(a, b)$-prior on $\boldsymbol{\eta} = (\eta_1, \eta_2)$, with $a >> b$. For a fixed small value of $b$, say $b = 1$, $a$ may be elicited from knowing a priori with high probability, say 95%, that at most $100\alpha\%$ outliers are present. Because for $b = 1$ the prior mode lies at 0, $a$ may be chosen in such a way that $\alpha$ is the upper limit of the 95% credibility interval $[0, \alpha]$ obtained for $\eta_2$ from the $\mathcal{D}(a, 1)$-prior:

$$a = \frac{\log(1 - 0.95)}{\log(1 - \alpha)}.$$

Table 7.6 shows various values of $a$ with the corresponding upper limit $\alpha$ for the proportion of outliers.

### 7.2.3 Outlier Modeling of Darwin's Data

Abraham and Box (1978) analyzed DARWIN'S DATA using the location shift model (7.25). Bayesian estimation was carried out under the assumption that the proportion of outliers $\eta_2$ is known. We reanalyze the data by the location shift model with $\eta_2$ being unknown. We combine a $\mathcal{D}(15, 1)$-prior on $\boldsymbol{\eta} = (\eta_1, \eta_2)$ with a noninformative independent prior $p(\mu, k, \sigma^2) \propto 1/\sigma^2$, and reject classifications where one group is left empty. Figure 7.3 shows paths of 3000 MCMC draws as well as posterior densities of $\mu$ and $\mu + k$ estimated from the last 2000 draws, whereas Figure 7.4 shows the posterior densities of $\sigma^2$ and $\eta_2$, with the posterior of $\eta_2$ being compared to the prior. Finally, outlier

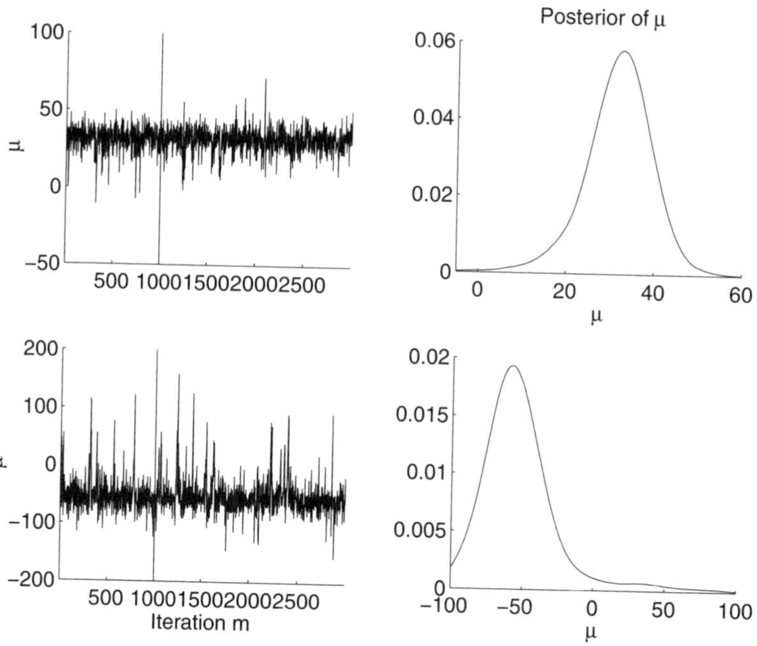


**Fig. 7.3.** DARWIN'S DATA, location shift model; MCMC paths and posterior densities for $\mu$ and $\mu + k$ under a $\mathcal{D}(15, 1)$-prior on $\boldsymbol{\eta}$

classification in the left-hand side of Figure 7.5 shows that the smallest two observations are classified as outliers.

We compare the results of the fully Bayesian analysis with the result obtained by assuming that $\eta_2$ is fixed as in Abraham and Box (1978). We select $\eta_2 = 0.18$ which corresponds to the upper limit of the 95%-H.P.D. interval obtained from the $\mathcal{D}(15, 1)$-prior. Figure 7.6 compares the posterior distribution of $\mu$, conditional on $\eta_2$ fixed, with the posterior obtained from the fully Bayesian analysis based on the $\mathcal{D}(15, 1)$-prior. Interestingly for this data set, there is little difference between these posterior densities, as well as between the classification probabilities; see Figure 7.5.

For illustration, we compare the nonsymmetric $\mathcal{D}(15, 1)$-prior with the invariant $\mathcal{D}(1, 1)$-prior. For DARWIN'S DATA, the invariant prior fails to identify the outliers. Label switching between $\mu$ and $\mu + k$ occurs frequently, causing classification probabilities that are extremely biased toward 0.5; see Figure 7.5.

### 7.2.4 Clustering Under Outliers and Noise

Clustering based on mixtures of multivariate normal distributions tends to suffer from sensitivity with respect to outlying values and misspecification of the component densities. Estimates of the component means and, even more,

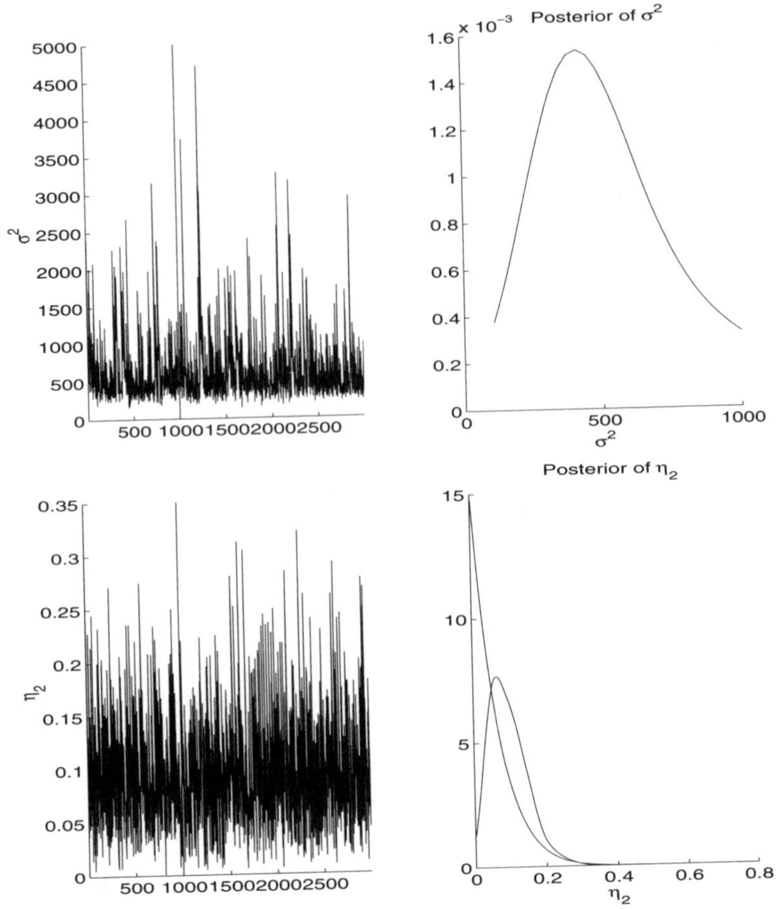


**Fig. 7.4.** DARWIN'S DATA, location shift model; MCMC paths and posterior densities for $\sigma^2$ and $\eta_2$ for the $\mathcal{D}(15,1)$-prior on $\boldsymbol{\eta}$; the posterior of $\eta_2$ is compared with the prior

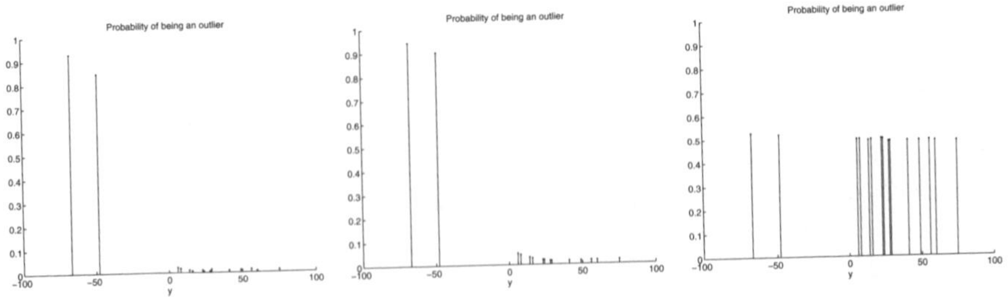


**Fig. 7.5.** DARWIN'S DATA, location shift model; outlier classification; left: $\mathcal{D}(15,1)$-prior on $\boldsymbol{\eta}$, middle: $\eta_2$ fixed at 0.18, right: $\mathcal{D}(1,1)$-prior on $\boldsymbol{\eta}$

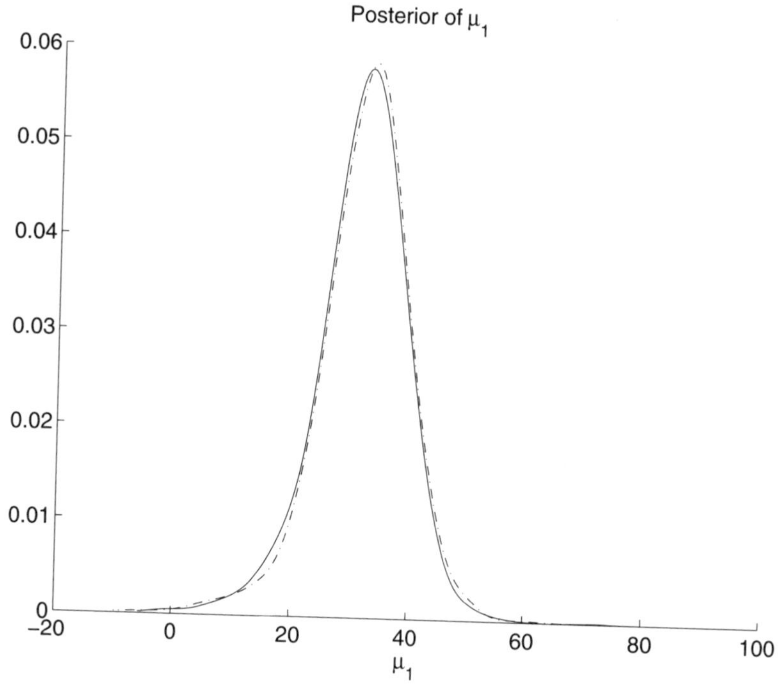


**Fig. 7.6.** DARWIN'S DATA, comparing the posterior density of $\mu$; full line: $\mathcal{D}(15,1)$-prior on $\boldsymbol{\eta}$, dashed line: $\eta_2$ fixed at 0.18

estimates of the component variance–covariance matrices, may be strongly affected by observations that are outlying with respect to the tails of the component-specific normal distribution.

Various suggestions have been made to achieve robustness of model-based clustering including distance-based estimation (Scott, 2004) which is less sensitive to outliers or clustering based on mixtures of multivariate Student-$t$ distribution (Peel and McLachlan, 2000); see also Subsection 7.3. A simple but efficient way to handle the presence of atypical observations or background noise is to add a noise component to the mixture (Banfield and Raftery, 1993):

$$p(\mathbf{y}_i|\boldsymbol{\vartheta}) = (1-w)\sum_{k=1}^{K} \eta_k f_N(\mathbf{y}_i;\boldsymbol{\mu}_k,\boldsymbol{\Sigma}_k) + w p_o(\mathbf{y}_i). \tag{7.26}$$

$p_o(\mathbf{y}_i)$ is an additional component that could have a uniform density over the support of the data (Peel and McLachlan, 2000).

For a Bayesian estimation of (7.26), a discrete latent indicator $J_i$ is introduced for each observation $\mathbf{y}_i$, taking the value 1 for an observation from the noise component $p_o(\mathbf{y}_i)$. The MCMC scheme introduced in *Algorithm 6.2* is extended by an additional step of data augmentation. Conditional on knowing

$\mathbf{J} = (J_1, \ldots, J_N)$, the data are grouped into noisy data, if $J_i = 1$, and into data from the multivariate normal mixture, if $J_i = 0$:

$$p(\mathbf{y}_i|\boldsymbol{\vartheta}, J_i = 0) = \sum_{k=1}^{K} \eta_k f_N(\mathbf{y}_i; \boldsymbol{\mu}_k, \boldsymbol{\Sigma}_k).$$

For all data that are not noise, steps (a) and (b) are carried out exactly as in *Algorithm 6.2* to update the component parameters $\boldsymbol{\mu}_k$ and $\boldsymbol{\Sigma}_k$, the weight distribution $\boldsymbol{\eta}$, and the allocations $S_i$. The allocations $S_i$ are not updated for observations that are currently considered to be noise.

An additional step is added at each sweep of the MCMC sampler, which performs classification of each observation either into the noise component or into the remaining components, according to the following classification rule,

$$\Pr(J_i = 1|\mathbf{y}, \boldsymbol{\mu}, \boldsymbol{\Sigma}, \boldsymbol{\eta}) = \frac{w p_o(\mathbf{y}_i)}{w p_o(\mathbf{y}_i) + (1-w)\sum_{k=1}^{K} \eta_k f_N(\mathbf{y}_i; \boldsymbol{\mu}_k, \boldsymbol{\Sigma}_k)}.$$

## 7.3 Robust Finite Mixtures Based on the Student-*t* Distribution

An important alternative to mixtures of normals is choosing the component densities from the family of $t_\nu(\boldsymbol{\mu}, \boldsymbol{\Sigma})$-distribution with $\nu$ degrees of freedom:

$$\mathbf{Y} \sim \eta_1 t_{\nu_1}(\boldsymbol{\mu}_1, \boldsymbol{\Sigma}_1) + \cdots + \eta_K t_{\nu_K}(\boldsymbol{\mu}_K, \boldsymbol{\Sigma}_K), \tag{7.27}$$

where $\mathbf{Y}$ is an $r$-dimensional random vector; see Appendix A.1.12 for a definition of the density of the $t_\nu(\boldsymbol{\mu}, \boldsymbol{\Sigma})$-distribution.

### 7.3.1 Parameter Estimation

Mixtures of Student-$t$ distributions have been estimated using the EM algorithm (Peel and McLachlan, 2000) and a Bayesian approach (Stephens, 1997a; Scott et al., 2005).

Bayesian estimation of multivariate mixtures of Student-$t$ distributions is based on the representation of the $t_\nu(\boldsymbol{\mu}, \boldsymbol{\Sigma})$-distribution as an infinite scale mixture of multivariate normal distributions, with the distribution of the scaling parameter being a Gamma distribution:

$$\mathbf{Y}_i|\omega_i \sim \mathcal{N}_r(\boldsymbol{\mu}, \boldsymbol{\Sigma}/\omega_i), \qquad \omega_i \sim \mathcal{G}(\nu/2, \nu/2).$$

For a mixture of Student-$t$ distributions, a similar representation exists:

$$\mathbf{Y}_i|S_i = k, \omega_i \sim \mathcal{N}_r(\boldsymbol{\mu}_k, \boldsymbol{\Sigma}_k/\omega_i), \qquad \omega_i \sim \mathcal{G}(\nu_k/2, \nu_k/2).$$

Thus a mixture of Student-$t$ distributions may be regarded as a mixture of normal distributions, where all group members have the same expectation $\boldsymbol{\mu}_k$, however, within each group there exists variance heterogeneity, captured by the scaling factor $\omega_i$, with smaller values of $\omega_i$ causing larger variances.

MCMC estimation of multivariate mixtures of Student-$t$ distributions is implemented through an additional step of data augmentation, which introduces for each observation $\mathbf{y}_i$ the scaling parameter $\omega_i$ as missing data. Conditional on knowing the scaling parameters $\boldsymbol{\omega} = (\omega_1, \dots, \omega_N)$ and the group indicators $\mathbf{S}$, the conditional posterior distribution of the location vector $\boldsymbol{\mu}_k$ and the scale matrix $\boldsymbol{\Sigma}_k$ are of a closed form similar to that of a mixture of Gaussian distributions, however, the relevant group-specific quantities $N_k(\mathbf{S})$, $\overline{\mathbf{y}}_k(\mathbf{S})$, and $\boldsymbol{W}_k(\mathbf{S})$ in *Algorithm 6.2* have to be substituted by the quantities $N_k(\mathbf{S}, \boldsymbol{\omega})$, $\overline{\mathbf{y}}_k(\mathbf{S}, \boldsymbol{\omega})$, and $\boldsymbol{W}_k(\mathbf{S}, \boldsymbol{\omega})$, which are adjusted by the scaling factor $\omega_i$:

$$N_k(\mathbf{S}, \boldsymbol{\omega}) = \sum_{i:S_i=k} \omega_i,$$

$$\overline{\mathbf{y}}_k(\mathbf{S}, \boldsymbol{\omega}) = \frac{1}{N_k(\mathbf{S}, \boldsymbol{\omega})} \sum_{i:S_i=k} \omega_i \mathbf{y}_i, \tag{7.28}$$

$$\boldsymbol{W}_k(\mathbf{S}, \boldsymbol{\omega}) = \sum_{i:S_i=k} \omega_i (\mathbf{y}_i - \overline{\mathbf{y}}_k(\mathbf{S}, \boldsymbol{\omega}))(\mathbf{y}_i - \overline{\mathbf{y}}_k(\mathbf{S}, \boldsymbol{\omega}))'. \tag{7.29}$$

Any of these quantities reduces to $N_k(\mathbf{S})$, $\overline{\mathbf{y}}_k(\mathbf{S})$, and $\boldsymbol{W}_k(\mathbf{S})$, when the Student-$t$ distribution approaches the normal distribution. In this case $\nu_k \rightarrow \infty$ and $\omega_i \equiv 1$ for all observations.

It is worth noting from (7.28) and (7.29) how robustness is achieved by introducing the Student-$t$ distribution. Observations with small scaling factors $\omega_i$, and consequently large individual variance, are down-weighted compared to observations with larger scaling factors and smaller individual variance.

The following algorithm demonstrates how MCMC estimation is carried out through Gibbs sampling if the degrees of freedom are known in each group and the same priors on $\boldsymbol{\mu}_k$ and $\boldsymbol{\Sigma}_k$ are adopted as used in Subsection 6.3.2 for multivariate mixtures of normals.

*Algorithm 7.1: Three-Block Gibbs Sampling for a Multivariate Mixture of Student-t Distributions*

(a) Parameter simulation conditional on the classification $\mathbf{S} = (S_1, \dots, S_N)$:
  (a1) Sample $\boldsymbol{\eta}$ from the conditional Dirichlet posterior $p(\boldsymbol{\eta}|\mathbf{S})$ as in *Algorithm 3.4*.
  (a2) Sample $\boldsymbol{\Sigma}_k^{-1}$ in each group $k$ from a $\mathcal{W}_r\left(c_k(\mathbf{S}), \mathbf{C}_k(\mathbf{S}, \boldsymbol{\omega})\right)$-distribution, where:

$$c_k(\mathbf{S}) = c_0 + \frac{N_k(\mathbf{S})}{2},$$

$$\mathbf{C}_k(\mathbf{S}, \boldsymbol{\omega}) = \mathbf{C}_0 + \frac{1}{2}\boldsymbol{W}_k(\mathbf{S}, \boldsymbol{\omega})$$

$$+\frac{1}{2}\left(\frac{N_k(\mathbf{S}, \boldsymbol{\omega})N_0}{N_k(\mathbf{S}, \boldsymbol{\omega}) + N_0}(\overline{\mathbf{y}}_k(\mathbf{S}, \boldsymbol{\omega}) - \mathbf{b}_0)(\overline{\mathbf{y}}_k(\mathbf{S}, \boldsymbol{\omega}) - \mathbf{b}_0)'\right).$$

(a3) Sample $\boldsymbol{\mu}_k$ in each group $k$ from an $\mathcal{N}_r\left(\mathbf{b}_k(\mathbf{S}, \boldsymbol{\omega}), \mathbf{B}_k(\mathbf{S}, \boldsymbol{\omega})\right)$-distribution, where:

$$\mathbf{B}_k(\mathbf{S}, \boldsymbol{\omega}) = \frac{1}{N_k(\mathbf{S}, \boldsymbol{\omega}) + N_0}\boldsymbol{\Sigma}_k,$$

$$\mathbf{b}_k(\mathbf{S}, \boldsymbol{\omega}) = \frac{1}{N_k(\mathbf{S}, \boldsymbol{\omega}) + N_0}\left(N_0\mathbf{b}_0 + N_k(\mathbf{S}, \boldsymbol{\omega})\overline{\mathbf{y}}_k(\mathbf{S}, \boldsymbol{\omega})\right).$$

(b) Classification of each observation $\mathbf{y}_i$, for $i = 1, \ldots, N$, conditional on knowing $\boldsymbol{\mu}, \boldsymbol{\Sigma}$, and $\boldsymbol{\eta}$:

$$p(S_i = k|\boldsymbol{\mu}_k, \boldsymbol{\Sigma}_k, \eta_k, \mathbf{y}_i) \propto \eta_k f_{t_{\nu_k}}(\mathbf{y}_i; \boldsymbol{\mu}_k, \boldsymbol{\Sigma}_k).$$

(c) Sample $\omega_i$ independently for each observation $\mathbf{y}_i$ for $i = 1, \ldots, N$ from the following Gamma posterior,

$$\omega_i|\boldsymbol{\mu}_{S_i}, \boldsymbol{\Sigma}_{S_i}, \nu_{S_i}, \mathbf{y} \sim \mathcal{G}\left(\frac{1}{2}(\nu_{S_i} + r), \frac{1}{2}(\nu_{S_i} + d_M(\mathbf{y}_i; \boldsymbol{\mu}_{S_i}, \boldsymbol{\Sigma}_{S_i})^2)\right),$$

where $d_M(\mathbf{y}_i; \boldsymbol{\mu}_{S_i}, \boldsymbol{\Sigma}_{S_i})$ is the Mahalanobis distance, defined in (1.17), between $\mathbf{y}_i$ and the cluster given by $\boldsymbol{\mu}_{S_i}$ and $\boldsymbol{\Sigma}_{S_i}$.

When sampling the indicators $S_i$ in step (b), a partially marginalized sampler has been used, where the unknown scaling factors are integrated out. Alternatively, full conditional Gibbs sampling would sample the indicators conditional on knowing the scaling parameters, by using the classification rule:

$$p(S_i = k|\boldsymbol{\mu}_k, \boldsymbol{\Sigma}_k, \eta_k, \omega_i, \mathbf{y}_i) \propto \eta_k f_N(\mathbf{y}_i; \boldsymbol{\mu}_k, \boldsymbol{\Sigma}_k/\omega_i).$$

Due to the results of Liu (1994), the partially marginalized sampler is likely to have better mixing properties than this full conditional Gibbs sampler.

### Dealing with Unknown Degrees of Freedom

An attractive alternative to fixing $\nu_k$ beforehand is estimating the degrees of freedom $\nu_k$ along with all other unknown quantities. For Bayesian estimation of the degrees of freedom $\nu_k$, the prior $p(\nu_k)$ has to be selected carefully, in order to avoid improper posteriors; see the discussion in Geweke (1993) and Bauwens and Lubrano (1998, Section 2). Proper priors that have been used for unknown degrees of freedom are the exponential prior (Geweke, 1993), uniform priors truncated to $[0, \nu_{\max}]$ (Hoek et al., 1995; Chib et al., 2002), and the uniform shrinkage prior (Scott et al., 2005):

$$p(\nu_k) = \frac{\bar{\nu}_k}{(\bar{\nu}_k + \nu_k)^2}. \tag{7.30}$$

Prior (7.30) is a density with median $\bar{\nu}_k$ but with no moments because of its heavy polynomial tail.

For sampling the degrees of freedom $\nu_k$ with an MCMC scheme, there are basically two ways to proceed. First, one could sample $\nu_k$ conditional on knowing the scaling parameters, by drawing from the full conditional posterior $p(\nu_k|\boldsymbol{\omega}, \mathbf{S}, \mathbf{y})$ given by

$$p(\nu_k|\boldsymbol{\omega}, \mathbf{S}, \mathbf{y}) \propto p(\nu_k) \frac{(\frac{\nu_k}{2})^{\nu_k N_k(\mathbf{S})/2}}{\Gamma(\frac{\nu_k}{2})^{N_k(\mathbf{S})}} \prod_{i:S_i=k} \omega_i^{\nu_k/2-1} \exp\{-\frac{\nu_k}{2}(\sum_{i:S_i=k} \omega_i)\},$$

by means of a Metropolis–Hastings algorithm. Alternatively, a partially marginalized sampler could be used, by drawing $\nu_k$ by means of a Metropolis–Hastings algorithm from the marginal distribution $p(\nu_k|\mathbf{S}, \mathbf{y})$,

$$p(\nu_k|\mathbf{S}, \mathbf{y}) \propto p(\nu_k) \prod_{i:S_i=k} f_{t_{\nu_k}}(\mathbf{y}_i; \boldsymbol{\mu}_k, \boldsymbol{\Sigma}_k),$$

where the scaling factors are integrated out. Again it is to be expected that the partially marginalized sampler is more efficient than full conditional sampling.

### 7.3.2 Dealing with Unknown Number of Components

Stephens (1997a, 2000a) applied birth and death MCMC methods to select the number of components for univariate and multivariate mixtures of $t$-distributions. Frühwirth-Schnatter et al. (2005) applied the bridge sampling estimator of the marginal likelihood to select the number of components in the more general framework of mixtures of regression models with errors from the Student-$t$ distributions, which encompass a multivariate mixture of Student-$t$ distributions as a special case.

## 7.4 Further Issues

### 7.4.1 Clustering High-Dimensional Data

When the data are of very high dimension relative to the sample size $N$, such as DNA micro-array data sets, some kind of reduction has to take place to enable clustering.

## Clustering Through Parsimonious Covariances

One successful strategy for model-based clustering of such data is to use parsimonious variance–covariance matrices. Yeung et al. (2001), for instance, use a high-dimensional but parsimonious normal mixture with $\mathbf{\Sigma}_k \equiv \sigma_k^2 \mathbf{I}_r$ for clustering gene expression data. To deal in a more flexible way with correlations within a cluster while remaining parsimonious, several authors (Banfield and Raftery, 1993; Celeux and Govaert, 1995; Bensmail et al., 1997) use parsimonious variance–covariance matrices derived from the eigenvalue decomposition already described in Subsection 6.4.1. Dasgupta and Raftery (1998) discuss an interesting application of these models for finding clusters that are highly concentrated around lines in a two-dimensional space such as mine fields or seismic faults.

Another approach of reducing the number of parameters of the component variance–covariance matrix $\mathbf{\Sigma}_k$ is to adopt the mixture of factor analyzer models suggested by McLachlan and Peel (2000, Chapter 8):

$$\mathbf{Y}_i = \boldsymbol{\mu}_{S_i} + \mathbf{B}_{S_i}\mathbf{z}_i + \boldsymbol{\varepsilon}_i, \qquad \boldsymbol{\varepsilon}_i \sim \mathcal{N}_r\left(\mathbf{0}, \mathbf{Q}_{S_i}\right),$$

where $S_i$ is the group indicator, $\mathbf{Q}_{S_i} = \text{Diag}\left(\sigma^2_{S_i,1}, \dots, \sigma^2_{S_i,r}\right)$, and $\mathbf{z}_i \sim \mathcal{N}_q\left(\mathbf{0}, \mathbf{I}_q\right)$ with $q < r$ are latent factors, which are assumed to be independent of $\boldsymbol{\varepsilon}_i$. $\mathbf{B}_{S_i}$ is an unknown $(r \times q)$-matrix of factor loadings. A related model called mixtures of probabilistic component analyzers (Tipping and Bishop, 1999) assumes the isotropic structure $\mathbf{Q}_{S_i} = \sigma^2_{S_i}\mathbf{I}_r$.

For both models unconditionally, with $\mathbf{z}_i$ being integrated out, a parsimonious mixture of normals results:

$$\mathbf{Y}_i|S_i = k \sim \mathcal{N}_r\left(\boldsymbol{\mu}_k, \mathbf{\Sigma}_k\right), \qquad \mathbf{\Sigma}_k = \mathbf{B}_k\mathbf{B}_k^{'} + \mathbf{Q}_k.$$

Estimations may be based on the EM algorithm (McLachlan et al., 2003) or a Bayesian approach (Fokoué and Titterington, 2003; Utsugi and Kumagai, 2001). Applications include analyzing handwritten digits (Hinton et al., 1997) and clustering of gene-expression data (McLachlan et al., 2003).

Finally, a time series covariance structure for $\mathbf{\Sigma}_k$ based on ARMA models is a very parsimonious and appropriate approach for handling high-dimensional data with a sequential nature, such as repeated measurements or panel data; see Frühwirth-Schnatter and Kaufmann (2006b) for applications in economics.

## Clustering and Variable Selection

In high-dimensional data sets the cluster structure is often confined to a small subset of variables. Because the inclusion of unnecessary variables may complicate or even hinder the recovery of the clusters (Fowlkes et al., 1988; Gnanadesikan et al., 1995; Brusco and Cradit, 2001) joint clustering and variable selection has become an important issue. Several approaches separate the two tasks and apply some dimension-reducing technique prior to applying a clustering procedure such as selecting variables based on preclustering of each variable separately using univariate mixtures of normals (McLachlan et al., 2002) or keeping only the leading components in a principal component analysis (Ghosh and Chinnaiyan, 2002). A related approach has been applied by Liu et al. (2003) within a Bayesian variable selection and clustering approach, by multiplying the original data by a dimension-reducing projection matrix obtained from a principal component analysis. However, as shown by Chang (1983), the leading principal components are not necessarily informative about the cluster structure; see also Yeung and Ruzzo (2001). Other approaches perform simultaneously clustering and variable selection using some heuristic criteria (Fowlkes et al., 1988; Brusco and Cradit, 2001).

Tadesse et al. (2005) use a more principled and promising approach by combining clustering of high-dimensional data via multivariate Gaussian mixtures with an unknown number of components with Bayesian variable selection (George and McCulloch, 1997). For each feature $Y_j$ of the observed random variable $\mathbf{Y} = (Y_1, \ldots, Y_r)$ a binary indicator $\gamma_j$ is introduced, which takes the value 1, iff for that particular feature unobserved heterogeneity is present. This defines a partition of the data into the discriminating variables $\mathbf{Y}_\gamma^D = \{Y_j : \gamma_j = 1\}$ and the nondiscriminating variables $\mathbf{Y}_\gamma^C = \{Y_j : \gamma_j = 0\}$. Conditional on $\boldsymbol{\gamma} = (\gamma_1, \ldots, \gamma_r)$, $\mathbf{Y}_\gamma^D$ and $\mathbf{Y}_\gamma^C$ are assumed to be independently and normally distributed:

$$\mathbf{Y}_\gamma^D \sim \sum_{k=1}^{K} \eta_k \mathcal{N}_d \left(\boldsymbol{\mu}_{k,\boldsymbol{\gamma}}, \boldsymbol{\Sigma}_{k,\boldsymbol{\gamma}}\right),$$
$$\mathbf{Y}_\gamma^C \sim \mathcal{N}_{r-d} \left(\boldsymbol{\beta}_{\boldsymbol{\gamma}}, \mathbf{R}_{\boldsymbol{\gamma}}\right),$$

where $d = \#\{\gamma_j = 1\}$. For MCMC estimation, first the allocations $\mathbf{S} = (S_1, \ldots, S_N)$ are introduced as missing data. Joint variable selection and clustering are carried out using a three-block Gibbs sampling scheme by drawing, respectively, the allocations $\mathbf{S}$, the variable selection indicators $\boldsymbol{\gamma}$, and the weight distribution $\boldsymbol{\eta}$ from the appropriate conditional distributions. Marginalizing over the component parameters is feasible under a conjugate prior, because it is possible to derive $p(\mathbf{y}|\mathbf{S}, \boldsymbol{\gamma})$ explicitly, up to a normalizing constant.

### 7.4.2 Discriminant Analysis

In discriminant analysis, classified observations, the so-called training sample $(\mathbf{S}^{(T)}, \mathbf{y}^{(T)})$, are used to derive a classification rule, whereby additional unclassified observations $\mathbf{y}_i$ could be assigned to one of $K$ classes, with the number of classes being known. This problem is also known as supervised classification. Discriminant analysis methods are often probabilistic, being based on the assumption that in the $k$th class the observations are generated by the class-specific probability distribution $p(\mathbf{y}_i|\boldsymbol{\theta}_k)$, most commonly assumed to arise from a multivariate normal distribution. Although it is not really necessary, it is helpful to keep in mind that marginally $\mathbf{y}_i$ arises from a multivariate finite mixture distribution.

Standard discriminant analysis assigns an observation $\mathbf{y}_i$ with unknown group membership $S_i$ to one of the groups based on the parameter estimates $\boldsymbol{\vartheta}$ obtained from the training sample $(\mathbf{S}^{(T)}, \mathbf{y}^{(T)})$, using the following classification rule,

$$\Pr(S_i = k|\boldsymbol{\vartheta}, \mathbf{y}_i) \propto \eta_k p(\mathbf{y}_i|\boldsymbol{\theta}_k), \tag{7.31}$$

where $\eta_k$ are the proportions of observations in the training sample that belong to class $k$. This choice will minimize the expected missclassification rate.

In discriminant analysis it is usual to study the classification rule (7.31) as a function of $\mathbf{y}_i$, in order to derive a discrimination rule. As discussed in Bensmail and Celeux (1996), classification based on Gaussian mixtures with homogeneous variance–covariance matrices corresponds to linear discriminant analysis, whereas Gaussian mixtures with heterogeneous variance–covariance matrices correspond to quadratic discriminant analysis.

Various suggestions have been made to achieve more flexibility concerning the classification boundaries. Bensmail and Celeux (1996) discuss eigenvalue discriminant analysis, based on parsimonious variance–covariance matrices derived from the eigenvalue decomposition as in Subsection 6.4.1. This approach is more flexible than linear discriminant analysis, while being more structured and parsimonious than quadratic discriminant analysis. Hastie and Tibshirani (1996) study mixture discriminant analysis by assuming that the components themselves are mixtures of normals:

$$p(\mathbf{y}_i|\boldsymbol{\theta}_k) = \sum_{j=1}^{G_k} \eta_{kj} f_N(\mathbf{y}_i; \boldsymbol{\mu}_{kj}, \boldsymbol{\Sigma}_{kj}),$$

where $\sum_{j=1}^{G_k} \eta_{kj} = 1$. This extension allows for nonlinear and nonmonotonic classification boundaries. For further details see the excellent review by Fraley and Raftery (2002, Section 6).

### 7.4.3 Combining Classified and Unclassified Observations

Hosmer (1973) showed that efficiency of joint parameter estimation and classification considerably improves if one has an additional small sample which is known to come from specific components. Lavine and West (1992) studied the problem of combining perfectly classified data $(\mathbf{S}^{(T)}, \mathbf{y}^{(T)})$ with unclassified observations $\mathbf{y}$ using a fully Bayesian viewpoint.

The Bayesian approach deals with this type of data by introducing the unknown class indicators $\mathbf{S}$ as missing data, and estimating $\mathbf{S}$ jointly with the component parameters $\boldsymbol{\theta}_1, \ldots, \boldsymbol{\theta}_K$ and the weight distribution $\boldsymbol{\eta}$ from the data $(\mathbf{S}^{(T)}, \mathbf{y}^{(T)}, \mathbf{y})$. For such data, data augmentation and MCMC as described in *Algorithm 6.2* are extended in an obvious way. In step (b), the unknown allocations $\mathbf{S}$ are sampled conditional on the most recent parameter $\boldsymbol{\vartheta} = (\boldsymbol{\theta}_1, \ldots, \boldsymbol{\theta}_K, \boldsymbol{\eta})$. In step (a), the parameter $\boldsymbol{\vartheta}$ is sampled conditional on knowing the complete sample $(\mathbf{S}^{(T)}, \mathbf{y}^{(T)}, \mathbf{S}, \mathbf{y})$.

Lavine and West (1992) emphasize that the assignment of $\mathbf{y}_i$ according to the classification rule (7.31) should account for the parameter uncertainty associated with estimating $\boldsymbol{\vartheta}$ from the data $(\mathbf{y}^{(T)}, \mathbf{S}^{(T)}, \mathbf{y})$ and suggest the following classification rule.

$$\Pr(S_i = k|\mathbf{y}^{(T)}, \mathbf{S}^{(T)}, \mathbf{y}) = \int \Pr(S_i = k|\boldsymbol{\vartheta}, \mathbf{y}_i) p(\boldsymbol{\vartheta}|\mathbf{y}^{(T)}, \mathbf{S}^{(T)}, \mathbf{y}) d\boldsymbol{\vartheta}.$$

Based on the MCMC sample $\boldsymbol{\vartheta}^{(1)}, \ldots, \boldsymbol{\vartheta}^{(M)}$, which draws from the posterior $p(\boldsymbol{\vartheta}|\mathbf{y}^{(T)}, \mathbf{S}^{(T)}, \mathbf{y})$, they approximate this integral by

$$\Pr(S_i = k|\mathbf{y}^{(T)}, \mathbf{S}^{(T)}, \mathbf{y}) \approx \frac{1}{M} \sum_{m=1}^{M} \Pr(S_i = k|\boldsymbol{\vartheta}^{(m)}, \mathbf{y}_i). \tag{7.32}$$

Lavine and West (1992) run a straightforward Gibbs sampler in the hope of achieving implicit labeling. Nevertheless, it is not guaranteed that the classification rule (7.32) is actually free from label switching.

### 7.4.4 Density Estimation Using Finite Mixtures

Consider the problem of estimating the density of an unknown probability distribution $p(\mathbf{y})$, given $N$ draws $\mathbf{y}_1, \ldots, \mathbf{y}_N$ from this distribution. Finite mixture distributions can be used to derive arbitrarily accurate approximations to practically any given probability distribution, provided that the number of components is not limited (Ferguson, 1973). Practically, density estimation has been based on univariate mixtures of normal distributions (Escobar and West, 1995; Roeder and Wasserman, 1997b; Scott and Szewczyk, 2001) and multivariate mixtures of normals (Scott, 1992; West, 1993; Li and Barron, 2000; Fraley and Raftery, 2002).

Finite mixtures of Gaussian distributions provide a (semi-)parametric alternative to nonparametric methods of density estimation, such as kernel density estimation. The usual nonparametric kernel density estimator results when fitting a mixture with $K = N$ components with $\eta_k = 1/N$, $\boldsymbol{\mu}_k = \mathbf{y}_k, k = 1, \ldots, N$, and $\boldsymbol{\Sigma} = h\boldsymbol{W}$:

$$p(\mathbf{y}|h) \approx \frac{1}{N} \sum_{i=1}^{N} f_N(\mathbf{y}; \mathbf{y}_i, h\boldsymbol{W}), \tag{7.33}$$

with $\boldsymbol{W}$ being an estimate of the variance–covariance matrix $\boldsymbol{\Sigma}$, for instance, the sample variance–covariance matrix. Note that in (7.33) the parameter

$h$, also known as the bandwidth, is the only unknown quantity. Bayesian inference on the parameter $h$ (West, 1993) is analogous to bandwidth selection in kernel density estimation (Silverman, 1999), where $h$ is chosen as a slowly decreasing function of the number of observations; for instance,

$$h = \left( \frac{4}{(1+2r)N} \right)^{1/(1+r)},$$

with $r$ being the dimension of $\mathbf{y}_i$.

The number $K$ of components in density approximation has no physical meaning, but is arbitrary in the sense that the choice of $K$ is guided by the desired degree of smoothness of the density estimate. Roeder and Wasserman (1997b) and Fraley and Raftery (2002) use BIC to select the number of components in mixture density approximation. Roeder and Wasserman (1997b) show that the density estimate based on BIC is a consistent estimator of the true density. Solka et al. (1998) use AIC-based pruning of the mixture density estimator. Scott and Szewczyk (2001) start from the standard kernel density estimate, which is then simplified by merging components that are similar in the sense of definition (1.18).

### 7.4.5 Finite Mixtures as an Auxiliary Computational Tool in Bayesian Analysis

Finite mixture distributions have been used in Bayesian analysis not only as a device of modeling certain nonstandard features of the data, but also as a useful computational tool to facilitate the necessary computations. Some work deals with approximating priors by mixtures of conjugate priors in a standard nonconjugate Bayesian analysis (Dalal and Hall, 1983). Other papers deal with approximating posterior distribution by mixtures. An early example is Alspach and Sorenson (1972), who used a multivariate Gaussian mixture in adaptive filtering of non-Gaussian data; see also West (1993) for further applications.

Several authors found it useful to approximate some nonnormal density appearing in the model definition by a mixture of normals, in order to facilitate Bayesian analysis. A prominent example is representing the Student-$t$ distribution as a scale mixture of normals, which reduces the handling of the somehow circumstantial Student-$t$ distribution to dealing with the more convenient normal distribution; see, for instance, Zellner (1971), Geweke (1993), and Shephard (1994).

Also in cases where a convenient theoretical mixture representation of a density, which is difficult to handle, does not exist, normal mixtures are helpful simply by approximating this density by a mixture of normals. Shephard (1994), Kim et al. (1998), and Chib et al. (2002) use a normal mixture approximation for the distribution of the log of a $\chi^2_1$-variable in the context of stochastic volatility models. Frühwirth-Schnatter and Wagner (2006) and

Frühwirth-Schnatter and Frühwirth (2006) use a normal mixture approximation for the distribution of the negative logarithm of an $\mathcal{E}(1)$-variable in the context of parameter-driven models for count, binary, and categorical data.

# 8

# Finite Mixtures of Regression Models

## 8.1 Introduction

In applied statistics as well as in econometrics a tremendous amount of applications deal with relating a random variable $Y_i$, which is observed on several occasions $i = 1, \ldots, N$, to a set of explanatory variables or covariates $(z_{i1}, \ldots, z_{i,d-1})$ through a regression-type model, where the conditional mean of $Y_i$ is assumed to depend on $\mathbf{x}_i = \begin{pmatrix} z_{i1} & \cdots & z_{i,d-1} & 1 \end{pmatrix}$ through $\mathrm{E}(Y_i|\boldsymbol{\beta}, \mathbf{x}_i) = \mathbf{x}_i\boldsymbol{\beta}$, where $\boldsymbol{\beta}$ is a vector of unknown regression coefficients of dimension $d$.

In many circumstances, however, the assumption that the regression coefficient is fixed over all possible realizations of $Y_1, \ldots, Y_N$ is inadequate, and models where the regression coefficient changes are of great practical importance. The most general alternative is to assume a different regression coefficient $\boldsymbol{\beta}_i^s$ for each realization $Y_i$, $\mathrm{E}(Y_i|\boldsymbol{\beta}, \mathbf{x}_i) = \mathbf{x}_i\boldsymbol{\beta}_i^s$, however, only in rare cases will it be possible to estimate $\boldsymbol{\beta}_i^s$ without imposing further structure, and modeling $\boldsymbol{\beta}^s = (\boldsymbol{\beta}_1^s, \ldots, \boldsymbol{\beta}_N^s)$ becomes an important issue.

For identifying a sensible model for $\boldsymbol{\beta}^s$, it is helpful to understand why the regression coefficients are different. For sequential observations the regression coefficient may change over time, whereas for cross-sectional data the regression coefficient may change between subgroups of observations. In both cases the model may be misspecified because of omitted variables and nonlinearities or the sample may contain outliers. Whatever information is available about the nature of heterogeneity for the problem at hand should be incorporated in an appropriate manner. Within a Bayesian approach, this information is included by choosing a specific probabilistic model for $\boldsymbol{\beta}^s$ which is specified in terms of the density $p(\boldsymbol{\beta}^s)$ of the joint distribution of $\boldsymbol{\beta}^s = (\boldsymbol{\beta}_1^s, \ldots, \boldsymbol{\beta}_N^s)$. $p(\boldsymbol{\beta}^s)$ plays the role of a prior distribution, imposing some model structure on the individual regression coefficients that may be overruled by the information in the data. Different such prior distributions defining different model structures may be compared in a principled way by Bayesian model comparison.

This chapter focuses on capturing parameter heterogeneity for cross-sectional data through finite mixtures of regression models where changes in $\boldsymbol{\beta}_i^s$ are driven by a hidden discrete indicator $S_i$, which is allowed to take one out of $K$ values for each observation $Y_i$. This model is formulated in Section 8.2, whereas statistical inference is discussed in Section 8.3.

Several useful extensions of this model are discussed in this chapter, such as mixed-effects finite mixtures of regression models in Section 8.4, which combine regression coefficients that are fixed across all realizations with regression coefficients that are allowed to change, and finite mixtures of random-effects models in Section 8.5, which are useful for longitudinal data and repeated measurements.

## 8.2 Finite Mixture of Multiple Regression Models

In this section focus lies on extending the standard multiple regression model with normally distributed errors by introducing a regression coefficient that changes between groups of otherwise homogeneous observations.

### 8.2.1 Model Definition

Let $(Y_i, \boldsymbol{z}_i)$ be a pair of a random variable $Y_i$ and a set of explanatory variables $\boldsymbol{z}_i = (z_{i1}, \dots, z_{i,d-1})$. Suppose that dependence of $Y_i$ on $\boldsymbol{z}_i$ is modeled by a multiple regression model:

$$Y_i = \mathbf{x}_i \boldsymbol{\beta} + \varepsilon_i, \qquad \varepsilon_i \sim \mathcal{N}\left(0, \sigma_\varepsilon^2\right), \tag{8.1}$$

where $\mathbf{x}_i = \left( z_{i1} \cdots z_{i,d-1} \; 1 \right)$ is a design point, and $\boldsymbol{\beta}$ and $\sigma_\varepsilon^2$ are unknown parameters. Assume that background information suggests that the regression coefficient $\boldsymbol{\beta}$ and the error variance $\sigma_\varepsilon^2$ are not homogeneous over all possible pairs $(Y_i, \boldsymbol{z}_i)$. One way to capture such changes in the parameter of a regression model is finite mixtures of regression models. A finite mixture regression model assumes that a set of $K$ regression models characterized by the parameters $(\boldsymbol{\beta}_1, \sigma_{\varepsilon,1}^2), \dots, (\boldsymbol{\beta}_K, \sigma_{\varepsilon,K}^2)$ exists, and that for each observation pair $(Y_i, \boldsymbol{z}_i)$ a hidden random indicator $S_i$ chooses one among these models to generate $Y_i$:

$$Y_i = \mathbf{x}_i \boldsymbol{\beta}_{S_i} + \varepsilon_i, \qquad \varepsilon_i \sim \mathcal{N}\left(0, \sigma_{\varepsilon,S_i}^2\right). \tag{8.2}$$

$\boldsymbol{\beta}_1, \dots, \boldsymbol{\beta}_K$ as well as $\sigma_{\varepsilon,1}^2, \dots, \sigma_{\varepsilon,K}^2$ are unknown parameters that need to be estimated from the data. The statistician applying a finite mixture of regression models has to specify how the random mechanism $S_i$ works. In the absence of any additional information it is usual to assume that $S_i$ and $S_{i'}$ are pairwise independent, and each $S_i$ is distributed according to an unknown probability distribution $\boldsymbol{\eta} = (\eta_1, \dots, \eta_K)$. In what follows, $\boldsymbol{\vartheta}$ summarizes all unknown model parameters, including the parameters $\boldsymbol{\eta}$ appearing in the definition of the distribution law of $\mathbf{S} = (S_1, \dots, S_N)$.

It is easy to verify that the marginal distribution of $Y_i$, when holding the design point $\mathbf{x}_i$ as well as $\boldsymbol{\vartheta}$ fixed, reads:

$$p(y_i|\mathbf{x}_i, \boldsymbol{\vartheta}) = \sum_{k=1}^{K} p(y_i|\mathbf{x}_i, S_i, \boldsymbol{\vartheta})\Pr(S_i = k|\boldsymbol{\vartheta}) = \sum_{k=1}^{K} \eta_k f_N(y_i; \mathbf{x}_i\boldsymbol{\beta}_k, \sigma^2_{\varepsilon,k}).$$

Thus for each value of the design point $\mathbf{x}_i$, the marginal distribution of $Y_i$ is a finite mixture of univariate normal distributions with mean $\mu_{k,i} = \mathbf{x}_i\boldsymbol{\beta}_k$ and variance $\sigma^2_{\varepsilon,k}$. Therefore a finite mixture of regression models may be seen as an extension of a finite mixture of univariate normal distributions where the mean in the mixture distribution depends on explanatory variables. On the other hand, a finite mixture of univariate normal distributions may be seen as that special case of finite mixtures of regression models where $\beta_k = \mu_k$ and $\mathbf{x}_i = 1$ for all $i = 1, \ldots, N$.

Various extensions of model (8.2) are useful. The mixture regression model defined in (8.2) is heteroscedastic because the variance of the error term $\varepsilon_i$ changes across the realizations. If the variance of the error term is unaffected by $S_i$, a homoscedastic finite mixture of regression models results:

$$Y_i = \mathbf{x}_i\boldsymbol{\beta}_{S_i} + \varepsilon_i, \qquad \varepsilon_i \sim \mathcal{N}\left(0, \sigma^2_\varepsilon\right). \tag{8.3}$$

The distributional law of $\mathbf{S}$ may be substituted by other structures, if more information about the nature of heterogeneity is available. As discussed in Subsection 8.6.2, the probability of belonging to a certain state may depend on a covariate. For random covariates, the covariate distribution may differ between the clusters, in which case a multivariate finite normal mixture model as discussed in Chapter 6 may be appropriate. Whenever data are collected sequentially, alternative probability structures on the hidden indicator turn out to be useful. Goldfeld and Quandt (1973) introduced a hidden Markov chain into a mixture regression model, in order to deal with time series data that depend on exogenous variables. This issue is discussed in Subsection 10.3.2.

### 8.2.2 Identifiability

Like any finite mixture model, finite mixtures of regression models suffer from nonidentifiability due to label switching and potential overfitting; see Section 1.3 for a general discussion of these issues. More importantly, generic identifiability of finite mixtures of regression models does not in general follow from the generic identifiability of Gaussian mixtures as falsely claimed, for instance, in DeSarbo and Cron (1988), despite the close relationship between these two model classes.

A necessary condition for identifiability of a standard regression model is that the matrix $\mathbf{X}'\mathbf{X}$, where

$$\mathbf{X} = \begin{pmatrix} \mathbf{x}_1 \\ \vdots \\ \mathbf{x}_N \end{pmatrix},$$

is of full rank. For finite mixtures of regression models nonidentifiability may occur, even if this condition is fulfilled. This was first noticed by Hennig (2000) who showed that the regression parameters are identifiable, iff the number $K$ of clusters is smaller than the number of distinct $(d-1)$-dimensional hyperplanes generated by the covariates (excluding the constant). Loosely speaking, identifiability problems occur for finite mixtures of regression models with covariates that show too little variability. Problems are to be expected, in particular, if covariates are dummy variables or reflect a few categories as in marketing research. In this section we provide more details on this important issue.

Consider the set of different covariates $\{\mathbf{x}_1, \dots, \mathbf{x}_p\}$. Assume that for each covariate $\mathbf{x}_i$ that the identity

$$\sum_{k=1}^{K} \eta_k f_N(y; \mu_{k,i}, \sigma^2_{\varepsilon,k}) = \sum_{k=1}^{K} \eta^\star_k f_N(y; \mu^\star_{k,i}, \sigma^{2,\star}_{\varepsilon,k}), \tag{8.4}$$

where $\mu_{k,i} = \mathbf{x}_i \boldsymbol{\beta}_k$ and $\mu^\star_{k,i} = \mathbf{x}_i \boldsymbol{\beta}^\star_k$, holds for all $y \in \Re$. If the model parameters $(\boldsymbol{\beta}_1, \dots, \boldsymbol{\beta}_K, \sigma^2_{\varepsilon,1}, \dots, \sigma^2_{\varepsilon,K}, \eta_1, \dots, \eta_K)$ and $(\boldsymbol{\beta}^\star_1, \dots, \boldsymbol{\beta}^\star_K, \sigma^{2,\star}_{\varepsilon,1}, \dots, \sigma^{2,\star}_{\varepsilon,K}, \eta^\star_1, \dots, \eta^\star_K)$ are related to each other by relabeling, then the finite mixture regression model is generically identifiable.

For a fixed covariate $\mathbf{x}_i$, (8.4) reduces to a Gaussian mixture, and generic identifiability of Gaussian mixtures implies the existence of a permutation $\rho_i(\cdot)$ of $\{1, \dots, K\}$ such that for all $k = 1, \dots, K$:

$$\eta^\star_k = \eta_{\rho_i(k)}, \qquad \mathbf{x}_i \boldsymbol{\beta}^\star_k = \mathbf{x}_i \boldsymbol{\beta}_{\rho_i(k)}, \qquad \sigma^{2,\star}_{\varepsilon,k} = \sigma^2_{\varepsilon,\rho_i(k)}. \tag{8.5}$$

A major cause for generic nonidentifiability is that the different permutations $\rho_i(\cdot)$ appearing in (8.5) are not necessarily the same for all design points $\mathbf{x}_i$, $i = 1, \dots, p$.

Nevertheless, let us assume for the moment that actually the same permutation $\rho_s(\cdot)$ has been applied for all design points $\mathbf{x}_i$, $i = 1, \dots, p$. Then (8.5) implies $\mathbf{x}_i \boldsymbol{\beta}^\star_k = \mathbf{x}_i \boldsymbol{\beta}_{\rho_s(k)}$ for all $i = 1, \dots, p$ and:

$$\mathbf{X} \boldsymbol{\beta}^\star_k = \mathbf{X} \boldsymbol{\beta}_{\rho_s(k)},$$

where the rows of the design matrix $\mathbf{X}$ are equal to $\mathbf{x}_1, \dots, \mathbf{x}_p$. If $\mathbf{X}'\mathbf{X}$ has full rank, then it follows immediately that the regression coefficients are determined up to relabeling:

$$\boldsymbol{\beta}^\star_k = \boldsymbol{\beta}_{\rho_s(k)}, \tag{8.6}$$

ensuring generic identifiability. The problem with this derivation is, that without further assumptions, the different permutations $\rho_i(\cdot)$ appearing in (8.5) are not necessarily the same for all $i = 1, \dots, p$.

It is possible to show that these permutations are necessarily the same, if any two regression models in the mixture differ at least in $\eta_k$ or $\sigma^2_{\varepsilon,k}$. Assume that (8.5) holds for two different permutations $\rho_s(\cdot)$ and $\rho_t(\cdot)$. Then $\eta_{k_1} = \eta_{k_2}$ and $\sigma^2_{\varepsilon,k_1} = \sigma^2_{\varepsilon,k_2}$ for regression model $k_1 = \rho_s(k)$ and $k_2 = \rho_t(k)$, which contradicts the assumption made above.

If $\eta_k$ and $\sigma^2_{\varepsilon,k}$ are the same for at least two regression models, then it is possible that (8.5) holds for two different permutations $\rho_s(\cdot)$ and $\rho_t(\cdot)$. Assume that $\eta_{k_1} = \eta_{k_2}$ and $\sigma^2_{\varepsilon,k_1} = \sigma^2_{\varepsilon,k_2}$. Then any two permutations where $\rho_s(k_1) = \rho_t(k_2)$, $\rho_s(k_2) = \rho_t(k_1)$, and $\rho_s(l) = \rho_t(l)$, for $l \neq k_1, k_2$, fulfill (8.5). In this case generic nonidentifiability may occur, even if the matrix $\mathbf{X}'\mathbf{X}$ has full rank.

Consider, for instance, a mixture of two regression models, where $\eta_1 = \eta_2$ and $\sigma^2_{\varepsilon,1} = \sigma^2_{\varepsilon,2}$. For each $i = 1, \ldots, p$, the permutation $\rho_i(\cdot)$ appearing in (8.5) is equal to one of the two possible permutations, namely the identity, $\rho_1(1) = 1$ and $\rho_1(2) = 2$, or the permutation $\rho_2(1) = 2$ and $\rho_2(2) = 1$, which interchanges the labeling. Reorder, for a given sequence of permutations, the equations in (8.5) according to the permutation applied to $k = 1$. Then:

$$\mathbf{X}_1\boldsymbol{\beta}_1^\star = \mathbf{X}_1\boldsymbol{\beta}_1, \tag{8.7}$$

$$\mathbf{X}_2\boldsymbol{\beta}_1^\star = \mathbf{X}_2\boldsymbol{\beta}_2, \tag{8.8}$$

where the rows of the design matrix $\mathbf{X}_1$ are built from all design points $\mathbf{x}_i$, where $\rho_i(1) = 1$, and the rows of the design matrix $\mathbf{X}_2$ are built from all design points $\mathbf{x}_i$, where $\rho_i(1) = 2$. If in (8.7) and (8.8) either $\text{rg}(\mathbf{X}_1'\mathbf{X}_1) = d$ or $\text{rg}(\mathbf{X}_2'\mathbf{X}_2) = d$ (or both), then (8.6) follows immediately.

Thus generic identifiability up to relabeling follows, if $\text{rg}(\mathbf{X}_1'\mathbf{X}_1) = d$ or $\text{rg}(\mathbf{X}_2'\mathbf{X}_2) = d$ holds for any partition of the set of different covariates $\{\mathbf{x}_1, \ldots, \mathbf{x}_p\}$ into two sets. This is essentially the same condition as the one given by Hennig (2000). Any partition that violates this condition defines an alternative solution. It follows that for $K = 2$ the minimum number of different design points is equal to $2\dim(\boldsymbol{\beta}_k) + 1$, which is sufficient to achieve identifiability, iff all subsets of size $\dim(\boldsymbol{\beta}_k)$ define a design matrix of full rank. To give an example, consider a mixture of two regression models where $\dim(\boldsymbol{\beta}_k) = 2$ where there are only two linear independent design points $\mathbf{x}_1 = (z_1\ 1)$ and $\mathbf{x}_2 = (z_2\ 1)$. A similar example appears in Hennig (2000). Evidently the partition $\{\mathbf{x}_1\} \cup \{\mathbf{x}_2\}$ violates the rank condition. Only if the two permutations in (8.5) are the same, do we obtain (8.6). However, if the two permutations in (8.5) are different, then another solution exists, which is given by

$$\mathbf{X}^\star = \begin{pmatrix} \mathbf{x}_1 \\ \mathbf{x}_2 \end{pmatrix}, \qquad \boldsymbol{\beta}_1^\star = (\mathbf{X}^\star)^{-1} \begin{pmatrix} \mathbf{x}_1\boldsymbol{\beta}_1 \\ \mathbf{x}_2\boldsymbol{\beta}_2 \end{pmatrix}, \qquad \boldsymbol{\beta}_2^\star = (\mathbf{X}^\star)^{-1} \begin{pmatrix} \mathbf{x}_1\boldsymbol{\beta}_2 \\ \mathbf{x}_2\boldsymbol{\beta}_1 \end{pmatrix}.$$

Consequently, this finite mixture regression model is generically unidentifiable. Whereas a single regression line is determined from two covariate points, for a mixture of two regressions this is not the case. Identifiability is achieved by adding a third design point $\mathbf{x}_3 = (z_3\ 1)$, with $z_3 \neq z_1, z_2$. Then any partition of the design points into two groups contains at least two different design points and the identifiability condition is fulfilled.

### Identifiability of Finite Mixtures of Regression Models

Consider a mixture of $K$ regression models, where $\eta_k$ and $\sigma^2_{\varepsilon,k}$ are the same in all groups. For each $k = 1, \dots, K$, reorder the equations in (8.5) according to the permutation applied to the label $k$. Then:

$$\mathbf{X}_1\boldsymbol{\beta}_k^\star = \mathbf{X}_1\boldsymbol{\beta}_1, \tag{8.9}$$
$$\mathbf{X}_2\boldsymbol{\beta}_k^\star = \mathbf{X}_2\boldsymbol{\beta}_2,$$
$$\vdots$$
$$\mathbf{X}_K\boldsymbol{\beta}_k^\star = \mathbf{X}_K\boldsymbol{\beta}_K, \tag{8.10}$$

where the rows of the design matrix $\mathbf{X}_j$ are built from all design points $\mathbf{x}_i$, where $\rho_i(k) = j$. If in (8.9) to (8.10) $\text{rg}(\mathbf{X}_j^{'}\mathbf{X}_j) = d$ for at least one $j = 1, \dots, K$, then (8.6) follows immediately. Thus generic identifiability up to relabeling follows, if $\text{rg}(\mathbf{X}_j^{'}\mathbf{X}_j) = d$ holds for any partition of the set of different covariates $\{\mathbf{x}_1, \dots, \mathbf{x}_p\}$ into $K$ subsets. This is essentially the same condition as the one given by Hennig (2000).

Any partition that violates this condition defines an alternative solution. It follows that the minimum number of different design points is equal to $K(\dim(\boldsymbol{\beta}_k) - 1) + 1$. If $p \leq K(\dim(\boldsymbol{\beta}_k) - 1)$, then evidently there exists a partition of the different design points into $K$ groups, where each set contains at most $\dim(\boldsymbol{\beta}_k) - 1$ design points and violates the rank condition. This minimum number of design points is sufficient to achieve identifiability, iff all subsets of size $\dim(\boldsymbol{\beta}_k)$ define a design matrix of full rank.

Grün and Leisch (2004) use bootstrap methods as a diagnostic tool for revealing identifiability problems in finite mixtures of normal and nonnormal regression models.

## 8.2.3 Statistical Modeling Based on Finite Mixture of Regression Models

In statistical modeling finite mixtures of regression models are also known as *switching regression models* in economics (Quandt, 1972), as *latent class regression models* in marketing (DeSarbo and Cron, 1988), as *mixture-of-expert models* in the machine-learning literature (Jacobs et al., 1991), and as *mixed models* in biology (Wang et al., 1996).

### The Switching Regression Model

For sequentially observed data, one source of heterogeneity is sudden changes in regression coefficients due to a structural break. A simple model to capture a sudden parameter change at a known breakpoint $\tau$ within the standard multiple regression model is the following,

$$Y_i = \begin{cases} \mathbf{x}_i\boldsymbol{\beta}_1 + \varepsilon_i, & \varepsilon_i \sim \mathcal{N}\left(0, \sigma^2_{\varepsilon,1}\right), & i < \tau, \\ \mathbf{x}_i\boldsymbol{\beta}_2 + \varepsilon_i, & \varepsilon_i \sim \mathcal{N}\left(0, \sigma^2_{\varepsilon,2}\right), & i \geq \tau. \end{cases} \tag{8.11}$$

It is useful to reparameterize model (8.11) as

$$Y_i = \mathbf{x}_i(1 - D_i)\boldsymbol{\beta}_1 + \mathbf{x}_i D_i \boldsymbol{\beta}_2 + \varepsilon_i, \qquad \varepsilon_i \sim \mathcal{N}\left(0, \sigma^2_i\right), \tag{8.12}$$

where $\sigma^2_i = \sigma^2_{\varepsilon,1}(1 - D_i) + \sigma^2_{\varepsilon,2}D_i$. $D_i$ is a dummy variable, taking the value 0 for $i < \tau$ and 1 otherwise. If the breakpoint $\tau$ is known, then $D_i$ is exogenous, and (8.12) is a regression model with heteroscedastic errors. If the exact position of the break point $\tau$ is unknown, $D_i$ is not observable, but a latent, discrete random variable, taking the values 0 and 1 according to some unknown probability law, and (8.11) turns out to be a finite mixture of regression models, also called a switching regression model.

An early example of a switching regression model with unknown breakpoint is considered in Quandt (1958) who studies the consumption function $Y = \beta X + \alpha$, where $X$ is the income and $Y$ is the consumption, and assumes that other factors, that are difficult to identify, affect the parameters of the consumption function. If this critical factor is below a threshold, then $Y = \beta_1 X + \alpha_1$, otherwise $Y = \beta_2 X + \alpha_2$. In general we are not able to identify the critical variable, and what we observe is a mixture of these two regression lines. Quandt (1958) considers a single shift between the two regimes at an unknown break point, mainly to make estimation feasible under the computational limitations of the 1950s.

A particularly important extension of this work is Quandt (1972), where for the first time a probability model is introduced, to model "that nature chooses between regimes with probability $\eta_1$ and $1 - \eta_1$"(Quandt, 1972, p.306).[1] Quandt (1972) starts directly from specifying the conditional density $p(y_i|\mathbf{x}_i, \boldsymbol{\beta}_1, \boldsymbol{\beta}_2, \sigma^2_{\varepsilon,1}, \sigma^2_{\varepsilon,2}, \eta_1)$ as a mixture of two normal distributions:

$$Y_i \sim \eta_1 \mathcal{N}\left(\mathbf{x}_i\boldsymbol{\beta}_1, \sigma^2_{\varepsilon,1}\right) + (1 - \eta_1)\mathcal{N}\left(\mathbf{x}_i\boldsymbol{\beta}_2, \sigma^2_{\varepsilon,2}\right). \tag{8.13}$$

In his summarizing remarks, Quandt (1972, p.310) concludes that "A notable disadvantage of the method is that it does not allow individual observations to be identified with particular regimes." The latent variable interpretation of his important contribution, which allows clustering observations into regimes, was discovered only later.

Further applications of switching regression models in econometrics are found in Fair and Jaffee (1972) and Quandt and Ramsey (1978), who consider the relation between wage bargains and unemployment rate through a Phillips curve which is expected to switch according to low and high changes on the consumer price index.

[1] Original notation of Quandt (1972) changed.

### Omitted Categorical Predictors

Mixtures of regression models arise whenever a categorical or dummy regressor is omitted. Hosmer (1974), which is an early reference in this area, considered a mixture of two regression lines with a nice application from fishery research. In commercial catches of halibut only age and length are measured, whereas the gender of the fish is unknown. For any particular age, the mean length of female fish exceeds that of male fish, and this difference increases with age. If gender were observed, length may be modeled in terms of gender $g_i$ and age $a_i$ in the following way.

$$Y_i = \beta_1 + a_i\beta_2 + g_i\beta_3 + g_ia_i\beta_4 + \varepsilon_i, \qquad \varepsilon_i \sim \mathcal{N}\left(0, \sigma_\varepsilon^2\right).$$

When coding gender as a 0/1 variable, this model may be written as

$$Y_i = \beta_{g_i,1} + a_i\beta_{g_i,2} + \varepsilon_i, \tag{8.14}$$

where $\beta_{g_i,1} = \beta_1 + g_i\beta_3$ and $\beta_{g_i,2} = \beta_2 + g_i\beta_4$. If gender is unobserved, then (8.14) is equal to a mixture of two regression models. In a scatter plot of $a_i$ versus the observed length $y_i$, the observations cluster around two regression lines, one corresponding to males, the other to females. When a switching regression model is fitted to the data, then both unknown regression lines have to be reconstructed from the data.

Note that the switching slope in (8.14) is caused by interaction between the observed and the omitted categorical variable. If such an interaction is not present, then $\beta_4 = 0$ and (8.14) reduces to a regression model with a shift in the intercept only:

$$Y_i = \beta_{g_i,1} + a_i\beta_2 + \varepsilon_i.$$

### Unknown Segments in the Population

Finite mixtures of regression models, introduced into marketing by DeSarbo and Cron (1988), found numerous applications in marketing research; see Wedel and DeSarbo (1993b) and Rossi et al. (2005) for a review. In marketing, consumers rate the quality of products or events. A regression model is built to describe the relation between the rating $Y_i$ of consumer $i$ and certain features of the product summarized in the design matrix $\mathbf{x}_i$. If unknown segments in the population are present, then the part-worths $\boldsymbol{\beta}_i^s$ of a certain consumer $i$ depend on membership in a certain segment. If we introduce a segment indicator $S_i$, then the market segmentation regression model reads:

$$Y_i = \mathbf{x}_i\boldsymbol{\beta}_{S_i} + \varepsilon_i, \qquad \varepsilon_i \sim \mathcal{N}\left(0, \sigma_\varepsilon^2\right). \tag{8.15}$$

Apart from estimating the regression coefficients in the different segments, the indicator $S_i$ itself is of interest, as it allows us to assign each consumer to a certain segment $k$.

### 8.2.4 Outliers in a Regression Model

The finite mixture model discussed in Section 7.2 for dealing with outliers in univariate data sets has been extended in several ways to deal with outliers in a linear regression model; see the review article by Peña and Guttman (1993).

Box and Tiao (1968), for instance, extend the variance inflation model (7.24) in the following way.

$$Y_i \sim (1-\eta_2)\mathcal{N}\left(\mathbf{x}_i\boldsymbol{\beta}, \sigma_\varepsilon^2\right) + \eta_2\mathcal{N}\left(\mathbf{x}_i\boldsymbol{\beta}, k\sigma_\varepsilon^2\right). \tag{8.16}$$

Model (8.16) is a regression model with switching variances, but a constant regression parameter $\boldsymbol{\beta}$. Abraham and Box (1978) extended the location shift model (7.25) to allow for outliers in a linear regression model:

$$Y_i \sim (1-\eta_2)\mathcal{N}\left(\mathbf{x}_i\boldsymbol{\beta}, \sigma_\varepsilon^2\right) + \eta_2\mathcal{N}\left(\mathbf{x}_i\boldsymbol{\beta}+k, \sigma_\varepsilon^2\right). \tag{8.17}$$

Model (8.17) allows for a switching intercept, while holding the variance fixed. Peña and Guttman (1993) show that these models are more effective in identifying outliers than methods which postulate a null model for the generation of the data with no alternative to the null model being entertained.

Various extensions to models (8.16) and (8.17) are worth mentioning. Guttman et al. (1978) combine a mixture of a normal regression models with a random-effects model to allow for a different shift for each outlier. Outlier modeling in nonnormal mixture regression models is considered in Pregibon (1981), Copas (1988), and Verdinelli and Wasserman (1991). West (1984, 1985) also studies more general scale mixtures of GLMs to deal with outliers.

## 8.3 Statistical Inference for Finite Mixtures of Multiple Regression Models

Parameter estimation for finite mixtures of regression models is usually based on ML estimation or Bayesian estimation, an exception being Quandt and Ramsey (1978) who used a method of moments estimator based on the moment-generating function.

### 8.3.1 Maximum Likelihood Estimation

Assume that $N$ observation pairs $(\mathbf{x}_1, y_1), \dots, (\mathbf{x}_N, y_N)$ are available. The appropriate likelihood function for parameter estimation for a finite mixture of an arbitrary number $K$ of regression models was derived for the first time by Quandt (1972). This function turns out to be the following extension of the mixture likelihood of a standard finite mixture model,

$$p(\mathbf{y}|\boldsymbol{\vartheta}) = \prod_{i=1}^{N}\left(\sum_{k=1}^{K} f_N(y_i; \mathbf{x}_i\boldsymbol{\beta}_k, \sigma_{\varepsilon,k}^2)\eta_k\right), \tag{8.18}$$

where $\boldsymbol{\vartheta} = (\boldsymbol{\beta}_1, \ldots, \boldsymbol{\beta}_K, \sigma^2_{\varepsilon,1}, \ldots, \sigma^2_{\varepsilon,K}, \boldsymbol{\eta})$. In contrast to this, Fair and Jaffee (1972) consider maximization of the classification likelihood $p(\mathbf{y}|\boldsymbol{\vartheta}, \mathbf{S})$ with respect to $\boldsymbol{\vartheta}$ and $\mathbf{S}$ for jointly solving the problem of parameter estimation and estimating the unknown allocations. However, Oberhofer (1980) showed that this approach leads in general to inconsistent estimators of $\boldsymbol{\beta}_1, \ldots, \boldsymbol{\beta}_K$.

In Quandt (1972), the mixture regression likelihood function $p(\mathbf{y}|\boldsymbol{\vartheta})$ is maximized numerically, and considerable convergence failures are reported for repeated experiments on artificially generated data. A mixture of two regression models, for instance, where $\boldsymbol{\beta}_1 = (1, 1)$, $\boldsymbol{\beta}_2 = (0.5, 1.5)$, $\sigma^2_{\varepsilon,1} = 2$, $\sigma^2_{\varepsilon,2} = 2.5$, $\eta_1 = \eta_2 = 0.5$, $N = 60$, and $\mathbf{x}_i = (1\, x_i)$, where $x_i \sim \mathcal{U}\,[0, 40]$, leads to a failure rates of 53 percent in 30 replications, where $x_i$ was kept fixed over the repetitions.

Later on, Hosmer (1974) realized that the problem of dealing with an unbounded likelihood function is of relevance not only for finite mixtures of normal distributions (see again Subsection 6.1.2), but also for heterogeneous mixtures of regression models, which include heterogeneous mixtures of normal distributions as a special case. Hosmer (1974) noted that any observation $y_i$ generates a singularity in the likelihood function if $\boldsymbol{\beta}_k$ is chosen such that $y_i = \mathbf{x}_i\boldsymbol{\beta}_k$, and $\sigma^2_{\varepsilon,k}$ goes to 0. More generally, each subgroup of $d$ observations generates a singularity in the likelihood function if $\boldsymbol{\beta}_k$ is chosen such that the regression plane provides a perfect fit to this subgroup.

Thus if the variances of a finite mixture of regression models are unconstrained, a global maximizer of the likelihood function does not exist. Nevertheless, Kiefer (1978) shows that a root of the log likelihood equations corresponding to a local maximizer in the interior of the parameter space is consistent, asymptotically normal, and efficient. In practice, however, it may be difficult to find the ML estimator numerically. An EM-type algorithm for finding the ML estimator was suggested by Hartigan (1977), whereas DeSarbo and Cron (1988) use the EM algorithm directly for this purpose.

As for mixtures of normal distributions, it is complete ignorance about the variance ratio $\sigma^2_{\varepsilon,k}/\sigma^2_{\varepsilon,l}$ that causes problems with maximum likelihood estimation, and again the Bayesian approach, discussed in the remaining subsections, is helpful in this respect, as it allows us to bound this ratio through choosing proper priors on $\sigma^2_{\varepsilon,k}$, $k = 1, \ldots, K$.

### 8.3.2 Bayesian Inference When the Allocations Are Known

If the allocations $\mathbf{S}$ are known, then Bayesian inference reduces to Bayesian analysis of the standard regression model as discussed first in Zellner (1971); see also Raftery et al. (1997) for a more recent review.

For each group, a separate regression model with parameters $\boldsymbol{\beta}_k$ and $\sigma^2_{\varepsilon,k}$ has to be estimated from all observations that fall into that group. In matrix notation, in each group the regression model reads:

$$\mathbf{y}_k = \mathbf{X}_k\boldsymbol{\beta}_k + \boldsymbol{\varepsilon}_k, \qquad \boldsymbol{\varepsilon}_k \sim \mathcal{N}_{N_k}\left(\mathbf{0}, \sigma^2_{\varepsilon,k}\mathbf{I}_{N_k}\right), \tag{8.19}$$

where $N_k = \#\{i : S_i = k\}$ is equal to the number of observations in group $k$, $\mathbf{y}_k$ is a vector containing all observations $y_i$ with $S_i = k$, and $\mathbf{X}_k$ is the corresponding design matrix, where each line contains the regressors $\mathbf{x}_i$ corresponding to $y_i$. The relevant group-specific data summaries are well known from the normal equations leading to the standard OLS estimator in econometrics:

$$\mathbf{X}_k^{'}\mathbf{y}_k = \sum_{i:S_i=k} \mathbf{x}_i^{'} y_i,$$
$$\mathbf{X}_k^{'}\mathbf{X}_k = \sum_{i:S_i=k} \mathbf{x}_i^{'}\mathbf{x}_i.$$

Note that $N_k$ as well as both group-specific data summaries depend on $\mathbf{S}$, however, as opposed to earlier chapters this dependence is not made explicit in this chapter.

Assume that observation $y_i$ is assigned to group $k$, $S_i = k$. Then the contribution of $y_i$ to the complete-data likelihood function $p(\mathbf{y}|\boldsymbol{\beta}, \boldsymbol{\sigma}^2, \mathbf{S})$ is equal to

$$p(y_i|\boldsymbol{\beta}_k, \sigma^2_{\varepsilon,k}, S_i) = \left(\frac{1}{2\pi\sigma^2_{\varepsilon,k}}\right)^{1/2} \exp\left(-\frac{1}{2\sigma^2_{\varepsilon,k}}(y_i - \mathbf{x}_i\boldsymbol{\beta}_k)^2\right).$$

The complete-data likelihood function $p(\mathbf{y}|\boldsymbol{\beta}, \boldsymbol{\sigma}^2, \mathbf{S})$ has $K$ independent factors, each carrying all information about the parameters in a certain group:

$$p(\mathbf{y}|\boldsymbol{\beta}, \boldsymbol{\sigma}^2, \mathbf{S}) = \prod_{k=1}^{K} \left(\frac{1}{2\pi\sigma^2_{\varepsilon,k}}\right)^{N_k/2} \times \exp\left(-\frac{1}{2\sigma^2_{\varepsilon,k}} \sum_{i:S_i=k} (y_i - \mathbf{x}_i\boldsymbol{\beta}_k)^2\right). \tag{8.20}$$

In a Bayesian analysis each of these factors is combined with a prior. When holding the variance $\sigma^2_{\varepsilon,k}$ fixed, the complete-data likelihood function, regarded as a function of $\boldsymbol{\beta}_k$, is the kernel of a multivariate normal distribution. Under the conjugate prior $\boldsymbol{\beta}_k \sim \mathcal{N}_d(\mathbf{b}_0, \mathbf{B}_0)$, the posterior density of $\boldsymbol{\beta}_k$ given $\sigma^2_{\varepsilon,k}$ and all observations assigned to group $k$, is again a density from the normal distribution, $\boldsymbol{\beta}_k|\sigma^2_{\varepsilon,k}, \mathbf{S}, \mathbf{y} \sim \mathcal{N}_d(\mathbf{b}_k, \mathbf{B}_k)$, where

$$\mathbf{B}_k = (\mathbf{B}_0^{-1} + \frac{1}{\sigma^2_{\varepsilon,k}}\mathbf{X}_k^{'}\mathbf{X}_k)^{-1}, \tag{8.21}$$
$$\mathbf{b}_k = \mathbf{B}_k(\mathbf{B}_0^{-1}\mathbf{b}_0 + \frac{1}{\sigma^2_{\varepsilon,k}}\mathbf{X}_k^{'}\mathbf{y}_k). \tag{8.22}$$

When holding the regression parameter $\boldsymbol{\beta}_k$ fixed, the complete-data likelihood function, regarded as a function of $\sigma^2_{\varepsilon,k}$, is the kernel of an inverted Gamma density. Under the conjugate inverted Gamma prior $\sigma^2_{\varepsilon,k} \sim \mathcal{G}^{-1}(c_0, C_0)$, the posterior density of $\sigma^2_{\varepsilon,k}$ given $\boldsymbol{\beta}_k$ and all observations assigned to this group, is again a density from the inverted Gamma distribution, $\sigma^2_{\varepsilon,k}|\boldsymbol{\beta}_k, \mathbf{S}, \mathbf{y} \sim \mathcal{G}^{-1}(c_k, C_k)$, where

$$c_k = c_0 + \frac{N_k}{2}, \qquad C_k = C_0 + \frac{1}{2}\boldsymbol{\varepsilon}_k^{'}\boldsymbol{\varepsilon}_k, \tag{8.23}$$

where $\boldsymbol{\varepsilon}_k = \mathbf{y}_k - \mathbf{X}_k\boldsymbol{\beta}_k$.

If both $\boldsymbol{\beta}_k$ and $\sigma^2_{\varepsilon,k}$ are unknown, a closed-form solution for the joint posterior $p(\boldsymbol{\beta}_k, \sigma^2_{\varepsilon,k}|\mathbf{S}, \mathbf{y})$ exists only if the prior of $\boldsymbol{\beta}_k$ is restricted by assuming that the prior covariance matrix depends on $\sigma^2_{\varepsilon,k}$ through $\mathbf{B}_{0,k} = \sigma^2_{\varepsilon,k}\tilde{\mathbf{B}}_0$. Then the joint posterior factors as $p(\boldsymbol{\beta}_k|\sigma^2_{\varepsilon,k}, \mathbf{y}, \mathbf{S})p(\sigma^2_{\varepsilon,k}|\mathbf{y}, \mathbf{S})$, where density of $\boldsymbol{\beta}_k$ given $\sigma^2_{\varepsilon,k}$ arises from an $\mathcal{N}_d(\mathbf{b}_k, \mathbf{B}_k)$ distribution with

$$\mathbf{B}_k = \sigma^2_{\varepsilon,k}\tilde{\mathbf{B}}_k, \qquad \tilde{\mathbf{B}}_k = (\tilde{\mathbf{B}}_0^{-1} + \mathbf{X}_k^{'}\mathbf{X}_k)^{-1}, \tag{8.24}$$

$$\mathbf{b}_k = \tilde{\mathbf{B}}_k(\tilde{\mathbf{B}}_0^{-1}\mathbf{b}_0 + \mathbf{X}_k^{'}\mathbf{y}_k), \tag{8.25}$$

whereas the marginal posterior of $\sigma^2_{\varepsilon,k}$ is a $\mathcal{G}^{-1}(c_k, C_k)$-distribution, where $c_k$ is the same as in (8.23), however,

$$C_k = C_0 + \frac{1}{2}\left(\mathbf{y}_k^{'}\mathbf{y}_k + \mathbf{b}_0^{'}\tilde{\mathbf{B}}_0^{-1}\mathbf{b}_0 - \mathbf{b}_k^{'}\tilde{\mathbf{B}}_k^{-1}\mathbf{b}_k\right). \tag{8.26}$$

### 8.3.3 Choosing Prior Distributions

The investigations of the previous subsection suggest choosing the following prior distributions for finite mixtures of regression models when the allocations are unknown, which were applied, for instance, in Hurn et al. (2003).

As a prior for the regression coefficient $\boldsymbol{\beta}_k$ one may use a conditionally conjugate prior:

$$\boldsymbol{\beta}_k|\sigma^2_{\varepsilon,k} \sim \mathcal{N}_d\left(\mathbf{b}_0, \sigma^2_{\varepsilon,k}\tilde{\mathbf{B}}_0\right), \tag{8.27}$$

which introduced prior dependence between $\boldsymbol{\beta}_k$ and $\sigma^2_{\varepsilon,k}$. Alternatively, a prior may be used, where $\boldsymbol{\beta}_k$ and $\sigma^2_{\varepsilon,k}$ are independent a priori:

$$\boldsymbol{\beta}_k \sim \mathcal{N}_d(\mathbf{b}_0, \mathbf{B}_0). \tag{8.28}$$

In both cases, the prior on $\sigma^2_{\varepsilon,k}$ is inverse Gamma, $\sigma^2_{\varepsilon,k} \sim \mathcal{G}^{-1}(c_0, C_0)$. As with for finite mixtures of normal distributions, $C_0$ may be considered as an unknown hyperparameter with a prior of its own, $C_0 \sim \mathcal{G}(g_0, G_0)$, in which case the resulting prior is called a hierarchical prior. The prior on the group sizes is the standard Dirichlet prior, $\boldsymbol{\eta} \sim \mathcal{D}(e_0, \ldots, e_0)$.

### 8.3.4 Bayesian Inference When the Allocations Are Unknown

MCMC estimation is usually carried out using data augmentation and Gibbs sampling, exceptions being Chen and Liu (1996) who discuss MCMC estimation of the allocations $\mathbf{S}$ without parameter estimation and Hurn et al. (2003) who discuss direct parameter estimation without data augmentation using the Metropolis–Hastings algorithm.

Albert and Chib (1993) consider Bayesian estimation using data augmentation and Gibbs sampling for the more general Markov mixture of regression model, however, their algorithm is also relevant for finite mixtures of regression models. They show that MCMC estimation along the lines indicated by *Algorithm 3.4* is feasible after introducing the group indicator $S_i$ for each observation pair $(\mathbf{x}_i, y_i)$ as missing data. Justel and Peña (1996) use a similar method and show that a false convergence of the Gibbs sampler may occur when one of the groups has a much smaller variance than the other. Otter et al. (2002) consider a Bayesian approach for more general finite mixtures of multivariate regression models and discuss an application in marketing. The following algorithm provides details for finite mixtures of heteroscedastic regression models.

*Algorithm 8.1: Unconstrained MCMC for a Multiple Normal Mixture Regression Model* Full conditional Gibbs sampling is carried out in two steps.

(a) Parameter simulation conditional on the allocations $\mathbf{S}$:
  (a1) Sample $\boldsymbol{\eta}$ from the conditional Dirichlet posterior $p(\boldsymbol{\eta}|\mathbf{S})$ as in *Algorithm 3.4*.
  (a2) Sample each regression coefficient $\boldsymbol{\beta}_k$, $k = 1, \dots, K$, from the posterior distribution $\boldsymbol{\beta}_k|\sigma^2_{\varepsilon,k}, \mathbf{S}, \mathbf{y} \sim \mathcal{N}_d(\mathbf{b}_k, \mathbf{B}_k)$.
  (a3) Sample each variance $\sigma^2_{\varepsilon,k}$, $k = 1, \dots, K$, from the posterior distribution $\sigma^2_{\varepsilon,k}|\boldsymbol{\beta}_k, \mathbf{S}, \mathbf{y} \sim \mathcal{G}^{-1}(c_k, C_k)$.

(b) Classification of each observation pair $(y_i, \mathbf{x}_i)$ conditional on $\boldsymbol{\vartheta}$: sample each element $S_i$ of $\mathbf{S}$ from the conditional posterior $p(S_i|\boldsymbol{\vartheta}, \mathbf{y})$ given by

$$\Pr(S_i = k|\boldsymbol{\vartheta}, \mathbf{y}) \propto \eta_k f_N(y_i; \mathbf{x}_i\boldsymbol{\beta}_k, \sigma^2_{\varepsilon,k}). \tag{8.29}$$

In step (a2), the posterior moments $\mathbf{b}_k$ and $\mathbf{B}_k$ are given by (8.21) and (8.22), whereas in step (a3) the posterior moments $c_k$ and $C_k$ are available from (8.23). These formulae could be applied for any prior. Under the conditionally conjugate prior or the hierarchical conditionally conjugate prior, computation of $\mathbf{b}_k$ and $\mathbf{B}_k$ may be simplified as in (8.24) and (8.25). Furthermore, under this prior, sampling of $\sigma^2_{\varepsilon,k}$ is possible from the marginal inverted Gamma posterior distribution $p(\sigma^2_{\varepsilon,k}|\mathbf{S}, \mathbf{y})$, where $c_k$ is the same as in (8.23) and $C_k$ is given by (8.26).

Under a hierarchical prior, where $C_0$ is a random hyperparameter with a prior of its own, $C_0 \sim \mathcal{G}(g_0, G_0)$, an additional step has to be added in

*Algorithm 8.1* to sample $C_0$ from $p(C_0|\mathbf{S}, \boldsymbol{\beta}, \boldsymbol{\sigma^2}, \mathbf{y})$, which is given by Bayes' theorem as $C_0|\mathbf{S}, \boldsymbol{\beta}, \boldsymbol{\sigma^2}, \mathbf{y} \sim \mathcal{G}(g_N, G_N)$, where:

$$g_N = g_0 + Kc_0, \qquad G_N = G_0 + \sum_{k=1}^{K} \frac{1}{\sigma^2_{\varepsilon,k}}.$$

### MCMC for Homoscedastic Mixtures of Regression Models

*Algorithm 8.1* could be applied for Bayesian estimation of a homoscedastic finite mixture regression model, where $\sigma^2_{\varepsilon,1} = \cdots = \sigma^2_{\varepsilon,K} = \sigma^2_{\varepsilon}$, however, step (a3) has to be modified by sampling $\sigma^2_{\varepsilon}$ from the appropriate posterior distribution. Under the inverted Gamma prior distribution $\sigma^2_{\varepsilon} \sim \mathcal{G}^{-1}(c_0, C_0)$, the posterior distribution is again inverted Gamma, $\sigma^2_{\varepsilon}|\boldsymbol{\beta}, \boldsymbol{\sigma^2}, \mathbf{S}, \mathbf{y} \sim \mathcal{G}^{-1}(c_N, C_N)$, where

$$c_N = c_0 + \frac{N}{2}, \qquad C_N = C_0 + \frac{1}{2}\sum_{i=1}^{N}(y_i - \mathbf{x}_i\boldsymbol{\beta}_{S_i})^2. \tag{8.30}$$

Under the conditionally conjugate prior (8.27) on $\boldsymbol{\beta}$, it is possible to sample $\sigma^2_{\varepsilon}$ from the marginal posterior $p(\sigma^2_{\varepsilon}|\boldsymbol{\sigma^2}, \mathbf{S}, \mathbf{y})$, where $\boldsymbol{\beta}$ is integrated out, as this density is available in closed form: $\sigma^2_{\varepsilon}|\boldsymbol{\sigma^2}, \mathbf{S}, \mathbf{y} \sim \mathcal{G}(c_N, C_N)$, with $c_N$ being the same as in (8.30), whereas $C_N$ is given by

$$C_N = C_0 + \frac{1}{2}\mathbf{b}_0'\tilde{\mathbf{B}}_0^{-1}\mathbf{b}_0 + \frac{1}{2}\sum_{k=1}^{K}\left(\mathbf{y}_k'\mathbf{y}_k - \mathbf{b}_k'\tilde{\mathbf{B}}_k^{-1}\mathbf{b}_k\right).$$

### Starting Values

Justel and Peña (1996) realized that for a finite mixture of regression models Gibbs sampling may be sensitive to choosing an appropriate initial classification. In particular under the presence of outliers that mask or swamp other observations, an erroneous initial classification of the observations will lead the algorithm to a wrong solution for thousands of iterations. As a remedy, Justel and Peña (2001) avoid random initial classification and search for a more sensible classification. They use an estimate of the covariance matrix of the allocations $\mathbf{S}$ and show that the eigenvectors associated with the nonzero eigenvalues provide information about which observations are possible outliers. The examples in Justel and Peña (2001) indicate considerable improvement of the Gibbs sampler based on these elaborated starting values.

## 8.3.5 Bayesian Inference Using Posterior Draws

As for a standard finite mixture model, label switching as discussed in detail in Subsection 3.5.5 is also an issue for finite mixtures of regression models.

Hurn et al. (2003) use the approach of Celeux et al. (2000) to deal with the labeling problem, by choosing that parameter for estimation which minimizes the symmetrized Kullback–Leibler distance measure, which is invariant to relabeling.

As noted by Hurn et al. (2003), a functional that is invariant to relabeling is the estimated regression hyperplane,

$$\mathrm{E}(Y_i|\mathbf{x}_i) = \sum_{k=1}^{K} \eta_k \mathbf{x}_i \boldsymbol{\beta}_k,$$

which reduces to the regression line

$$\mathrm{E}(Y_i|x_i) = \sum_{k=1}^{K} \eta_k (x_i \beta_{k,1} + \beta_{k,2})$$

for simple regression problems. In the latter case, the regression line may be visualized by showing for each MCMC draw several points from this regression line for selected values of $x_i$ (either sampled randomly from $[x_{\min}, x_{\max}]$ for continuous covariates, or sampled randomly from the set of observed covariates).

Finding identifiability constraints is not trivial, particularly in higher dimensions, however, producing scatter plots of $\beta_{k,j}$ against $\beta_{k',j'}$ for all pairs of coefficients of $\boldsymbol{\beta}$ may be helpful, as shown, for instance, in Frühwirth-Schnatter and Kaufmann (2006a). The predicted points on the regression line could also help to identify groups. If for a certain $\mathbf{x}_i$, all simulated points obey $\mathbf{x}_i\boldsymbol{\beta}_1 < \cdots < \mathbf{x}_i\boldsymbol{\beta}_K$, then this constraint could be used for identification. Thus for a switching regression model constraints need not be simple order constraints on the regression parameter, but could also be linear constraints as applied, for instance, in Otter et al. (2002).

### 8.3.6 Dealing with Model Specification Uncertainty

Testing for the presence of switching regression parameters was already considered by Quandt (1958), who performed an F-Test involving the ratio of variances under a switching and a nonswitching regression model, and by Quandt (1960) who considered a likelihood ratio test.

Bayes factors for testing a switching regression model with $K = 2$ against homogeneity are considered by Peña and Tiao (1992) who investigate the relation between the Bayes factor and the Chow test introduced by Chow (1960). Otter et al. (2002) and Frühwirth-Schnatter et al. (2004) use the bridge sampling estimator of the marginal likelihoods (see also Subsection 5.4.6 for more detail on this estimator) to select the number of groups in mixtures of regression models.

Hurn et al. (2003) use the birth and death process method of Stephens (2000a), discussed in Subsection 5.2.3 in detail, to select the number of groups in a finite mixture regression model.

## 8.4 Mixed-Effects Finite Mixtures of Regression Models

A mixed-effects model allows us to combine regression coefficients that are fixed across all realizations $(Y_i, \mathbf{x}_i)$ with regression coefficients that are allowed to change.

### 8.4.1 Model Definition

A mixed-effects finite mixture of regression models results if only some regression coefficients are different among the hidden groups:

$$Y_i = \mathbf{x}_i^f \boldsymbol{\alpha} + \mathbf{x}_i^r \boldsymbol{\beta}_{S_i} + \varepsilon_i, \qquad \varepsilon_i \sim \mathcal{N}\left(0, \sigma^2_{\varepsilon,S_i}\right), \tag{8.31}$$

where $\mathbf{x}_i^f$ are the fixed effects, whereas $\mathbf{x}_i^r$ are the random effects. A necessary condition for identifiability is that the columns of the design matrix defined by

$$\mathbf{X} = \begin{pmatrix} \mathbf{x}_1^f & \mathbf{x}_1^r \\ \vdots & \vdots \\ \mathbf{x}_N^f & \mathbf{x}_N^r \end{pmatrix}$$

are linearly independent.

Considering certain effects as being fixed may help to avoid generic identifiability, in particular for categorical covariates. For a regression model, where only the intercept is switching,

$$Y_i = \mathbf{x}_i \boldsymbol{\alpha} + \beta_{S_i} + \varepsilon_i, \qquad \varepsilon_i \sim \mathcal{N}\left(0, \sigma^2_{\varepsilon,S_i}\right), \tag{8.32}$$

generic identifiability follows immediately from pointwise identifiability, given by (8.5):

$$\eta_k^\star = \eta_{\rho_i(k)}, \qquad \beta_k^\star + \mathbf{x}_i \boldsymbol{\alpha} = \beta_{\rho_i(k)} + \mathbf{x}_i \boldsymbol{\alpha}, \qquad \sigma^{2,\star}_{\varepsilon,k} = \sigma^2_{\varepsilon,\rho_i(k)},$$

hence $\beta_k^\star = \beta_{\rho_i(k)}$. For the general mixed-effects model defined in (8.31) pointwise identifiability, given by (8.5),

$$\eta_k^\star = \eta_{\rho_i(k)}, \qquad \mathbf{x}_i^f \boldsymbol{\alpha} + \mathbf{x}_i^r \boldsymbol{\beta}_k = \mathbf{x}_i^f \boldsymbol{\alpha} + \mathbf{x}_i^r \boldsymbol{\beta}_{\rho_i(k)}, \qquad \sigma^{2,\star}_{\varepsilon,k} = \sigma^2_{\varepsilon,\rho_i(k)}, \tag{8.33}$$

implies $\mathbf{x}_i^r \boldsymbol{\beta}_k = \mathbf{x}_i^r \boldsymbol{\beta}_{\rho_i(k)}$, and generic identifiability holds if the identifiability condition discussed in Section 8.2.2 is applied to the design points defining only the random effects $\mathbf{x}_i^r$.

### 8.4.2 Choosing Priors for Bayesian Estimation

It is assumed that the priors of all parameters but $\boldsymbol{\alpha}$ are the same as in Subsection 8.3.3, whereas $\boldsymbol{\alpha} \sim \mathcal{N}_r(\mathbf{a}_0, \mathbf{A}_0)$. If $\boldsymbol{\alpha}$ and $\boldsymbol{\beta}_k$ are pairwise independent a priori, then the joint prior on $\boldsymbol{\alpha}^* = (\boldsymbol{\alpha}, \boldsymbol{\beta}_1, \dots, \boldsymbol{\beta}_K)$ is a normal prior, $\boldsymbol{\alpha}^* \sim \mathcal{N}_{r^*}(\mathbf{a}_0^*, \mathbf{A}_0^*)$, where $r^* = r + Kd$ and $\mathbf{a}_0^*$ and $\mathbf{A}_0^*$ are derived from $\mathbf{a}_0, \mathbf{A}_0, \mathbf{b}_0$, and $\mathbf{B}_0$ in an obvious way.

### 8.4.3 Bayesian Parameter Estimation When the Allocations Are Known

In matrix notation, in each group the regression model reads:

$$\mathbf{y}_k = \mathbf{X}_k^f \boldsymbol{\alpha} + \mathbf{X}_k^r \boldsymbol{\beta}_k + \boldsymbol{\varepsilon}_k, \qquad \boldsymbol{\varepsilon}_k \sim \mathcal{N}_{N_k}\left(\mathbf{0}, \sigma_{\varepsilon,k}^2 \mathbf{I}_{N_k}\right),$$

where $N_k = \#\{i : S_i = k\}$ is equal to the number of observations in group $k$, $\mathbf{y}_k$ is a vector containing all observations $y_i$ with $S_i = k$, and $\mathbf{X}_k^f$ and $\mathbf{X}_k^r$ are the corresponding design matrices, where each line contains the regressors $\mathbf{x}_i^f$ and $\mathbf{x}_i^r$ corresponding to $y_i$.

Due to the presence of the common regression parameter $\boldsymbol{\alpha}$ in each group, conditional independence across the groups as in Subsection 8.3.2 is lost, even conditional on known allocations $\mathbf{S}$, and inference is carried out simultaneously for all regression coefficients $\boldsymbol{\alpha}^\star = (\boldsymbol{\alpha}, \boldsymbol{\beta}_1, \ldots, \boldsymbol{\beta}_K)$. This inference problem is closely related to Bayesian inference for a single regression model. By introducing a dummy coding for $S_i$ through $K$ binary variables $D_{ik}, k = 1, \ldots, K$, where $D_{ik} = 1$, iff $S_i = k$, and 0 otherwise, model (8.31) is written as a heteroscedastic regression model with regression parameter $\boldsymbol{\alpha}^\star$:

$$\begin{aligned} y_i &= \mathbf{x}_i^f \boldsymbol{\alpha} + \mathbf{x}_i^r D_{i1} \boldsymbol{\beta}_1 + \cdots + \mathbf{x}_i^r D_{iK} \boldsymbol{\beta}_K + \varepsilon_i, \\ \varepsilon_i &\sim \mathcal{N}\left(0, \sigma_i^2\right), \qquad \sigma_i^2 = D_{i1}\sigma_{\varepsilon,1}^2 + \cdots + D_{iK}\sigma_{\varepsilon,K}^2. \end{aligned} \tag{8.34}$$

Normalization yields a regression model with homoscedastic errors:

$$\frac{y_i}{\sigma_i} = \frac{1}{\sigma_i}\mathbf{x}_i^f \boldsymbol{\alpha} + \frac{1}{\sigma_i}\mathbf{x}_i^r D_{i1}\boldsymbol{\beta}_1 + \cdots + \frac{1}{\sigma_i}\mathbf{x}_i^r D_{iK}\boldsymbol{\beta}_K + \tilde{\varepsilon}_i, \tag{8.35}$$

where $\tilde{\varepsilon}_i \sim \mathcal{N}(0,1)$. Under a normal prior on the regression coefficients $\boldsymbol{\alpha}^\star$, $\boldsymbol{\alpha}^\star \sim \mathcal{N}_{r^\star}(\mathbf{a}_0^\star, \mathbf{A}_0^\star)$, the joint posterior of $\boldsymbol{\alpha}^\star$, conditional on knowing the variance parameters $\sigma_{\varepsilon,1}^2, \ldots, \sigma_{\varepsilon,K}^2$, is again a normal distribution: $\boldsymbol{\alpha}^\star | \sigma_{\varepsilon,1}^2, \ldots, \sigma_{\varepsilon,K}^2, \mathbf{y}, \mathbf{S} \sim \mathcal{N}_{r^\star}(\mathbf{a}_N^\star, \mathbf{A}_N^\star)$. $\mathbf{a}_N^\star$ and $\mathbf{A}_N^\star$ are given by:

$$(\mathbf{A}_N^\star)^{-1} = (\mathbf{A}_0^\star)^{-1} + \sum_{i=1}^N \frac{1}{\sigma_{\varepsilon,S_i}^2} \boldsymbol{Z}_i' \boldsymbol{Z}_i, \tag{8.36}$$

$$\mathbf{a}_N^\star = \mathbf{A}_N^\star \left( (\mathbf{A}_0^\star)^{-1}\mathbf{a}_0^\star + \sum_{i=1}^N \frac{1}{\sigma_{\varepsilon,S_i}^2} \boldsymbol{Z}_i' y_i \right), \tag{8.37}$$

where $\boldsymbol{Z}_i = (\mathbf{x}_i^f \; \mathbf{x}_i^r D_{i1} \; \cdots \; \mathbf{x}_i^r D_{iK})$. If $N$ is not too large, these moments could be determined from a single matrix manipulation:

$$\begin{aligned} (\mathbf{A}_N^\star)^{-1} &= (\mathbf{A}_0^\star)^{-1} + \mathbf{X}'\mathbf{X} \\ \mathbf{a}_N^\star &= \mathbf{A}_N^\star \left( (\mathbf{A}_0^\star)^{-1}\mathbf{a}_0^\star + \mathbf{X}'\tilde{\mathbf{y}} \right), \end{aligned}$$

where

$$\mathbf{X} = \begin{pmatrix} \boldsymbol{Z}_1/\sigma_{\varepsilon,S_1} \\ \vdots \\ \boldsymbol{Z}_N/\sigma_{\varepsilon,S_N} \end{pmatrix}, \qquad \tilde{\mathbf{y}} = \begin{pmatrix} y_1/\sigma_{\varepsilon,S_1} \\ \vdots \\ y_N/\sigma_{\varepsilon,S_N} \end{pmatrix}.$$

In contrast to the regression parameters, the variance parameters $\sigma^2_{\varepsilon,1}, \ldots, \sigma^2_{\varepsilon,K}$ are independent, conditional on knowing $\boldsymbol{\alpha}, \boldsymbol{\beta}_1, \ldots, \boldsymbol{\beta}_K$. Under the conjugate inverted Gamma prior $\sigma^2_{\varepsilon,k} \sim \mathcal{G}^{-1}(c_0, C_0)$, the posterior density of $\sigma^2_{\varepsilon,k}$ given $\boldsymbol{\alpha}, \boldsymbol{\beta}_k$, and all observations assigned to this group, is again a density from the inverted Gamma distribution, $\sigma^2_{\varepsilon,k}|\boldsymbol{\alpha}, \boldsymbol{\beta}_k, \mathbf{S}, \mathbf{y} \sim \mathcal{G}^{-1}(c_k, C_k)$, where

$$c_k = c_0 + \frac{N_k}{2}, \qquad C_k = C_0 + \frac{1}{2}\boldsymbol{\varepsilon}_k'\boldsymbol{\varepsilon}_k, \tag{8.38}$$

where $\boldsymbol{\varepsilon}_k = \mathbf{y}_k - \mathbf{X}_k^f\boldsymbol{\alpha} - \mathbf{X}_k^r\boldsymbol{\beta}_k$.

### 8.4.4 Bayesian Parameter Estimation When the Allocations Are Unknown

Bayesian parameter estimation using data augmentation and MCMC as in *Algorithm 8.1* is easily adapted to deal with mixed-effects finite mixtures of regression models.

*Algorithm 8.2: Unconstrained MCMC for a Mixed-Effects Normal Mixture Regression Model* Full conditional Gibbs sampling is carried out in two steps.

(a) Parameter simulation conditional on the allocations $\mathbf{S}$:
  (a1) Sample $\boldsymbol{\eta}$ from the conditional Dirichlet posterior $p(\boldsymbol{\eta}|\mathbf{S})$ as in *Algorithm 3.4*.
  (a2) Sample all regression coefficients $\boldsymbol{\alpha}^* = (\boldsymbol{\alpha}, \boldsymbol{\beta}_1, \ldots, \boldsymbol{\beta}_K)$ jointly from the posterior distribution $\boldsymbol{\alpha}^*|\sigma^2_{\varepsilon,1}, \ldots, \sigma^2_{\varepsilon,K}, \mathbf{y}, \mathbf{S} \sim \mathcal{N}_{r^\star}(\mathbf{a}_N^*, \mathbf{A}_N^*)$.
  (a3) Sample each variance $\sigma^2_{\varepsilon,k}$, $k = 1, \ldots, K$, from the posterior distribution $\sigma^2_{\varepsilon,k}|\boldsymbol{\alpha}, \boldsymbol{\beta}_k, \mathbf{S}, \mathbf{y} \sim \mathcal{G}^{-1}(c_k, C_k)$.
(b) Classification of each observation $(y_i, \mathbf{x}_i)$ conditional on $\boldsymbol{\vartheta}$: sample each element $S_i$ of $\mathbf{S}$ from the conditional posterior $p(S_i|\boldsymbol{\vartheta}, \mathbf{y})$ given by

$$\Pr(S_i = k|\boldsymbol{\vartheta}, \mathbf{y}) \propto \eta_k f_N(y_i; \mathbf{x}_i^f\boldsymbol{\alpha} + \mathbf{x}_i^r\boldsymbol{\beta}_k, \sigma^2_{\varepsilon,k}). \tag{8.39}$$

In step (a3), the posterior moments $c_k$ and $C_k$ are available from (8.38). In step (a2), joint sampling of all regression parameters $(\boldsymbol{\alpha}, \boldsymbol{\beta}_1, \ldots, \boldsymbol{\beta}_K)$ is easily carried out from the conditional posterior $\mathcal{N}_{r^\star}(\mathbf{a}_N^*, \mathbf{A}_N^*)$, where the moments are given by (8.36) and (8.37). With increasing number $K$ of groups joint sampling may be rather timeconsuming, especially for regression models with high-dimensional parameter vectors. Then one of the following variants may be useful.

### Variants of Sampling the Regression Parameters for a Mixed-Effects Model

As $\boldsymbol{\beta}_1, \dots, \boldsymbol{\beta}_K$ are independent conditional on $\boldsymbol{\alpha}$, sampling in step (a2) of *Algorithm 8.2* may be carried out in two subblocks as in Albert and Chib (1993):

(a2-1) Conditional on $\boldsymbol{\alpha}$, sample $\boldsymbol{\beta}_1, \dots, \boldsymbol{\beta}_K$ independently for each group from the regression model:

$$\mathbf{y}_k - \mathbf{X}_k^f \boldsymbol{\alpha} = \mathbf{X}_k^r \boldsymbol{\beta}_k + \boldsymbol{\varepsilon}_k, \qquad \boldsymbol{\varepsilon}_k \sim \mathcal{N}_{N_k}\left(\mathbf{0}, \sigma^2_{\varepsilon,k} \mathbf{I}_{N_k}\right),$$

where only observations with $S_i = k$ are considered. This is exactly the same situation as in Subsection 8.3.2, with a slight modification of the left-hand side variable.

(a2-2) Conditional on $\boldsymbol{\beta}_1, \dots, \boldsymbol{\beta}_K$, sample $\boldsymbol{\alpha}$ from the posterior obtained from the regression model:

$$y_i - \mathbf{x}_i^r \boldsymbol{\beta}_{S_i} = \mathbf{x}_i^f \boldsymbol{\alpha} + \varepsilon_i, \qquad \varepsilon_i \sim \mathcal{N}\left(0, \sigma^2_{\varepsilon,S_i}\right),$$

where $i = 1, \dots, N$.

This sampler may be less efficient than joint sampling of all regression coefficients as in step (a2) of *Algorithm 8.2*, in particular if posterior correlations are high among parameters appearing in different blocks.

The following variant which has been suggested by Frühwirth-Schnatter et al. (2004) is equivalent to joint sampling of all parameters as in step (a2) of *Algorithm 8.2* and is based on decomposing the joint posterior as

$$p(\boldsymbol{\beta}_1, \dots, \boldsymbol{\beta}_K, \boldsymbol{\alpha} | \mathbf{S}, \sigma^2_{\varepsilon,1}, \dots, \sigma^2_{\varepsilon,K}, \mathbf{y}) = \prod_{k=1}^{K} p(\boldsymbol{\beta}_k | \mathbf{S}, \sigma^2_{\varepsilon,k}, \mathbf{y}) p(\boldsymbol{\alpha} | \mathbf{S}, \sigma^2_{\varepsilon,1}, \dots, \sigma^2_{\varepsilon,K}, \mathbf{y}).$$

The group-specific parameters $\boldsymbol{\beta}_1, \dots, \boldsymbol{\beta}_K$ are sampled conditional on $\boldsymbol{\alpha}$ as in step (a2-1) above. To sample $\boldsymbol{\alpha}$, however, the marginal posterior density $p(\boldsymbol{\alpha} | \mathbf{S}, \sigma^2_{\varepsilon,1}, \dots, \sigma^2_{\varepsilon,K}, \mathbf{y})$ is considered. The moments of this density are derived in Frühwirth-Schnatter et al. (2004).

## 8.5 Finite Mixture Models for Repeated Measurements

An often occurring problem in applied statistics is simultaneous inference on a set of parameters for similar units such as schools from a certain region, firms from the same branch, or consumers in a market. In economics, for instance, data may be available for many countries for several years, whereas in marketing the purchase behavior of many consumers may be observed on several occasions. In econometrics such data are referred to as panel data (Baltagi, 1995), whereas in statistics they are more commonly called longitudinal data (Verbeke and Molenberghs, 2000) or repeated measurements (Crowder and Hand, 1990; Davidian and Giltinan, 1998). In this section we discuss some finite mixture models that are useful for such data.

### 8.5.1 Pooling Information Across Similar Units

Assume that for $N$ units $i$, $i = 1, \ldots, N$, outcomes $y_{it}$ are observed on several occasions $t = 1, \ldots T_i$ where $T_i$ may vary between units. In each unit $i$, the outcomes $y_{it}$ are assumed to be generated by a probability law $p(y_{it}|\boldsymbol{\beta}_i^s)$ that is governed by a unit-specific parameter $\boldsymbol{\beta}_i^s$ of dimension $d$. It is to be expected that the parameters $\boldsymbol{\beta}_1^s, \ldots, \boldsymbol{\beta}_N^s$ albeit being different across the units are related to each other. One way to model such a relation is to assume that $\boldsymbol{\beta}_i^s$ is drawn from some distribution $p(\boldsymbol{\beta}_i^s|\boldsymbol{\vartheta})$ which may depend on some unknown hyperparameter $\boldsymbol{\vartheta}$. Note, however, that the distribution $p(\boldsymbol{\beta}_i^s|\boldsymbol{\vartheta})$ is unknown and needs to be estimated from the data. This problem is known as unobserved heterogeneity in marketing and economics, as residual heterogeneity in the social sciences, and as frailty in medical statistics.

One way to capture unobserved heterogeneity is to assume the existence of $K$ subpopulations of size $\eta_1, \ldots, \eta_K$ with $\boldsymbol{\beta}_i^s$ being equal to a group-specific parameter $\boldsymbol{\beta}_k$ within subpopulation $k$. The distribution $p(\boldsymbol{\beta}_i^s|\boldsymbol{\vartheta})$ is a discrete distribution with $K$ unknown support points $\boldsymbol{\beta}_1, \ldots, \boldsymbol{\beta}_K$, where $\Pr(\boldsymbol{\beta}_i^s = \boldsymbol{\beta}_k) = \eta_k$. Alternatively, it is common to assume random deviation of $\boldsymbol{\beta}_i^s$ from a population mean $\boldsymbol{\beta}$ following a normal distribution, $\boldsymbol{\beta}_i^s \sim \mathcal{N}_d(\boldsymbol{\beta}, \mathbf{Q})$, with $\boldsymbol{\beta}$ and $\mathbf{Q}$ being unknown parameters. Without much thought the normality assumption is almost automatically taken for granted, however, as shown by Heckman and Singer (1984), the distribution of heterogeneity is rather influential and quite small changes may lead to substantial changes in the estimated parameters. The effect of misspecifying the distribution of heterogeneity is also discussed in Verbeke and Lesaffre (1997).

To achieve some robustness against the misspecification of this distribution, West (1985) chooses Student-$t$ distributions of heterogeneity instead of normal ones, whereas Verbeke and Lesaffre (1996) choose a mixture of multivariate normal distributions to capture unobserved heterogeneity:

$$\boldsymbol{\beta}_i^s \sim \sum_{k=1}^{K} \eta_k \mathcal{N}_d(\boldsymbol{\beta}_k, \mathbf{Q}_k).$$

This distribution of heterogeneity has been called shrinkage within clusters by Frühwirth-Schnatter and Kaufmann (2006b).

### 8.5.2 Finite Mixtures of Random-Effects Models

The linear mixed-effects model for modeling longitudinal data was introduced by Laird and Ware (1982) and reads for each unit $i$:

$$y_{it} = \mathbf{x}^f_{it}\boldsymbol{\alpha} + \mathbf{x}^r_{it}\boldsymbol{\beta}^s_i + \varepsilon_{it}, \qquad \varepsilon_{it} \sim \mathcal{N}\left(0, \sigma^2_\varepsilon\right), \tag{8.40}$$

for $t = 1, \ldots, T_i$. $\mathbf{x}^f_{it}$ is the $(1 \times r)$ design matrix for the unknown coefficient $\boldsymbol{\alpha}$, where $r = \dim(\boldsymbol{\alpha})$. $\mathbf{x}^r_{it}$ is a $(1 \times d)$ design matrix for the unknown coefficient $\boldsymbol{\beta}^s_i$, where $d = \dim(\boldsymbol{\beta}^s_i)$. $\mathbf{x}^f_{it}$ are called the fixed effects, because changing $\mathbf{x}^f_{it}$ by the same $(1 \times r)$ vector $\Delta$ changes the mean of $y_{it}$ by the same constant $\Delta\boldsymbol{\alpha}$ for all units $i$. $\mathbf{x}^r_{it}$ are called the random effects, because changing $\mathbf{x}^r_{it}$ by the same $(1 \times d)$ vector $\Delta$ changes the mean of $y_{it}$ by $\Delta\boldsymbol{\beta}^s_i$, which is different across units. Textbooks dealing with this model are Baltagi (1995), Verbeke and Molenberghs (2000), and Diggle et al. (2002).

In the standard mixed-effects model the errors $\varepsilon_{it}$ are assumed to be homogeneous across the units. To deal with unit-specific variance heterogeneity, model (8.40) has been extended in the following way,

$$y_{it} = \mathbf{x}^f_{it}\boldsymbol{\alpha} + \mathbf{x}^r_{it}\boldsymbol{\beta}^s_i + \varepsilon_{it}, \qquad \varepsilon_{it} \sim \mathcal{N}\left(0, \sigma^2_\varepsilon/\omega_i\right), \tag{8.41}$$

which reduces to (8.40), if $\omega_i \equiv 1$ for all $i = 1, \ldots, N$. Unit-specific scaling factors $\omega_i$ different from 1 are included to capture variance heterogeneity across the units. Like the unit-specific regression coefficients $\boldsymbol{\beta}^s_i$, the scaling factors are also assumed to arise from some distribution of variance heterogeneity, a common choice being a Gamma distribution:

$$\omega_i \sim \mathcal{G}\left(\nu/2, \nu/2\right). \tag{8.42}$$

For a fixed unit $i$, model (8.41) could be written as a multivariate regression model,

$$\mathbf{y}_i = \mathbf{X}^f_i\boldsymbol{\alpha} + \mathbf{X}^r_i\boldsymbol{\beta}^s_i + \boldsymbol{\varepsilon}_i, \qquad \boldsymbol{\varepsilon}_i \sim \mathcal{N}_{T_i}\left(\mathbf{0}, \sigma^2_\varepsilon/\omega_i\mathbf{I}_{T_i}\right), \tag{8.43}$$

with regression parameter $(\boldsymbol{\alpha}, \boldsymbol{\beta}^s_i)$ using the matrix notation

$$\mathbf{y}_i = \begin{pmatrix} y_{i1} \\ \vdots \\ y_{i,T_i} \end{pmatrix}, \qquad \mathbf{X}^f_i = \begin{pmatrix} \mathbf{x}^f_{i1} \\ \vdots \\ \mathbf{x}^f_{i,T_i} \end{pmatrix}, \qquad \mathbf{X}^r_i = \begin{pmatrix} \mathbf{x}^r_{i1} \\ \vdots \\ \mathbf{x}^r_{i,T_i} \end{pmatrix}.$$

Note that unit-specific variances introduced through the variance model (8.42) imply the following marginal distribution for $\mathbf{y}_i$,

$$\mathbf{y}_i = \mathbf{X}^f_i\boldsymbol{\alpha} + \mathbf{X}^r_i\boldsymbol{\beta}^s_i + \boldsymbol{\varepsilon}_i, \qquad \boldsymbol{\varepsilon}_i \sim t_\nu\left(\mathbf{0}, \sigma^2_\varepsilon\mathbf{I}_{T_i}\right). \tag{8.44}$$

Unobserved heterogeneity caused by omitted variables may be summarized by a regression intercept $\alpha_i$ that varies between the units:

$$\mathbf{y}_i = \mathbf{1}_{T_i}\alpha_i + \mathbf{X}_i\boldsymbol{\beta} + \boldsymbol{\varepsilon}_i, \qquad \boldsymbol{\varepsilon}_i \sim \mathcal{N}_{T_i}\left(\mathbf{0}, \sigma^2_\varepsilon/\omega_i\mathbf{I}_{T_i}\right);$$

in other cases it will make sense to assume that all effects are random, in which case the random coefficient model results:

$$\mathbf{y}_i = \mathbf{X}_i \boldsymbol{\beta}_i^s + \boldsymbol{\varepsilon}_i, \qquad \boldsymbol{\varepsilon}_i \sim \mathcal{N}_{T_i}\left(\mathbf{0}, \sigma_\varepsilon^2/\omega_i \mathbf{I}_{T_i}\right).$$

If $T_i \geq d$ and $\sum T_i \geq r+Kd$, then it would be possible to combine the information from all units to estimate one large regression vector $\boldsymbol{\alpha}, \boldsymbol{\beta}_1^s, \ldots, \boldsymbol{\beta}_N^s$ without imposing further assumptions. This so-called fixed-effects approach estimates $\boldsymbol{\alpha}, \boldsymbol{\beta}_1^s, \ldots, \boldsymbol{\beta}_N^s$ from the complete-data likelihood $p(\mathbf{y}|\boldsymbol{\alpha}, \boldsymbol{\beta}_1^s, \ldots, \boldsymbol{\beta}_N^s)$, which reduces to estimating $\boldsymbol{\beta}_i^s$ separately for each unit, if no common coefficient $\boldsymbol{\alpha}$ is present. The fixed-effects approach leads to estimates that are more dispersed than the set of parameters one is estimating. Think, for instance, of the extreme case that all $\boldsymbol{\beta}_i^s$s are actually equal. Nevertheless the individual ML estimators of $\boldsymbol{\beta}_1^s, \ldots, \boldsymbol{\beta}_N^s$ will be dispersed, with the dispersion disappearing only for $T_i$ going to infinity.

Thus even for a likelihood-based approach it has been long recommended to consider the so-called random-effects approach where it is assumed that $\boldsymbol{\beta}_1^s, \ldots, \boldsymbol{\beta}_N^s$ are drawn independently from an underlying distribution $p(\boldsymbol{\beta}_i^s|\boldsymbol{\vartheta})$, which may depend on some hyperparameter $\boldsymbol{\vartheta}$, therefore:

$$p(\boldsymbol{\beta}_1^s, \ldots, \boldsymbol{\beta}_N^s|\boldsymbol{\vartheta}) = \prod_{i=1}^N p(\boldsymbol{\beta}_i^s|\boldsymbol{\vartheta}).$$

By combining model (8.43) with one the distributions $p(\boldsymbol{\beta}_i^s|\boldsymbol{\vartheta})$ discussed earlier in Subsection 8.5.1 different useful models emerge. An early reference that shows how pooling helps in problems of simultaneous inference on a set of related parameters $\boldsymbol{\beta}_1^s, \ldots, \boldsymbol{\beta}_N^s$ is Rao (1975); see also Efron and Morris (1977) for some enlightening discussion.

In a Bayesian approach, the distribution $p(\boldsymbol{\beta}_1^s, \ldots, \boldsymbol{\beta}_N^s|\boldsymbol{\vartheta})$ takes the role of a prior distribution which is combined with observations arising from model (8.43) through Bayes' theorem; see Lindley and Smith (1972).

### The Hierarchical Bayes Model

The standard mixed-effects model introduced in Laird and Ware (1982), and applied in many subsequent papers, results from combining model (8.43) with the normal distribution of heterogeneity

$$\boldsymbol{\beta}_i^s \sim \mathcal{N}_d\left(\boldsymbol{\beta}, \mathbf{Q}\right), \tag{8.45}$$

where $\boldsymbol{\beta}$ and $\mathbf{Q}$ are unknown parameters. Morris (1983) discusses that such a prior allows borrowing strength from the ensemble, when estimating $\boldsymbol{\beta}_i^s$ which is shrunken toward the population mean $\boldsymbol{\beta}$. In marketing research this model is also known as the hierarchical Bayes model; see, for instance, Rossi et al. (2005, Chapter 5). If we rewrite (8.45) as $\boldsymbol{\beta}_i^s = \boldsymbol{\beta} + \mathbf{w}_i$, $\mathbf{w}_i \sim \mathcal{N}_d\left(\mathbf{0}, \mathbf{Q}\right)$, and substitute into (8.43), we obtain:

$$\mathbf{y}_i = \mathbf{X}_i^f \boldsymbol{\alpha} + \mathbf{X}_i^r \boldsymbol{\beta} + \mathbf{X}_i^r \mathbf{w}_i + \boldsymbol{\varepsilon}_i.$$

Under the common assumption that $\mathbf{w}_i$ and $\boldsymbol{\varepsilon}_i$ are independent, the hierarchical Bayes model corresponds to the following multivariate regression model,

$$\mathbf{y}_i = \mathbf{X}_i^f \boldsymbol{\alpha} + \mathbf{X}_i^r \boldsymbol{\beta} + \tilde{\boldsymbol{\varepsilon}}_i, \qquad \tilde{\boldsymbol{\varepsilon}}_i \sim \mathcal{N}_{T_i}\left(\mathbf{0}, \mathbf{V}_i\right), \tag{8.46}$$

with constrained error variance–covariance matrix

$$\mathbf{V}_i = \mathbf{X}_i^r \mathbf{Q} (\mathbf{X}_i^r)' + \sigma_\varepsilon^2 / \omega_i \mathbf{I}_{T_i}.$$

Subsequently, model (8.46) is referred to as the *marginal model*, because the random coefficients $\boldsymbol{\beta}_i^s$ no longer appear in this specification. The marginal model clearly indicates that despite allowing for heterogeneity the hierarchical Bayes model implies the rather inflexible normal distribution as a marginal distribution for $\mathbf{y}_i$. Further issues, in particular estimation of this widely used model, are well discussed in the many excellent monographs mentioned at the beginning of this section.

Verbeke and Lesaffre (1997) study the effect of misspecifying the random effect distribution in the linear mixed-effects model. They show that the normal shrinkage prior (8.45) yields consistent estimates of $\boldsymbol{\alpha}, \boldsymbol{\beta}, \mathbf{Q}$, and $\sigma_\varepsilon^2$ even if the random effects are not normal, however, standard errors need to be corrected.

### The Latent Class Regression Model

More flexibility in the marginal distribution of $\mathbf{y}_i$ is achieved by assuming that the distribution $p(\boldsymbol{\beta}_i^s|\boldsymbol{\vartheta})$ is a discrete distribution with $K$ unknown support points $\boldsymbol{\beta}_1, \ldots, \boldsymbol{\beta}_K$ with $\Pr(\boldsymbol{\beta}_i^s = \boldsymbol{\beta}_k) = \eta_k$. In this case, the marginal distribution of $\mathbf{y}_i$ is the following finite mixture distribution,

$$p(\mathbf{y}_i|\omega_i, \boldsymbol{\vartheta}) = \sum_{k=1}^K \eta_k f_N(\mathbf{y}_i; \mathbf{X}_i^f \boldsymbol{\alpha} + \mathbf{X}_i^r \boldsymbol{\beta}_k, \sigma_\varepsilon^2 / \omega_i \mathbf{I}_{T_i}).$$

By introducing the hidden allocation variable $S_i$, which takes the value $k$, iff $\boldsymbol{\beta}_i^s = \boldsymbol{\beta}_k$, the model may be written as the following finite mixture of multivariate mixed-effects regression models,

$$\mathbf{y}_i = \mathbf{X}_i^f \boldsymbol{\alpha} + \mathbf{X}_i^r \boldsymbol{\beta}_{S_i} + \boldsymbol{\varepsilon}_i, \qquad \boldsymbol{\varepsilon}_i \sim \mathcal{N}_{T_i}\left(\mathbf{0}, \sigma_\varepsilon^2 / \omega_i \mathbf{I}_{T_i}\right), \tag{8.47}$$

which is an extension of the finite mixture regression model discussed in Section 8.4 to multivariate observations $\mathbf{y}_i$. This model is also called the latent class regression model, as conditional on knowing $S_i$ and $\omega_i$ the observations $y_{i1}, \ldots, y_{i,T_t}$ are independent.

Many interesting applications of this model are found in marketing research; see, for instance, DeSarbo et al. (1992) for metric conjoint analysis, Ramaswamy et al. (1993) for latent pooling of marketing mix elasticities, as well as Wedel and Steenkamp (1991) and the review in Wedel and DeSarbo (1993b).

### The Heterogeneity Model

A very general model results if the observation model (8.43) is combined with a heterogeneity distribution assumed to be a mixture of multivariate normal distributions:

$$\boldsymbol{\beta}_i^s \sim \sum_{k=1}^{K} \eta_k \mathcal{N}_d\left(\boldsymbol{\beta}_k, \mathbf{Q}_k\right), \tag{8.48}$$

with unknown component means $\boldsymbol{\beta}_1, \ldots, \boldsymbol{\beta}_K$, unknown component variance–covariance matrices $\mathbf{Q}_1, \ldots, \mathbf{Q}_K$, and unknown weight distribution $\boldsymbol{\eta} = (\eta_1, \ldots, \eta_K)$. A constrained version of this model with $\mathbf{Q}_1, \ldots, \mathbf{Q}_K$ being the same for all components was introduced by Verbeke and Lesaffre (1996) for homogeneous error variances. A similar model is discussed in Allenby et al. (1998), however, without considering fixed effects. Lenk and DeSarbo (2000) extend this model to observations from distributions from general exponential families; see also Section 9.6.2. Verbeke and Molenberghs (2000) introduced the terminology *heterogeneity model* for this model.

The heterogeneity model encompasses the other models discussed above. If $\mathbf{Q}_k$ is equal to a null matrix in all groups, the latent class regression model results, whereas the hierarchical Bayes model results as that special case where $K = 1$. After introducing the allocation variable $S_i$ in the finite mixture distribution (8.48), the following distribution of heterogeneity results conditional on holding $S_i$ fixed,

$$\boldsymbol{\beta}_i^s | S_i \sim \mathcal{N}_d\left(\boldsymbol{\beta}_{S_i}, \mathbf{Q}_{S_i}\right).$$

Because the $N$ units form $K$ groups, where within each group heterogeneity is described by a group-specific normal distribution, the heterogeneity model may be regarded as a mixture of random-effects models.

The marginal model where the random effects are integrated out, while still conditioning on $S_i$ and $\omega_i$, reads:

$$\mathbf{y}_i = \mathbf{X}_i^f \boldsymbol{\alpha} + \mathbf{X}_i^r \boldsymbol{\beta}_{S_i} + \tilde{\varepsilon}_i, \qquad \tilde{\varepsilon}_i \sim \mathcal{N}_{T_i}\left(\mathbf{0}, \mathbf{V}_i\right), \tag{8.49}$$

where

$$\mathbf{V}_i = \mathbf{X}_i^r \mathbf{Q}_{S_i} (\mathbf{X}_i^r)' + \sigma_\varepsilon^2 / \omega_i \mathbf{I}_{T_i}. \tag{8.50}$$

Therefore the heterogeneity model may also be regarded as a finite mixture of multivariate mixed-effects regression models, where the errors within each unit are correlated, as opposed to the latent class regression model, where these errors are uncorrelated.

The model found applications in marketing to deal with preference heterogeneity of consumers (Allenby et al., 1998; Otter et al., 2004), in economics to analyze individual records of work and life history data (Oskrochi and Davies, 1997) and to find convergence clubs in a macroeconomic panel (Canova, 2004;

Frühwirth-Schnatter and Kaufmann, 2006b), and in biology to analyze micro-array data (Lopes et al., 2003). An extension of this model which includes a dynamic linear trend model is studied in Gamerman and Smith (1996). Nobile and Green (2000) apply a modification of this model with separate random effects, each following a mixture of normal distributions, to estimate main and interaction effects in a factorial experiment.

### 8.5.3 Choosing the Prior for Bayesian Estimation

For Bayesian estimation, a prior on $\boldsymbol{\vartheta} = (\boldsymbol{\beta}_1, \ldots, \boldsymbol{\beta}_K, \mathbf{Q}_1, \ldots, \mathbf{Q}_K, \boldsymbol{\eta}, \boldsymbol{\alpha}, \sigma_\varepsilon^2)$ has to be chosen. Because $(\boldsymbol{\beta}_1, \ldots, \boldsymbol{\beta}_K, \mathbf{Q}_1, \ldots, \mathbf{Q}_K, \boldsymbol{\eta})$ are unknown parameters in a finite mixture of multivariate normal distributions, the same priors as in Subsection 6.3.2 may be applied.

One could choose a conditionally conjugate prior for $\boldsymbol{\beta}_k$ where the prior variance depends on $\mathbf{Q}_k$, $\mathbf{B}_{0,k} = \mathbf{Q}_k/N_0$. On the other hand, in the marginal model (8.49), where the random effects are integrated out, $\boldsymbol{\beta}_k$ appears as a regression coefficient in a finite mixture of regression models, where no conditionally conjugate prior variance exists due to the correlation in the errors. This suggests choosing $\mathbf{B}_0$ independent of $\mathbf{Q}_k$.

$\boldsymbol{\alpha}$ and $\sigma_\varepsilon^2$ have a similar meaning as for a finite mixture of mixed-effects regression models, therefore the prior is chosen as in Subsection 8.4.2. The joint prior reads:

$$\begin{aligned} &\boldsymbol{\beta}_k \sim \mathcal{N}_d(\mathbf{b}_0, \mathbf{B}_0), && \mathbf{Q}_k^{-1} \sim \mathcal{W}_d\left(c_0^Q, \mathbf{C}_0^Q\right), \\ &\boldsymbol{\alpha} \sim \mathcal{N}_r(\mathbf{a}_0, \mathbf{A}_0), && \sigma_\varepsilon^2 \sim \mathcal{G}^{-1}(c_0^\varepsilon, C_0^\varepsilon), \\ &\boldsymbol{\eta} \sim \mathcal{D}(e_0, \ldots, e_0). && \end{aligned} \tag{8.51}$$

### 8.5.4 Bayesian Parameter Estimation When the Allocations Are Known

For a general Bayesian analysis of the heterogeneity model it is helpful to start with parameter estimation, when the allocations $\mathbf{S} = (S_1, \ldots, S_N)$ as well as the variance parameters $\mathbf{Q}_1, \ldots, \mathbf{Q}_K, \sigma_\varepsilon^2$ and $\boldsymbol{\omega}$ are known.

Then the joint posterior of the regression parameters $\boldsymbol{\alpha}^\star = (\boldsymbol{\alpha}, \boldsymbol{\beta}_1, \ldots, \boldsymbol{\beta}_K)$ and the random coefficients $\boldsymbol{\beta}^s = (\boldsymbol{\beta}_1^s, \ldots, \boldsymbol{\beta}_N^s)$ partitions as follows,

$$\begin{aligned} &p(\boldsymbol{\alpha}^\star, \boldsymbol{\beta}^s | \mathbf{y}, \mathbf{Q}_1, \ldots, \mathbf{Q}_K, \sigma_\varepsilon^2, \boldsymbol{\omega}, \mathbf{S}) \\ &= p(\boldsymbol{\alpha}^\star | \mathbf{y}, \mathbf{Q}_1, \ldots, \mathbf{Q}_K, \sigma_\varepsilon^2, \boldsymbol{\omega}, \mathbf{S}) \prod_{i=1}^N p(\boldsymbol{\beta}_i^s | \mathbf{y}_i, \boldsymbol{\alpha}, \boldsymbol{\beta}_{S_i}, \mathbf{Q}_{S_i}, \omega_i) . \end{aligned}$$

Conditional on knowing the fixed effects, the random coefficients $\boldsymbol{\beta}_i^s$ are independent. Because the allocations $\mathbf{S}$ are known, the prior of $\boldsymbol{\beta}_i^s$ is normal,

$$\boldsymbol{\beta}_i^s \sim \mathcal{N}_d\left(\boldsymbol{\beta}_{S_i}, \mathbf{Q}_{S_i}\right),$$

whereas the complete-data likelihood results from:

$$\mathbf{y}_i - \mathbf{X}_i^f \boldsymbol{\alpha} = \mathbf{X}_i^r \boldsymbol{\beta}_i^s + \boldsymbol{\varepsilon}_i, \qquad \boldsymbol{\varepsilon}_i \sim \mathcal{N}_{T_i}\left(\mathbf{0}, \sigma_\varepsilon^2/\omega_i \mathbf{I}_{T_i}\right).$$

Combining these two sources of information yields the following posterior of $\boldsymbol{\beta}_i^s$,

$$\boldsymbol{\beta}_i^s|\mathbf{y}_i, \boldsymbol{\alpha}, \boldsymbol{\beta}_{S_i}, \mathbf{Q}_{S_i}, \omega_i \sim \mathcal{N}_d\left(\mathbf{b}_i^s, \mathbf{B}_i^s\right),$$

where the moments are given in terms of an information filter:

$$\begin{aligned} \mathbf{B}_i^s &= (\mathbf{Q}_{S_i}^{-1} + (\mathbf{X}_i^r)' \mathbf{X}_i^r \omega_i/\sigma_\varepsilon^2)^{-1}, \\ \mathbf{b}_i^s &= \mathbf{B}_i^s(\mathbf{Q}_{S_i}^{-1}\boldsymbol{\beta}_{S_i} + (\mathbf{X}_i^r)'(\mathbf{y}_i - \mathbf{X}_i^f\boldsymbol{\alpha})\omega_i/\sigma_\varepsilon^2). \end{aligned} \tag{8.52}$$

If $T_i < d$, it is more efficient to work with the following filter form which is derived in Subsection 13.3.2,

$$\begin{aligned} \mathbf{b}_i^s &= \boldsymbol{\beta}_{S_i} + \mathbf{K}_i(\mathbf{y}_i - \mathbf{X}_i^f\boldsymbol{\alpha} - \mathbf{X}_i^r\boldsymbol{\beta}_{S_i}), \\ \mathbf{B}_i^s &= (\mathbf{I}_{T_i} - \mathbf{K}_i\mathbf{X}_i^r)\mathbf{Q}_{S_i}, \\ \mathbf{K}_i &= \mathbf{Q}_{S_i}(\mathbf{X}_i^r)'\mathbf{V}_i^{-1}, \end{aligned} \tag{8.53}$$

with $\mathbf{V}_i$ being the error variance–covariance matrix of the marginal model defined in (8.50).

The posterior $p(\boldsymbol{\alpha}^*|\mathbf{y}, \mathbf{Q}_1, \ldots, \mathbf{Q}_K, \sigma_\varepsilon^2, \boldsymbol{\omega}, \mathbf{S})$ is a conditional distribution, where the allocations are known, whereas the random coefficients $\boldsymbol{\beta}_1^s, \ldots, \boldsymbol{\beta}_N^s$ are unknown. The prior of $\boldsymbol{\alpha}^*$ is a normal distribution, $\boldsymbol{\alpha}^* \sim \mathcal{N}_{r^*}(\mathbf{a}_0^*, \mathbf{A}_0^*)$, where $r^* = r + Kd$ and $\mathbf{a}_0^*$ and $\mathbf{A}_0^*$ are derived in an obvious way from the parameters $\mathbf{a}_0, \mathbf{A}_0$, $\mathbf{b}_0$, and $\mathbf{B}_0$ of the prior defined in (8.51). This prior is combined with the likelihood function $p(\mathbf{y}|\boldsymbol{\alpha}^*, \mathbf{y}, \mathbf{Q}_1, \ldots, \mathbf{Q}_K, \sigma_\varepsilon^2, \boldsymbol{\omega}, \mathbf{S})$ of the marginal model (8.49), where the random effects are integrated out.

The posterior distribution $p(\boldsymbol{\alpha}^*|\mathbf{y}, \mathbf{Q}_1, \ldots, \mathbf{Q}_K, \sigma_\varepsilon^2, \boldsymbol{\omega}, \mathbf{S})$ is derived in a similar way as in Subsection 8.4.3, which concerned finite mixtures of multiple mixed-effects models, whereas in the present case we are dealing with a multivariate one. By introducing a dummy coding for $S_i$ through $K$ binary variables $D_{ik}, k = 1, \ldots, K$, where $D_{ik} = 1$, iff $S_i = k$, and 0 otherwise, we rewrite the marginal model (8.49) as

$$\mathbf{y}_i = \boldsymbol{Z}_i^*\boldsymbol{\alpha}^* + \tilde{\boldsymbol{\varepsilon}}_i, \qquad \tilde{\boldsymbol{\varepsilon}}_i \sim \mathcal{N}_{T_i}\left(\mathbf{0}, \mathbf{V}_i\right), \tag{8.54}$$

where the design matrix $\boldsymbol{Z}_i^*$ is defined as

$$\boldsymbol{Z}_i^* = \left(\mathbf{X}_i^f \ \mathbf{X}_i^r D_{i1} \ \ldots \ \mathbf{X}_i^r D_{iK}\right).$$

Because model (8.54) is a multivariate regression model, the posterior of $\boldsymbol{\alpha}^*$ arises from a normal distribution:

$$\boldsymbol{\alpha}^*|\mathbf{y}, \mathbf{Q}_1, \ldots, \mathbf{Q}_K, \sigma_\varepsilon^2, \boldsymbol{\omega}, \mathbf{S} \sim \mathcal{N}_{r^*}\left(\mathbf{a}_N^*, \mathbf{A}_N^*\right),$$

where

$$(\mathbf{A}_N^*)^{-1} = \sum_{i=1}^{N} (\mathbf{Z}_i^*)' \mathbf{V}_i^{-1} \mathbf{Z}_i^* + (\mathbf{A}_0^*)^{-1},$$

$$\mathbf{a}_N^* = (\mathbf{A}_N^*)^{-1} \left( \sum_{i=1}^{N} (\mathbf{Z}_i^*)' \mathbf{V}_i^{-1} \mathbf{y}_i + (\mathbf{A}_0^*)^{-1} \mathbf{a}_0^* \right).$$

### 8.5.5 Practical Bayesian Estimation Using MCMC

Empirical Bayesian estimation of the heterogeneity model, including classification, is discussed in Verbeke and Lesaffre (1996). A fully Bayesian analysis of the heterogeneity model for a fixed number $K$ of groups via MCMC methods is discussed by Allenby et al. (1998), Lenk and DeSarbo (2000), and Frühwirth-Schnatter et al. (2004).

Let $\mathbf{y} = (\mathbf{y}_1, \dots, \mathbf{y}_N)$ denote all observations. MCMC estimation of the most general model is based on three levels of data augmentation. First, one introduces the discrete latent group indicators $\mathbf{S} = (S_1, \dots, S_N)$, with $S_i$ taking values in $\{1, \dots, K\}$ and thereby indicating to which group unit $i$ belongs. Second, the vector of unknowns is augmented by the random effects $\boldsymbol{\beta}^s = (\boldsymbol{\beta}_1^s, \dots, \boldsymbol{\beta}_N^s)$. And finally, under heterogeneous error variances the scale factors $\boldsymbol{\omega} = (\omega_1, \dots, \omega_N)$ are added in a third data augmentation step. The joint posterior distribution of all unknowns reads:

$$p(\boldsymbol{\vartheta}, \boldsymbol{\beta}^s, \mathbf{S}, \boldsymbol{\omega}|\mathbf{y}) \propto p(\mathbf{y}|\boldsymbol{\omega}, \boldsymbol{\beta}^s, \boldsymbol{\alpha}, \sigma_\varepsilon^2) p(\boldsymbol{\omega}) p(\boldsymbol{\alpha}, \sigma_\varepsilon^2)$$
$$\times p(\boldsymbol{\beta}^s|\mathbf{S}, \boldsymbol{\beta}_1, \dots, \boldsymbol{\beta}_K, \mathbf{Q}_1, \dots, \mathbf{Q}_K) p(\boldsymbol{\beta}_1, \dots, \boldsymbol{\beta}_K, \mathbf{Q}_1, \dots, \mathbf{Q}_K) p(\mathbf{S}|\boldsymbol{\eta}) p(\boldsymbol{\eta}).$$

A straightforward way of Bayesian estimation of the heterogeneity model via MCMC methods is Gibbs sampling from full conditional distributions. The sampler is discussed in Allenby et al. (1998) and Lenk and DeSarbo (2000) for a heterogeneity model with homogeneous error variances and draws in turn $\boldsymbol{\alpha}$, $\sigma_\varepsilon^2$, $\boldsymbol{\eta}$, $\boldsymbol{\beta}_k$, and $\mathbf{Q}_k$ for $k = 1, \dots, K$, and $S_i$ and $\boldsymbol{\beta}_i^s$ for $i = 1, \dots, N$, from the appropriate full conditional distributions given the remaining parameters and the data $\mathbf{y}$.

It has been demonstrated in Frühwirth-Schnatter et al. (2004) that the full conditional Gibbs sampler is sensitive to the way model (8.43) is parameterized, depending on whether $\mathbf{X}_i^f$ and $\mathbf{X}_i^r$ have common columns. Sensitivity of Gibbs sampling with respect to parameterizing the standard mixed-effects model was noted earlier by Gelfand et al. (1995), and several papers show that marginalization helps in improving the performance of the Gibbs sampler; see, for instance, Meng and Van Dyk (1997, 1999), Chib and Carlin (1999), and van Dyk and Meng (2001).

The partly marginalized Gibbs sampler suggested in Frühwirth-Schnatter et al. (2004) for homogeneous error variances draws $\mathbf{S}$, $\boldsymbol{\alpha}$, and $\boldsymbol{\beta}_1, \dots, \boldsymbol{\beta}_K$ from conditional distributions where the random effects $\boldsymbol{\beta}^s$ are integrated out. This sampler is shown to be less sensitive to the parameterization and has been extended in Frühwirth-Schnatter et al. (2005) to deal with heterogeneous error variances. It is summarized in the following algorithm.

*Algorithm 8.3: MCMC Estimation of the Heterogeneity Model*

(a) Parameter simulation conditional on the allocations $\mathbf{S}$, the random effects $\boldsymbol{\beta}^s$ and the scaling factors $\boldsymbol{\omega}$.
  (a1) Sample $\boldsymbol{\eta}$ from the conditional Dirichlet posterior $p(\boldsymbol{\eta}|\mathbf{S})$ as in *Algorithm 3.4*.
  (a2) Sample all regression coefficients $\boldsymbol{\alpha}^* = (\boldsymbol{\alpha}, \boldsymbol{\beta}_1, \dots, \boldsymbol{\beta}_K)$ jointly from the posterior distribution $\boldsymbol{\alpha}^* \sim \mathcal{N}_{r^*}(\mathbf{a}_N^*, \mathbf{A}_N^*)$, derived conditional on $\mathbf{y}, \mathbf{S}, \mathbf{Q}_1, \dots, \mathbf{Q}_K, \sigma_\varepsilon^2$, and $\boldsymbol{\omega}$.
  (a3) Sample each variance–covariance matrix $\mathbf{Q}_k$, $k = 1, \dots, K$, from the posterior distribution $\mathbf{Q}_k^{-1} \sim \mathcal{W}_d\left(c_k^Q, \mathbf{C}_k^Q\right)$, derived conditional on $\boldsymbol{\beta}_1, \dots, \boldsymbol{\beta}_K, \boldsymbol{\beta}^s$, and $\mathbf{S}$.
  (a4) Sample $\sigma_\varepsilon^2$ from $\mathcal{G}^{-1}(c_N^\varepsilon, C_N^\varepsilon)$, derived conditional on $\mathbf{y}, \boldsymbol{\alpha}, \boldsymbol{\beta}^s$, and $\boldsymbol{\omega}$.

(b) Classification of each unit based on $\mathbf{y}_i$, $\boldsymbol{\omega}$, and $\boldsymbol{\vartheta}$: sample each element $S_i$ of $\mathbf{S}$ from the conditional posterior $p(S_i|\boldsymbol{\vartheta}, \mathbf{y}_i, \omega_i)$ given by

$$\Pr(S_i = k|\boldsymbol{\vartheta}, \mathbf{y}_i, \omega_i) \propto \eta_k f_N(\mathbf{y}_i; \mathbf{X}_i^f \boldsymbol{\alpha} + \mathbf{X}_i^r \boldsymbol{\beta}_k, \mathbf{V}_i), \tag{8.55}$$

where $\mathbf{V}_i$ has been defined in (8.50).

(c) Dealing with parameter heterogeneity: sample each random coefficient $\boldsymbol{\beta}_i^s$ for $i = 1, \dots, N$ from the $\mathcal{N}_d(\mathbf{b}_i^s, \mathbf{B}_i^s)$-distribution, derived conditional on $\mathbf{y}, \boldsymbol{\alpha}, \boldsymbol{\beta}_1, \dots, \boldsymbol{\beta}_K, \mathbf{S}, \mathbf{Q}_1, \dots, \mathbf{Q}_K, \sigma_\varepsilon^2$, and $\boldsymbol{\omega}$.

(d) Dealing with variance heterogeneity: sample each scaling factor $\omega_i$ from the $\mathcal{G}(c_i^\omega, C_i^\omega)$ distribution, derived conditional on $\mathbf{y}, \boldsymbol{\alpha}, \boldsymbol{\beta}^s$, and $\sigma_\varepsilon^2$.

Estimation of the regression coefficients $\boldsymbol{\alpha}^* = (\boldsymbol{\alpha}, \boldsymbol{\beta}_1, \dots, \boldsymbol{\beta}_K)$ in step (a2) is based on the marginal model (8.49) where the random effects are integrated out in order to improve the mixing properties of the sampler. The appropriate moments were derived in Subsection 8.5.4.

Sampling of the covariance matrices $\mathbf{Q}_1, \dots, \mathbf{Q}_K$ in step (a3) follows immediately from *Algorithm 6.2*, dealing with mixtures of normal distributions, because the random effects $\boldsymbol{\beta}_i^s$ are assumed to be known in this step. The precise form of $c_k^Q$ and $\mathbf{C}_k^Q$ depends upon the chosen prior covariance matrix $\mathbf{B}_0$. If $\mathbf{B}_0$ is independent of $\mathbf{Q}_k$, then

$$c_k^Q = c_0^Q + \frac{N_k}{2},$$

$$\mathbf{C}_k^Q = \mathbf{C}_0^Q + \frac{1}{2} \sum_{i:S_i=k} (\boldsymbol{\beta}_i^s - \boldsymbol{\beta}_k)(\boldsymbol{\beta}_i^s - \boldsymbol{\beta}_k)',$$

where $N_k = \#\{S_i = k\}$.

The appropriate posterior distribution in step (a4) is easily derived from the complete-data likelihood function, which reads:

$$p(\mathbf{y}|\boldsymbol{\alpha}, \boldsymbol{\beta}^s, \boldsymbol{\omega}) = \prod_{i=1}^{N} \left( \frac{\omega_i}{2\pi\sigma_\varepsilon^2} \right)^{T_i/2} \times \exp\left( -\frac{1}{2\sigma_\varepsilon^2} \sum_{i=1}^{N} \omega_i \|\mathbf{y}_i - \mathbf{X}_i^f \boldsymbol{\alpha} - \mathbf{X}_i^r \boldsymbol{\beta}_i^s \|_2^2 \right). \tag{8.56}$$

Therefore:

$$c_N^\varepsilon = c_0^\varepsilon + \frac{1}{2} \left( \sum_{i=1}^{N} T_i \right),$$

$$C_N^\varepsilon = C_0^\varepsilon + \frac{1}{2} \left( \sum_{i=1}^{N} \omega_i \|\mathbf{y}_i - \mathbf{X}_i^f \boldsymbol{\alpha} - \mathbf{X}_i^r \boldsymbol{\beta}_i^s \|_2^2 \right).$$

In step (b), the indicators $S_1, \ldots, S_N$ are conditionally independent given $\mathbf{y}$, $\boldsymbol{\omega}$, and $\boldsymbol{\vartheta}$, as it is assumed that the units are drawn randomly from the underlying population. The classification rule (8.55) is based on the marginal model (8.49), where the random effects are integrated out, in order to improve the mixing properties of the sampler.

In step (c), the moments of the $\mathcal{N}_d(\mathbf{b}_i^s, \mathbf{B}_i^s)$ distribution to sample the random effects are given by (8.52) or (8.53).

Finally, the posterior in step (d) follows immediately from the complete-data likelihood given in (8.56) in combination with the prior (8.42):

$$c_i^\omega = \frac{\nu}{2} + \frac{T_i}{2}, \qquad C_i^\omega = \frac{\nu}{2} + \frac{1}{2\sigma_\varepsilon^2} \|\mathbf{y}_i - \mathbf{X}_i^f \boldsymbol{\alpha} - \mathbf{X}_i^r \boldsymbol{\beta}_i^s \|_2^2.$$

### 8.5.6 Dealing with Model Specification Uncertainty

BIC or Schwarz criterion is quite popular for model selection in random-effect models, however, problems are reported in Stone (1974) and McCulloch and Rossi (1992) for few repeated measurements with large heterogeneity, where the number of parameters actually grows with $N$.

Watier et al. (1999) and Nobile and Green (2000) extend the reversible jump MCMC method of Richardson and Green (1997) to select the unknown number of components in a finite mixture of random-effects models.

Marginal likelihoods for selecting between the different models were considered by Lenk and DeSarbo (2000) and Frühwirth-Schnatter et al. (2004, 2005). Marginal likelihoods allow not only choosing the number of components, but also a comparison between the different types of heterogeneity distributions; see also the case study in Subsection 8.5.7.

**Table 8.1.** MARKETING DATA, logarithm of marginal likelihoods $p(\mathbf{y}|K, \nu)$ (from Frühwirth-Schnatter et al. (2005) with permission granted by Springer-Verlag, Wien)

| | $\log p(\mathbf{y}\|K, \nu = \infty)$ | | $\log p(\mathbf{y}\|K, \nu = 4)$ | |
|---|---|---|---|---|
| $K$ | $\mathbf{Q}_k \neq \mathbf{O}$ | $\mathbf{Q}_k = \mathbf{O}$ | $\mathbf{Q}_k \neq \mathbf{O}$ | $\mathbf{Q}_k = \mathbf{O}$ |
| 1 | -9222.36 | -10077.31 | -9101.52 | -9980.21 |
| 2 | -9165.66 | -9881.49 | **-9028.81** | -9727.13 |
| 3 | **-9161.27** | -9733.98 | -9043.96 | -9576.97 |
| 4 | -9165.73 | -9669.98 | -9045.86 | -9522.18 |
| 5 | — | -9596.61 | — | -9453.22 |
| ⋮ | | | | |
| 12 | — | — | — | -9332.96 |
| 13 | — | — | — | -9329.49 |
| 14 | — | — | — | **-9326.26** |
| 15 | — | — | — | -9327.27 |
| 16 | — | -9464.77 | — | — |
| 17 | — | **-9460.61** | — | — |
| 18 | — | -9465.79 | — | — |

### 8.5.7 Application to the Marketing Data

This application concerns conjoint analysis in marketing, a procedure that is focused on obtaining the importance of certain product attributes and their significance in motivating a consumer toward purchase from a holistic appraisal of attribute combinations.

The MARKETING DATA come from a brand–price trade-off study in the mineral water market. Each of 213 Austrian consumers evaluated their likelihood of purchasing 15 different product-profiles offering five different brands of mineral water at different prices on 20-point rating scales. The goal of the modeling exercise is to find a model describing consumers' heterogeneous preferences toward the different brands of mineral water and their brand–price trade-offs. These data were analyzed in several studies based on homogeneous errors using a random coefficient model (Frühwirth-Schnatter and Otter, 1999), the latent class regression model (Otter et al., 2002), and the heterogeneity model (Otter et al., 2004). The material in this subsection is based on Frühwirth-Schnatter et al. (2005), where these models are compared to models based on unit-specific variance heterogeneity.

The design matrix consists of 15 columns corresponding to the constant, the four brands Römerquelle (*RQ*), Vöslauer (*VOE*), Juvina (*JU*), and Waldquelle (*WA*), a linear and a quadratic price effect, and four brand by linear price and four brand by quadratic price interaction effects. A dummy coding is used for the brands, hence the fifth brand Kronsteiner (*KR*) was chosen as the baseline. The smallest price is subtracted from the linear price column; the quadratic price is a contrast from the centered linear price. Therefore, the constant corresponds to the purchase likelihood of Kronsteiner at the lowest price level, if quadratic price effects are not present. The investigations of these data in Otter et al. (2002) indicated that a specification with fixed brand by quadratic price interactions is preferable, therefore the dimension of $\boldsymbol{\beta}_k$ is equal to $d = 11$, whereas the dimension of $\boldsymbol{\alpha}$ is equal to $r = 4$.

The prior is chosen as in Subsection 8.5.3. $\mathbf{a}_0$ and $\mathbf{b}_0$ are equal to the population mean of the random coefficient model reported in Frühwirth-Schnatter and Otter (1999), whereas $\mathbf{A}_0^{-1} = 0.04 \times \mathbf{I}_4$ and $\mathbf{B}_0^{-1} = 0.04 \times \mathbf{I}_{11}$. In the prior of $\mathbf{Q}_k$, $c_0^Q = 10$ whereas $\mathbf{C}_0^Q$ is derived by matching the prior mean, $E(\mathbf{Q}_k) = (c_0^Q - (d+1)/2)^{-1}\mathbf{C}_0^Q$, to a sample estimate computed from individual OLS estimation. In the prior of $\sigma_\varepsilon^2$, $c_0^\varepsilon = C_0^\varepsilon = 0$, whereas the prior on $\boldsymbol{\eta}$ is a $\mathcal{D}(1, \dots, 1)$ distribution.

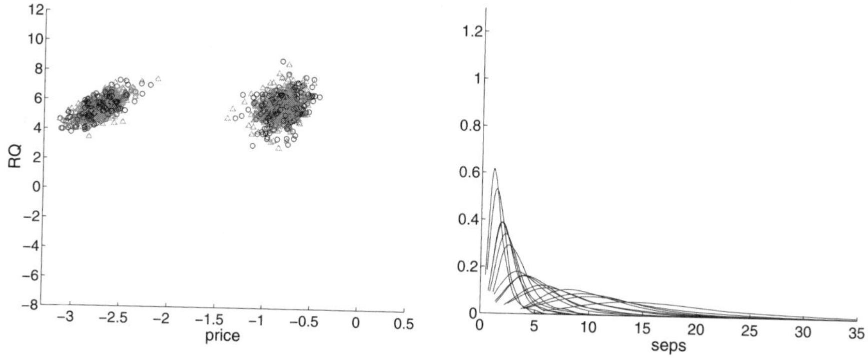


**Fig. 8.1.** MARKETING DATA; heterogeneity model with $K = 2$ and heterogeneous variances with $\nu = 4$, scatter plot of price against brand $RQ$ (left-hand side) and posterior distribution of individual variances $\sigma_\varepsilon^2/\omega_i$ for 15 randomly selected consumers (from Frühwirth-Schnatter et al. (2005) with permission granted by Springer-Verlag, Wien)

The following finite mixture models with $K > 1$ were fitted to these data with varying the number $K$ of groups: the general heterogeneity model, where $\mathbf{Q}_k \neq \mathbf{O}$ for all $k = 1, \dots, K$ and the latent class regression model, where $\mathbf{Q}_k = \mathbf{O}$ for all $k = 1, \dots, K$. These models were compared to the hierarchical Bayes model, which formally corresponds to a heterogeneity model with $K = 1$ and $\mathbf{Q}_1 \neq \mathbf{O}$. Each of these models was fitted with heterogeneous variances with $\nu = 4$ as well as with homogeneous variances that correspond to $\nu = \infty$. Estimation was carried through 30,000 MCMC iterations, with the last 6000 draws being kept for inference.

Table 8.1 shows estimates of the logarithm of the marginal likelihood $p(\mathbf{y}|K, \nu)$ for various models obtained by bridge sampling. The hierarchical Bayes model (column $\mathbf{Q}_k \neq \mathbf{O}$, line $K = 1$) is clearly preferred to all latent class regression models (column $\mathbf{Q}_k = \mathbf{O}$), but is outperformed by the heterogeneity model (column $\mathbf{Q}_k \neq \mathbf{O}$, lines with $K > 1$), regardless of the assumption made concerning the variances.

The specification chosen for the variance exercises a considerable influence on the number of optimal classes. Under the assumption of homogeneous variances the optimal latent class regression model has seventeen classes, whereas the number reduces to fourteen under heterogeneous errors. Also the heterogeneity model has a different number of optimal classes, namely two under heterogeneous errors and three under homogeneous errors. The optimal model of all models under consideration is a heterogeneity model with heterogeneous error variances and $K = 2$ classes. The preference of a model with heterogeneous variances is also supported by Figure 8.1, which shows considerable differences in the posterior distribution of the individual variances $\sigma_\varepsilon^2/\omega_i$ for 15 randomly selected consumers.

**Table 8.2.** MARKETING DATA, heterogeneity model with $K = 2$ and heterogeneous variances with $\nu = 4$; posterior expectation of the group-specific parameters $\boldsymbol{\beta}_k$ and the group-specific weights $\eta_k$; posterior standard deviations in parentheses (from Frühwirth-Schnatter et al. (2005) with permission granted by Springer-Verlag, Wien)

| | $\beta_{k,j}$ | | | $\beta_{k,j}$ | |
|---|---|---|---|---|---|
| | $k=1$ | $k=2$ | | $k=1$ | $k=2$ |
| $const$ | 14.78<br>(0.67) | 12.43<br>(0.75) | $RQ \times p$ | -0.71<br>(0.16) | -0.04<br>(0.15) |
| $RQ$ | 5.44<br>(0.65) | 5.65<br>(0.84) | $VOE \times p$ | -0.85<br>(0.16) | -0.02<br>(0.16) |
| $VOE$ | 5.30<br>(0.65) | 5.17<br>(0.97) | $JU \times p$ | -0.38<br>(0.16) | 0.07<br>(0.16) |
| $JU$ | 1.28<br>(0.66) | 0.38<br>(0.97) | $WA \times p$ | -0.58<br>(0.15) | -0.10<br>(0.13) |
| $WA$ | 2.24<br>(0.68) | 1.10<br>(0.78) | | | |
| $p$ | -2.72<br>(0.15) | -0.82<br>(0.15) | | $\eta_k$<br>$k=1$ | <br>$k=2$ |
| $p^2$ | -0.03<br>(0.07) | 0<br>(0.06) | | 0.58<br>(0.04) | 0.42<br>(0.04) |

We proceed with estimating the group-specific parameters for this model. The posterior draws in Figure 8.1 are the point process representation of the projection onto the coefficients $\beta_{k,2}$ and $\beta_{k,6}$ which correspond to the effect of the brand $RQ$ and the price effect. We find two clearly separated simulation clusters, with one group collecting very price-sensitive consumers whereas the consumers of the other group are less price sensitive. Therefore it is possible to identify the model through putting the constraint $\beta_{1,6} < \beta_{2,6}$

on the group-specific price coefficient. Table 8.2 gives the resulting estimates for the group-specific parameters and the group weights.

## 8.6 Further Issues

### 8.6.1 Regression Modeling Based on Multivariate Mixtures of Normals

Müller et al. (1996) show that for stochastic regressor variables finite mixtures of multivariate normal distributions could be used as an alternative tool for flexible regression modeling. Consider, for example, a bivariate random variable $(X, Y)$, modeled by a mixture of bivariate normal distributions with component means $\boldsymbol{\mu}_k$, component covariance matrices $\boldsymbol{\Sigma}_k$, and weight distribution $\boldsymbol{\eta} = (\eta_1, \dots, \eta_K)$.

The conditional density $p(y|X = x_i, \boldsymbol{\vartheta})$ of $Y$ given $X = x_i$ is easily found to be equal to the following univariate mixture of normal distributions,

$$p(y|X = x_i, \boldsymbol{\vartheta}) = \sum_{k=1}^{K} w_k(x_i, \boldsymbol{\vartheta}) f_N(y; \beta_{k,1} x_i + \beta_{k,2}, \Sigma_{k,22}(1 - \rho_k^2)), \tag{8.57}$$

where

$$\beta_{k,1} = \rho_k \sqrt{\frac{\Sigma_{k,11}}{\Sigma_{k,22}}}, \qquad \beta_{k,2} = \mu_{k,2} - \beta_{k,1}\mu_{k,1},$$

with $\rho_k$ being the group-specific correlation coefficient

$$\rho_k = \frac{\Sigma_{k,12}}{\sqrt{\Sigma_{k,11}\Sigma_{k,22}}},$$

and

$$w_k(x_i, \boldsymbol{\vartheta}) \propto \eta_k f_N(x_i; \mu_{k,1}, \Sigma_{k,11}).$$

Density (8.57) is closely related to the density of a finite mixture of regression models, where the slope, the intercept, and the error variance of the regression model switch among the different components. The component weights $w_k(x_i, \boldsymbol{\vartheta})$, however, are not fixed, but vary with $x_i$, and will be higher for components that are closer to $x_i$ than others.

The dependence of the weights on observations is implicit in this application of a multivariate mixture distribution to a regression type analysis. Several extensions of standard finite mixture models and finite mixtures of a regression model that are based on explicitly modeling such a dependence of the weights on observations are discussed in Subsections 8.6.2 and 8.6.3.

### 8.6.2 Modeling the Weight Distribution

For a standard finite mixture regression model the joint distribution $p(\mathbf{y}_i, S_i|\boldsymbol{\vartheta})$ factors as $p(\mathbf{y}_i, S_i|\boldsymbol{\vartheta}) = p(\mathbf{y}_i|S_i, \boldsymbol{\beta}_1, \ldots, \boldsymbol{\beta}_K)p(S_i|\boldsymbol{\eta})$, where the prior classification probabilities $\Pr(S_i = k|\boldsymbol{\eta}) = \eta_k$ are modeled as being independent of any data. In the marginal mixture distribution of $\mathbf{y}_i$ this leads to mixture density with fixed weight distribution $\boldsymbol{\eta} = (\eta_1, \ldots, \eta_K)$.

Various authors suggested modeling the prior classification probabilities $\Pr(S_i = k|\boldsymbol{\eta})$ in terms of covariates $\mathbf{z}_i$; see Fair and Jaffee (1972) for an early application. This is sensible whenever the span of the covariates is different between the different clusters. A typical example is a change-point regression, where the covariate $z_i = i$ is likely to determine cluster membership.

To include covariate information, $\Pr(S_i = k|\boldsymbol{\eta})$ is first reparameterized for $k = 1, \ldots, K-1$ in terms of an unconstrained parameter $\boldsymbol{\alpha} = (\alpha_1, \ldots, \alpha_{K-1})$ using the logistic transformation:

$$\log \frac{\Pr(S_i = k|\boldsymbol{\alpha})}{\Pr(S_i = K|\boldsymbol{\alpha})} = \log \frac{\eta_k}{1 - \sum_{j=1}^{K-1} \eta_j} = \alpha_k. \tag{8.58}$$

If for each unit $i$ a subject-specific variable $\mathbf{z}_i$ is observed additionally to $\mathbf{y}_i$, that might help to classify the subjects, then this information could be included through a multinomial logistic regression model:

$$\log \frac{\Pr(S_i = k|\boldsymbol{\alpha}, \boldsymbol{\gamma})}{\Pr(S_i = K|\boldsymbol{\alpha}, \boldsymbol{\gamma})} = \alpha_k + \mathbf{z}_i \boldsymbol{\gamma}_k, \tag{8.59}$$

where $\boldsymbol{\gamma} = (\boldsymbol{\gamma}_1, \ldots, \boldsymbol{\gamma}_{K-1})$ is an unknown regression parameter. If $\mathbf{z}_i$ fails to improve the resulting classification, then all components of $\boldsymbol{\gamma}$ are zero and model (8.59) reduces to (8.58). Frühwirth-Schnatter and Kaufmann (2006b) assume dependence of $\Pr(S_i = k|\boldsymbol{\alpha})$ on the initial income in an economic study involving panels of income data and use marginal likelihoods to test the more general model against a model where $\eta_k$ is fixed. Scaccia and Green (2003) use time and age in a growth curve analysis to model the weight distribution in a mixture of normal distributions.

### 8.6.3 Mixtures-of-Experts Models

Mixtures-of-experts models have been proposed in the neural network literature by Jacobs et al. (1991), and have found widespread application for modeling relationships among variables. They are defined as the following mixture distribution,

$$p(y_i) = \sum_{k=1}^{K} \eta_{k,i} f_N(y_i; \mathbf{x}_i \boldsymbol{\beta}_k, \sigma^2_{\varepsilon,k}),$$

where

$$\text{logit } \eta_{k,i} = \alpha_k + \mathbf{x}_i \boldsymbol{\gamma}_k.$$

From a statistical point of view, such a model is a finite mixture of regression model with observation-dependent weight distribution; see again Subsection 8.6.2. Note that the mixture weights may depend on the same covariates as the mean of the regression model. This may lead to identifiability problems (Jiang and Tanner, 1999).

Jacobs et al. (1996) and Peng et al. (1996) consider Bayesian parameter estimation using MCMC. Jacobs et al. (1997) discuss Bayesian methods for model selection in mixtures-of-experts models.

Hierarchical mixtures-of-experts result if the component densities themselves are mixtures-of-experts models; see Jordan and Jacobs (1994) for estimation based on the EM algorithm and Peng et al. (1996) for a Bayesian approach.

# 9

# Finite Mixture Models with Nonnormal Components

## 9.1 Finite Mixtures of Exponential Distributions

### 9.1.1 Model Formulation and Parameter Estimation

It is often assumed that nonnegative observations are realizations of a random variable $Y$ arising from a finite mixture of exponential distributions:

$$Y \sim \eta_1 \mathcal{E}(\lambda_1) + \cdots + \eta_K \mathcal{E}(\lambda_K), \tag{9.1}$$

where $\mathcal{E}(\lambda_k)$ is parameterized as in Appendix A.1.4. This mixture distribution is parameterized in terms of $\boldsymbol{\vartheta} = (\lambda_1, \dots, \lambda_K, \boldsymbol{\eta})$. Teicher (1963) showed that mixtures of exponential distributions are identifiable.

Following Farewell (1982), various mixture survival models, based on the exponential or more general distributions, were suggested and studied by many authors; see, for instance, Morbiducci et al. (2003), who studied such models with special focus on cure-rate models, to estimate the unknown rate of cured patients and the survival function of uncured patients in a clinical trial. The popularity of these models in duration or survival analysis is explained by their ability to explain the frequently observed fact that hazards decline with the length of spells (Heckman et al., 1990). Another interesting application of mixtures of exponential distributions appears in failure analysis, where failure often occurs for more than one reason (Everitt and Hand, 1981; Taylor, 1995). Slud (1997) proposes a two-component exponential mixture model to test imperfect debugging in software reliability.

Heckman et al. (1990) consider a consistent method of moments estimator and present Bayesian and classical tests for testing the hypothesis of dealing with mixtures of exponential distributions. Taylor (1995) uses the EM algorithm. Gruet et al. (1999) use MCMC methods for Bayesian estimation and apply reversible jump MCMC to select the number of components.

### 9.1.2 Bayesian Inference

Bayesian estimation using data augmentation and MCMC as in *Algorithm 3.4* is easily implemented for a mixture of exponential distributions. Based on the prior $\lambda_k \sim \mathcal{G}(a_0, b_0)$, the complete-data posterior $p(\lambda_k|\mathbf{y}, \mathbf{S})$ is given by $\lambda_k|\mathbf{y}, \mathbf{S} \sim \mathcal{G}(a_k(\mathbf{S}), b_k(\mathbf{S}))$, where:

$$a_k(\mathbf{S}) = a_0 + N_k(\mathbf{S}),$$
$$b_k(\mathbf{S}) = b_0 + \sum_{i:S_i=k} y_i.$$

Gruet et al. (1999) show how a reparameterization of the exponential mixture model (9.1) can allow for noninformative priors. They count the mixture components starting from $k = 0$, rather than $k = 1$. They leave $\lambda_0$ and $\eta_0 = \omega_0$ unchanged, whereas each $\lambda_k$ and $\eta_k$ is expressed for $k = 1, \ldots, K-1$ as

$$\lambda_k = \lambda_0 \prod_{j=1}^{k} \tau_j,$$
$$\eta_k = (1 - \omega_0) \cdots (1 - \omega_{k-1})\omega_k.$$

This parameterization allows us to select a partially proper prior distribution, based on the improper $\mathcal{G}(0, 0)$-prior for $\lambda_0$, whereas $\tau_1, \ldots, \tau_{K-1}$ are assumed to be uniform on [0,1]. As $\lambda_0$ appears as a common parameter in all component densities, this leads to a proper posterior density, as shown in the appendix of Gruet et al. (1999). This prior implies the order constraint $\lambda_0 > \cdots > \lambda_{K-1}$ on the component parameters, leading to an automatic identification of the model.

Casella et al. (2002) illustrate how perfect slice sampling may be implemented for mixtures of exponential distributions.

#### Reversible Jump MCMC

Gruet et al. (1999) apply reversible jump MCMC to select the number of components for an exponential mixture. Their parameterization introduces quite a natural strategy for carrying out split and merge moves, because in a mixture with $K-1$ components, the last component $\mathcal{E}(\lambda_0\tau_1 \cdots \tau_{K-2})$ is replaced by a two-component exponential mixture:

$$\omega_{K-2}\mathcal{E}(\lambda_0\tau_1 \cdots \tau_{K-2}) + (1 - \omega_{K-2})\mathcal{E}(\lambda_0\tau_1 \cdots \tau_{K-1})$$

to obtain a mixture with $K$ components.

To perform a split move in a mixture with $K$ components, first a component $k$ is chosen randomly. The index of all components from $k+1, \ldots, K-1$ is shifted by one. To split the old component $k$ into the two new components $k$ and $k+1$, the new parameters $\tau_k^{new}$ and $\tau_{k+1}^{new}$ satisfy:

$$\tau_k^{new}\tau_{k+1}^{new} = \tau_k,$$

whereas the weights satisfy:

$$(1-\omega_k^{new})(1-\omega_{k+1}^{new}) = (1-\omega_k).$$

To perform the split move, two random numbers $u_1$ and $u_2$ are introduced:

$$\begin{aligned}
\tau_k^{new} &= u_1 + \tau_k(1-u_1),\\
\tau_{k+1}^{new} &= \frac{\tau_k}{\tau_k^{new}} = \frac{\tau_k}{u_1+\tau_k(1-u_1)},\\
\omega_k^{new} &= u_2\omega_k,\\
\omega_{k+1}^{new} &= \frac{\omega_k(1-u_2)}{1-\omega_k u_2}.
\end{aligned}$$

If $k > 0$, then $u_1, u_2 \sim \mathcal{U}[0,1]$, whereas $u_1 \sim \mathcal{U}[0,.5]$ for $k=0$, in which case

$$\lambda_0^{new} = \lambda_0/u_1, \qquad \tau_1^{new} = u_1.$$

The determinant of the Jacobian is given by

$$|\text{Jacobian}| = \begin{cases} \dfrac{\omega_0(1-\tau_k)}{(1-\omega_k^{new})(\tau_k^{new})^2}, & \text{if } k>0,\\ \dfrac{\omega_0}{(1-\omega_0^{new})u_1}, & \text{if } k=0. \end{cases}$$

Gruet et al. (1999) report no improvement in refining reversible jumps by adding a move that introduces empty components.

## 9.2 Finite Mixtures of Poisson Distributions

### 9.2.1 Model Formulation and Estimation

A popular model for describing the distribution of count data is the Poisson mixture model, where it is assumed that $y_1, \ldots, y_N$ are independent realization of a random variable $Y$ arising from a mixture of Poisson distributions:

$$Y \sim \eta_1 \mathcal{P}(\mu_1) + \cdots + \eta_K \mathcal{P}(\mu_K),$$

with $\mathcal{P}(\mu_k)$ being a Poisson distribution with mean $\mu_k$; see Appendix A.1.11. This distribution is parameterized in terms of $2K-1$ distinct model parameters $\boldsymbol{\vartheta} = (\mu_1, \ldots, \mu_K, \boldsymbol{\eta})$. Mixtures of Poisson distributions are identifiable; see Feller (1943) and Teicher (1960).

Applications of mixtures of Poisson distributions appear in particular in biology and medicine; see, for example, Farewell and Sprott (1988) and Pauler et al. (1996). Applications to disease mapping are briefly discussed in Subsection 9.4.1. Karlis and Xekalaki (2005) provide a recent review of Poisson mixtures.

Mixtures of Poisson distributions served to illustrate statistical inference for finite mixtures throughout Chapter 2 to Chapter 5. Reversible jump MCMC has been used for finite mixtures of Poisson distributions by Dellaportas et al. (2002) and Viallefont et al. (2002); see also Subsection 5.2.2.

For Bayesian estimation, we add only comments on choosing the hyperparameters $a_0$ and $b_0$ of the prior of the group means, $\mu_k \sim \mathcal{G}(a_0, b_0)$. Viallefont et al. (2002) suggest fixing $a_0$ around 1 and choosing $b_0$ in such a way that the prior mean $\mathrm{E}(Y|\boldsymbol{\vartheta}) = a_0/b_0$ is matched to the midrange of the data, for example, the mean:

$$b_0 = \frac{a_0}{\overline{y}}. \tag{9.2}$$

For data where overdispersion is actually present, meaning that $s_y^2 - \overline{y} > 0$, it is possible to choose $a_0$ in such a way that the expectation of the second factorial moment with respect to the $\mathcal{G}(a_0, b_0)$-prior, which is by $\mathrm{E}(Y(Y-1)|\boldsymbol{\vartheta}) = a_0/b_0^2(1 + 1/a_0)$, is matched to the second factorial moment of the data, $v_2$, defined earlier in (2.26):

$$a_0 = \frac{\overline{y}^2}{v_2 - \overline{y}^2}, \tag{9.3}$$

where due to (9.6) $v_2 - \overline{y}^2$ could be substituted by $s_y^2 - \overline{y}$. Thus the larger the overdispersion in the data is, the smaller $a_0$ should be chosen.

If overdispersion is small, then $a_0$ is large and $\mu_k$ is strongly shrunken toward $a_0/b_0$. In this case it useful to assume a hierarchical prior as defined in Subsection 3.2.4, where $b_0 \sim \mathcal{G}(g_0, G_0)$. Estimation and model selection are rather insensitive to the parameter $g_0$ and could be chosen as $g_0 = 0.5$, whereas matching $\mathrm{E}(b_0) = g_0/G_0$ to $a_0/\overline{y}$ yields:

$$G_0 = \frac{g_0 \overline{y}}{a_0}.$$

### 9.2.2 Capturing Overdispersion in Count Data

Overdispersion occurs for a random variable $Y$, if the variance is bigger than the mean, whereas mean and variance are identical for a Poisson distribution. Overdispersion is present in many data sets involving counts. For illustration, consider the EYE TRACKING DATA counting eye anomalies in 101 schizophrenic patients studied by Pauler et al. (1996) and Escobar and West (1998), where the sample variance $s_y^2 = 35.89$ shows overdispersion in comparison to the sample mean $\overline{y} = 3.5248$; see also the histogram of the data in Figure 9.1.

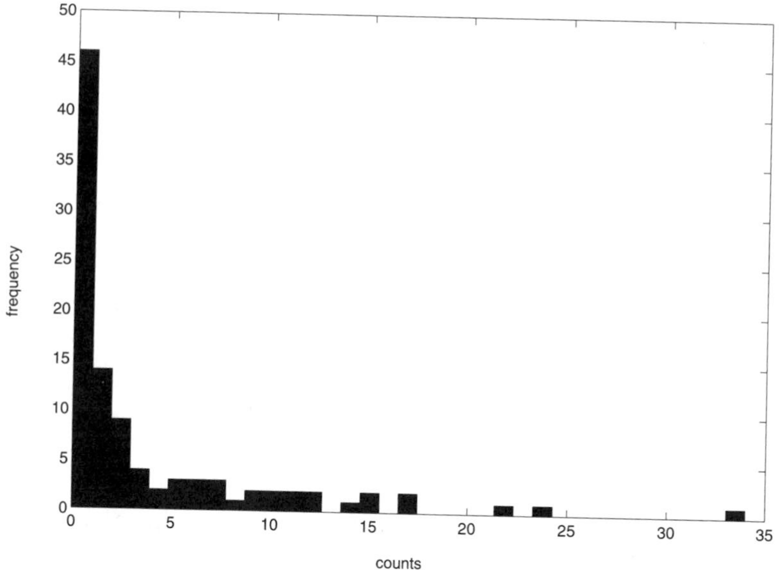


**Fig. 9.1.** Eye Tracking Data, empirical distribution of the observations

Many authors have studied the effect of overdispersion; see Wang et al. (1996) for some review. One possible reason for overdispersion is unobserved heterogeneity in the sample, causing the mean to be different among the observed subjects. A model commonly used in this context and discussed already in Feller (1943), is the Poisson–Gamma model which is a continuous mixture of Poisson distributions:

$$Y \sim \mathcal{P}\left(\mu_i^s\right), \qquad \mu_i^s \sim \mathcal{G}\left(\alpha, \alpha/\mu\right). \tag{9.4}$$

Marginally, $Y$ arises from the NegBin $(\alpha, \alpha/\mu)$-distribution, with $\mathrm{E}(Y|\boldsymbol{\vartheta}) = \mu$ and

$$\mathrm{Var}(Y|\boldsymbol{\vartheta}) = \mathrm{E}(Y|\boldsymbol{\vartheta})\frac{\alpha + \mathrm{E}(Y|\boldsymbol{\vartheta})}{\alpha} \geq \mathrm{E}(Y|\boldsymbol{\vartheta}),$$

where $\boldsymbol{\vartheta} = (\alpha, \mu)$. As long as $\alpha$ is not too large, this distribution actually captures overdispersion.

Overdispersion of a random variable $Y$ drawn from a Poisson mixture is evident from the first two moments of this mixture given by (1.19):

$$\mathrm{E}(Y|\boldsymbol{\vartheta}) = \sum_{k=1}^{K} \mu_k \eta_k,$$

$$\mathrm{Var}(Y|\boldsymbol{\vartheta}) = \sum_{k=1}^{K} \mu_k(1 + \mu_k)\eta_k - \mathrm{E}(Y|\boldsymbol{\vartheta})^2 = \mathrm{E}(Y|\boldsymbol{\vartheta}) + B(\boldsymbol{\vartheta}),$$

where $B(\boldsymbol{\vartheta})$ is the between-group heterogeneity:

$$B(\boldsymbol{\vartheta}) = \sum_{k=1}^{K} (\mu_k - \mu(\boldsymbol{\vartheta}))^2 \eta_k, \tag{9.5}$$

with $\mu(\boldsymbol{\vartheta}) = \mathrm{E}(Y|\boldsymbol{\vartheta})$. As $\mathrm{Var}(Y|\boldsymbol{\vartheta}) - \mathrm{E}(Y|\boldsymbol{\vartheta}) = B(\boldsymbol{\vartheta})$, finite mixtures of Poisson distributions explain overdispersion through unobserved heterogeneity in the sample, causing the mean to be different among the observed subjects. For $K = 2$, for instance, $B(\boldsymbol{\vartheta}) = 2\eta_1\eta_2(\mu_2 - \mu_1)^2$. Overdispersion occurs as long as the means of at least two components are different. Overdispersion could also be determined from the difference of the second factorial moment of the Poisson mixture, $\mathrm{E}(Y(Y-1)|\boldsymbol{\vartheta})$, and $\mathrm{E}(Y|\boldsymbol{\vartheta})^2$, as $\mathrm{E}(Y(Y-1)|\boldsymbol{\vartheta}) = \mathrm{E}(Y^2|\boldsymbol{\vartheta}) - \mathrm{E}(Y|\boldsymbol{\vartheta})$, and therefore:

$$\mathrm{E}(Y(Y-1)|\boldsymbol{\vartheta}) - \mathrm{E}(Y|\boldsymbol{\vartheta})^2 = B(\boldsymbol{\vartheta}). \tag{9.6}$$

The use of finite mixtures of Poisson distributions, rather than the more commonly used Poisson–Gamma model, to account for overdispersion has attracted several researchers, among them Simar (1976), Manton et al. (1981), Lawless (1987), Leroux (1992a), Leroux and Puterman (1992), Wang et al. (1996), and Viallefont et al. (2002).

It is possible to include observed covariates to explain part of the unobserved heterogeneity as discussed in Subsection 9.4.1, dealing with mixtures of Poisson regression models.

### 9.2.3 Modeling Excess Zeros

Count data often contain more zeros than expected under the Poisson distribution. In medical data excess zeros occur if the zero-class is inflated by the inclusion of observations that belong to a noninfected group. The EYE TRACKING DATA, for instance, contain 46 zeros, whereas under the $\mathcal{P}(\mu)$-distribution, the number of zeros in a sample of size $N$ follows a $\mathrm{BiNom}(N, e^{-\mu})$-distribution. For $N = 101$ and $\mu = \overline{y} = 3.5248$, the expected number of zero counts is roughly equal to 3, whereas the probability to observe at least 46 zero counts in a sample from the $\mathcal{P}(3.5248)$-distribution is as small as $1.9 \cdot 10^{-14}$, clearly indicating the presence of excess zeros.

Analyzing count data with excess zeros, sometimes also called inflated zeros, has a long tradition in applied statistics; see Meng (1997) for an interesting review. Feller (1943) proves that the number of zeros in a Poisson mixture is always larger than the number of zeros in a single Poisson distribution $\mathcal{P}(\mu)$ with the same mean $\mu = \mathrm{E}(Y|\boldsymbol{\vartheta})$ as the mixture distribution. This follows immediately from:

$$\Pr(Y=0|\boldsymbol{\vartheta}) = \sum_{k=1}^{K} \eta_k e^{-\mu_k} = e^{-\mu} \sum_{k=1}^{K} \eta_k e^{\mu-\mu_k} \tag{9.7}$$

$$\geq e^{-\mu} \sum_{k=1}^{K} \eta_k (1+\mu-\mu_k) \geq e^{-\mu}.$$

Cohen (1966) considers the following two-component mixture:

$$Y \sim \eta_1 I_{\{0\}}(y_i) + \eta_2 \mathcal{P}(\mu_2), \tag{9.8}$$

where $I_{\{0\}}(y_i)$ is 1 iff $y_i = 0$. A limitation of (9.8) is that the second group is assumed to be homogeneous. To capture overdispersion among nonzero individuals, it is sensible to substitute the Poisson distribution by a more general distribution, such as a finite mixture of $K-1$ Poisson distributions as in Cohen (1960) or a negative binomial distribution as in Cohen (1966). Such models are known as hurdle models; see, for instance, Cameron and Trivedi (1998) and Dalrymple et al. (2003) for an application to sudden infant death syndrome.

### 9.2.4 Application to the Eye Tracking Data

For illustration, consider the count data on eye tracking anomalies in 101 schizophrenic patients studied by Escobar and West (1998). To capture overdispersion and excess zeros for this data set, diagnosed in Subsection 9.2.2, we model the data by a finite mixture of $K$ Poisson distributions as in Congdon (2001), with increasing number $K$ of potential groups. We use the hierarchical prior (3.12) with $a_0 = 0.1$, $g_0 = 0.5$, and $G_0 = g_0\overline{y}/a_0$, and a $\mathcal{D}(4,\ldots,4)$-prior for $\boldsymbol{\eta}$. We use *Algorithm 3.3* for MCMC estimation, and store 8000 MCMC draws after a burn-in-phase of 3000.

Figure 9.2 shows, for an increasing number of components $K = 1,\ldots,7$, the posterior distribution of the probability $p_0(\boldsymbol{\vartheta})$ to observe 0, which is given by (9.7), of the overdispersion $B(\boldsymbol{\vartheta})$ defined in (9.5), and of the $l$th factorial moment, $\sum_{k=1}^{K} \eta_k \mu_k^l$ for $l = 3$ and $l = 4$. A comparison of these posterior distributions to the corresponding sample moments indicates that either four or five components are sufficient to capture the moments under investigation. Adding additional components hardly changes the posterior distribution of these moments.

Formal model selection, either using marginal likelihoods or reversible jump MCMC, is not really conclusive. Table 9.1 shows the log of the marginal likelihood for an increasing number of components, estimated through various simulation-based approximations, that were discussed in Section 5.4, namely bridge sampling, importance sampling, and reciprocal importance sampling. The importance density is constructed from the MCMC draws as in (5.36) with $S = \min(50K!, 5000)$, and the estimators are based on $M = 5000$ MCMC draws and $L = 5000$ draws from the importance density. Up to $K = 4$, these

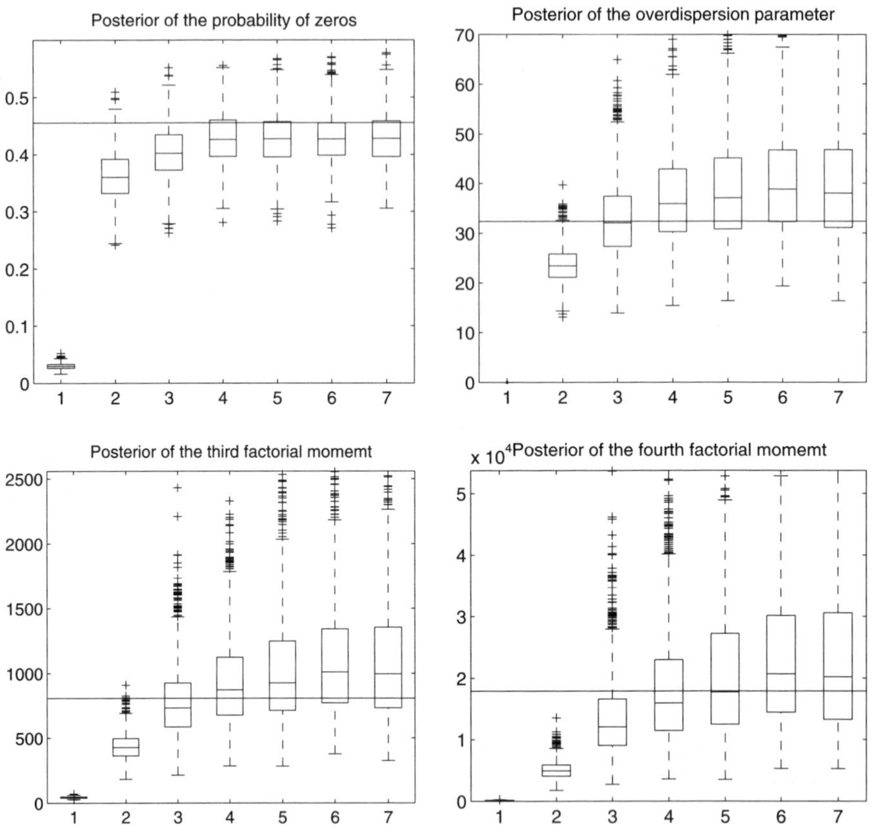


**Fig. 9.2.** EYE TRACKING DATA, finite Poisson mixtures with increasing numbers $K$ of potential groups; posterior distribution of the probability $p_0(\boldsymbol{\vartheta})$ to observe 0 (top left), of the overdispersion $B(\boldsymbol{\vartheta})$ (top right), the third (bottom left), and the fourth (bottom right) factorial moment in comparison to the corresponding sample moments (black horizontal line) for $K = 1, \dots, 7$

**Table 9.1.** EYE TRACKING DATA, various estimators of the marginal likelihood $p(\mathbf{y}|K)$ for finite mixtures of Poisson distributions with $K = 1$ to $K = 7$ components

| $p(\mathbf{y}\|K)$ | $K=1$ | $K=2$ | $K=3$ | $K=4$ | $K=5$ | $K=6$ | $K=7$ |
|---|---|---|---|---|---|---|---|
| $\log \hat{p}_{BS}(\mathbf{y}\|K)$ | –472.02 | –252.61 | –237.29 | –232.81 | –232.55 | –234.07 | –235.68 |
| $\log \hat{p}_{IS}(\mathbf{y}\|K)$ | –472.02 | –252.62 | –237.28 | –232.67 | –231.08 | –230.37 | –231.53 |
| $\log \hat{p}_{RI}(\mathbf{y}\|K)$ | –472.02 | –252.61 | –237.32 | –233.40 | –234.44 | –236.74 | –238.28 |

estimators are rather similar, but from $K = 5$ onwards, the estimators tend to be rather unstable. This leads to quite different estimators for the model posterior probabilities $\Pr(K|\mathbf{y})$, displayed in Table 9.2, which were computed under the truncated Poisson prior $p(K) \propto f_P(K;1)$. Although all estimators suggest choosing $K = 4$, the estimated posterior probabilities are quite different, and differ substantially from the posterior probabilities obtained from reversible jump MCMC, which are given in the same table. By considering a different importance density, namely a full permutation of single draw $\mathbf{S}^\star$, we obtained estimators of the model probabilities that are rather close to the estimators obtained from reversible jump MCMC. This suggests that simulation-based approximations to the marginal likelihood are sensitive for $K$ larger than 3 or 4, and reversible jump MCMC is preferable for mixtures with a medium to large number of components.

The same table shows that AIC and BIC also lead to the conclusion to choose $K = 4$, however, again evidence in favor of this model is very fragile, as AIC for $K = 4$ is only slightly larger than AIC for $K = 5$, whereas BIC for $K = 4$ is only slightly larger than BIC for $K = 3$.

**Table 9.2.** EYE TRACKING DATA, posterior probabilities $\Pr(K|\mathbf{y})$ based on the prior $p(K) \propto f_P(K;1)$, obtained from $\log \hat{p}_{BS}(\mathbf{y}|K)$ and reversible jump MCMC (RJMCM), AIC and BIC for $K = 1$ to $K = 7$ number of components

| $\Pr(K\|\mathbf{y})$ | $K=1$ | $K=2$ | $K=3$ | $K=4$ | $K=5$ | $K=6$ | $K=7$ |
|---|---|---|---|---|---|---|---|
| Based on $\hat{p}_{BS}(\mathbf{y}\|K)$ | 0.00 | 0.00 | 0.03 | 0.76 | 0.20 | 0.01 | 0.00 |
| Based on $\hat{p}_{IS}(\mathbf{y}\|K)$ | 0.00 | 0.00 | 0.02 | 0.42 | 0.41 | 0.14 | 0.01 |
| Based on $\hat{p}_{RI}(\mathbf{y}\|K)$ | 0.00 | 0.00 | 0.07 | 0.87 | 0.06 | 0.00 | 0.00 |
| RJMCMC | 0.00 | 0.00 | 0.01 | 0.33 | 0.40 | 0.20 | 0.06 |
| Based on $\hat{p}_{BS,2}(\mathbf{y}\|K)$ | 0.00 | 0.00 | 0.02 | 0.36 | 0.32 | 0.22 | 0.09 |
| AIC | -472.02 | -247.48 | -230.22 | -227.60 | -227.94 | -229.94 | -231.94 |
| BIC | -472.02 | -251.40 | -236.76 | -236.76 | -239.71 | -244.32 | -248.94 |

To obtain estimates of the group means and group sizes for a mixture of $K = 4$ groups, we need to identify a unique labeling among the MCMC draws. We first ran Gibbs sampling for an unconstrained mixture model with $K = 4$, and found that label switching took place between the two groups with the smallest means. For this data set, we could not achieve a unique labeling through unsupervised clustering. Next we imposed the constraint $\mu_1 < \cdots < \mu_K$ using the permutation sampler. Whenever the constraint was violated, we reordered the MCMC output in such a way that the constraint is fulfilled. Imposing the constraint eliminated label switching. Table 9.3 summarizes estimates of all group means and group sizes for $K = 4$, respectively. Evidently, the first group is practically a zero-movement group.

**Table 9.3.** EYE TRACKING DATA, posterior inference based on a Poisson mixture with $K = 4$ groups (hierarchical prior with $a_0 = 0.1$, $g_0 = 0.5$, and $G_0 = g_0\overline{y}/a_0$); identification obtained through imposing the constraint $\mu_1 < \cdots < \mu_4$

| $k$ | $\mathrm{E}(\mu_k\|\mathbf{y})$ | (95% Confidence Region) | $\mathrm{E}(\eta_k\|\mathbf{y})$ | (95% Confidence Region) |
|---|---|---|---|---|
| 1 | 0.09 | (0.00,0.38) | 0.36 | (0.19,0.55) |
| 2 | 1.38 | (0.61,2.89) | 0.33 | (0.17,0.49) |
| 3 | 7.95 | (5.07,10.83) | 0.20 | (0.12,0.30) |
| 4 | 20.17 | (15.16,27.55) | 0.10 | (0.04,0.18) |

## 9.3 Finite Mixture Models for Binary and Categorical Data

### 9.3.1 Finite Mixtures of Binomial Distributions

For binomial mixtures the component densities arise from $\mathrm{BiNom}(T, \pi)$-distributions, where $T$ is commonly assumed to be known, whereas the component-specific probabilities $\pi$ are unknown and heterogeneous:

$$Y \sim \eta_1 \mathrm{BiNom}(T, \pi_1) + \cdots + \eta_K \mathrm{BiNom}(T, \pi_K) .$$

The density of this mixture is given by

$$p(y|\boldsymbol{\vartheta}) = \sum_{k=1}^{K} \eta_k \binom{T}{y} \pi_k^y (1 - \pi_k)^{T-y}, \tag{9.9}$$

with $\boldsymbol{\vartheta} = (\pi_1, \ldots, \pi_K, \eta_1, \ldots, \eta_K)$. Binomial mixtures are not necessarily identifiable, as discussed already in Subsection 1.3.4. A necessary and sufficient condition is $T \geq 2K - 1$; see Teicher (1961).

Ever since Pearson (1915) employed a mixture of two binomial distributions to model yeast cell count data, discrete as well as continuous binomial mixtures have been suggested as overdispersed alternatives to the binomial distribution. Farewell and Sprott (1988), for instance, discuss an application in medicine to model the effect of a drug on patients who experience frequent premature ventricular contraction, whereas Brooks et al. (1997) and Brooks (2001) apply finite mixtures of binomials to dominant-lethal testing in a biological experiment.

Finite mixtures of binomial distributions may be extended to the case where $T_i$ varies between the realizations $y_1, \ldots, y_N$:

$$p(y_i|\boldsymbol{\vartheta}) = \sum_{k=1}^{K} \eta_k \binom{T_i}{y_i} \pi_k^{y_i} (1 - \pi_k)^{T_i - y_i} .$$

For identifiability of the corresponding mixture we refer to Teicher (1963, p.1268). Böhning et al. (1998) discuss an application of this model to a prevalence study in veterinary science.

### Unobserved Heterogeneity in Occurrence Probabilities

Assume that $K$ hidden groups are present in a population with heterogeneity in the occurrence probability of a certain event, such as the choice probability for a certain product. Let $\pi_1, \dots, \pi_K$ denote the different probabilities. Assume that for $N$ randomly selected subjects it is observed if the event under investigation has occurred or not, with $Y_i = 1$ indicating occurrence.

The identifiability condition discussed above becomes essential if we want to use this information to estimate the unknown probabilities $\pi_1, \dots, \pi_K$ as well as the unknown group sizes $\eta_1, \dots, \eta_K$. Evidently $\Pr(Y_i = 1|\boldsymbol{\vartheta}) = \pi_k$, if subject $i$ belongs to class $k$. As we observed for each subject only once whether the event has occurred, the marginal distribution of $Y_i$ is a mixture of binomial distributions with $T = 1$, which is not identified. Hence it is not possible to estimate $\boldsymbol{\vartheta}$.

To this aim, it is necessary to observe the event under investigation more than once, thus we need repeated measurements $Y_{it}, t = 1, \dots, T$ for each subject. In this case, the distribution of the number of successes, $Y_i = \sum_{t=1}^{T} Y_{it}$, is a realization from a mixture of binomial distributions $\text{BiNom}(T, \pi_k)$, which is identifiable only if $T \geq 2K - 1$. Thus even two repeated measurements are not sufficient to estimate the unknown parameter $\boldsymbol{\vartheta}$. For $K = 2$, for instance, the identifiability condition implies that we need for each subject $i$ at least $T = 3$ repeated measurements on occurrence/nonoccurrence of the event in order to identify the group sizes and the probabilities. With increasing number of hidden groups, the number of repeated measurement increases.

It is possible to include observed covariates to explain part of the heterogeneity in the occurrence probabilities $\pi_1, \dots, \pi_K$, as discussed in Subsection 9.4.2.

### Extra-Binomial Variation

Extra-binomial variation, meaning that $\text{Var}(Y|\boldsymbol{\vartheta}) > \text{E}(Y|\boldsymbol{\vartheta})(1 - \text{E}(Y|\boldsymbol{\vartheta})/T)$, is present in many data sets involving binary data. Extra-binomial variation is often due to unobserved heterogeneity in the population, for example, if an important covariate is omitted.

A common way of dealing with extra-binomial variation is the Beta-binomial model, which is a continuous mixture of binomial distributions, where $Y \sim \text{BiNom}(T, \pi_i^s)$ and $\pi_i^s \sim \mathcal{B}(\alpha, \beta)$. Marginally, this leads to the Beta-binomial distribution:

$$p(y|\boldsymbol{\vartheta}) = \binom{T}{y} \frac{B(\alpha + y, \beta + T - y)}{B(\alpha, \beta)},$$

with $\boldsymbol{\vartheta} = (\alpha, \beta)$. The first two moments of this distribution read with $\pi = \alpha/\beta$:

$$\begin{aligned} \text{E}(Y|\boldsymbol{\vartheta}) &= T\pi, \\ \text{Var}(Y|\boldsymbol{\vartheta}) &= T\pi(1 - \pi) + (T - 1)T\frac{\pi(1 - \pi)}{\alpha + \beta + 1}. \end{aligned}$$

A finite mixture of binomial distributions is an interesting alternative to the Beta-binomial distribution. Overdispersion of a random variable $Y$, drawn from the binomial mixture (9.9) is evident from the first two moments of this mixture, which are easily derived from (1.19):

$$\mathrm{E}(Y|\boldsymbol{\vartheta}) = T\pi, \qquad \pi = \sum_{k=1}^{K} \eta_k \pi_k,$$

$$\mathrm{Var}(Y|\boldsymbol{\vartheta}) = T\pi(1-\pi) + (T-1)T\left(\sum_{k=1}^{K} \eta_k \pi_k^2 - \pi^2\right). \tag{9.10}$$

For $T > 1$ extra variation due to the second term in (9.10) is present for any mixture with at least two different occurrence probabilities.

### Bayesian Estimation of Binomial Finite Mixture Models

Bayesian inference for mixtures of binomial distributions is considered in Brooks (2001), who applied a Metropolis–Hastings algorithm. Bayesian estimation using data augmentation and MCMC as in *Algorithm 3.4* is easily implemented for a mixture of binomial distributions; see again Brooks (2001). Based on the conjugate Beta prior $\pi_k \sim \mathcal{B}(a_0, b_0)$, the posterior $p(\pi_k|\mathbf{y}, \mathbf{S})$ is again a Beta distribution, $\pi_k|\mathbf{y}, \mathbf{S} \sim \mathcal{B}(a_k(\mathbf{S}), b_k(\mathbf{S}))$, where:

$$a_k(\mathbf{S}) = a_0 + \sum_{i:S_i=k} y_i,$$

$$b_k(\mathbf{S}) = b_0 + \sum_{i:S_i=k} (T - y_i).$$

Brooks (2001) applies the reversible jump MCMC method to jump between mixtures with different number of components and between mixtures of binomial distributions and mixtures of Beta-binomial distributions, where the number of components is left unchanged.

## 9.3.2 Finite Mixtures of Multinomial Distributions

Consider a categorical variable of more than two categories $\{1, \dots, D\}$. Let $Y_l$, for $l = 1, \dots, D$, be the number of occurrences of category $l$ among $T$ trials. If the occurrence probability distribution $\boldsymbol{\pi} = (\pi_1, \dots, \pi_D)$ of each category is homogeneous among the observed subjects, then $\mathbf{Y} = (Y_1, \dots, Y_D) \sim \mathrm{MulNom}(T, \boldsymbol{\pi})$; see also Appendix A.1.8 for a definition of the multinomial distribution.

To deal with unobserved heterogeneity in the occurrence probability of the various categories, $\mathbf{Y} = (Y_1, \dots, Y_D)$ is assumed to follow a finite mixture of multinomial distributions,

$$Y \sim \eta_1 \text{MulNom}(T, \pi_1) + \cdots + \eta_K \text{MulNom}(T, \pi_K),$$

with $\boldsymbol{\pi}_k = (\pi_{k,1}, \ldots, \pi_{k,D})$ being the unknown occurrence probability in group $k$. The density is given by

$$p(\mathbf{y}|\boldsymbol{\vartheta}) = \sum_{k=1}^{K} \eta_k \binom{T}{y_1 \ldots y_D} \prod_{l=1}^{D} \pi_{k,l}^{y_l}, \tag{9.11}$$

where $\boldsymbol{\vartheta} = (\boldsymbol{\pi}_1, \ldots, \boldsymbol{\pi}_K, \boldsymbol{\eta})$,

Morel and Nagaraj (1993) use such a model for capturing multinomial extra variation. In this respect, the finite mixture distribution (9.11) is an interesting alternative to the more commonly applied Dirichlet-multinomial distribution, where $\mathbf{Y} \sim \text{MulNom}(T, \boldsymbol{\pi}^s)$ and $\boldsymbol{\pi}^s \sim \mathcal{D}(\alpha_1, \ldots, \alpha_D)$; see, for instance, Paul et al. (1989) and Kim and Margolin (1992).

Further applications are found in clustering Internet traffic (Jorgensen, 2004) and developmental psychology (Cruz-Medina et al., 2004). Banjeree and Paul (1999) extend (9.11) to deal with multinomial clustered data. Further extensions are finite mixtures of multinomial logit models that are discussed in Subsection 9.4.1.

### Bayesian Estimation of Multinomial Finite Mixture Models

Let $\mathbf{y}_i = (y_{i1}, \ldots, y_{iD})$, $i = 1, \ldots, N$, be $N$ observations, where each $y_{il}, l = 1, \ldots, D$, counts the number of occurrences of category $l$ in a series of $T$ independent Bernoulli trials. Assume that a finite mixture of multinomial distributions should be fitted to these data.

Bayesian estimation using data augmentation and MCMC as in *Algorithm 3.4* is easily implemented for a mixture of multinomial distributions. Let $\boldsymbol{\pi}_k = (\pi_{k,1}, \ldots, \pi_{k,D})$ be the unknown discrete probability distribution in group $k$. Based on the Dirichlet prior $\boldsymbol{\pi}_k \sim \mathcal{D}(a_{0,1}, \ldots, a_{0,D})$, the posterior $p(\boldsymbol{\pi}_k|\mathbf{y}, \mathbf{S})$ is again a Dirichlet distribution $\boldsymbol{\pi}_k|\mathbf{y}, \mathbf{S} \sim \mathcal{D}(a_{k,1}(\mathbf{S}), \ldots, a_{k,D}(\mathbf{S}))$, where:

$$a_{k,l}(\mathbf{S}) = a_{0,l} + \sum_{i:S_i=k} y_{il}, \qquad l = 1, \ldots, D.$$

## 9.4 Finite Mixtures of Generalized Linear Models

Any of the finite mixture models discussed earlier in this chapter may be extended by assuming that in each group the underlying discrete distribution depends on some covariates. A common way to accommodate dependence of a nonnormal distribution on covariates is the generalized linear model (Nelder and Wedderburn, 1972; McCullagh and Nelder, 1999). Finite mixtures of generalized linear models extend the finite mixture of regression models discussed in Chapter 8 to nonnormal data and find numerous applications in particular in biology, medicine, and marketing. A very useful taxonomical review of numerous applications of mixture regression models for various types of data may be found in Wedel and DeSarbo (1993b, Table 10.1), which includes a lot of additional references.

After reviewing in Subsections 9.4.1 and 9.4.2 some specific examples for count, binary, and multinomial data, estimation of such models is discussed in detail in Subsection 9.4.3.

### 9.4.1 Finite Mixture Regression Models for Count Data

Finite mixture regression models for count data are either based on the Poisson or the negative binomial distribution.

#### Finite Mixtures of Poisson Regression Models

Let $Y_i$ denote the $i$th response variable, observed in reaction to a covariate $\mathbf{x}_i$, where the last element of $\mathbf{x}_i$ is 1, corresponding to an intercept. It is assumed that the marginal distribution of $Y_i$ follows a mixture of Poisson distributions,

$$Y_i \sim \sum_{k=1}^{K} \eta_k \mathcal{P}\left(\mu_{k,i}\right), \tag{9.12}$$

where $\mu_{k,i} = \exp(\mathbf{x}_i \boldsymbol{\beta}_k)$. If exposure data $e_i$ are available for each subject, then $\mu_{k,i} = e_i \exp(\mathbf{x}_i \boldsymbol{\beta}_k)$. If $\mathbf{x}_i = 1$, a finite mixture of Poisson distributions with $\mu_k = \exp(\boldsymbol{\beta}_k)$ results; if $K = 1$, the standard Poisson regression model results.

For a standard Poisson regression model, conditional on a given covariate, the data often exhibit overdispersion. Wang et al. (1996) showed that a mixture of Poisson regression models is able to capture overdispersion. For a fixed covariate, the mean and variance of $Y_i$ are easily obtained as in Subsection 9.2.2:

$$\mathrm{E}(Y_i|\boldsymbol{\vartheta}) = \sum_{k=1}^{K} \eta_k \mu_{k,i},$$

$$\mathrm{Var}(Y_i|\boldsymbol{\vartheta}) = \mathrm{E}(Y_i|\boldsymbol{\vartheta}) + \left(\sum_{k=1}^{K} \eta_k \mu_{k,i}^2 - \mathrm{E}(Y_i|\boldsymbol{\vartheta})^2\right).$$

Wang et al. (1996) falsely claim that a mixture of Poisson regression models is identifiable if the regressor matrix is of full rank. However, as discussed in Subsection 8.2.2 for mixtures of normal regression models, this condition is in general not sufficient.

Pointwise identifiability for a fixed covariate $\mathbf{x}_i$ follows from the generic identifiability of Poisson mixtures. Thus

$$\sum_{k=1}^{K} \eta_k f_P(y; \mu_{k,i}) = \sum_{k=1}^{K} \eta_k^\star f_P(y; \mu_{k,i}^\star),$$

where $\log \mu_{k,i} = \mathbf{x}_i \boldsymbol{\beta}_k$ and $\log \mu_{k,i}^\star = \mathbf{x}_i \boldsymbol{\beta}_k^\star$, implies that the condition

$$\eta_k^\star = \eta_{\rho_i(k)}, \qquad \mathbf{x}_i \boldsymbol{\beta}_k^\star = \mathbf{x}_i \boldsymbol{\beta}_{\rho_i(k)}, \tag{9.13}$$

holds for some permutation $\rho_i(\cdot)$. This is exactly what resulted for finite mixtures of the standard regression model studied earlier in Section 8.2.2. It follows immediately that a mixture of Poisson regressions, where only the intercept is switching, is identifiable. In all other cases, evidently the same conditions as for a Gaussian mixture of regression models hold.

Many applications of this model appear in medicine such as modeling epileptic seizure frequency data in a clinical trial (Wang et al., 1996; Wang and Puterman, 1999) or modeling the length of hospital stay (Lu et al., 2003). Wedel et al. (1993) and Wedel and DeSarbo (1995) discuss applications in marketing, such as modeling the number of coupons used by a household and evaluating direct marketing strategies. Applications in road safety appear in Viallefont et al. (2002) and Hurn et al. (2003), who relate the number of accidents to covariates.

## Disease Mapping

An area where mixtures of Poisson regression models are applied frequently is the study of disease distributions. The analysis of the geographic variation of disease and the representation of a disease distribution on a map is one of the oldest applications of statistics in epidemiology; see Schlattmann and Böhning (1993) for a review. Simple probabilistic models are based on the assumption that the number $Y_i$ of cases observed in region $i$ follows a $\mathcal{P}(\lambda e_i)$-distribution, where $\lambda$ is the relative risk and $e_i$ are the exposures. Rather than assuming that the risk is the same in all areas, Schlattmann and Böhning (1993) consider the case where the relative risk differs among the different areas, and takes one out of $K$ values $\lambda_1, \dots, \lambda_K$; see also Viallefont et al. (2002). This model is extended in Schlattmann et al. (1996) to accommodate dependence of covariates $\mathbf{x}_i = \begin{pmatrix} z_{i1} & \cdots & z_{i,d} \end{pmatrix}$ measured in each area:

$$Y_i | S_i \sim \mathcal{P}\left(\lambda_{S_i} \exp(\mathbf{x}_i \boldsymbol{\beta}) e_i\right),$$

whereas Viallefont et al. (2002) also consider heterogeneous covariate effects $\boldsymbol{\beta}_{S_i}$. Marginally, these models are finite mixtures of regression models that allow the detection of disease clusters, that is, areas of high or low risk. It provides an alternative to hierarchical Bayesian models for disease mapping; see, for instance, Bernardinelli et al. (1995).

Extensions of these models which substitute the unrealistic independence assumption among the indicators $S_1, \dots, S_N$ by a spatial dependence model are discussed, among others, by Fernández and Green (2002) and Green and Richardson (2002).

### Zero-Inflated Poisson Regressions

Lambert (1992) proposed the zero-inflated Poisson mixture regression model for dealing with zero-inflated count data with covariates and discussed an application where a production system switches between a perfect state where defects are extremely rare and an imperfect state where defects are possible. Both $\eta_1$, the probability of the perfect state as well as $\mu_2$, the mean of the imperfect state depend on covariates through a logit-type model. Further applications are disease mapping (Böhning, 1998) and the analysis of sudden infant death syndrome in relation to climate (Dalrymple et al., 2003).

### Finite Mixtures of Negative Binomial Regression Models

Ramaswamy et al. (1994) apply a finite mixture of negative binomial regression models in marketing research to model the purchase behavior of consumers.

## 9.4.2 Finite Mixtures of Logit and Probit Regression Models

### Finite Mixture Regression Models for Binary Data

Let $Y_{i,t}$ denote a binary variable, observed for $T_i$ times in reaction to a covariate $\mathbf{x}_i$, where the last element of $\mathbf{x}_i$ is 1 corresponding to an intercept. Define $Y_i = \sum_{t=1}^{T_i} Y_{i,t}$. It is assumed that the marginal distribution of $Y_i$ follows a mixture of binomial distributions,

$$Y_i \sim \sum_{k=1}^{K} \eta_k \text{BiNom}\left(T_i, \pi_{k,i}\right), \tag{9.14}$$

where logit $\pi_{k,i} = \mathbf{x}_i \boldsymbol{\beta}_k$ in finite mixtures of logit regression models, whereas $\pi_{k,i} = \Phi(\mathbf{x}_i \boldsymbol{\beta}_k)$ in finite mixtures of probit regression models.

Both models capture extra-binomial variation due to unobserved heterogeneity in the population, for example, if an important covariate is omitted. It follows

$$\begin{aligned} \mathrm{E}(Y_i|\boldsymbol{\vartheta}) &= \pi_i = \sum_{k=1}^{K} \eta_k \pi_{k,i}, \\ \mathrm{Var}(Y_i|\boldsymbol{\vartheta}) &= T_i \pi_i (1-\pi_i) + \frac{(T_i-1)T_i^2}{T_i} \left( \sum_{k=1}^{K} \eta_k \pi_{k,i}^2 - \pi_i^2 \right). \end{aligned} \tag{9.15}$$

For $T_i > 1$, extra-binomial variation due to the second term in (9.15) is present.

Identifiability is rather evolved for mixtures of logistic and probit regression models, the reason being that for $\mathbf{x}_i = 1$ such a mixture reduces to a finite mixture of binomial distributions, which is not necessarily identifiable; see Subsection 9.3.1.

Pointwise identifiability for a fixed covariate $\mathbf{x}_i$ follows from the identifiability of a binomial mixture only, if $T_i \geq 2K - 1$. In this case

$$\sum_{k=1}^{K} \eta_k f_{BN}(y; T_i, \pi_{k,i}) = \sum_{k=1}^{K} \eta_k^\star f_{BN}(y; T_i, \pi_{k,i}^\star),$$

where logit $\pi_{k,i} = \mathbf{x}_i \boldsymbol{\beta}_k$ and logit $\pi_{k,i}^\star = \mathbf{x}_i \boldsymbol{\beta}_k^\star$, implies

$$\eta_k^\star = \eta_{\rho_i(k)}, \qquad \mathbf{x}_i \boldsymbol{\beta}_k^\star = \mathbf{x}_i \boldsymbol{\beta}_{\rho_i(k)}, \tag{9.16}$$

which is exactly what resulted for a Gaussian mixture of regression models. It follows immediately that a mixture of logistic regressions where only the intercept is switching is identifiable, if $T_i \geq 2K - 1$ for at least one covariate $\mathbf{x}_i$; see also Follmann and Lambert (1991).

Applications of finite mixtures of logistic regression models appear in biology to analyze the effect of a drug on the death rate of a protozoan trypanosome and to study the effect of salinity and temperature on the hatch rate of English sole eggs (Follmann and Lambert, 1989), in medicine to determine the risk factors of preterm delivery (Zhu and Zhang, 2004), in genetics to detect inheritance patterns of a binary trait such as alcoholism (Zhang and Merikangas, 2000), in marketing research to deal with the analysis of paired comparison choice data (Wedel and DeSarbo, 1993a), and in agriculture (Wang and Puterman, 1998).

Finite mixtures of probit regression models are applied in medical research to analyze the resistance to treatment of parasites in sheep (Lwin and Martin, 1989), in marketing research to analyze pick and/N data (De Soete and DeSarbo, 1991), and in the economics of labor markets (Geweke and Keane, 1999).

#### Finite Mixture Regression Models for Categorical Data

Extensions to multinomial mixtures are considered by Paul et al. (1989), Kim and Margolin (1992), and Morel and Nagaraj (1993). Kamakura and Russell (1989) applied a multinomial logit mixture regression model in marketing research to model consumers' choices among a set of brands and identified segments of consumers that differ in price sensitivity. Kamakura (1991) proposed a multinomial probit finite mixture regression model. Identifiability for multinomial mixture regression models is investigated in Grün (2002).

### 9.4.3 Parameter Estimation for Finite Mixtures of GLMs

ML estimation for finite mixtures of generalized linear models is considered, for instance, by Jansen (1993), Wedel and DeSarbo (1995), and Aitkin (1996).

Wedel et al. (1993), Lambert (1992), and Wang et al. (1996) use the EM algorithm for the estimation of mixtures of Poisson regression models, whereas Wedel and DeSarbo (1995) consider more general mixtures of GLMs.

For Bayesian estimation of mixtures of GLMs, Hurn et al. (2003) use the same prior as for a normal regression model, namely $\boldsymbol{\beta}_k \sim \mathcal{N}_d(\mathbf{b}_0, \mathbf{B}_0)$. Various suggestions have been put forward of how to estimate mixtures of nonnormal regression models using MCMC. Viallefont et al. (2002) use a single-move random walk Metropolis–Hastings algorithm, whereas Hurn et al. (2003) use a multivariate random walk Metropolis–Hastings algorithm for sampling directly from the marginal posterior distribution $p(\boldsymbol{\vartheta}|\mathbf{y})$ for mixtures of logistic and Poisson regressions; see also *Algorithm 3.6*. This is feasible, as the likelihood $p(\mathbf{y}|\boldsymbol{\vartheta})$ is available in closed form.

Alternatively, one could use data augmentation by introducing a group indicator $S_i$ for each observation pair $(\mathbf{x}_i, \mathbf{y}_i)$ as missing data to obtain a sampling scheme comparable to *Algorithm 8.1*, which was derived in the context of finite mixtures of standard regression models. The resulting scheme, however, is not a Gibbs sampling scheme. Difficulties arise when drawing the group-specific parameters $\boldsymbol{\beta}_k$ in group $k$, because the conditional posterior distribution $p(\boldsymbol{\beta}_k|\mathbf{y}, \mathbf{S})$ has to be derived from a nonnormal regression model and does not belong to a well-known distribution family. To sample from this distribution, usually a Metropolis–Hastings step is applied; alternatively Hurn et al. (2003) mention the possibility of using the slice sampler (Damien et al., 1999). Classification, however, does not cause any problem, as it is sufficient to know the conditional distribution $p(\mathbf{y}_i|\boldsymbol{\beta}_k)$ for each $k = 1, \dots, K$ for each observation $\mathbf{y}_i$.

### 9.4.4 Model Selection for Finite Mixtures of GLMs

Wang et al. (1996) use AIC and BIC for model selection of the number of mixture regressions, while including all possible covariates. Covariates are selected in a second step, after having chosen the number of components. In their simulation study BIC always selected the correct model.

Viallefont et al. (2002) use the reversible jump MCMC for mixtures of Poisson regression models to determine the number of mixture regressions, whereas Hurn et al. (2003) use the birth and death MCMC of Stephens (2000a). Both papers illustrate that the prior on $K$, which is usually assumed to be $\mathcal{P}(\lambda_0)$, is not without effects on the resulting inference.

## 9.5 Finite Mixture Models for Multivariate Binary and Categorical Data

In this section we consider finite mixture modeling of multivariate binary or categorical data $\{\mathbf{y}_1, \dots, \mathbf{y}_N\}$, where $\mathbf{y}_i = (y_{i1}, \dots, y_{ir})$ is the realization of an $r$-dimensional discrete random variable $\mathbf{Y} = (Y_1, \dots, Y_r)$. Mixture models for multivariate discrete data, usually called latent class models, or latent structure analysis, have long been recognized as a useful tool in the behavioral and biomedical sciences, as exemplified by Lazarsfeld and Henry (1968), Goodman (1974b, 1978), Clogg and Goodman (1984), among many others; see Formann and Kohlmann (1996) and Clogg (1995) for a review.

In latent structure analysis the correlation between the elements $Y_1, \dots, Y_r$ of $\mathbf{Y}$ is assumed to be caused by a discrete latent variable $S_i$, also called the latent class. It is then assumed that the variables $Y_1, \dots, Y_r$, which are also called manifest variables, are stochastically independent conditional on the latent variable. Latent structure analysis is closely related to multivariate mixture modeling, as marginally the distribution of $\mathbf{Y}$ is a multivariate discrete mixture with density:

$$p(\mathbf{y}_i|\boldsymbol{\vartheta}) = \sum_{k=1}^{K} \eta_k \prod_{j=1}^{r} p(y_{ij}|\boldsymbol{\pi}_{k,j}),$$

where $\boldsymbol{\pi}_{k,j}$ is a parameter modeling the discrete probability distribution of $Y_j$ in class $k$.

### 9.5.1 The Basic Latent Class Model

In this section we consider a collection of multivariate binary observations $\mathbf{y}_1, \dots, \mathbf{y}_N$, where $\mathbf{y}_i = (y_{i1}, \dots, y_{ir})'$ is an $r$-dimensional vector of 0s and 1s, assumed to be the realization of a binary multivariate random variable $\mathbf{Y} = (Y_1, \dots, Y_r)$. The latent class model assumes that associations between the manifest variables $Y_j$ are caused by the presence of "latent classes" within which the features are independent. These latent classes may be seen as arising from an unobserved categorical variable $S_i$, which causes differences in occurrence probabilities $\pi_{k,j} = \Pr(Y_j = 1|S_i = k)$ of the manifest variable $Y_j$ in the different classes $k$.

The marginal distribution of $\mathbf{Y}$ is equal to a mixture of $r$ independent Bernoulli distributions, with density:

$$p(\mathbf{y}_i|\boldsymbol{\vartheta}) = \sum_{k=1}^{K} \eta_k \prod_{j=1}^{r} \pi_{k,j}^{y_{ij}} (1 - \pi_{k,j})^{1-y_{ij}}, \qquad (9.17)$$

where the $K$ components of the mixture correspond to the $K$ latent classes.

It is possible to verify that differences in the occurrence probabilities $\pi_{k,j}$ between the latent classes cause associations between the components of $\mathbf{Y}$ in the corresponding cell with respect to the marginal distribution, where the latent class is integrated out. For $K = 2$ and $r = 2$, for instance, Gilula (1979) shows that the marginal probability $\Pr(Y_1 = 1, Y_2 = 1|\boldsymbol{\vartheta})$ may be expressed as

$$\Pr(Y_1 = 1, Y_2 = 1|\boldsymbol{\vartheta}) =$$
$$\Pr(Y_1 = 1|\boldsymbol{\vartheta})\Pr(Y_2 = 1|\boldsymbol{\vartheta}) + \eta_1(1 - \eta_1)(\pi_{1,1} - \pi_{2,1})(\pi_{1,2} - \pi_{2,2}).$$

Thus associations between the components of $\mathbf{Y}$ will be observed, whenever both occurrence probabilities are different. Bartholomew (1980) regards the latent class models as factor analysis for categorical data.

Historically seen, model (9.17) was originated by psychometricians and sociologists, and goes back to Lazarsfeld (1950). The main purpose was to study hypothetical constructs such as "intelligence" or "attitude." There is a large body of literature with many applications of these models to problems in behavioral, medical, and social sciences, such as finding associations between teaching style and pupil performance (Aitkin et al., 1981), tumor diagnostics based on a sequence of binary test results (Albert and Dodd, 2004), analyzing historical household data (Liao, 2004), and texture analysis (Grim and Haindl, 2003), just to mention a few.

Celeux and Govaert (1991) discuss the application of latent class models for clustering discrete data and, using the classification likelihood approach discussed earlier in Subsection 7.1.3, show that clustering based on the latent class model is closely related to clustering based on minimizing entropy-type criteria.

### 9.5.2 Identification and Parameter Estimation

A difficult problem with the latent class model is verifying if the model is identifiable for a given number of classes, given a certain collection of the data; see Goodman (1974b) and Clogg (1995). If $\pi_{k,1} = \cdots = \pi_{k,r} = \pi_k$, then a binomial finite mixture with component density BiNom $(r, \pi_k)$ results; consequently the more general latent class models could be applied only to at least three manifest variables ($r \geq 3$). As outlined by Formann and Kohlmann (1996, p.194), "In general it is not possible to say a priori whether these models may be identifiable or not." Statements about identifiability are usually made after having estimated the parameters under a certain model by considering the rank of the observed information matrix evaluated at the ML estimator as in Catchpole and Morgan (1997) to prove local identifiability (Rothenberg, 1971); see also Carreira-Perpiñán and Renals (2000).

In (9.17), the $Kr$ unknown probabilities $(\pi_{1,1}, \ldots, \pi_{K,r})$ as well as the weight distribution $\boldsymbol{\eta}$ are unknown parameters that need to be estimated from the data. Pioneering work on ML estimation for the latent class model is found in Wolfe (1970). The basic latent class model is usually formulated as a generalized linear model and fitted by some iterative method, for instance, the proportional fitting algorithm of Goodman (1974a, 1974b), which later on turned out to be a variant of the EM algorithm.

An early reference on Bayesian estimation of latent class models is Evans et al. (1989), where the practical implementation was carried out using adaptive importance sampling. Again it is surprising to see how easily the marginal posterior density $p(\pi_{1,1}, \ldots, \pi_{K,r}, \boldsymbol{\eta}|\mathbf{y})$, which is quite complicated, is obtained, using data augmentation and MCMC as in *Algorithm 3.4*. Assume that all probabilities $\pi_{k,j}$ are independent a priori, with $\pi_{k,j} \sim \mathcal{B}(a_{0,j}, b_{0,j})$. Conditional on the class indicator $S_i$, the conditional posterior $p(\boldsymbol{\pi}_1, \ldots, \boldsymbol{\pi}_K|\mathbf{S}, \mathbf{y})$ is the product of $Kr$ independent Beta distributions with

$$\pi_{k,j}|\mathbf{S}, \mathbf{y} \sim \mathcal{B}\left(a_{0,j} + N_{k,j}(\mathbf{S}), b_{0,j} + N_k(\mathbf{S}) - N_{k,j}(\mathbf{S})\right),$$

where $N_{k,j}(\mathbf{S})$ is the number of ones observed for feature $Y_j$ in latent class $k$, and $N_k(\mathbf{S})$ is the total number of observations in latent class $k$:

$$N_{k,j}(\mathbf{S}) = \sum_{i:S_i=k} y_{ij}, \qquad N_k(\mathbf{S}) = \#\{i : S_i = k\}.$$

### 9.5.3 Extensions of the Basic Latent Class Model

Over the years, many variants and extensions of the basic latent class model have been considered. One particularly useful extension deals with multivariate categorical data $\mathbf{y}_1, \ldots, \mathbf{y}_N$, where $\mathbf{y}_i = (y_{i1}, \ldots, y_{ir})$ is the realization of an $r$-dimensional categorical random variable $\mathbf{Y} = (Y_1, \ldots, Y_r)$, with each element $Y_j$ taking one value out of $D_j$ categories $\{1, \ldots, D_j\}$. Again, a mixture density results:

$$p(\mathbf{y}_i|\boldsymbol{\vartheta}) = \sum_{k=1}^{K} \eta_k \prod_{j=1}^{r} \prod_{l=1}^{D_j} \pi_{k,jl}^{I_{\{y_{ij}=l\}}}, \tag{9.18}$$

where $\pi_{k,jl} = \Pr(Y_j = l|S_i = k)$ is the probability of category $l$ for feature $Y_j$ in class $k$.

The unknown parameter $\boldsymbol{\vartheta}$ appearing in (9.18) contains the unknown weight distribution $\boldsymbol{\eta}$ as well as the $Kr$ unknown probability distributions $\boldsymbol{\pi}_{k,j} = (\pi_{k,j1}, \cdots, \pi_{k,jD_j})$ of feature $Y_j$ in class $k$. Again Bayesian estimation is easy implemented, by sampling from the marginal posterior density $p(\boldsymbol{\pi}_{1,1}, \ldots, \boldsymbol{\pi}_{K,r}|\mathbf{y})$, using data augmentation and MCMC as in *Algorithm 3.4*. Assume that all probability distributions $\boldsymbol{\pi}_{k,j}$ are independent a priori, with $\boldsymbol{\pi}_{k,j} \sim \mathcal{D}(e_{0,j}, \ldots, e_{0,j})$. Conditional on the class indicator $S_i$, the conditional posterior $p(\boldsymbol{\pi}_{1,1}, \ldots, \boldsymbol{\pi}_{K,r}|\mathbf{S}, \mathbf{y})$ is the product of $Kr$ independent Dirichlet distributions with

$$\boldsymbol{\pi}_{k,j}|\mathbf{S}, \mathbf{y} \sim \mathcal{D}\left(e_{0,j} + N_{k,j1}(\mathbf{S}), \ldots, e_{0,j} + N_{k,jD_j}(\mathbf{S})\right),$$

where, for each class $k$, $N_{k,jl}(\mathbf{S})$ counts how often category $l$ is observed for feature $j$:

$$N_{k,jl}(\mathbf{S}) = \sum_{i:S_i=k} I_{\{y_{ij}=l\}}.$$

In a series of papers, Formann (1982, 1992, 1993, 1994a, 1994b) considered further extensions of (9.18) such as linear logistic latent class analysis, where both $\eta_k$ as well as the probabilities $\pi_{k,jl}$ depend on some covariates through a linear logistic model; see also Formann and Kohlmann (1996) for a review.

Clogg and Goodman (1984) consider simultaneously a latent structure analysis of a whole set of multinomial contingency tables and discuss methods for testing complete or partial homogeneity across tables.

## 9.6 Further Issues

### 9.6.1 Finite Mixture Modeling of Mixed-Mode Data

Often data are realizations of a mixed random variable $\mathbf{Y} = (\mathbf{Y}^C, \mathbf{Y}^D)$ with $\mathbf{Y}^C$ containing metric features and $\mathbf{Y}^D$ containing categorical features. Within a latent class analysis of such mixed-mode data, it is assumed that the distribution of $\mathbf{Y}$ depends on a latent unknown variable, which again leads to a finite mixture model.

Everitt (1988) and Everitt and Merette (1990) deal with mixed-mode data by incorporating the use of thresholds for the categorical data, however, the resulting model is difficult to estimate. Muthén and Shedden (1999) suggest combining features of Gaussian multivariate mixtures with a latent class model. The density of this mixture model reads:

$$p(\mathbf{y}_i|\boldsymbol{\vartheta}) = \sum_{k=1}^{K} \eta_k f_N(\mathbf{y}_i^C; \boldsymbol{\mu}_k, \boldsymbol{\Sigma}_k) p(\mathbf{y}_i^D|\boldsymbol{\theta}_k^D), \tag{9.19}$$

$$p(\mathbf{y}_i^D|\boldsymbol{\theta}_k^D) = \prod_{j=1}^{r} \prod_{l=1}^{D_j} \pi_{k,jl}^{I_{\{y_{ij}^D = l\}}}.$$

Bacher (2000) discusses an application of this model in sociology. Clustering multivariate data through probabilistic models based on finite mixtures is particularly useful for mixed continuous and categorical data; see Bock (1996) for some review.

Muthén and Shedden (1999) use the EM algorithm for estimation. Bayesian estimation of model (9.19) using data augmentation and MCMC as in *Algorithm 3.4* is easily implemented, as conditionally on a known classification we only need to combine the sampling step for multivariate mixtures of normals discussed in Subsection 6.3.3 with those obtained for the latent class model in Subsection 9.5.3.

A disadvantage of model (9.19) is stochastic independence of the categorical and the continuous variables within each class; more refined models, that allow for association between both types of variables are discussed in Lawrence and Krzanowski (1996). The idea is to replace the multivariate categorical variable $\mathbf{Y}^D = (Y_1^D, \ldots, Y_r^D)$, each of which is assumed to have $D_j$ different categories by a single multinomial variable $\mathbf{Y}^M$, which has $D_M = \prod_{j=1}^r D_j$ different cells, corresponding to the number of distinct patterns produced by $\mathbf{Y}^D$. Furthermore the group-specific mean $\boldsymbol{\mu}_{k,l}$ of the continuous variable is allowed to be different for all patterns $l = 1, \ldots, D_M$.

Willse and Boik (1999) show that the model in its unrestricted form is not identifiable. There exist $(K!)^{D_M-1}$ distinct parameters that define the same mixture distribution. Identifiability is achieved by imposing the restrictions $\boldsymbol{\mu}_{k,l} = \boldsymbol{\mu}_k + \boldsymbol{\beta}_l$ on the group-specific mean of the continuous variable.

Hunt and Jorgensen (2003) extended model (9.19) to mixed-mode data with missing observations. The modeling of mixed-mode data in a time series context is discussed in Cosslett and Lee (1985).

### 9.6.2 Finite Mixtures of GLMs with Random Effects

As discussed in Section 9.4, finite mixtures of GLMs are able to deal with overdispersion and extra-binomial or multinomial variation in regression models for discrete valued data. An alternative popular approach is based on GLMs with random-effects models (Schall, 1991; Breslow and Clayton, 1993; Aitkin, 1996), which regard overdispersion and extra-binomial or multinomial variation as a nuisance factor that needs to be accounted for in order to obtain consistent estimates of the other parameters. GLMs with random effects are also applied to pool information across similar units as in Section 8.5 for repeated measurements where the dependent variable is a discrete rather than a normally distributed random variable.

Usually the distribution of the random effects is chosen to be normal. Neuhaus et al. (1992) studied the effect of misspecifying the distribution of the random effects for logistic mixed-effects models and found cases of inconsistency both for the fixed and the random effects. Much more flexibility is achieved by assuming that the random effects follow a mixture of normal distributions as in Section 8.5, in which case a finite mixture of GLMs with random effects results. Such a model has been applied by Lenk and DeSarbo (2000) in marketing research, who discuss Bayesian estimation using data augmentation and MCMC. Yau et al. (2003) apply a two-component mixture of binary logit models with random effects to the analysis of hospital length of stay. Bottolo et al. (2003) apply a mixture of Poisson models with random effects to modeling extreme values in a data set of large insurance claims, using reversible jump MCMC.

# 10

# Finite Markov Mixture Modeling

## 10.1 Introduction

In this and the following chapters, finite mixture models are extended to deal with time series data that exhibit dependence over time. Broadly speaking, this is achieved by substituting the discrete latent indicator $S_i$ introduced as an allocation variable for finite mixture models by a hidden Markov chain. This leads to a surprisingly rich class of nonlinear time series models that solve a variety of interesting problems in applied time series analysis, as demonstrated in Chapter 12.

Section 10.2 starts with the definition of a finite Markov mixture distribution, whose properties are studied in some detail. Section 10.3 introduces the basic Markov switching model and deals with its extensions. The problem of econometric estimation of a Markov switching model from an observed time series is then discussed in Chapter 11.

## 10.2 Finite Markov Mixture Distributions

Let $\{y_t, t = 1, \ldots, T\}$ denote a time series of $T$ univariate observations taking values in a sampling space $\mathcal{Y}$ which may be either discrete or continuous. As common in time series analysis, $\{y_t, t = 1, \ldots, T\}$ is considered to be the realization of a stochastic process $\{Y_t\}_{t=1}^T$. Modeling is based on special cases from the class of doubly stochastic time series models (Tjøstheim, 1986) that have been found to be very useful for applied time series analysis.

It is assumed that the probability distribution of the stochastic process $Y_t$ depends on the realizations of a hidden discrete stochastic process $S_t$. The stochastic process $Y_t$ is directly observable, whereas $S_t$ is a latent random process that is observable only indirectly through the effect it has on the realizations of $Y_t$. A simple example is the hidden Markov chain model $Y_t = \mu_{S_t} + \varepsilon_t$, where $\varepsilon_t$ is a zero-mean white noise process with variance $\sigma^2$.

### 10.2.1 Basic Definitions

We start with specifying the properties of the hidden process $\{S_t\}_{t=0}^T$, which is assumed to be a discrete-time process with finite state space $\{1,\dots,K\}$ that obeys the following condition **S4**.

**S4** $S_t$ is an irreducible, aperiodic Markov chain starting from its ergodic distribution $\boldsymbol{\eta} = (\eta_1,\dots,\eta_K)$:

$$\Pr(S_0 = k|\boldsymbol{\xi}) = \eta_k.$$

The stochastic properties of $S_t$ are sufficiently described by the $(K \times K)$ transition matrix $\boldsymbol{\xi}$, where each element $\xi_{jk}$ of $\boldsymbol{\xi}$ is equal to the transition probability from state $j$ to state $k$:

$$\xi_{jk} = \Pr(S_t = k|S_{t-1} = j), \qquad \forall j,k \in \{1,\dots,K\}.$$

Evidently, the $j$th row of the transition matrix $\boldsymbol{\xi}$ defines, for all $t = 1,\dots,T$, the conditional distribution of $S_t$ given the information that $S_{t-1}$ is in state $j$. We sometimes use the notation $\boldsymbol{\xi}_{j\cdot}$ to refer to row $j$. All elements of $\boldsymbol{\xi}$ are nonnegative and the elements of each row sum to 1:

$$\begin{aligned} &\xi_{jk} \geq 0, \qquad \forall j,k \in \{1,\dots,K\},\\ &\sum_{k=1}^{K} \xi_{jk} = 1, \qquad \forall j = 1,\dots,K. \end{aligned} \tag{10.1}$$

$\boldsymbol{\xi} = (\boldsymbol{\xi}_{1\cdot},\dots,\boldsymbol{\xi}_{K\cdot})$ takes values in the product space $(\mathcal{E}_K)^K$, where $\mathcal{E}_K$ is the unit simplex defined in Subsection 1.2.1. Further assumptions about $\boldsymbol{\xi}$ are necessary to fulfill condition **S4**; see Subsection 10.2.2 for more details.

We continue with describing how the distribution of $Y_t$ depends on $S_t$. Let $\mathcal{T}(\boldsymbol{\theta})$ be a parametric distribution family, defined over a sampling space $\mathcal{Y}$ which may be either discrete or continuous, with density $p(y|\boldsymbol{\theta})$, indexed by a parameter $\boldsymbol{\theta} \in \Theta$. Let $\{Y_t\}_{t=1}^T$ be a sequence of random variables that depend on $\{S_t\}_{t=0}^T$ in the following way.

**Y4** Conditional on knowing $\mathbf{S} = (S_0,\dots,S_T)$, the random variables $Y_1,\dots,Y_T$ are stochastically independent. For each $t \geq 1$, the distribution of $Y_t$ arises from one out of $K$ distributions $\mathcal{T}(\boldsymbol{\theta}_1),\dots,\mathcal{T}(\boldsymbol{\theta}_K)$, depending on the state of $S_t$:

$$Y_t|S_t = k \sim \mathcal{T}(\boldsymbol{\theta}_k).$$

Hidden indicators comparable to $S_t$ have been introduced also for a finite mixture model, using the symbol $S_i$. The original definition of a mixture distribution in Section 1.2, however, started with the marginal distribution of $Y_i$ without introducing the latent indicator $S_i$ right from the beginning.

For the doubly stochastic process $\{S_t, Y_t\}_{t=1}^T$ obeying conditions **S4** and **Y4** it is rather easy to derive the marginal distribution of $Y_t$:

$$p(y_t|\boldsymbol{\vartheta}) = \sum_{k=1}^{K} p(y_t|S_t = k, \boldsymbol{\vartheta})\Pr(S_t = k|\boldsymbol{\vartheta}).$$

Because $S_t$ is a stationary Markov chain and the conditional distribution of $Y_t$ given $S_t = k$ has density $p(y_t|\boldsymbol{\theta}_k)$, one obtains that the unconditional distribution of $Y_t$ is a finite mixture of $\mathcal{T}(\boldsymbol{\theta})$ distribution with the ergodic probabilities $\boldsymbol{\eta} = (\eta_1, \dots, \eta_K)$ acting as weight distribution (Baum et al., 1970):

$$p(y_t|\boldsymbol{\vartheta}) = \sum_{k=1}^{K} p(y_t|\boldsymbol{\theta}_k)\eta_k. \tag{10.2}$$

Hence the process $Y_t$ is said to be generated by a finite Markov mixture of $\mathcal{T}(\boldsymbol{\theta})$ distributions. Stationarity of $Y_t$ is evident from (10.2). Furthermore such a process is autocorrelated (see Subsection 10.2.4), which is an important difference to a (standard) finite mixture of $\mathcal{T}(\boldsymbol{\theta})$ distributions, which produces sequences of independent random variables.

One early example of a finite Markov mixture distribution is the hidden Markov chain model (Baum and Petrie, 1966), where $Y_t$ is a discrete random signal taking one out of $D$ values $\{1, \dots, D\}$ according to a discrete probability distribution, which depends on the state of $S_t$:

$$\Pr(Y_t = l|S_t = k) = \pi_{k,l},$$

for $k = 1, \dots, K$ and $l = 1, \dots, D$. The transition matrix $\boldsymbol{\xi}$ as well as the matrix $\boldsymbol{\Pi} = (\pi_{k,l})$ is assumed to be unknown and has to be recovered from observations $\mathbf{y} = (y_1, \dots, y_T)$ of the process $\{Y_t\}_{t=1}^T$, whereas $S_t$ is unobserved.

Another early example is a Markov mixture of normal distributions (Baum et al., 1970), where $Y_t$ is a discrete signal observed with noise:

$$Y_t = \begin{cases} \mu_1 + \varepsilon_t, \ \varepsilon_t \sim \mathcal{N}\left(0, \sigma_1^2\right), & S_t = 1, \\ \vdots & \\ \mu_K + \varepsilon_t, \varepsilon_t \sim \mathcal{N}\left(0, \sigma_K^2\right), & S_t = K. \end{cases}$$

Many more examples appear throughout the remaining chapters.

The mathematical properties of a process generated by a finite Markov mixture distribution have been studied for specific processes obeying conditions **Y4** and **S4** such as hidden Markov chain models (Blackwell and Koopmans, 1957; Heller, 1965), white noise driven by a hidden Markov chain (Francq and Roussignol, 1997), discrete-valued time series generated by a hidden Markov chain (MacDonald and Zucchini, 1997), Markov mixtures of normal distributions (Krolzig, 1997), and Markov mixtures of more general location-scale families, where $Y_t = \mu_{S_t} + \sigma_{S_t}\varepsilon_t$, with $\varepsilon_t$ being an i.i.d. process (Timmermann, 2000).

After a short introduction into irreducible, aperiodic Markov chains in Subsection 10.2.2 these results are summarized for arbitrary processes $Y_t$ being generated by a Markov mixture obeying conditions **Y4** and **S4**.

### 10.2.2 Irreducible Aperiodic Markov Chains

In this subsection we briefly review properties of irreducible aperiodic Markov chains, focusing on results that are needed later on. For a more detailed survey we refer to Karlin and Taylor (1975).

Let $S_t$ be a homogeneous first-order Markov chain with transition matrix $\boldsymbol{\xi}$, where the elements of $\boldsymbol{\xi}$ are unconstrained apart from the natural constraints defined in (10.1). The transition matrix $\boldsymbol{\xi}$ plays a prominent role in understanding the properties of the corresponding Markov chain.

Any probability distribution $\boldsymbol{\eta} = (\eta_1, \ldots, \eta_K)$ that fulfills the invariance property

$$\boldsymbol{\xi}^{'}\boldsymbol{\eta} = \boldsymbol{\eta}, \tag{10.3}$$

is called an invariant distribution of $S_t$. The practical importance of the invariant distribution for the Markov chain $S_t$ is the following. Assume that at time $t-1$ the states of $S_{t-1}$ are drawn from an invariant distribution of $\boldsymbol{\xi}$. Then the following holds $\forall k = 1, \ldots, K$,

$$\Pr(S_t = k|\boldsymbol{\xi}) = \sum_{j=1}^{K} \Pr(S_t = k|S_{t-1} = j, \boldsymbol{\xi})\Pr(S_{t-1} = j|\boldsymbol{\xi})$$

$$= \sum_{j=1}^{K} \xi_{jk}\eta_j = \eta_k.$$

Therefore the states of $S_t$ are again drawn from $\boldsymbol{\eta}$, and so on for $S_{t+1}, \ldots$.

It is possible to show that such an invariant distribution exists for any finite Markov chain. By rewriting the constraint (10.1) as

$$\boldsymbol{\xi}_{j\cdot}\mathbf{1}_{K\times 1} = 1, \qquad \forall j = 1, \ldots, K,$$

where $\boldsymbol{\xi}_{j\cdot}$ refers to row $j$ of $\boldsymbol{\xi}$, and $\mathbf{1}_{K\times 1}$ is a column vector of ones, it becomes apparent, that for any transition matrix $\boldsymbol{\xi}$ one of the eigenvalues is equal to 1:

$$\boldsymbol{\xi}\mathbf{1}_{K\times 1} = 1 \times \mathbf{1}_{K\times 1}.$$

By rewriting (10.3) as

$$\boldsymbol{\eta}^{'}\boldsymbol{\xi} = \boldsymbol{\eta}^{'} \times 1,$$

it becomes apparent that, formally, $\boldsymbol{\eta}$ is the (suitably normalized) left-hand eigenvector of $\boldsymbol{\xi}$, associated with the eigenvalue 1.

The invariant distribution, however, is not unique for arbitrary transition matrices $\boldsymbol{\xi} \in (\mathcal{E}_K)^K$; consider, for instance, the transition matrix $\boldsymbol{\xi} = \mathbf{I}_K$, for which any arbitrary probability distribution will be invariant. An outstanding subset in the class $(\mathcal{E}_K)^K$ contains transition matrices for which this invariant distribution is unique and, additionally, the distribution of $S_t$ converges to this invariant distribution, regardless of the state of $S_0$. Such a Markov chain is called an ergodic Markov chain, and the invariant distribution $\boldsymbol{\eta}$ is called the ergodic distribution of the Markov chain.

Necessary restrictions on $\boldsymbol{\xi}$ to achieve ergodicity may be defined in terms of properties of $\boldsymbol{\xi}^h = \boldsymbol{\xi} \cdots \boldsymbol{\xi}$, the $h$th power of the transition matrix $\boldsymbol{\xi}$. $\boldsymbol{\xi}^h$ determines the long-run behavior of the Markov chain in terms of the $h$-step ahead predictive distribution $\Pr(S_{t+h} = l|S_t = k, \boldsymbol{\xi})$ of $S_{t+h}$ given $S_t = k$:

$$\Pr(S_{t+h} = l|S_t = k, \boldsymbol{\xi}) = (\boldsymbol{\xi}^h)_{kl}, \tag{10.4}$$

where $(\boldsymbol{\xi}^h)_{kl}$ is the element $(k,l)$ of $\boldsymbol{\xi}^h$. (10.4) is obvious for $h = 1$ from the definition of $\boldsymbol{\xi}$. For $h > 1$, (10.4) is easily derived by induction:

$$\begin{aligned} &\Pr(S_{t+h} = l|S_t = k, \boldsymbol{\xi}) \\ &= \sum_{j=1}^K \Pr(S_{t+h} = l|S_{t+h-1} = j, \boldsymbol{\xi})\Pr(S_{t+h-1} = j|S_t = k, \boldsymbol{\xi}) \\ &= \sum_{j=1}^K \xi_{jl}(\boldsymbol{\xi}^{h-1})_{kj} = (\boldsymbol{\xi}^h)_{kl}. \end{aligned}$$

Uniqueness of the invariant distribution follows for any transition matrix that leads to an irreducible Markov chain. Irreducibility means that starting $S_t$ from an arbitrary state $k \in \{1, \ldots, K\}$ any state $l \in \{1, \ldots, K\}$ must be reachable in finite time, or in terms of $(\boldsymbol{\xi}^h)_{kl}$:

$$\forall (k,l) \in \{1, \ldots, K\} \quad \Rightarrow \quad \exists h(k,l) : (\boldsymbol{\xi}^{h(k,l)})_{kl} > 0. \tag{10.5}$$

It follows that any transition matrix $\boldsymbol{\xi}$ where all elements $\xi_{jk}$ are positive leads to irreducibility and uniqueness of the invariant distribution. More generally, irreducibility follows if $(\boldsymbol{\xi}^h)_{kl} > 0$ for some $h \geq 1$, independent of $k, l$. If any element $(\boldsymbol{\xi}^h)_{kl}$ is 0 for all $h \geq 1$, then the Markov chain is reducible; consider, for instance, the following transition matrix

$$\boldsymbol{\xi} = \begin{pmatrix} \xi_{11} & 1 - \xi_{11} \\ 0 & 1 \end{pmatrix}, \tag{10.6}$$

which reappears in Subsection 10.3.3 in the context of change-point modeling. It is easily verified that this transition matrix leads to a reducible Markov chain, in as much as for all $h \geq 1$:

$$\boldsymbol{\xi}^h = \begin{pmatrix} \xi_{11}^h & 1 - \xi_{11}^h \\ 0 & 1 \end{pmatrix}.$$

Solving (10.3) for $K = 2$ leads to the following invariant probabilities,

$$\begin{aligned} \eta_1 &= \frac{1 - \xi_{22}}{(1 - \xi_{11}) + (1 - \xi_{22})} = \frac{\xi_{21}}{\xi_{12} + \xi_{21}}, \\ \eta_2 &= \frac{1 - \xi_{11}}{(1 - \xi_{11}) + (1 - \xi_{22})} = \frac{\xi_{12}}{\xi_{12} + \xi_{21}}. \end{aligned} \tag{10.7}$$

For a Markov chain with $\xi_{11} = \xi_{22}$, the invariant probability distribution is uniform: $\eta_1 = \eta_2 = 0.5$; $\xi_{11} > \xi_{22}$ favors state 1: $\eta_1 > \eta_2$, whereas $\xi_{11} < \xi_{22}$ favors state 2: $\eta_1 < \eta_2$.

For $K > 2$, some numerical method has to be used for solving (10.3). A closed-form expression for the invariant probability distribution $\boldsymbol{\eta}$ in terms of the transition matrix $\boldsymbol{\xi}$ is derived in Hamilton (1994b, Section 22.2). Define a matrix $\mathbf{A}$ as

$$\mathbf{A} = \begin{pmatrix} \mathbf{I}_K - \boldsymbol{\xi}' \\ \mathbf{1}_{1 \times K} \end{pmatrix}, \tag{10.8}$$

with $\mathbf{I}_K$ being the identity matrix with $K$ rows and $\mathbf{1}_{1\times K}$ being a row vector of ones. Then $\boldsymbol{\eta}$ is given as the $(K+1)$th column of the matrix $(\mathbf{A}'\mathbf{A})^{-1}\mathbf{A}'$:

$$\boldsymbol{\eta} = \left( (\mathbf{A}'\mathbf{A})^{-1}\mathbf{A}' \right)_{\cdot, K+1}. \tag{10.9}$$

Now let us turn to the distribution $\Pr(S_t|\boldsymbol{\xi})$ of a Markov chain $S_t$, starting with $S_0$ being drawn from a certain probability distribution. If the states of $S_0$ are drawn from the invariant distribution $\boldsymbol{\eta}$ of $\boldsymbol{\xi}$, then by the invariance property $\Pr(S_t|\boldsymbol{\xi})$ is equal $\boldsymbol{\eta}$ for all $t \geq 1$, but what happens if $S_0$ is drawn from a different distribution or is assumed to be a fixed starting value? Consider, for instance, the following irreducible transition matrix

$$\boldsymbol{\xi} = \begin{pmatrix} 0 & 1 & 0 \\ 0 & 0 & 1 \\ 1 & 0 & 0 \end{pmatrix}. \tag{10.10}$$

This matrix is an example of a doubly stochastic matrix where both the row and the column sums are equal to 1:

$$\sum_{k=1}^{K} \xi_{jk} = 1, \qquad \sum_{j=1}^{K} \xi_{jk} = 1.$$

For such matrices the uniform distribution, $\eta_k = 1/K$, is an invariant distribution:

$$\sum_{j=1}^{K} \xi_{jk}\eta_j = \frac{1}{K}\sum_{j=1}^{K} \xi_{jk} = \frac{1}{K}, \qquad \forall k = 1, \dots, K.$$

Because (10.10) is irreducible, the uniform distribution is the unique invariant distribution; the distribution $\Pr(S_t|\boldsymbol{\xi})$, however, does not converge to the invariant distribution if $S_0$ is started with a different distribution, such as the degenerate distribution $\Pr(S_0 = 1) = 1$, because

$$\begin{aligned}
\Pr(S_t = 1|S_0 = 1, \boldsymbol{\xi}) &= 1, \qquad \text{iff } t = 3m+1, m \in \{1, 2, 3, \dots\},\\
\Pr(S_t = 2|S_0 = 1, \boldsymbol{\xi}) &= 1, \qquad \text{iff } t = 3m+2, m \in \{1, 2, 3, \dots\},\\
\Pr(S_t = 3|S_0 = 1, \boldsymbol{\xi}) &= 1, \qquad \text{iff } t = 3m, m \in \{1, 2, 3, \dots\}.
\end{aligned}$$

The main reason for this failure of convergence is that the transition matrix (10.10) is periodic and captures a kind of seasonal pattern.

Ergodicity of a Markov chain with transition matrix $\boldsymbol{\xi}$ holds, if the Markov chain is aperiodic. Aperiodicity is defined as the absence of periodicity such as the one observed in the transition matrix (10.10). Consider, for each state $k$, all periods $n$ for which the transition probability $\Pr(S_{t+n} = k|S_t = k, \boldsymbol{\xi}) = (\boldsymbol{\xi}^n)_{kk}$ is positive. The period of a state is the greatest common divisor (GCD) of all periods $n$. A Markov chain is aperiodic, if the period of each state is equal to one:

$$\text{GCD}\{n \geq 1 : (\boldsymbol{\xi}^n)_{kk} > 0\} = 1, \qquad \forall k \in \{1, \dots, K\}.$$

A Markov chain is aperiodic if all diagonal elements of $\boldsymbol{\xi}$ are positive.

Ergodicity of a Markov chain implies that the distribution $\Pr(S_t|\boldsymbol{\xi}, S_0 = k)$ which is equal to the $k$th row $(\boldsymbol{\xi}^h)_{k\cdot}$ of $\boldsymbol{\xi}^h$ converges to the ergodic distribution, regardless of the state $k$ of $S_0$:

$$\lim_{h\to\infty} (\boldsymbol{\xi}^h)_{k\cdot} = \boldsymbol{\eta}'.$$

For understanding Markov mixture models it is helpful to know if this convergence is fast or if the Markov chain $S_t$ is persistent, meaning that the state of $S_t$ is mainly defined by the state of $S_{t-1}$. It turns out that the second largest eigenvalues of $\boldsymbol{\xi}$ play a crucial role in this respect.

Consider, for instance, a two-state Markov chain, where

$$\boldsymbol{\xi} = \begin{pmatrix} \xi_{11} & 1-\xi_{11} \\ 1-\xi_{22} & \xi_{22} \end{pmatrix}.$$

A two-state Markov chain is ergodic if $0 < \xi_{11} + \xi_{22} < 2$. The eigenvalues are obtained from

$$\begin{vmatrix} \xi_{11} - \lambda & 1 - \xi_{11} \\ 1 - \xi_{22} & \xi_{22} - \lambda \end{vmatrix} = (\lambda - 1)(\lambda - (\xi_{11} + \xi_{22} - 1)) = 0.$$

Apart from $\lambda = 1$, the other eigenvalue is equal to:

$$\lambda = \xi_{11} + \xi_{22} - 1 = \xi_{11} - \xi_{21}. \tag{10.11}$$

For $K = 2$ a simple representation of $\boldsymbol{\xi}^h$ in terms of the ergodic probability distribution is possible (Hamilton, 1994a, p.683):

$$\boldsymbol{\xi}^h = \begin{pmatrix} \eta_1 & \eta_2 \\ \eta_1 & \eta_2 \end{pmatrix} + \lambda^h \begin{pmatrix} \eta_2 & -\eta_2 \\ -\eta_1 & \eta_1 \end{pmatrix}, \tag{10.12}$$

with $\lambda$ being the second eigenvalue derived in (10.11) which demonstrates that persistence of $S_t$ is higher, the closer $\lambda$ is to 1.

Persistence is also related to the issue of duration of a certain state. Given that the Markov chain $S_t$ is currently in state $j$, the duration $D_j$ of that state is a random variable following a geometric distribution with parameter $1-\xi_{jj}$ (see Appendix A.1.7),

$$\begin{aligned}
\Pr(D_j = l|S_t = j) &= \Pr(S_{t+1} = j, \dots, S_{t+l-1} = j, S_{t+l} \neq j|S_t = j) \\
&= \prod_{m=1}^{l-1} \Pr(S_{t+m} = j|S_{t+m-1} = j)\Pr(S_{t+l} \neq j|S_{t+l-1} = j) \\
&= \xi_{jj}^{l-1}(1 - \xi_{jj}).
\end{aligned}$$

Therefore the expected duration of state $j$ is given by

$$\mathrm{E}(D_j) = \frac{1}{1 - \xi_{jj}}. \tag{10.13}$$

Two interesting conclusions may be drawn from (10.13). First, the expected duration of state $j$ is longer the closer the persistence probability $\xi_{jj}$ is to 1. Second, if the persistence probabilities differ in the various states, then also the expected duration of the state differs across states. Therefore Markov mixture distributions are able to capture asymmetry over time as observed for economic time series such as unemployment (Neftçi, 1984) and GDP, investment, and industrial production (Falk, 1986) over the business cycle.

### 10.2.3 Moments of a Markov Mixture Distribution

Because the unconditional distribution of a random process $Y_t$, being generated by a Markov mixture of $\mathcal{T}(\boldsymbol{\theta})$-distribution is a standard finite mixture of $\mathcal{T}(\boldsymbol{\theta})$-distribution with the ergodic probabilities acting as weights, the expectation of any function $h(Y_t)$ of $Y_t$ is given by the results of Subsection 1.2.4, where $\boldsymbol{\eta}$ is substituted by the ergodic distribution of $S_t$.

From Subsection 1.2.4 it is known that standard finite mixture distributions are able to generate probability distributions with asymmetry and fat tails. Timmermann (2000) studied finite Markov mixture distributions taken from a location-scale family and demonstrated that the introduction of Markovian dependence into the hidden indicator $S_t$ even increases the scope for asymmetry and fat tails in the generated process.

### Moments of a Markov Mixture of Two Normal Distributions

More explicit results are given for a Markov mixture of two normal distributions:

$$Y_t = \begin{cases} \mu_1 + \varepsilon_t,\ \varepsilon_t \sim \mathcal{N}\left(0, \sigma_1^2\right), & S_t = 1, \\ \mu_2 + \varepsilon_t,\ \varepsilon_t \sim \mathcal{N}\left(0, \sigma_2^2\right), & S_t = 2. \end{cases}$$

The unconditional distribution of $Y_t$ is given by a mixture of two normal distributions:

$$p(y_t|\boldsymbol{\vartheta}) = \eta_1 f_N(y_t; \mu_1, \sigma_1^2) + \eta_2 f_N(y_t; \mu_2, \sigma_2^2), \tag{10.14}$$

where the ergodic probabilities $\eta_1$ and $\eta_2$ are given by (10.7). The marginal distribution (10.14) exhibits nonnormality as long as either $\mu_1 \neq \mu_2$ or $\sigma_1^2 \neq \sigma_2^2$. Multimodality of the marginal distribution is possible for appropriate choices of $(\mu_1, \mu_2, \sigma_1^2, \sigma_2^2, \xi_{11}, \xi_{21})$ and could be checked for a given parameter using the results of Subsection 1.2.2.

From Subsection 1.2.4 the following coefficient of skewness results,

$$\frac{\mathrm{E}((Y_t - \mu)^3|\boldsymbol{\vartheta})}{\mathrm{E}((Y_t - \mu)^2|\boldsymbol{\vartheta})^{3/2}} = \eta_1\eta_2(\mu_1 - \mu_2)\frac{3(\sigma_2^2 - \sigma_1^2)^2 + (\eta_2 - \eta_1)(\mu_2 - \mu_1)^2}{\sigma^3},$$

with $\mu = \mathrm{E}(Y_t|\boldsymbol{\vartheta})$ and $\sigma^2 = \mathrm{Var}(Y_t|\boldsymbol{\vartheta})$ being the mean and variance of the mixture distribution (10.14):

$$\mu = \eta_1\mu_1 + \eta_2\mu_2,$$
$$\sigma^2 = \eta_1\sigma_1^2 + \eta_2\sigma_2^2 + \eta_1\eta_2(\mu_2 - \mu_1)^2.$$

Skewness in the marginal distribution will be present whenever both the means and the variances are different. For a model where the means are the same, no skewness is present. If the variances are the same and the means are different, skewness is possible only iff $\eta_1 \neq \eta_2$. Thus, for a Markov mixture model with different means but equal variances, asymmetry is introduced into the marginal distribution only through asymmetry in the persistence probabilities, namely $\xi_{11} \neq \xi_{22}$.

From Subsection 1.2.4, excess kurtosis is given by

$$\frac{\mathrm{E}((Y_t - \mu)^4|\boldsymbol{\vartheta})}{\mathrm{E}((Y_t - \mu)^2|\boldsymbol{\vartheta})^2} - 3 = \eta_1\eta_2\frac{3(\sigma_2^2 - \sigma_1^2)^2 + c(\mu_1, \mu_2)}{\sigma^4}, \tag{10.15}$$

where $c(\mu_1, \mu_2) = 6(\eta_1 - \eta_2)(\sigma_2^2 - \sigma_1^2)(\mu_2 - \mu_1)^2 + (\mu_2 - \mu_1)^4(1 - 6\eta_1\eta_2)$; see also Timmermann (2000, Corollary 1). Therefore if $\mu_1 = \mu_2$, the marginal distribution has fatter tails than a normal distribution as long as $\sigma_1^2 \neq \sigma_2^2$.

### 10.2.4 The Autocorrelation Function of a Process Generated by a Markov Mixture Distribution

A finite Markov mixture distribution generates an autocorrelated process $Y_t$ where the autocorrelation strongly depends on the persistence of $S_t$. The autocorrelation function of $Y_t$ is defined in the usual way as

$$\rho_{Y_t}(h|\boldsymbol{\vartheta}) = \frac{\mathrm{E}(Y_t Y_{t+h}|\boldsymbol{\vartheta}) - \mu^2}{\sigma^2}, \tag{10.16}$$

with $\mu = \mathrm{E}(Y_t|\boldsymbol{\vartheta})$ and $\sigma^2 = \mathrm{Var}(Y_t|\boldsymbol{\vartheta})$ being the unconditional moments and

$$\mathrm{E}(Y_t Y_{t+h}|\boldsymbol{\vartheta}) = \int y_t y_{t+h} p(y_t, y_{t+h}|\boldsymbol{\vartheta}) dy_t dy_{t+h}.$$

MacDonald and Zucchini (1997) derive the autocorrelation function for hidden Markov chain models for time series of counts, whereas Krolzig (1997), Rydén et al. (1998), and Timmermann (2000) consider continuous data. In the following we provide results for arbitrary processes obeying conditions **S4** and **Y4**.

To this aim it is useful to give an explicit form for the density $p(y_t, y_{t+h}|\boldsymbol{\vartheta})$ of the joint unconditional distribution of $Y_t$ and $Y_{t+h}$:

$$\begin{aligned} p(y_t, y_{t+h}|\boldsymbol{\vartheta}) = \sum_{k,l=1}^{K} & p(y_t|S_t = k, \boldsymbol{\vartheta}) p(y_{t+h}|S_{t+h} = l, \boldsymbol{\vartheta}) \\ & \times \Pr(S_{t+h} = l|S_t = k, \boldsymbol{\xi}) \Pr(S_t = k|\boldsymbol{\xi}). \end{aligned} \tag{10.17}$$

The predictive distribution $\Pr(S_{t+h} = l|S_t = k, \boldsymbol{\xi})$ is given by (10.4), and (10.17) reduces to

$$p(y_t, y_{t+h}|\boldsymbol{\vartheta}) = \sum_{k=1}^{K} p(y_t|\boldsymbol{\theta}_k)\eta_k \sum_{l=1}^{K} p(y_{t+h}|\boldsymbol{\theta}_l)(\boldsymbol{\xi}^h)_{kl}, \tag{10.18}$$

where $(\boldsymbol{\xi}^h)_{kl}$ is the element $(k, l)$ of the $h$th power of the transition matrix $\boldsymbol{\xi}$. Therefore $\mathrm{E}(Y_t Y_{t+h}|\boldsymbol{\vartheta})$ is given by

$$\mathrm{E}(Y_t Y_{t+h}|\boldsymbol{\vartheta}) = \sum_{k=1}^{K} \eta_k \mu_k \sum_{l=1}^{K} (\boldsymbol{\xi}^h)_{kl} \mu_l, \tag{10.19}$$

and the autocorrelation function results from (10.16):

$$\rho_{Y_t}(h|\boldsymbol{\vartheta}) = \frac{\sum_{k=1}^{K} \mu_k \eta_k \sum_{l=1}^{K} \mu_l (\boldsymbol{\xi}^h)_{kl} - \mu^2}{\sigma^2}.$$

Because the process $Y_t$ is uncorrelated conditional on knowing $S_t$, the autocorrelation function depends on $h$ only through $\boldsymbol{\xi}^h$, and autocorrelation in the marginal process $Y_t$, where $S_t$ is unknown, enters through persistence in $S_t$, only. Note that $Y_t$, in contrast to $S_t$, is no longer a Markov process of first order.

### Autocorrelation for a Two-State Model

From the specific form of $\boldsymbol{\xi}^h$ given in (10.12), the following autocorrelation function results for any two-state finite Markov mixture model,

$$\rho_{Y_t}(h|\boldsymbol{\vartheta}) = \frac{\eta_1\eta_2(\mu_1 - \mu_2)^2}{\sigma^2}\lambda^h, \tag{10.20}$$

with $\lambda = \xi_{11} - \xi_{21}$ being the second eigenvalue of $\boldsymbol{\xi}$.

No autocorrelation in $Y_t$ is present if $\mu_1 = \mu_2$. Otherwise, autocorrelation of $Y_t$ is caused through the hidden Markov chain $S_t$, whenever $\xi_{11} \neq \xi_{21}$. The process $Y_t$ exhibits positive autocorrelation provided that $\xi_{11} > \xi_{21}$, otherwise negative autocorrelation results. An equivalent criterion is to check if $\xi_{11}+\xi_{22}$ is larger or smaller than 1.

### Relation to ARMA Models

There exists a close relationship between Markov mixture models and nonnormal ARMA models. For a two-state Markov mixture model, for instance, the autocorrelation function of $Y_t$ given in (10.20) fulfills, for $h > 1$, the following recursion,

$$\rho_{Y_t}(h|\boldsymbol{\vartheta}) = \lambda\rho_{Y_t}(h-1|\boldsymbol{\vartheta}),$$

and corresponds to the autocorrelation function of an ARMA$(1,1)$ process, whereas the nonnormality of the unconditional distribution of $Y_t$ is preserved through the mixture distribution. In general, Poskitt and Chung (1996) proved for a univariate $K$-state hidden Markov chain $Y_t = \mu_{S_t} + u_t$ the existence of an ARMA$(K-1, K-1)$ representation with a homogeneous zero-mean white noise process.

## 10.2.5 The Autocorrelation Function of the Squared Process

An interesting feature of any finite Markov mixture model is that it generates processes $Y_t$, with $Y_t^2$ being autocorrelated. This is of particular interest when Markov mixture models are applied to financial time series; see Section 12.5. Timmermann (2000, Proposition 5) derived the autocorrelation function for a Markov mixture based on the continuous location-scale family. It is quite easy to generalize these results to any process obeying conditions **Y4** and **S4**.

The autocorrelation function of $Y_t^2$ is defined as

$$\rho_{Y_t^2}(h|\boldsymbol{\vartheta}) = \frac{\mathrm{E}(Y_t^2Y_{t+h}^2|\boldsymbol{\vartheta}) - \mathrm{E}(Y_t^2|\boldsymbol{\vartheta})^2}{\mathrm{E}(Y_t^4|\boldsymbol{\vartheta}) - \mathrm{E}(Y_t^2|\boldsymbol{\vartheta})^2}, \tag{10.21}$$

where $\mathrm{E}(Y_t^2|\boldsymbol{\vartheta}) = \sum_{k=1}^K \mathrm{E}(Y_t^2|\boldsymbol{\theta}_k)\eta_k$, and $\mathrm{E}(Y_t^4|\boldsymbol{\vartheta}) = \sum_{k=1}^K \mathrm{E}(Y_t^4|\boldsymbol{\theta}_k)\eta_k$, and $\mathrm{E}(Y_t^2Y_{t+h}^2|\boldsymbol{\vartheta})$ is obtained from (10.18) as

$$\mathrm{E}(Y_t^2 Y_{t+h}^2|\boldsymbol{\vartheta}) = \sum_{k=1}^{K} \eta_k \mathrm{E}(Y_t^2|\boldsymbol{\theta}_k) \sum_{l=1}^{K} \mathrm{E}(Y_{t+h}^2|\boldsymbol{\theta}_l)(\boldsymbol{\xi}^h)_{kl}, \tag{10.22}$$

with $(\boldsymbol{\xi}^h)_{kl}$ being the element $(k,l)$ of the $h$th power of the transition matrix $\boldsymbol{\xi}$. Although the process $Y_t^2$ is uncorrelated conditional on knowing $S_t$, autocorrelation in $Y_t^2$ enters through persistence in $S_t$.

### Autocorrelation in the Squared Process for a Two-State Model

We provide here further details for a Markov mixture of two normal distributions. From the general autocorrelation function of $Y_t^2$ given by (10.21), together with the representation of the transition matrix $\boldsymbol{\xi}$ of a two-state Markov model as in (10.12), one obtains:

$$\mathrm{E}(Y_t^2 Y_{t+h}^2|\boldsymbol{\vartheta}) = \mathrm{E}(Y_t^2|\boldsymbol{\vartheta}) + \eta_1\eta_2(\mu_1^2 - \mu_2^2 + \sigma_1^2 - \sigma_2^2)^2\lambda^h, \tag{10.23}$$

with $\lambda = \xi_{11} - \xi_{21}$ being the second eigenvalue of $\boldsymbol{\xi}$. Therefore:

$$\rho_{Y_t^2}(h|\boldsymbol{\vartheta}) = \frac{\eta_1\eta_2(\mu_1^2 - \mu_2^2 + \sigma_1^2 - \sigma_2^2)^2}{\mathrm{E}(Y_t^4|\boldsymbol{\vartheta}) - \mathrm{E}(Y_t^2|\boldsymbol{\vartheta})^2}\lambda^h. \tag{10.24}$$

The squared process exhibits positive autocorrelation provided that $\xi_{11} > \xi_{21}$, otherwise if $\xi_{11} < \xi_{21}$ negative autocorrelation will result. An equivalent criterion is to check if $\xi_{11} + \xi_{22}$ is larger or smaller than 1. Interestingly, state dependent variances are neither necessary nor sufficient for autocorrelation in the squared process. Even if $\sigma_1^2 = \sigma_2^2$, the marginal process shows conditional heteroscedasticity, as long as $S_t$ does not degenerate to an i.i.d. process. On the other hand, if $\xi_{11} = \xi_{21}$, no autocorrelation in the squared returns is present, even if $\sigma_1^2 \neq \sigma_2^2$.

By comparing the autocorrelation of $Y_t^2$, given by (10.24), with the autocorrelation of $Y_t$, given by (10.20), we find that a Markov mixture of two normal distributions with $\mu_1 = \mu_2$ will produce an uncorrelated process without skewness in the marginal distribution, whereas $Y_t^2$ is correlated and the marginal distribution has fat tails, as long as $\sigma_1^2 \neq \sigma_2^2$. As for other models that capture autocorrelation in the squared process, such as the GARCH model (Bollerslev, 1986), differences in the variances alone are insufficient to capture asymmetry in the marginal distribution.

### 10.2.6 The Standard Finite Mixture Distribution as a Limiting Case

Any standard finite mixture of $\mathcal{T}(\boldsymbol{\theta})$-distributions defined in Chapter 1 may be thought of as that limiting case of a finite Markov mixture of $\mathcal{T}(\boldsymbol{\theta})$-distribution where $S_t$ is an i.i.d. random sequence, in which case the transition probabilities from state $j$ to state $k$ are equal to $\Pr(S_t = k|S_{t-1} = j) = \Pr(S_t = k) = \eta_k$.

Thus a random variable $Y_t$ drawn from a standard finite mixture of $\mathcal{T}(\boldsymbol{\theta})$-distribution with weight distribution $\boldsymbol{\eta}$ is observationally equivalent with a process $Y_t$ generated by a finite Markov mixture of $\mathcal{T}(\boldsymbol{\theta})$-distributions where all rows of the transition matrix of $S_t$ are identical to $\boldsymbol{\eta}$:

$$\boldsymbol{\xi} = \begin{pmatrix} \eta_1 & \cdots & \eta_K \\ \vdots & & \vdots \\ \eta_1 & \cdots & \eta_K \end{pmatrix}.$$

In this case the transition matrix $\boldsymbol{\xi}$ is idempotent, $\boldsymbol{\xi}^h = \boldsymbol{\xi}$ for all $h \geq 1$, and (10.19) reduces to

$$\mathrm{E}(Y_t Y_{t+h}|\boldsymbol{\vartheta}) = \sum_{k=1}^{K} \eta_k \mu_k \sum_{l=1}^{K} \xi_{kl} \mu_l = \mu^2.$$

Thus the autocorrelation $\rho_{Y_t}(h|\boldsymbol{\vartheta})$ of $Y_t$, given by (10.16), is equal to 0 for $h > 1$. Similarly, (10.22) reduces to

$$\mathrm{E}(Y_t^2 Y_{t+h}^2|\boldsymbol{\vartheta}) = \sum_{k=1}^{K} \eta_k \mathrm{E}(Y_t^2|\boldsymbol{\theta}_k) \sum_{l=1}^{K} \mathrm{E}(Y_{t+h}^2|\boldsymbol{\theta}_l)\xi_{kl} = \mathrm{E}(Y_t^2|\boldsymbol{\vartheta})^2,$$

and the autocorrelation $\rho_{Y_t^2}(h|\boldsymbol{\vartheta})$ of $Y_t^2$, given by (10.21), is equal to 0 for $h > 1$.

### 10.2.7 Identifiability of a Finite Markov Mixture Distribution

For a finite Markov mixture distribution one has to distinguish between the same three types of nonidentifiability that have been discussed for a standard finite mixture distribution in Section 1.3. There exists nonidentifiability due to invariance to relabeling the states of the hidden Markov chain as well as generic nonidentifiability.

Consider all $s = 1, \ldots, K!$ different permutations $\rho_s : \{1, \ldots, K\} \rightarrow \{1, \ldots, K\}$, where the value $\rho_s(k)$ is assigned to each value $k \in \{1, \ldots, K\}$. Let $\boldsymbol{\vartheta} = (\boldsymbol{\theta}_1, \ldots, \boldsymbol{\theta}_K, \boldsymbol{\xi})$ be an arbitrary point in the parameter space $\Theta_K = \Theta^K \times (\mathcal{E}_K)^K$, and define a subset $\mathcal{U}^P(\boldsymbol{\vartheta}) \subset \Theta_K$ by

$$\mathcal{U}^P(\boldsymbol{\vartheta}) = \bigcup_{s=1}^{K!} \{\boldsymbol{\vartheta}^\star \in \Theta_K : \boldsymbol{\vartheta}^\star = (\boldsymbol{\theta}_{\rho_s(1)}, \ldots, \boldsymbol{\theta}_{\rho_s(K)}, \boldsymbol{\xi}^{\rho_s})\}, \qquad (10.25)$$

where $\boldsymbol{\xi}^{\rho_s}$ is related to $\boldsymbol{\xi}$ by permuting the rows and the column in the same fashion:

$$\xi_{jk}^{\rho_s} = \xi_{\rho_s(j),\rho_s(k)}, \qquad \forall j, k \in \{1, \ldots, K\}. \qquad (10.26)$$

Then evidently, all points in $\mathcal{U}^P(\boldsymbol{\vartheta})$ generate the same Markov mixture distribution, however, with a different labeling of the states of the hidden Markov chain.

A weak inequality constraint, similar to the one discussed for finite mixtures in Subsection 1.3.3 requiring that the state-specific parameters $\boldsymbol{\theta}_k$ and $\boldsymbol{\theta}_l$ differ in *at least one* element, which need not be the same for all states, will rule out these identifiability problems.

Blackwell and Koopmans (1957) is an early reference addressing generic identifiability problems for some special hidden Markov chain models, where $Y_t$ is a discrete signal. Petrie (1969) proved generic identifiability for hidden Markov chain models, where the observed process $Y_t$ takes values in a finite set. Identifiability for rather general finite Markov mixtures is addressed in Leroux (1992b).

One necessary condition for generic identifiability of a Markov mixture of $\mathcal{T}(\boldsymbol{\theta})$-distributions is that a standard finite mixture of $\mathcal{T}(\boldsymbol{\theta})$-distributions is generically identifiable; see again Subsection 1.3.4. A second necessary condition is that the hidden Markov chain is irreducible and aperiodic; it is, however, not necessary to assume that $S_0$ started from the invariant distribution.

## 10.3 Statistical Modeling Based on Finite Markov Mixture Distributions

Researchers have found Markov mixture models increasingly useful in applied time series analysis.

### 10.3.1 The Basic Markov Switching Model

Assume that a time series $\{y_1, \ldots, y_T\}$ is observed as a single realization of a stochastic process $\{Y_1, \ldots, Y_T\}$. In the basic Markov switching model the time series $\{y_1, \ldots, y_T\}$ is assumed to be a realization of a stochastic process $Y_t$ generated by a finite Markov mixture from a specific distribution family:

$$Y_t|S_t \sim \mathcal{T}(\boldsymbol{\theta}_{S_t}),$$

where $S_t$ is an unobservable (hidden) $K$ state ergodic Markov chain, and $Y_t$ fulfills assumption **Y4**.

The basic Markov switching model found widespread applications in many practical areas including bioinformatics, biology, economics, finance, hydrology, marketing, medicine, and speech recognition. Various terminology became usual to denote models based on hidden Markov chains. The term Markov mixture models is preferred by biologists (Albert, 1991). Markov mixture models are usually called hidden Markov models in engineering applications (Zucchini and Guttorp, 1991; Thyer and Kuczera, 2000) and in speech recognition

(Levison et al., 1983; Rabiner, 1989). The terms Markov switching models or regime-switching models are preferred by economists who used Markov switching models to analyze stock market returns (Pagan and Schwert, 1990; Engel and Hamilton, 1990), interest rates (Ang and Bekaert, 2002) and asymmetries over the business cycle (Neftçi, 1984; Hamilton, 1989); see the monographs by Bhar and Hamori (2004), Krolzig (1997) and Kim and Nelson (1999) and Chapter 12 for further references and more details.

An interesting special case of the basic Markov switching model arises if $\{y_1, \ldots, y_T\}$ is a discrete-valued time series (MacDonald and Zucchini, 1997). Because one may choose Markov mixtures of any discrete distribution, it is possible to model many different types of discrete valued time series data, for example, binary time series by

$$\Pr(Y_t = 1|S_t) = \pi_{S_t}, \tag{10.27}$$

time series of bounded counts by a Markov mixture of binomial distributions,

$$Y_t|S_t \sim \text{BiNom}\left(n_t, \pi_{S_t}\right), \tag{10.28}$$

or time series of unbounded counts by a Markov mixture of Poisson distributions,

$$Y_t|S_t \sim \mathcal{P}\left(\mu_{S_t}\right); \tag{10.29}$$

see also Section 11.7. An important feature of applying Markov mixture models to discrete-valued time series is the ease with which autocorrelation is introduced, and the properties of the marginal distribution are easily analyzed.

Similarly, the basic Markov switching model could be applied to deal with autoregression in positive-valued time series (Lawrance and Lewis, 1985) simply by choosing the observation density $p(y_t|\boldsymbol{\theta})$ from any density on $\Re^+$, such as the exponential, the Gamma, or the Weibull distribution.

The basic Markov switching model has been generalized in several ways as outlined in the following subsections as well as in Chapter 12.

### 10.3.2 The Markov Switching Regression Model

An early attempt at introducing Markov switching models into econometrics in order to deal with time series data that depends on exogenous variables is the switching regression model of Goldfeld and Quandt (1973), which extends the switching regression model (Quandt, 1972) described earlier in Section 8.2. Whereas Quandt (1972) assumes that $S_t$ is an i.i.d. random sequence, Goldfeld and Quandt (1973) allow explicitly for dependence between the states by modeling $S_t$ as a two-state hidden Markov chain.

The general Markov switching regression model reads,

$$Y_t = \mathbf{x}_t \boldsymbol{\beta}_{S_t} + \varepsilon_t, \qquad \varepsilon_t \sim \mathcal{N}\left(0, \sigma^2_{\varepsilon,S_t}\right), \tag{10.30}$$

where $S_t$ is a hidden Markov chain and $\mathbf{x}_t$ is a row vector of explanatory variables including the constant (Lindgren, 1978; Cosslett and Lee, 1985). For discrete-valued explanatory variables, the Markov switching regression model will suffer from the same identifiability problems as the standard finite mixture of regression models studied in Subsection 8.2.2, a fact that has remained unnoted in the literature.

### 10.3.3 Nonergodic Markov Chains

In certain applications it makes sense to consider Markov switching models driven by a nonergodic Markov chain. An important example is a model driven by a Markov chain with transition matrix $\boldsymbol{\xi}$ defined in (10.6) which captures a single structural break or change-point. Assume that the Markov chain starts in $S_0 = 1$. The Markov chain will stay in state 1 for $h$ periods; that is, $S_1 = \cdots = S_h = 1$ with probability $\xi_{11}^h$. Once state 2 is reached for the first time, the process remains there. An important aspect of this model is that the time of change-point occurrence is random.

A multiple change-point model with $K$ change-points may be modeled through a Markov switching model with the following transition matrix (Chib, 1998),

$$\boldsymbol{\xi} = \begin{pmatrix} \xi_{11} & 1-\xi_{11} & 0 & \cdots & & 0 \\ 0 & \xi_{22} & 1-\xi_{22} & \cdots & & 0 \\ & \ddots & \ddots & \ddots & & \\ & & 0 & \xi_{K-1,K-1} & 1-\xi_{K-1,K-1} & \\ & & & 0 & 1 & \end{pmatrix}. \qquad (10.31)$$

A more general Bayesian time series model of multiple structural changes in level, trend, and variance is studied in Wang and Zivot (2000). For a review of other methods of testing for the presence of unknown breakpoints in normal linear regression see Ploberger et al. (1989) and Andrews et al. (1996).

### 10.3.4 Relaxing the Assumptions of the Basic Markov Switching Model

The basic Markov switching model has been extended by many authors with the aim of formulating even more flexible models for a wide range of time series data.

Let $\{S_t\}_{t=0}^T$ be a finite-state Markov process with state space $\{1, \ldots, K\}$, and let $\{Y_t\}_{t=1}^T$ be a sequence of random variables with sampling space $\mathcal{Y}$. A general Markov switching model is obtained by specifying the density $p(\mathbf{S}, \mathbf{y}|\boldsymbol{\vartheta})$ of the joint distribution of $\mathbf{S} = \{S_t\}_{t=0}^T$ and $\mathbf{Y} = \{Y_t\}_{t=1}^T$, which is equal to:

$$p(\mathbf{S}, \mathbf{y}|\boldsymbol{\vartheta}) = p(S_0|\boldsymbol{\vartheta}) \prod_{t=1}^T p(y_t|\mathbf{y}^{t-1}, \mathbf{S}^t, \boldsymbol{\vartheta}) p(S_t|\mathbf{S}^{t-1}, \mathbf{y}^{t-1}, \boldsymbol{\vartheta}). \qquad (10.32)$$

$p(y_t|\mathbf{y}^{t-1}, \mathbf{S}^t, \boldsymbol{\vartheta})$ is the one-step ahead predictive density of the conditional distribution of $Y_t$, knowing the past realizations $\mathbf{y}^{t-1} = (y_1, \dots, y_{t-1})$ of $\mathbf{Y}^{t-1}$ and knowing the states $\mathbf{S}^t = (S_0, \dots, S_t)$. $p(S_t|\mathbf{y}^{t-1}, \mathbf{S}^{t-1}, \boldsymbol{\vartheta})$ is the density of the conditional distribution of $S_t$, knowing all past states $\mathbf{S}^{t-1} = (S_0, \dots, S_{t-1})$ and the past realizations $\mathbf{y}^{t-1}$. The parameter $\boldsymbol{\vartheta}$ contains unknown model parameters such as the transition matrix $\boldsymbol{\xi}$, and other parameters indexing the densities $p(y_t|\mathbf{y}^{t-1}, \mathbf{S}^t, \boldsymbol{\vartheta})$ and $p(S_t|\mathbf{S}^{t-1}, \boldsymbol{\vartheta})$.

The basic Markov switching model, formulated in Subsection 10.2.1, results under rather strong assumptions concerning the densities $p(y_t|\mathbf{y}^{t-1}, \mathbf{S}^t, \boldsymbol{\vartheta})$ and $p(S_t|\mathbf{S}^{t-1}, \mathbf{y}^{t-1}, \boldsymbol{\vartheta})$. Under assumption **Y4**, the density $p(y_t|\mathbf{y}^{t-1}, \mathbf{S}^t, \boldsymbol{\vartheta})$ is not allowed to depend on past realizations $\mathbf{y}^{t-1}$ nor on the previous states of $\mathbf{S}^{t-1}$: $p(y_t|\mathbf{y}^{t-1}, \mathbf{S}^t, \boldsymbol{\vartheta}) = p(y_t|\boldsymbol{\theta}_{S_t})$. Assumption **S4** implies that the conditional distribution $p(S_t|\mathbf{S}^{t-1}, \mathbf{y}^{t-1}, \boldsymbol{\vartheta})$ is influenced by the state of $S_{t-1}$, only, and is independent of $t$. More general Markov switching models result by considering more general observation densities $p(y_t|\mathbf{y}^{t-1}, \mathbf{S}^t, \boldsymbol{\vartheta})$ or more general probability models of the hidden Markov chain.

### More General Observation Densities

First of all, the conditional distribution of $Y_t$ given $S_t$ may be allowed to depend on past realizations $\mathbf{y}^{t-1} = (y_1, \dots, y_{t-1})$ of $Y_1, \dots, Y_{t-1}$, leading to assumption

**Y3** Only the present value of $S_t$ influences the density $p(y_t|\mathbf{y}^{t-1}, \mathbf{S}^t, \boldsymbol{\vartheta})$ and dependence on past values of $S_t$ is not allowed:

$$p(y_t|\mathbf{y}^{t-1}, \mathbf{S}^t, \boldsymbol{\vartheta}) = p(y_t|\mathbf{y}^{t-1}, S_t, \boldsymbol{\vartheta}), \tag{10.33}$$

for $t = 1, \dots, T$. Furthermore, $p(y_t|\mathbf{y}^{t-1}, S_t, \boldsymbol{\vartheta})$ is allowed to depend on exogenous variables $\boldsymbol{z}_t$.

The Markov switching regression model discussed in Subsection 10.3.2 results as that special case where $p(y_t|S_t, \boldsymbol{\vartheta})$ is independent of $\mathbf{y}^{t-1}$ while depending on exogenous variables $\boldsymbol{z}_t$. Further examples are the Markov switching autoregressive model suggested by McCulloch and Tsay (1994b), which is discussed in Section 12.2, and the Markov switching dynamic regression model, discussed in Section 12.3.

Assumption **Y3** is not fulfilled by the original Markov switching autoregressive model suggested by Hamilton (1989), which fulfills the more general condition

**Y2** The present value of $S_t$, as well as a limited number of past values $S_{t-1}, \dots, S_{t-p}$ influences the observation density $p(y_t|\mathbf{y}^{t-1}, \mathbf{S}^t, \boldsymbol{\vartheta})$:

$$p(y_t|\mathbf{y}^{t-1}, \mathbf{S}^t, \boldsymbol{\vartheta}) = p(y_t|\mathbf{y}^{t-1}, S_t, \dots, S_{t-p}, \boldsymbol{\vartheta}). \tag{10.34}$$

Assumption **Y2** is still too restrictive for switching ARMA models (Billio and Monfort, 1998) and switching GARCH models (Francq et al., 2001); see also Subsection 12.5.5. These models fulfill only the most general assumption

**Y1** The observation density $p(y_t|\mathbf{y}^{t-1}, \mathbf{S}^t, \boldsymbol{\vartheta})$ depends on $\mathbf{y}^{t-1}$ and all past states of $\mathbf{S}^t$.

### More General Models for the Hidden Markov Chain

The change-point model discussed in Subsection 10.3.3 shows that sensible Markov switching models result, when assumption **S4** is relaxed in the following way.

**S3** $S_t$ is a first-order homogeneous Markov chain with arbitrary transition matrix $\boldsymbol{\xi}$, which need not be irreducible or aperiodic, and starts from an arbitrary distribution $\mathbf{p}_0 = (p_{0,1}, \dots, p_{0,K})$, where

$$p_{0,k} = \Pr(S_0 = k). \tag{10.35}$$

Furthermore it is possible to relax the assumption of homogeneity of the hidden Markov chain $S_t$ as done in Subsection 12.6.1 for models with time-varying transition probabilities:

**S2** $S_t$ is a first-order inhomogeneous Markov chain, with the conditional distribution of $S_t$ being independent of $\mathbf{y}^{t-1}$ and depending on the most recent value $S_{t-1}$ and on some exogenous variables $\boldsymbol{z}_t$:

$$\Pr(S_t = k|\mathbf{S}^{t-1}, \mathbf{y}^{t-1}) = \Pr(S_t = k|S_{t-1}, \boldsymbol{z}_t), \qquad \forall k \in \{1, \dots, K\}.$$

Some Markov switching models with time-varying transition matrices also allow for dependence of the transition matrix on previous realizations $\mathbf{y}^{t-1}$.

**S1** $S_t$ is a first-order Markov chain, and the conditional distribution of $S_t$ depends on the history $\mathbf{y}^{t-1}$ of $Y_t$:

$$\Pr(S_t = k|\mathbf{S}^{t-1}, \mathbf{y}^{t-1}) = \Pr(S_t = k|S_{t-1}, \mathbf{y}^{t-1}), \qquad \forall k \in \{1, \dots, K\},$$

for $t = 1, \dots, T$.

### The Initial Distribution of $S_0$

To complete the model specification for the process $S_t$, the distribution $\mathbf{p}_0$ needs to be specified. Under assumption **S4**, $S_t$ starts from the ergodic probability distribution, hence $\mathbf{p}_0 = \boldsymbol{\eta}$. This assumption could be relaxed by assuming that $S_t$ starts from an arbitrary discrete probability distribution $\mathbf{p}_0$, independent of $\boldsymbol{\xi}$. Note that the resulting Markov chain is no longer stationary.

The initial distribution $\mathbf{p}_0$ could either be a uniform distribution over $\{1, \dots, K\}$ (Frühwirth-Schnatter, 2001b), or could be treated as an unknown parameter to be estimated from the data (Goldfeld and Quandt, 1973; Leroux and Puterman, 1992).

For certain reducible Markov chains it is sensible to assume that the starting value $S_0$ is a known value. Consider, for instance, the transition matrix (10.31), which captures structural breaks at unknown time points when starting with $S_0 = 1$.

# 11

# Statistical Inference for Markov Switching Models

## 11.1 Introduction

For a Markov switching model there are three key problems of statistical inference that must be solved to render them useful for applied time series analysis. First, modeling a time series by a Markov switching model requires some specification of $K$, the number of states of the hidden Markov chain. Second, the state-specific parameters and the transition matrix of the hidden Markov chain are unknown and need to be estimated from the data. Finally, estimates of the hidden Markov chain $\mathbf{S} = (S_1, \ldots, S_T)$ are of interest.

We start in Section 11.2 with state estimation for known state parameters and introduce the filtering and the smoothing problem. In Section 11.3 parameter estimation when the states of the hidden Markov chains are known is discussed. Section 11.4 deals with parameter estimation when the states are unknown and the Markov mixture likelihood function is derived. Practical Bayesian parameter estimation for a known number of states is discussed in Section 11.5, making use of the principle of data augmentation, by introducing the latent Markov chain as missing data, and running a Gibbs sampler. Section 11.6 deals with model specification uncertainty for finite Markov mixture models. Finally, Section 11.7 applies the methods of this chapter to modeling overdispersion and autocorrelation in a time series of count data.

## 11.2 State Estimation for Known Parameters

In this section, statistical inference on the states of the hidden Markov chain $\mathbf{S}$ for fixed state parameters and a known transition matrix, that is, inference on $\boldsymbol{\vartheta} = (\boldsymbol{\theta}_1, \ldots, \boldsymbol{\theta}_K, \boldsymbol{\xi})$, is considered. In many applications statistical inference about the hidden Markov chain is of interest in its own right, because the states or regimes, as economists prefer to call them, may be given some substantive meaning such as being the state of the economy in terms of boom and recession (Hamilton, 1989) or being a climatic state (Zucchini and Guttorp, 1991).

### 11.2.1 Statistical Inference About the States

Inference about the state of $S_t$ at time $t$, given information $\mathbf{y}^\tau = (y_1, \dots, y_\tau)$ about the observable process $Y_t$ for all $t \leq \tau$, is expressed in terms of the probability distribution $\Pr(S_t = l|\mathbf{y}^\tau, \boldsymbol{\vartheta}), l = 1, \dots, K$. The precise meaning of these probabilities depends on the relation between $t$ and $\tau$. The probabilities $\Pr(S_t = l|\mathbf{y}^\tau, \boldsymbol{\vartheta})$ with $t > \tau$ are the predictive state probabilities, with the one-step ahead predictive probabilities $\Pr(S_t = l|\mathbf{y}^{t-1}, \boldsymbol{\vartheta})$ being the most important ones. The probabilities $\Pr(S_t = l|\mathbf{y}^t, \boldsymbol{\vartheta})$ with $t = \tau$ are the filtered state probabilities, and their derivation is also known as the filtering problem. The probabilities $\Pr(S_t = l|\mathbf{y}^\tau, \boldsymbol{\vartheta})$ with $t < \tau$ are the smoothed state probabilities with the full-sample smoothed probabilities $\Pr(S_t = l|\mathbf{y}, \boldsymbol{\vartheta})$, with $\tau = T$ being the most important ones. Related estimation problems known as Kalman filtering and Kalman smoothing occur for state space models with continuous state space (see Chapter 13). Filter and smoother formulae developed for a hidden Markov chain may be regarded as a discrete state space version of Kalman filtering because a hidden Markov chain model may be thought of as being a state space model with discrete state space (Hamilton, 1994a; Cappé et al., 2005).

Many researchers have contributed to the development of efficient methods for filtering and smoothing the unknown states of $S_t$ for specific finite Markov switching models such as hidden Markov chain models for discrete-valued processes (Baum and Petrie, 1966), Markov mixtures of more general distribution families (Baum et al., 1970), Markov switching regression models (Lindgren, 1978), Markov switching autoregressive models (Hamilton, 1989; Chib, 1996), and for the more general switching state space model; see Section 13.5.

There exists a unifying algorithm for filtering and smoothing of all Markov switching models fulfilling at least assumption **Y3**, whereas $S_t$ only needs to be a first-order Markov chain fulfilling at least assumption **S1**; see Subsections 11.2.2 and 11.2.4. Subsection 11.2.5 briefly treats the case of Markov switching models violating assumption **Y3**.

### 11.2.2 Filtered State Probabilities

Speaking in general, the filtering problem for a state space model means statistical inference about the state variable given observations up to $t$. The discreteness of the support of the state variable $S_t$ allows us to derive the complete filtering distribution $\Pr(S_t = l|\mathbf{y}^t, \boldsymbol{\vartheta})$ for all possible realizations $l \in \{1, \dots, K\}$ of $S_t$.

*Algorithm 11.1: Filtering the States* The following steps are carried out recursively for $t = 1, 2, \dots, T$.

(a) One-step ahead prediction of $S_t$:

$$\Pr(S_t = l|\mathbf{y}^{t-1}, \boldsymbol{\vartheta}) = \sum_{k=1}^{K} \xi^\star_{kl}(t-1) \Pr(S_{t-1} = k|\mathbf{y}^{t-1}, \boldsymbol{\vartheta}), \tag{11.1}$$

for $l = 1, \ldots, K$, where $\xi^\star_{kl}(t-1) = \Pr(S_t = l|S_{t-1} = k, \mathbf{y}^{t-1}, \boldsymbol{\vartheta})$ simplifies to the transition probability $\xi_{kl}$ for homogeneous Markov chains.

(b) Filtering for $S_t$:

$$\Pr(S_t = l|\mathbf{y}^t, \boldsymbol{\vartheta}) = \frac{p(y_t|S_t = l, \mathbf{y}^{t-1}, \boldsymbol{\vartheta})\Pr(S_t = l|\mathbf{y}^{t-1}, \boldsymbol{\vartheta})}{p(y_t|\mathbf{y}^{t-1}, \boldsymbol{\vartheta})}, \tag{11.2}$$

where

$$p(y_t|\mathbf{y}^{t-1}, \boldsymbol{\vartheta}) = \sum_{k=1}^{K} p(y_t|S_t = k, \mathbf{y}^{t-1}, \boldsymbol{\vartheta})\Pr(S_t = k|\mathbf{y}^{t-1}, \boldsymbol{\vartheta}). \tag{11.3}$$

At $t = 1$, the filter is started with the initial distribution $\Pr(S_0 = k|\boldsymbol{\xi})$; see Subsection 10.3.4 for various choices of this distribution. Therefore:

$$\Pr(S_1 = l|\mathbf{y}^0, \boldsymbol{\vartheta}) = \sum_{k=1}^{K} \xi^\star_{kl}(0)\Pr(S_0 = k|\boldsymbol{\xi}), \tag{11.4}$$

where $\xi^\star_{kl}(0) = \Pr(S_1 = l|S_0 = k, \boldsymbol{\vartheta})$ simplifies to the transition probability $\xi_{kl}$ for homogeneous Markov chains.

As is typical for any filter, the discrete filter described in *Algorithm 11.1* is an adaptive inference tool. At time $t-1$, the filtered probabilities $\Pr(S_{t-1} = l|\mathbf{y}^{t-1}, \boldsymbol{\vartheta})$ summarize, for a fixed value of $\boldsymbol{\vartheta}$, all information the observations $y_1, \ldots, y_{t-1}$ contain about $S_{t-1}$. To obtain inference about $S_t$ at time $t$ in terms of the "posterior" distribution $\Pr(S_t = l|\mathbf{y}^t, \boldsymbol{\vartheta})$, knowledge of the posterior distribution $\Pr(S_{t-1} = l|\mathbf{y}^{t-1}, \boldsymbol{\vartheta})$ at time $t-1$ and only the *actual* value of $y_t$ are sufficient; see also Figure 11.1. Formula (11.1) delivers the "prior" distribution of $S_t$ given information up to $t-1$, whereas formula (11.2) corrects the prediction through the information contained in the actual observation $y_t$.

Filter at $t-1$: $\Pr(S_{t-1} = l|\mathbf{y}^{t-1}, \boldsymbol{\vartheta})$

$\Downarrow$

Prediction for $t$: $\Pr(S_t = l|\mathbf{y}^{t-1}, \boldsymbol{\vartheta})$
Data at $t$: $y_t$

$\Downarrow$ $\Downarrow$

Filter at $t$: $\Pr(S_t = l|\mathbf{y}^t, \boldsymbol{\vartheta})$

**Fig. 11.1.** Filtering the states

For each run of the discrete filter conditional on a fixed value of $\boldsymbol{\vartheta}$, the one-step ahead predictive densities $p(y_t|\boldsymbol{\vartheta}, \mathbf{y}^{t-1})$ are available from (11.3) as the normalizing constant of the nonnormalized discrete posterior. These densities are useful for computing the likelihood function in Subsection 11.4.1.

### Numerical Stabilization

A straightforward implementation of (11.2) may lead to numerical problems if $y_t$ is very unlikely compared to the prediction from the model. Then $p(y_t|S_t = l, \mathbf{y}^{t-1}, \boldsymbol{\vartheta})$ may be (numerically) 0 for all $l \in \{1, \dots, K\}$ and the filter breaks down as the denominator in (11.2) is 0. To avoid numerical problems, it is usually better to work on the logarithmic scale. To be more specific, compute first $\log p(y_t|S_t = l, \mathbf{y}^{t-1}, \boldsymbol{\vartheta})$ for all $l = 1, \dots, K$ by taking the analytical logarithm of the predictive density. For a basic Markov mixture of Poisson distributions, for instance, this reads:

$$\log p(y_t|S_t = l, \mathbf{y}^{t-1}, \boldsymbol{\vartheta}) = \log f_P(y_t; \mu_l) = y_t \log \mu_l - \log \Gamma(y_t + 1) - \mu_l.$$

Next define

$$L_{t,\max} = \max_{l \in \{1,\dots,K\}} \log p(y_t|S_t = l, \mathbf{y}^{t-1}, \boldsymbol{\vartheta}),$$

and compute for each $l = 1, \dots, K$:

$$p^\star(y_t|S_t = l, \mathbf{y}^{t-1}, \boldsymbol{\vartheta}) = \exp(\log p(y_t|S_t = l, \mathbf{y}^{t-1}, \boldsymbol{\vartheta}) - L_{t,\max}),$$

as well as the following sum,

$$p^\star(y_t|\mathbf{y}^{t-1}, \boldsymbol{\vartheta}) = \sum_{k=1}^{K} p^\star(y_t|S_t = k, \mathbf{y}^{t-1}, \boldsymbol{\vartheta}) \Pr(S_t = k|\mathbf{y}^{t-1}, \boldsymbol{\vartheta}).$$

Then the filtered state probabilities are given by

$$\Pr(S_t = l|\mathbf{y}^t, \boldsymbol{\vartheta}) = \frac{p^\star(y_t|S_t = l, \mathbf{y}^{t-1}, \boldsymbol{\vartheta}) \Pr(S_t = l|\mathbf{y}^{t-1}, \boldsymbol{\vartheta})}{p^\star(y_t|\mathbf{y}^{t-1}, \boldsymbol{\vartheta})}.$$

If $\log p(y_t|\mathbf{y}^{t-1}, \boldsymbol{\vartheta})$ is also needed as part of evaluating the Markov mixture likelihood function $p(\mathbf{y}|\boldsymbol{\vartheta})$, then this numerical value is obtained from:

$$\log p(y_t|\mathbf{y}^{t-1}, \boldsymbol{\vartheta}) = L_{t,\max} + \log p^\star(y_t|\mathbf{y}^{t-1}, \boldsymbol{\vartheta}).$$

### Derivation of the Filtered State Probabilities

It is instructive to verify that the formulae given in *Algorithm 11.1* actually hold. The filter distribution is most conveniently derived by the help of Bayes' theorem, however, other derivations are possible. By Bayes' theorem:

$$\begin{aligned}\Pr(S_t = l|\mathbf{y}^t, \boldsymbol{\vartheta}) &= \Pr(S_t = l|y_t, \mathbf{y}^{t-1}, \boldsymbol{\vartheta}) \\ &= \frac{p(y_t|S_t = l, \mathbf{y}^{t-1}, \boldsymbol{\vartheta}) \Pr(S_t = l|\mathbf{y}^{t-1}, \boldsymbol{\vartheta})}{p(y_t|\mathbf{y}^{t-1}, \boldsymbol{\vartheta})},\end{aligned}$$

where $p(y_t|\mathbf{y}^{t-1}, \boldsymbol{\vartheta})$ is the normalizing constant of the nonnormalized filter probability distribution:

$$p(y_t|\mathbf{y}^{t-1}, \boldsymbol{\vartheta}) = \sum_{k=1}^{K} p(y_t|S_t = k, \mathbf{y}^{t-1}, \boldsymbol{\vartheta})\mathrm{Pr}(S_t = k|\mathbf{y}^{t-1}, \boldsymbol{\vartheta}).$$

$p(y_t|S_t = l, \mathbf{y}^{t-1}, \boldsymbol{\vartheta})$ is the sampling density of $Y_t$ assuming that $S_t$ takes the value $l$, whereas $\mathrm{Pr}(S_t = l|\mathbf{y}^{t-1}, \boldsymbol{\vartheta})$ are the one-step ahead predictive or "prior" probabilities of $S_t$ given information about $Y_s$ for $s \leq t-1$. These one-step ahead predictive probabilities may be derived as marginal probabilities in the following way.

$$\mathrm{Pr}(S_t = l|\mathbf{y}^{t-1}, \boldsymbol{\vartheta}) = \sum_{k=1}^{K} \mathrm{Pr}(S_t = l, S_{t-1} = k|\mathbf{y}^{t-1}, \boldsymbol{\vartheta})$$

$$= \sum_{k=1}^{K} \mathrm{Pr}(S_t = l|S_{t-1} = k, \mathbf{y}^{t-1}, \boldsymbol{\vartheta})\mathrm{Pr}(S_{t-1} = k|\mathbf{y}^{t-1}, \boldsymbol{\vartheta}),$$

$$= \sum_{k=1}^{K} \xi_{kl}^{\star}(t-1)\mathrm{Pr}(S_{t-1} = k|\mathbf{y}^{t-1}, \boldsymbol{\vartheta}).$$

### 11.2.3 Filtering for Special Cases

It is illustrative to apply Algorithm 11.2 to a model where $S_t$ is a hidden i.i.d. process with marginal distribution $\mathrm{Pr}(S_t = k) = \eta_k$, rather than a hidden Markov chain. Because a hidden i.i.d. process is observationally equivalent with a hidden Markov chain with a transition matrix $\xi_{kl} = \eta_l, \forall k, l$ (see Subsection 10.2.6), one-step ahead prediction through (11.1) yields:

$$\mathrm{Pr}(S_t = l|\mathbf{y}^{t-1}, \boldsymbol{\vartheta}) = \eta_l \sum_{k=1}^{K} \mathrm{Pr}(S_{t-1} = k|\mathbf{y}^{t-1}, \boldsymbol{\vartheta}) = \eta_l.$$

As expected, the predictive distribution of an i.i.d. process is independent of the past, and the filter distribution in (11.2) depends on $y_t$ only:

$$\mathrm{Pr}(S_t = l|\mathbf{y}^{t}, \boldsymbol{\vartheta}) = \mathrm{Pr}(S_t = l|y_t, \boldsymbol{\vartheta}).$$

Thus for an i.i.d. process only $y_t$ is informative about $S_t$.

How much the filtered state probabilities depend for a general hidden Markov chain model on the past observations is strongly influenced by the persistence of the transition matrix $\boldsymbol{\xi}$. This should be made more explicit for a homogeneous two-state hidden Markov chain. For a two-state Markov chain, one-step ahead prediction through (11.1) yields the following predictive probability for $S_t = 1$,

$$\Pr(S_t = 1|\mathbf{y}^{t-1}, \boldsymbol{\vartheta})$$
$$= \eta_1 + \lambda\eta_2 \Pr(S_{t-1} = 1|\mathbf{y}^{t-1}, \boldsymbol{\vartheta}) - \lambda\eta_1 \Pr(S_{t-1} = 2|\mathbf{y}^{t-1}, \boldsymbol{\vartheta}),$$

where the relationship between the transition matrix $\boldsymbol{\xi}$ and the ergodic probabilities $\boldsymbol{\eta}$ given in (10.12) have been exploited for $h = 1$ and $\lambda = \xi_{11} - \xi_{12}$ is equal to the second eigenvalue of $\boldsymbol{\xi}$. The predictive probability may be expressed as a weighted mean of the ergodic probability $\eta_1$ and filtered state probability $\Pr(S_{t-1} = 1|\mathbf{y}^{t-1}, \boldsymbol{\vartheta})$ at $t-1$:

$$\Pr(S_t = 1|\mathbf{y}^{t-1}, \boldsymbol{\vartheta}) = (1-\lambda)\eta_1 + \lambda \Pr(S_{t-1} = 1|\mathbf{y}^{t-1}, \boldsymbol{\vartheta}), \qquad (11.5)$$

with the weights being equal to $(1-\lambda)$ and $\lambda$. For chains that are not very persistent ($\lambda$ close to 0), the predictive distribution for $S_t$ will be dominated by the ergodic probabilities, and the filter distribution in (11.2) will barely depend on the past observations $\mathbf{y}^{t-1}$ of the time series. For highly persistent chains ($\lambda$ close to 1), the predictive distribution for $S_t$ will be dominated by the filtered state probability $\Pr(S_{t-1} = k|\mathbf{y}^{t-1}, \boldsymbol{\vartheta})$ obtained for $t-1$, and the filter distribution in (11.2) will strongly depend on all observations $\mathbf{y}^t$.

### 11.2.4 Smoothing the States

The filter described in Subsection 11.2.2 yields the probability distribution of $S_t$ given information $\mathbf{y}^t$ up to $t$. When analyzing a time series in a post-processing manner, probability statements about $S_t$ that incorporate the whole information $\mathbf{y} = (y_1, \ldots, y_T)$ may be preferable. Such probability statements are given by the full-sample smoothed probabilities $\Pr(S_t = l|\mathbf{y}, \boldsymbol{\vartheta})$. Different approaches have been suggested to derive these probabilities.

The smoother suggested in Hamilton (1988, 1989) expresses these probabilities as marginal probabilities from the joint distribution of $S_t$ and $S_T$ conditional on $\mathbf{y}$:

$$\Pr(S_t = l|\mathbf{y}, \boldsymbol{\vartheta}) = \sum_{k=1}^{K} \Pr(S_t = l, S_T = k|\mathbf{y}, \boldsymbol{\vartheta}),$$

where $\Pr(S_t = l, S_T = k|\mathbf{y}, \boldsymbol{\vartheta})$ may be obtained recursively from the filter probabilities. A more efficient smoother, described in *Algorithm 11.2*, operates as a backward smoothing algorithm, running for $t = T, T-1, \ldots, 0$, after having carried out filtering using *Algorithm 11.1*. This smoother has been derived independently by Chib (1996) for Markov switching models and by Shephard (1994) and Kim (1994) for the more general dynamic linear model with switching.

*Algorithm 11.2: Smoothing the States* The smoother is implemented in a forward-filtering-backward-smoothing manner.

(a) First the filter described in *Algorithm 11.1* is carried out to obtain the filtered probabilities $\Pr(S_t = l|\mathbf{y}^t, \boldsymbol{\vartheta}), l = 1, \ldots, K$, for each $t = 1, \ldots, T$.

(b) Then the smoother operates as a backward algorithm, starting from $t = T$ and running backwards in time. The recursions are initialized at $t = T$ with the distribution $\Pr(S_T = l|\mathbf{y}, \boldsymbol{\vartheta})$ which is equal to the filter distribution at $t = T$.
(c) For each $t = T-1, T-2, \ldots, t_0$ the smoothed probability distribution $\Pr(S_t = l|\mathbf{y}, \boldsymbol{\vartheta}), l = 1, \ldots, K$, is derived from

$$\Pr(S_t = l|\mathbf{y}, \boldsymbol{\vartheta}) = \sum_{k=1}^{K} \frac{\xi^\star_{lk}(t)\Pr(S_t = l|\mathbf{y}^t, \boldsymbol{\vartheta})\Pr(S_{t+1} = k|\mathbf{y}, \boldsymbol{\vartheta})}{\displaystyle\sum_{j=1}^{K} \xi^\star_{jk}(t)\Pr(S_t = j|\mathbf{y}^t, \boldsymbol{\vartheta})}, \quad (11.6)$$

where $\xi^\star_{lk}(t) = \Pr(S_{t+1} = k|S_t = l, \mathbf{y}^t, \boldsymbol{\vartheta})$ simplifies to the transition probability $\xi_{lk}$ for homogeneous Markov chains.

To obtain the smoothed probability distribution $\Pr(S_t = l|\mathbf{y}, \boldsymbol{\vartheta})$ for a certain time point $t$ from (11.6), one only needs to know the filtered probabilities $\Pr(S_t = l|\mathbf{y}^t, \boldsymbol{\vartheta})$, and the smoothed probability distribution $\Pr(S_{t+1} = l|\mathbf{y}, \boldsymbol{\vartheta})$ at time point $t + 1$. The recursions of *Algorithm 11.2* can be nicely expressed in terms of matrix operations; see Scott (2002).

The smoother stops at $t_0 = 1$ if $S_0$ is deterministic. For a random initial value $S_0$, the smoother stops at $t_0 = 0$ and delivers updated probability statements $\Pr(S_0 = l|\mathbf{y}, \boldsymbol{\vartheta})$ about the starting value in the light of the observed time series. For this final step, (11.6) reduces to

$$\Pr(S_0 = l|\mathbf{y}, \boldsymbol{\vartheta}) = \sum_{k=1}^{K} \frac{\xi^\star_{lk}(0)\Pr(S_0 = l|\boldsymbol{\xi})}{\displaystyle\sum_{j=1}^{K} \xi^\star_{jk}(0)\Pr(S_0 = j|\boldsymbol{\xi})},$$

with $\Pr(S_0 = l|\boldsymbol{\xi})$ being the initial distribution; see Subsection 10.3.4 for various choices of this distribution. $\xi^\star_{lk}(0) = \Pr(S_1 = k|S_0 = l, \boldsymbol{\vartheta})$ simplifies to the transition probability $\xi_{kl}$ for homogeneous Markov chains.

## Derivation of Forward-Filtering-Backward-Smoothing

To derive the smoother in *Algorithm 11.2*, the full-sample smoothed probabilities $\Pr(S_t = l|\mathbf{y}, \boldsymbol{\vartheta})$ are expressed as marginal probabilities in the following way,

$$\Pr(S_t = l|\mathbf{y}, \boldsymbol{\vartheta}) = \sum_{k=1}^{K} \Pr(S_t = l, S_{t+1} = k|\mathbf{y}, \boldsymbol{\vartheta}) =$$

$$\sum_{k=1}^{K} \Pr(S_t = l|S_{t+1} = k, \mathbf{y}, \boldsymbol{\vartheta})\Pr(S_{t+1} = k|\mathbf{y}, \boldsymbol{\vartheta}). \quad (11.7)$$

Therefore the smoothed probabilities $\Pr(S_t = l|\mathbf{y}, \boldsymbol{\vartheta})$ may be obtained recursively from the smoothed probabilities $\Pr(S_{t+1} = k|\mathbf{y}, \boldsymbol{\vartheta})$ at $t+1$, if the probabilities $\Pr(S_t = l|S_{t+1} = k, \mathbf{y}, \boldsymbol{\vartheta})$ were available. These are the smoothed probabilities for $S_t$ obtained under the assumption that the state of $S_{t+1}$ is known to be equal to $k$. Bayes' theorem could be applied to obtain these probabilities:

$$\begin{aligned}\Pr(S_t = l|S_{t+1} = k, \mathbf{y}, \boldsymbol{\vartheta}) &\propto p(y_{t+1}, \ldots, y_T|S_t = l, S_{t+1} = k, \mathbf{y}^t, \boldsymbol{\vartheta}) \\ &\quad \times \Pr(S_t = l|S_{t+1} = k, \mathbf{y}^t, \boldsymbol{\vartheta}). \end{aligned} \tag{11.8}$$

As $Y_{t+1}, \ldots, Y_T$ are independent of $S_t$, given $S_{t+1}$, the term $p(y_{t+1}, \ldots, y_T|S_t = l, S_{t+1} = k, \mathbf{y}^t, \boldsymbol{\vartheta})$ is independent of $S_t$ and cancels from (11.8):

$$\Pr(S_t = l|S_{t+1} = k, \mathbf{y}, \boldsymbol{\vartheta}) \propto \Pr(S_t = l|S_{t+1} = k, \mathbf{y}^t, \boldsymbol{\vartheta}).$$

To derive $\Pr(S_t = l|S_{t+1} = k, \mathbf{y}^t, \boldsymbol{\vartheta})$, Bayes' theorem is applied once more:

$$\begin{aligned}&\Pr(S_t = l|S_{t+1} = k, \mathbf{y}^t, \boldsymbol{\vartheta}) \\ &\propto \Pr(S_{t+1} = k|S_t = l, \mathbf{y}^t, \boldsymbol{\vartheta})\Pr(S_t = l|\mathbf{y}^t, \boldsymbol{\vartheta}) = \xi^\star_{lk}(t)\Pr(S_t = l|\mathbf{y}^t, \boldsymbol{\vartheta}).\end{aligned} \tag{11.9}$$

The right-hand side of (11.9) has to be normalized to obtain the desired full-sample smoothing probability from (11.7):

$$\Pr(S_t = l|S_{t+1} = k, \mathbf{y}, \boldsymbol{\vartheta}) = \frac{\xi^\star_{lk}(t)\Pr(S_t = l|\mathbf{y}^t, \boldsymbol{\vartheta})}{\sum_{j=1}^K \xi^\star_{jk}(t)\Pr(S_t = j|\mathbf{y}^t, \boldsymbol{\vartheta})}. \tag{11.10}$$

### 11.2.5 Filtering and Smoothing for More General Models

Filtering and smoothing may be extended to Markov switching models violating assumption **Y3**.

#### Models with Longer Memory of the Indicators

For various Markov switching models the conditional density $p(y_t|\mathbf{y}^{t-1}, \mathbf{S}^t, \boldsymbol{\vartheta})$ depends on $p \geq 1$ past values of $S_t$, formulated as condition **Y2**. Examples are certain Markov switching AR models (Hamilton, 1989; see Subsection 12.2.2), and the switching ARCH model (Hamilton and Susmel, 1994; see Subsection 12.5.3). A complication for models of this kind is that no simple filter for deriving the probabilities $\Pr(S_t = k|\mathbf{y}^t, \boldsymbol{\vartheta})$ is available. A filter, however, may be derived for the multivariate state vector $\mathbf{S}_t$, defined as

$$\mathbf{S}_t = \begin{pmatrix} S_t \\ \vdots \\ S_{t-p} \end{pmatrix}.$$

We refer to Hamilton (1994a) for more details on how to implement the multivariate filter. As discussed in Chapter 12, the need to implement a multivariate filter may often be avoided by introducing the hidden Markov model into an alternative parameterization of the model.

#### Models with Infinite Memory of the Indicators

For some Markov switching models the conditional density $p(y_t|\mathbf{y}^{t-1}, \mathbf{S}^t, \boldsymbol{\vartheta})$ depends on the whole history of $S_t$, formulated as condition **Y1**. Examples are the switching ARMA model (Billio and Monfort, 1998; Billio et al., 1999) and the switching GARCH model (Francq et al., 2001); see also Subsection 12.5.5. No exact finite-dimensional filter is available for this type of model, an exception to this rule being a GARCH model based on the $t_\nu$-distribution, where only the degree of freedom changes according to a Markov switching model (Dueker, 1997). Various approximate filters have been derived (Lam, 1990; Kim, 1994; Gray, 1996; Klaasen, 2002) based on writing such models as switching state space models; see Chapter 13.

## 11.3 Parameter Estimation for Known States

Assume that a single realization $\mathbf{y} = (y_1, \ldots, y_T)$ from a finite Markov mixture distribution with $K$ states has been observed. In this section, attention is shifted toward estimating the state-specific parameters $\boldsymbol{\theta}_1, \ldots, \boldsymbol{\theta}_K$ and the transition matrix $\boldsymbol{\xi}$ of a hidden Markov chain under the assumption that the states $\mathbf{S} = (S_0, S_1, \ldots, S_T)$ of the hidden Markov chain are observed as well. Although this is rarely the situation in practice, it is an inference problem that occurs as part of the more general problem of joint parameter and state estimation for hidden Markov models discussed in Section 11.4. Once we condition on the state of the hidden Markov chain $S_t$, there is a close relationship between the present inference problem and parameter estimation for a standard finite mixture model considered in Section 2.3. Indeed, it turns out that estimation of the state-specific parameters $\boldsymbol{\theta}_1, \ldots, \boldsymbol{\theta}_K$ is essentially the same problem as in Section 2.3; the only new challenge is estimating the transition matrix $\boldsymbol{\xi}$ of the hidden Markov chain.

### 11.3.1 The Complete-Data Likelihood Function

Let $\boldsymbol{\vartheta}$ be a vector containing all different elements in the state-specific parameters $\boldsymbol{\theta}_1, \ldots, \boldsymbol{\theta}_K$, and in the transition matrix $\boldsymbol{\xi}$. The complete-data likelihood function is equal to the joint sampling distribution $p(\mathbf{y}, \mathbf{S}|\boldsymbol{\vartheta})$ for the complete-data $(\mathbf{S}, \mathbf{y})$ given $\boldsymbol{\vartheta}$, which is then regarded as a function of $\boldsymbol{\vartheta}$ in order to estimate the unknown parameters $\boldsymbol{\vartheta}$. The joint sampling distribution $p(\mathbf{y}, \mathbf{S}|\boldsymbol{\vartheta})$ is immediately available from the model definitions, given in Subsection 10.3.4:

$$p(\mathbf{S}, \mathbf{y}|\boldsymbol{\vartheta}) = p(\mathbf{y}|\mathbf{S}, \boldsymbol{\vartheta})p(\mathbf{S}|\boldsymbol{\vartheta}), \tag{11.11}$$

where $p(\mathbf{y}|\mathbf{S}, \boldsymbol{\vartheta})$ is the density of the sampling distribution of the time series $\mathbf{y}$ given $\mathbf{S}$ and $p(\mathbf{S}|\boldsymbol{\vartheta})$ is the density of the sampling distribution on the state process $\mathbf{S}$. Under model assumptions **S3** or **S4**, the density $p(\mathbf{S}|\boldsymbol{\vartheta})$ depends on $\boldsymbol{\xi}$ only, and reads:

$$\begin{aligned} p(\mathbf{S}|\boldsymbol{\xi}) &= \prod_{t=1}^{T} p(S_t|S_{t-1}, \boldsymbol{\xi})p(S_0|\boldsymbol{\xi}) = \prod_{t=1}^{T} \xi_{S_{t-1},S_t} p(S_0|\boldsymbol{\xi}) \\ &= p(S_0|\boldsymbol{\xi}) \prod_{j=1}^{K} \prod_{k=1}^{K} \xi_{jk}^{N_{jk}(\mathbf{S})}, \end{aligned} \tag{11.12}$$

where $N_{jk}(\mathbf{S})$ counts the numbers of transitions from $j$ to $k$:

$$N_{jk}(\mathbf{S}) = \# \left\{ S_{t-1} = j, S_t = k \right\}, \qquad \forall j,k \in \{1, \ldots, K\}. \tag{11.13}$$

For the derivation of this density it is not necessary to assume that $S_t$ starts from the ergodic distribution.

Under assumption **Y3** or **Y4**, the complete-data likelihood $p(\mathbf{y}|\mathbf{S}, \boldsymbol{\vartheta})$ is the product of the one-step ahead predictive densities $p(y_t|S_t, \mathbf{y}^{t-1}, \boldsymbol{\vartheta})$ given $S_t$:

$$p(\mathbf{y}|\mathbf{S}, \boldsymbol{\vartheta}) = \prod_{t=1}^{T} p(y_t|S_t, \mathbf{y}^{t-1}, \boldsymbol{\vartheta}).$$

For many Markov mixture models, in particular for the basic Markov switching model, the complete-data likelihood function, when regarded as a function of $\boldsymbol{\vartheta}$, has a rather convenient structure that highly facilitates parameter estimation. It is usually the product of $K+1$ factors, where each of the first $K$ factors depends on a single state-specific parameter $\boldsymbol{\theta}_k$, whereas the last factor depends only on the transition matrix $\boldsymbol{\xi}$:

$$p(\mathbf{y}, \mathbf{S}|\boldsymbol{\vartheta}) = \prod_{k=1}^{K} \left( \prod_{t:S_t=k} p(y_t|\boldsymbol{\theta}_k, \mathbf{y}^{t-1}) \right) \prod_{j=1}^{K} \prod_{k=1}^{K} \xi_{jk}^{N_{jk}(\mathbf{S})} p(S_0|\boldsymbol{\xi}). \tag{11.14}$$

For a Markov mixture of Poisson distributions with unknown state-specific means $\mu_1, \ldots, \mu_K$ and unknown transition matrix $\boldsymbol{\xi}$, for instance, the complete-data likelihood $p(\mathbf{y}, \mathbf{S}|\boldsymbol{\vartheta})$, after dropping factors independent of $\boldsymbol{\vartheta} = (\mu_1, \ldots, \mu_K, \boldsymbol{\xi})$, reads:

$$p(\mathbf{y}, \mathbf{S}|\mu_1, \ldots, \mu_K, \boldsymbol{\xi}) \propto \prod_{k=1}^{K} \mu_k^{N_k(\mathbf{S})\overline{y}_k(\mathbf{S})} e^{-N_k(\mathbf{S})\mu_k} p(\mathbf{S}|\boldsymbol{\xi}),$$

where $N_k(\mathbf{S}) = \#\{S_t = k\}$ and $\overline{y}_k(\mathbf{S})$ is the mean of all observations, where $S_t = k$. Apart from the factor $p(\mathbf{S}|\boldsymbol{\xi})$, this is exactly the same complete-data likelihood function as occurred for finite mixtures of Poisson distributions in Subsection 2.3.3.

### 11.3.2 Complete-Data Bayesian Parameter Estimation

As in Subsection 2.3.3, the complete-data likelihood $p(\mathbf{y}, \mathbf{S}|\boldsymbol{\vartheta})$, regarded as a function of $\boldsymbol{\vartheta}$, is combined through Bayes' theorem with a prior distribution $p(\boldsymbol{\vartheta})$ on the parameter $\boldsymbol{\vartheta}$, to obtain the complete-data posterior distribution $p(\boldsymbol{\vartheta}|\mathbf{y}, \mathbf{S})$:

$$p(\boldsymbol{\vartheta}|\mathbf{y}, \mathbf{S}) \propto p(\mathbf{y}, \mathbf{S}|\boldsymbol{\vartheta})p(\boldsymbol{\vartheta}).$$

If the prior were improper, $p(\boldsymbol{\vartheta}) \propto$ constant, the complete-data posterior distribution $p(\boldsymbol{\vartheta}|\mathbf{y}, \mathbf{S})$ would factor in the same ways as the complete-data likelihood $p(\mathbf{y}, \mathbf{S}|\boldsymbol{\vartheta})$ factors in (11.14). This convenient structure is preserved by choosing the prior

$$p(\boldsymbol{\vartheta}) = \prod_{k=1}^{K} p(\boldsymbol{\theta}_k)p(\boldsymbol{\xi}),$$

in which case the posterior reads:

$$p(\boldsymbol{\vartheta}|\mathbf{S}, \mathbf{y}) = \prod_{k=1}^{K} p(\boldsymbol{\theta}_k|\mathbf{y}, \mathbf{S})p(\boldsymbol{\xi}|\mathbf{S}),$$

where

$$p(\boldsymbol{\theta}_k|\mathbf{y}, \mathbf{S}) \propto \prod_{t:S_t=k} p(y_t|\boldsymbol{\theta}_k, \mathbf{y}^{t-1})p(\boldsymbol{\theta}_k),$$

$$p(\boldsymbol{\xi}|\mathbf{S}) \propto p(S_0|\boldsymbol{\xi}) \prod_{j=1}^{K}\prod_{k=1}^{K} \xi_{jk}^{N_{jk}(\mathbf{S})} p(\boldsymbol{\xi}).$$

### 11.3.3 Complete-Data Bayesian Estimation of the Transition Matrix

Each row $\boldsymbol{\xi}_{j\cdot}$ of the transition matrix $\boldsymbol{\xi}$ of the hidden Markov chain is an unknown probability distribution that has to be estimated from the data. This inference problem is discussed, for instance, in Chib (1996). We consider in this subsection complete-data estimation, when a realization $\mathbf{S}$ of the hidden chain is available; the more general case where both $\mathbf{S}$ and $\boldsymbol{\xi}$ are unknown is treated in Section 11.4. Furthermore we assume that $S_0$ is independent of $\boldsymbol{\xi}$.

The complete-data likelihood $p(\mathbf{S}|\boldsymbol{\xi})$, given in (11.12), factors in the following way,

$$p(\mathbf{S}|\boldsymbol{\xi}) = \prod_{j=1}^{K}\prod_{k=1}^{K} \xi_{jk}^{N_{jk}(\mathbf{S})}, \tag{11.15}$$

where $N_{jk}(\mathbf{S})$ counts the number of transitions from $j$ to $k$; see also (11.13). The right-hand side of (11.15), when regarded as a function of the unknown transition matrix, is the product of $K$ factors that are proportional to densities from the Dirichlet distribution. Each factor involves just one row $\boldsymbol{\xi}_{j\cdot}$ of the transition matrix. This convenient structure of the complete-data posterior distribution of $\boldsymbol{\xi}$ is preserved under a prior distribution with similar structure. In particular, if it is assumed that the rows of $\boldsymbol{\xi}$ are independent a priori, each following a Dirichlet distribution,

$$\boldsymbol{\xi}_{j\cdot} \sim \mathcal{D}\left(e_{j1}, \ldots, e_{jK}\right), \qquad j = 1, \ldots, K,$$

then the rows $\boldsymbol{\xi}_{j\cdot}$ of $\boldsymbol{\xi}$ remain independent also a posteriori, each following again a Dirichlet distribution:

$$\boldsymbol{\xi}_{j\cdot}|\mathbf{S} \sim \mathcal{D}\left(e_{j1} + N_{j1}(\mathbf{S}), \ldots, e_{jK} + N_{jK}(\mathbf{S})\right), \qquad j = 1, \ldots, K,$$

where $N_{jk}(\mathbf{S})$ counts the number of transitions from $j$ to $k$. For $K = 2$, the transition matrix $\boldsymbol{\xi}$ consists only of two distinct elements, for instance, the persistence probabilities $\xi_{11}$ and $\xi_{22}$. Given $\mathbf{S}$, $\xi_{11}$ and $\xi_{22}$ are independent, each following a Beta distribution:

$$\begin{aligned}\xi_{11} &\sim \mathcal{B}\left(e_{11} + N_{11}(\mathbf{S}), e_{12} + N_{12}(\mathbf{S})\right),\\ \xi_{22} &\sim \mathcal{B}\left(e_{21} + N_{21}(\mathbf{S}), e_{22} + N_{22}(\mathbf{S})\right).\end{aligned}$$

Chib (1996) realized that these densities generalize to the Dirichlet distribution for $K > 2$.

## 11.4 Parameter Estimation When the States are Unknown

In this section we consider estimation of the unknown parameters $\boldsymbol{\vartheta}$ of a Markov switching model for given time series data $\mathbf{y} = (y_1, \ldots, y_T)$ for models where the observation density $p(y_t|y^{t-1}, S_t, \boldsymbol{\vartheta})$ is allowed to depend on past values of $Y_t$ (assumption **Y3** or **Y4**) and $S_t$ fulfills assumption **S3** or **S4**. $\boldsymbol{\vartheta}$ summarizes all unknown parameters of the observation density $p(y_t|\mathbf{y}^{t-1}, S_t, \boldsymbol{\vartheta})$, such as location and scale parameters for the normal distributions, and all distinct parameters appearing in the unknown transition matrix $\boldsymbol{\xi}$.

The most commonly applied estimation methods are maximum likelihood estimation and Bayesian estimation based on the Markov mixture likelihood function $p(\mathbf{y}|\boldsymbol{\vartheta})$.

### 11.4.1 The Markov Mixture Likelihood Function

The derivation of the likelihood function is much more involved for finite Markov mixture models than it has been for standard finite mixture models.

Consider the following representation of the Markov mixture likelihood function as a mixture over the complete-data likelihood (11.11),

$$\begin{aligned} p(\mathbf{y}|\boldsymbol{\vartheta}) &= \sum_{\mathbf{S}\in\mathcal{S}_K} p(\mathbf{y}|\mathbf{S},\boldsymbol{\theta}_1,\ldots,\boldsymbol{\theta}_K)p(\mathbf{S}|\boldsymbol{\xi}) \\ &= \sum_{\mathbf{S}\in\mathcal{S}_K} \prod_{t=p+1}^{T} p(y_t|\mathbf{y}^{t-1},\boldsymbol{\theta}_{S_t}) \prod_{j=1}^{K}\prod_{k=1}^{K} \xi_{jk}^{N_{jk}(\mathbf{S})} p(S_0|\boldsymbol{\xi}), \end{aligned} \tag{11.16}$$

where $\mathcal{S}_K = \{1,\ldots K\}^{T+1}$ is the space of all possible realizations of $\mathbf{S}$, whereas $N_{jk}(\mathbf{S})$ counts the number of transitions from $j$ to $k$; see also (11.13). The sum in (11.16) is over $K^{T+1}$ elements and quickly becomes infeasible for practical evaluation of the Markov mixture likelihood function.

An early solution to the problem of computing the likelihood function of a Markov mixture model is provided by the forward–backward recursions of Baum et al. (1970) which were designed for reconstructing a hidden Markov chain from a discrete signal observed with noise. It took quite a while before it became common knowledge that the work of Baum et al. (1970) is easily extended to more general Markov mixture models such as Markov switching regression models (Lindgren, 1978; Cosslett and Lee, 1985), Markov switching autoregressive models (Hamilton, 1989), and discrete Markov mixture models (Le et al., 1992).

To understand some of the difficulties associated with deriving the likelihood function of a finite Markov mixture model, it is illuminating to study once more the likelihood function of the standard finite mixture regression model introduced by Quandt (1972):

$$p(\mathbf{y}|\boldsymbol{\vartheta}) = \prod_{t=1}^{T} \left( \sum_{k=1}^{K} p(y_t|\boldsymbol{\vartheta}, S_t = k)\Pr(S_t = k|\boldsymbol{\vartheta}) \right). \tag{11.17}$$

Because this model corresponds to a model with $S_t$ being independent over time, $\Pr(S_t = k|\boldsymbol{\vartheta}) = \eta_k$ and the standard mixture likelihood results.

Goldfeld and Quandt (1973), although allowing for Markov dependence between the states of $S_t$, maximize the same objective function, with $\Pr(S_t = k|\boldsymbol{\vartheta})$ being computed as the prior probability $p_{t,k}$ of being in state $k$ at time $t$, when $S_0$ is distributed according to the probability distribution $\mathbf{p}_0$ defined in (10.35). The vector $\mathbf{p}_t = (p_{t,1} \cdots p_{t,K})'$ containing these probabilities is given by

$$\mathbf{p}_t = (\boldsymbol{\xi}')^t \mathbf{p}_0.$$

Although this method yields consistent parameter estimates, (11.17) is not the likelihood function of this model, as noted by Cosslett and Lee (1985). In order to obtain the correct Markov mixture likelihood function $p(\mathbf{y}|\boldsymbol{\vartheta})$, the prior probabilities $\Pr(S_t = k|\boldsymbol{\vartheta})$ have to be substituted by the one-step ahead predictive probabilities $\Pr(S_t = k|\mathbf{y}^{t-1}, \boldsymbol{\vartheta})$:

$$p(\mathbf{y}|\boldsymbol{\vartheta}) = \prod_{t=1}^{T} \left( \sum_{k=1}^{K} p(y_t|S_t = k, \boldsymbol{\vartheta}) \Pr(S_t = k|\mathbf{y}^{t-1}, \boldsymbol{\vartheta}) \right). \tag{11.18}$$

The easiest way to derive this likelihood function is through the product of the one-step ahead predictive densities, a method commonly applied for time series models; see, for instance, Hamilton (1994b):

$$p(\mathbf{y}|\boldsymbol{\vartheta}) = \prod_{t=1}^{T} p(y_t|\mathbf{y}^{t-1}, \boldsymbol{\vartheta}).$$

For a Markov mixture model the predictive density $p(y_t|\mathbf{y}^{t-1}, \boldsymbol{\vartheta})$ is, for each $t = 1, \ldots, T$, the normalizing constant of the filtered probability distribution $\Pr(S_t = k|\mathbf{y}^t, \boldsymbol{\vartheta})$ (see again (11.3)), and therefore directly available as a by-product of running *Algorithm 11.1* conditional on $\boldsymbol{\vartheta}$. Hence (11.18) follows immediately. Computational complexity is of order $\mathcal{O}(TK^2)$.

An alternative evaluation of the Markov mixture likelihood function is based on the following likelihood recursion (Cosslett and Lee, 1985; MacDonald and Zucchini, 1997; Rydén et al., 1998),

$$\begin{aligned} p(\mathbf{y}^t, S_t = k|\boldsymbol{\vartheta}) &= \sum_{j=1}^{K} p(y_t, \mathbf{y}^{t-1}, S_t = k, S_{t-1} = j|\boldsymbol{\vartheta}) \\ &= p(y_t|\mathbf{y}^{t-1}, S_t = k, \boldsymbol{\vartheta}) \sum_{j=1}^{K} \xi_{jk} p(\mathbf{y}^{t-1}, S_{t-1} = j|\boldsymbol{\vartheta}). \end{aligned}$$

This leads to the following iteration for $t = 1, \ldots, T$, starting with $\mathbf{l}_0 = \mathbf{p}_0$,

$$\mathbf{l}_t = \text{Diag}\big( p(y_t|\mathbf{y}^{t-1}, \boldsymbol{\theta}_1) \cdots p(y_t|\mathbf{y}^{t-1}, \boldsymbol{\theta}_K) \big) \boldsymbol{\xi}' \mathbf{l}_{t-1}, \tag{11.19}$$

which finally yields the Markov mixture likelihood function:

$$p(\mathbf{y}|\boldsymbol{\vartheta}) = \sum_{k=1}^{K} p(\mathbf{y}^T, S_T = k|\boldsymbol{\vartheta}) = \mathbf{1}_{1\times K} \mathbf{l}_T,$$

where $\mathbf{1}_{1\times K}$ is a row vector of ones. Computational complexity is of order $\mathcal{O}(TK^2)$, however, recursion (11.19) is sensitive to numerical underflow; see Scott (2002) for how it may be stabilized.

A certain complication arises with the definition of the Markov mixture likelihood function for Markov switching models, which do not obey condition **Y4**, as the predictive density $p(y_t|\mathbf{y}^{t-1}, \boldsymbol{\vartheta}, S_t)$ depends on the past values $y_{t-1}, \ldots, y_{t-p}$. As for the standard AR model (Hamilton, 1994b), there exist two likelihood functions, namely the conditional and the unconditional Markov mixture likelihood function. The conditional Markov mixture likelihood function is obtained by conditioning on the first $p$ observations, therefore filtering and evaluating of the likelihood function starts at $t = p + 1$:

$$p(\mathbf{y}|\boldsymbol{\vartheta}) \approx p(y_{p+1}, \ldots, y_T|\boldsymbol{\vartheta}, \mathbf{y}^p) = \prod_{t=p+1}^{T} p(y_t|\mathbf{y}^{t-1}, \boldsymbol{\vartheta}).$$

Most papers use this likelihood function for parameter estimation, a notable exception being Albert and Chib (1993) who consider the unconditional Markov mixture likelihood for a Markov switching autoregressive model with state-independent autoregressive parameters.

Finally, like the mixture likelihood function studied in Subsection 2.4.2, the Markov mixture likelihood function usually has $K!$ different, but equivalent modes corresponding to all different ways of labeling the states of the hidden Markov chain. Consider again the representation of the Markov mixture likelihood function as a mixture over the complete-data likelihood as in (11.16). By relabeling the states of $S_t$ according to an arbitrary permutation $\rho_s(\cdot)$, one finds that the parameter $\boldsymbol{\vartheta}^\star = (\boldsymbol{\theta}_{\rho_s(1)}, \ldots, \boldsymbol{\theta}_{\rho_s(K)}, \boldsymbol{\xi}^{\rho_s})$, where the elements of $\boldsymbol{\xi}^{\rho_s}$ have been defined in (10.26), yields the same Markov mixture likelihood as $\boldsymbol{\vartheta}$ for all possible observations $\mathbf{y}$:

$$p(\mathbf{y}|\boldsymbol{\vartheta}^\star) = \sum_{\mathbf{S} \in \mathcal{S}_K} \prod_{t=p+1}^{T} p(y_t|\mathbf{y}^{t-1}, \boldsymbol{\theta}_{\rho_s(S_t)})$$
$$\times \prod_{j=1}^{K} \prod_{k=1}^{K} \xi_{\rho_s(j),\rho_s(k)}^{N_{\rho_s(j),\rho_s(k)}(\mathbf{S})} p(\rho_s(S_0)|\boldsymbol{\xi}^{\rho_s}) = p(\mathbf{y}|\boldsymbol{\vartheta}).$$

### 11.4.2 Maximum Likelihood Estimation

Maximization of the likelihood function may be carried out numerically (Hamilton, 1989; Rydén et al., 1998), through the EM algorithm (Baum et al., 1970; Hamilton, 1990), or may be considered as an optimization problem under $K$ linear constraints, as the rows of $\boldsymbol{\xi}$ have to sum up to 1 (Levison et al., 1983).

An excellent review of the asymptotic properties of ML estimators for hidden Markov models appears in Cappé et al. (2005, Chapter 12); we mention here only some seminal papers. Consistency and asymptotic normality of the maximum likelihood for hidden Markov chain models, where $Y_t$ is a discrete signal, are established in Baum and Petrie (1966) and, with a more careful discussion of identifiability issues, in Petrie (1969). Lindgren (1978) established similar results for Markov switching regression models. Leroux (1992b) proved consistency of the ML estimator for rather general hidden Markov models under certain regularity condition. Bickel et al. (1998) verified that the ML estimator is asymptotically normal with the observed information matrix being a consistent estimator of the expected information matrix under certain regularity conditions.

For Markov mixtures of normal distributions with switching variances the likelihood function is unbounded (Lindgren, 1978; Hamilton, 1988) as it is for finite mixtures of normal distributions with heterogeneous variances; see again Subsection 6.1.2. This is easily seen by choosing the unknown parameter is such a way that under state 1 $\mu_{t,1} = y_t$ for an arbitrary observation $y_t$. Then as $\sigma_1$ approaches 0, the Markov mixture likelihood function goes to infinity. This problem may be avoided by bounding $\sigma_k^2$ away from 0 (Lindgren, 1978) or choosing proper priors (Hamilton, 1988).

### 11.4.3 Bayesian Estimation

Within the Bayesian framework, the Markov mixture likelihood $p(\mathbf{y}|\boldsymbol{\vartheta})$ is combined with a prior $p(\boldsymbol{\vartheta})$ using Bayes' theorem:

$$p(\boldsymbol{\vartheta}|\mathbf{y}) \propto p(\mathbf{y}|\boldsymbol{\vartheta})p(\boldsymbol{\vartheta}). \tag{11.20}$$

As for the standard finite mixture model, various comments on this posterior are in order. First, the prior has to be proper to obtain a proper posterior distribution. Second, for the Markov mixture likelihood appearing in (11.20) no conjugate analysis is possible, meaning that whatever (proper) prior $p(\boldsymbol{\vartheta})$ one chooses, the posterior density obtained from (11.20) does not belong to any tractable distribution family. Finally, through Bayes' theorem (11.20), the posterior distribution of a Markov switching model inherits the invariance properties of the Markov mixture likelihood function discussed above.

For practical Bayesian estimation of Markov switching models it is nowadays common to use MCMC methods for Bayesian estimation, following the pioneering work by Robert et al. (1993), Albert and Chib (1993), and McCulloch and Tsay (1994b). For the most part, practical MCMC estimation makes use of the principle of data augmentation by choosing the latent Markov chain as missing data. The gain of introducing the latent Markov chain as missing data is evident: with regard to the complete-data posterior distribution $p(\boldsymbol{\vartheta}|\mathbf{S}, \mathbf{y})$ we are often back in the conjugate setting of Section 11.3. Then it is rather straightforward to sample from the posterior (11.20) using Gibbs sampling. Practical Bayesian estimation is described in detail in Section 11.5.

### 11.4.4 Alternative Estimation Methods

Alternative estimation methods for Markov mixture models include the method of moments (Quandt and Ramsey, 1978), a classification likelihood approach based on maximizing the complete-data likelihood function $p(\mathbf{y}|\mathbf{S}, \boldsymbol{\vartheta})p(\mathbf{S}|\boldsymbol{\vartheta})$ jointly with respect to $(\mathbf{S}, \boldsymbol{\vartheta})$ (Sclove, 1983) and dynamic programming (Kim, 1993c). Recently, Francq and Zakoian (1999) considered the so-called linear-representation-based estimator, where the parameters of the model are estimated from minimizing the weighted mean-squared prediction error obtained from the linear ARMA representation of powers of $Y_t$.

## 11.5 Bayesian Parameter Estimation with Known Number of States

### 11.5.1 Choosing the Prior for the Parameters of a Markov Mixture Model

A standard prior assumption is that the state-specific parameters $\boldsymbol{\theta}_1, \ldots, \boldsymbol{\theta}_K$ are a priori independent of the transition matrix $\boldsymbol{\xi}$:

$$p(\boldsymbol{\vartheta}) = p(\boldsymbol{\xi})p(\boldsymbol{\theta}_1, \ldots, \boldsymbol{\theta}_K). \tag{11.21}$$

It is convenient to choose a prior $p(\boldsymbol{\theta}_1, \ldots, \boldsymbol{\theta}_K|\boldsymbol{\delta})$ that is conjugate with respect to the complete-data likelihood $p(\mathbf{y}|\mathbf{S}, \boldsymbol{\theta}_1, \ldots, \boldsymbol{\theta}_K)$, especially if Bayesian parameter estimation is based on data augmentation. This usually leads to choosing the following prior,

$$p(\boldsymbol{\theta}_1, \ldots, \boldsymbol{\theta}_K|\boldsymbol{\delta}) = \prod_{k=1}^{K} p(\boldsymbol{\theta}_k|\boldsymbol{\delta}),$$

where $p(\boldsymbol{\theta}_k|\boldsymbol{\delta})$ is the density of some distribution family with some fixed hyperparameter $\boldsymbol{\delta}$. This prior is obviously invariant to relabeling the states of the hidden Markov chain. When working with this prior, choosing improper priors $p(\boldsymbol{\theta}_k|\boldsymbol{\delta})$ should be avoided, as it may be shown in a similar way as was done for finite mixture models in Subsection 3.2.2 that the Markov mixture posterior density $p(\boldsymbol{\theta}_1, \ldots, \boldsymbol{\theta}_K, \boldsymbol{\xi}|\mathbf{y})$ is improper in this case.

As discussed for finite mixture models in Subsection 3.2.4, it is also possible for Markov switching models to treat $\boldsymbol{\delta}$ as an unknown hyperparameter with prior $p(\boldsymbol{\delta})$:

$$p(\boldsymbol{\theta}_1, \ldots, \boldsymbol{\theta}_K, \boldsymbol{\delta}) = p(\boldsymbol{\delta}) \prod_{k=1}^{K} p(\boldsymbol{\theta}_k|\boldsymbol{\delta}).$$

To choose the prior of the transition matrix $\boldsymbol{\xi}$, we recall that each row of this matrix is a discrete probability distribution. As in Subsection 11.3, where we treated the case of an observed state process $S_t$, rather than a hidden one, we assume that the rows of $\boldsymbol{\xi}$ are independent a priori, each following a Dirichlet distribution:

$$\boldsymbol{\xi}_{k\cdot} \sim \mathcal{D}\left(e_{k1}, \ldots, e_{kK}\right), \qquad k = 1, \ldots, K. \tag{11.22}$$

To obtain a prior that is invariant to relabeling, Frühwirth-Schnatter (2001b) suggested choosing $e_{kk} = e^P$ and $e_{kk'} = e^T$, if $k \neq k'$. By choosing $e^P > e^T$, the Markov switching model is bounded away from a finite mixture model.

### 11.5.2 Some Properties of the Posterior Distribution of a Markov Switching Model

The posterior density $p(\boldsymbol{\vartheta}|\mathbf{y})$ of a Markov switching model has similar properties as the posterior density of a finite mixture model, discussed in depth in Section 3.3. Most important, the posterior is invariant to relabeling the states of the hidden Markov chain if the same is true for the prior:

$$p(\boldsymbol{\theta}_1, \ldots, \boldsymbol{\theta}_K, \boldsymbol{\xi}|\mathbf{y}) = p(\boldsymbol{\theta}_{\rho_s(1)}, \ldots, \boldsymbol{\theta}_{\rho_s(K)}, \boldsymbol{\xi}^{\rho_s}|\mathbf{y}), \tag{11.23}$$

where $\boldsymbol{\xi}^{\rho_s}$ is related to $\boldsymbol{\xi}$ by permuting the rows and the columns; see also definition (10.26). This causes multimodality of the posterior of a Markov switching model, and potential label switching when sampling from this density. Furthermore many functionals derived from the posterior, which are seemingly state dependent, are actually state invariant.

#### Marginal Distributions of State-Specific Parameters

As in Section 3.3, it is possible to show that the marginal posterior $p(\boldsymbol{\theta}_k|\mathbf{y})$ of any state-specific parameter is actually state invariant:

$$p(\boldsymbol{\theta}_k|\mathbf{y}) = p(\boldsymbol{\theta}_{k'}|\mathbf{y}), \qquad \forall k, k' = 1, \ldots, K, k \neq k'.$$

It could be proven that the bivariate marginal distribution of the parameters $\boldsymbol{\theta}_k$ and $\boldsymbol{\theta}_{k'}$ of different states $k$ and $k'$ is the same for all possible pairs $k$ and $k'$, and therefore symmetric:

$$p(\boldsymbol{\theta}_k, \boldsymbol{\theta}_{k'}|\mathbf{y}) = p(\boldsymbol{\theta}_{\rho_s(k)}, \boldsymbol{\theta}_{\rho_s(k')}|\mathbf{y}) = p(\boldsymbol{\theta}_{k'}, \boldsymbol{\theta}_k|\mathbf{y}), \quad \forall k, k' = 1, \ldots, K, k \neq k'.$$

The marginal posterior of each persistence probability is state invariant,

$$p(\xi_{kk}|\mathbf{y}) = p(\xi_{k',k'}|\mathbf{y}), \qquad \forall k, k' = 1, \ldots, K, k \neq k',$$

as is the posterior of each transition probability:

$$p(\xi_{kk'}|\mathbf{y}) = p(\xi_{\rho_s(k),\rho_s(k')}|\mathbf{y}), \quad \forall k, k' = 1, \ldots, K, k \neq k'.$$

#### The Joint Posterior of the Hidden Markov Chain

For any arbitrary permutation $\rho_s(\cdot)$ of $\{1, \ldots, K\}$, the conditional posterior $p(\mathbf{S}|\boldsymbol{\vartheta}, \mathbf{y})$ is invariant to relabeling the states,

$$p(S_1, \ldots, S_T|\boldsymbol{\vartheta}, \mathbf{y}) = p(\rho_s(S_1), \ldots, \rho_s(S_T)|\boldsymbol{\vartheta}^\star, \mathbf{y}),$$

with $\boldsymbol{\vartheta}^\star = (\boldsymbol{\theta}_{\rho_s(1)}, \ldots, \boldsymbol{\theta}_{\rho_s(K)}, \boldsymbol{\xi}^{\rho_s})$, where the elements of $\boldsymbol{\xi}^{\rho_s}$ have been defined in (10.26). The proof of this invariance property is exactly the same as the proof for finite mixture models appearing in Subsection 3.3.4. It follows that any two sequences $\mathbf{S}$ and $\mathbf{S}'$ that imply the same partition of the observed time series obtain the same posterior probability.

### The Marginal Posterior of a Single State

When a Markov mixture model is fitted to a time series with the aim of performing joint posterior parameter and state estimation, one would hope to infer how likely the event $\{S_t = k\}$ is in light of the data. A natural candidate appears to be the posterior probability $\Pr(S_t = k|\mathbf{y})$. However, as already observed for finite mixture models in Subsection 3.3.4, this marginal posterior probability does not contain any information if it is derived from the Markov mixture posterior distribution but is equal to $1/K$ regardless of the observed time series:

$$\Pr(S_t = k|\mathbf{y}) = \frac{1}{K}. \tag{11.24}$$

The proof is the same as for finite mixture models.

## 11.5.3 Parameter Estimation Through Data Augmentation and MCMC

Early papers realizing the importance of Gibbs sampling for the Bayesian analysis of Markov switching models are Robert et al. (1993), Albert and Chib (1993), and McCulloch and Tsay (1994b). Improvements and modifications suggested later on concern multi-move sampling of the hidden Markov chain $\mathbf{S}$ rather than single-move sampling (Shephard, 1994; Chib, 1996), improving MCMC performance through reparameterization (Robert and Titterington, 1998), and dealing with the label switching problem (Frühwirth-Schnatter, 2001b).

### MCMC Estimation for a Markov Mixture of Poisson Distributions

Consider a Markov mixture of Poisson distributions, defined as $Y_t \sim \mathcal{P}(\mu_{S_t})$, with $S_t$ being a $K$-state hidden Markov chain with transition matrix $\boldsymbol{\xi}$. Assume that a realization $\mathbf{y} = (y_1, \ldots, y_T)$ of the process $Y_t$ is available, which should be used for inference on the parameter $\boldsymbol{\vartheta} = (\mu_1, \ldots, \mu_K, \boldsymbol{\xi})$ and the hidden states $\mathbf{S} = (S_0, S_1, \ldots, S_T)$. For a Bayesian analysis, choose the conditional conjugate Gamma priors $\mu_k \sim \mathcal{G}(a_0, b_0)$ for $\mu_k$, and assume prior independence of the means.

If for each observation $y_t, t = 1, \ldots, T$, the state $S_t$ of the hidden Markov chain is introduced as missing datum, data augmentation and MCMC estimation may be carried out for Markov mixtures of Poisson distribution in a similar way as discussed for finite mixtures of Poisson distributions in Subsection 3.5.2, by iterating between state estimation for known parameters and parameter estimation for known states. Part of this algorithm is closely related to the one developed in *Algorithm 3.3*; part of it is essentially different.

We start with parameter estimation conditional on knowing the states of $\mathbf{S} = (S_0, S_1, \ldots, S_T)$. The complete-data posterior is given by the results of Subsection 11.3.2 as

$$p(\mu_1, \ldots, \mu_K, \boldsymbol{\xi}|\mathbf{S}, \mathbf{y}) = \prod_{k=1}^{K} p(\mu_k|\mathbf{y}, \mathbf{S}) p(\boldsymbol{\xi}|\mathbf{S}),$$

where $p(\boldsymbol{\xi}|\mathbf{S}) \propto p(\mathbf{S}|\boldsymbol{\xi})p(\boldsymbol{\xi})$. Each of the first $K$ factors depends only on $\mu_k$, whereas the last factor depends only on $\boldsymbol{\xi}$. Under the $\mathcal{G}(a_0, b_0)$-prior, the posterior distribution of $p(\mu_k|\mathbf{S}, \mathbf{y})$ is a $\mathcal{G}(a_k(\mathbf{S}), b_k(\mathbf{S}))$-distribution, where:

$$a_k(\mathbf{S}) = a_0 + N_k(\mathbf{S})\overline{y}_k(\mathbf{S}), \qquad b_k(\mathbf{S}) = b_0 + N_k(\mathbf{S}), \tag{11.25}$$

and $N_k(\mathbf{S}) = \#\{S_t = k\}$ is the number of observations in state $k$. Sampling of the parameters $\mu_1, \ldots, \mu_K$ is straightforward, as $\mu_1, \ldots, \mu_K$ are conditionally independent, and we simply have to draw $\mu_k$ from the $\mathcal{G}(a_k(\mathbf{S}), b_k(\mathbf{S}))$ posterior.

It is important to realize that this is exactly the same sampling step as for a standard finite mixture of Poisson distributions, because for state parameter estimation only the number of observations in state $k$ are relevant but not the number of transitions. The number of transitions are relevant only for sampling of the transition matrix $\boldsymbol{\xi}$, as discussed in Subsection 11.5.5.

The second step in this MCMC procedure, namely sampling $\mathbf{S}$, is much more involved for a Markov mixture than is the corresponding step for a standard finite mixture model. The main reason is that $\mathbf{S}$ is a path of a stochastic process with dependence among successive values of $S_t$, even if the parameters are known, whereas for a finite mixture model the indicators are independent conditional on $\mathbf{y}$ and $\boldsymbol{\vartheta}$. Efficient methods for sampling a path of $\mathbf{S}$ are developed in Subsection 11.5.6.

### MCMC Estimation for General Markov Switching Models

Bayesian estimation of a general Markov mixture model through data augmentation estimates the augmented parameter $(\mathbf{S}, \boldsymbol{\vartheta})$ by sampling from the posterior distribution $p(\mathbf{S}, \boldsymbol{\vartheta}|\mathbf{y})$,

$$p(\mathbf{S}, \boldsymbol{\vartheta}|\mathbf{y}) \propto p(\mathbf{y}|\mathbf{S}, \boldsymbol{\vartheta}) p(\mathbf{S}|\boldsymbol{\vartheta}) p(\boldsymbol{\vartheta}). \tag{11.26}$$

Sampling from the posterior (11.26) is most commonly carried out by the Markov chain Monte Carlo sampling scheme described below in *Algorithm 11.3*, where $\boldsymbol{\vartheta}$ is sampled conditional on knowing $\mathbf{S}$, whereas $\mathbf{S}$ is sampled conditional on knowing $\boldsymbol{\vartheta}$.

*Algorithm 11.3: Unconstrained MCMC for a Markov Switching Model* Start with some state process $\mathbf{S}^{(0)}$ and repeat the following steps for $m = 1, \ldots, M_0, \ldots, M + M_0$.

(a) Parameter simulation conditional on the states $\mathbf{S}^{(m-1)}$:
   (a1) Sample the transition matrix $\boldsymbol{\xi}$ from the complete-data posterior distribution $p(\boldsymbol{\xi}|\mathbf{S}^{(m-1)})$.

(a2) Sample the model parameter $\boldsymbol{\theta}_1, \ldots, \boldsymbol{\theta}_K$ from the complete-data posterior $p(\boldsymbol{\theta}_1, \ldots, \boldsymbol{\theta}_K|\mathbf{y}, \mathbf{S}^{(m-1)})$.

Store the actual values of all parameters as $\boldsymbol{\vartheta}^{(m)} = (\boldsymbol{\theta}_1^{(m)}, \ldots, \boldsymbol{\theta}_K^{(m)}, \boldsymbol{\xi}^{(m)})$.

(b) State simulation conditional on knowing $\boldsymbol{\vartheta}^{(m)}$ by sampling a path $\mathbf{S}$ of the hidden Markov chain from the conditional posterior $p(\mathbf{S}|\boldsymbol{\vartheta}^{(m)}, \mathbf{y})$ using *Algorithm 11.5*, discussed below.

Store the actual values of all states as $\mathbf{S}^{(m)}$, increase $m$ by one, and return to step (a). Finally, the first $M_0$ draws are discarded.

More details on how to sample the unknown transition matrix $\boldsymbol{\xi}$ in step (a1) are provided in Subsection 11.5.5. For many important Markov switching models, sampling of the state parameters in step (a2) is straightforward, as the relevant posterior densities are of closed form. For other models, further blocking of the elements of $\boldsymbol{\theta}_k$ will lead to closed-form conditional densities. In cases where this is not feasible, a Metropolis–Hastings algorithm may be implemented, as demonstrated in Subsection 12.5.4 for a switching ARCH model.

The precise form of the posterior $p(\boldsymbol{\theta}_1, \ldots, \boldsymbol{\theta}_K|\mathbf{S}, \mathbf{y})$ appearing in step (a2) crucially depends on the chosen parametric family, but also on the prior $p(\boldsymbol{\theta}_1, \ldots, \boldsymbol{\theta}_K)$. It is important to emphasize that the structure of the conditional posterior $p(\boldsymbol{\theta}_1, \ldots, \boldsymbol{\theta}_K|\mathbf{S}, \mathbf{y})$ where $\mathbf{S}$ is a Markov chain, is usually the same as for a standard finite mixture model, where $\mathbf{S}$ is an i.i.d. sequence, because the complete-data posterior $p(\boldsymbol{\theta}_1, \ldots, \boldsymbol{\theta}_K|\mathbf{S}, \mathbf{y})$ factors in the following way,

$$p(\boldsymbol{\vartheta}|\mathbf{S}, \mathbf{y}) = p(\boldsymbol{\xi}|\mathbf{S}) \prod_{k=1}^{K} p(\boldsymbol{\theta}_k|\mathbf{y}, \mathbf{S}),$$

under the conditionally conjugate prior discussed in Subsection 11.5.1. As only the observations $y_t$ in a certain state $k$ are relevant to draw the state-specific parameters, step (a2) reduces to exactly the same procedure as would be applied for a finite mixture model. Thus many results on MCMC estimation of finite mixtures derived in earlier chapters, in particular MCMC estimation of finite mixtures of regression models discussed in Subsection 8.3.4, are useful for Markov mixture models.

Step (b), however, is much more involved for a Markov switching model than the corresponding step of sampling the hidden indicator $\mathbf{S}$ in a standard mixture model, the main reason being that $\mathbf{S}$ is a path of a stochastic process with dependence among successive values of $S_t$, whereas for a mixture model the indicators are independent conditional on $\mathbf{y}$ and $\boldsymbol{\vartheta}$. Subsection 11.5.6 is devoted to a detailed discussion of various algorithms for implementing step (b).

### 11.5.4 Permutation MCMC Sampling

As for finite mixture models, the behavior of the sampler described in *Algorithm 11.3* is somewhat unpredictable, and may be trapped at one modal region of the Markov mixture posterior distribution or may jump occasionally between different modal regions causing label switching. In most cases the sampler does not explore the full Markov mixture posterior distribution which matters in particular when estimating marginal densities. A simple but efficient solution to obtain a sampler that explores the full Markov mixture posterior distribution is to extend random permutation MCMC sampling (Frühwirth-Schnatter, 2001b), discussed for finite mixture models in *Algorithm 3.5*, to Markov switching models.

*Algorithm 11.4: Random Permutation MCMC Sampling for a Markov Switching Model* Start as described in *Algorithm 11.3.*

(a) and (b) are the same steps as in *Algorithm 11.3.*

(c) Conclude each draw by selecting randomly one of the $K!$ possible permutations $\rho_s(1), \ldots, \rho_s(K)$ of the current labeling. This permutation is applied to $\boldsymbol{\xi}^{(m)}$, the state-specific parameters $\boldsymbol{\theta}_1^{(m)}, \ldots, \boldsymbol{\theta}_K^{(m)}$, and the states $\mathbf{S}^{(m)}$:
 - (c1) Each element $\xi_{jk}^{(m)}$ of the simulated transition matrix is substituted by $\xi_{\rho_s(j),\rho_s(k)}^{(m)}$ for $j, k = 1, \ldots, K$.
 - (c2) The state-specific parameter $\boldsymbol{\theta}_k^{(m)}$ is substituted by $\boldsymbol{\theta}_{\rho_s(k)}^{(m)}$ for $k = 1, \ldots, K$.
 - (c3) The states $S_t^{(m)}$ are substituted by $\rho_s(S_t^{(m)})$ for $t = 0, \ldots, T$.

### 11.5.5 Sampling the Unknown Transition Matrix

The precise algorithm for sampling the transition matrix $\boldsymbol{\xi}$ from the conditional posterior $p(\boldsymbol{\xi}|\mathbf{S})$ for a given trajectory $\mathbf{S}$ of the hidden Markov chain depends on the assumptions made concerning the distribution $\mathbf{p}_0$ of the initial value $S_0$; see Subsection 10.3.4 for various choices of $\mathbf{p}_0$. As in Subsection 11.5.1, it is assumed that the rows of $\boldsymbol{\xi}$ are independent a priori, each following a Dirichlet distribution, $\boldsymbol{\xi}_{j\cdot} \sim \mathcal{D}(e_{j1}, \ldots, e_{jK})$, $j = 1, \ldots, K$.

#### Gibbs Sampling for Nonstationary Markov Chains

If the initial distribution $\mathbf{p}_0$ is independent of $\boldsymbol{\xi}$, Gibbs sampling from the conditional posterior $p(\boldsymbol{\xi}|\mathbf{S})$ is feasible, as shown by Chib (1995), as the problem reduces to the classical Bayesian inference problem discussed in Subsection 11.3.3. The rows $\boldsymbol{\xi}_{j\cdot}$ of $\boldsymbol{\xi}$ are independent a posteriori, and are drawn from the following Dirichlet distribution,

$$\boldsymbol{\xi}_{j\cdot} \sim \mathcal{D}(e_{j1} + N_{j1}(\mathbf{S}), \ldots, e_{jK} + N_{jK}(\mathbf{S})), \qquad j = 1, \ldots, K, \quad (11.27)$$

where $N_{jk}(\mathbf{S}) = \#\{S_{t-1} = j, S_t = k\}$ counts the numbers of transitions from $j$ to $k$ for the actual draw of $\mathbf{S}$.

Interestingly, the usefulness of the Dirichlet distribution for handling the case $K > 2$ has not been recognized by all authors implementing the Gibbs sampler for more than two states. For $K = 3$, for instance, Kim and Nelson (1998, 1999), sample for each row the two distinct transition probabilities in a two-step procedure by first sampling the persistence probability $\xi_{jj}$ from the appropriate marginal distribution, namely

$$\xi_{jj} \sim \mathcal{B}\left(e_{jj} + N_{jj}(\mathbf{S}), \sum_{k \neq j} e_{jk} + N_{jk}(\mathbf{S})\right).$$

Then $\tilde{\xi}_{jk} = \Pr(S_t = k|S_{t-1} = j, S_t \neq j)$, with $k \neq j$, is sampled from another Beta distribution. A generalization of this somewhat circumstantial scheme to arbitrary $K > 3$ is discussed in Krolzig (1997, p.156).

### Metropolis–Hastings Algorithm for Stationary Markov Chains

For a stationary Markov chain, the initial distribution is equal to the ergodic distribution, $\mathbf{p}_0 = \boldsymbol{\eta}$, and depends on the transition matrix $\boldsymbol{\xi}$ through formula (10.9), discussed in Subsection 10.2.2. Gibbs sampling from the conditional posterior $p(\boldsymbol{\xi}|\mathbf{S})$ is no longer feasible, as due to the presence of $\mathbf{p}_0$ the rows of $\boldsymbol{\xi}$ are no longer independent a posteriori. The joint posterior $p(\boldsymbol{\xi}|\mathbf{S})$ of all rows takes the form:

$$p(\boldsymbol{\xi}|\mathbf{S}) \propto \prod_{j=1}^{K} g_j(\boldsymbol{\xi}_{j\cdot})\eta_{S_0},$$

where $g_j(\boldsymbol{\xi}_{j\cdot})$ is equal to the density of the Dirichlet distribution given in (11.27). To sample the rows of $\boldsymbol{\xi}$ one could use a Metropolis–Hastings algorithm, with $g_j(\boldsymbol{\xi}_{j\cdot})$ being the proposal density for the $j$th row. Starting from the old transition matrix $\boldsymbol{\xi}^{old}$, a new transition matrix $\boldsymbol{\xi}^{new}$ is proposed by drawing some or all rows of $\boldsymbol{\xi}^{old}$ from the Dirichlet proposal density $g_j(\boldsymbol{\xi}_{j\cdot})$, given in (11.27). The acceptance rate for the Metropolis–Hastings algorithm is equal to $\min(1, A)$, where

$$A = \frac{p(\boldsymbol{\xi}^{new}|\mathbf{S}) \prod_{j=1}^{K} g_j(\boldsymbol{\xi}_{j\cdot}^{old})}{p(\boldsymbol{\xi}^{old}|\mathbf{S}) \prod_{j=1}^{K} g_j(\boldsymbol{\xi}_{j\cdot}^{new})} = \frac{\eta_{S_0}^{new}}{\eta_{S_0}^{old}}.$$

Therefore $\boldsymbol{\xi}^{new}$ is accepted, if

$$U \leq \frac{\eta_{S_0}^{new}}{\eta_{S_0}^{old}},$$

where $U$ is a random draw from $\mathcal{U}[0, 1]$.

### 11.5.6 Sampling Posterior Paths of the Hidden Markov Chain

In this section, sampling a path of the hidden Markov chain $\mathbf{S} = (S_0, \dots, S_T)$ from the conditional posterior distribution $p(\mathbf{S}|\mathbf{y}, \boldsymbol{\vartheta})$ is discussed in full detail; see also Scott (2002) for a recent review.

#### Multi-Move Sampling

Many of the early papers on Gibbs sampling for Markov switching models (Robert et al., 1993; Albert and Chib, 1993; McCulloch and Tsay, 1994b) use single-move Gibbs sampling, meaning that the state of $S_t$ is sampled conditional on all other states of the hidden Markov chain. A more efficient way to sample $\mathbf{S}$, however, is multi-move sampling, meaning joint sampling of the states of the whole path $\mathbf{S}$ from the conditional posterior distribution $p(\mathbf{S}|\mathbf{y}, \boldsymbol{\vartheta})$, discussed earlier in Section 11.2. Multi-move sampling of the hidden Markov chain has been suggested independently for the Markov switching autoregressive model (Chib, 1996; Krolzig, 1997) and switching state space models (Carter and Kohn, 1994; Shephard, 1994). There exists a unifying algorithm for any Markov switching model that fulfills at least assumption **Y3**, whereas $S_t$ only needs to fulfill the most general assumption **S1**.

Multi-move sampling is based on writing the joint posterior $p(\mathbf{S}|\mathbf{y}, \boldsymbol{\vartheta})$ as

$$p(\mathbf{S}|\mathbf{y}, \boldsymbol{\vartheta}) = \left[\prod_{t=0}^{T-1} p(S_t|S_{t+1}, \dots, S_T, \boldsymbol{\vartheta}, \mathbf{y})\right] p(S_T|\mathbf{y}, \boldsymbol{\vartheta}).$$

$p(S_T|\mathbf{y}, \boldsymbol{\vartheta})$ is the filtered probability distribution at $t = T$. The conditional probability distribution $p(S_t|S_{t+1}, \dots, S_T, \boldsymbol{\vartheta}, \mathbf{y})$ has been found in Subsection 11.2.4 to be proportional to:

$$p(S_t|S_{t+1}, \dots, S_T, \boldsymbol{\vartheta}, \mathbf{y}) \propto \xi^\star_{S_t,S_{t+1}}(t) p(S_t|\mathbf{y}^t, \boldsymbol{\vartheta}),$$

where $p(S_t|\mathbf{y}^t, \boldsymbol{\vartheta})$ is the filtered probability distribution at $t$ and $\xi^\star_{S_t,S_{t+1}}(t) = p(S_{t+1}|S_t, \boldsymbol{\vartheta}, \mathbf{y}^t)$ reduces to $\xi_{S_t,S_{t+1}}$ for a homogeneous Markov chain $S_t$.

It is therefore rather clear how to carry out a forward-filtering-backward-sampling algorithm which is summarized in *Algorithm 11.5*. A similar forward-filtering-backward-sampling algorithm has been derived for Gibbs sampling for normal Gaussian state space models (Frühwirth-Schnatter, 1994; Carter and Kohn, 1994; De Jong and Shephard, 1995); see also Subsection 13.5.2.

*Algorithm 11.5: Multi-Move Sampling of the States* To sample a path $\mathbf{S}^{(m)}$ of the hidden Markov chain, while holding $\boldsymbol{\vartheta}$ fixed, perform the following steps.

(a) Run the filter described in *Algorithm 11.1* conditional on $\boldsymbol{\vartheta}$ and store the filtered state probability distribution $\Pr(S_t = j|\mathbf{y}^t, \boldsymbol{\vartheta}), j = 1, \dots, K$, for $t = 1, \dots, T$.

(b) Sample $S_T^{(m)}$ from the filtered state probability distribution $\Pr(S_T = j|\mathbf{y}^T, \boldsymbol{\vartheta})$.
(c) For $t = T-1, T-2, \ldots, 0$ sample $S_t^{(m)}$ from the conditional distribution $\Pr(S_t = j|S_{t+1}^{(m)}, \mathbf{y}^t, \boldsymbol{\vartheta})$ given by

$$\Pr(S_t = j|S_{t+1}^{(m)}, \mathbf{y}^t, \boldsymbol{\vartheta}) = \frac{\xi_{j,l_m}^\star(t)\Pr(S_t = j|\mathbf{y}^t, \boldsymbol{\vartheta})}{\sum_{k=1}^{K} \xi_{k,l_m}^\star(t)\Pr(S_t = k|\mathbf{y}^t, \boldsymbol{\vartheta})},$$

where $\xi_{j,l_m}^\star(t) = \Pr(S_{t+1} = l_m|S_t = j, \boldsymbol{\vartheta}, \mathbf{y}^t)$ reduces to $\xi_{j,l_m}$ for a homogeneous Markov chain $S_t$ and $l_m$ is equal to the state of $S_{t+1}^{(m)}$.

For each $t$, $\Pr(S_t = j|S_{t+1}^{(m)}, \mathbf{y}^t, \boldsymbol{\vartheta})$ needs to be evaluated for all $j = 1, \ldots, K$. This requires knowledge of the filtered state probabilities $\Pr(S_t = j|\mathbf{y}^t, \boldsymbol{\vartheta})$, which were stored in step (a) of the algorithm.

### Single-Move Sampling

An alternative method of sampling a path $\mathbf{S}$ from the conditional posterior distribution $p(\mathbf{S}|\mathbf{y}, \boldsymbol{\vartheta})$ is single-move sampling of the state of $S_t$ conditional on all other states of the hidden Markov chain from the conditional posterior probability distribution $\Pr(S_t = j|\mathbf{S}_{-t}, \mathbf{y}, \boldsymbol{\vartheta})$ as discussed in Robert et al. (1993), Albert and Chib (1993), and McCulloch and Tsay (1994b). Here $\mathbf{S}_{-t}$ is a commonly used abbreviation to denote the whole path of $\mathbf{S}$ but the element $S_t$.

A computational advantage of single-move sampling over multi-move sampling is that running a time-consuming filter is avoided, because the latter method is $\mathcal{O}(TK)$, whereas the former was $\mathcal{O}(TK^2)$. Furthermore, single-move sampling is also possible for any Markov switching models, even if the predictive density $p(y_t|\mathbf{S}^t, \mathbf{y}^{t-1}, \boldsymbol{\vartheta})$ depends on the whole history of $S_t$ (assumption **Y1**). A theoretical disadvantage of single-move sampling, however, is that the autocovariance function of any complete-data sufficient statistics drawn is equal to the autocovariance function of the same statistics under multi-move sampling plus a penalty term that increases with the posterior covariance of the hidden states (Scott, 2002). Hence the single-move sampler may be poorly mixing for highly correlated hidden Markov chains (Albert and Chib, 1993; McCulloch and Tsay, 1994b).

The precise form of the conditional posterior probability distribution $\Pr(S_t = j|\mathbf{S}_{-t}, \mathbf{y}, \boldsymbol{\vartheta})$ is obtained from Bayes' theorem:

$$p(S_t|\mathbf{S}_{-t}, \mathbf{y}, \boldsymbol{\vartheta}) \propto p(\mathbf{y}|\mathbf{S}, \boldsymbol{\vartheta})p(\mathbf{S}|\boldsymbol{\xi}) \propto$$
$$\prod_{t=1}^{T} p(y_t|\mathbf{y}^{t-1}, \mathbf{S}^t, \boldsymbol{\vartheta}) \prod_{t=1}^{T} p(S_t|S_{t-1}, \boldsymbol{\vartheta}, \mathbf{y}^{t-1})p(S_0|\boldsymbol{\xi}).$$

After dropping from the right-hand side all factors that are independent of $S_t$, we obtain for $1 \leq t \leq T-1$:

$$\Pr(S_t = j|\mathbf{S}_{-t}, \mathbf{y}, \boldsymbol{\vartheta}) \propto p(y_t|\mathbf{y}^{t-1}, \mathbf{S}^{t-1}, S_t = j, \boldsymbol{\vartheta})\xi^\star_{S_{t-1},j}(t-1)\xi^\star_{j,S_{t+1}}(t).$$

Obvious modifications for $t = 0$ and $t = T$ are:

$$\begin{aligned}
&\Pr(S_0 = j|S_1, \ldots, S_T, \mathbf{y}, \boldsymbol{\vartheta}) \propto \xi^\star_{j,S_1}(0)\Pr(S_0 = j|\boldsymbol{\xi}), \\
&\Pr(S_T = j|\mathbf{S}^{T-1}, \mathbf{y}, \boldsymbol{\vartheta}) \propto p(y_T|\mathbf{y}^{T-1}, \mathbf{S}^{T-1}, S_T = j, \boldsymbol{\vartheta})\xi^\star_{S_{T-1},j}(T-1).
\end{aligned}$$

The states of $S_t$ are then sampled for each $t = 0, 1, \ldots, T$ from these discrete probability distributions with the most recent draw for all other states of $\mathbf{S}$ being used in the conditioning argument. In any of these formulae

$$\xi^\star_{S_{t-1},S_t}(t-1) = p(S_t|S_{t-1}, \boldsymbol{\vartheta}, \mathbf{y}^{t-1})$$

reduces to $\xi_{S_{t-1},S_t}$ for a homogeneous Markov chain.

#### Blocked Sampling

To improve mixing, Albert and Chib (1993) mention the possibility of using blocked sampling by updating $b$ consecutive elements $\{S_t, \ldots, S_{t+b-1}\}$ of $\mathbf{S}$ at a time by sampling from the appropriate conditional density

$$p(S_t, \ldots, S_{t+b-1}|S_0, \ldots, S_{t-1}, S_{t+b}, \ldots, S_T, \mathbf{y}, \boldsymbol{\vartheta}),$$

however, without providing any details. Slightly more details on sampling from this posterior were provided by McCulloch and Tsay (1994b) for $K = 2$, who consider evaluating the posterior over all $2^b$ possible realizations of $\{S_t, \ldots, S_{t+b-1}\}$.

A more efficient sampling method for sampling $\{S_t, \ldots, S_{t+b-1}\}$ that easily is applied to models with more than two states is to apply the multi-move sampler as in *Algorithm 11.5* with deterministic starting value $\tilde{S}_0 = S_{t-1}$ and $b$ observations $\tilde{\mathbf{y}} = \{y_t, \ldots, y_{t+b-1}\}$. One could follow Shephard (1994) in choosing the blocks.

#### Marginal Single-Move Sampling

Chen and Liu (1996) demonstrated that it is possible to marginalize over the parameters of a Markov mixture model, if the observation density comes from the exponential family as in (1.11) and the complete-data likelihood is combined with a conditionally conjugate prior $p(\boldsymbol{\theta}_k)$.

This allows state estimation without parameter estimation. A related approach has been discussed in detail in Section 3.4 for a standard finite mixture model, and is easily modified to deal with Markov switching models. The application of the Gibbs sampling algorithm described in Subsection 3.4.1 or the

Metropolis–Hastings algorithm described in Subsection 3.4.2 to sample from $p(\mathbf{S}|\mathbf{y})$ only requires marginalizing the prior $p(\mathbf{S}|\boldsymbol{\xi})$ of a hidden Markov chain with respect to the prior on $\boldsymbol{\xi}$:

$$p(\mathbf{S}) = \int p(\mathbf{S}|\boldsymbol{\xi})p(\boldsymbol{\xi})d\boldsymbol{\xi};$$

all other steps are similar. A closed form of this prior is available under the Dirichlet prior (11.22), if $p(S_0)$ is independent of the transition matrix $\boldsymbol{\xi}$:

$$p(\mathbf{S}) \propto p(S_0) \prod_{j=1}^{K} \frac{\prod_{k=1}^{K} \Gamma(N_{jk}(\mathbf{S}) + e_{jk})}{\Gamma(\sum_{k=1}^{K}(N_{jk}(\mathbf{S}) + e_{jk}))},$$

where $N_{jk}(\mathbf{S}) = \#\{S_{t-1} = j, S_t = k\}$ and $e_{jk}$ are the prior parameters.

### 11.5.7 Other Sampling-Based Approaches

Reparameterization techniques that were discussed for finite mixture models in Subsection 6.2.6 have been extended to Markov mixtures by Robert and Titterington (1998). For Markov mixtures of normal distributions the component parameters are parameterized as in (6.26) with the weights being equal to the ergodic distribution of the hidden Markov chain $S_t$, and the standard parameterization is kept for the transition matrix. Reparameterization is also discussed for Markov mixtures of Poisson distributions.

Particle filter methods which were briefly mentioned in Subsection 6.2.7 have been applied to Markov mixture models by Fearnhead and Clifford (2003).

### 11.5.8 Bayesian Inference Using Posterior Draws

As discussed in Section 3.7 for finite mixture models, posterior draws may be used for statistical inference, provided that a sufficiently large number $M_0$ of draws are discarded. Convergence diagnostics for MCMC methods with applications to hidden Markov models are discussed in Robert et al. (1999). Because most of the issues and methods discussed in Section 3.7 are adapted to Markov mixture models in a straightforward manner, we focus only on the question of how to obtain point estimates for the hidden Markov chain.

Point estimators of the hidden Markov chain $\mathbf{S}$ may be obtained by minimizing a loss function in a similar way as was done in Subsection 7.1.7 for finite mixture models for the purpose of clustering observations. There exists indeed a close connection between clustering observations into groups and segmenting time series observations into states or regimes, and one could effectively use the same loss functions as in Subsection 7.1.7.

Minimizing the 0/1 loss function requires finding the most likely sequence $\mathbf{S}$ from the joint posterior $p(\mathbf{S}|\mathbf{y})$. Such a sequence could be found by applying a Viterbi algorithm, which is a special form of dynamic programming, utilizing the Markovian structure of $S_t$. Details are found in MacDonald and Zucchini (1997, p.64ff), Chib (1996, p.85ff), Bhar and Hamori (2004, Section 1.9), and Cappé et al. (2005, Section 5.1.2).

Minimizing the loss function based on the misclassification risk leads to finding that sequence $\mathbf{S} = (S_1, \ldots, S_T)$, where each $S_t$ maximizes the smoothed probability distribution $\Pr(S_t^{\mathcal{L}} = l|\mathbf{y})$ obtained from an identified Markov mixture model. Finally, state estimation could be based on minimizing the loss function based on the posterior similarity matrix, as this may lead to smoother paths of $S_t$ than individual state estimation.

## 11.6 Statistical Inference Under Model Specification Uncertainty

The most commonly occurring model selection problem for finite Markov mixture models is selecting the number of states of the hidden Markov chain, although a couple of other model selection problems, such as order selection for Markov switching autoregressive model (see Subsection 12.2.5), and order selection for switching ARCH models (see Subsection 12.5.4), are also relevant. Another important model selection problem is choosing the appropriate structure for $\mathbf{S}$; see Subsection 11.6.1.

### 11.6.1 Diagnosing Markov Switching Models

As discussed in Subsections 4.3.3 and 4.3.4 for finite mixtures, diagnosing the goodness-of-fit for Markov switching models may be based on studying the posterior distribution of certain moments implied by the Markov mixture and studying the predictive posterior distribution of certain statistics (Scott, 2002; MacKay Altman, 2004). A particularly useful statistic for assessing goodness-of-fit for a Markov switching model is the predictive posterior distribution of the implied autocorrelation function $\rho_{Y_t}(h|\boldsymbol{\vartheta})$ in comparison to the observed autocorrelation function; see also Subsection 11.7.3.

### 11.6.2 Likelihood-Based Methods

Testing a Markov switching model against homogeneity through the likelihood ratio statistics has to cope with similar problems as for finite mixture models; see again Subsection 4.4.1. The limiting distribution of the LR statistic has been approximated using empirical process theory (Hansen, 1992) and bootstrapping technique (Rydén et al., 1998).

AIC and BIC as defined in Subsection 4.4.2 have been used by several authors to deal with model selection problems for Markov mixtures (Sclove, 1983; Leroux and Puterman, 1992; Wang and Puterman, 1999):

$$\begin{aligned}\mathrm{AIC}_K &= -2\log p(\mathbf{y}|\hat{\boldsymbol{\vartheta}}_K, \mathcal{M}_K) + 2d_K,\\ \mathrm{BIC}_K &= -2\log p(\mathbf{y}|\hat{\boldsymbol{\vartheta}}_K, \mathcal{M}_K) + \log(N)d_K,\end{aligned}$$

where $\log p(\mathbf{y}|\hat{\boldsymbol{\vartheta}}_K, \mathcal{M}_K)$ is the log of the Markov mixture likelihood function, evaluated at the ML estimator, and $d_K$ is the number of distinct parameters in $\boldsymbol{\vartheta}_K$. For a $K$-state Markov mixture model with state-specific parameters $\boldsymbol{\theta}_k$ where all elements are different among all states, and unconstrained transition matrix, $d_K$ is equal to $K\dim(\boldsymbol{\theta}_k) + K(K-1)$.

### 11.6.3 Marginal Likelihoods for Markov Switching Models

The definition of the marginal likelihoods, given for finite mixture models in Section 5.3, is practically of the same form for Markov switching models:

$$p(\mathbf{y}|\mathcal{M}_K) = \int_{\mathcal{S}_K \times \Theta_K} p(\mathbf{y}|\mathbf{S}, \boldsymbol{\vartheta}_K)p(\mathbf{S}|\boldsymbol{\xi})p(\boldsymbol{\vartheta}_K)d(\mathbf{S}, \boldsymbol{\vartheta}_K), \tag{11.28}$$

because only the prior of $\mathbf{S}$ has to be substituted. Also for Markov switching models the dimensionality of (11.28) can often be reduced by solving the integration with respect to the hidden Markov chain $\mathbf{S}$ analytically, and using numerical methods only for the remaining parameters:

$$p(\mathbf{y}) = \int p(\mathbf{y}|\boldsymbol{\theta}_1, \ldots, \boldsymbol{\theta}_K, \boldsymbol{\xi})p(\boldsymbol{\theta}_1, \ldots, \boldsymbol{\theta}_K, \boldsymbol{\xi})d(\boldsymbol{\theta}_1, \ldots, \boldsymbol{\theta}_K, \boldsymbol{\xi}).$$

Marginalizing over $\mathbf{S}$ is possible for the Markov switching model, whenever the memory of $y_t$ with respect to $S_t$ is finite (assumption **Y2**). This class encompasses many important Markov switching models such as the Markov switching autoregressive model (McCulloch and Tsay, 1994b), but also non-conjugate models such as switching ARCH models (Hamilton and Susmel, 1994; Kaufmann and Frühwirth-Schnatter, 2002).

Chib (1995, 1996) estimates the marginal likelihood for simple Markov switching models using Chib's estimator as discussed in Subsection 5.5.2, however, Frühwirth-Schnatter (2004) showed that this estimator may be biased; see also Subsection 11.7.3.

The sampling-based estimators of the marginal likelihood, discussed in Section 5.4 for standard finite mixture models, are easily extended to Markov switching models (Frühwirth-Schnatter, 2004). Based on a random subsequence $\mathbf{S}^{(s)}, s = 1, \ldots, S$ of the MCMC draws $\mathbf{S}^{(m)}, m = 1, \ldots, M$, an importance density comparable to (5.36) is available through:

$$q(\boldsymbol{\vartheta}_K) = \frac{1}{S}\sum_{s=1}^{S} p(\boldsymbol{\xi}|\mathbf{S}^{(s)}) \prod_{k=1}^{K} p(\boldsymbol{\theta}_k|\mathbf{S}^{(s)}, \mathbf{y}), \tag{11.29}$$

with $p(\boldsymbol{\xi}|\mathbf{S})$ and $p(\boldsymbol{\theta}_k|\mathbf{y}, \mathbf{S})$ being the complete-data posterior densities under the assumption that $\mathbf{p}_0$ is independent of $\boldsymbol{\xi}$. Again it is important that the states $\mathbf{S}$ are sampled from the unconstrained Markov mixture posterior distribution with balanced label switching using random permutation sampling as in *Algorithm 11.4*. Frühwirth-Schnatter (2004) argued that in particular the bridge sampling technique, discussed in detail in Subsection 5.4.6, is useful for estimating the marginal likelihood of a Markov switching model; see also Kaufmann and Frühwirth-Schnatter (2002) and Frühwirth-Schnatter and Kaufmann (2006a, 2006b) for further applications.

### 11.6.4 Model Space MCMC

Robert et al. (2000) and Cappé et al. (2003) applied model space MCMC methods for Markov switching models in order to select the number of states. Birth and death methods are considered by Cappé et al. (2003); reversible jump MCMC methods and birth and death MCMC methods are explored by Robert et al. (2000).

### 11.6.5 Further Issues

A further important issue is the effect of misspecifying the dependence structure of the hidden Markov chain. Lindgren (1978) argues that the assumption of first-order Markovian dependence for the hidden Markov chain $S_t$ is not crucial, but is merely the simplest structure to transfer information between successive states and proves robustness of the ML estimator against misspecifying the dependence structure of the hidden Markov chain. From a Bayesian perspective, the Markovian dependence structure is a prior distribution on $\mathbf{S}$, which may be overruled by the information contained in the observed time series. Chen and Liu (1996) used a Bayesian approach based on marginal likelihoods to test the Markov dependence priors on $\mathbf{S}$ against the i.i.d. assumption.

## 11.7 Modeling Overdispersion and Autocorrelation in Time Series of Count Data

### 11.7.1 Motivating Example

We consider the LAMB DATA, a time series of count data analyzed originally in Leroux and Puterman (1992), and reanalyzed by Chib (1996) and Frühwirth-Schnatter (2004). The data plotted in Figure 11.2 are the number $y_t$ of movements by a fetal lamb in $T = 240$ consecutive five-second intervals. Assuming that the counts are i.i.d. realization from a Poisson distribution implies first, that the mean is equal to the variance and, second, that the realizations are independent over time. Both assumptions, however, are violated for these data. First overdispersion with $\mathrm{Var}(Y_t) > \mathrm{E}(Y_t)$ is present, because

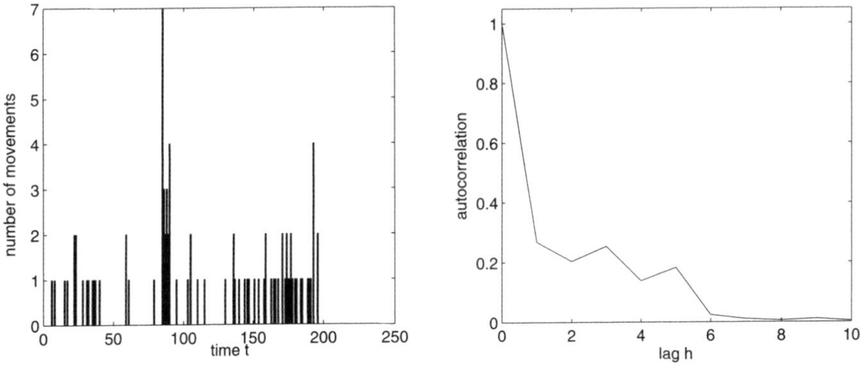


**Fig. 11.2.** LAMB DATA, left: time series plot of $y_t$, right: empirical autocorrelogram

the sample variance $s_y^2 = 0.658$ is nearly twice the sample mean $\overline{y} = 0.358$. Second, the empirical autocorrelogram in Figure 11.2 indicates stochastic dependence between $Y_{t-1}$ and $Y_t$.

To capture overdispersion, a finite Poisson mixture distribution could be applied as was done in Subsection 9.2.4 for the EYE TRACKING DATA. A standard Poisson mixture with $K = 3$ components is actually able to capture the overdispersion in the marginal distribution of this time series; see Figure 11.3. Autocorrelation, however, is not captured by this model, as by the results of Subsection 10.2.6 the autocorrelation of any standard finite mixture model is 0. To capture both overdispersion and autocorrelation, a Poisson Markov mixture model is applied in Subsection 11.7.3.

### 11.7.2 Capturing Overdispersion and Autocorrelation Using Poisson Markov Mixture Models

To capture autocorrelation and overdispersion for time series of counts, often observation-driven models in the sense of Cox (1981) are applied. Models that introduce autocorrelation for discrete-valued time series by allowing direct dependence of the predictive distribution $p(y_t|\mathbf{y}^{t-1}, \boldsymbol{\vartheta})$ on past values, have been considered by, among others, Zeger and Qaqish (1988) and Chan and Ledolter (1995).

Parameter-driven models introduce autocorrelation through a hidden structure. Parameter-driven models for count data, where autocorrelation is introduced through a latent continuous state process were considered by Harvey and Fernandes (1989) and Mayer (1999). Alternatively, the latent structure could be modeled as a discrete Markov chain $S_t$ with transition matrix $\boldsymbol{\xi}$, leading to a Markov mixture model. Markov mixture models for count data are typically based on the Poisson distribution:

$$Y_t|S_t \sim \begin{cases} \mathcal{P}(\mu_1), & S_t = 1, \\ \vdots & \\ \mathcal{P}(\mu_K), & S_t = K, \end{cases} \tag{11.30}$$

and were considered, among many others, by MacRae Keenan (1982), Leroux and Puterman (1992), Wang and Puterman (1999), and Wang et al. (1996). Albert (1991) considered a two-state Markov mixture model for a time series of epileptic seizure counts.

#### Capturing Autocorrelation

The presence of autocorrelation in a discrete-valued process $Y_t$ generated by a Markov mixture is evident from the results of Subsection 10.2.4, where the autocorrelation function is derived in (10.20). Autocorrelation in the observed counts is introduced through autocorrelation in the hidden Markov chain.

For a Poisson Markov mixture with $K = 2$ states, for instance, the autocorrelation function simplifies to

$$\rho_{Y_t}(h|\boldsymbol{\vartheta}) = \left( \frac{\eta_1\eta_2(\mu_2 - \mu_1)^2}{\mu + \eta_1\eta_2(\mu_2 - \mu_1)^2} \right) (\xi_{11} - \xi_{21})^h, \tag{11.31}$$

with $\eta_1, \eta_2$ being the ergodic probabilities of the Markov chain $S_t$; see also Subsection 10.2.2. The values of the generated time series are correlated, as long as $S_t$ does not degenerate to an i.i.d. process. As discussed in Subsection 10.2.4, the autocorrelation function (11.31) is the same for an ARMA(1, 1) model, without requiring that the data be drawn from a normal distribution.

#### Capturing Overdispersion

An additional important feature of applying Poisson Markov mixture models to time series of counts is to capture overdispersion in the marginal distribution. By the results of Subsection 10.2.3, the first two moments of a process generated by a Poisson Markov mixture are given by

$$\mathrm{E}(Y_t|\boldsymbol{\vartheta}) = \sum_{k=1}^{K} \mu_k \eta_k,$$

$$\mathrm{Var}(Y_t|\boldsymbol{\vartheta}) = \sum_{k=1}^{K} \mu_k(1 + \mu_k)\eta_k - \mathrm{E}(Y_t|\boldsymbol{\vartheta})^2 = \mathrm{E}(Y_t|\boldsymbol{\vartheta}) + B(\boldsymbol{\vartheta}),$$

where $\eta_1, \ldots, \eta_K$ is the ergodic probability distribution of the hidden Markov chain and $B(\boldsymbol{\vartheta})$ is the between-group heterogeneity:

$$B(\boldsymbol{\vartheta}) = \sum_{k=1}^{K} (\mu_k - \mu(\boldsymbol{\vartheta}))^2 \eta_k,$$

with $\mu(\boldsymbol{\vartheta}) = \mathrm{E}(Y|\boldsymbol{\vartheta})$. As $B(\boldsymbol{\vartheta}) > 0$, if at least two means of the Poisson mixture are different, the marginal distribution of the process $Y_t$ is actually overdispersed.

Note that the Poisson distribution might be substituted by another discrete distribution. MacDonald and Zucchini (1997, p.68) note that a particularly useful model to capture overdispersion is a negative binomial distribution based on a hidden Markov chain, because such a model introduces overdispersion via the conditional distribution $p(y_t|S_t, \boldsymbol{\vartheta})$ as well as via the hidden Markov chain $S_t$.

### 11.7.3 Application to the Lamb Data

We return to the LAMB DATA, introduced in Subsection 11.7.1. To capture overdispersion and autocorrelation in this time series, $Y_t$ is modeled as a Poisson process where the intensity changes according to a $K$-state hidden Markov process $S_t$ as in (11.30). The unknown model parameter is $\boldsymbol{\vartheta}_K = (\mu_1, \dots, \mu_K, \boldsymbol{\xi})$.

For several values of $K$, Bayesian estimation of $\boldsymbol{\vartheta}_K$ is carried out by MCMC sampling based on data augmentation through *Algorithm 11.3.* The priors are chosen as $\boldsymbol{\xi}_{k\cdot} \sim \mathcal{D}(e_{k1}, \dots, e_{kK})$ with $e_{kk} = 2$ and $e_{kk'} = 1/(K-1)$, if $k \neq k'$ and $\mu_k \sim \mathcal{G}(1, 0.5)$. The sampler is run for $M = 5000$ draws, with a burn-in of 1000 draws.

#### Capturing Autocorrelation and Overdispersion

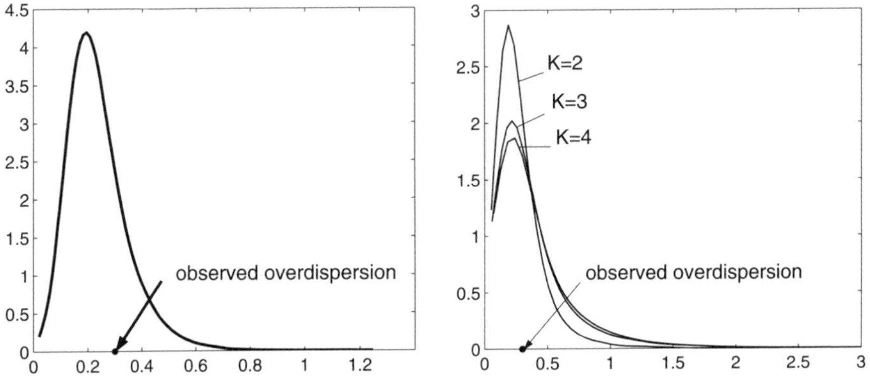


**Fig. 11.3.** LAMB DATA, posterior distribution of the overdispersion parameter in comparison to the observed value; left-hand side: finite Poisson mixture with $K = 3$ components; right-hand side: Poisson Markov mixture with $K = 2$, $K = 3$, and $K = 4$ states

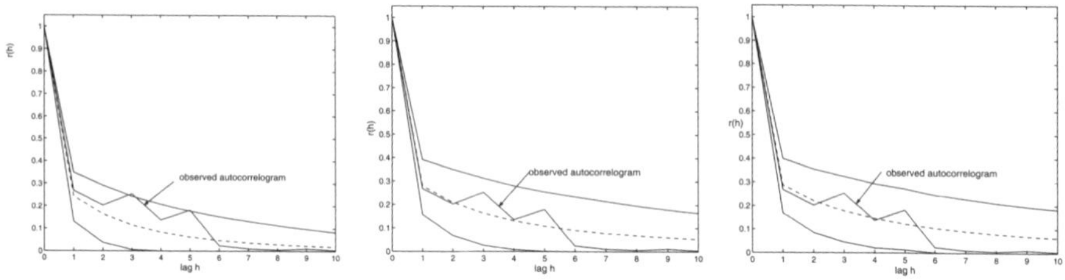


**Fig. 11.4.** LAMB DATA, modeled by a Poisson Markov mixture with $K = 2$, $K = 3$, and $K = 4$ states; posterior confidence bands obtained from the posterior distribution $p(\rho_{Y_t}(h|\boldsymbol{\vartheta}_K)|\mathbf{y}, K)$ of the autocorrelation $\rho_{Y_t}(h|\boldsymbol{\vartheta}_K)$ plotted against $h$, in comparison to the empirical autocorrelogram

To assess model fit as in Subsection 4.3.4, we consider overdispersion and autocorrelation. Figure 11.3 shows that any of the Markov Poisson mixture models is able to produce the observed overdispersion, as did the standard Poisson mixture model fitted in Subsection 11.7.1.

To assess model fit with respect to autocorrelation, the posterior of the implied autocorrelation function $\rho_{Y_t}(h|\boldsymbol{\vartheta}_K)$ is compared in Figure 11.4 with the observed autocorrelation for various values of $K$. From (10.20) it is evident that the implied autocorrelation function depends on the unknown model parameters $\boldsymbol{\vartheta}_K$. In a classical framework one would compute $\rho_{Y_t}(h|\hat{\boldsymbol{\vartheta}}_K)$ for some estimator $\hat{\boldsymbol{\vartheta}}_K$. Within the Bayesian framework, however, one could take full account of the uncertainty associated with estimating $\rho_{Y_t}(h|\boldsymbol{\vartheta}_K)$. Although $\rho_{Y_t}(h|\boldsymbol{\vartheta}_K)$ is a nonlinear function of $\boldsymbol{\vartheta}_K$, posterior draws from $p(\rho_{Y_t}(h|\boldsymbol{\vartheta}_K)|\mathbf{y}, K)$ are available simply by transforming the draws $\boldsymbol{\vartheta}_K^{(m)}$ from the Markov mixture posterior $p(\boldsymbol{\vartheta}_K|\mathbf{y}, K)$. Note that $\rho_{Y_t}(h|\boldsymbol{\vartheta}_K)$ is invariant to relabeling the states of the Markov chain and may be estimated without identifying the states.

In Figure 11.4, the posterior mean $\mathrm{E}(\rho_{Y_t}(h|\boldsymbol{\vartheta}_K)|\mathbf{y}, K)$ is systematically different from the empirical autocorrelogram for $K = 2$ and some of the empirical autocorrelations lie outside the confidence band derived from the posterior. For $K = 3$ and $K = 4$ the observed autocorrelogram lies within these bands. Introducing the fourth component hardly changes the implied autocorrelation function. To sum up, a Markov mixture of three components seems to be sufficient to capture both autocorrelation and overdispersion.

### Selecting the Number of Components Through Marginal Likelihoods

Leroux and Puterman (1992) when using AIC and BIC found conflicting evidence concerning the number of states, as AIC selects a model with three states, whereas BIC selects a model with two states. The application of the Bayesian model selection technique to choose the number of states requires

**Table 11.1.** LAMB DATA, modeled by a single Poisson distribution ($K = 1$) and Poisson Markov mixture with $K = 2$, $K = 3$, and $K = 4$ states; log marginal likelihood $\log p(\mathbf{y}|K)$ for $K = 1$ and various estimates of $\log p(\mathbf{y}|K)$ for $K = 2$, $K = 3$, and $K = 4$; relative standard errors are given in parentheses (from Frühwirth-Schnatter (2004) with permission granted by the Royal Economic Society)

| $K$ | 1 | 2 | 3 | 4 |
|---|---|---|---|---|
| $\log p(\mathbf{y}\|K)$ | -204.25 | — | — | — |
| $\log \hat{p}_{BS}(\mathbf{y}\|K)$ | — | -185.08 (.004) | -179.14 (.002) | -179.17 (.003) |
| $\log \hat{p}_{IS}(\mathbf{y}\|K)$ | — | -185.08 (.003) | -179.11 (.003) | -178.33 (.059) |
| $\log \hat{p}_{RI}(\mathbf{y}\|K)$ | — | -185.08 (.005) | -179.99 (.013) | -184.83 (.040) |
| $\log \hat{p}_{CH}(\mathbf{y}\|K)$ | -204.25 | -185.74 (.013) | -181.37 (.088) | -177.84 (.057) |

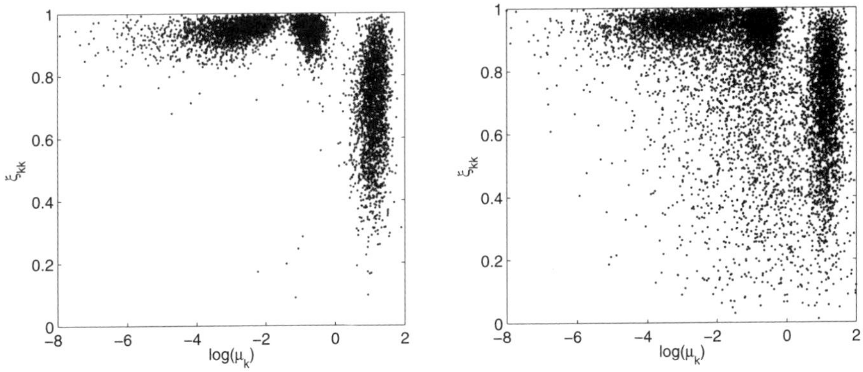


**Fig. 11.5.** LAMB DATA, modeled by a Poisson Markov mixture with $K = 3$ (left-hand side) and $K = 4$ (right-hand side); scatter plot of the MCMC draws $\log(\mu_k^{(m)})$ versus $\xi_{kk}^{(m)}$ (from Frühwirth-Schnatter (2001b) with permission granted by the American Statistical Association)

the computation of the marginal likelihood $p(\mathbf{y}|K)$ for different number $K$ of states. For $K = 1$ the marginal likelihood $p(\mathbf{y}|K)$ is known analytically from (5.62), leading to $\log p(\mathbf{y}|K = 1) = -204.25$. For $K > 1$ we use the various estimators of $p(\mathbf{y}|K)$ discussed in Section 5.4.2. We computed the "optimal" bridge sampling estimator, the importance sampling estimator, and the reciprocal importance sampling estimator based on the unsupervised importance density (11.29) which is constructed from the MCMC output of a random permutation sampler with $M = L = 5000$ and $S = 100 \cdot K!$. The resulting marginal likelihoods are summarized in Table 11.1.

Concerning the number of states, there is overwhelming evidence against the hypothesis that the process is homogeneous. There is clear evidence in favor of the three-state Markov mixture model compared to the two-state Markov mixture model. When adding a fourth state, the various estimators grow sensitive to the estimation method. The bridge sampling estimator has the smallest standard error and indicates no increase in the marginal likelihood. The scatter plots in Figure 11.5 indicate that the four-state Markov mixture seems to be overfitting.

### Identifying a Markov Mixture Model with Three States

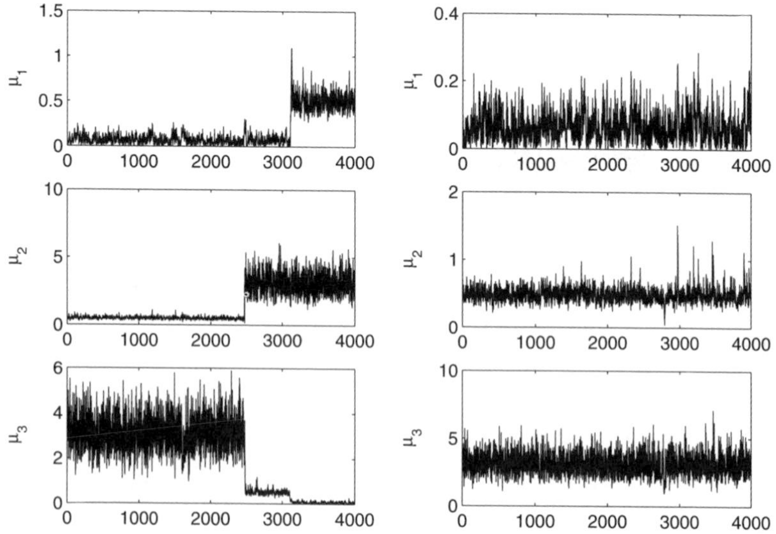


**Fig. 11.6.** LAMB DATA, modeled by a Poisson Markov mixture with $K = 3$ states; MCMC draws of $\mu_1$ (top), $\mu_2$ (middle), and $\mu_3$ (bottom); unconstrained Gibbs sampling (left-hand side) and permutation sampling under the constraint $\mu_1 < \mu_2 < \mu_3$ (right-hand side) (from Frühwirth-Schnatter (2001b) with permission granted by the American Statistical Association)

As in Frühwirth-Schnatter (2004), we discuss identification of a three-state Markov mixture model. Because of the biological background of the example, we expect that the states differ in the intensity of movement. The direct connection between the intensity and the state-specific parameter $\mu_k$ suggest the identifiability constraint

$$\mu_1 < \mu_2 < \mu_3. \tag{11.32}$$

To check the constraint we exploit the draws from random permutation sampling for $K = 3$. Figure 11.5 shows a scatter plot of $\log(\mu_k^{(m)})$ versus $\xi_{kk}^{(m)}$ from the random permutation sampler. The various states actually differ in the intensity, supporting constraint (11.32). This constraint is included in the

**Table 11.2.** LAMB DATA, modeled by a Poisson Markov mixture with $K = 3$ states; parameter estimates and 95% confidence regions (from Frühwirth-Schnatter (2001b) with permission granted by the American Statistical Association)

| | Unrestricted Gibbs Sampling | | | Permutation Sampling | | |
|---|---|---|---|---|---|---|
| | Mean | Lower | Upper | Mean | Lower | Upper |
| $\mu_1$ | 0.145 | 9.43e–005 | 0.545 | 0.0648 | 4.84e–005 | 0.149 |
| $\mu_2$ | 1.28 | 0.27 | 3.85 | 0.488 | 0.299 | 0.69 |
| $\mu_3$ | 2.2 | 0.000444 | 4.21 | 3.08 | 1.65 | 4.51 |
| $\xi_{11}$ | 0.943 | 0.871 | 0.996 | 0.944 | 0.875 | 0.994 |
| $\xi_{12}$ | 0.031 | 2.11e–010 | 0.093 | 0.039 | 1.52e–007 | 0.105 |
| $\xi_{13}$ | 0.026 | 2.44e–010 | 0.0743 | 0.018 | 4.43e–010 | 0.049 |
| $\xi_{21}$ | 0.076 | 8.38e–010 | 0.296 | 0.043 | 9.56e–008 | 0.107 |
| $\xi_{22}$ | 0.871 | 0.543 | 0.999 | 0.944 | 0.873 | 0.996 |
| $\xi_{23}$ | 0.053 | 6.03e–009 | 0.279 | 0.013 | 4.95e–011 | 0.0434 |
| $\xi_{31}$ | 0.137 | 4.29e–008 | 0.427 | 0.172 | 9.88e–008 | 0.449 |
| $\xi_{32}$ | 0.087 | 3.66e–009 | 0.349 | 0.129 | 7.35e–010 | 0.398 |
| $\xi_{33}$ | 0.777 | 0.463 | 0.994 | 0.699 | 0.41 | 0.964 |

permutation sampler to obtain estimates of the state-specific intensities $\mu_k$ and the transition matrix $\boldsymbol{\xi}$; see Table 11.2.

### Consequences of Undetected Label Switching

The three-state Markov mixture model has been estimated by Chib (1996) using unconstrained Gibbs sampling based on the nonsymmetric priors $\mu_1 \sim \mathcal{G}(1,2)$, $\mu_2 \sim \mathcal{G}(2,1)$, and $\mu_3 \sim \mathcal{G}(3,1)$. Although the choice of the prior means obviously reflects the belief that the first state has the lowest and the last state has the highest intensity, the prior does not prevent label switching, as is evident from Figure 11.6. Table 11.2 illustrates the effect of undetected label switching on parameter estimates in comparison to estimation based on the MCMC output of an identified model using constraint (11.32). For the sake of comparison unconstrained sampling is based on the vague, symmetric prior used above, rather than on the prior used by Chib (1996). For unrestricted Gibbs sampling, label switching pulls the estimates $\hat{\mu}_k$ of all state-specific parameters $\mu_k$ toward the mean $\mathrm{E}(\mu_k|\mathbf{y}) = 1.23$ of the Markov mixture posterior distribution. Similarly, all estimates of the persistence probabilities $\xi_{kk}$ are pulled toward $\mathrm{E}(\xi_{kk}|\mathbf{y}) = 0.8641$ and all estimates of the transition probabilities $\xi_{kk'}, k \neq k'$ are pulled toward $\mathrm{E}(\xi_{kk'}|\mathbf{y}) = 0.0680$. All means with respect to the Markov mixture posterior distribution have been estimated from the output of the random permutation sampler.

Finally, we study the effect of label switching on Chib's estimator (Chib, 1995) given by (5.63). $\boldsymbol{\vartheta}_K^\star = (\mu_1^\star, \ldots, \mu_K^\star, \boldsymbol{\xi}^\star)$ is estimated as that MCMC draw $\boldsymbol{\vartheta}_K^{(m)}$ which maximizes the nonnormalized posterior $p(\mathbf{y}|\boldsymbol{\vartheta}_K^{(m)})p(\boldsymbol{\vartheta}_K^{(m)})$. $\hat{p}(\boldsymbol{\vartheta}_K^\star|\mathbf{y})$ is estimated from the MCMC output of the unconstrained Gibbs sampler by:

$$\hat{p}(\boldsymbol{\vartheta}_K^\star|\mathbf{y}) = M^{-1}\sum_{m=1}^{M} p(\boldsymbol{\xi}^\star|(\mathbf{S})^{(m)})p(\mu_1^\star,\ldots,\mu_K^\star|(\mathbf{S})^{(m)},\mathbf{y}). \qquad (11.33)$$

From Table 11.1 we see that Chib's estimator is biased in comparison to the other methods. As for $K = 2$ no label switching occurred, it is possible to correct for this bias as in Section 4.5.2: $\log \hat{p}_{CH}^\star(\mathbf{y}|K) = \log \hat{p}_{CH}(\mathbf{y}|K) + \log(2) = -185.048$. Bias correction is not possible for $K = 3$. From Figure 11.6 it is clear that for $K = 3$ the unconstrained Gibbs sampler shows frequent label switching causing a switching behavior in the functional values of the conditional densities $p(\boldsymbol{\xi}^\star|(\mathbf{S})^{(m)})p(\mu_1^\star,\mu_2^\star,\mu_3^\star|(\mathbf{S})^{(m)},\mathbf{y})$ in (11.33). This causes an uncontrollable bias in estimating the functional value of the posterior and subsequently in estimating the marginal likelihood.

# 12

# Nonlinear Time Series Analysis Based on Markov Switching Models

## 12.1 Introduction

In practical time series analysis, an important aspect is properties of the marginal distribution of $Y_t$ as well as properties of the one-step ahead predictive density $p(y_t|\mathbf{y}^{t-1}, \boldsymbol{\vartheta})$, implied by the chosen time series model. Typical stylized facts of the marginal distribution of practical time series are asymmetry and nonnormality with rather fat tails, and autocorrelation not only in the level $Y_t$, but also in the squared process $Y_t^2$. Properties of the predictive distribution are nonlinear effects of past observation on the mean and conditional heteroscedasticity.

It is well known that standard ARMA models (Box and Jenkins, 1970) often are not able to capture stylized facts of practical time series. Some unrealistic features of ARMA models based on normal errors are normality of the predictive as well as the marginal density, linearity of the expectation $\mathrm{E}(Y_t|\mathbf{y}^{t-1}, \boldsymbol{\vartheta})$ in the past observation $y_1, \dots, y_{t-1}$, and homoscedasticity of $\mathrm{Var}(Y_t|\mathbf{y}^{t-1}, \boldsymbol{\vartheta})$ (Brockwell and Davis, 1991; Hamilton, 1994b). Numerous nonlinear time series models such as GARCH models, threshold autoregressive models, and many others have been designed to reproduce empirical features of practical time series (Tong, 1990; Granger and Teräsvirta, 1993; Franses and van Dijk, 2000).

This chapter discusses Markov switching models that constitute another very flexible class of nonlinear time series models and are able to capture many features of practical time series by appropriate modifications of the basic Markov switching model introduced in Subsection 10.3.1. Section 12.2 deals with the Markov switching autoregressive model and Section 12.3 considers the related Markov switching dynamic regression model. Section 12.4 shows that Markov switching models give rise to very flexible predictive distributions. Section 12.5 deals with Markov switching conditional heteroscedasticity and switching ARCH models are introduced. Section 12.6 studies further extensions, namely hidden Markov chains with time-varying transition probabilities and hidden Markov models for longitudinal data and multivariate time series.

## 12.2 The Markov Switching Autoregressive Model

It has been discussed in Subsection 10.2.4 that a Markov mixture model introduces autocorrelation in the process $Y_t$ even for the basic Markov switching model, where conditionally on knowing the states, the process $Y_t$ is uncorrelated. In this section the Markov switching autoregressive model is introduced that deals with autocorrelation in a more flexible way than the basic Markov switching model.

### 12.2.1 Motivating Example

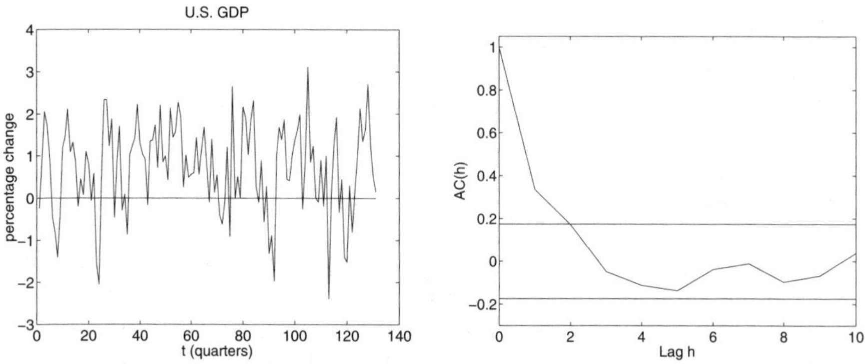


**Fig. 12.1.** GDP DATA, quarterly data 1951.II to 1984.IV, left: time series plot of $y_t$ in comparison to level 0, right: empirical autocorrelogram

A standard time series that has been analyzed in numerous papers is the percentage growth rate of the U.S. quarterly real GDP series:

$$Y_t = 100(\log(\mathrm{GDP}_t) - \log(\mathrm{GDP}_{t-1})), \tag{12.1}$$

$t = 1, \ldots, T$. Figure 12.1 shows a time series plot of the data for the period 1951.II to 1984.IV, together with empirical autocorrelation. First, we fit various AR($p$) models to these data to capture autocorrelation in this time series. Figure 12.2, comparing the unconditional distribution of $Y_t$, implied by each of the fitted AR($p$) models with the empirical histogram of $y_t$, reveals a striking difference between the empirical histogram which evidently shows bimodality, and any of the implied marginal distributions which are unimodal and, by the way, show surprisingly little difference for the different model orders.

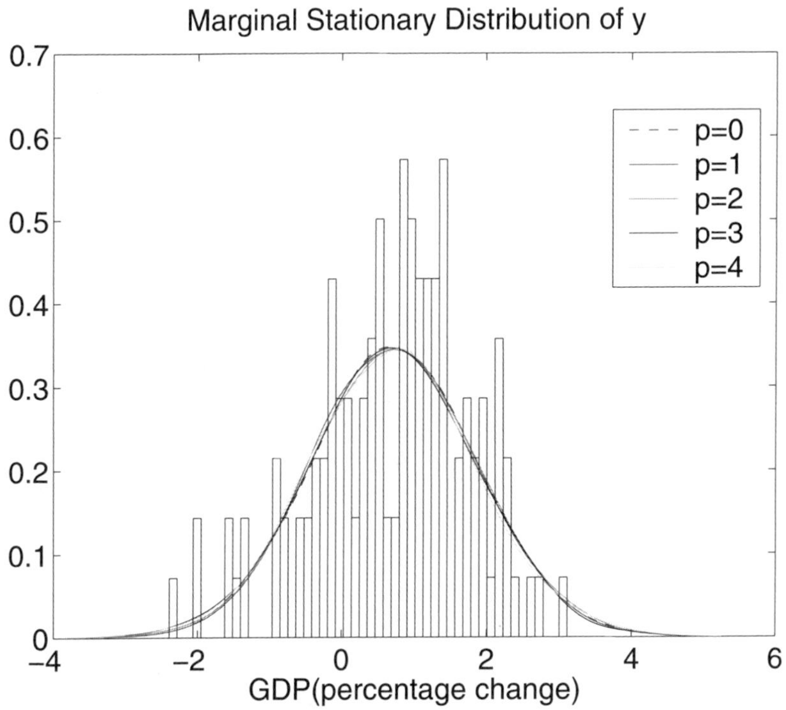


**Fig. 12.2.** GDP DATA, modeled by an AR($p$) model with $p = 0, 1, \ldots, 4$; implied unconditional distribution of $Y_t$ (full line) in comparison to the empirical marginal distribution of $Y_t$ (histogram)

From where does this bimodality in the empirical time series come? Figure 12.1 displays the growth rate of the U.S. GDP series in comparison to the zero line. Evidently periods of positive growth rate, where $y_t > 0$, are followed by periods of negative growth rate, where $y_t < 0$. What we find here is known by economists as the business cycle. Macro-economic variables such as the GDP are influenced by the state of the economy and follow different processes, depending on whether the economy is in a boom or in a recession. Figure 12.1 suggests that the marginal distribution of $Y_t$ is a mixture distribution with different means and possibly different variances. If we fitted a standard mixture of two normal distributions, the implied marginal distribution of $Y_t$ would be in fact multimodal, but marginally $Y_t$ would be a process that is uncorrelated over time. To capture both multimodality and autocorrelation for such time series, Hamilton (1989) introduced the Markov switching autoregressive model.

### 12.2.2 Model Definition

The standard model to capture autocorrelation is the AR($p$) model,

$$Y_t - \mu = \delta_1(Y_{t-1} - \mu) + \cdots + \delta_p(Y_{t-p} - \mu) + \varepsilon_t, \tag{12.2}$$

where $\varepsilon_t \sim \mathcal{N}(0, \sigma_\varepsilon^2)$, which is equivalent to model

$$Y_t = \delta_1 Y_{t-1} + \cdots + \delta_p Y_{t-p} + \zeta + \varepsilon_t, \tag{12.3}$$

with $\zeta = \mu(1 - \delta_1 - \cdots - \delta_p)$.

An important extension of the basic Markov switching model is the Markov switching autoregressive (MSAR) model, where a hidden Markov chain is introduced into model (12.2). This model was used independently in the work of Neftçi (1984) and Sclove (1983), and became popular in econometrics for analyzing economic time series such as the GDP data introduced in Subsection 12.2.1 through the work of Hamilton (1989) who allowed for a random shift in the mean level of process (12.2) through a two-state hidden Markov chain:

$$Y_t - \mu_{S_t} = \delta_1(Y_{t-1} - \mu_{S_{t-1}}) + \cdots + \delta_p(Y_{t-p} - \mu_{S_{t-p}}) + \varepsilon_t. \tag{12.4}$$

An important alternative to model (12.4) was suggested by McCulloch and Tsay (1994b), who introduced the hidden Markov chain into (12.3) rather than into (12.2), by assuming that the intercept is driven by the hidden Markov chain rather than the mean level:

$$Y_t = \delta_1 Y_{t-1} + \cdots + \delta_p Y_{t-p} + \zeta_{S_t} + \varepsilon_t. \tag{12.5}$$

Although the parameterization (12.2) and (12.3) are equivalent for the standard AR model, a model with a Markov switching intercept turns out to be different from a model with a Markov switching mean level. In (12.4), after a one-time change from $S_{t-1}$ to $S_t \neq S_{t-1}$, an immediate mean level shift from $\mu_{S_{t-1}}$ to $\mu_{S_t}$ occurs. In (12.5), however, the mean level approaches the new value smoothly over several time periods.

Both models violate assumption **Y4** as the one-step ahead predictive density $p(y_t|\mathbf{y}^{t-1}, \mathbf{S}^t, \boldsymbol{\vartheta})$ depends on past values $\mathbf{y}^{t-1}$. For a model with switching mean level it is evident from (12.4) that the predictive density $p(y_t|\mathbf{y}^{t-1}, \mathbf{S}^t, \boldsymbol{\vartheta})$ depends not only on $S_t$, but also on the past values $S_{t-1}, \ldots, S_{t-p}$ of the hidden Markov chain fulfilling only assumption **Y2** stated in Subsection 10.3.4. On the other hand for a model with switching intercept the predictive density $p(y_t|\mathbf{y}^{t-1}, \mathbf{S}^t, \boldsymbol{\vartheta})$ depends only on $S_t$ and such a process fulfills the stronger condition **Y3**. As discussed in Subsection 11.2.5, condition **Y3** essentially influences the complexity of econometric inference about the hidden Markov chain $S_t$. As a result, econometric inference for an MSAR model with switching intercept is not more complicated than for the basic Markov switching model, whereas for an MSAR model with switching mean inference on the hidden Markov chain $S_t$ is far more involved.

In its most general form the MSAR model allows that the autoregressive coefficients are also affected by $S_t$ (Sclove, 1983; Holst et al., 1994; McCulloch and Tsay, 1994b):

$$Y_t = \delta_{S_t,1} Y_{t-1} + \cdots + \delta_{S_t,p} Y_{t-p} + \zeta_{S_t} + \varepsilon_t. \tag{12.6}$$

The assumption that the autoregressive parameters switch between the two states implies different dynamic patterns in the various states, and introduces asymmetry over time. Asymmetry over time between the states is introduced also through the hidden Markov chain as different persistence probabilities imply different state duration; see (10.13). This combined asymmetry leads to a rather flexible model that is able to capture asymmetric patterns observed in economics time series, such as the fast rise and the slow decay in the U.S. quarterly unemployment rate.

In any of these models the variance may be assumed to be constant, irrespective of the state of $S_t$, or it is possible to assume a shift in the variance, $\varepsilon_t \sim \mathcal{N}\left(0, \sigma^2_{\varepsilon,S_t}\right)$.

Subsequently the notation MS($K$)-AR($p$) is used occasionally to denote a Markov switching autoregressive model with $K$ states and autoregressive order $p$. A more subtle notation that also differentiates between homo- and heteroscedastic variances, switching in the mean level or in the intercept as well as between invariant and switching autoregressive parameters is introduced in Krolzig (1997).

## Related Models

The mixture autoregressive model (Juang and Rabiner, 1985; Wong and Li, 2000) defines the one-step ahead predictive $p(y_t|\mathbf{y}^{t-1}, \boldsymbol{\vartheta})$ directly as a mixture of normal distributions with an AR structure in the mean:

$$p(y_t|\mathbf{y}^{t-1}, \boldsymbol{\vartheta}) = \sum_{k=1}^{K} \eta_k f_N(y_t; \mu_{k,t}, \sigma^2_k), \tag{12.7}$$

where $\mu_{k,t} = \mathrm{E}(Y_t|\mathbf{y}^{t-1}, \boldsymbol{\theta}_k) = \delta_{k,1} y_{t-1} + \cdots + \delta_{k,p} y_{t-p} + \zeta_k$. This model results as that special of an MSAR model, where $S_t$ is an i.i.d. process, with each row of the transition matrix $\boldsymbol{\xi}$ being equal to the weight distribution in (12.7). Because autocorrelation in $Y_t$ is introduced only through the observation equation this model is not able to capture spurious autocorrelation that disappears once we condition on the state of $S_t$.

MSAR models are related to the self-exciting threshold autoregressive (SETAR) models (Jalali and Pemberton, 1995; Clements and Krolzig, 1998) which are themselves that special case of a threshold autoregressive (TAR) model (Tong, 1990), where the mean and the autoregressive parameters switch according to the level of the threshold variable $z_t = Y_{t-d}$:

$$Y_t = \begin{cases} \delta_{1,1}Y_{t-1} + \cdots + \delta_{1,p}Y_{t-p} + \zeta_1 + \varepsilon_t, & Y_{t-d} \leq r, \\ \delta_{2,1}Y_{t-1} + \cdots + \delta_{2,p}Y_{t-p} + \zeta_2 + \varepsilon_t, & Y_{t-d} > r, \end{cases}$$

with $\varepsilon_t \sim \mathcal{N}\left(0, \sigma_\varepsilon^2\right)$. Consider, for instance, a first-order SETAR model, where $p = 1$ and $d = 1$ and define an indicator $S_t$ such that

$$S_t = \begin{cases} 1, & Y_{t-1} \leq r, \\ 2, & Y_{t-1} > r. \end{cases}$$

Then $S_t$ follows a first-order Markov process with transition matrix $\boldsymbol{\xi}$ given by

$$\boldsymbol{\xi} = \begin{pmatrix} \Phi(r_1) & 1 - \Phi(r_1) \\ \Phi(r_2) & 1 - \Phi(r_2) \end{pmatrix},$$

with $\Phi(\cdot)$ being the standard normal distribution, and $r_k = (r - \mu_k)/\sigma_\varepsilon$. Therefore the first-order SETAR model with $d = 1$ corresponds to a two-state Markov switching autoregressive model with a restricted transition matrix, which has a single free parameter, once $\mu_1, \mu_2$, and $\sigma_\varepsilon^2$ are known.

### 12.2.3 Features of the MSAR Model

The Markov switching autoregressive model is a special case of a dynamic stochastic system with stochastic autoregressive parameters for which it is not straightforward to find conditions under which the process $Y_t$ is strictly stationary and certain moments exist (Tjøstheim, 1986; Karlsen, 1990; Bougerol and Picard, 1992b). Results on the stationarity of Markov switching autoregressive models can be found in Holst et al. (1994), Krolzig (1997), Yao and Attali (2000), and Francq and Zakoian (2001). Timmermann (2000) illustrates how the variance and higher-order moments of a process generated by an MSAR model may be computed explicitly provided the process is stationary.

The Markov switching autoregressive model introduces autocorrelation both through the hidden Markov chain as well as through the observation equation, leading to rather flexible autocorrelation structures. The autocorrelation function may be computed explicitly provided that the process is second-order stationary (Timmermann, 2000, Proposition 4). For an MS(2)-AR(1) model with switching mean, fixed variance, and fixed AR coefficients, for instance, the autocorrelation function of $Y_t$ reads:

$$\rho_{Y_t}(h|\boldsymbol{\vartheta}) = \frac{1}{\mathrm{Var}(Y_t|\boldsymbol{\vartheta})} \left( \lambda^h (\mu_1 - \mu_2)^2 \eta_1 \eta_2 + \delta_1^h \frac{\sigma_\varepsilon^2}{1 - \delta_1^2} \right), \tag{12.8}$$

with $\lambda = \xi_{11} - \xi_{21}$ being the second eigenvalue of the transition matrix $\boldsymbol{\xi}$ and the unconditional variance $\mathrm{Var}(Y_t|\boldsymbol{\vartheta})$ being equal to

$$\mathrm{Var}(Y_t|\boldsymbol{\vartheta}) = (\mu_1 - \mu_2)^2 \eta_1 \eta_2 + \frac{\sigma_\varepsilon^2}{1 - \delta_1^2}.$$

The autocorrelation function fulfills, for $h > 2$, the following recursion,

$$\rho_{Y_t}(h|\boldsymbol{\vartheta}) = (\delta_1 + \lambda)\rho_{Y_t}(h-1|\boldsymbol{\vartheta}) - \delta_1\lambda\rho_{Y_t}(h-2|\boldsymbol{\vartheta}), \tag{12.9}$$

and corresponds to the autocorrelation function of an ARMA(2, 1) model, but has a nonnormal unconditional distribution.

Krolzig (1997) derived general results on the relation between Markov switching autoregressive models and nonnormal ARMA models. A process generated by an MS($K$)-AR($p$) model with switching intercept, but fixed variances and AR coefficients, for instance, possesses an ARMA($K+p-1, K-1$) representation (Krolzig, 1997, Proposition 3), whereas an ARMA($K+p-1, K+p-2$) representation results, if a switching mean is considered, rather than a switching intercept (Krolzig, 1997, Proposition 4).

### 12.2.4 Markov Switching Models for Nonstationary Time Series

The work of Nelson and Plosser (1982) started a discussion in econometrics, as to whether macro-economic time series contain a deterministic or a stochastic trend, the latter typically being a unit root in the autoregressive representation of the time series. This is tested by applying a unit root test to $Y_t$ which often leads to nonrejection of the unit root null hypothesis. Perron (1989, 1990) found evidence for spurious unit roots in real interest rates under structural breaks in the trend level and the growth rate.

Markov switching models are to a certain degree able to deal with spurious unit roots caused by structural breaks. To illustrate this point consider a process $Y_t$, generated by a two-state Markov mixture of normal distributions with $\mu_2 \neq \mu_1$ and a highly persistent transition matrix where $\xi_{11}$ and $\xi_{22}$ are close to one, pushing the second eigenvalue $\lambda = \xi_{11} - \xi_{21}$ toward 1. It is evident from the autocorrelation function of $Y_t$, derived in (10.20), that high autocorrelation in the marginal process $Y_t$ is present, although there exists no autocorrelation within the two regimes. Furthermore the autocorrelation increases as the size $|\mu_2 - \mu_1|$ of the shift in the mean increases. This may lead to detecting a spurious unit root because a unit root test applied to $Y_t$ is biased toward nonrejection of the unit root hypothesis under a sudden change in the mean with increasing rate of nonrejection as the size $|\mu_2 - \mu_1|$ of the break increases. Garcia and Perron (1996), by modeling interest rates by a three-state MSAR model with state-invariant autocorrelation and heteroscedastic variances, show that the autocorrelation actually nearly disappears in the various regimes.

This raises the question as to whether a Markov switching model should be applied to the level or to the growth rate of a nonstationary time series. Hamilton (1989), following the standard ARIMA modeling approach, which is based on autoregressive modeling of the growth rate, applied the MSAR model to the growth rate of a nonstationary time series such as the GDP. In terms of the (log) level $Y_t$ the model reads:

$$
\begin{aligned}
Y_t &= \mu_t + Z_t, \\
\mu_t &= \mu_{t-1} + \zeta_{S_t}, \\
\boldsymbol{\delta}(L)\triangle Z_t &= \boldsymbol{\delta}(L)(1-L)Z_t = \varepsilon_t,
\end{aligned}
\tag{12.10}
$$

where $L$ is the lag operator, $\boldsymbol{\delta}(L) = 1 - \delta_1 L - \cdots - \delta_p L^p$ and all roots of $\boldsymbol{\delta}(L)$ lie outside the unit circle. This model is also called the Markov switching trend model, because the untransformed time series $Y_t$ has a stochastic trend with a drift that is switching according to a hidden Markov chain.

Specification (12.10) assumes that $Y_t$ has a unit root, however, as noted by Lam (1990), the results of Perron (1989, 1990) suggest that the unit root in $Y_t$ disappears once occasional shifts in the deterministic trend are allowed for. Lam (1990) assumes that $Y_t$ is trend stationary around a Markov switching trend:

$$
\begin{aligned}
Y_t &= \mu_t + Z_t, \\
\mu_t &= \mu_{t-1} + \zeta_{S_t}, \\
\boldsymbol{\delta}(L) Z_t &= \varepsilon_t,
\end{aligned}
\tag{12.11}
$$

where all roots of $\boldsymbol{\delta}(L)$ lie outside the unit circle. In this model the predictive density $p(y_t|\mathbf{S}, \mathbf{y}^{t-1})$ depends on the whole history of $S_t$ (assumption **Y1**) and estimation has to be carried within the framework of switching state space models; see Chapter 13.

As a compromise between these two models, Hall et al. (1999) consider a model based on the Dickey–Fuller regression (Dickey and Fuller, 1981) and allow for regression parameter switching according to a two-state hidden Markov chain:

$$
\triangle Y_t = \zeta_{S_t} + \psi_{S_t} Y_{t-1} + \sum_{j=1}^{p} \delta_{S_t,j} \triangle Y_{t-j} + \varepsilon_t. \tag{12.12}
$$

In (12.12), $Y_t$ is the (log) level of the observed process, whereas $\triangle Y_t = Y_t - Y_{t-1}$ is the growth rate. If $\psi_1 = \psi_2 = 0$ in both regimes then a unit root is present in $Y_t$, and the Markov switching trend model of Hamilton (1989) results. On the other hand, if $\psi_1 \neq 0$ and $\psi_2 \neq 0$, then $Y_t$ is stationary around a trend with Markov switching slope, leading to the model of Lam (1990).

Model (12.12) allows that $Y_t$ has a unit root in one state ($\psi_1 = 0$), whereas $Y_t$ is stationary in the other state ($\psi_2 \neq 0$). This model has been found useful in applied time series analysis, for instance, in economics for modeling the GDP (McCulloch and Tsay, 1994a), in finance for modeling interest rates (Ang and Bekaert, 2002), as well as in geophysics (Karlsen and Tjøstheim, 1990).

Several authors investigate the power of unit root tests when the data arise from particular Markov switching alternatives (Nelson et al., 2001; Psaradakis, 2001, 2002).

### 12.2.5 Parameter Estimation and Model Selection

ML estimation is usually carried out through the EM algorithm (Hamilton, 1990; Holst et al., 1994). Asymptotic properties of the ML estimator for MSAR models are established in Francq and Roussignol (1998), Krishnamurthy and Rydén (1998), and Douc et al. (2004).

Bayesian estimation of the MSAR model relies on data augmentation and MCMC (Albert and Chib, 1993; McCulloch and Tsay, 1994b; Chib, 1996; Frühwirth-Schnatter, 2001b). For an MSAR model where all coefficients, including the intercept and the variance, are switching, MCMC estimation is carried out along the lines indicated in *Algorithm 11.3*, with step (a2) being the only model-specific part. Sampling the model parameters $\boldsymbol{\vartheta} = (\boldsymbol{\beta}_1, \ldots, \boldsymbol{\beta}_K, \sigma^2_{\varepsilon,1}, \ldots, \sigma^2_{\varepsilon,K})$, with $\boldsymbol{\beta}_k = (\delta_{k,1}, \ldots, \delta_{k,p}, \zeta_k)$, in combination with the conjugate priors

$$\boldsymbol{\beta}_k \sim \mathcal{N}_{p+1}(\mathbf{b}_0, \mathbf{B}_0), \qquad \sigma^2_{\varepsilon,k} \sim \mathcal{G}(c_0, C_0),$$

is closely related to sampling these parameters for a finite mixture regression model as in steps (a2) and (a3) of *Algorithm 8.1.* An MSAR model, where only some parameters are switching, may be considered as a special case of a Markov switching dynamic regression model, which is introduced in Section 12.3, where Bayesian estimation is discussed in Subsection 12.3.2.

The presence of the lagged values $y_{t-1}, \ldots, y_{t-p}$, however, causes certain technical problems that are avoided if inference is carried out conditional on the first $p$ values. For an unconditional analysis as in Albert and Chib (1993), the first $p$ values are considered to be random draws from the stationary distribution $p(y_1, \ldots, y_p|\boldsymbol{\vartheta})$. An undesirable effect of an unconditional analysis is that the posterior of $\boldsymbol{\vartheta} = (\boldsymbol{\beta}_1, \ldots, \boldsymbol{\beta}_K, \sigma^2_{\varepsilon,1}, \ldots, \sigma^2_{\varepsilon,K})$ no longer has a standard form, as the stationary distribution depends on these parameters in a nonconjugate manner. Albert and Chib (1993) suggest using rejection sampling to sample from this posterior.

The most commonly occurring model selection problems for MSAR models is selecting the number of states of the hidden Markov chains well as order selection. Frühwirth-Schnatter (2004) shows that it is important to consider these model selection problems jointly in order to avoid underfitting the number of states while overfitting the AR order; see also Subsection 12.2.6.

### 12.2.6 Application to Business Cycle Analysis of the U.S. GDP Data

The motivating example studied in Subsection 12.2.1 demonstrated one of the key features of the business cycle, namely that periods of expansion and contraction are quite different. Whereas in expansion periods the output growth rate is high and the economy is booming, growth rates are typically negative in contraction periods, where the economy is in a recession. An important feature of macro-economic time series such as the GDP or industrial production is persistence of the respective states. Once the economy is in a certain state it tends to remain there for more than one period. Furthermore there is some asymmetry in this persistency, as longer periods of positive growth rates are followed by shorter periods of negative growth rates. This asymmetry over the business cycle has been captured using a basic Markov switching model for unemployment rates (Neftçi, 1984) and GDP, investment, and productivity (Falk, 1986).

Markov switching autoregressive models, often also called regime switching models by economists, became extremely popular in business cycle analysis since Hamilton's (1989) paper, and further applications include Goodwin (1993), Sichel (1994), Clements and Krolzig (1998), and Kaufmann (2000), among many others. For a theoretical justification of why Markov switching might be sensible models for the economy we refer to Hamilton and Raj (2002) and Raj (2002) and the references therein.

### Model Selection for the GDP Data

We return to modeling the U.S. quarterly GDP series introduced in Subsection 12.2.1 within the framework of MSAR models, by comparing different Markov switching models. The first model is the K-state MSAR model with switching intercept, but state-independent AR parameters and state-independent variances, defined in (12.5), which has been applied by Chib (1996). The second is the K-state MSAR model with switching intercept, switching AR parameters, and switching error variance ("totally switching"), defined in (12.6) which was applied by McCulloch and Tsay (1994b). The priors are selected to be rather vague and state-independent. We assume no prior correlation among the regression parameters. The prior on the switching intercept is $\mathcal{N}(0,1)$; the prior both on switching and state-independent AR parameters is $\mathcal{N}(0,0.25)$. The prior both on switching and state-independent variances is $\mathcal{G}^{-1}(2,0.5)$. The prior on the rows $\boldsymbol{\xi}_{k\cdot}$ of the transition matrix is, for all $k$, $\mathcal{D}(e_{k1},\ldots,e_{kK})$ with $e_{kk}=2$ and $e_{kk'}=1/(K-1)$, if $k\neq k'$.

We compare the Markov switching models (12.5) and (12.6), where $K$ is equal to 2 or 3, with the classical AR($p$) model, which corresponds to $K=1$, using marginal likelihoods. We assume that $p$ varies between 0 and 4, leading to a total of 25 different models. The marginal likelihoods are estimated from the MCMC output of a random permutation sampler ($M=6000$ after a burn-in phase of 1000 simulations) using the "optimal" bridge sampling estimator described in Subsection 5.4.6, where the construction of the importance density $q(\boldsymbol{\vartheta})$ according to (11.29) is based on $S=100\cdot K!$ components.

From Table 12.1, reporting the log of the estimated marginal likelihoods, we find that the two-state totally switching MSAR model of order $p=2$ has the highest marginal likelihood. This result is interesting for various reasons: first, we were able to produce evidence in favor of Markov switching heterogeneity from univariate time series observations of the GDP alone, without the need to include other time series as in Kim and Nelson (2001). Second, the

**Table 12.1.** GDP DATA, modeled by an AR($p$) model ($\mathcal{M}_1$) and different Markov switching models ($\mathcal{M}_2$ ... switching intercept, $\mathcal{M}_3$ ... totally switching) with different order $p$ and different number of states $K$; log of marginal likelihoods $\log p(\mathbf{y}|\mathcal{M}_j, K, p)$ (from Frühwirth-Schnatter (2004) with permission granted by the Royal Economic Society)

| | $\mathcal{M}_1$ | $\mathcal{M}_2$ | | $\mathcal{M}_3$ | |
|---|---|---|---|---|---|
| $p$ | $K = 1$ | $K = 2$ | $K = 3$ | $K = 2$ | $K = 3$ |
| 0 | −199.71 | −193.54 | −192.25 | −194.25 | −193.10 |
| 1 | −194.22 | −192.54 | −192.75 | −193.58 | −194.71 |
| 2 | −196.30 | −194.15 | −194.38 | **−191.62** | −194.33 |
| 3 | −197.26 | −194.59 | −194.74 | −193.67 | −196.78 |
| 4 | −199.18 | −195.70 | −195.72 | −195.34 | −199.88 |

evidence in favor of the hypothesis that the dynamic pattern of the economy is different between contraction and expansion periods confirms the empirical results of McCulloch and Tsay (1994b).

Testing for Markov switching heterogeneity is highly influenced by selecting the appropriate model order. If we compare in Table 12.1 the AR(1) model, which has highest marginal likelihood among all AR($p$) models considered, with a two-state totally switching model of order four, which is the model considered by McCulloch and Tsay (1994b), we end up with evidence in favor of *no* Markov switching heterogeneity. For a two-state totally switching MSAR model, however, the optimal model order is $p = 2$ rather than $p = 4$. Only if we compare the AR(1) model with a two-state switching model with $p$ close to the optimal order, will we end up with evidence *in favor of* Markov switching heterogeneity. These results indicate the importance of simultaneously testing for Markov switching heterogeneity and selecting the appropriate model order and might explain why other studies, reviewed in Kim and Nelson (2001), have produced somewhat conflicting evidence concerning the presence or absence of Markov switching heterogeneity in this time series.

### Exploratory Bayesian Analysis

A number of exploratory cues with regard to model selection are available from the point process representations of the MCMC output of the various models. We start with the point process representations of various bivariate marginal distributions for the three-state totally switching MSAR model of order four. Although we allowed for three states, the scatter plots in Figure 12.3 indicate that a model with three states is overfitting. If we compare this figure with the simulations of a two-state totally switching MSAR model of order four in Figure 12.4, we obtain a similar picture, with fuzziness being reduced due to the smaller number of parameters; nevertheless the two states are not very clearly separated. The bivariate marginal density of the autoregressive parameters $\delta_{k,3}$ and $\delta_{k,4}$ clusters around 0 for all states, suggesting reducing

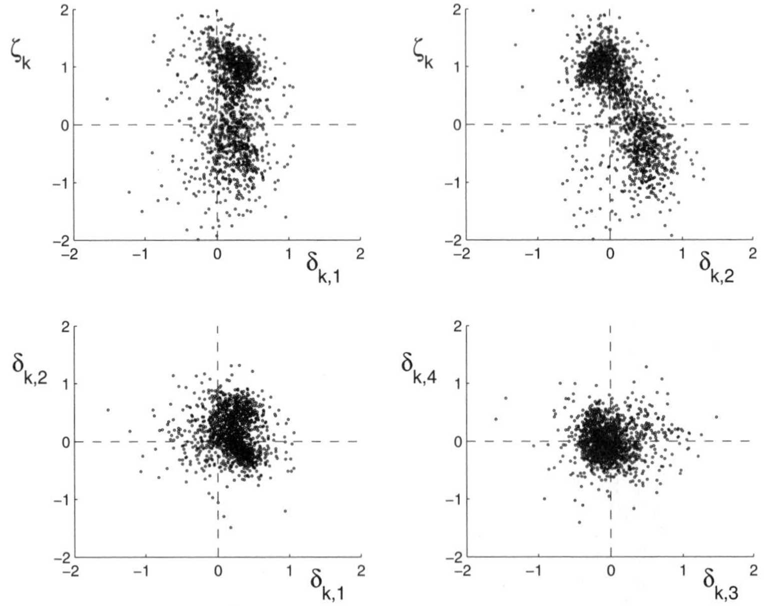


**Fig. 12.3.** GDP DATA, totally Markov switching model with $K = 3$ and $p = 4$; MCMC simulations from various bivariate marginal densities obtained from random permutation sampling

the model order $p$ to 2. The point process representations MCMC simulations for the two-state totally switching MSAR model of order 2 in Figure 12.5 show a much clearer picture. As $\delta_{k,2}$ has two simulation clusters, one of which is shifted away from 0, there is no exploratory evidence that we should reduce the model order further. Furthermore the two simulation clusters provide evidence in favor of a totally switching rather than a switching intercept MSAR model.

On the whole, exploratory Bayesian analysis using projections of the point process representations of the MCMC draws supports the findings from formal model selection using marginal likelihood.

### Parameter Estimation for the "Best" Model

To identify the two-state totally switching MSAR model of order two, we use the identifiability constraint $\zeta_1 < \zeta_2$, as the growth rate in the two states is expected to be different. This choice is supported by point process representation in Figure 12.5, showing that the simulations of $\zeta_k$ cluster around two points, one having an intercept bigger, the other having an intercept smaller than zero.

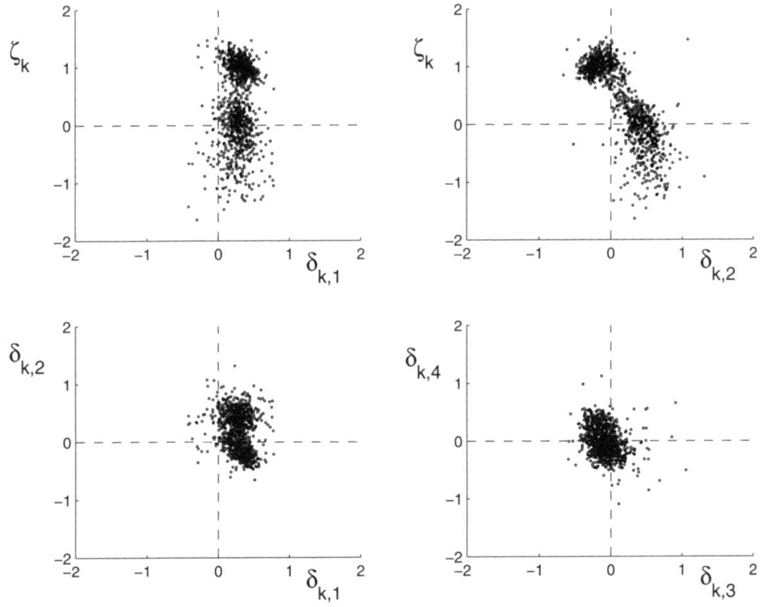


**Fig. 12.4.** GDP DATA, totally Markov switching model with $K = 2$ and $p = 4$; MCMC simulations from various bivariate marginal densities obtained from random permutation sampling

To produce simulations under the identifiability constraint we apply the permutation sampler by reordering the MCMC output according to the constraint $\zeta_1 < \zeta_2$. If the constraint is violated for any MCMC draw, $\zeta_1^{(m)} > \zeta_2^{(m)}$, we permute the labels of all state-dependent parameters with $\rho(1) = 2$ and $\rho(2) = 1$. This is the basic idea behind permutation sampling under an identifiability constraint. It has been proven in Frühwirth-Schnatter (2001b) that due to the invariance of the posterior distribution to relabeling the states, this is a valid strategy to produce a sample from the constrained Markov mixture posterior distribution.

The resulting parameter estimates are summarized in Table 12.2. Positive growth in expansion is followed by negative growth in contraction. The dynamic behavior of the U.S. GDP growth rate is different between contraction and expansion with reaction to a percentage change of the GDP growth being faster in expansion than in contraction. The expected duration of expansion is longer than that of contraction.

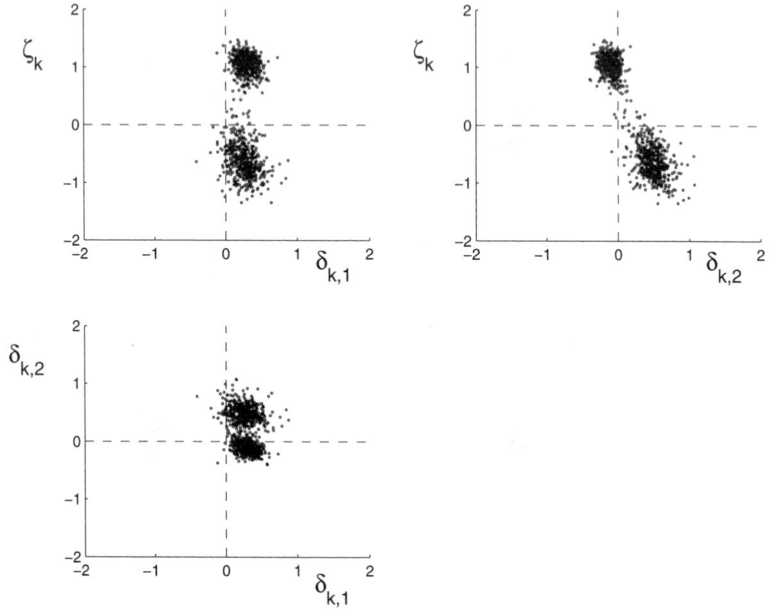


**Fig. 12.5.** GDP Data, totally Markov switching model with $K = 2$ and $p = 2$; MCMC simulations from various bivariate marginal densities obtained from random permutation sampling (from Frühwirth-Schnatter (2001b) with permission granted by the American Statistical Association)

**Table 12.2.** GDP Data, totally Markov switching model with $K = 2$ and $p = 2$, identified through $\zeta_1 < \zeta_2$; parameters estimated by posterior means; standard errors given by posterior standard deviations in parentheses

| Parameter | Contraction ($k = 1$) | Expansion ($k = 2$) |
|---|---|---|
| $\delta_{k,1}$ | 0.249 (0.164) | 0.295 (0.116) |
| $\delta_{k,2}$ | 0.462 (0.164) | –0.114 (0.098) |
| $\zeta_k$ | –0.557 (0.322) | 1.060 (0.175) |
| $\sigma_{\varepsilon,k}$ | 0.768 (0.161) | 0.692 (0.115) |
| $\xi_{kk'}$ | 0.489 (0.165) | 0.337 (0.145) |

## 12.3 Markov Switching Dynamic Regression Models

An important extension both of Markov switching autoregressive models and the Markov switching regression model, discussed in Subsection 10.3.2, is the Markov switching dynamic regression model.

### 12.3.1 Model Definition

The MSAR model (12.4) has been extended in the following way to deal with the presence of exogenous variables $\boldsymbol{z}_t = \begin{pmatrix} z_{t1} & \cdots & z_{td} \end{pmatrix}$ (Cosslett and Lee, 1985; Albert and Chib, 1993),

$$\begin{aligned} Y_t - \mu_{S_t} - \boldsymbol{z}_t\boldsymbol{\beta} = \delta_1(Y_{t-1} - \mu_{S_{t-1}} - \boldsymbol{z}_{t-1}\boldsymbol{\beta}) + \cdots \\ + \delta_p(Y_{t-p} - \mu_{S_{t-p}} - \boldsymbol{z}_{t-p}\boldsymbol{\beta}) + \varepsilon_t, \end{aligned}$$

where the regression coefficient $\boldsymbol{\beta}$ is considered to be unaffected by $S_t$. In the following dynamic regression model all parameters, including the regression coefficient $\boldsymbol{\beta}$, are affected by endogenous regime shifts following a hidden Markov chain (McCulloch and Tsay, 1994b),

$$Y_t = \delta_{S_t,1}Y_{t-1} + \cdots + \delta_{S_t,p}Y_{t-p} + \boldsymbol{z}_t\boldsymbol{\beta}_{S_t} + \zeta_{S_t} + \varepsilon_t.$$

For estimation it is useful to view this model as a Markov switching regression model as in Subsection 10.3.2, without distinguishing between endogenous variables, exogenous variables, and the intercept:

$$Y_t = \mathbf{x}_t\boldsymbol{\beta}_{S_t} + \varepsilon_t, \tag{12.13}$$

where $\mathbf{x}_t = \begin{pmatrix} y_{t-1} & \cdots & y_{t-p} & z_{t1} & \cdots & z_{td} & 1 \end{pmatrix}$. In the mixed-effects Markov switching dynamic regression model only certain elements of the parameter $\boldsymbol{\beta}_{S_t}$ in (12.13) actually depend on the state of the hidden Markov chain, and others are state independent (McCulloch and Tsay, 1994b):

$$Y_t = \mathbf{x}_t^f\boldsymbol{\alpha} + \mathbf{x}_t^r\boldsymbol{\beta}_{S_t} + \varepsilon_t, \tag{12.14}$$

where $\mathbf{x}_t^f$ are those columns of $\mathbf{x}_t$ that correspond to the state-independent parameters $\boldsymbol{\alpha}$ whereas the columns of $\mathbf{x}_t^r$ correspond to the state-dependent parameters. Any of these models may be combined with homoscedastic variances, $\varepsilon_t \sim \mathcal{N}\left(0, \sigma_\varepsilon^2\right)$, or heteroscedastic variances, where the error variances are different in the various states, $\varepsilon_t \sim \mathcal{N}\left(0, \sigma_{\varepsilon,S_t}^2\right)$.

### 12.3.2 Bayesian Estimation

Bayesian estimation of the Markov switching dynamic regression model along the lines indicated in *Algorithm 11.3* is closely related to Bayesian estimation of finite mixtures of regression models. Sampling the parameters

$(\boldsymbol{\alpha}, \boldsymbol{\beta}_1, \ldots, \boldsymbol{\beta}_K, \sigma^2_{\varepsilon,1}, \ldots, \sigma^2_{\varepsilon,K})$ conditional on a known trajectory $\mathbf{S}$ of the hidden Markov chain in step (a2) of *Algorithm 11.3* is exactly the same as for a mixed-effects finite mixture regression model and may be implemented as in *Algorithm 8.2.* As the Markov switching dynamic regression model includes lagged values of $Y_t$, inference is usually carried out conditional on the first $p$ observations $y_1, \ldots, y_p$ and $t$ runs from $t_0 = p+1$ to $T$. To adapt the formulae of Subsection 8.4.4, in particular (8.36) and (8.37), to the slightly different notation used here, note that $i$ corresponds to $t-p$, whereas $N$ corresponds to $T-p$.

Usually an independence prior is applied where location and scale parameters are assumed to be independent a priori (Albert and Chib, 1993; McCulloch and Tsay, 1994b):

$$p(\boldsymbol{\alpha}, \boldsymbol{\beta}_1, \ldots, \boldsymbol{\beta}_K, \sigma^2_{\varepsilon,1}, \ldots, \sigma^2_{\varepsilon,K}) = p(\boldsymbol{\alpha}) \prod_{k=1}^{K} p(\boldsymbol{\beta}_k) p(\sigma^2_{\varepsilon,k}), \tag{12.15}$$

$$\boldsymbol{\alpha} \sim \mathcal{N}_r(\mathbf{a}_0, \mathbf{A}_0), \qquad \boldsymbol{\beta}_k \sim \mathcal{N}_d(\mathbf{b}_0, \mathbf{B}_0), \qquad \sigma^2_{\varepsilon,k} \sim \mathcal{G}(c_0, C_0).$$

Conditionally conjugate priors exist only for two special cases of model (12.14); first, for a model with homoscedastic variances, namely

$$p(\boldsymbol{\alpha}, \boldsymbol{\beta}_1, \ldots, \boldsymbol{\beta}_K, \sigma^2_{\varepsilon}) = p(\sigma^2_{\varepsilon}) p(\boldsymbol{\alpha}|\sigma^2_{\varepsilon}) \prod_{k=1}^{K} p(\boldsymbol{\beta}_k|\sigma^2_{\varepsilon}), \tag{12.16}$$

$$\boldsymbol{\alpha}|\sigma^2_{\varepsilon} \sim \mathcal{N}_r\left(\mathbf{a}_0, \sigma^2_{\varepsilon}\mathbf{A}_0\right), \qquad \boldsymbol{\beta}_k|\sigma^2_{\varepsilon} \sim \mathcal{N}_d\left(\mathbf{b}_0, \sigma^2_{\varepsilon}\mathbf{B}_0\right), \qquad \sigma^2_{\varepsilon} \sim \mathcal{G}(c_0, C_0),$$

and, second, for a model with heteroscedastic variances and no common parameters, where $\mathbf{x}^f_t \boldsymbol{\alpha}$ vanishes in (12.14), namely

$$p(\boldsymbol{\beta}_1, \ldots, \boldsymbol{\beta}_K, \sigma^2_{\varepsilon,1}, \ldots, \sigma^2_{\varepsilon,K}) = \prod_{k=1}^{K} p(\boldsymbol{\beta}_k|\sigma^2_{\varepsilon,k}) p(\sigma^2_{\varepsilon,k}), \tag{12.17}$$

$$\boldsymbol{\beta}_k|\sigma^2_{\varepsilon,k} \sim \mathcal{N}_d\left(\mathbf{b}_0, \sigma^2_{\varepsilon,k}\mathbf{B}_0\right), \qquad \sigma^2_{\varepsilon,k} \sim \mathcal{G}(c_0, C_0).$$

With increasing number $K$ of states joint sampling of all regression parameters $(\boldsymbol{\alpha}, \boldsymbol{\beta}_1, \ldots, \boldsymbol{\beta}_K)$ may be rather time consuming, especially for regression models with high-dimensional parameter vectors, and further blocking may be applied (Albert and Chib, 1993; McCulloch and Tsay, 1994b; Kim and Nelson, 1999).

## 12.4 Prediction of Time Series Based on Markov Switching Models

### 12.4.1 Flexible Predictive Distributions

The predictive distribution of a Markov switching model is much more flexible than the predictive distribution of more traditional time series models. Consider the one-step ahead predictive density $p(y_t|\mathbf{y}^{t-1}, \boldsymbol{\vartheta})$ of a Markov switching model that reads

$$p(y_t|\mathbf{y}^{t-1}, \boldsymbol{\vartheta}) = \sum_{k=1}^{K} p(y_t|\mathbf{y}^{t-1}, \boldsymbol{\theta}_k)\Pr(S_t = k|\mathbf{y}^{t-1}, \boldsymbol{\vartheta}), \tag{12.18}$$

if at least assumption **Y2** holds. Various features of (12.18) are worth mentioning.

First, the one-step ahead predictive density $p(y_t|\mathbf{y}^{t-1}, \boldsymbol{\vartheta})$ is a finite mixture distribution and potentially nonnormal, even if the component densities $p(y_t|\mathbf{y}^{t-1}, \boldsymbol{\theta}_k)$ are normal. The weights of this mixture density are given by the one-step ahead predictive probabilities $\Pr(S_t = k|\mathbf{y}^{t-1}, \boldsymbol{\vartheta}), k = 1, \ldots, K$, which are determined recursively by the filter derived in Subsection 11.2.2, and are dynamic, depending on the past values of $\mathbf{y}^{t-1}$, as long as $S_t$ does not reduce to an i.i.d. process. Additional important features of the predictive density are nonlinearity of $\mathrm{E}(Y_t|\mathbf{y}^{t-1}, \boldsymbol{\vartheta})$ in the past values $\mathbf{y}^{t-1}$, and conditional heteroscedasticity, meaning that $\mathrm{Var}(Y_t|\mathbf{y}^{t-1}, \boldsymbol{\vartheta})$ depends on the past. Also higher-order moments of $p(y_t|\mathbf{y}^{t-1}, \boldsymbol{\vartheta})$ are dynamic and depend on the past.

These features are made more explicit for a two-state Markov switching model with normal component densities, $p(y_t|\mathbf{y}^{t-1}, \boldsymbol{\theta}_k) = f_N(y_t; \mu_{k,t}, \sigma^2_{k,t})$, with $\mu_{k,t} = \mathrm{E}(Y_t|\mathbf{y}^{t-1}, \boldsymbol{\theta}_k)$ and $\sigma^2_{k,t} = \mathrm{Var}(Y_t|\mathbf{y}^{t-1}, \boldsymbol{\theta}_k)$ being the conditional mean and the conditional variance. Obviously from (12.18), the predictive density $p(y_t|\mathbf{y}^{t-1}, \boldsymbol{\vartheta})$ is a mixture of two normal distributions,

$$\begin{aligned} p(y_t|\mathbf{y}^{t-1}, \boldsymbol{\vartheta}) &= w_{t-1}(\mathbf{y}^{t-1}) f_N(y_t; \mu_{1,t}, \sigma^2_{1,t}) \\ &\quad + (1 - w_{t-1}(\mathbf{y}^{t-1})) f_N(y_t; \mu_{2,t}, \sigma^2_{2,t}), \end{aligned} \tag{12.19}$$

where $w_{t-1}(\mathbf{y}^{t-1}) = \Pr(S_t = 1|\mathbf{y}^{t-1}, \boldsymbol{\vartheta})$ is the predictive probability of $S_t = 1$ given time series observations up to $t-1$. Using the filter equations given in Subsection 11.2.3, it is possible to show that $w_{t-1}(\mathbf{y}^{t-1})$ is a nonlinear function of the past values $\mathbf{y}^{t-1}$. From (11.5) follows

$$w_{t-1}(\mathbf{y}^{t-1}) = (1 - \lambda)\eta_1 + \lambda \Pr(S_{t-1} = 1|\mathbf{y}^{t-1}, \boldsymbol{\vartheta}), \tag{12.20}$$

where the filter equation (11.2) implies that the odds ratio for the filter probability $\Pr(S_{t-1} = 1|\mathbf{y}^{t-1}, \boldsymbol{\vartheta})$ is given by

$$\begin{aligned} &\operatorname{logit} \Pr(S_{t-1} = 1|\mathbf{y}^{t-1}, \boldsymbol{\vartheta}) = -.5 \\ &\times \left( \frac{(y_{t-1} - \mu_{1,t-1})^2}{\sigma^2_{1,t-1}} - \frac{(y_{t-1} - \mu_{2,t-1})^2}{\sigma^2_{2,t-1}} + \log \frac{\sigma^2_{1,t-1}}{\sigma^2_{2,t-1}} \right) + \operatorname{logit} w_{t-2}(\mathbf{y}^{t-1}). \end{aligned}$$

Consequently, the right-hand side of (12.20) is a nonlinear function of $y_{t-1}$, and by recursion, of all other previous values $\mathbf{y}^{t-2}$. Hence, the mean of the one-step ahead predictive distribution $p(y_t|\mathbf{y}^{t-1}, \boldsymbol{\vartheta})$ of a two-state hidden Markov model, which is given by

$$\mathrm{E}(Y_t|\mathbf{y}^{t-1},\boldsymbol{\vartheta}) = \mu_{1,t}w_{t-1}(\mathbf{y}^{t-1}) + \mu_{2,t}(1 - w_{t-1}(\mathbf{y}^{t-1})),$$

is nonlinear in the past $\mathbf{y}^{t-1}$, even if the conditional means $\mu_{1,t}$ and $\mu_{2,t}$ are linear as for the MSAR model.

Furthermore, the dependence of the weights $w_{t-1}(\mathbf{y}^{t-1})$ on past observations through the nonlinear function (12.20) introduces conditional heteroscedasticity, even if the predictive densities are homoscedastic. For a two-state hidden Markov model the variance of the one-step ahead predictive distribution $p(y_t|\mathbf{y}^{t-1},\boldsymbol{\vartheta})$ is given by

$$\begin{aligned}\mathrm{Var}(Y_t|\mathbf{y}^{t-1},\boldsymbol{\vartheta}) &= \sigma^2_{1,t}w_{t-1}(\mathbf{y}^{t-1}) + \sigma^2_{2,t}(1 - w_{t-1}(\mathbf{y}^{t-1})) \\ &\quad + 2\mu_{1,t}\mu_{2,t}w_{t-1}(\mathbf{y}^{t-1})(1 - w_{t-1}(\mathbf{y}^{t-1})),\end{aligned}$$

$w_{t-1}(\mathbf{y}^{t-1})$ depends on past observations through the nonlinear function (12.20). Thus the conditional variance $\mathrm{Var}(Y_t|\mathbf{y}^{t-1},\boldsymbol{\vartheta})$ of a Markov switching model is in general a nonlinear function of past squared errors and able to capture conditional heteroscedasticy observed in financial time series; see Section 12.5.

### 12.4.2 Forecasting of Markov Switching Models via Sampling-Based Methods

Predictors $\hat{y}_{T+1},\ldots,\hat{y}_{T+h}$ of a time series $\mathbf{y} = (y_1,\ldots,y_T)$ which are optimal with respect to the mean-squared prediction error criterion may be computed recursively for most Markov switching models (Krolzig, 1997; Clements and Krolzig, 1998).

Bayesian forecasting of future observations $y_{T+1},\ldots,y_{T+h}$ of a time series $\mathbf{y} = (y_1,\ldots,y_T)$ is based on the predictive density $p(y_{T+1},\ldots,y_{T+h}|\mathbf{y})$ which is not available in closed form for most time series models, even for simple AR($p$) models (Schnatter, 1988a). Sampling-based forecasting procedures that have been applied to AR models (Thompson and Miller, 1986) and to ARCH models (Geweke, 1992) were extended to deal with Markov switching autoregressive models (Albert and Chib, 1993).

The following algorithm shows how forecasting by a sampling-based approach is implemented for arbitrary Markov switching models fulfilling at least assumption **Y3** whereas $S_t$ only needs to fulfill **S1**.

*Algorithm 12.1: Forecasting of a Markov Switching Time Series Model* For each MCMC draw $(\boldsymbol{\vartheta}^{(m)}, S_1^{(m)},\ldots,S_T^{(m)})$ from the joint posterior $p(\mathbf{S},\boldsymbol{\vartheta}|\mathbf{y})$ carry out the following steps to sample from the posterior predictive density $p(y_{T+1},\ldots,y_{T+h}|\mathbf{y})$.

(a) Starting with $S_T^{(m)}$, sample a future path of the hidden Markov chain by sampling $S_{T+s}^{(m)}$ recursively for $s = 1,\ldots,h$ from the discrete distribution $p(S_{T+s}|S_{T+s-1}^{(m)},\boldsymbol{\vartheta}^{(m)})$. For a homogeneous Markov chain, this distribution is equal to the $k$th row of $\boldsymbol{\xi}^{(m)}$, if $S_{T+s-1}^{(m)}$ takes the value $k$.

(b) Given $\boldsymbol{\vartheta}^{(m)}$ and $S_{T+1}^{(m)}, \dots, S_{T+h}^{(m)}$, sample $y_{T+1}^{(m)}$ from the predictive density $p(y_{T+1}|\mathbf{y}, \boldsymbol{\vartheta}, S_{T+1}^{(m)})$, and for $s = 2, \dots, h$, sample $y_{T+s}^{(m)}$ recursively from the predictive density $p(y_{T+s}|y_{T+s-1}^{(m)}, \dots, y_{T+1}^{(m)}, \mathbf{y}, \boldsymbol{\vartheta}, S_{T+s}^{(m)})$.

To implement step (b) for the MSAR model, for instance, one samples future paths $y_{T+1}^{(m)}, \dots, y_{T+h}^{(m)}$ recursively from:

$$\begin{aligned}
&y_{T+1}|S_{T+1}^{(m)} = k_1, \mathbf{y}, \boldsymbol{\vartheta}^{(m)} \sim \\
&\qquad \mathcal{N}\left(\zeta_{k_1}^{(m)} + \delta_{k_1,1}^{(m)} y_T + \dots + \delta_{k_1,p}^{(m)} y_{T-p}, \sigma_{\varepsilon,k_1}^{(2,m)}\right) \\
&y_{T+2}|S_{T+2}^{(m)} = k_2, y_{T+1}^{(m)}, \mathbf{y}, \boldsymbol{\vartheta}^{(m)} \sim \\
&\qquad \mathcal{N}\left(\zeta_{k_2}^{(m)} + \delta_{k_2,1}^{(m)} y_{T+1}^{(m)} + \delta_{k_2,2}^{(m)} y_T + \dots, \sigma_{\varepsilon,k_2}^{(2,m)}\right) \\
&y_{T+3}|S_{T+3}^{(m)} = k_3, y_{T+2}^{(m)}, y_{T+1}^{(m)}, \mathbf{y}, \boldsymbol{\vartheta}^{(m)} \sim \\
&\qquad \mathcal{N}\left(\zeta_{k_3}^{(m)} + \delta_{k_3,1}^{(m)} y_{T+2}^{(m)} + \delta_{k_3,2}^{(m)} y_{T+1}^{(m)} + \delta_{k_3,3}^{(m)} y_T + \dots, \sigma_{\varepsilon,k_3}^{(2,m)}\right), \\
&\dots
\end{aligned}$$

where $\delta_{k,l}^{(m)} = 0$ for $l > p$.

## 12.5 Markov Switching Conditional Heteroscedasticity

### 12.5.1 Motivating Example

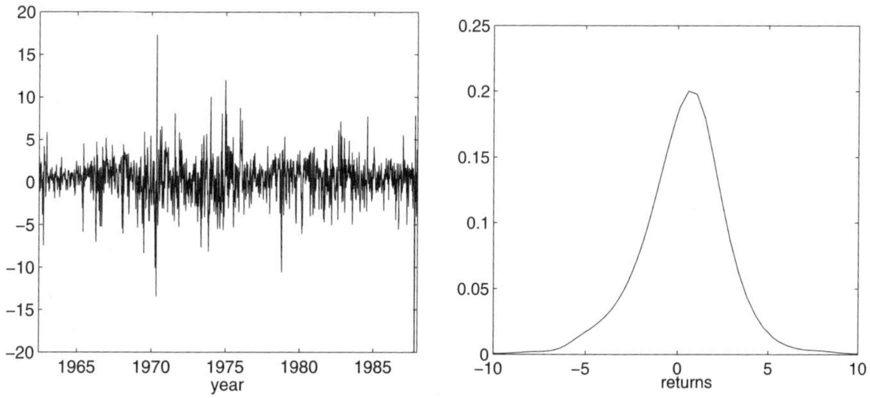


**Fig. 12.6.** New York Stock Exchange Data, left: time series plot; right: smoothed histogram of the marginal distribution

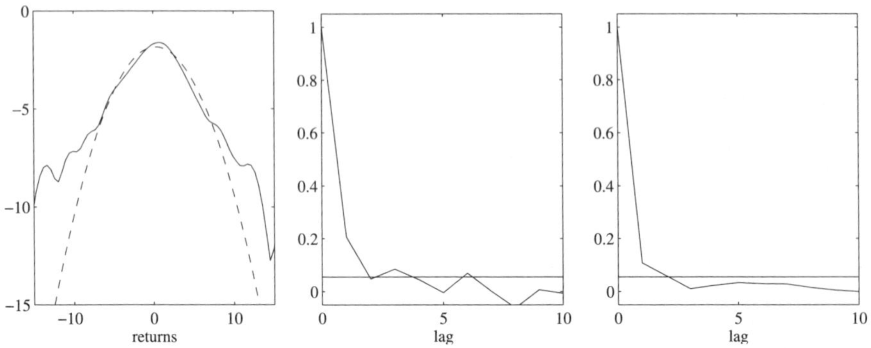


**Fig. 12.7.** NEW YORK STOCK EXCHANGE DATA, left: log of the smoothed histogram (solid line) in comparison to the log of a normal distribution with same mean and variance (dashed line); middle: empirical autocorrelogram of the returns; right: empirical autocorrelogram of the squared returns

Figure 12.6 shows the weekly NEW YORK STOCK EXCHANGE DATA investigated in Hamilton and Susmel (1994). The series originates from the CRISP data tapes and consists of a value-weighted portfolio of stocks traded on the New York Stock Exchange and starts with the week ending Tuesday, July 3, 1962 and ends with the week ending Tuesday, December 29, 1987, making in total 1330 observations. The smoothed histogram of the marginal distribution indicates asymmetry and fat tails. The empirical skewness coefficient and excess kurtosis are given by $-1.2923$ and 17.6394, respectively.

A central topic of econometrics of financial markets is the question of how to model the distribution of such returns, and how to estimate the variability, usually termed volatility, of financial time series. The returns are in general defined as $y_t = \log p_t - \log p_{t-1}$, where $p_t$ is the price of a financial asset or a stock index. Two important stylized facts of financial time series, known as fat tails and volatility clustering, were discovered in the 1960s. Fama (1965, p.48), when studying 30 stocks from the Dow Jones industrial average index, summarized:

> *In any case the empirical distributions are more peaked than the normal in the centre and have longer tails than the normal distribution.*

Departure from normality also occurs for the returns of the NEW YORK STOCK EXCHANGE DATA. Nonnormal tail behavior is evident in particular from the left plot in Figure 12.7, comparing the log of the smoothed histogram with the log of a normal distribution with the same mean and the same variance.

Concerning the second stylized fact, Mandelbrot (1963, p.418) states,

> *Large changes tends to be followed by large changes — of either sign — and small changes tend to be followed by small changes.*

This kind of volatility clustering is also evident for the returns of the New York Stock Exchange Data from the time series plot in Figure 12.6. The presence of volatility clusters if often tested by analyzing the autocorrelation in the squared process. Many studies find significant serial correlation in the squared values of financial time series; see also the right-hand side plot in Figure 12.7 for the returns of the New York Stock Exchange Data.

Later on researchers realized various asymmetric effects also called leverage effects. As Engle and Ng (1995, p.173) summarize,

> *Overall, these results show a greater impact on volatility of negative, rather than positive return shocks.*

### 12.5.2 Capturing Features of Financial Time Series Through Markov Switching Models

Markov switching models are often used by researchers to account for specific features of financial time series such as asymmetries, fat tails, and volatility clusters.

To deal with skewness and excess kurtosis in the unconditional distribution of daily stock returns standard finite mixtures of normal distributions have been applied quite frequently (Fama, 1965; Granger and Orr, 1972; Kon, 1984; Tucker, 1992). Such a modeling approach, however, is appropriate for time series data only if the processes $Y_t$ and $Y_t^2$ do not exhibit autocorrelation, as by the results of Subsections 10.2.4 and 10.2.5 a standard finite mixture model implies zero autocorrelation in $Y_t$ and $Y_t^2$.

Volatility clustering implies persistence of states of high volatility and leads to the rejection of standard time series models in favor of time series models that allow the conditional variance $\mathrm{Var}(Y_t|\mathbf{y}^{t-1}, \boldsymbol{\vartheta})$ to depend on the history $y_{t-1}, y_{t-2}, \ldots$ of the observed process such as the autoregressive conditionally heteroscedastic (ARCH) model (Engle, 1982), where

$$\mathrm{Var}(Y_t|\mathbf{y}^{t-1}, \boldsymbol{\vartheta}) = \gamma_t + \alpha_1 y_{t-1}^2 + \cdots + \alpha_m y_{t-m}^2,$$

and the generalized autoregressive conditionally heteroscedastic (GARCH) model (Bollerslev, 1986). The popularity of ARCH models, in particular if they are based on the $t_\nu$-error distributions (Bollerslev et al., 1992), can certainly be explained by their ability to generate processes with serial correlation in $Y_t^2$, whereas the introduction of $t_\nu$-error helps to capture the tail behavior appropriately. Tsay (1987) considered random coefficient autoregressive models which are another example of a conditional heteroscedastic time series model and showed that the ARCH process is a special case of this model class. More recently, stochastic volatility models have been increasingly applied to financial time series (Shephard, 1996; Kim et al., 1998; Chib et al., 2002).

As an alternative to these models, Markov mixture models where the variance of a location-scale family is driven by a hidden Markov chain have been applied to financial time series (Engel and Hamilton, 1990; McQueen and

Thorely, 1991; Rydén et al., 1998). Although these models introduce autocorrelation in $Y_t^2$, while preserving nonnormality of the marginal distribution (see again Subsection 10.2.5), a closer inspection of the autocorrelation function of $Y_t^2$ reveals that the basic Markov switching model generates only limited persistence in the squared process. For a two-state model, where $\mu_1 = \mu_2$, for instance, there exists a strong relationship between the fatness of the tails, measured by the excess kurtosis, and autocorrelation of the squared process.

Far more general autocorrelation functions of $Y_t^2$ are possible if $Y_t$ is generated by an MSAR model with or without switching AR coefficients; see Timmermann (2000, Proposition 5). Hence the MSAR model has been applied to a number of financial time series (Hamilton, 1988; Turner et al., 1989; Cecchetti et al., 1990; Engel, 1994; Gray, 1996; Ang and Bekaert, 2002).

To obtain even more flexibility in the autocorrelation of $Y_t^2$, for a given marginal distribution of $Y_t$, Hamilton and Susmel (1994), Cai (1994), and Gray (1996) proposed to combine ARCH and Markov switching effects to formulate the switching ARCH model, which is defined in Subsection 12.5.3 as a highly flexible, nonlinear time series model. Bekaert and Harvey (1995) introduced a model that combines Markov switching models with multivariate ARCH models to allow for time-dependence in the integration of emerging markets. Francq et al. (2001) considered the switching GARCH model; see Subsection 12.5.5.

Smith (2002) extends Markov switching models further, by incorporating a regime-dependent variance parameter, when modeling stochastic volatility in interest rates.

### 12.5.3 Switching ARCH Models

A simple model to capture volatility clusters in financial time series is the ARCH model (Engle, 1982) which may be written as

$$\begin{aligned} Y_t &= \sigma_t \varepsilon_t, \qquad \varepsilon_t \sim \mathcal{N}(0,1), \\ \sigma_t^2 &= \gamma_t + \alpha_1 Y_{t-1}^2 + \cdots + \alpha_m Y_{t-m}^2 \end{aligned} \tag{12.21}$$

with $\gamma_t \equiv \gamma$. An alternative parameterization of this model reads:

$$\begin{aligned} Y_t &= \sqrt{\gamma_t} h_t \varepsilon_t, \\ h_t^2 &= 1 + \frac{\alpha_1}{\gamma_{t-1}} Y_{t-1}^2 + \cdots + \frac{\alpha_m}{\gamma_{t-m}} Y_{t-m}^2. \end{aligned} \tag{12.22}$$

The two parameterizations are equivalent if $\gamma_t \equiv \gamma$, however, they generate different processes if $\gamma_t$ is time dependent. The switching ARCH model results by allowing time dependence of $\gamma_t$ through a hidden $K$-state Markov chain $S_t$: $\gamma_t = \gamma_{S_t}$.

Such a switching parameter was introduced by Hamilton and Susmel (1994) into parameterization (12.22):

$$\begin{aligned} Y_t &= \sqrt{\gamma_{S_t}} h_t \varepsilon_t, \\ h_t^2 &= 1 + \frac{\alpha_1}{\gamma_{S_{t-1}}} Y_{t-1}^2 + \cdots + \frac{\alpha_m}{\gamma_{S_{t-m}}} Y_{t-m}^2, \end{aligned}$$

whereas Cai (1994) introduced a two-state and Kaufmann and Frühwirth-Schnatter (2002) a $K$-state switching parameter into parameterization (12.21):

$$\begin{aligned} Y_t &= \sigma_t \varepsilon_t, \\ \sigma_t^2 &= \gamma_{S_t} + \alpha_1 Y_{t-1}^2 + \cdots + \alpha_m Y_{t-m}^2. \end{aligned}$$

Gray (1996) introduced switching into all coefficients of the ARCH process, represented by (12.21):

$$\begin{aligned} Y_t &= \sigma_t \varepsilon_t, \\ \sigma_t^2 &= \gamma_{S_t} + \alpha_{S_t,1} Y_{t-1}^2 + \cdots + \alpha_{S_t,m} Y_{t-m}^2. \end{aligned} \tag{12.23}$$

A special case of model (12.23) is the mixture autoregressive conditional heteroscedastic model (Wong and Li, 2001), where $S_t$ is an i.i.d. process rather than a Markov process. Francq et al. (2001) provide conditions under which model (12.23) is second-order stationary; see also the discussion in Subsection 12.5.5.

The switching ARCH model may be combined with a Markov switching autoregressive model for the mean equation that includes the same hidden Markov chain (Gray, 1996):

$$\begin{aligned} Y_t &= \zeta_{S_t} + \delta_{S_t,1} Y_{t-1} + u_t, \\ u_t &= \sigma_t \varepsilon_t, \qquad \varepsilon_t \sim \mathcal{N}(0,1), \\ \sigma_t^2 &= \gamma_{S_t} + \alpha_{S_t,1} u_{t-1}^2 + \cdots + \alpha_{S_t,m} u_{t-m}^2. \end{aligned} \tag{12.24}$$

The switching ARCH model has been extended by including a leverage effect into the ARCH specification (Hamilton and Susmel, 1994; Kaufmann and Frühwirth-Schnatter, 2002) to deal with asymmetries in the marginal distribution:

$$\begin{aligned} Y_t &= \sigma_t \varepsilon_t, \qquad \varepsilon_t \sim \mathcal{N}(0,1), \\ \sigma_t^2 &= \gamma_{S_t} + \alpha_1 y_{t-1}^2 + \cdots + \alpha_m y_{t-m}^2 + \varrho d_{t-1} y_{t-1}^2, \end{aligned} \tag{12.25}$$

where $d_t = 1$ if $y_t \leq 0$, $d_t = 0$ if $y_t > 0$ and $\varrho > 0$.

Further applications of switching ARCH models in financial econometrics include modeling of stock market returns (Hamilton and Lin, 1996; Fong, 1997), interest rates (Cai, 1994; Gray, 1996; Ang and Bekaert, 2002), and exchange rate data (Klaasen, 2002).

### Spurious Persistency in Squared Returns

A common finding when fitting GARCH models to high-frequency financial data is the somewhat unexpected persistence of shocks to the variance implied by the estimated coefficients which led to the development of the class of integrated generalized autoregressive conditional heteroscedasticity (IGARCH) models (Engle and Bollerslev, 1986). Lamoureux and Lastrapes (1990) investigated the possibility that the appearance of a unit root in the GARCH model may be due to time-varying GARCH parameters. They show that a deterministic structural shift in the unconditional variance, caused by exogenous shocks such as changes in the monetary policy, will increase persistency of squared residuals, however, when the structural break is accounted for, persistency often decreases dramatically.

Introducing a hidden Markov chain into a variance model helps to explain spurious persistence in squared returns. Consider, for illustration, a simple Markov mixture of two normal distributions with $\mu_1 = \mu_2$ and $\sigma_1^2 \neq \sigma_2^2$ driven by a highly persistent transition matrix $\boldsymbol{\xi}$ with $\lambda = \xi_{11} - \xi_{21}$ being close to 1. Together with $\sigma_2^2 - \sigma_1^2$ being large this leads to slowly decaying persistence in $Y_t^2$:

$$\rho_{Y_t^2}(h|\boldsymbol{\vartheta}) = \frac{\eta_1\eta_2(\sigma_1^2 - \sigma_2^2)^2}{\mathrm{E}(Y_t^4|\boldsymbol{\vartheta}) - \mathrm{E}(Y_t^2|\boldsymbol{\vartheta})^2}\lambda^h$$

(see again (10.24)), although the squared returns are uncorrelated within each regime. Also for the more general switching ARCH model, Hamilton and Susmel (1994) attribute part of the high marginal persistence in $Y_t^2$, which is typically much larger than autocorrelation of $Y_t^2$ in the various regimes, to this effect.

### 12.5.4 Statistical Inference for Switching ARCH Models

Parameter estimation for switching ARCH models may be carried out by ML estimation (Hamilton and Susmel, 1994; Francq et al., 2001). Hamilton and Susmel (1994) report extreme difficulties with maximizing the likelihood function for the New York Stock Exchange Data, and only by restricting seven transition probabilities to 0 were they able to run the optimization procedure and to report standard errors for their final model.

Bayesian estimation of the switching ARCH model as exemplified in Kaufmann and Frühwirth-Schnatter (2002) has the advantage of coping with the near boundary space problem by imposing a proper prior on the transition matrix $\boldsymbol{\xi}$ as discussed in Subsection 11.5.1, in which case the posterior density is proper also for unobserved transitions, and standard errors and confidence regions are directly available. MCMC sampling may be carried out along the lines indicated in *Algorithm 11.3*, however, the Metropolis–Hastings algorithm is needed to implement step (a2) due to the nonlinear structure of the underlying model which does not lead to simple conditional densities.

Consider, as an example, the following special case of the switching AR-ARCH model,

$$y_t = \zeta + \delta_1 y_{t-1} + u_t, \tag{12.26}$$
$$u_t = \sigma_t \varepsilon_t, \qquad \varepsilon_t \sim \mathcal{N}(0,1), \tag{12.27}$$
$$\sigma_t^2 = \gamma_{S_t} + \alpha_1 u_{t-1}^2 + \cdots + \alpha_m u_{t-m}^2,$$

where step (a2) in *Algorithm 11.3* requires sampling the AR parameters $\boldsymbol{\phi}_1 = (\zeta, \delta_1)$ and the ARCH parameters $\boldsymbol{\phi}_2 = (\gamma_1, \ldots, \gamma_K, \alpha_1, \ldots, \alpha_m)$ from the appropriate conditional densities. To this aim, Kaufmann and Frühwirth-Schnatter (2002) developed the following two-block Metropolis–Hastings step, building on Nakatsuma (2000).

(a2-1) Sample the AR parameters $\boldsymbol{\phi}_1 = (\zeta, \delta_1)$ from the conditional posterior $p(\boldsymbol{\phi}_1|\mathbf{S}, \boldsymbol{\phi}_2, \mathbf{y})$ using a Metropolis–Hastings algorithm with proposal density $q(\boldsymbol{\phi}_1^{new}|\boldsymbol{\phi}_1^{old})$.
(a2-2) Sample the ARCH parameters $\boldsymbol{\phi}_2 = (\gamma_1, \ldots, \gamma_K, \alpha_1, \ldots, \alpha_m)$ from the conditional posterior $p(\boldsymbol{\phi}_2|\boldsymbol{\phi}_1, \mathbf{S}, \mathbf{y})$ using a Metropolis–Hastings algorithm with proposal density $q(\boldsymbol{\phi}_2^{new}|\boldsymbol{\phi}_2^{old})$.

Due to the presence of ARCH errors in regression model (12.26) no direct method of sampling the AR parameters $\boldsymbol{\phi}_1$ is available even if the ARCH parameters $\boldsymbol{\phi}_2$ are known and a normal prior $\boldsymbol{\phi}_1 \sim \mathcal{N}(\mathbf{b}_0, \mathbf{B}_0)$ is assumed (Bauwens and Lubrano, 1998; Kim et al., 1998; Nakatsuma, 2000). The crucial point is that the error variance $\sigma_t^2$ depends on $\boldsymbol{\phi}_1 = (\zeta, \delta_1)$ through the lagged residuals $u_{t-1}, \ldots, u_{t-m}$:

$$\sigma_t^2(\boldsymbol{\phi}_1, \boldsymbol{\phi}_2) = \gamma_{S_t} + \alpha_1 (y_{t-1} - \zeta - \delta_1 y_{t-2})^2 + \cdots \\ + \alpha_m (y_{t-m} - \zeta - \delta_1 y_{t-m-1})^2.$$

Because model (12.26) is a standard regression model with heteroscedastic errors, if $\sigma_t^2(\boldsymbol{\phi}_1, \boldsymbol{\phi}_2)$ is independent of $\boldsymbol{\phi}_1$, the following normal proposal density results when substituting $\sigma_t^2(\boldsymbol{\phi}_1, \boldsymbol{\phi}_2)$ by $\sigma_t^2(\boldsymbol{\phi}_1^{old}, \boldsymbol{\phi}_2)$ (Kaufmann and Frühwirth-Schnatter, 2002),

$$q(\boldsymbol{\phi}_1^{new}|\boldsymbol{\phi}_1^{old}) = f_N(\boldsymbol{\phi}_1^{new}; \mathbf{b}_N(\boldsymbol{\phi}_1^{old}), \mathbf{B}_N(\boldsymbol{\phi}_1^{old})),$$
$$\mathbf{b}_N(\boldsymbol{\phi}_1) = \mathbf{B}_N(\boldsymbol{\phi}_1) \left( \sum_{t=m+2}^{T} \frac{1}{\sigma_t^2(\boldsymbol{\phi}_1, \boldsymbol{\phi}_2)} \mathbf{x}_t' y_t + \mathbf{B}_0^{-1} \mathbf{b}_0 \right),$$
$$\mathbf{B}_N(\boldsymbol{\phi}_1) = \left( \sum_{t=m+2}^{T} \frac{1}{\sigma_t^2(\boldsymbol{\phi}_1, \boldsymbol{\phi}_2)} \mathbf{x}_t' \mathbf{x}_t + \mathbf{B}_0^{-1} \right)^{-1},$$

where $\mathbf{b}_0$ and $\mathbf{B}_0$ are the prior parameters and $\mathbf{x}_t = \begin{pmatrix} 1 & y_{t-1} \end{pmatrix}$.

Also the conditional posterior of the ARCH parameters $\boldsymbol{\phi}_2$ is not of any closed form. To derive a proposal density $q(\boldsymbol{\phi}_2^{new}|\boldsymbol{\phi}_2^{old})$, the switching ARCH model is reformulated in Kaufmann and Frühwirth-Schnatter (2002) as a generalized linear model. From (12.27), $u_t^2 = \sigma_t^2(\boldsymbol{\phi}_1, \boldsymbol{\phi}_2)\varepsilon_t^2$, where $\varepsilon_t^2$ is a $\chi_1^2$ random variable that may be expressed as $\varepsilon_t^2 = 1 + \tilde{\varepsilon}_t$ with $\mathrm{E}(\tilde{\varepsilon}_t) = 0$ and $\mathrm{Var}(\tilde{\varepsilon}_t) = 2$. Therefore:

$$u_t^2 = \gamma_1 D_t^1 + \cdots + \gamma_K D_t^K + u_{t-1}^2 \alpha_1 + \cdots \\ + u_{t-m}^2 \alpha_m + \sigma_t^2(\phi_1, \phi_2)\tilde{\varepsilon}_t, \tag{12.28}$$

where $D_t^k = 1$ iff $S_t = k$. A normal proposal for $\phi_2$ has been derived in Kaufmann and Frühwirth-Schnatter (2002) from model (12.28) by substituting the nonnormal errors by normal ones with variance $2(\sigma_t^2(\phi_1, \phi_2^{old}))^2$.

Note that both the basic Markov switching model with heterogeneous variances as well as the ARCH model are nested within the switching ARCH model. Therefore model selection may be used to test for the usefulness of the combined model as well as the correct model order. Francq et al. (2001) show that the AIC and Schwarz criteria do not underestimate the correct order of the switching ARCH model. Kaufmann and Frühwirth-Schnatter (2002) use marginal likelihoods to select both the number of states as well as the model order of a switching ARCH model.

### Application to the New York Stock Exchange Data

For illustration, we return to the NEW YORK STOCK EXCHANGE DATA. To account for the autocorrelation found in $y_t$ and $y_t^2$, as well as for the fat tails and the asymmetry observed in the marginal distribution, Kaufmann and Frühwirth-Schnatter (2002) fitted the following switching AR-ARCH model to these data, which includes a leverage term,

$$y_t = \zeta + \delta_1 y_{t-1} + u_t, \tag{12.29}$$

$$u_t = \sigma_t \varepsilon_t, \qquad \varepsilon_t \sim \mathcal{N}(0,1),$$

$$\sigma_t^2 = \gamma_{S_t} + \alpha_1 u_{t-1}^2 + \cdots + \alpha_m u_{t-m}^2 + \varrho d_{t-1} y_{t-1}^2. \tag{12.30}$$

**Table 12.3.** NEW YORK STOCK EXCHANGE DATA, modeled by a switching AR-ARCH model with leverage with different numbers of states $K$ and different model orders $m$; log of the marginal likelihoods under different priors on the switching ARCH intercept (from Kaufmann and Frühwirth-Schnatter (2002) with permission granted by Blackwell Publisher Ltd.)

| | | $\log p(\mathbf{y}\|K,m)$ | |
|---|---|---|---|
| $K$ | $m$ | (prior 1) | (prior 2) |
| 3 | 2 | –2858.5 | –2858.0 |
| 3 | 3 | –2858.2 | –2857.7 |
| 3 | 4 | **–2857.1** | –2856.4 |
| 4 | 2 | –2861.0 | –2859.7 |
| 4 | 3 | –2860.7 | –2859.4 |
| 4 | 4 | –2859.1 | **–2855.9** |

Table 12.3 summarizes the marginal likelihoods $p(\mathbf{y}|K, m)$ for different numbers of states $K$ and different model orders $m$. The marginal likelihoods

are estimated using bridge sampling as described in Section 5.4.6. Kaufmann and Frühwirth-Schnatter (2002) noted sensitivity of the model selection procedure with respect to the prior on $\gamma_k$. Selecting the model order $m$ is unaffected by this prior and yields $m = 4$. Depending on the prior, the marginal likelihood would favor either a model with three or four states; see Table 12.3. This sensitivity may be explained by the fact that for a four-state model one of the states corresponds to a single outlier. Thus little information on the parameters of the fourth state is available from the likelihood and the prior dominates the posterior distribution.

### 12.5.5 Switching GARCH Models

Francq et al. (2001) consider the following switching GARCH$(m, n)$ model, where all coefficients are switching,

$$\begin{aligned} &Y_t = \sigma_t \varepsilon_t, \qquad \varepsilon_t \sim \mathcal{N}(0,1), \\ &\sigma_t^2 = \gamma_{S_t} + \alpha_{S_t,1} y_{t-1}^2 + \cdots + \alpha_{S_t,m} y_{t-m}^2 + \delta_{S_t,1}\sigma_{t-1}^2 + \cdots + \delta_{S_t,n}\sigma_{t-n}^2. \end{aligned} \tag{12.31}$$

By recursive substitution it becomes evident that the predictive density $p(y_t|\mathbf{y}^{t-1}, \mathbf{S}^t, \boldsymbol{\vartheta})$ depends on the whole history of $S_t$. For the switching GARCH$(1, 1)$ model, for instance, the variance of the predictive density reads:

$$\begin{aligned} \sigma_t^2 = {} & \gamma_{S_t} + \alpha_{S_t,1} y_{t-1}^2 + \delta_{S_t,1}(\gamma_{S_{t-1}} + \alpha_{S_{t-1},1} y_{t-2}^2) \\ & + \delta_{S_t,1}\gamma_{S_{t-1}}(\gamma_{S_{t-2}} + \alpha_{S_{t-2},1} y_{t-3}^2) + \cdots . \end{aligned}$$

Thus the model obeys only the weakest assumption **Y1** defined in Subsection 10.3.4. Due to the work of Francq et al. (2001), the theoretical properties of the switching GARCH models are well understood.

First, Francq et al. (2001) establish necessary and sufficient conditions ensuring the existence of a strictly stationary solution by rewriting (12.31) as a stochastic dynamic system and considering the Lyapunov exponent of this system as in Bougerol and Picard (1992a). For the switching GARCH$(1, 1)$ model, for instance, this condition reads:

$$\sum_{k=1}^{K} \eta_k \mathrm{E}(\log\left(\alpha_{k,1}\varepsilon_t^2 + \delta_{k,1}\right)) < 0,$$

which reduces for $K = 1$ to the result given by Nelson (1990) for the standard GARCH$(1, 1)$ model. This condition, however, does not guarantee the existence of the unconditional variance of $Y_t$.

Francq et al. (2001) establish necessary and sufficient conditions for the existence of second-order stationary solutions, which reduce to the requirement that the spectral radius of a matrix derived from the stochastic dynamic system mentioned above is strictly less than one. For a GARCH$(m, n)$ model where only the intercept is switching, this condition reduces to:

$$\sum_{j=1}^{m} \alpha_j + \sum_{j=1}^{n} \delta_j < 1,$$

which is equal to the condition given by Bollerslev (1986) for a standard GARCH$(m, n)$ model.

Finally, Francq and Zakoian (1999) establish necessary and sufficient conditions for the existence of higher-order moments of $Y_t^2$. They show that $Y_t^2$ admits a linear ARMA representation where the orders depend on $m$, $n$, and the model coefficients, extending the well-known result that a GARCH$(m, n)$ process has the same autocorrelation as an ARMA$(\max(m, n), m)$ process. Similar ARMA representations are also derived for powers of $Y_t^2$.

Practical application of switching GARCH models include stock market returns (Dueker, 1997) and exchange rate data (Klaasen, 2002).

## 12.6 Some Extensions

### 12.6.1 Time-Varying Transition Matrices

Whereas the transition matrix $\boldsymbol{\xi}$ of the hidden process $S_t$ is time invariant under assumption **S3** or **S4**, the transition probability from $S_{t-1}$ to $S_t$ may depend on exogenous variables under assumption **S1** or **S2**, as suggested by Goldfeld and Quandt (1973).

For a two-state Markov switching model, the transition probabilities $\xi_{S_{t-1},S_t}$ may be reparameterized through a logit model in the following way,

$$\xi_{S_{t-1},S_t} = \frac{\exp(\kappa_{S_{t-1},1})}{1 + \exp(\kappa_{S_{t-1},1})}, \qquad S_t \neq S_{t-1}.$$

A univariate exogenous variable $z_t$ may then be included as in Subsection 8.6.2:

$$\xi_{S_{t-1},S_t} = \frac{\exp(\kappa_{S_{t-1},1} + z_t\kappa_{S_{t-1},2})}{1 + \exp(\kappa_{S_{t-1},1} + z_t\kappa_{S_{t-1},2})}, \qquad S_t \neq S_{t-1}, \qquad (12.32)$$

with $\kappa_{j,1}$ and $\kappa_{j,2}$, $j = 1, 2$ being unknown parameters. Note that the transition probability $\xi_{S_{t-1},S_t}$ not only depends on $z_t$, but also on the state of $S_{t-1}$. The logit transform could be substituted by another increasing function $F(\cdot)$,

$$\xi_{S_{t-1},S_t} = F(\kappa_{S_{t-1},1} + z_t\kappa_{S_{t-1},2}), \qquad S_t \neq S_{t-1}, \qquad (12.33)$$

for instance, the standard normal distribution. If $z_t$ is equal to a lagged value of $Y_t$, $z_t = y_{t-d}$ for some $d > 0$, then the so-called endogenous selection MSAR model (Krolzig, 1997, Subsection 10.3.2) results. If, in addition, the parameters of model (12.32) or (12.33) are independent of the state of $S_{t-1}$, $\kappa_{1,1} = \kappa_{2,1}$, $\kappa_{1,2} = \kappa_{2,2}$, then the resulting model is closely related to the smooth transition autoregressive model (Teräsvirta and Anderson, 1992; Granger and Teräsvirta, 1993). Extensions to multiple exogenous variables $\boldsymbol{z}_t = (z_{t1}, \dots, z_{tr})$ and to more than two states are possible.

Models with time-varying transition matrices found applications in hydrology (Zucchini and Guttorp, 1991; Pfeiffer and Jeffries, 1999), in financial econometrics (Diebold et al., 1994; Peria, 2002; Schaller and van Norden, 2002; Ang and Bekaert, 2002), and in business cycle analysis to capture duration dependence, meaning that the transition probability between recession and boom depends on how long the economy remained within the same regime (Durland and McCurdy, 1994; Filardo, 1994; Filardo and Gordon, 1998).

A model with time-varying transition matrices may be estimated through the EM algorithm (Diebold et al., 1994) or through MCMC methods (Filardo and Gordon, 1998).

### 12.6.2 Markov Switching Models for Longitudinal and Panel Data

Some recent papers combine clustering methods and longitudinal analysis using hidden Markov models. In a health state model comparing the effectiveness of two different medications for schizophrenia, Scott et al. (2005) assume that the observed response $\mathbf{y}_{it}$ for patient $i$ at time $t$ follows a multivariate Student-$t$ distribution,

$$\mathbf{y}_{it} | S_{it} = k \sim t_{\nu_k} (\boldsymbol{\mu}_k, \boldsymbol{\Sigma}_k), \tag{12.34}$$

depending on a latent health state $S_{it}$. The health state is assumed to be a hidden Markov chain with treatment-dependent transition matrix. Estimation of this model is carried out using MCMC and BIC was used to select the number of health states.

Frühwirth-Schnatter and Kaufmann (2006b) combine clustering and Markov switching models in economic panel data analysis by assuming that $K$ hidden groups are present in a panel and that within each group the parameters may switch according to a hidden Markov chain. Consider, for example, the mixed-effects model defined in Subsection 8.5.2,

$$y_{it} = \mathbf{x}_{it}^f \boldsymbol{\alpha} + \mathbf{x}_{it}^r \boldsymbol{\beta}_{it}^s + \varepsilon_{it}, \qquad \varepsilon_{it} \sim \mathcal{N}\left(0, \sigma_\varepsilon^2\right). \tag{12.35}$$

$\boldsymbol{\beta}_{it}^s$ depends on two latent discrete indicators, first on a group indicator $S_i$. Second, within each group the regression coefficient corresponding to $\mathbf{x}_{it}^r$ may switch between two states, commonly thought of as the state of the economy, depending on a group-specific hidden Markov chain $I_{t,k}$ with group-specific transition matrix $\boldsymbol{\xi}_k$:

$$\boldsymbol{\beta}_{it}^s = \boldsymbol{\beta}_k + (I_{t,k} - 1)\boldsymbol{\gamma}_k, \quad S_i = k.$$

This model allows pooling all time series within each group and is robust against structural changes through including the hidden Markov chain. A simplified version where the hidden Markov chain is group independent has been considered by Frühwirth-Schnatter and Kaufmann (2006a). Estimation is carried out using MCMC methods.

For $K = 1$, this model reduces to the panel data Markov switching model that has been applied to analyze the lending behavior of banks over the business cycle (Asea and Blomberg, 1998; Kaufmann, 2002):

$$y_{it} = \mathbf{x}_{it}^f \boldsymbol{\alpha} + \mathbf{x}_{it}^r \boldsymbol{\beta}_{S_t} + \varepsilon_{it}, \qquad \varepsilon_{it} \sim \mathcal{N}\left(0, \sigma_\varepsilon^2\right), \tag{12.36}$$

which allows a shift in the regression coefficient corresponding to $\mathbf{x}_{it}^r$ between the two states of $S_t$, commonly thought of as the state of the economy. Estimation of model (12.36) may be carried out using the EM algorithm (Asea and Blomberg, 1998) or MCMC methods (Kaufmann, 2002; Frühwirth-Schnatter and Kaufmann, 2006a).

### 12.6.3 Markov Switching Models for Multivariate Time Series

Hidden Markov models have been extended in several ways to deal with multivariate time series $\{\mathbf{Y}_t, t = 1, \ldots, T\}$, where $\mathbf{Y}_t$ is random vector of $r$ different variables, for instance, the GDP from different countries. Common multivariate time series models are the vector autoregressive (VAR) model (Sims, 1980) and cointegration models (Engle and Granger, 1987); see also Shumway and Stoffer (2000, Chapter 4) for a review of multivariate time series analysis.

To analyze the growth rate of GDP in a two-country set-up, Phillips (1991) generalized the univariate MSAR model (Hamilton, 1989) by introducing a hidden Markov chain into a bivariate VAR(1) model:

$$\mathbf{Y}_t - \boldsymbol{\mu}_{\mathbf{S}_t} = \boldsymbol{\Phi}(\mathbf{Y}_{t-1} - \boldsymbol{\mu}_{\mathbf{S}_{t-1}}) + \boldsymbol{\varepsilon}_t, \qquad \boldsymbol{\varepsilon}_t \sim \mathcal{N}_r\left(\mathbf{0}, \boldsymbol{\Sigma}\right),$$

with $\boldsymbol{\Phi}$ and $\boldsymbol{\Sigma}$ being $(2 \times 2)$ matrices, and $\boldsymbol{\mu}_{\mathbf{S}_t}$ being a vector of length 2. $\mathbf{S}_t = (S_{t,1}, S_{t,2})$ is a bivariate two-state hidden Markov chain, with $S_{t,j}$ describing the state of the economy in country $j$, which could be coded as a single Markov chain with four states. The $(4 \times 4)$-transition matrix $\boldsymbol{\xi}$ of $\mathbf{S}_t$ is unrestricted if the states of the two economies are correlated, a restricted transition matrix results if $S_{t,1}$ and $S_{t,2}$ are assumed to be independent. Hamilton and Lin (1996) apply a related model to analyze jointly growth in industrial production and volatility in stock returns, and discuss restricted transition matrices where one indicator is leading the other.

Krolzig (1997) considered multivariate MS-VAR models, where a single hidden Markov chain $S_t$ may affect the intercept (or the mean level), the matrix containing the AR coefficients as well as the error covariance matrix:

$$\mathbf{Y}_t = \boldsymbol{\Phi}_{S_t}\mathbf{Y}_{t-1} + \boldsymbol{\zeta}_{S_t} + \boldsymbol{\varepsilon}_t, \qquad \boldsymbol{\varepsilon}_t \sim \mathcal{N}_r\left(\mathbf{0}, \boldsymbol{\Sigma}_{S_t}\right),$$

with $\boldsymbol{\Phi}_{S_t}$ and $\boldsymbol{\Sigma}_{S_t}$ being $(r \times r)$ matrices and $\boldsymbol{\zeta}_{S_t}$ being a vector of length $r$. Krolzig (1997) discusses ML estimation of this model using the EM algorithm as well as Bayesian estimation using Gibbs sampling.

A related model is applied in Ang and Bekaert (2002) to model interest rates from three different countries, however, $\mathbf{S}_t = (S_{t,1}, S_{t,2}, S_{t,3})$ is a trivariate two-state hidden Markov chain, with $S_{t,j}$ describing the hidden state in country $j$, coded as a single Markov chain with eight states. Because the states in different countries are assumed to be independent, a restricted transition matrix $\boldsymbol{\xi}$ results for $\mathbf{S}_t$.

Economic theory implies a long-run relationship between certain integrated time series such as consumption and disposable income, implying that the time series are cointegrated. As with unit root tests, discussed in Subsection 12.2.4, common cointegration tests are affected by shifts in the growth rate of the underlying time series (Hall et al., 1997). For this reason several authors considered the introduction of a hidden Markov chain into cointegration models to account for unexpected shifts.

Paap and van Dyck (2003) introduce a multivariate Markov switching trend model that accounts for different growth rates in a bivariate time series $\mathbf{Y}_t$, containing the log of per capita consumption and disposable income:

$$\mathbf{Y}_t = \boldsymbol{\mu}_t + (S_t - 1)\begin{pmatrix} \delta \\ 0 \end{pmatrix} + \mathbf{Z}_t,$$

$$\boldsymbol{\mu}_t = \boldsymbol{\mu}_{t-1} + \boldsymbol{\beta}_{S_t},$$

where $S_t$ is a two-state hidden Markov chain. $\boldsymbol{\beta}_1$ is a vector containing the slopes of the trend function of both time series, if $S_t = 1$ (expansion) and $\boldsymbol{\beta}_2$ contains the slopes if $S_t = 2$ (recession). $\delta$ accounts for possible level shifts in the first time series during recession. $\mathbf{Z}_t$ is assumed to follow a standard VAR($p$) process. Cointegration analysis based on the vector error correction model is then carried out for $\mathbf{Z}_t$:

$$\triangle \mathbf{Z}_t = \boldsymbol{\Pi} \mathbf{Z}_t + \sum_{j=1}^{p-1} \tilde{\boldsymbol{\Phi}}_j \triangle \mathbf{Z}_{t-j} + \boldsymbol{\varepsilon}_t, \qquad \boldsymbol{\varepsilon}_t \sim \mathcal{N}_2\left(\mathbf{0}, \boldsymbol{\Sigma}\right). \tag{12.37}$$

Depending on the rank of $\boldsymbol{\Pi}$ three cases arise. If $\boldsymbol{\Pi}$ has rank zero, then the bivariate MS-VAR model for the growth rates results; if $\boldsymbol{\Pi}$ has rank two, then $\mathbf{Z}_t$ is stationary and a generalization of the model of Lam (1990) results; and finally, if $\boldsymbol{\Pi}$ has rank one, then the two time series are cointegrated. Bayesian estimation of this model is carried out in Paap and van Dyck (2003) using MCMC and the Bayes factor is used to test for the cointegration rank (Kleibergen and Paap, 2002). The empirical results of Paap and van Dyck (2003) suggest the existence of a cointegration relationship between U.S. per capita disposable income and consumption.

Related approaches are a single equation cointegration analysis where the parameters are allowed to undergo changes driven by a hidden Markov chain (Hall et al., 1997) and an alternative Markov switching vector error correction (MS-VEC) model (Krolzig and Sensier, 2000; Krolzig, 2001) where a Markov switching intercept is introduced directly into a vector error correction model for $\mathbf{Y}_t$; see also Krolzig (1997, Chapter 13).

Further Markov switching models for multivariate time series, in particular the Markov switching model dynamic factor model (Diebold and Rudebusch, 1996; Kim and Nelson, 1998), are special cases of switching Gaussian state space models which are studied in Chapter 13.

# 13

# Switching State Space Models

## 13.1 State Space Modeling

As an introduction into the vast area of state space modeling, we start in Subsection 13.1.1 with the local level model which is a simple but characteristic example of the linear Gaussian state space form that is discussed in full generality in Section 13.1.2.

### 13.1.1 The Local Level Model with and Without Switching

In a local level model, a random process $\{Y_1, \dots, Y_T\}$ is generated by the following stochastic difference equation,

$$\mu_t = \mu_{t-1} + w_t, \qquad w_t \sim \mathcal{N}\left(0, \sigma_\mu^2\right), \tag{13.1}$$

$$Y_t = \mu_t + \varepsilon_t, \qquad \varepsilon_t \sim \mathcal{N}\left(0, \sigma_\varepsilon^2\right), \tag{13.2}$$

where all error terms $w_t$ and $\varepsilon_t$ are mutually independent and independent of $\mu_0$; see, for instance, Durbin and Koopman (2001) for an excellent introduction. The random process $\{Y_1, \dots, Y_T\}$ is assumed to be observable and the realizations are denoted by $\{y_1, \dots, y_T\}$. The distribution of $Y_t$ is allowed to depend on an unobservable latent variable, in this case the level $\mu_t$, which follows a random walk. Because $\mu_t$ is a hidden Markov process, this model is related to the hidden Markov chain model considered in Chapter 10; the latent process, however, does not live on a discrete state space, but on a continuous one.

The process $Y_t$ is nonstationary as long as the variance $\sigma_\mu^2$ is positive. There is a close relationship between the local level model and more classical time series models; see, for instance, Abraham and Ledolter (1986). By taking first differences, we obtain:

$$\Delta Y_t = w_t + \varepsilon_t - \varepsilon_{t-1}.$$

The lag 1 autocorrelation is given by

$$\rho_{\Delta Y_t}(1) = -\sigma_\varepsilon^2/(2\sigma_\varepsilon^2 + \sigma_\mu^2),$$

whereas higher autocorrelations are zero. Because this autocorrelation function is the same as that of an MA(1) process, the local level model has an ARIMA(0, 1, 1) representation, where the MA(1) coefficient $\theta_1$ is constrained to the interval $[0, 1]$ and results from equating the lag 1 autocorrelation in both models:

$$\rho_{\Delta Y_t}(1) = \frac{\theta_1}{1+\theta_1^2} \Rightarrow \theta_1 = \frac{1 - \sqrt{1 - 4\rho_{\Delta Y_t}(1)^2}}{2\rho_{\Delta Y_t}(1)}.$$

The advantage of the state space form (13.1) and (13.2) as compared to the ARIMA(0, 1, 1) representation is manifold as discussed extensively in the monograph of West and Harrison (1997). First, one may extract much more information from the observed time series as it is possible to estimate the level $\mu_t$ for each $t$ using the Kalman filter; see Section 13.3. Second, further components capturing seasonal patterns in the time series or trend behavior are easily added, as discussed in Subsection 13.2.1. Third, it is much easier to deal with time series irregularities such as outliers or structural breaks in the state space form.

Consider, for instance, a process generated by a local level model that is disrupted by occasional observation outliers. Based on the finite mixture approach to outlier modeling discussed in Section 7.2, the local level model may be modified in the following way,

$$\begin{aligned} \mu_t &= \mu_{t-1} + w_t, & w_t &\sim \mathcal{N}\left(0, \sigma_\mu^2\right), \\ Y_t &= \mu_t + \varepsilon_t, & \varepsilon_t &\sim \eta_1 \mathcal{N}\left(0, \sigma_{\varepsilon,1}^2\right) + \eta_2 \mathcal{N}\left(0, \sigma_{\varepsilon,2}^2\right). \end{aligned}$$

After introducing an i.i.d. binary indicator $S_t$ with $\Pr(S_t = 1) = \eta_1$ as in earlier chapters, conditional on knowing $S_t$, this model is a local linear model as defined in (13.1) and (13.2), however, the observation variance $\sigma_\varepsilon^2$ is switching between two values:

$$\begin{aligned} \mu_t &= \mu_{t-1} + w_t, & w_t &\sim \mathcal{N}\left(0, \sigma_\mu^2\right), \\ Y_t &= \mu_t + \varepsilon_t, & \varepsilon_t &\sim \mathcal{N}\left(0, \sigma_{\varepsilon,S_t}^2\right). \end{aligned}$$

This is a first example of a switching state space model, which is commonly applied to deal with outliers in time series; see Subsection 13.2.3 for a more detailed discussion. Another useful switching state space model results if $S_t$ follows a hidden Markov chain as in Chapter 10, rather than an i.i.d. process, because this introduces conditional heteroscedasticity in the error term $\varepsilon_t$; see also Subsection 13.2.2.

Apart from the observation variance, other model parameters may be switching in a state space model. Consider, for instance, the variance $\sigma_\mu^2$ in the local level model, which determines how much $\mu_t$ changes over time. The smaller $\sigma_\mu^2$, the less flexibility of $\mu_t$ is allowed a priori. To distinguish between periods of smaller and greater variability of $\mu_t$ it may be assumed that the variance $\sigma_\mu^2$ switches between two values:

$$\mu_t = \mu_{t-1} + w_t, \qquad w_t \sim \mathcal{N}\left(0, \sigma_{\mu,S_t}^2\right).$$

A special case of this model is one where $\sigma_{\mu,1}^2$ is 0, whereas $\sigma_{\mu,2}^2 > 0$. Such a model allows for an occasional level shift:

$$\mu_t = \begin{cases} \mu_{t-1}, & S_t = 1 \\ \mu_{t-1} + w_t, \quad w_t \sim \mathcal{N}\left(0, \sigma_\mu^2\right), & S_t = 2. \end{cases} \tag{13.3}$$

Finally, two independent indicators $S_t^1$ and $S_t^2$ may be introduced to combine heteroscedasticity in $w_t$ with observation outliers:

$$\begin{aligned} \mu_t &= \mu_{t-1} + w_t, \qquad w_t \sim \mathcal{N}\left(0, \sigma_{\mu,S_t^1}^2\right), \\ Y_t &= \mu_t + \varepsilon_t, \qquad \varepsilon_t \sim \mathcal{N}\left(0, \sigma_{\varepsilon,S_t^2}^2\right). \end{aligned}$$

### 13.1.2 The Linear Gaussian State Space Form

The local level model introduced in the previous subsection is a special case of a linear Gaussian state space model, which is a dynamic stochastic system defined in the following way,

$$\mathbf{x}_t = \mathbf{F}_t \mathbf{x}_{t-1} + \mathbf{w}_t, \qquad \mathbf{w}_t \sim \mathcal{N}_d\left(\mathbf{0}, \mathbf{Q}_t\right), \tag{13.4}$$

$$\mathbf{Y}_t = \mathbf{H}_t \mathbf{x}_t + \boldsymbol{\varepsilon}_t, \qquad \boldsymbol{\varepsilon}_t \sim \mathcal{N}_r\left(\mathbf{0}, \mathbf{R}_t\right), \tag{13.5}$$

where $t = 1, \dots, T$. The key variables in these formulations are the state variable $\mathbf{x}_t$ and the observation variable $\mathbf{Y}_t$.

The state variable $\mathbf{x}_t$ is a latent $d$-dimensional random vector, which is observed only indirectly through the effect it has on the distribution of $\mathbf{Y}_t$. The transition equation (13.4), also called the state equation, specifies for each $t \geq 1$ how $\mathbf{x}_t$ is generated from the previous state variable $\mathbf{x}_{t-1}$. The linear relationship between $\mathbf{x}_t$ and $\mathbf{x}_{t-1}$ which depends on the $(d \times d)$ matrix $\mathbf{F}_t$ is disturbed by a zero-mean error $\mathbf{w}_t$ following a normal distribution with variance–covariance matrix $\mathbf{Q}_t$. To complete the model formulation, the distribution of $\mathbf{x}_0$ is specified as $\mathbf{x}_0 \sim \mathcal{N}_d\left(\hat{\mathbf{x}}_{0|0}, \mathbf{P}_{0|0}\right)$.

The observation variable $\mathbf{Y}_t$ is a random vector of dimension $r$, which is assumed to be observable for all time points $t = 1, \dots, T$. A single realization of this process is denoted by $\{\mathbf{y}_1, \dots, \mathbf{y}_T\}$. The dimension of $\mathbf{Y}_t$ may be smaller, larger or equal to the dimension of $\mathbf{x}_t$. For a scalar observation variable with $r = 1$ we write $Y_t$ and denote the observed time series by $\{y_1, \dots, y_T\}$. The observation equation (13.5), also called the measurement equation, specifies how the distribution of $\mathbf{Y}_t$ is influenced by the state variable $\mathbf{x}_t$. The linear relationship between $\mathbf{Y}_t$ and $\mathbf{x}_t$ which depends on the $(r \times d)$ matrix $\mathbf{H}_t$ is disturbed by a zero-mean random observation error $\boldsymbol{\varepsilon}_t$ following a normal distribution with variance–covariance matrix $\mathbf{R}_t$, which reduces to a scalar variance for $r = 1$.

Often the matrices $\mathbf{F}_t$, $\mathbf{H}_t$, $\mathbf{Q}_t$, and $\mathbf{R}_t$ emerge from putting a specific time series model into a state space form; see Subsection 13.2.1. The local level model introduced in Subsection 13.1.1, for instance, results with $\mathbf{x}_t = \mu_t$, $\mathbf{F}_t = \mathbf{H}_t = 1$, $\mathbf{Q}_t = \sigma^2_\mu$, and $\mathbf{R}_t = \sigma^2_\varepsilon$. The matrices $\mathbf{F}_t$, $\mathbf{H}_t$, $\mathbf{Q}_t$, and $\mathbf{R}_t$ need not depend on time, in which case the notation $\mathbf{F}$, $\mathbf{H}$, $\mathbf{Q}$, and $\mathbf{R}$ will be used.

Some elements of the matrices $\mathbf{F}_t$, $\mathbf{H}_t$, $\mathbf{Q}_t$, and $\mathbf{R}_t$ may depend on unknown model parameters $\boldsymbol{\vartheta}$, such as for the local level model, where $\boldsymbol{\vartheta} = (\sigma^2_\mu, \sigma^2_\varepsilon)$. The notations $\mathbf{F}_t(\boldsymbol{\vartheta})$, $\mathbf{H}_t(\boldsymbol{\vartheta})$, $\mathbf{Q}_t(\boldsymbol{\vartheta})$, and $\mathbf{R}_t(\boldsymbol{\vartheta})$ are used whenever it is necessary to make this dependence explicit. Identification becomes an important issue whenever part of the system matrices $\mathbf{F}_t(\boldsymbol{\vartheta})$ and $\mathbf{H}_t(\boldsymbol{\vartheta})$ are unknown; see Hannan and Deistler (1988) for an extensive treatment of this issue.

Further assumptions are necessary to complete the model definition. Most important, $\mathbf{w}_t$ is uncorrelated with $\mathbf{x}_{t-1}$ for all $t$:

$$\mathrm{E}\left(\mathbf{w}_t\mathbf{x}_{t-1}^{'}\right) = \mathbf{0}, \qquad t = 1, \ldots, T.$$

Second, the observation error $\boldsymbol{\varepsilon}_t$ as well as $\mathbf{w}_t$ is uncorrelated over time:

$$\mathrm{E}\left(\boldsymbol{\varepsilon}_t\boldsymbol{\varepsilon}_s^{'}\right) = \mathbf{0}, \qquad \mathrm{E}\left(\mathbf{w}_t\mathbf{w}_s^{'}\right) = \mathbf{0}, \qquad \forall s, t \in \{1, \ldots, T\}, t \neq s.$$

Finally, the two error sequences $\boldsymbol{\varepsilon}_t$ and $\mathbf{w}_s$ are uncorrelated for all $t, s$:

$$\mathrm{E}\left(\boldsymbol{\varepsilon}_t\mathbf{w}_s^{'}\right) = \mathbf{0}, \qquad \forall s, t \in \{1, \ldots, T\}.$$

On various occasions, it is useful to introduce additional terms that influence the mean of the transition as well as the observation equation:

$$\mathbf{x}_t = \mathbf{F}_t\mathbf{x}_{t-1} + \mathbf{G}_t\mathbf{u}_t + \mathbf{w}_t, \tag{13.6}$$

$$\mathbf{Y}_t = \mathbf{H}_t\mathbf{x}_t + \mathbf{A}_t\mathbf{z}_t + \boldsymbol{\varepsilon}_t. \tag{13.7}$$

In (13.6), $\mathbf{u}_t$ is a vector of dimension $n$, which may be smaller, larger, or equal to the dimension of $\mathbf{x}_t$. In engineering applications $\mathbf{u}_t$ often is a controllable input vector (Anderson and Moore, 1979). In econometric problems $\mathbf{u}_t$ often is a vector of $n$ exogenous variables being observable at time $t$. The expected value of $\mathbf{x}_t$ given $\mathbf{u}_t$ and $\mathbf{x}_{t-1}$ is a linear function in $\mathbf{u}_t$, depending on the $(d \times n)$ matrix $\mathbf{G}_t$. In (13.7), $\mathbf{z}_t$ is a vector of $m$ variables being observable at time $t$, which could be exogenous variables or past values of $\mathbf{Y}_t$. The expected value of $\mathbf{Y}_t$ given $\mathbf{x}_t$ and $\mathbf{z}_t$ is a linear function in $\mathbf{z}_t$, depending on the $(r \times m)$ matrix $\mathbf{A}_t$. For a further review of various aspects of state space modeling we refer to the monographs of Aoki (1990), Harvey (1993), West and Harrison (1997), and Durbin and Koopman (2001).

Originally, the state space model was developed by Kalman (1960, 1961) in aerospace research for tracking some target such as an aircraft. In this application the transition equation is derived from physical laws describing the motion of the target, whereas the observation vector measures properties of this target that are observable through some device such as a radar, subject to some measurement error; see also related tracking problems in high-energy physics (Frühwirth, 1987). Due to their flexibility and generality state space models found applications in many research areas in engineering such as hydrology (Schnatter et al., 1987) and speech recognition (Juang and Rabiner, 1985; Rabiner and Juang, 1986), just to mention two; see Anderson and Moore (1979) for further references.

The application of state space models in the econometric literature started in the 1970s with the time-varying coefficient model; see the review of Nicholls and Pagan (1985). In the 1980s, it was recognized that econometric models that rely on unobservable quantities could be cast into a state space form, and state space models found wide applications in economics and finance, for instance, to estimate the ex ante real interest rate (Fama and Gibbons, 1982), unobserved expected inflation (Burmeister et al., 1986), or the potential real GDP (Kuttner, 1994); see Granger and Teräsvirta (1993) and Kim and Nelson (1999) for further applications of state space models in econometrics.

### 13.1.3 Multiprocess Models

The simplest way of introducing a latent discrete indicator into the linear Gaussian state space form is multiprocess models. A multiprocess model is a collection of $K$ state space models, indexed by a hidden random indicator $S$ taking values in a discrete space $\{1, \dots, K\}$. Conditional on knowing the state of $S$, the model for $\mathbf{Y}_t$ is a linear Gaussian state space form:

$$\mathbf{x}_t = \mathbf{F}_t^{[S]}\mathbf{x}_{t-1} + \mathbf{G}_t^{[S]}\mathbf{u}_t + \mathbf{w}_t, \qquad \mathbf{w}_t \sim \mathcal{N}_d\left(\mathbf{0}, \mathbf{Q}_t^{[S]}\right), \tag{13.8}$$

$$\mathbf{Y}_t = \mathbf{H}_t^{[S]}\mathbf{x}_t + \mathbf{A}_t^{[S]}\mathbf{z}_t + \boldsymbol{\varepsilon}_t, \qquad \boldsymbol{\varepsilon}_t \sim \mathcal{N}_r\left(\mathbf{0}, \mathbf{R}_t^{[S]}\right). \tag{13.9}$$

Multiprocess models were well known in the control engineering literature for many years (see, for instance, Magill, 1965) before they were introduced into the statistics literature by Harrison and Stevens (1976). Multiprocess models were applied to forecasting multiple time series (Schnatter et al., 1987), to deal with unobserved heterogeneity in longitudinal studies (Gamerman and Smith, 1996), or to cluster time series in panel data (Frühwirth-Schnatter and Kaufmann, 2006b).

### 13.1.4 Switching Linear Gaussian State Space Models

The basic idea of a switching state space model is that a priori no single model is expected to hold for all time points $t$, rather the possibility that different models hold at different times points is explicitly recognized by modeling the hidden model indicator $S_t$ as being dynamic over time.

A switching linear Gaussian state space model is based on the state space form introduced in Subsection 13.1.2, however, some (or all) system matrices are driven by a hidden model indicator $S_t$:

$$\mathbf{x}_t = \mathbf{F}_t^{[S_t]}\mathbf{x}_{t-1} + \mathbf{G}_t^{[S_t]}\mathbf{u}_t + \mathbf{w}_t, \quad \mathbf{w}_t \sim \mathcal{N}_d\left(\mathbf{0}, \mathbf{Q}_t^{[S_t]}\right), \tag{13.10}$$

$$\mathbf{Y}_t = \mathbf{H}_t^{[S_t]}\mathbf{x}_t + \mathbf{A}_t^{[S_t]}\mathbf{z}_t + \boldsymbol{\varepsilon}_t, \quad \boldsymbol{\varepsilon}_t \sim \mathcal{N}_r\left(\mathbf{0}, \mathbf{R}_t^{[S_t]}\right). \tag{13.11}$$

$\{S_t, t = 1, \ldots, T\}$ is a sequence of random variables, allowed to take values in the discrete space $\{1, \ldots, K\}$. The degenerate case $S_t \equiv S_{t-1} \equiv S$ reduces to the multiprocess model introduced in Subsection 13.1.3.

To complete the model specification, some probabilistic structure has to be imposed on $S_t$. We distinguish two cases of switching state space models, namely finite mixtures of state space models, if $S_t$ is an i.i.d. sequence with probability distribution $\boldsymbol{\eta} = (\eta_1, \ldots, \eta_K)$, and Markov switching state space models, if $S_t$ is a hidden Markov chain with transition matrix $\boldsymbol{\xi}$ as introduced in Chapter 10. The first structure may be regarded as a special case of the second structure with restricted transition matrix; see Subsection 10.2.6. Finite mixtures of state space models are sometimes called "multi process models" (Harrison and Stevens, 1976; Smith and West, 1983), whereas Markov switching state space models are sometimes called "state space models with regime switching" (Kim and Nelson, 1999).

The engineering literature has seen several pioneering works on switching state space models since the 1960s. For target tracking problems, for instance, Nahi (1969) assumes a nonzero probability that any observation consists of noise only, leading to a state space model where $\mathbf{H}_t$ is switching between a zero and a nonzero value; see Bar-Shalom and Tse (1975) and Shumway and Stoffer (1991) for a related application. In control engineering research Ackerson and Fu (1970) consider a linear Gaussian state space model, where the covariance matrices in the transition and in the observation equation are allowed to depend on a hidden Markov chain.

Harrison and Stevens (1976) introduced finite mixtures of linear Gaussian state space models into the statistics literature; further applications are found in medicine (Smith and West, 1983; Gordon and Smith, 1990), speech recognition (Juang and Rabiner, 1985; Rabiner and Juang, 1986), and hydrology (Schnatter, 1988b). The Markov switching linear Gaussian state space model became popular in econometrics through the work of Kim (1993a, 1993b, 1994) and Shephard (1994); see also the monograph of Kim and Nelson (1999) for further applications and references.

### 13.1.5 The General State Space Form

A useful way of thinking of a state space model is in terms of a hierarchical model, where on a first level the model specifies the conditional distribution

$p(\mathbf{y}_1, \dots, \mathbf{y}_T|\mathbf{x}_1, \dots, \mathbf{x}_T)$ of the process $\mathbf{Y}_1, \dots, \mathbf{Y}_T$ given the whole state process $\mathbf{x}_1, \dots, \mathbf{x}_T$. On a second level the model characterizes the distribution $p(\mathbf{x}_1, \dots, \mathbf{x}_T)$ of the state process. The following structure is characteristic of a state space model. The random variables $\mathbf{Y}_1, \dots, \mathbf{Y}_T$ are independent of each other given the state process $\mathbf{x}_1, \dots, \mathbf{x}_T$ and $\mathbf{Y}_t$ is independent of $\mathbf{x}_1, \dots, \mathbf{x}_{t-1}$ given $\mathbf{x}_t$:

$$p(\mathbf{y}_1, \dots, \mathbf{y}_T|\mathbf{x}_1, \dots, \mathbf{x}_T) = \prod_{t=1}^{T} p(\mathbf{y}_t|\mathbf{x}_t). \tag{13.12}$$

The state variable $\mathbf{x}_t$ is a first-order hidden Markov process, hence independent of $\mathbf{x}_1, \dots, \mathbf{x}_{t-2}$ given $\mathbf{x}_{t-1}$:

$$p(\mathbf{x}_1, \dots, \mathbf{x}_T) = \prod_{t=1}^{T} p(\mathbf{x}_t|\mathbf{x}_{t-1}). \tag{13.13}$$

Thus to define a state space model, one could directly specify for each $t = 1, \dots, T$ the observation density $p(\mathbf{y}_t|\mathbf{x}_t)$ and the transition density $p(\mathbf{x}_t|\mathbf{x}_{t-1})$. Because these densities are in principle arbitrary, the hierarchical formulation is very useful, as it allows us to introduce nonlinearities in the relationship between $\mathbf{y}_t$ and $\mathbf{x}_t$ and $\mathbf{x}_t$ and $\mathbf{x}_{t-1}$ as well as nonnormality by densities that are intrinsically nonnormal.

For a linear Gaussian state space model the observation and the transition density evidently are given by

$$\begin{aligned} \mathbf{Y}_t|\mathbf{x}_t &\sim \mathcal{N}_r\left(\mathbf{H}_t\mathbf{x}_t + \mathbf{A}_t\mathbf{z}_t, \mathbf{R}_t\right), \\ \mathbf{x}_t|\mathbf{x}_{t-1} &\sim \mathcal{N}_d\left(\mathbf{F}_t\mathbf{x}_{t-1} + \mathbf{G}_t\mathbf{u}_t, \mathbf{Q}_t\right). \end{aligned}$$

For a switching state space model the distributions (13.12) and (13.13) are formulated conditional on knowing the hidden indicators $\mathbf{S} = (S_0, S_1, \dots, S_T)$, whereas a third level is added by describing the probability law for $\mathbf{S}$:

$$\begin{aligned} p(\mathbf{y}_1, \dots, \mathbf{y}_T|\mathbf{x}_1, \dots, \mathbf{x}_T, \mathbf{S}) &= \prod_{t=1}^{T} p(\mathbf{y}_t|\mathbf{x}_t, S_t), \\ p(\mathbf{x}_1, \dots, \mathbf{x}_T|\mathbf{S}) &= \prod_{t=1}^{T} p(\mathbf{x}_t|\mathbf{x}_{t-1}, S_t), \\ p(\mathbf{S}) &= \prod_{t=1}^{T} p(S_t|S_{t-1}). \end{aligned}$$

As in Chapter 10, the transition density as well as the observation density is assumed to depend on $S_t$ only.

## 13.2 Nonlinear Time Series Analysis Based on Switching State Space Models

### 13.2.1 ARMA Models with and Without Switching

The linear Gaussian state space form introduced in Subsection 13.1.2 subsumes many models that are popular in time series analysis, including regression models and ARMA models (Harvey, 1989; Shumway and Stoffer, 2000; Durbin and Koopman, 2001). Consider the ARMA$(p, q)$ process,

$$\boldsymbol{\delta}(L)(Y_t - \mu) = \boldsymbol{\theta}(L)\varepsilon_t, \qquad \varepsilon_t \sim \mathcal{N}\left(0, \sigma_\varepsilon^2\right),$$

where $L$ is the lag operator, $\boldsymbol{\delta}(L) = 1 - \delta_1 L - \cdots - \delta_p L^p$, and $\boldsymbol{\theta}(L) = 1 - \theta_1 L - \cdots - \theta_q L^q$, with $\delta_1, \ldots, \delta_p$ being the AR coefficients, and $\theta_1, \ldots, \theta_q$ being the MA coefficients. This model possesses for $q = p - 1$ the following state space representation with $\mathbf{x}_t \in \Re^p$,

$$Y_t = \mu + \mathbf{H}(\boldsymbol{\theta})\mathbf{x}_t, \tag{13.14}$$

$$\mathbf{x}_t = \mathbf{F}(\boldsymbol{\delta})\mathbf{x}_{t-1} + \begin{pmatrix} 1 \\ 0 \\ \vdots \\ 0 \end{pmatrix} \varepsilon_t, \qquad \varepsilon_t \sim \mathcal{N}\left(0, \sigma_\varepsilon^2\right), \tag{13.15}$$

with

$$\mathbf{F}(\boldsymbol{\delta}) = \begin{pmatrix} \delta_1 \ldots \delta_{p-1} & \delta_p \\ \mathbf{I}_{p-1} & \mathbf{0}_{(p-1)\times 1} \end{pmatrix}, \qquad \mathbf{H}(\boldsymbol{\theta}) = \begin{pmatrix} 1 & -\theta_1 & \cdots & -\theta_q \end{pmatrix}. \tag{13.16}$$

The same state space representation could be used if $q < p - 1$ after adding $(p - 1 - q)$ MA coefficients equal to 0: $\boldsymbol{\theta} = (\theta_1, \ldots, \theta_q, 0, \ldots, 0)$. If $q \geq p$, a similar state space representation with $\mathbf{x}_t \in \Re^{q+1}$ could be used, with the matrix $\mathbf{F}(\boldsymbol{\delta})$ in (13.15) being defined after adding $(q - p + 1)$ AR coefficients equal to 0: $\boldsymbol{\delta} = (\delta_1, \ldots, \delta_p, 0, \ldots, 0)$.

In complex modeling situations it is often easier to work with the state space representation, in particular when dealing with outliers, missing data, interventions, mixed-effects, and structural changes (Kohn and Ansley, 1986; Harvey et al., 1998).

This is, for instance, the case if a hidden Markov chain $S_t$ is introduced into an ARMA model, one example being the switching ARMA model (Billio and Monfort, 1998; Billio et al., 1999), for which the one-step ahead predictive density depends on the whole history of $S_t$. This long-range dependence disappears conditional on the latent variables $\mathbf{x}_t$ and $S_t$, if an ARMA$(p, q)$ process with switching mean is represented by the following switching state space model,

$$Y_t = \mu_{S_t} + \mathbf{H}(\boldsymbol{\theta})\mathbf{x}_t,$$

with the state equation being the same as in (13.15). This facilitates statistical inference in Section 13.3 and 13.4.

A similar result holds for the Markov switching autoregressive model of Lam (1990), defined earlier in (12.11), which has the following state space form,

$$Y_t = (1\ 1)\begin{pmatrix} \mu_t \\ \mathbf{x}_t \end{pmatrix},$$

$$\begin{pmatrix} \mu_t \\ \mathbf{x}_t \end{pmatrix} = \begin{pmatrix} 1 & 0 \\ 0 & \mathbf{F}(\boldsymbol{\delta}) \end{pmatrix} \begin{pmatrix} \mu_{t-1} \\ \mathbf{x}_{t-1} \end{pmatrix} + \begin{pmatrix} \beta_{S_t} \\ 0 \\ \vdots \\ 0 \end{pmatrix} + \begin{pmatrix} 0 \\ 1 \\ \vdots \\ 0 \end{pmatrix} \varepsilon_t,$$

with $\mathbf{x}_t$ and $\mathbf{F}(\boldsymbol{\delta})$ being the same as in (13.15).

### 13.2.2 Unobserved Component Time Series Models

The state space approach is also useful for decomposing a time series into unobserved components such as trend, cycles, seasonal, and irregular components (Harvey, 1989). A simple example of such a model is the local level model discussed in Subsection 13.1.1; a more flexible one is the basic structural model (Harvey and Todd, 1983):

$$\begin{aligned}
\mu_t &= \mu_{t-1} + \beta_{t-1} + w_{t,1}, \qquad w_{t,1} \sim \mathcal{N}\left(0, \sigma^2_\mu\right), \\
\beta_t &= \beta_{t-1} + w_{t,2}, \qquad w_{t,2} \sim \mathcal{N}\left(0, \sigma^2_\beta\right), \\
\gamma_t &= -\sum_{j=1}^{s-1} \gamma_{t-j} + w_{t,3}, \qquad w_{t,3} \sim \mathcal{N}\left(0, \sigma^2_\gamma\right), \\
Y_t &= \mu_t + \gamma_t + \varepsilon_t, \qquad \varepsilon_t \sim \mathcal{N}\left(0, \sigma^2_\varepsilon\right).
\end{aligned} \tag{13.17}$$

$\mu_t$ is the slowly varying trend of the time series, $\gamma_t$ is a periodic seasonal component, and $\varepsilon_t$ is a random disturbance term. If no seasonal component is present in (13.17) then the resulting model is called the local linear trend model.

Decomposing a time series into a stochastic and stationary component may lead to identification problems (Nelson, 1988). The local level model discussed in Subsection 13.1.1, for instance, is not identified if the two noise terms $\varepsilon_t$ and $w_t$ are allowed to be correlated. For this reason, it is assumed in the basic structural model that all error terms are uncorrelated.

Unobserved component models found numerous applications in economics and have been extended in several ways by including a hidden indicator. Many applications of this model typically are based on the assumption that the error terms in the state and in the observation equation are homoscedastic. Heteroscedasticity may be caused by outliers as discussed in Subsection 13.2.3.

In addition, it is reasonable to assume there exists some kind of conditional heteroscedasticity in that errors with large variances tend to be followed by errors with large variances and similarly errors with small variances tend to be followed by errors with small variances.

To capture heteroscedasticity, Harvey et al. (1992) consider unobserved component models with ARCH disturbances both in the transition as well as in the observation equation. As an alternative, Kim (1993b) introduced unobserved component time series models with Markov switching heteroscedasticity, by assuming that the variances depend on a hidden Markov chain:

$$\begin{aligned}\mathbf{x}_t &= \mathbf{F}\mathbf{x}_{t-1} + \mathbf{w}_t, \qquad \mathbf{w}_t \sim \mathcal{N}_d\left(\mathbf{0}, \mathbf{Q}_t^{[S_t]}\right),\\ Y_t &= \mathbf{H}\mathbf{x}_t + \varepsilon_t, \qquad \varepsilon_t \sim \mathcal{N}\left(0, \sigma^2_{\varepsilon,S_t}\right).\end{aligned}$$

Applications of this model include modeling the link between inflation rates and inflation uncertainty (Kim, 1993b) and analyzing the U.S. stock market with focus on the 1987 crash (Kim and Kim, 1996).

Alternatively, hidden indicators have been introduced into the structural part of unobserved component models. Whittaker and Frühwirth-Schnatter (1994) define a dynamic change-point model, where in a local level model, a random walk drift is added after a structural break:

$$\begin{aligned}\mu_t &= \mu_{t-1} + (S_t - 1)\beta_{t-1} + w_{t,1}, \qquad w_{t,1} \sim \mathcal{N}\left(0, \sigma^2_\mu\right),\\ \beta_t &= \beta_{t-1} + w_{t,2}, \qquad w_{t,2} \sim \mathcal{N}\left(0, \sigma^2_\beta\right),\end{aligned}$$

where $S_t$ is allowed a one-time change between state 1 and 2 at an unknown change-point $\tau$.

To capture different growth behavior in boom and recession, Luginbuhl and de Vos (1999) model the log gross domestic product by a switching local linear trend model. Two different drift components $\alpha_t$ and $\beta_t$ are assumed to be present, each of which follows a random walk, but only one of them contributes to the trend:

$$\begin{aligned}\mu_t &= \mu_{t-1} + (1 - S_t)\alpha_{t-1} + S_t\beta_{t-1} + w_{t,1},\\ \alpha_t &= \alpha_{t-1} + w_{t,2},\\ \beta_t &= \beta_{t-1} + w_{t,3},\\ Y_t &= \mu_t + \varepsilon_t, \qquad \varepsilon_t \sim \mathcal{N}\left(0, \sigma^2_\varepsilon\right).\end{aligned}$$

The indicator $S_t$, selecting one of the two trend components, is assumed to follow a hidden Markov chain with state space $\{0, 1\}$.

### 13.2.3 Capturing Sudden Changes in Time Series

Detecting sudden changes, outliers, and level shifts is an important aspect of practical time series analysis, often called intervention analysis (Tsay, 1988).

Many authors generalized the linear Gaussian state space model with the aim of establishing recursive estimation procedures that are robust to outliers (Masreliez, 1975; Masreliez and Martin, 1975; West, 1981, 1984; Tsai and Kurz, 1983; Peña and Guttman, 1988; Meinhold and Singpurwalla, 1989). Peña and Guttman (1988) generalized the scale-contaminated model (Tukey, 1960; Box and Tiao, 1968), already discussed in Subsection 7.2.1, to the framework of robust linear Gaussian state space models with univariate observation vector $Y_t$, by assuming that the noise $\varepsilon_t$ in the observation equation (13.5) follows a mixture of two normal distributions with mean zero, but different variances:

$$\varepsilon_t \sim (1-\eta_2)\mathcal{N}\left(0, \sigma_\varepsilon^2\right) + \eta_2 \mathcal{N}\left(0, k\sigma_\varepsilon^2\right),$$

where typically $\eta_2$ is a small fraction of outliers, whereas $k >> 1$. For estimation, however, it is useful to view such a robust state space model as a switching Gaussian state space model, where the distribution of the observation noise is driven by a hidden i.i.d. sequence $S_t$:

$$\sigma_{\varepsilon,S_t}^2 = \begin{cases} \sigma_\varepsilon^2, & S_t = 1, \\ k\sigma_\varepsilon^2, & S_t = 2, \end{cases}$$

with probability $\Pr(S_t = 2) = \eta_2$.

In Meinhold and Singpurwalla (1989) robustness is achieved by assuming that both $\mathbf{w}_t$ and $\varepsilon_t$ have a marginal $t$-distribution of differing degree of freedom $\nu_1$ and $\nu_2$, which may be written as

$$\begin{aligned} \mathbf{w}_t &\sim \mathcal{N}_d\left(\mathbf{0}, \mathbf{Q}/\omega_t^1\right), & \omega_t^1 &\sim \mathcal{G}\left(\nu_1/2, \nu_1/2\right), \\ \varepsilon_t &\sim \mathcal{N}\left(0, \sigma_\varepsilon^2/\omega_t^2\right), & \omega_t^2 &\sim \mathcal{G}\left(\nu_2/2, \nu_2/2\right). \end{aligned}$$

A combination of these two robust state space models appears in Godsill and Rayner (1998) for the reconstruction of signals that are degraded by an impulsive noise:

$$Y_t = \mathbf{x}_t + (S_t - 1)v_t, \qquad v_t \sim \mathcal{N}\left(0, \sigma_v^2/\omega_t\right), \qquad \omega_t \sim \mathcal{G}\left(\nu/2, \nu/2\right),$$

where $S_t$ is a hidden Markov chain taking the value 1, if no noise is present, and 2 otherwise. $\mathbf{x}_t$ is an AR($p$) process modeled through a state space model as in (13.15).

A more general model, where outliers may be observational as well as innovational is considered in Godsill (1997) in the context of reconstructing acoustically recorded signals, such as speech and music. The statistical model is an ARMA($p, q$) process observed with noise, which possesses the following state space representation with observation equation,

$$Y_t = \mu + \mathbf{H}(\boldsymbol{\theta})\mathbf{x}_t + v_t,$$

with $\mathbf{x}_t$ and $\mathbf{H}(\boldsymbol{\theta})$ being the same as in (13.15) and (13.16). Both $v_t$ as well as the error term $\varepsilon_t$ appearing in (13.15) are assumed to follow a mixture of a normal and a $t$-distribution:

$$\begin{aligned}
&\upsilon_t \sim \mathcal{N}\left(0, \sigma^2_{\upsilon,S_t^1}\right), \\
&\quad \sigma^2_{\upsilon,S_t^1} = (2 - S_t^1)\sigma^2_\upsilon + (S_t^1 - 1)\sigma^2_\upsilon/\omega_t^1, \qquad \omega_t^1 \sim \mathcal{G}\left(\nu_1/2, \nu_1/2\right), \\
&\varepsilon_t \sim \mathcal{N}\left(0, \sigma^2_{\varepsilon,S_t^2}\right), \\
&\quad \sigma^2_{\varepsilon,S_t^2} = (2 - S_t^2)\sigma^2_\varepsilon + (S_t^2 - 1)\sigma^2_\varepsilon/\omega_t^2, \qquad \omega_t^2 \sim \mathcal{G}\left(\nu_2/2, \nu_2/2\right).
\end{aligned}$$

$S_t^1$ and $S_t^2$ are two independent two-state hidden Markov chains with unknown transition matrices $\boldsymbol{\xi}^1$ and $\boldsymbol{\xi}^2$.

Another useful model to deal with structural or innovation outliers is the random level shift time series model (Chen and Tiao, 1990; McCulloch and Tsay, 1993):

$$\begin{aligned}
&Y_t = \mu_t + Z_t, \\
&\mu_t = \mu_{t-1} + (S_t - 1)\beta_t, \qquad \beta_t \sim \mathcal{N}\left(0, k\sigma^2_\varepsilon\right), \\
&\boldsymbol{\delta}(L) Z_t = \boldsymbol{\theta}(L)\varepsilon_t, \qquad \varepsilon_t \sim \mathcal{N}\left(0, \sigma^2_\varepsilon\right),
\end{aligned}$$

where $S_t$ is a two-state hidden i.i.d. indicator with $S_t = 2$ corresponding to a shift that occurs a priori with probability $\Pr(S_t = 2) = \eta_2$. If $\eta_2 = 1$, then the level changes all the time and the model is related to the local trend model (13.17). Gerlach and Kohn (2000) show how intervention analysis may be treated through a switching state space model including both a hidden Markov indicator as well as a second i.i.d. indicator to deal with outliers.

For any of these models traditional likelihood estimation is rather involved. The Bayesian framework discussed in Section 13.4 offers the possibility of locating the position and the size of outlier and shifts simultaneously with parameter estimation.

### 13.2.4 Switching Dynamic Factor Models

Dynamic factor models, in which a large number of observed time series are assumed to be influenced by a common unobserved component, are a special case of a state space model which found various applications in economics, for instance, to estimate wage rates (Engle and Watson, 1981) and to analyze economic indicators that move together (Stock and Watson, 2002).

Diebold and Rudebusch (1996) combine the dynamic factor model with the Markov switching model, one example being the following model,

$$\begin{aligned}
&\triangle \mathbf{Y}_t = \boldsymbol{\beta} + \boldsymbol{\lambda} f_t + \boldsymbol{\varepsilon}_t, \qquad \boldsymbol{\varepsilon}_t \sim \mathcal{N}_r\left(\mathbf{0}, \boldsymbol{\Sigma}\right), \\
&\boldsymbol{\delta}(L)(f_t - \mu_{S_t}) = w_t, \qquad w_t \sim \mathcal{N}\left(0, 1\right),
\end{aligned}$$

where $\boldsymbol{\Sigma}$ is a diagonal matrix, and $w_t$ and $\boldsymbol{\varepsilon}_t$ are pairwise independent. $f_t$ is the latent dynamic factor, $\boldsymbol{\beta}$ and the factor loadings $\boldsymbol{\lambda}$ are unknown parameters. Diebold and Rudebusch (1996) extended this model by considering more general structures for the error process $\boldsymbol{\varepsilon}_t$ such as a VAR model. Kim and

Nelson (1998) generalize this model by introducing time-varying transition matrices.

Application appeared mainly in business cycle analysis (Kim and Nelson, 1998, 2001; Kaufmann, 2000).

### 13.2.5 Switching State Space Models as a Semi-Parametric Smoothing Device

State space models are a useful device for smoothing and interpolating time series (Wecker and Ansley, 1983; Kohn and Ansley, 1987) which are closely related to semiparametric optimal smoothing methods based on the roughness penalty approach.

Kitagawa (1981), for instance, considers the following smoothness prior approach for smoothing nonstationary time series,

$$\begin{aligned} y_t &= \mu_t + \varepsilon_t, & \varepsilon_t &\sim \mathcal{N}\left(0, \sigma_\varepsilon^2\right), \\ \mu_t - 2\mu_{t-1} + \mu_{t-2} &= w_t, & w_t &\sim \mathcal{N}\left(0, \sigma_\mu^2\right), \end{aligned} \tag{13.18}$$

which is closely related to basic structural model (13.17) without a seasonal component $\gamma_t$ and has a very simple state space form. A model that is similar to (13.17) was introduced by Kitagawa and Gersch (1984) for smoothing time series with trends and seasonal components.

Posterior mode estimation for model (13.18) under diffuse priors on $\mu_{-1}$ and $\mu_0$ corresponds to minimizing the penalized least square criterion

$$\sum_{t=1}^{T} (y_t - \mu_t)^2 + \lambda \sum_{t=3}^{T} (\mu_t - 2\mu_{t-1} + \mu_{t-2})^2, \tag{13.19}$$

with respect to $\mu_1, \ldots, \mu_T$, where the smoothness parameter $\lambda$ is related to the variances of the error terms through $\lambda = \sigma_\varepsilon^2/\sigma_\mu^2$ (Fahrmeir and Knorr-Held, 2000). If in (13.18), the fixed variance $\sigma_\mu^2$ is substituted by the switching variance $\sigma_{\mu,S_t}^2$, then the smoothness parameter itself depends on the hidden Markov chain $S_t$: $\lambda_t = \sigma_\varepsilon^2/\sigma_{\mu,S_t}^2$. In this respect, switching state space models with heteroscedastic variances $\sigma_{\mu,S_t}^2$ may be seen as a device for smoothing time series where the smoothness parameter changes over time.

## 13.3 Filtering for Switching Linear Gaussian State Space Models

Filtering aims at deriving the posterior density $p(\mathbf{x}_t|\mathbf{y}^t, \boldsymbol{\vartheta})$ of $\mathbf{x}_t$ given observations $\mathbf{y}^t = (\mathbf{y}_1, \ldots, \mathbf{y}_t)$ up to $t$ in an efficient manner for a fixed model parameter $\boldsymbol{\vartheta}$. To keep notation simple, dependence on $\boldsymbol{\vartheta}$ is not made explicit.

### 13.3.1 The Filtering Problem

Regrettably, the posterior density $p(\mathbf{x}_t|\mathbf{y}^t)$ is of closed form only for very restricted state space models with the linear Gaussian state space model being the most prominent one. For this model class, the posterior density $p(\mathbf{x}_t|\mathbf{y}^t)$ is a normal distribution, where the first two moments are given by the Kalman filter (Kalman, 1960, 1961); see also *Algorithm 13.1* below.

Long before the Bayesian community became aware of the Kalman filter, the importance of the Bayesian approach for solving the filtering problem was realized in the engineering literature (Magill, 1965; Alspach and Sorenson, 1972). As pointed out by Alspach and Sorenson (1972, p.439) regarding $p(\mathbf{x}_t|\mathbf{y}^t)$,

> *If this posterior density function were known, an estimate of the state for any performance criterion could be determined.*

Also for a nonlinear non-Gaussian state space model the filter problem is solved by recursions similar in structure, but not in complexity, to the Kalman filter. Let $p(\mathbf{x}_{t-1}|\mathbf{y}^{t-1})$ be the posterior density of the state $\mathbf{x}_{t-1}$ given information up to $t-1$. The first part of the filtering step is to propagate this information into the future, by deriving the density $p(\mathbf{x}_t|\mathbf{y}^{t-1})$ which may be obtained from integrating the density $p(\mathbf{x}_t, \mathbf{x}_{t-1}|\mathbf{y}^{t-1})$ with respect to $\mathbf{x}_{t-1}$. By assumption (13.13), the propagation step reads:

$$p(\mathbf{x}_t|\mathbf{y}^{t-1}) = \int p(\mathbf{x}_t|\mathbf{x}_{t-1})p(\mathbf{x}_{t-1}|\mathbf{y}^{t-1})d\mathbf{x}_{t-1}.$$

Once an observation $\mathbf{y}_t$ is available, Bayes' theorem plays a crucial role in finding a coherent way of combining information propagated from the past with the information contained in $\mathbf{y}_t$. The updated posterior density $p(\mathbf{x}_t|\mathbf{y}^t)$ is obtained from Bayes' theorem as

$$p(\mathbf{x}_t|\mathbf{y}^t) = \frac{p(\mathbf{y}_t|\mathbf{x}_t)p(\mathbf{x}_t|\mathbf{y}^{t-1})}{p(\mathbf{y}_t|\mathbf{y}^{t-1})},$$

with the normalizing constant being identical to the one-step ahead predictive density $p(\mathbf{y}_t|\mathbf{y}^{t-1})$:

$$p(\mathbf{y}_t|\mathbf{y}^{t-1}) = \int p(\mathbf{y}_t|\mathbf{x}_t)p(\mathbf{x}_t|\mathbf{y}^{t-1})d\mathbf{x}_t.$$

### 13.3.2 Bayesian Inference for a General Linear Regression Model

It is useful to discuss the filtering problem first for a multivariate regression model with general error variance–covariance matrix:

$$\mathbf{Y} = \mathbf{X}\boldsymbol{\beta} + \boldsymbol{\varepsilon}, \qquad \boldsymbol{\varepsilon} \sim \mathcal{N}_r\left(\mathbf{0}, \mathbf{R}\right), \tag{13.20}$$

where $\mathbf{Y}$ is a vector-valued random variable of dimension $r$, $\boldsymbol{\beta}$ is an unknown regression coefficient of dimension $d$, $\mathbf{X}$ is a known $(r \times d)$ design matrix, and $\mathbf{R}$ is a known variance–covariance matrix. In this context filtering refers to inference on $\boldsymbol{\beta}$ through combining of the information contained in a single observation $\mathbf{y}$ from model (13.20) with prior information on $\boldsymbol{\beta}$ expressed through a prior distribution $p(\boldsymbol{\beta})$. Bayes' theorem provides a coherent way of combining these two sources of information by deriving the posterior distribution $p(\boldsymbol{\beta}|\mathbf{R}, \mathbf{y})$:

$$p(\boldsymbol{\beta}|\mathbf{R}, \mathbf{y}) \propto p(\mathbf{y}|\boldsymbol{\beta}, \mathbf{R})p(\boldsymbol{\beta}), \tag{13.21}$$

where the likelihood function $p(\mathbf{y}|\boldsymbol{\beta}, \mathbf{R})$ is equal to:

$$p(\mathbf{y}|\boldsymbol{\beta}, \mathbf{R}) = (2\pi)^{-r/2}|\mathbf{R}|^{-1/2}\exp\left(-\frac{1}{2}(\mathbf{y} - \mathbf{X}\boldsymbol{\beta})'\mathbf{R}^{-1}(\mathbf{y} - \mathbf{X}\boldsymbol{\beta})\right).$$

For a known variance–covariance matrix $\mathbf{R}$, the likelihood function $p(\mathbf{y}|\boldsymbol{\beta}, \mathbf{R})$ is a quadratic form in $\boldsymbol{\beta}$, hence the conjugate prior $p(\boldsymbol{\beta})$ for the regression coefficient $\boldsymbol{\beta}$ is a normal distribution, $\boldsymbol{\beta} \sim \mathcal{N}_d(\mathbf{b}_0, \mathbf{B}_0)$, as is the resulting posterior distribution:

$$\boldsymbol{\beta}|\mathbf{R}, \mathbf{y} \sim \mathcal{N}_d(\mathbf{b}_1, \mathbf{B}_1). \tag{13.22}$$

If $\mathbf{R}^{-1}$ and $\mathbf{B}_0^{-1}$ exist, then the moments of the posterior density are given in terms of the following information filter,

$$\begin{aligned}\mathbf{b}_1 &= \mathbf{B}_1(\mathbf{B}_0^{-1}\mathbf{b}_0 + \mathbf{X}'\mathbf{R}^{-1}\mathbf{y}), \\ \mathbf{B}_1 &= (\mathbf{B}_0^{-1} + \mathbf{X}'\mathbf{R}^{-1}\mathbf{X})^{-1}.\end{aligned} \tag{13.23}$$

The information filter expresses the posterior mean $\mathbf{b}_1$ as a weighted average of the prior mean $\mathbf{b}_0$ and an estimator that is based entirely on the observation $\mathbf{y}$, with the weights depending on the information obtained in the prior distribution and the likelihood function. If $\mathbf{X}'\mathbf{R}^{-1}\mathbf{X}$ is invertible, the data-based estimator is equal to the weighted least square estimator $(\mathbf{X}'\mathbf{R}^{-1}\mathbf{X})^{-1}\mathbf{X}'\mathbf{R}^{-1}\mathbf{y}$, and the weight matrices are equal to $\mathbf{B}_1\mathbf{B}_0^{-1}$ and $\mathbf{B}_1\mathbf{X}'\mathbf{R}^{-1}\mathbf{X}$, respectively.

The information filter involves the inversion of a $(d \times d)$ matrix to obtain the posterior variance–covariance matrix $\mathbf{B}_1$. If the dimension of $\boldsymbol{\beta}$ is larger than the dimension of the observation $\mathbf{y}$ (i.e., $r < d$), or if $\mathbf{R}$ or $\mathbf{B}_0$ are not invertible, it is preferable to work with the following prediction-correction filter which involves the inversion of an $(r \times r)$ matrix, only,

$$\begin{aligned}\mathbf{b}_1 &= \mathbf{b}_0 + \mathbf{K}_1(\mathbf{y} - \mathbf{X}\mathbf{b}_0), &(13.24)\\ \mathbf{B}_1 &= (\mathbf{I}_d - \mathbf{K}_1\mathbf{X})\mathbf{B}_0, \\ \mathbf{K}_1 &= \mathbf{B}_0\mathbf{X}'\mathbf{C}^{-1}, \\ \mathbf{C} &= \mathbf{X}\mathbf{B}_0\mathbf{X}' + \mathbf{R}. &(13.25)\end{aligned}$$

The prediction-correction filter expresses the posterior mean $\mathbf{b}_1$ as a correction of the prior mean $\mathbf{b}_0$, which is based on the prediction error $\mathbf{y}-\mathbf{X}\mathbf{b}_0$, resulting from using the prior mean $\mathbf{b}_0$ as an estimator of $\boldsymbol{\beta}$.

It is useful to have an explicit form of the marginal likelihood $p(\mathbf{y}|\mathbf{R})$, that is equal to the normalizing constant of the nonnormalized posterior $p(\boldsymbol{\beta}|\mathbf{R},\mathbf{y})$, given by (13.21):

$$p(\mathbf{y}|\mathbf{R}) = \int p(\mathbf{y}|\boldsymbol{\beta},\mathbf{R})p(\boldsymbol{\beta})d\boldsymbol{\beta}.$$

The marginal likelihood $p(\mathbf{y}|\mathbf{R})$ is obtained from evaluating the following ratio for an arbitrary value of $\boldsymbol{\beta}$,

$$p(\mathbf{y}|\mathbf{R}) = \frac{p(\mathbf{y}|\boldsymbol{\beta},\mathbf{R})p(\boldsymbol{\beta})}{p(\boldsymbol{\beta}|\mathbf{R},\mathbf{y})}.$$

Choosing $\boldsymbol{\beta} = \mathbf{b}_0$ yields

$$p(\mathbf{y}|\mathbf{R}) = (2\pi)^{-r/2}|\mathbf{C}|^{-1/2}\exp\left(-\frac{1}{2}(\mathbf{y}-\mathbf{X}\mathbf{b}_0)'\mathbf{C}^{-1}(\mathbf{y}-\mathbf{X}\mathbf{b}_0)\right), \quad (13.26)$$

which is the density of a multivariate normal distribution with mean $\mathbf{X}\mathbf{b}_0$ and variance–covariance matrix $\mathbf{C}$, when regarded as a function of $\mathbf{y}$.

### 13.3.3 Filtering for the Linear Gaussian State Space Model

For the linear Gaussian state space model defined in (13.6) and (13.7) the posterior density $p(\mathbf{x}_t|\mathbf{y}^t)$ is a normal distribution, where the first two moments are given by the Kalman filter recursions, derived for the first time in Kalman (1960) and Kalman (1961).

*Algorithm 13.1: Kalman Filter* Assume that the filter density $p(\mathbf{x}_{t-1}|\mathbf{y}^{t-1})$ is the density of a normal distribution:

$$\mathbf{x}_{t-1}|\mathbf{y}^{t-1} \sim \mathcal{N}_d\left(\hat{\mathbf{x}}_{t-1|t-1}, \mathbf{P}_{t-1|t-1}\right). \quad (13.27)$$

Then for a linear Gaussian state space model, the filter density $p(\mathbf{x}_t|\mathbf{y}^t)$ at time $t$ is again the density of a normal distribution obtained from $p(\mathbf{x}_{t-1}|\mathbf{y}^{t-1})$ and $\mathbf{y}_t$ through the following steps.

(a) Propagation — determine the density $p(\mathbf{x}_t|\mathbf{y}^{t-1})$:

$$\mathbf{x}_t|\mathbf{y}^{t-1} \sim \mathcal{N}_d\left(\hat{\mathbf{x}}_{t|t-1}, \mathbf{P}_{t|t-1}\right), \quad (13.28)$$
$$\hat{\mathbf{x}}_{t|t-1} = \mathbf{F}_t\hat{\mathbf{x}}_{t-1|t-1} + \mathbf{G}_t\mathbf{u}_t,$$
$$\mathbf{P}_{t|t-1} = \mathbf{F}_t\mathbf{P}_{t-1|t-1}\mathbf{F}_t' + \mathbf{Q}_t.$$

(b) Prediction — determine the predictive density $p(\mathbf{y}_t|\mathbf{y}^{t-1})$:

$$\begin{aligned}\mathbf{y}_t|\mathbf{y}^{t-1} &\sim \mathcal{N}_r\left(\hat{\mathbf{y}}_{t|t-1}, \mathbf{C}_{t|t-1}\right), \\ \hat{\mathbf{y}}_{t|t-1} &= \mathbf{H}_t\hat{\mathbf{x}}_{t|t-1} + \mathbf{A}_t\mathbf{z}_t, \\ \mathbf{C}_{t|t-1} &= \mathbf{H}_t\mathbf{P}_{t|t-1}\mathbf{H}_t^{'} + \mathbf{R}_t.\end{aligned} \tag{13.29}$$

(c) Correction — determine the filter density $p(\mathbf{x}_t|\mathbf{y}^t)$:

$$\begin{aligned}\mathbf{x}_t|\mathbf{y}^t &\sim \mathcal{N}_d\left(\hat{\mathbf{x}}_{t|t}, \mathbf{P}_{t|t}\right), \\ \hat{\mathbf{x}}_{t|t} &= \mathbf{H}_t\hat{\mathbf{x}}_{t|t-1} + \mathbf{K}_t(\mathbf{y}_t - \hat{\mathbf{y}}_{t|t-1}), \\ \mathbf{K}_t &= \mathbf{P}_{t|t-1}\mathbf{H}_t^{'}\mathbf{C}_{t|t-1}^{-1}, \\ \mathbf{P}_{t|t} &= (\mathbf{I} - \mathbf{K}_t\mathbf{H}_t)\mathbf{P}_{t|t-1}.\end{aligned} \tag{13.30}$$

To start the Kalman filter, one has to choose the normal prior $\mathcal{N}_d\left(\hat{\mathbf{x}}_{0|0}, \mathbf{P}_{0|0}\right)$. It is often recommended to start with a diffuse prior with $\mathbf{P}_{0|0} = \kappa\mathbf{I}_d$ with $\kappa$ being a large value. For state vectors containing both nonstationary and stationary components, De Jong and Chu-Chun-Lin (1994) suggest combining a vague prior with a stationary prior. On the whole, the correct initialization of the Kalman filter is a very subtle issue, and we refer to Koopman (1997) and Durbin and Koopman (2001, Chapter 5) for a very concise and excellent discussion of this issue.

### Derivation of the Kalman Filter

The Kalman filter is easily derived using filtering for a general linear model as in Subsection 13.3.2, as exemplified in Harrison and Stevens (1976) and Meinhold and Singpurwalla (1983).

The density $p(\mathbf{x}_t|\mathbf{y}^{t-1})$ appearing in the propagation step is the normalizing constant of the posterior density $p(\mathbf{x}_{t-1}|\mathbf{x}_t, \mathbf{y}^{t-1})$, given by Bayes' theorem as

$$p(\mathbf{x}_{t-1}|\mathbf{x}_t, \mathbf{y}^{t-1}) \propto p(\mathbf{x}_t|\mathbf{x}_{t-1})p(\mathbf{x}_{t-1}|\mathbf{y}^{t-1}). \tag{13.31}$$

In (13.31), the transition density $p(\mathbf{x}_t|\mathbf{x}_{t-1})$ is the likelihood of a general linear model with error variance-covariance matrix $\mathbf{Q}_t$, where the unknown regression parameter $\mathbf{x}_{t-1}$ follows the prior $p(\mathbf{x}_{t-1}|\mathbf{y}^{t-1})$, being equal to the filtering density (13.27). The marginal likelihood for this problem is given by (13.26) and takes the form of a normal density in $\mathbf{x}_t$ with the moments being given exactly as in (13.28).

The predictive density $p(\mathbf{y}_t|\mathbf{y}^{t-1})$ is the normalizing constant of the filter density $p(\mathbf{x}_t|\mathbf{y}^t)$ which is given by Bayes' theorem:

$$p(\mathbf{x}_t|\mathbf{y}^t) \propto p(\mathbf{y}_t|\mathbf{x}_t)p(\mathbf{x}_t|\mathbf{y}^{t-1}). \tag{13.32}$$

In (13.32), the observation density $p(\mathbf{y}_t|\mathbf{x}_t)$ is the likelihood of a general linear model with error variance–covariance matrix $\mathbf{R}_t$, where the unknown regression parameter $\mathbf{x}_t$ follows the prior $p(\mathbf{x}_t|\mathbf{y}^{t-1})$ being equal to the propagated density (13.28). Again from Subsection 13.3.2, the posterior $p(\mathbf{x}_t|\mathbf{y}^t)$ is normal with the moments given by (13.30), whereas the marginal likelihood $p(\mathbf{y}_t|\mathbf{y}^{t-1})$ takes the form of a normal density in $\mathbf{y}_t$ with the moments given exactly by (13.29).

For alternative derivations of the Kalman filter based on the concept of projection and minimum mean-squared estimation, see Jazwinski (1970), Anderson and Moore (1979), and Harvey (1989).

### 13.3.4 Filtering for Multiprocess Models

In his pioneering work, Magill (1965) used Bayesian methods to show that for a multiprocess model an explicit solution for the filtering problem is available. If the hidden model indicator $S$ takes $K$ values, then the filter density is a mixture of $K$ normal distributions:

$$p(\mathbf{x}_t|\mathbf{y}^t) = \sum_{k=1}^{K} f_N(\mathbf{x}_t; \hat{\mathbf{x}}_{t|t}^{[k]}, \mathbf{P}_{t|t}^{[k]}) \Pr(S = k|\mathbf{y}^t), \qquad (13.33)$$

where the number of components remains fixed for all $t = 1, \dots, T$. The moments of the various components are obtained by running $K$ parallel Kalman filters as in *Algorithm 13.1*, each conditional on assuming that the state of $S$ is equal to $k$, for $k = 1, \dots, K$. The component weights are dynamically changing over time and Sims and Lainiotis (1969) showed how they may be updated recursively using Bayes' theorem:

$$\Pr(S = k|\mathbf{y}^t) \propto f_N(\mathbf{y}_t; \hat{\mathbf{y}}_{t|t-1}^{[k]}, \mathbf{C}_{t|t-1}^{[k]}) \Pr(S = k|\mathbf{y}^{t-1}),$$

where the moments of the predictive density $p(\mathbf{y}_t|S = k, \mathbf{y}^{t-1})$ are obtained from the Kalman filter corresponding to $S = k$.

### 13.3.5 Approximate Filtering for Switching Linear Gaussian State Space Models

For a switching linear Gaussian state space model the filter density is a mixture of normal distributions:

$$p(\mathbf{x}_t|\mathbf{y}^t) = \qquad (13.34)$$
$$\sum_{(k_1,\dots,k_t)\in\mathcal{S}_t} f_N(\mathbf{x}_t; \hat{\mathbf{x}}_{t|t}^{[k_1,\dots,k_t]}, \mathbf{P}_{t|t}^{[k_1,\dots,k_t]}) \Pr(\mathbf{S}^t = (k_1, \dots, k_t)|\mathbf{y}^t),$$

where $\mathcal{S}_t = \{1, \dots, K\}^t$ is the space of all paths $\mathbf{S}^t = (S_1, \dots, S_t)$ up to $t$. This representation holds both for finite mixture as well as Markov switching state space models. In contrast to the multiprocess model, the number of components in the filtering density is increasing exponentially fast. Running an exact recursive filter requires combining all $K^{t-1}$ normal posterior densities $f_N(\mathbf{x}_{t-1}; \hat{\mathbf{x}}_{t-1|t-1}^{[k_1,\ldots,k_{t-1}]}, \mathbf{P}_{t-1|t-1}^{[k_1,\ldots,k_{t-1}]})$ with each of the $K$ states of $S_t$, running in total $K^t$ parallel Kalman filters as in *Algorithm 13.1*. This is operational only if the total number $T$ of observations is not too large; see, for instance, Schervish and Tsay (1988) for an empirical application of this filter.

In most cases some approximate filter has to be applied. Approximate filters for switching Gaussian state space models were studied rather early in the engineering literature; we mention here in particular Ackerson and Fu (1970), Bar-Shalom and Tse (1975), Akashi and Kumamoto (1977), Tugnait (1982), and Blom and Bar-Shalom (1988). Approximations in the statistical and econometric literature were suggested by Harrison and Stevens (1976), Cosslett and Lee (1985), Peña and Guttman (1988), Lam (1990), Gordon and Smith (1990), Shumway and Stoffer (1991), and Kim (1994). To keep the filter operational, the number of components of the filtering density has to be limited, usually by merging components at each filter step. Other techniques are trimming by removing unlikely components with small probability and combining similar components into a single component.

A useful starting point for discussing the various approximate filters is writing the filter density $p(\mathbf{x}_t|\mathbf{y}^t)$ as

$$p(\mathbf{x}_t|\mathbf{y}^t) = \sum_{k=1}^{K} p(\mathbf{x}_t|\mathbf{y}^t, S_t = k)\Pr(S_t = k|\mathbf{y}^t). \qquad (13.35)$$

In (13.35) we identify two filtering problems. First, we need to derive the discrete filter probabilities $\Pr(S_t = k|\mathbf{y}^t)$ for $k = 1, \ldots, K$ without conditioning on the continuous state vector $\mathbf{x}_t$; second, we need to derive filter recursion for the continuous state $\mathbf{x}_t$ conditional on knowing only the present state of $S_t$.

For a hidden Markov chain $S_t$ with transition matrix $\boldsymbol{\xi}$, the discrete filter is derived through Bayes' theorem in a similar way as was done for Markov switching models in Section 11.2:

$$\Pr(S_t = k|\mathbf{y}^t) \propto p(\mathbf{y}_t|S_t = k, \mathbf{y}^{t-1})\Pr(S_t = k|\mathbf{y}^{t-1}). \qquad (13.36)$$

The propagated probabilities $\Pr(S_t = k|\mathbf{y}^{t-1})$ are essentially the same as in Section 11.2 and read:

$$\Pr(S_t = k|\mathbf{y}^{t-1}) = \sum_{j=1}^{K} \xi_{jk}\Pr(S_{t-1} = j|\mathbf{y}^{t-1}).$$

For a hidden i.i.d. indicator this reduces to $\Pr(S_t = k|\mathbf{y}^{t-1}) = \eta_k$ as $\xi_{jk} = \eta_k$. Because the likelihood $p(\mathbf{y}_t|S_t = k, \mathbf{y}^{t-1})$ in (13.36) will also appear in the prediction step of the second filtering problem, both filtering problems are related.

To solve the second filtering problem, a recursion between the filter densities $p(\mathbf{x}_{t-1}|\mathbf{y}^{t-1}, S_{t-1} = j)$ and $p(\mathbf{x}_t|\mathbf{y}^t, S_t = k)$ has to be established. One could, in principle, proceed as in Subsection 13.3.3, using the propagation step

$$p(\mathbf{x}_t|\mathbf{y}^{t-1}, S_t = k) = \int p(\mathbf{x}_t|\mathbf{x}_{t-1}, S_t = k)p(\mathbf{x}_{t-1}|\mathbf{y}^{t-1}, S_t = k)d\mathbf{x}_{t-1}, \tag{13.37}$$

the prediction step

$$p(\mathbf{y}_t|S_t = k, \mathbf{y}^{t-1}) = \int p(\mathbf{y}_t|\mathbf{x}_t, S_t = k)p(\mathbf{x}_t|\mathbf{y}^{t-1}, S_t = k)d\mathbf{x}_t, \tag{13.38}$$

and the correction step

$$p(\mathbf{x}_t|\mathbf{y}^t, S_t = k) \propto p(\mathbf{y}_t|\mathbf{x}_t, S_t = k)p(\mathbf{x}_t|\mathbf{y}^{t-1}, S_t = k). \tag{13.39}$$

Because we are dealing with a finite or Markov mixture of linear Gaussian state space models, the transition density $p(\mathbf{x}_t|\mathbf{x}_{t-1}, S_t = k)$ and the observation density $p(\mathbf{y}_t|\mathbf{x}_t, S_t = k)$ are normal, however, $p(\mathbf{x}_{t-1}|\mathbf{y}^{t-1}, S_t = k)$ does not have the required form of a conjugate normal prior. Nonnormality of $p(\mathbf{x}_{t-1}|\mathbf{y}^{t-1}, S_t = k)$ arises due to possible changes in the states of $S_{t-1}$ and $S_t$ between $t-1$ and $t$, which may occur both for finite mixture as well as Markov switching state space models. $p(\mathbf{x}_{t-1}|\mathbf{y}^{t-1}, S_t = k)$ is a finite mixture of the filtering densities at $t-1$:

$$p(\mathbf{x}_{t-1}|\mathbf{y}^{t-1}, S_t = k) = \sum_{j=1}^{K} p(\mathbf{x}_{t-1}|\mathbf{y}^{t-1}, S_{t-1} = j)w_{jk}, \tag{13.40}$$

where the weights are given by

$$w_{jk} = \Pr(S_{t-1} = j|\mathbf{y}^{t-1}, S_t = k) \propto \xi_{jk}\Pr(S_{t-1} = j|\mathbf{y}^{t-1}).$$

For a Markov switching state space model, the weights read:

$$w_{jk} = \frac{\xi_{jk}\Pr(S_{t-1} = j|\mathbf{y}^{t-1})}{\sum_{l=1}^{K} \xi_{lk}\Pr(S_{t-1} = l|\mathbf{y}^{t-1})}. \tag{13.41}$$

For a finite mixture state space model, the weights are identical with the discrete filter probabilities:

$$w_{jk} = \Pr(S_{t-1} = j|\mathbf{y}^{t-1}). \tag{13.42}$$

In principle, these formulae provide a recursion comparable to the Kalman filter. However, because the filter density $p(\mathbf{x}_t|\mathbf{y}^t, S_t = k)$ is given by a mixture of $K_t = KK_{t-1}$ components, where $K_{t-1}$ is the number of components at $t-1$, some method of limiting the number of components must be found to make this filter operational. As pointed out by Blom and Bar-Shalom (1988), different algorithms emerge, depending on the precise density and the precise time point chosen for this simplification.

**Kim's Algorithm**

This algorithm was suggested independently by Tugnait (1982) and Kim (1994). Assume that the filter density $p(\mathbf{x}_{t-1}|\mathbf{y}^{t-1}, S_{t-1} = j)$ is a normal distribution:

$$p(\mathbf{x}_{t-1}|\mathbf{y}^{t-1}, S_{t-1} = j) = f_N(\mathbf{x}_{t-1}; \hat{\mathbf{x}}^{[j]}_{t-1|t-1}, \mathbf{P}^{[j]}_{t-1|t-1}). \qquad (13.43)$$

Then the prior $p(\mathbf{x}_{t-1}|\mathbf{y}^{t-1}, S_t = k)$ in (13.40) is a mixture of $K$ normal distributions as is the filter density $p(\mathbf{x}_t|\mathbf{y}^t, S_t = k)$ in (13.39):

$$p(\mathbf{x}_t|\mathbf{y}^t, S_t = k) = \sum_{j=1}^{K} f_N(\mathbf{x}_t; \hat{\mathbf{x}}^{[j,k]}_{t|t}, \mathbf{P}^{[j,k]}_{t|t}) \Pr(S_{t-1} = j|\mathbf{y}^t, S_t = k). \qquad (13.44)$$

The component densities in the filter density are obtained by running in total $K^2$ Kalman filters, combining each normal density $p(\mathbf{x}_{t-1}|\mathbf{y}^{t-1}, S_{t-1} = j)$ with each possible value for $S_t = k$. Each Kalman filter delivers the normal one-step ahead predictive density

$$p(\mathbf{y}_t|\mathbf{y}^{t-1}, S_{t-1} = j, S_t = k) = f_N(\mathbf{y}_t; \hat{\mathbf{y}}^{[j,k]}_{t|t-1}, \mathbf{C}^{[j,k]}_{t|t-1}),$$

which could be used to compute the weights $\Pr(S_{t-1} = j|\mathbf{y}^t, S_t = k)$ in (13.44) through Bayes' theorem:

$$\Pr(S_{t-1} = j|\mathbf{y}^t, S_t = k) \propto p(\mathbf{y}_t|\mathbf{y}^{t-1}, S_{t-1} = j, S_t = k) w_{jk}, \qquad (13.45)$$

where $w_{jk}$ were defined in (13.41) and (13.42), respectively. For each value of $k$, the normalizing constant of the right-hand side of (13.45) is equal to the one-step ahead predictive density $p(\mathbf{y}_t|S_t = k, \mathbf{y}^{t-1})$,

$$p(\mathbf{y}_t|S_t = k, \mathbf{y}^{t-1}) = \sum_{j=1}^{K} p(\mathbf{y}_t|\mathbf{y}^{t-1}, S_{t-1} = j, S_t = k) w_{jk},$$

which is necessary for the computation of the discrete filter probabilities $\Pr(S_t = k|\mathbf{y}^t)$ through (13.36).

To keep the filter operational, Kim (1994) collapses the mixture (13.44) to a single normal density after having finished filtering at time $t$, which it is then used as a prior density for the next filtering step:

$$p(\mathbf{x}_t|\mathbf{y}^t, S_t = k) \approx f_N(\mathbf{x}_t; \hat{\mathbf{x}}^{[k]}_{t|t}, \mathbf{P}^{[k]}_{t|t}),$$

$$\hat{\mathbf{x}}^{[k]}_{t|t} = \sum_{j=1}^{K} \hat{\mathbf{x}}^{[j,k]}_{t|t} \Pr(S_{t-1} = j|\mathbf{y}^t, S_t = k),$$

$$\mathbf{P}^{[k]}_{t|t} = \sum_{j=1}^{K} (\hat{\mathbf{x}}^{[j,k]}_{t|t} (\hat{\mathbf{x}}^{[j,k]}_{t|t})' + \mathbf{P}^{[j,k]}_{t|t}) \Pr(S_{t-1} = j|\mathbf{y}^t, S_t = k) - \hat{\mathbf{x}}^{[k]}_{t|t} (\hat{\mathbf{x}}^{[k]}_{t|t})'.$$

A comparison of this approximate filter with exact inference in Kim (1994) for the model of Lam (1990) indicates that this approximate filter is quite accurate.

Tugnait (1982) extended this method by updating a whole sequence $(S_{t-h}, \ldots, S_t)$ with $h > 1$.

### Other Approximations

Several other approximations also assume that the prior $p(\mathbf{x}_{t-1}|\mathbf{y}^{t-1}, S_{t-1} = j)$ is a normal distribution as in (13.43), reduction of filter complexity, however, is carried out in a different manner. Blom and Bar-Shalom (1988) suggest collapsing the mixture density $p(\mathbf{x}_{t-1}|\mathbf{y}^{t-1}, S_t = k)$ given by (13.40) to a single normal density with the same moments *prior* to running through the filter steps (13.37) to (13.38) at time $t$:

$$p(\mathbf{x}_{t-1}|\mathbf{y}^{t-1}, S_t = k) \approx f_N(\mathbf{x}_t; \hat{\mathbf{x}}^{[k]}_{t-1|t-1}, \mathbf{P}^{[k]}_{t-1|t-1})$$

$$\hat{\mathbf{x}}^{[k]}_{t-1|t-1} = \sum_{j=1}^{K} w_{jk} \mathrm{E}(\mathbf{x}_{t-1}|\mathbf{y}^{t-1}, S_{t-1} = j), \tag{13.46}$$

with a similar formula for the variance–covariance matrix. Filtering then reduces to running $K$ Kalman filters, however, this filter is less precise than Kim's algorithm.

For finite mixture of state space models, the weights in (13.46) are independent of $k$, $w_{jk} = \Pr(S_{t-1} = j|\mathbf{y}^{t-1})$ (see again (13.42)), and all moments in (13.46) reduce to the moments $\hat{\mathbf{x}}_{t-1|t-1}$ and $\mathbf{P}_{t-1|t-1}$ of the marginal posterior $p(\mathbf{x}_{t-1}|\mathbf{y}^{t-1})$,

$$p(\mathbf{x}_{t-1}|\mathbf{y}^{t-1}, S_t = k) \approx f_N(\mathbf{x}_t; \hat{\mathbf{x}}_{t-1|t-1}, \mathbf{P}_{t-1|t-1}).$$

Such a filter is running through the filter steps (13.37) to (13.38) with the same prior $p(\mathbf{x}_{t-1}|\mathbf{y}^{t-1}, S_t = k)$ for all $k$ and reduces the collapsing procedures suggested by Harrison and Stevens (1976), Peña and Guttman (1988), and Shumway and Stoffer (1991) for finite mixtures of state space models.

Ackerson and Fu (1970) and Bar-Shalom and Tse (1975) use the same collapsing technique, where the unconditional posterior $p(\mathbf{x}_{t-1}|\mathbf{y}^{t-1})$ is approximated by a single normal density prior to filtering also for Markov switching state space models. This procedure, however, is likely to be less optimal than the collapsing method of Blom and Bar-Shalom (1988), especially for highly persistent Markov chains, whereas there is little computational gain.

## 13.4 Parameter Estimation for Switching State Space Models

Let $\boldsymbol{\vartheta}$ summarize all unknown distinct parameters appearing in the definition of a switching state space model that should be fitted to a univariate or multivariate time series $\mathbf{y} = (\mathbf{y}_1, \ldots, \mathbf{y}_T)$. In various applications of switching state space models, the parameters of the probability law of $S_t$ and the covariances $\mathbf{Q}_t$ and $\mathbf{R}_t$ are assumed to be known, often based by choosing somewhat arbitrary values (Harrison and Stevens, 1976; Carter and Kohn, 1994), but in general these parameters may be estimated from the data as well.

### 13.4.1 The Likelihood Function of a State Space Model

The likelihood function $p(\mathbf{y}|\boldsymbol{\vartheta})$ is defined as the density $p(\mathbf{y}_1, \ldots, \mathbf{y}_T|\boldsymbol{\vartheta})$ of the joint distribution of $\mathbf{Y}_1, \ldots, \mathbf{Y}_T$ where all latent variables, in particular the state process $\mathbf{x} = (\mathbf{x}_0, \ldots, \mathbf{x}_T)$ and the indicator process $\mathbf{S} = (S_0, \ldots, S_T)$, are integrated out. In general, the likelihood of a state space model is derived by using the following decomposition into one-step ahead predictive densities (Schweppe, 1965; Kashyap, 1970),

$$p(\mathbf{y}|\boldsymbol{\vartheta}) = p(\mathbf{y}_1, \ldots, \mathbf{y}_T|\boldsymbol{\vartheta}) = \prod_{t=1}^{T} p(\mathbf{y}_t|\mathbf{y}^{t-1}, \boldsymbol{\vartheta}).$$

For a linear Gaussian state space model the predictive density $p(\mathbf{y}_t|\mathbf{y}^{t-1}, \boldsymbol{\vartheta})$ appears as part of the Kalman filter (see *Algorithm 13.1*), and the likelihood function is obtained from a single run of the Kalman filter conditional on $\boldsymbol{\vartheta}$, if the initial moments $\hat{\mathbf{x}}_{0|0}$ and $\mathbf{P}_{0|0}$ are known:

$$\begin{aligned}
&-2 \log p(\mathbf{y}_1, \ldots, \mathbf{y}_T|\boldsymbol{\vartheta}) \\
&= \sum_{t=1}^{T} \left( \log |\mathbf{C}_{t|t-1}(\boldsymbol{\vartheta})| + (\mathbf{y}_t - \hat{\mathbf{y}}_{t|t-1}(\boldsymbol{\vartheta}))' \mathbf{C}_{t|t-1}(\boldsymbol{\vartheta})^{-1} (\mathbf{y}_t - \hat{\mathbf{y}}_{t|t-1}(\boldsymbol{\vartheta})) \right),
\end{aligned}$$

where $\hat{\mathbf{y}}_{t|t-1}(\boldsymbol{\vartheta})$ and $\mathbf{C}_{t|t-1}(\boldsymbol{\vartheta})$ are given by (13.29). Some care needs to be exercised if the initial moments $\hat{\mathbf{x}}_{0|0}$ and $\mathbf{P}_{0|0}$ are unknown, and we refer to Durbin and Koopman (2001, Section 7.2) for further discussion.

For a switching linear Gaussian state space model, the likelihood $p(\mathbf{y}|\boldsymbol{\vartheta})$ where both sets of latent variables are integrated out is not available in closed form. Like the filter density $p(\mathbf{x}_t|\mathbf{y}^t, \boldsymbol{\vartheta})$, the one-step ahead predictive density $p(\mathbf{y}_t|\mathbf{y}^{t-1}, \boldsymbol{\vartheta})$ is a mixture of normal densities with an increasing number of components. However, any of the approximate filters discussed in Subsection 13.3.5 leads immediately to an approximation to the log likelihood function. By rewriting the predictive density as

$$p(\mathbf{y}_t|\mathbf{y}^{t-1}, \boldsymbol{\vartheta}) = \sum_{k=1}^{K} p(\mathbf{y}_t|\mathbf{y}^{t-1}, S_t = k, \boldsymbol{\vartheta}) \Pr(S_t = k|\mathbf{y}^{t-1}, \boldsymbol{\vartheta}),$$

it becomes evident that $p(\mathbf{y}_t|\mathbf{y}^{t-1}, \boldsymbol{\vartheta})$ is the normalizing constant of the right-hand side of discrete filter distribution $\Pr(S_t = k|\mathbf{y}, \boldsymbol{\vartheta})$, given in (13.36). Approximate ML estimation based on approximate filters has been applied by Shumway and Stoffer (1991) and Kim (1994), among others.

It is worth noting that certain partial likelihood functions are available in closed form. When holding $\mathbf{S}$ fixed, one is dealing with a standard state space model, and the likelihood $p(\mathbf{y}|\boldsymbol{\vartheta}, \mathbf{S})$ is obtained by running a Kalman filter conditional on $\boldsymbol{\vartheta}$ and $\mathbf{S}$.

### 13.4.2 Maximum Likelihood Estimation

A straightforward method of obtaining the ML estimator is direct maximization of the exact or approximate log likelihood function $\log p(\mathbf{y}_1, \dots, \mathbf{y}_T|\boldsymbol{\vartheta})$ using some numerical technique such as Newton–Raphson methods; see, for instance, Hamilton (1994b, Section 5.7) for a review of these methods.

It was realized by Shumway and Stoffer (1982) and Watson and Engle (1983) that the EM algorithm of Dempster et al. (1977) may be applied to linear Gaussian state space models without switching, because the complete-data likelihood function $p(\mathbf{y}|\mathbf{x}, \boldsymbol{\vartheta})p(\mathbf{x}|\boldsymbol{\vartheta})$ turns out to be of simple form. Koopman (1993) proposed a very simple and efficient EM algorithm for unknown parameters inside the variance–covariance matrices $\mathbf{Q}_t$ and $\mathbf{R}_t$ of a linear Gaussian state space form.

For a switching state space model, the presence of two sets of latent variables hinders a straightforward application of the EM algorithm, because the required smoothed probabilities $\Pr(S_t = k|\mathbf{y})$ are not available in closed form. Shumway and Stoffer (1991) substitute these probabilities by $\Pr(S_t = k|\mathbf{y}^t)$ which are available from any approximate filter discussed in Subsection 13.3.5 and report that this pseudo EM algorithm works well.

Consistency and asymptotic normality of the ML estimator of the parameters of a state space model hold under fairly general conditions; see Shumway and Stoffer (1982), Schneider (1988), Hamilton (1994b, Section 13.4), Jensen and Petersen (1999), and Shumway and Stoffer (2000, p.326ff). The observed time series, however, needs to be fairly long in order to achieve asymptotic normality. Moreover, problems occur if some of the parameters are close to the boundary of the parameter space. For this reason it seems sensible to consider a Bayesian approach.

### 13.4.3 Bayesian Inference

Bayesian inference for switching state space models is based on deriving the joint posterior density $p(\mathbf{x}, \mathbf{S}, \boldsymbol{\vartheta}|\mathbf{y})$ of all continuous states $\mathbf{x} = (\mathbf{x}_0, \dots, \mathbf{x}_T)$, all discrete states $\mathbf{S} = (S_0, \dots, S_T)$, and unknown model parameters $\boldsymbol{\vartheta}$, including unknown parameters in the probability law of $\mathbf{S}$, if any are present. Due to the hierarchical structure of a switching state space model, this density is proportional to:

$$p(\mathbf{x}, \mathbf{S}, \boldsymbol{\vartheta}|\mathbf{y}) \propto p(\mathbf{y}|\mathbf{x}, \mathbf{S}, \boldsymbol{\vartheta})p(\mathbf{x}|\mathbf{S}, \boldsymbol{\vartheta})p(\mathbf{S}|\boldsymbol{\vartheta})p(\boldsymbol{\vartheta}),$$

which simplifies to:

$$p(\mathbf{x}, \mathbf{S}, \boldsymbol{\vartheta}|\mathbf{y}) \propto p(\mathbf{x}_0|\boldsymbol{\vartheta})p(\boldsymbol{\vartheta}) \times \prod_{t=1}^{N} p(\mathbf{y}_t|S_t, \mathbf{x}_t, \boldsymbol{\vartheta})p(\mathbf{x}_t|S_t, \mathbf{x}_{t-1}, \boldsymbol{\vartheta})p(\mathbf{S}|\boldsymbol{\vartheta}). \tag{13.47}$$

The densities $p(\mathbf{y}_t|S_t, \mathbf{x}_t, \boldsymbol{\vartheta})$ and $p(\mathbf{x}_t|S_t, \mathbf{x}_{t-1}, \boldsymbol{\vartheta})$ result directly from the definition of the state space model, where $p(\mathbf{x}_0|\boldsymbol{\vartheta})$ is the prior of $\mathbf{x}_0$. $p(\boldsymbol{\vartheta})$ is the prior density of all model parameters. The density $p(\mathbf{S}|\boldsymbol{\vartheta})$ results directly from the definition of the probability law of $S_t$. If $S_t$ is a hidden Markov chain, then

$$p(\mathbf{S}|\boldsymbol{\vartheta}) = p(S_0|\boldsymbol{\vartheta}) \prod_{t=1}^{N} p(S_t|S_{t-1}, \boldsymbol{\vartheta}).$$

If $S_t$ is a hidden i.i.d. indicator, then

$$p(\mathbf{S}|\boldsymbol{\vartheta}) = \prod_{t=1}^{N} p(S_t|\boldsymbol{\vartheta}).$$

Note that the derivation of the posterior density in (13.47) is not limited to switching linear Gaussian state space models, but is valid for any switching state space model.

The posterior density $p(\mathbf{x}, \mathbf{S}, \boldsymbol{\vartheta}|\mathbf{y})$, however, is not of any closed form, even for linear Gaussian state space models without switching and simulation-based methods are usually applied for Bayesian estimation. Durbin and Koopman (2000) propagate the application of importance sampling, several other authors explored MCMC methods; see Section 13.5.

### Choosing the Priors for Bayesian Estimation

If $S_t$ is a hidden Markov chain with transition matrix $\boldsymbol{\xi}$, then the joint prior reads

$$p(\mathbf{x}_0|\boldsymbol{\vartheta}, S_0)p(S_0|\boldsymbol{\xi})p(\boldsymbol{\vartheta})p(\boldsymbol{\xi}), \tag{13.48}$$

where each row $\boldsymbol{\xi}_{j\cdot}$ of the transition matrix $\boldsymbol{\xi}$ is chosen from a Dirichlet distribution as in Chapter 11:

$$\boldsymbol{\xi}_{k\cdot} \sim \mathcal{D}\left(e_{k1}, \ldots, e_{kK}\right), \qquad k = 1, \ldots, K. \tag{13.49}$$

To obtain a prior that is invariant to relabeling, Frühwirth-Schnatter (2001a) suggested choosing $e_{kk} = e^P$ and $e_{kk'} = e^T$, if $k \neq k'$. By choosing $e^P > e^T$, a Markov switching state space model is bounded away from a finite mixture state space model. Choosing the prior $p(S_0|\boldsymbol{\xi})$ of the discrete-valued state variable $S_0$ is closely related to choosing the same prior for finite Markov mixture models; see Subsection 10.3.4 for various choices of this distribution.

If $S_t$ is a hidden i.i.d. indicator with probability distribution $\boldsymbol{\eta}$, then the joint prior reduces to

$$p(\mathbf{x}_0|\boldsymbol{\vartheta})p(\boldsymbol{\vartheta})p(\boldsymbol{\eta}), \tag{13.50}$$

where the prior for $\boldsymbol{\eta}$ is chosen from the Dirichlet distribution as in Chapter 2:

$$\boldsymbol{\eta} \sim \mathcal{D}\left(e_0, \ldots, e_0\right). \tag{13.51}$$

In both cases, $p(\mathbf{x}_0|\boldsymbol{\vartheta}, S_0)$ is the prior for the continuous state variable $\mathbf{x}_0$ used for initialization in the Kalman filter, and is allowed to depend on $S_0$ for a Markov switching state space model.

The prior for the remaining parameters $\boldsymbol{\vartheta}$ is usually chosen to be conditionally conjugate to the complete-data likelihood $p(\mathbf{y}|\mathbf{x}, \mathbf{S}, \boldsymbol{\vartheta})p(\mathbf{x}|\mathbf{S}, \boldsymbol{\vartheta})$. To give an example, consider a local level model where both variances are switching,

$$\begin{aligned} \mu_t &= \mu_{t-1} + w_t, \qquad & w_t &\sim \mathcal{N}\left(0, \sigma^2_{\mu,S_t}\right), \\ Y_t &= \mu_t + \varepsilon_t, \qquad & \varepsilon_t &\sim \mathcal{N}\left(0, \sigma^2_{\varepsilon,S_t}\right). \end{aligned} \tag{13.52}$$

The complete-data likelihood reads with $\mathbf{x} = (\mu_0, \ldots, \mu_T)$ and $\boldsymbol{\vartheta} = (\sigma^2_{\mu,1}, \ldots, \sigma^2_{\mu,K}, \sigma^2_{\varepsilon,1}, \ldots, \sigma^2_{\varepsilon,K})$:

$$p(\mathbf{y}|\mathbf{x}, \mathbf{S}, \boldsymbol{\vartheta})p(\mathbf{x}|\mathbf{S}, \boldsymbol{\vartheta}) \propto \prod_{k=1}^{K} \left(\frac{1}{\sigma^2_{\varepsilon,k}}\right)^{N_k(\mathbf{S})/2} \exp\left\{-\frac{\sum_{t:S_t=k} (y_t - \mu_t)^2}{2\sigma^2_{\varepsilon,k}}\right\}$$
$$\times \left(\frac{1}{\sigma^2_{\mu,k}}\right)^{N_k(\mathbf{S})/2} \exp\left\{-\frac{\sum_{t:S_t=k} (\mu_t - \mu_{t-1})^2}{2\sigma^2_{\mu,k}}\right\},$$

where $N_k(\mathbf{S}) = \#\{S_t = k\}$. Considered as a function of $\sigma^2_{\varepsilon,k}$, this is an inverted Gamma density. Therefore the conditionally conjugate prior for $\sigma^2_{\varepsilon,k}$ is an inverted Gamma density $\mathcal{G}^{-1}\left(c_{\varepsilon,0}, C_{\varepsilon,0}\right)$. Similarly, the complete-data likelihood is an inverted Gamma density, when considered as a function of $\sigma^2_{\mu,k}$. Thus the conditionally conjugate prior for $\sigma^2_{\mu,k}$ is again an inverted Gamma density $\mathcal{G}^{-1}\left(c_{\mu,0}, C_{\mu,0}\right)$.

### Complete-Data Bayesian Estimation

Estimation of the unknown model parameters $\boldsymbol{\vartheta}$ conditional on the complete data $\mathbf{S}$, $\mathbf{x}$, and $\mathbf{y}$ is closely related to various Bayesian inference problems discussed earlier. If parameters appearing in the definition of the probability law of $S_t$ are a priori independent of parameters appearing in the definition of the transition and observation densities, then this independence is preserved a posteriori. If $S_t$ is an i.i.d. indicator with unknown probability distribution $\boldsymbol{\eta}$, then $\boldsymbol{\eta}|\mathbf{S},\mathbf{x},\mathbf{y}$ follows a Dirichlet distribution as discussed for finite mixture models in Subsection 3.5.3, whereas the posterior of $\boldsymbol{\xi}|\mathbf{S},\mathbf{x},\mathbf{y}$ under a hidden Markov chain $S_t$ with unknown transition matrix $\boldsymbol{\xi}$ is the same as in Subsection 11.5.5. For unknown parameters appearing in the definition of the observation and the transition density, the complete-data likelihood $p(\mathbf{y}|\mathbf{x},\mathbf{S},\boldsymbol{\vartheta})p(\mathbf{x}|\mathbf{S},\boldsymbol{\vartheta})$ in combination with a conditionally conjugate prior $p(\boldsymbol{\vartheta})$ often leads to a posterior density $p(\boldsymbol{\vartheta}|\mathbf{S},\mathbf{x},\mathbf{y})$ that is of closed form.

To give an example, consider a local level model where both variances are switching as in (13.52) and $S_t$ is a hidden Markov chain. Then $\boldsymbol{\vartheta} = (\sigma^2_{\mu,1},\ldots,\sigma^2_{\mu,K},\sigma^2_{\varepsilon,1},\ldots,\sigma^2_{\varepsilon,K},\boldsymbol{\xi})$ and the complete-data posterior $p(\boldsymbol{\vartheta}|\mathbf{x},\mathbf{S},\mathbf{y})$ reads:

$$
\begin{aligned}
p(\boldsymbol{\vartheta}|\mathbf{x},\mathbf{S},\mathbf{y}) &\propto p(\mathbf{y}|\mathbf{x},\mathbf{S},\boldsymbol{\vartheta})p(\mathbf{x}|\mathbf{S},\boldsymbol{\vartheta})p(\mathbf{S}|\boldsymbol{\xi})p(\boldsymbol{\vartheta}) \propto p(S_0|\boldsymbol{\vartheta}) \prod_{j=1}^{K}\prod_{k=1}^{K} \xi_{jk}^{N_{jk}(\mathbf{S})} \\
&\times \prod_{k=1}^{K} \left(\frac{1}{\sigma^2_{\varepsilon,k}}\right)^{N_k(\mathbf{S})/2+c_{\varepsilon,0}+1} \exp\left\{ -\frac{\sum_{t:S_t=k} (y_t-\mu_t)^2}{2\sigma^2_{\varepsilon,k}} - \frac{C_{\varepsilon,0}}{\sigma^2_{\varepsilon,k}} \right\} \\
&\times \prod_{k=1}^{K} \left(\frac{1}{\sigma^2_{\mu,k}}\right)^{N_k(\mathbf{S})/2+c_{\mu,0}+1} \exp\left\{ -\frac{\sum_{t:S_t=k} (\mu_t-\mu_{t-1})^2}{2\sigma^2_{\mu,k}} - \frac{C_{\mu,0}}{\sigma^2_{\mu,k}} \right\},
\end{aligned}
$$

where $N_{jk}(\mathbf{S}) = \#\{S_{t-1}=j, S_t=k\}$ counts the numbers of transitions from $j$ to $k$ and $N_k(\mathbf{S}) = \#\{S_t=k\} = \sum_{j=1}^K N_{jk}(\mathbf{S})$. The transition matrix $\boldsymbol{\xi}$, as well as all variances $\sigma^2_{\mu,k}$ and $\sigma^2_{\varepsilon,k}$ are conditionally independent. The precise form of the posterior of $\boldsymbol{\xi}$ and the method used for sampling from this density depend on the assumptions concerning $p(S_0|\boldsymbol{\vartheta})$, as has been discussed earlier in Subsection 11.5.5. The variances $\sigma^2_{\mu,k}$ and $\sigma^2_{\varepsilon,k}$ each follow an inverted Gamma density $\mathcal{G}^{-1}(c_{\mu,k}(\mathbf{S}),C_{\mu,k}(\mathbf{S}))$ and $\mathcal{G}^{-1}(c_{\varepsilon,k}(\mathbf{S}),C_{\varepsilon,k}(\mathbf{S}))$, where

$$
\begin{aligned}
c_{\varepsilon,k}(\mathbf{S}) &= c_{\varepsilon,0} + 0.5N_k(\mathbf{S}), \qquad C_{\varepsilon,k}(\mathbf{S}) = C_{\varepsilon,0} + 0.5\sum_{t:S_t=k}(y_t-\mu_t)^2, \\
c_{\mu,k}(\mathbf{S}) &= c_{\mu,0} + 0.5N_k(\mathbf{S}), \qquad C_{\mu,k}(\mathbf{S}) = C_{\mu,0} + 0.5\sum_{t:S_t=k}(\mu_t-\mu_{t-1})^2.
\end{aligned}
$$

## 13.5 Practical Bayesian Estimation Using MCMC

Practical Bayesian estimation of switching state space models usually relies on MCMC estimation and was implemented for specific models discussed in Section 13.2 such as the state space model with Markov switching conditional heteroscedasticity (Carlin et al., 1992; Carter and Kohn, 1994, 1996), the random level shift model (McCulloch and Tsay, 1993), partial Gaussian state space model (Shephard, 1994), robust state space model (Godsill, 1997; Godsill and Rayner, 1998), dynamic factor model with regime switching (Kim and Nelson, 1998; Kaufmann, 2000), and various unobserved component models with Markov switching (Luginbuhl and de Vos, 1999; Engel and Kim, 1999). Frühwirth-Schnatter (2001a) provides a general discussion of MCMC methods for switching linear Gaussian state space models.

### 13.5.1 Various Data Augmentation Schemes

Various MCMC schemes have been suggested to implement data augmentation and Gibbs sampling for switching linear Gaussian state space models. The following three-block Gibbs sampler has been applied in Shephard (1994), Carter and Kohn (1994), and Frühwirth-Schnatter (2001a).

*Algorithm 13.2: MCMC for a Switching Linear Gaussian State Space Model — Full Conditional Gibbs Sampling* Sampling is carried out in three steps.

(a) Sample a path $\mathbf{x} = (\mathbf{x}_0, \dots, \mathbf{x}_T)$ of the continuous state variable conditional on $\boldsymbol{\vartheta}$ and $\mathbf{S}$ from the density $p(\mathbf{x}|\boldsymbol{\vartheta}, \mathbf{S}, \mathbf{y})$, preferably using forward-filtering-backward-sampling; see *Algorithm 13.4*.
(b) Sample a path $\mathbf{S} = (S_0, \dots, S_T)$ of the discrete state variable conditional on $\boldsymbol{\vartheta}$ and $\mathbf{x}$ from the density $p(\mathbf{S}|\boldsymbol{\vartheta}, \mathbf{x}, \mathbf{y})$.
(c) Sample $\boldsymbol{\vartheta}$ conditional on $\mathbf{x}$ and $\mathbf{S}$ from the complete-data posterior density $p(\boldsymbol{\vartheta}|\mathbf{x}, \mathbf{S}, \mathbf{y})$.

Sampling a path of the state process $\mathbf{x}_0, \dots, \mathbf{x}_T$ in step (a) is discussed in full detail in Subsection 13.5.2. Sampling the indicators in step (b) is straightforward, if $S_t$ is a hidden i.i.d. sequence with probability distribution $\boldsymbol{\eta} = (\eta_1, \dots, \eta_K)$. In this case, $S_t$ is independent of all other indicators $\mathbf{S}_{-t}$ given $\mathbf{x}$, and step (b) could be carried out in one sweep by sampling $S_t$ for each $t = 1, \dots, T$ from

$$\begin{aligned}&\Pr(S_t = j|\mathbf{y}, \mathbf{x}, \boldsymbol{\vartheta}) \\ &\propto p(\mathbf{y}_t|S_t = j, \mathbf{x}_t, \boldsymbol{\vartheta}) p(\mathbf{x}_t|S_t = j, \mathbf{x}_{t-1}, \boldsymbol{\vartheta})\eta_k.\end{aligned} \tag{13.53}$$

If $S_t$ is a hidden Markov chain, then the results derived earlier for sampling hidden Markov chains are extended to deal with switching state space models; see *Algorithm 13.5* for more details. Sampling the unknown model parameters in step (c) conditional on $\mathbf{S}$, $\mathbf{x}$, and $\mathbf{y}$ has been discussed earlier in Subsection 13.4.3.

Carter and Kohn (1996, Lemma 2.2) prove that full conditional Gibbs sampling may lead to a reducible sampler for certain state space models. This is the case, for instance, if one of the variances, say $\mathbf{Q}_t^{[k]}$, is assumed to be exactly 0, if $S_t = k$. As a remedy, Carter and Kohn (1996) substitute step

(b) in *Algorithm 13.2* by a step that samples $S_t$ without conditioning on the continuous states $\mathbf{x}$.

*Algorithm 13.3: MCMC for a Switching Linear Gaussian State Space Model — Marginal Sampling of the Indicators* Whereas sampling of $\mathbf{x}$ and $\boldsymbol{\vartheta}$ is the same as in step (a) and (c) in *Algorithm 13.2*, marginal sampling of the indicators is carried out in the following way.

(b) For $t = 1, \ldots, T$, sample $S_t$ from $p(S_t|\mathbf{S}_{-t}, \boldsymbol{\vartheta}, \mathbf{y})$ without conditioning on $\mathbf{x}$.

Generating the indicators $S_t$ in step (b) of this algorithm in an efficient way is far from straightforward. Carter and Kohn (1996) and Gerlach and Kohn (2000) discuss various samplers, that are reviewed in Subsection 13.5.3. The results of Liu et al. (1994) suggest that *Algorithm 13.3* is more efficient than *Algorithm 13.2*, because the indicators are conditioned on fewer variables when they are generated. This is supported by a small simulation study in Gerlach and Kohn (2000).

Another modification of *Algorithm 13.2* is a partially marginalized sampler (McCulloch and Tsay, 1993; Godsill, 1997; Godsill and Rayner, 1998), where sampling of the indicators and the states is carried out in a different manner.

### 13.5.2 Sampling the Continuous State Process from the Smoother Density

In this section, sampling a path of the state process $\mathbf{x}_0, \ldots, \mathbf{x}_T$ from the conditional posterior $p(\mathbf{x}_0, \ldots, \mathbf{x}_T|\mathbf{y}, \mathbf{S}, \boldsymbol{\vartheta})$, also called smoother density, is discussed in full detail. The transition density $p(\mathbf{x}_t|\mathbf{x}_{t-1})$ as well as the observation density $p(\mathbf{y}_t|\mathbf{x}_t)$ depends on unknown parameters $\boldsymbol{\vartheta}$ and the latent processes $\mathbf{S}$. This dependence, however, is dropped for the remainder of this subsection for notational convenience.

#### Single-Move Sampling of the Continuous State Process

Carlin et al. (1992) used a single-move Gibbs sampler based on sampling the state $\mathbf{x}_t$ for each $t = 1, \ldots, T$ from the conditional posterior $\mathbf{x}_t \sim p(\mathbf{x}_t|\mathbf{x}_{-t}, \mathbf{y})$, where $\mathbf{x}_{-t}$ is the collection all state vectors $\mathbf{x}_0, \ldots, \mathbf{x}_T$ excluding $\mathbf{x}_t$. The posterior $p(\mathbf{x}_t|\mathbf{x}_{-t}, \mathbf{y})$ is given by

$$\begin{aligned} p(\mathbf{x}_t|\mathbf{x}_{-t}, \mathbf{y}) &\propto p(\mathbf{y}|\mathbf{x})p(\mathbf{x}) \\ &\propto \prod_{t=1}^{T} p(\mathbf{y}_t|\mathbf{x}_t) \prod_{t=1}^{T} p(\mathbf{x}_t|\mathbf{x}_{t-1})p(\mathbf{x}_0). \end{aligned}$$

Dropping all quantities that are independent of $\mathbf{x}_t$ yields for $t = 1, \ldots, T-1$:

$$p(\mathbf{x}_t|\mathbf{x}_{-t}, \mathbf{y}) \propto p(\mathbf{y}_t|\mathbf{x}_t)p(\mathbf{x}_{t+1}|\mathbf{x}_t)p(\mathbf{x}_t|\mathbf{x}_{t-1}), \tag{13.54}$$

with obvious simplifications for $t = 0$ and $t = T$:

$$p(\mathbf{x}_0|\mathbf{x}_1,\ldots,\mathbf{x}_T,\mathbf{y}) \propto p(\mathbf{x}_1|\mathbf{x}_0)p(\mathbf{x}_0),$$
$$p(\mathbf{x}_T|\mathbf{x}_0,\ldots,\mathbf{x}_{T-1},\mathbf{y}) \propto p(\mathbf{y}_T|\mathbf{x}_T)p(\mathbf{x}_T|\mathbf{x}_{T-1}).$$

For a linear Gaussian state space model the first two densities in (13.54) may be considered as the likelihood of a linear model with general, but known, error covariance matrices and independent observations $\mathbf{y}_t$ and $\mathbf{x}_{t+1}$,

$$\begin{pmatrix} \mathbf{y}_t \\ \mathbf{x}_{t+1} \end{pmatrix} = \begin{pmatrix} \mathbf{H}_t \\ \mathbf{F}_{t+1} \end{pmatrix} \mathbf{x}_t + \begin{pmatrix} \varepsilon_t \\ \mathbf{w}_{t+1} \end{pmatrix},$$
$$\varepsilon_t \sim \mathcal{N}_r(\mathbf{0}, \mathbf{R}_t), \qquad \mathbf{w}_{t+1} \sim \mathcal{N}_d(\mathbf{0}, \mathbf{Q}_{t+1}),$$

where the unknown regression parameter $\mathbf{x}_t$ follows the conjugate normal prior $p(\mathbf{x}_t|\mathbf{x}_{t-1})$ as in Subsection 13.3.2. Thus for $t = 1, \ldots, T-1$ the density $p(\mathbf{x}_t|\mathbf{x}_{-t},\mathbf{y})$ is normal with

$$\mathbf{x}_t|\mathbf{x}_{-t},\mathbf{y} \sim \mathcal{N}_d\left(\hat{\mathbf{x}}_{t|-t}, \mathbf{P}_{t|-t}\right),$$
$$\mathbf{P}_{t|-t}^{-1} = \mathbf{H}_t^{'}\mathbf{R}_t^{-1}\mathbf{H}_t + \mathbf{F}_{t+1}^{'}\mathbf{Q}_{t+1}^{-1}\mathbf{F}_{t+1} + \mathbf{Q}_t^{-1},$$
$$\hat{\mathbf{x}}_{t|-t} = \mathbf{P}_{t|-t}(\mathbf{H}_t^{'}\mathbf{R}_t^{-1}\mathbf{y}_t + \mathbf{F}_{t+1}^{'}\mathbf{Q}_{t+1}^{-1}\mathbf{x}_{t+1} + \mathbf{Q}_t^{-1}\mathbf{F}_t\mathbf{x}_{t-1}),$$

a result that allows direct sampling. For more general state space models, $p(\mathbf{x}_t|\mathbf{x}_{-t},\mathbf{y})$ is no longer a normal density, but it is possible to draw from this density using a Metropolis–Hastings step (Carlin et al., 1992; Jacquier et al., 1994).

As noted by Carter and Kohn (1994), this sampler converges rather slowly when $\mathbf{Q}_t$ approaches singularity and breaks down to a reducible sampler; see also Pitt and Shephard (1999) for a theoretical investigation of this issue.

### Multi-Move Sampling of the Continuous State Process

A more efficient way to sample $\mathbf{x}_0, \ldots, \mathbf{x}_T$ for the linear Gaussian state space model is joint or multi-move sampling of the states (Carter and Kohn, 1994; Frühwirth-Schnatter, 1994; De Jong and Shephard, 1995; Koopman and Durbin, 2000). In contrast to single-move sampling, multi-move sampling draws the whole path $\mathbf{x} = (\mathbf{x}_0, \ldots, \mathbf{x}_T)$ from the joint posterior of all states: $(\mathbf{x}_0, \ldots, \mathbf{x}_T) \sim p(\mathbf{x}_0, \ldots, \mathbf{x}_T|\mathbf{y})$. The multi-move sampler starts by representing the joint density $p(\mathbf{x}|\mathbf{y})$ as the product of $T+1$ conditional densities:

$$p(\mathbf{x}|\mathbf{y}) = p(\mathbf{x}_T|\mathbf{y}) \prod_{t=0}^{T-1} p(\mathbf{x}_t|\mathbf{x}_{t+1}, \ldots, \mathbf{x}_T, \mathbf{y}). \tag{13.55}$$

The densities $p(\mathbf{x}_t|\mathbf{x}_{t+1}, \ldots, \mathbf{x}_T, \mathbf{y})$ are the posterior densities of $\mathbf{x}_t$ knowing not only all observations $\mathbf{y}$, but also all future values $\mathbf{x}_{t+1}, \ldots, \mathbf{x}_T$. This posterior is obtained by Bayes' theorem as

$$p(\mathbf{x}_t|\mathbf{x}_{t+1},\ldots,\mathbf{x}_T,\mathbf{y}) \propto p(\mathbf{y}_{t+1},\ldots,\mathbf{y}_T,\mathbf{x}_{t+1},\ldots,\mathbf{x}_T|\mathbf{x}_t,\mathbf{y}^t)p(\mathbf{x}_t|\mathbf{y}^t)$$
$$\propto \prod_{s=t+1}^{T} p(\mathbf{y}_s|\mathbf{x}_s) \prod_{s=t}^{T-1} p(\mathbf{x}_{s+1}|\mathbf{x}_s)p(\mathbf{x}_t|\mathbf{y}^t).$$

Dropping terms that are independent of $\mathbf{x}_t$ we find that this density is obtained by combining the filter density $p(\mathbf{x}_t|\mathbf{y}^t)$ with the likelihood of $\mathbf{x}_{t+1}$ measured in terms of the transition density $p(\mathbf{x}_{t+1}|\mathbf{x}_t)$:

$$p(\mathbf{x}_t|\mathbf{x}_{t+1},\ldots,\mathbf{x}_T,\mathbf{y}) \propto p(\mathbf{x}_{t+1}|\mathbf{x}_t)p(\mathbf{x}_t|\mathbf{y}^t). \tag{13.56}$$

Equations (13.55) and (13.56) motivate was has been called forward-filtering-backward-sampling (Frühwirth-Schnatter, 1994).

*Algorithm 13.4: Forward-Filtering-Backward-Sampling (FFBS)*

(a) Determine and store the moments $\hat{\mathbf{x}}_{t|t}$ and $\mathbf{P}_{t|t}$ of the filtering density $p(\mathbf{x}_t|\mathbf{y}^t)$ by running a Kalman filter from $t = 1,\ldots,T$ as described in *Algorithm 13.1.*
(b) Start sampling of the path $\mathbf{x}_0,\ldots,\mathbf{x}_T$ by sampling the latest state vector $\mathbf{x}_T$ from the most recent filter density $p(\mathbf{x}_T|\mathbf{y}^T)$.
(c) Sample the remaining states $\mathbf{x}_t$ from $p(\mathbf{x}_t|\mathbf{x}_{t+1},\ldots,\mathbf{x}_T,\mathbf{y})$ backward in time for $t = T-1,\ldots,0$.

There exist various ways to implement step (c). Following Carter and Kohn (1994), $p(\mathbf{x}_{t+1}|\mathbf{x}_t)$ may be considered as the likelihood of a general linear model with known error covariance matrices as in Subsection 13.3.2, with observations $\mathbf{x}_{t+1}$ and regression parameter $\mathbf{x}_t$ following the conjugate normal prior $p(\mathbf{x}_t|\mathbf{y}^t)$. From Subsection 13.3.2, the density $p(\mathbf{x}_t|\mathbf{x}_{t+1},\ldots,\mathbf{x}_T,\mathbf{y})$ is normal with

$$\begin{aligned}
&\mathbf{x}_t|\mathbf{x}_{t+1},\ldots,\mathbf{x}_T,\mathbf{y} \sim \mathcal{N}_d\left(\hat{\mathbf{x}}_{t|T}(\mathbf{x}_{t+1}),\mathbf{P}_{t|T}\right), \qquad (13.57)\\
&\hat{\mathbf{x}}_{t|T}(\mathbf{x}_{t+1}) = (\mathbf{I}-\mathbf{B}_{t+1}\mathbf{F}_{t+1})\hat{\mathbf{x}}_{t|t} + \mathbf{B}_{t+1}(\mathbf{x}_{t+1}-\mathbf{G}_{t+1}\mathbf{u}_{t+1}),\\
&\mathbf{P}_{t|T} = (\mathbf{I}-\mathbf{B}_{t+1}\mathbf{F}_{t+1})\mathbf{P}_{t|t},\\
&\mathbf{B}_{t+1} = \mathbf{P}_{t|t}\mathbf{F}'_{t+1}\left(\mathbf{F}_{t+1}\mathbf{P}_{t|t}\mathbf{F}'_{t+1}+\mathbf{Q}_{t+1}\right)^{-1}.
\end{aligned}$$

If $\mathbf{Q}_{t+1}$ is positive definite, one could also use the information form of updating the posterior in a general linear model. If $\mathbf{Q}_{t+1}$ is singular, then the conditional density $p(\mathbf{x}_t|\mathbf{x}_{t+1},\ldots,\mathbf{x}_T,\mathbf{y})$ is degenerate because part of $\mathbf{x}_t$ is deterministic given $\mathbf{x}_{t+1}$. Sampling from (13.57) based on a Cholesky decomposition of $\mathbf{P}_{t|T}$ will lead to numerical problems. Furthermore the recursions in (13.57) are inefficient, as they involve the inversion of a $(d \times d)$ matrix, with $d = \dim \mathbf{x}_t$, whereas $\mathbf{x}_t$ only has $s = \text{rg}(\mathbf{Q}_t) < d$ random components. Frühwirth-Schnatter (1994) suggested transforming the state vector $\mathbf{x}_t$ to a new state variable with only $s$ random components. Another efficient sampler is to simulate the disturbances $\mathbf{w}_t$ rather than $\mathbf{x}_t$ using a disturbance smoother (De Jong and Shephard, 1995; Durbin and Koopman, 2002).

For more general state space models, such a multi-move sampler does not exist. Shephard and Pitt (1997) designed a blocked sampler, where an entire subblock $\mathbf{x}_t, \ldots, \mathbf{x}_{t+h}$ is sampled from the appropriate density using a Metropolis–Hastings step.

### 13.5.3 Sampling the Discrete States for a Switching State Space Model

The notation $\mathbf{S}^t = (S_0, \ldots, S_t)$ is used to denote a whole path of the hidden Markov chain $S_t$ up to $t$, with $S_0$ being dropped for finite mixtures of state space models.

#### Full Conditional Sampling of a Hidden Markov Chain

Full conditional sampling of the states $\mathbf{S}$ of a hidden Markov chain is not restricted to linear Gaussian state space models, but may be applied also to more general models with nonnormal or nonlinear densities $p(\mathbf{y}_t|S_t = j, \mathbf{x}_t, \boldsymbol{\vartheta})$ and $p(\mathbf{x}_t|S_t = j, \mathbf{x}_{t-1}, \boldsymbol{\vartheta})$.

Single-move sampling of $p(S_t|\mathbf{S}_{-t}, \mathbf{x}, \mathbf{y}, \boldsymbol{\vartheta})$ could be used as in Subsection 11.5.6, however, it is much more efficient to use a multi-move sampler (Carter and Kohn, 1994; Shephard, 1994) that samples the whole path $\mathbf{S} = (S_0, \ldots, S_T)$ jointly from $p(\mathbf{S}|\mathbf{x}, \mathbf{y}, \boldsymbol{\vartheta})$. This multi-move sampler is closely related to the sampler discussed in *Algorithm 11.5* for finite Markov mixture models.

*Algorithm 13.5: Multi-Move Sampling of the Discrete States of a Switching State Space Model*

(a) Run a filter conditional on $\boldsymbol{\vartheta}$ and $\mathbf{x}$ to obtain the filtered probability distribution $\Pr(S_t = j|\mathbf{y}^t, \mathbf{x}^t, \boldsymbol{\vartheta})$ for $t = 1, \ldots, T$. The filter is started at $t = 1$ with the initial distribution $\Pr(S_0 = k|\boldsymbol{\xi})$. For each $t \geq 1$, perform one-step ahead prediction,

$$\Pr(S_t = j|\mathbf{y}^{t-1}, \mathbf{x}^{t-1}, \boldsymbol{\vartheta}) = \sum_{k=1}^{K} \xi_{kj} \Pr(S_{t-1} = k|\mathbf{y}^{t-1}, \mathbf{x}^{t-1}, \boldsymbol{\vartheta}),$$

and filtering for each possible value $j = 1, \ldots, K$ of $S_t$:

$$\begin{aligned} &\Pr(S_t = j|\mathbf{y}^t, \mathbf{x}^t, \boldsymbol{\vartheta}) \qquad (13.58)\\ &\propto p(\mathbf{y}_t|S_t = j, \mathbf{x}_t, \boldsymbol{\vartheta}) p(\mathbf{x}_t|S_t = j, \mathbf{x}_{t-1}, \boldsymbol{\vartheta}) \Pr(S_t = j|\mathbf{y}^{t-1}, \mathbf{x}^{t-1}, \boldsymbol{\vartheta}). \end{aligned}$$

The probabilities in (13.58) need to be normalized to obtain a proper filter distribution.

(b) Sample $S_T$ from the discrete probability distribution $\Pr(S_T = j|\mathbf{y}^T, \mathbf{x}^T, \boldsymbol{\vartheta})$.

(c) For $t = T-1, T-2, \ldots, 0$ sample $S_t$ from the conditional distribution $\Pr(S_t = j|S_{t+1}, \mathbf{y}^t, \mathbf{x}^t, \boldsymbol{\vartheta})$ given by

$$\Pr(S_t = j|S_{t+1}, \mathbf{y}^t, \mathbf{x}^t, \boldsymbol{\vartheta}) = \frac{\xi_{j,S_{t+1}} \Pr(S_t = j|\mathbf{y}^t, \mathbf{x}^t, \boldsymbol{\vartheta})}{\sum_{k=1}^{K} \xi_{k,S_{t+1}} \Pr(S_t = k|\mathbf{y}^t, \mathbf{x}^t, \boldsymbol{\vartheta})}.$$

Here $S_{t+1}$ is the most recent value sampled for the hidden Markov chain at $t+1$.

### Marginal Sampling of the Indicators

Both Carter and Kohn (1996) and Gerlach and Kohn (2000) generate $S_t$ from the discrete density $p(S_t|\mathbf{S}_{-t}, \mathbf{y}, \boldsymbol{\vartheta})$ without conditioning on the continuous states $\mathbf{x}$. Marginalization over $\mathbf{x}$, however, leads to dependence among all the values of $S_t$, even if the indicators are i.i.d., and generating $S_t$ in an efficient way is far from straightforward.

Suppose that $\mathbf{S}^{t-1}$ has already been updated and that the first two moments of the normal density $p(\mathbf{x}_{t-1}|\mathbf{y}^{t-1}, \mathbf{S}^{t-1}, \boldsymbol{\vartheta})$ are known. Bayes' theorem is used to obtain the density $p(S_t|\mathbf{S}_{-t}, \mathbf{y}, \boldsymbol{\vartheta})$:

$$p(S_t|\mathbf{S}_{-t}, \mathbf{y}, \boldsymbol{\vartheta}) \propto p(S_t|\mathbf{S}_{-t}, \boldsymbol{\vartheta}) p(\mathbf{y}_t|\mathbf{y}^{t-1}, \mathbf{S}^t, \boldsymbol{\vartheta}) p(\mathbf{y}_{t+1}, \ldots, \mathbf{y}_T|\mathbf{y}^t, \mathbf{S}_{-t}, S_t, \boldsymbol{\vartheta}).$$

For each of the $K$ values of $S_t$, the predictive density $p(\mathbf{y}_t|\mathbf{y}^{t-1}, \mathbf{S}^t, \boldsymbol{\vartheta})$ as well as the filtering density $p(\mathbf{x}_t|\mathbf{y}^t, \mathbf{S}^t, \boldsymbol{\vartheta})$, is obtained from a single step of the Kalman filter. A direct but inefficient method to evaluate the predictive density $p(\mathbf{y}_{t+1}, \ldots, \mathbf{y}_T|\mathbf{y}^t, \mathbf{S}_{-t}, S_t)$ for the $K$ different values of $S_t$ is to use $T-t+1$ forecasting steps of the Kalman filter, which requires $\mathcal{O}(T)$ steps to generate $S_t$, and hence $\mathcal{O}(T^2)$ steps to generate the whole path $\mathbf{S}$. Gerlach and Kohn (2000) show how to obtain the term $p(\mathbf{y}_{t+1}, \ldots, \mathbf{y}_T|\mathbf{y}^t, \mathbf{S}_{-t}, S_t)$ in one step after an initial set of backward recursions, requiring $\mathcal{O}(T)$ steps to generate the whole path $\mathbf{S}$. We refer to Gerlach and Kohn (2000) for more details.

Finally, Gerlach and Kohn (2000) discuss an efficient way of sampling a binary indicator $S_t$ which takes one of two values most of the time, for instance, an indicator corresponding to an outlier or to an intervention variable.

## 13.6 Further Issues

### 13.6.1 Model Specification Uncertainty in Switching State Space Modeling

The application of the state space approach to socioeconomic or biological sciences is complicated by the need of model identification, because often little a priori information about the dynamics of the system is available. One approach toward this model specification uncertainty is to fit several state space models to a given time series and to apply some method of model selection.

AIC was used in the context of model selection for state space models by, among others, Kitagawa (1981) and Harvey (1989). AIC and BIC are defined for state space models in the usual way as

$$\mathrm{AIC} = -2\log p(\mathbf{y}|\hat{\boldsymbol{\vartheta}}) + 2\dim(\boldsymbol{\vartheta}), \tag{13.59}$$

$$\mathrm{BIC} = -2\log p(\mathbf{y}|\hat{\boldsymbol{\vartheta}}) + \log(T)\dim(\boldsymbol{\vartheta}), \tag{13.60}$$

where $p(\mathbf{y}|\hat{\boldsymbol{\vartheta}})$ is the (approximate) likelihood of a (switching) state space model evaluated at the ML estimator $\hat{\boldsymbol{\vartheta}}$. Durbin and Koopman (2001, p.152) provide a corrected AIC and BIC for state space models with diffuse initial conditions. Harvey (1989) and Durbin and Koopman (2001) prefer a definition where the right-hand side of (13.59) and (13.60) is divided by $T$.

The marginal likelihood has been applied to model selection problems involving state space models by, among many others, Frühwirth-Schnatter (1995), Shively and Kohn (1997), and Koop and van Dijk (2000). Frühwirth-Schnatter (2001a) discusses model comparison based on marginal likelihoods for switching linear Gaussian state space models and uses the bridge sampling techniques discussed in Subsection 5.4.6 to obtain a numerical approximation of the marginal likelihood.

A Bayesian variable selection approach (Carlin and Chib, 1995) has been applied to switching dynamic factor models by Kim and Nelson (2001).

### 13.6.2 Auxiliary Mixture Sampling for Nonlinear and Nonnormal State Space Models

To deal with non-Gaussian or nonlinear state space models it is useful to approximate nonnormal densities by a finite mixture of common distributions. Sorenson and Alspach (1971) and Alspach and Sorenson (1972) are pioneering works using a Gaussian sum approximation to derive an approximate filter for nonlinear and non-Gaussian state space models. Meinhold and Singpurwalla (1989) represent the posterior density $p(\mathbf{x}_{t-1}|\mathbf{y}^{t-1})$ by a mixture of $t$-distributions and suggest some approximate recursive scheme to obtain a similar mixture approximation to $p(\mathbf{x}_t|\mathbf{y}^t)$.

To facilitate statistical inference, Shephard (1994) introduced the concept of partially Gaussian state space models and suggested approximating nonnormal densities appearing in the definition of the state space model by mixtures of normal distributions. This allows MCMC estimation through efficient multi-move sampling of the state process as in *Algorithm 13.4* also for non-Gaussian state space models, where usually single-move sampling has to be applied.

MCMC methods based on a finite mixture approximation have been developed in particular for stochastic volatility models (Shephard, 1994; Kim et al., 1998; Chib et al., 2002; Omori et al., 2004). A stochastic volatility model is a state space model with state vector $h_t$, usually assumed to follow an AR(1)-process, where the observation equation is nonlinear, because the variance of the observation error is a nonlinear function of $h_t$:

$$\begin{aligned} h_t &= \delta h_{t-1} + \zeta + w_t, \qquad w_t \sim \mathcal{N}\left(0, \sigma_\mu^2\right), \\ Y_t &= e^{h_t/2} z_t, \qquad z_t \sim \mathcal{N}(0,1). \end{aligned}$$

This model may be transformed into a linear state space model with nonnormal errors in the following way,

$$\log Y_t^2 = h_t + \varepsilon_t,$$

where $\varepsilon_t$ is equal to the log of a $\chi_1^2$ random variable. The density of the $\log \chi_1^2$ is approximated in Shephard (1994) by a mixture of univariate normal distributions,

$$p(\varepsilon_t) = \sum_{k=1}^{K} w_k f_N(\varepsilon_t; m_k, s_k^2).$$

Shephard (1994) derived appropriate parameters $(w_k, m_k, s_k^2), k = 1, \ldots, K$, for mixtures up to $K = 7$ components, whereas a more accurate approximation with $K = 10$ components appears in Omori et al. (2004). By introducing i.i.d. hidden indicators $S_t$ for each $t$, the following finite mixture of linear Gaussian state space models results,

$$\begin{aligned} h_t &= \delta h_{t-1} + \zeta + w_t, \qquad w_t \sim \mathcal{N}\left(0, \sigma_\mu^2\right), \\ \log Y_t^2 &= h_t + m_{S_t} + \varepsilon_t, \qquad \varepsilon_t \sim \mathcal{N}\left(0, s_{S_t}^2\right), \end{aligned}$$

with $\Pr(S_t = k) = w_k$. Filtering and parameter estimation as discussed in Sections 13.3 to 13.5 may be applied.

Recently, Frühwirth-Schnatter and Wagner (2006) developed a similar auxiliary mixture sampler for state space modeling of count data, based on a finite mixture approximation to the type I extreme value distribution. Frühwirth-Schnatter and Frühwirth (2006) show that this sampler may be extended to deal with state space modeling of binary and multinomial data.

## 13.7 Illustrative Application to Modeling Exchange Rate Data

For illustration we reanalyze the U.S./U.K. real exchange rate from January 1885 to November 1995, originally published in Grilli and Kaminsky (1991)

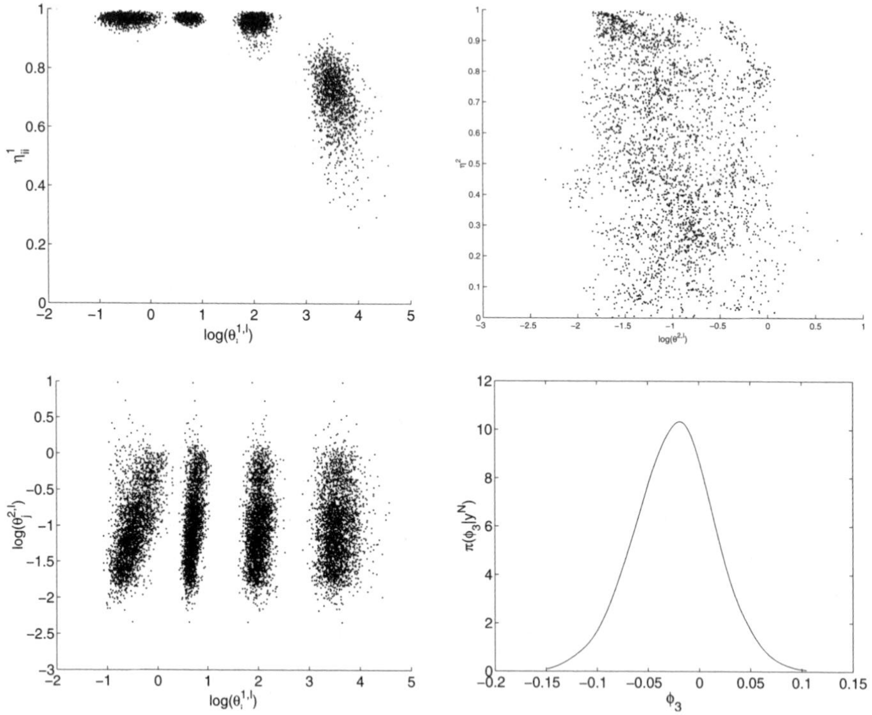


**Fig. 13.1.** U.S./U.K. Real Exchange Rate Data, exploratory Bayesian analysis for a switching model with $K_1 = 4, K_2 = 2, p = 3$; top left-hand side: $\log(\sigma^2_{1,k})$ versus $\boldsymbol{\xi}^1_{kk}$ for all possible $k$; top right-hand side: $\log(\sigma^2_{2,k})$ versus $\boldsymbol{\xi}^2_{kk}$ for all possible $k$; bottom left-hand side: $\log(\sigma^2_{1,k})$ versus $\log(\sigma^2_{2,k})$ for all possible $k$; bottom right-hand side: posterior of $\delta_3$ (from Frühwirth-Schnatter (2001a) with permission granted by The Institute of Statistical Mathematics)

and reanalyzed by Engel and Kim (1999) and Frühwirth-Schnatter (2001a). The real exchange rate is defined as the relative price of U.K. to U.S. producer goods; that is, U.S./U.K. nominal exchange rate times the U.K. producer price index divided by the U.S. producer price index. Engel and Kim (1999) suggested decomposing the log of the real exchange rate $Y_t$ into a permanent component $\mu_t$ and a transitory component $c_t$:

$$\log Y_t = \mu_t + c_t,$$

where $c_t$ is assumed to follow an AR($p$) process:

$$c_t = \delta_1 c_{t-1} + \cdots + \delta_p c_{t-p} + w_{t,1},$$

and $\mu_t$ follows a random walk process:

$$\mu_t = \mu_{t-1} + w_{t,2}.$$

The conditional variance of the transitory component $c_t$ is assumed to switch between $K_1$ values according to a Markov chain $S_t^1$ with transition matrix $\boldsymbol{\xi}^1$, whereas the conditional variance of the permanent component $\mu_t$ is assumed to switch between $K_2$ values according to a Markov chain $S_t^2$ with transition matrix $\boldsymbol{\xi}^2$:

$$w_{t,1} \sim \mathcal{N}\left(0, \sigma^2_{1,S_t^1}\right), \qquad w_{t,2} \sim \mathcal{N}\left(0, \sigma^2_{2,S_t^2}\right).$$

The model can be put into state space form with the following state vector $\mathbf{x}_t$ and matrix $\mathbf{F}$,

$$\mathbf{x}_t = \begin{pmatrix} \mu_t \\ c_t \\ \vdots \\ c_{t-p+1} \end{pmatrix}, \qquad \mathbf{F} = \begin{pmatrix} 1 & \mathbf{0}_{1\times p} \\ \mathbf{0}_{p\times 1} & \mathbf{F}(\boldsymbol{\delta}) \end{pmatrix},$$

and $\mathbf{F}(\boldsymbol{\delta})$ being the same as in (13.15).

This model is a switching linear Gaussian state space model with two hidden indicators. The estimation method used by Frühwirth-Schnatter (2001a) is an extension of *Algorithm 13.2* to the case of two hidden switching variables. Frühwirth-Schnatter (2001a) did not condition on the first values of the state process as in Engel and Kim (1999), but sample in step (a) the whole processes $c_{1-p}, \dots, c_0, \dots, c_T$ and $\mu_0, \dots, \mu_T$ including the starting values by applying the multi-move sampler of Frühwirth-Schnatter (1994). The filter is initialized with the prior $\mathbf{x}_0 \sim \mathcal{N}\left(\hat{\mathbf{x}}_{0|0}, \mathbf{P}_{0|0}\right)$, where

$$\hat{\mathbf{x}}_{0|0} = \begin{pmatrix} \log y_1 \\ 0 \\ \vdots \\ 0 \end{pmatrix}, \qquad \mathbf{P}_{0|0} = \begin{pmatrix} 1000 & \mathbf{0}_{1\times(p-1)} \\ \mathbf{0}_{(p-1)\times 1} & \mathbf{M} \end{pmatrix}$$

with

$$\text{vec}(\mathbf{M}) = (\mathbf{I}_{p^2} - \mathbf{F}(\boldsymbol{\delta}) \otimes \mathbf{F}(\boldsymbol{\delta}))^{-1} \begin{pmatrix} \sigma^2_{1,S_0^1} \\ \mathbf{0}_{(p-1)\times 1} \end{pmatrix},$$

and $\otimes$ is the Kronecker product of two matrices. This choice is based on the suggestion of De Jong and Chu-Chun-Lin (1994) for combining a vague prior with a stationary prior for state vectors containing both nonstationary and stationary components.

As the Markov processes $S_t^1$ and $S_t^2$ are independent a posteriori, sampling in step (b) is carried out independently for both indicators using *Algorithm 13.5*. For $S_t^1$ the filter step (13.58) is based on

$$\begin{aligned} &\Pr(S_t^1 = j|\mathbf{y}^t, \mathbf{x}^t, \boldsymbol{\vartheta}) \\ &\propto f_N(c_t; \delta_1 c_{t-1} + \cdots + \delta_p c_{t-p}, \sigma^2_{1,j}) \Pr(S_t^1 = j|\mathbf{y}^{t-1}, \mathbf{x}^{t-1}, \boldsymbol{\vartheta}), \end{aligned}$$

whereas for $S_t^2$ this step reads:

$$\Pr(S_t^2 = j|\mathbf{y}^t, \mathbf{x}^t, \boldsymbol{\vartheta}) \propto f_N(\mu_t; \mu_{t-1}, \sigma_{2,j}^2)\Pr(S_t^2 = j|\mathbf{y}^{t-1}, \mathbf{x}^{t-1}, \boldsymbol{\vartheta}).$$

Parameter estimation is based on the priors $\sigma_{1,k}^2 \sim \mathcal{G}^{-1}(3,8)$, $k = 1, \dots, K_1$, and $\sigma_{2,k}^2 \sim \mathcal{G}^{-1}(3,2)$, $k = 1, \dots, K_2$. The prior for all rows of the transition matrices $\boldsymbol{\xi}_1$ and $\boldsymbol{\xi}_2$ is chosen to be $\mathcal{D}(1, \dots, 1)$.

All variances $\sigma_{1,k}^2, k = 1, \dots, K_1$ and $\sigma_{2,k}^2, k = 1, \dots, K_2$ are sampled at the same time, as they are conditionally independent, inverted Gamma distributed. This is different from Engel and Kim (1999) who impose a priori an identifiability constraint on the variances and sample the variances in a single-move manner from the constrained posterior.

Sampling of the AR($p$) parameters $\delta_1, \dots, \delta_p$ is carried out from the regression model $c_t = \delta_1 c_{t-1} + \dots + \delta_p c_{t-p} + \sigma_{1,S_t^1}\varepsilon_t$, where $\varepsilon_t$ is i.i.d. standard normal. As samples of $c_0, \dots, c_{1-r}$ are available from step (a), $t$ is running from 1 to $T$. Within one iteration, sampling of the AR($p$) parameters $\delta_1, \dots, \delta_p$ is repeated until the stationarity condition on the AR($p$) process is fulfilled.

**Table 13.1.** U.S./U.K. REAL EXCHANGE RATE DATA, model selection using marginal likelihoods (from Frühwirth-Schnatter (2001a) with permission granted by The Institute of Statistical Mathematics)

| Model | $\log p(\mathbf{y}\|\text{Model})$ |
|---|---|
| $K_1 = 4$, $K_2 = 2$, $p = 3$ | -2562.4 |
| $K_1 = 4$, $K_2 = 1$, $p = 2$ | **-2515.5** |
| $K_1 = 4$, $K_2 = 1$, $p = 1$ | -2612.5 |
| $K_1 = 3$, $K_2 = 1$, $p = 2$ | -2605.9 |
| $K_1 = 5$, $K_2 = 1$, $p = 2$ | -2880.2 |
| No switching, $p = 2$ | -2914.4 |

Engel and Kim (1999) selected a model where the variance of the transitory component is driven by a three-state Markov switching process, the variance of the permanent component is constant, and the order of the AR process is equal to two, that is, $K_1 = 3, K_2 = 1, p = 2$. They adopt this specification by exploring the posterior distributions without formal Bayesian model selection.

We proceed with an exploratory Bayesian analysis of a model with $K_1 = 4$, $K_2 = 2$, and $p = 3$, using the MCMC output of a random permutation sampler. Parts (a) and (b) of Figure 13.1 show a point process representation of $(\sigma_{1,k}^2)^{(m)}$ versus $(\boldsymbol{\xi}_{kk}^1)^{(m)}$ and $(\sigma_{2,j}^2)^{(m)}$ versus $(\boldsymbol{\xi}_{jj}^2)^{(m)}$ for all possible states $k \in \{1, \dots, K_1\}$ and $j \in \{1, \dots, K_2\}$, respectively. For $S_t^1$ we have allowed for four states and there are actually four simulation clusters; for $S_t^2$, however, we have allowed for two states, but there is just one simulation cluster. This provides empirical evidence in favor of a homogeneous rather than a switching variance of the permanent component. This hypothesis is further supported by part (c) of the figure where the point process representation of $(\sigma^2_{1,k})^{(m)}$ versus $(\sigma^2_{2,k})^{(m)}$ is plotted. Finally, part (d) of the same figure plots the posterior of the AR parameter $\delta_3$ which may be estimated directly from the output of the random permutation sampler as $\delta_3$ is state independent. The mode of the posterior is close to 0 providing evidence for the hypothesis that $\delta_3$ is equal to zero. To sum up, the exploratory analysis provides evidence in favor of a model with $K_1 = 4$, $K_2 = 1$, and $p = 2$ rather than $K_1 = 3$, $K_2 = 2$, and $p = 2$.

In Frühwirth-Schnatter (2001a) the marginal likelihood, based on a bridge sampling estimator, was used for model selection; see Table 13.1. For the best model the variance of the transitory component is driven by a four-state Markov switching process, the variance of permanent component is constant, and the order of the AR process is equal to two; that is, $K_1 = 4$, $K_2 = 1$, $p = 2$.

The marginal likelihoods reported in Table 13.1, however, clearly favor the model with $K_1 = 4$, $K_2 = 1$, and $p = 2$, which differs from the one selected in Engel and Kim (1999) by the number of states of the variance of the transitory component. Increasing the number of states from four to five, however, reduces the marginal likelihood drastically. For completeness, the marginal likelihood for a model without switching is reported, showing that this model is the most unlikely of all.

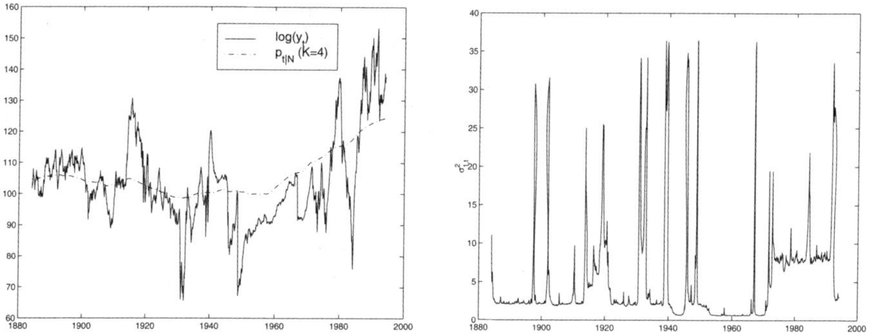


**Fig. 13.2.** U.S./U.K. REAL EXCHANGE RATE DATA, four-state model ($K_1 = 4$, $K_2 = 1$, $p = 2$); left-hand side: smoothed real exchange rate $\hat{p}_{t|T}$; right-hand side: estimated time-varying variance $\hat{\sigma}^2_{1,t}$ ($K_1 = 3$, $K_2 = 1$, $p = 2$) (from Frühwirth-Schnatter (2001a) with permission granted by The Institute of Statistical Mathematics)

We can draw further interesting inferences from the output of the random permutation sampler without the need to identify the model. This is especially true for the smoothed permanent component $\hat{p}_{t|T}$ which is compared in Figure 13.2 with the observed time series. The resulting estimator of the permanent component is much smoother than the rather noisy estimate published in Engel and Kim (1999), being nearly constant until the end of the fifties and increasing afterwards. Another interesting picture is obtained if we plot the time-varying variance $\sigma^2_{1,t}$ estimated from:

$$\hat{\sigma}^2_{1,t} = \frac{1}{M} \sum_{m=1}^{M} \left(\sigma^2_{1,s}\right)^{(m)},$$

where $s = (S_t^1)^{(m)}$ over time $t$ as in Figure 13.2.

**Table 13.2.** U.S./U.K. Real Exchange Rate Data, estimation results for $K_1 = 4, K_2 = 1, p = 2$ (from Frühwirth-Schnatter (2001a) with permission granted by The Institute of Statistical Mathematics)

| Parameter | Mean | Std.Dev. | 95%-H.P.D. Regions | |
|---|---|---|---|---|
| $\sigma^2_{1,1}$ | 0.634 | 0.151 | 0.371 | 0.93 |
| $\sigma^2_{1,2}$ | 2.05 | 0.196 | 1.67 | 2.42 |
| $\sigma^2_{1,3}$ | 7.63 | 1.07 | 5.9 | 9.88 |
| $\sigma^2_{1,4}$ | 36.4 | 9.13 | 20.7 | 53.9 |
| $\sigma^2_2$ | 0.366 | 0.132 | 0.121 | 0.608 |
| $\delta_1$ | 1.06 | 0.0474 | 0.967 | 1.14 |
| $\delta_2$ | −0.0729 | 0.046 | −0.158 | 0.0139 |
| $\xi_{11}$ | 0.968 | 0.0132 | 0.943 | 0.991 |
| $\xi_{12}$ | 0.0091 | 0.00861 | 2.84e−006 | 0.0256 |
| $\xi_{13}$ | 0.00639 | 0.00586 | 2.87e−006 | 0.0189 |
| $\xi_{14}$ | 0.0162 | 0.00987 | 0.000231 | 0.0341 |
| $\xi_{21}$ | 0.00855 | 0.00576 | 0.000165 | 0.0205 |
| $\xi_{22}$ | 0.973 | 0.00853 | 0.957 | 0.988 |
| $\xi_{23}$ | 0.00587 | 0.0057 | 6.19e−006 | 0.0155 |
| $\xi_{24}$ | 0.0123 | 0.00697 | 0.000484 | 0.0246 |
| $\xi_{31}$ | 0.00498 | 0.00489 | 1.24e−005 | 0.0144 |
| $\xi_{32}$ | 0.0139 | 0.0123 | 9.59e−006 | 0.0373 |
| $\xi_{33}$ | 0.956 | 0.0222 | 0.916 | 0.992 |
| $\xi_{34}$ | 0.0248 | 0.0161 | 0.00129 | 0.0562 |
| $\xi_{41}$ | 0.039 | 0.0338 | 0.000159 | 0.103 |
| $\xi_{42}$ | 0.147 | 0.0691 | 0.024 | 0.288 |
| $\xi_{43}$ | 0.123 | 0.0934 | 0.00108 | 0.309 |
| $\xi_{44}$ | 0.691 | 0.116 | 0.438 | 0.865 |

The selected model has to be identified to draw inference on the variances of the different states as well as to obtain state estimates over the whole observation period. The identifiability constraint $\sigma^2_{1,1} < \sigma^2_{1,2} < \sigma^2_{1,3} < \sigma^2_{1,4}$ is suggested by the point process representation in Figure 13.1, showing that the states of $S_t^1$ differ in the variance of the transitory component. If this constraint is included in the permutation sampler, no label switching occurs.

Table 13.2 reports point estimates as well as 95%-H.P.D.-regions for all model parameters, including estimates of the state-specific variances as well as estimates of the transition probabilities.

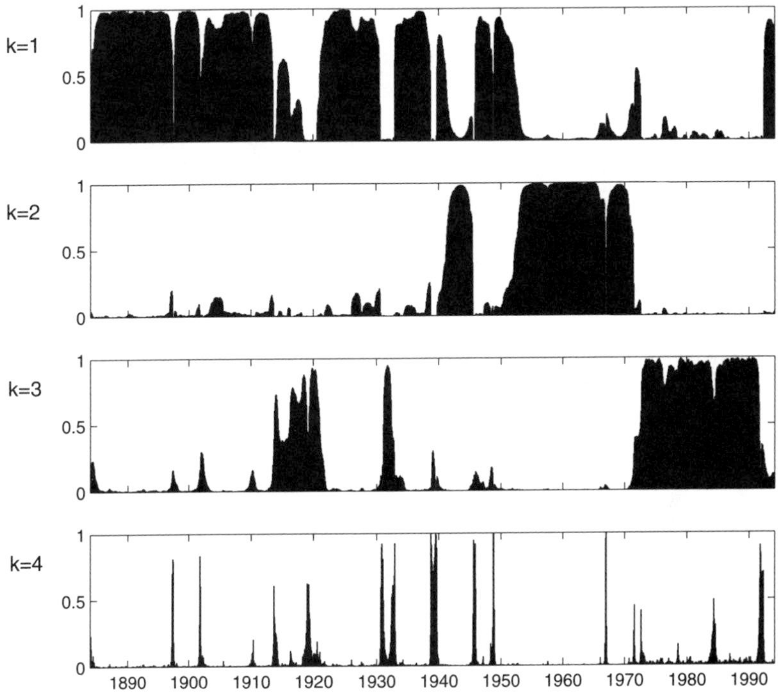


**Fig. 13.3.** U.S./U.K. REAL EXCHANGE RATE DATA, smoothed state probabilities for $S_t^1$ for a switching state space model with $K_1 = 4$, $K_2 = 1$, and $p = 2$ (from Frühwirth-Schnatter (2001a) with permission granted by The Institute of Statistical Mathematics)

Figure 13.3 plots the smoothed posterior state probabilities $\Pr(S_t^{1,\mathcal{L}} = k|\mathbf{y})$ of being in a certain state $k \in \{1, 2, 3, 4\}$ over time $t$, for a four-state switching model, and compares them with the probabilities obtained from the three-state model. The probabilities $\Pr(S_t^{1,\mathcal{L}} = k|\mathbf{y})$ are estimated from the constrained MCMC output by

$$\Pr(S_t^{1,\mathcal{L}} = k|\mathbf{y}) = \frac{1}{M} \#\{(S_t^{1,\mathcal{L}})^{(m)} = k\}.$$

Engel and Kim (1999) found the following interpretation of these probabilities. The quietest state occurred during the first half of the forties and then from about 1952 to the end of the seventies, which are periods in which the nominal exchange rate was fixed. The two medium-state variances correspond to periods of floating nominal exchange rates. Periods of high-state variance are rather singular events and can be identified with specific historical events.

# A

# Appendix

## A.1 Summary of Probability Distributions

Here we briefly summarize all probability distributions used in this book. For an exhaustive review of probability distributions see Johnson et al. (1993, 1994, 1995). The parameterization of the densities closely follows Bernardo and Smith (1994).

*Notation*

$Y$ denotes a univariate random variable; $y$ refers to a realization of $Y$. $f(y)$ refers to the density of the probability distribution of $Y$ with respect to an appropriate measure (Lebesgue measure or counting measure, depending on the context). $\mathrm{E}(Y)$ denotes the expectation of $Y$, whereas $\mathrm{Var}(Y)$ denotes the variance of $Y$. For multivariate random variables $\mathbf{Y}$ and $\mathbf{y}$ are used, whereas $Y_j$ and $y_j$ refer to a certain element of $\mathbf{Y}$ and $\mathbf{y}$, respectively. $\mathrm{E}(\mathbf{Y})$ denotes the mean vector; $\mathrm{Var}(Y)$ denotes the variance–covariance matrix of $\mathbf{Y}$.

### A.1.1 The Beta Distribution

The Beta distribution $Y \sim \mathcal{B}(\alpha, \beta)$ with $\alpha, \beta \in \Re^+$, is a univariate distribution defined on the unit interval $y \in [0, 1]$. For mixture models the Beta distribution appears mainly as a posterior distribution of an unknown probability. Density, mean, and variance are given by

$$f_B(y; \alpha, \beta) = \frac{1}{B(\alpha, \beta)} y^{\alpha-1}(1-y)^{\beta-1}, \tag{A.1}$$

$$\mathrm{E}(Y) = \frac{\alpha}{\alpha+\beta}, \qquad \mathrm{Var}(Y) = \frac{\alpha\beta}{(\alpha+\beta)^2(\alpha+\beta+1)},$$

where $B(\alpha, \beta)$ is the Beta function:

$$B(\alpha, \beta) = \frac{\Gamma(\alpha)\Gamma(\beta)}{\Gamma(\alpha + \beta)}. \tag{A.2}$$

For a proper density both $\alpha$ and $\beta$ need to be positive. If $0 < \alpha < 1$, the density is unbounded at 0; if $0 < \beta < 1$, the density is unbounded at 1. If $\alpha = 1$ and $\beta = 1$, the density is equal to the density of a uniform distribution. If $\alpha = 1$ and $\beta > 1$, the mode of the density lies at 0; if $\alpha > 1$ and $\beta = 1$, the mode of the density lies at 1. For $\alpha, \beta > 1$ the mode of the density lies in the interior of $[0, 1]$ at $(\alpha - 1)/(\alpha + \beta - 2)$.

### A.1.2 The Binomial Distribution

The binomial distribution, $Y \sim \text{BiNom}(n, p)$, with $n = 1, 2, \ldots$ and $p \in [0, 1]$, is frequently chosen to model the outcome of repeated measurements. The density is defined for $y \in \{0, 1, 2, \ldots, n\}$,

$$f_{BN}(y; n, p) = \binom{n}{y} p^y (1 - p)^{n-y}, \tag{A.3}$$

and mean and variance are given by

$$\mathrm{E}(Y) = np, \qquad \mathrm{Var}(Y) = np(1 - p). \tag{A.4}$$

### A.1.3 The Dirichlet Distribution

The Dirichlet distribution, $\mathbf{Y} \sim \mathcal{D}(\alpha_1, \ldots, \alpha_K)$, is a standard choice in the context of modeling an unknown discrete probability distribution $\mathbf{Y} = (Y_1, \ldots, Y_K)$ where $\sum_{j=1}^K Y_j = 1$ and therefore is of great importance for mixture and switching models. The Dirichlet distribution is a distribution on the unit simplex $\mathcal{E}_K \subset (\Re^+)^K$, defined by the following constraint,

$$\mathcal{E}_K = \left\{ \mathbf{y} = (y_1, \ldots, y_K) \in (\Re^+)^K : \sum_{j=1}^K y_j = 1 \right\}.$$

The density is given by

$$f_D(\mathbf{y}; \alpha_1, \ldots, \alpha_K) = y_1^{\alpha_1 - 1} \cdots y_K^{\alpha_K - 1} c, \qquad c = \frac{\Gamma(\Sigma_\alpha)}{\prod_{i=1}^K \Gamma(\alpha_i)}, \tag{A.5}$$

where $\Sigma_\alpha = \sum_{j=1}^K \alpha_j$. As each $y_j$ is completely determined given the remaining elements of $\mathbf{y}$, one element $y_j$ has to be substituted in (A.5) by $1 - \sum_{k \neq j} y_k$ to obtain a proper density. The density is proper if all $\alpha_j$ are positive; the density is bounded if all $\alpha_j$ are greater than or equal to 1. The density of a $\mathcal{D}(\alpha_1, \ldots, \alpha_K)$-distribution is improper whenever any $\alpha_j$ is equal to 0.

For $\alpha_1 = \cdots = \alpha_K = 1$ the uniform distribution over the unit simplex results. For $K = 2$ the Dirichlet distribution is equal to $Y_1 \sim \mathcal{B}(\alpha_1, \alpha_2)$.

The marginal distribution of $Y_j$ is a $\mathcal{B}(\alpha_j, \Sigma_\alpha - \alpha_j)$-distribution, therefore mean and variance of $Y_j$ are given by

$$\mathrm{E}(Y_j) = \frac{\alpha_j}{\Sigma_\alpha}, \qquad \mathrm{Var}(Y_j) = \frac{\alpha_j(\Sigma_\alpha - \alpha_j)}{\Sigma_\alpha^2(\Sigma_\alpha + 1)}. \tag{A.6}$$

The mode of the marginal density of $Y_j$ is given by $(\alpha_j - 1)/(\Sigma_\alpha - K)$.

The easiest way to sample a random variable $\mathbf{Y} = (Y_1, \dots, Y_K)$ from the $\mathcal{D}(\alpha_1, \dots, \alpha_K)$-distribution is to sample $K$ independent random variables $Y_1^\star$, $\dots$, $Y_K^\star$ from the following Gamma distributions, $Y_j^\star \sim \mathcal{G}(\alpha_j, 1)$, $j = 1, \dots, K$, and to normalize: $Y_j = Y_j^\star / \sum_{k=1}^K Y_k^\star$.

### A.1.4 The Exponential Distribution

The exponential distribution, $Y \sim \mathcal{E}(\beta)$ with $\beta > 0$, is a univariate distribution defined on the positive real line $y \geq 0$ and is mainly used as a sampling distribution in the context of finite mixture models. There exist different ways to parameterize this distribution and we follow Bernardo and Smith (1994), by defining the density as

$$f_E(y; \beta) = \beta e^{-\beta y}. \tag{A.7}$$

Mean and variance are given by

$$\mathrm{E}(Y) = \mathrm{Var}(Y) = 1/\beta. \tag{A.8}$$

In this parameterization, the exponential distribution is equal to the $\mathcal{G}(1, \beta)$-distribution.

### A.1.5 The F-Distribution

The F-distribution, $Y \sim \mathrm{F}(\alpha_1, \alpha_2)$, is a univariate distribution defined on the positive real line $y \geq 0$. For finite mixture models based on the normal distribution, it appears as part of modeling the prior of heterogeneous variances. For $\alpha_1 > 0, \alpha_2 > 0$ the density is given by

$$f_F(y; \alpha_1, \alpha_2) = \frac{\Gamma((\alpha_1 + \alpha_2)/2)\alpha_1}{\Gamma(\alpha_1/2)\Gamma(\alpha_2/2)\alpha_2} \left(\frac{\alpha_1}{\alpha_2} y\right)^{\alpha_1/2-1} \left(1 + \frac{\alpha_1}{\alpha_2} y\right)^{-(\alpha_1+\alpha_2)/2} \tag{A.9}$$

For $y \to \infty$, the density behaves as $y^{-\alpha_2/2-1}$, whereas for $y \to 0$, the density behaves as $y^{\alpha_1/2-1}$. The density is an improper density, whenever $\alpha_1$ or $\alpha_2$ are 0. If $\alpha_1 > 0$ and $\alpha_2 > 0$, the density is proper, but unbounded at $y = 0$ for $0 < \alpha_1 < 2$.

### Relation to the $\chi^2$-Distribution

Assume that $Y_1$ and $Y_2$ are independent random variables, each following a $\chi^2$-distribution: $Y_1 \sim \chi^2_{\alpha_1}$ and $Y_2 \sim \chi^2_{\alpha_2}$. Then

$$Y = \frac{Y_1/\alpha_1}{Y_2/\alpha_2} \sim \mathrm{F}\left(\alpha_1, \alpha_2\right). \tag{A.10}$$

## A.1.6 The Gamma Distribution

The Gamma distribution, $Y \sim \mathcal{G}\left(\alpha, \beta\right)$, is a univariate distribution defined on the positive real line $y \geq 0$. It is encountered in finite mixture models as a posterior density within a Bayesian analysis for certain nonnormal models, in particular for observations from the Poisson and the exponential distribution. There exist various ways to parameterize this distribution and we follow Bernardo and Smith (1994), where the density is given by

$$f_G(y; \alpha, \beta) = \frac{\beta^\alpha}{\Gamma(\alpha)} y^{\alpha-1} e^{-\beta y}. \tag{A.11}$$

The density is an improper density whenever $\beta = 0$. If $\beta > 0$, the density is proper, but unbounded at $y = 0$ for $0 < \alpha < 1$. For $\alpha > 0, \beta > 0$, mean and variance are given by

$$\mathrm{E}(Y) = \frac{\alpha}{\beta}, \qquad \mathrm{Var}(Y) = \frac{\alpha}{\beta^2}. \tag{A.12}$$

The mode is given by $(\alpha - 1)/\beta$. If $Y \sim \mathcal{G}\left(\alpha, \beta\right)$, then $\omega Y \sim \mathcal{G}\left(\alpha, \beta/\omega\right)$.

### The $\chi^2$-Distribution

The $\mathcal{G}\left(\nu/2, 1/2\right)$-distribution with $\nu = 1, 2, \ldots$ is called the $\chi^2_\nu$-distribution, which is the distribution of the sum of squares of $\nu$ independent standard normal random variables, $Y = \sum_{j=1}^{\nu} Y_j^2$, $Y_j \sim \mathcal{N}\left(0, 1\right)$. Therefore $\mathrm{E}(Y) = \nu$, $\mathrm{Var}(Y) = 2\nu$.

### Inverted Gamma Distribution

A random variable $Y$ follows an inverted Gamma distribution, $Y \sim \mathcal{G}^{-1}\left(\alpha, \beta\right)$, if $Y^{-1}$ has a Gamma distribution: $Y^{-1} \sim \mathcal{G}\left(\alpha, \beta\right)$. The inverted Gamma distribution often appears as a posterior distribution of an unknown variance within a Bayesian analysis of finite mixture models based on the normal distribution. The density is given by

$$f_{IG}(y; \alpha, \beta) = \frac{\beta^\alpha}{\Gamma(\alpha)} \left(\frac{1}{y}\right)^{\alpha+1} e^{-\beta/y}. \tag{A.13}$$

$Y$ has finite expectation, iff $\alpha > 1$, and finite variance, iff $\alpha > 2$:

$$\mathrm{E}(Y) = \frac{\beta}{\alpha - 1}, \qquad \mathrm{Var}(Y) = \frac{\beta^2}{(\alpha-1)^2(\alpha-2)}. \tag{A.14}$$

The mode is given by $\beta/(\alpha+1)$. $\mathcal{G}^{-1}(0,0)$ has an improper density proportional to $1/y$. If $Y \sim \mathcal{G}^{-1}(\alpha,\beta)$, then $\omega Y \sim \mathcal{G}^{-1}(\alpha, \omega\beta)$.

### A.1.7 The Geometric Distribution

The geometric distribution is defined for $y \in \{1, 2, \ldots\}$. Density, mean and variance are given by

$$f(y; p) = p(1-p)^{y-1}, \tag{A.15}$$

$$\mathrm{E}(Y) = \frac{1}{p}, \qquad \mathrm{Var}(Y) = \frac{1-p}{p^2}.$$

### A.1.8 The Multinomial Distribution

Let $\mathbf{Y}$ be a categorical random variable with $D$ categories coded as 1 to $D$. $\mathbf{Y} \sim \mathrm{MulNom}(n, p_1, \ldots, p_D)$, iff for all $\mathbf{y} = (y_1, \ldots, y_D)$, where $\sum_{j=1}^D y_j = n$ and $y_j \geq 0$,

$$\Pr(\mathbf{Y} = \mathbf{y}) = f_{MN}(\mathbf{y}; n, p_1, \ldots, p_D) = \binom{n}{y_1 \ldots y_D} p_1^{y_1} \ldots p_D^{y_D}, \tag{A.16}$$

with the general binomial coefficient being defined as:

$$\binom{n}{y_1 \ldots y_D} = \frac{n!}{y_1! \ldots y_D!}$$

The mean and variance of $Y_j$ are given by

$$\mathrm{E}(Y_j) = np_j, \qquad \mathrm{Var}(Y_j) = np_j(1-p_j). \tag{A.17}$$

The binomial distribution is that special case of the multinomial distribution where $D = 2$.

### A.1.9 The Negative Binomial Distribution

The negative binomial distribution is defined for $y \in \{0, 1, 2, \ldots\}$: $Y \sim \mathrm{NegBin}(\alpha, \beta)$. For $\alpha > 0, \beta > 0$ density, mean, and variance of $Y$ are given by

$$f_{NB}(y; \alpha, \beta) = \binom{\alpha + y - 1}{\alpha - 1} \left(\frac{\beta}{\beta+1}\right)^{\alpha} \left(\frac{1}{\beta+1}\right)^{y},$$

$$\mathrm{E}(Y) = \frac{\alpha}{\beta}, \qquad \mathrm{Var}(Y) = \frac{\alpha}{\beta^2}(\beta+1). \tag{A.18}$$

The $\mathrm{NegBin}(\alpha, \beta)$-distribution is an infinite mixture of $\mathcal{P}(\mu)$-distributions, where $\mu \sim \mathcal{G}(\alpha, \beta)$:

$$f_{NB}(y; \alpha, \beta) = \int f_P(y; \mu) f_G(\mu; \alpha, \beta) d\mu.$$

### A.1.10 The Normal Distribution

The normal distribution is the most important density in finite mixture modeling, both as a common choice as sampling density as well as posterior density in a Bayesian analysis of such models.

#### The Univariate Normal Distribution

The univariate normal distribution, $Y \sim \mathcal{N}(\mu, \sigma^2)$, with $\mu \in \Re$ and $\sigma > 0$, is defined on $\Re$. Density, mean, and variance of $Y$ are given by

$$f_N(y;\mu,\sigma^2) = \frac{1}{\sqrt{2\pi\sigma^2}} e^{-(y-\mu)^2/(2\sigma^2)},$$
$$\mathrm{E}(Y) = \mu, \qquad \mathrm{Var}(Y) = \sigma^2. \tag{A.19}$$

$1/\sigma^2$ is often called precision. A normal distribution with zero mean and zero precision has an improper density proportional to a constant. Higher-order moments $\mathrm{E}((Y-\mu)^m)$ are zero for $m$ odd, otherwise:

$$\mathrm{E}((Y-\mu)^m) = \sigma^{2m} \prod_{n=1}^{m/2} (2n-1). \tag{A.20}$$

#### The Multivariate Normal Distribution

The multivariate normal distribution, $\mathbf{Y} \sim \mathcal{N}_d(\boldsymbol{\mu}, \boldsymbol{\Sigma})$, where $\boldsymbol{\mu} \in \Re^d$ and $\boldsymbol{\Sigma} \in \Re^{d\times d}$ is a symmetric, positive definite matrix, is defined for $\mathbf{y} \in \Re^d$. Density, mean, and covariance matrix of $\mathbf{Y}$ are given by

$$f_N(\mathbf{y};\boldsymbol{\mu},\boldsymbol{\Sigma}) = (2\pi)^{-d/2}|\boldsymbol{\Sigma}|^{-1/2}\exp\left(-\tfrac{1}{2}(\mathbf{y}-\boldsymbol{\mu})'\boldsymbol{\Sigma}^{-1}(\mathbf{y}-\boldsymbol{\mu})\right),$$
$$\mathrm{E}(Y) = \boldsymbol{\mu}, \quad \mathrm{Var}(\mathbf{Y}) = \boldsymbol{\Sigma}. \tag{A.21}$$

$\boldsymbol{\Sigma}^{-1}$ is the information matrix. A normal distribution with $\boldsymbol{\mu} = \mathbf{0}$ and $\boldsymbol{\Sigma}^{-1} = \mathbf{O}$ has an improper density proportional to a constant.

To sample from a multivariate normal distribution the following result is useful. Let $\mathbf{Z} = (Z_1 \dots Z_d)'$ be $d$ independent copies of an $\mathcal{N}(0,1)$-distribution and let $\boldsymbol{\Sigma} = \mathbf{A}\mathbf{A}'$ be the Cholesky decomposition of $\boldsymbol{\Sigma}$. Then $\mathbf{Y} = \boldsymbol{\mu} + \mathbf{A}\mathbf{Z} \sim \mathcal{N}_d(\boldsymbol{\mu}, \boldsymbol{\Sigma})$.

#### Marginal and Conditional Distributions

The quadratic form $(\mathbf{Y}-\boldsymbol{\mu})'\boldsymbol{\Sigma}^{-1}(\mathbf{Y}-\boldsymbol{\mu}) \sim \chi^2_d$. Assume that $\mathbf{Y}$ is divided into two blocks, $\mathbf{Y}_1$ containing the first $d_1$ components of $\mathbf{Y}$, and $\mathbf{Y}_2$ containing the remaining $d_2 = d - d_1$ components. Apply a similar partition on $\boldsymbol{\mu}$ and $\boldsymbol{\Sigma}$:

$$\boldsymbol{\mu} = \begin{pmatrix} \boldsymbol{\mu}_1 \\ \boldsymbol{\mu}_2 \end{pmatrix}, \qquad \boldsymbol{\Sigma} = \begin{pmatrix} \boldsymbol{\Sigma}_{11} & \boldsymbol{\Sigma}_{12} \\ \boldsymbol{\Sigma}_{21} & \boldsymbol{\Sigma}_{22} \end{pmatrix}.$$

Then, marginally, $\mathbf{Y}_2 \sim \mathcal{N}_{d_2}(\boldsymbol{\mu}_2, \boldsymbol{\Sigma}_{22})$ and the conditional distribution of $\mathbf{Y}_1|\mathbf{Y}_2$ is $\mathcal{N}_{d_1}\left(\boldsymbol{\mu}_{1|2}, \boldsymbol{\Sigma}_{11|2}\right)$, where

$$\begin{aligned} \boldsymbol{\mu}_{1|2} &= \boldsymbol{\mu}_1 + \boldsymbol{\Sigma}_{12}\boldsymbol{\Sigma}_{22}^{-1}(\mathbf{Y}_2 - \boldsymbol{\mu}_2), \\ \boldsymbol{\Sigma}_{11|2} &= \boldsymbol{\Sigma}_{11} - \boldsymbol{\Sigma}_{12}\boldsymbol{\Sigma}_{22}^{-1}\boldsymbol{\Sigma}_{12}'. \end{aligned}$$

### The Normal Gamma Distribution

A pair $(\mathbf{Y}_1, Y_2)$ of random variables $\mathbf{Y}_1$ and $Y_2$ with $\mathbf{Y}_1|Y_2 \sim \mathcal{N}_d(\boldsymbol{\mu}, Y_2\boldsymbol{\Sigma})$ and $1/Y_2 \sim \mathcal{G}(\nu/2, \nu/2)$ follows the so-called normal Gamma distribution. The marginal distribution of $\mathbf{Y}_1$ is the $d$-variate $t_\nu(\boldsymbol{\mu}, \boldsymbol{\Sigma})$-distribution, whereas the marginal distribution of $Y_2$ is $\mathcal{G}^{-1}(\nu/2, \nu/2)$.

## A.1.11 The Poisson Distribution

The Poisson distribution, $Y \sim \mathcal{P}(\mu)$ with $\mu > 0$, is defined for $y \in \{0, 1, 2, \ldots\}$ making it a standard distribution to model a random count variable. Density, mean, and variance are given by

$$\begin{aligned} f_P(y; \mu) &= \frac{\mu^y}{y!} e^{-\mu}, \qquad \text{(A.22)} \\ \mathrm{E}(Y) &= \mathrm{Var}(Y) = \mu. \end{aligned}$$

## A.1.12 The Student-$t$ Distribution

The Student-$t$ distribution is useful as a sampling distribution for robust modeling.

### The Univariate Student-$t$ Distribution

The univariate Student-$t$ distribution, $Y \sim t_\nu(\mu, \sigma^2)$ with $\mu \in \Re$, $\sigma > 0$, and $\nu > 0$, is defined for $y \in \Re$. The density is given by

$$f_{t_\nu}(y; \mu, \sigma^2) = \frac{\Gamma((\nu+1)/2)}{\Gamma(\nu/2)\sqrt{\nu\pi\sigma^2}} \left(1 + \frac{(y-\mu)^2}{\nu\sigma^2}\right)^{-(\nu+1)/2}.$$

If $\nu > 1$, then $\mathrm{E}(Y) = \mu$; if $\nu > 2$, then

$$\mathrm{Var}(Y) = \sigma^2 \frac{\nu}{\nu - 2}.$$

The univariate $t_\nu(\mu, \sigma^2)$-distribution is an infinite mixture of $\mathcal{N}(\mu, \sigma^2/\omega)$-distributions, where $\omega \sim \mathcal{G}(\nu/2, \nu/2)$:

$$f_{t_\nu}(y; \mu, \sigma^2) = \int f_N(y; \mu, \sigma^2/\omega) f_G(\omega; \nu/2, \nu/2) d\omega.$$

### The Multivariate Student-$t$ Distribution

The multivariate Student-$t$ distribution, $\mathbf{Y} \sim t_{\nu}(\boldsymbol{\mu}, \boldsymbol{\Sigma})$, where $\boldsymbol{\mu} \in \Re^d$ and $\boldsymbol{\Sigma} \in \Re^{d \times d}$ is a symmetric, positive definite matrix, is defined for $\mathbf{y} \in \Re^d$. The density is given by

$$f_{t_\nu}(\mathbf{y}; \boldsymbol{\mu}, \boldsymbol{\Sigma}) = \tag{A.23}$$
$$\frac{\Gamma((\nu+d)/2)}{\Gamma(\nu/2)}(\nu\pi)^{-d/2}|\boldsymbol{\Sigma}|^{-1/2}\left(1+\frac{1}{\nu}(\mathbf{y}-\boldsymbol{\mu})'\boldsymbol{\Sigma}^{-1}(\mathbf{y}-\boldsymbol{\mu})\right)^{-(\nu+d)/2}.$$

If $\nu > 1$, then $\mathrm{E}(\mathbf{Y}) = \boldsymbol{\mu}$; if $\nu > 2$, then

$$\mathrm{Var}(\mathbf{Y}) = \frac{\nu}{\nu-2}\boldsymbol{\Sigma}.$$

The multivariate $t_{\nu}(\boldsymbol{\mu}, \boldsymbol{\Sigma})$-distribution is an infinite mixture of $\mathcal{N}_d(\boldsymbol{\mu}, \boldsymbol{\Sigma}/\omega)$-distributions, where $\omega \sim \mathcal{G}(\nu/2, \nu/2)$:

$$f_{t_\nu}(\mathbf{y}; \boldsymbol{\mu}, \boldsymbol{\Sigma}) = \int f_N(\mathbf{y}; \boldsymbol{\mu}, \boldsymbol{\Sigma}/\omega) f_G(\omega; \nu/2, \nu/2) d\omega.$$

### A.1.13 The Uniform Distribution

The uniform distribution, $Y \sim \mathcal{U}[0,1]$, is defined over the unit interval $y \in [0,1]$. Density, mean, and variance are given by

$$f_U(y) = 1, \qquad \mathrm{E}(Y) = 1/2, \qquad \mathrm{Var}(Y) = 1/12. \tag{A.24}$$

### A.1.14 The Wishart Distribution

The Wishart distribution, $\mathbf{Y} \sim \mathcal{W}_d(\alpha, \mathbf{S})$, with $\mathbf{S}$ being a $(d \times d)$ symmetric, nonsingular matrix ($|\mathbf{S}| > 0$), is a standard distribution law used for a random $(d \times d)$ symmetric, positive-definite matrix $\mathbf{Y}$. Many parameterizations are in use for this density; the following one, introduced by Bernardo and Smith (1994), is used in this book, because in this parameterization the $\mathcal{W}_1(\alpha, S)$ distribution reduces the $\mathcal{G}(\alpha, S)$ distribution.

For $\alpha > (d-1)/2$ the density of the $d(d+1)/2$-dimensional random vector of distinct elements of $\mathbf{Y}$ is given by

$$f_W(\mathbf{Y}; \alpha, \mathbf{S}) = \frac{|\mathbf{S}|^{\alpha}}{\Gamma_d(\alpha)}|\mathbf{Y}|^{\alpha-(d+1)/2}\exp\{-\mathrm{tr}(\mathbf{SY})\}, \tag{A.25}$$

where

$$\Gamma_d(\alpha) = \pi^{d(d-1)/4}\prod_{j=1}^{d}\Gamma\left(\frac{2\alpha+1-j}{2}\right) \tag{A.26}$$

is the generalized Gamma function. The mean of $\mathbf{Y}$ is given by

$$\mathrm{E}(\mathbf{Y}) = \alpha \mathbf{S}^{-1};$$

the mode reads:

$$(\alpha - \frac{d+1}{2})\mathbf{S}^{-1}.$$

The variance of the elements $Y_{ij}$ of $\mathbf{Y}$ is equal to

$$\mathrm{Var}(Y_{ij}) = \alpha \left( (S_{ij}^{-1})^2 + S_{ii}^{-1} S_{jj}^{-1} \right),$$

where $S_{ij}$ is the $(i,j)$th element of $\mathbf{S}$. If $\mathbf{Y} \sim \mathcal{W}_d(\alpha, \mathbf{S})$ and $\mathbf{A}$ is an $(m \times d)$ matrix with $m \leq d$, then

$$\mathbf{AYA}' \sim \mathcal{W}_m \left( \alpha, (\mathbf{AS}^{-1}\mathbf{A}')^{-1} \right),$$

provided that the scale matrix exists.

### Inverted Wishart Distribution

A random $(d \times d)$ symmetric, positive-definite matrix $\mathbf{Y}$ follows an inverted Wishart distribution, $\mathbf{Y} \sim \mathcal{W}_d^{-1}(\alpha, \mathbf{S})$, if $\mathbf{Y}^{-1} \sim \mathcal{W}_d(\alpha, \mathbf{S})$. In the parameterization used for the Wishart density in (A.25), the $\mathcal{W}_1^{-1}(\alpha, S)$ distribution reduces to the $\mathcal{G}^{-1}(\alpha, S)$ distribution. The density of the inverted Wishart distribution is given by

$$f_{IW}(\mathbf{Y}; \alpha, \mathbf{S}) = \frac{|\mathbf{S}|^{\alpha}}{\Gamma_d(\alpha)} |\mathbf{Y}^{-1}|^{\alpha+(d+1)/2} \exp\{-\mathrm{tr}\left(\mathbf{SY}^{-1}\right)\}, \qquad \text{(A.27)}$$

with $\Gamma_d(\alpha)$ being the same as in (A.26). $\mathbf{Y}$ has finite expectation, iff $\alpha > (d+1)/2$:

$$\mathrm{E}(\mathbf{Y}) = \mathbf{S}/(\alpha - (d+1)/2);$$

the mode is given by

$$\mathbf{S}/(\alpha + (d+1)/2).$$

## A.2 Software

A toolbox of MATLAB version 6 scripts and functions has been written by the author with the purpose of carrying out practical statistical modeling based on finite mixture and Markov switching models. The package is available at `http://www.ifas.jku.at/personal/fruehwirth/fruehwirthspringer06`.

You will need a valid MATLAB license including the STATISTICS toolbox to run functions from this package. The MATLAB software is available from MathWorks, Inc. (`http://www.mathworks.com/`).

# References


Abraham, B. and G. E. P. Box (1978). Linear models and spurious observations. *Applied Statistics 27*, 131–138.

Abraham, B. and J. Ledolter (1986). Forecast functions implied by autoregressive integrated moving average models and other related forecast procedures. *International Statistical Review 54*, 51–66.

Ackerson, G. A. and K. S. Fu (1970). On state estimation in switching environments. *IEEE Transactions on Automatic Control 15*, 10–17.

Aitkin, M. (1996). A general maximum likelihood analysis of overdispersion in generalized linear models. *Statistics and Computing 6*, 251–262.

Aitkin, M. and I. Aitkin (1996). A hybrid EM/Gauss-Newton algorithm for maximum likelihood in mixture distributions. *Statistics and Computing 6*, 127–130.

Aitkin, M., D. Anderson, and J. Hinde (1981). Statistical modelling of data on teaching styles. *Journal of the Royal Statistical Society, Ser. A 144*, 419–461.

Aitkin, M. and D. B. Rubin (1985). Estimation and hypothesis testing in finite mixture models. *Journal of the Royal Statistical Society, Ser. B 47*, 67–75.

Aitkin, M. and G. T. Wilson (1980). Mixture models, outliers, and the EM algorithm. *Technometrics 22*, 325–331.

Akaike, H. (1974). A new look at statistical model identification. *IEEE Transactions on Automatic Control 19*, 716–723.

Akashi, H. and H. Kumamoto (1977). Random sampling approach to state estimation in switching environments. *Automatica 13*, 429–434.

Albert, J. H. and S. Chib (1993). Bayes inference via Gibbs sampling of autoregressive time series subject to Markov mean and variance shifts. *Journal of Business & Economic Statistics 11*, 1–15.

Albert, P. S. (1991). A two-stage Markov mixture model for a time series of epileptic seizure counts. *Biometrics 47*, 1371–1381.

Albert, P. S. and L. E. Dodd (2004). A cautionary note on the robustness of latent class models for estimating diagnostic error without a gold standard. *Biometrics 60*, 427–435.

Allenby, G. M., N. Arora, and J. L. Ginter (1998). On the heterogeneity of demand. *Journal of Marketing Research 35*, 384–389.

Alspach, D. L. and H. W. Sorenson (1972). Nonlinear Bayesian estimation using Gaussian sum approximations. *IEEE Transactions on Automatic Control 17*, 439–448.

Anderson, B. D. O. and J. B. Moore (1979). *Optimal Filtering*. Englewood Cliffs, NJ: Prentice-Hall.

Anderson, T. W. (1984). *An Introduction to Multivariate Statistical Analysis* (2 ed.). Chichester: Wiley.

Andrews, D. W. K., I. Lee, and W. Ploberger (1996). Optimal changepoint tests for normal linear regression. *Journal of Econometrics 70*, 9–38.

Ang, A. and G. Bekaert (2002). Regime switches in interest rates. *Journal of Business & Economic Statistics 20*, 163–182.

Antelman, G. (1997). *Elementary Bayesian Statistics*. Cheltenham: Edward Elgar.

Aoki, M. (1990). *State Space Modeling of Time Series*. New York: Springer.

Asea, P. K. and B. Blomberg (1998). Lending cycles. *Journal of Econometrics 83*, 89–128.

Atkinson, A. C. (1978). Posterior probabilities for choosing a regression model. *Biometrika 65*, 39–48.

Bacher, J. (2000). A probabilistic clustering model for variables of mixed type. *Quality & Quantity 34*, 223–235.

Baltagi, B. H. (1995). *Econometric Analysis of Panel Data*. Chichester: Wiley.

Banfield, J. D. and A. E. Raftery (1993). Model-based Gaussian and non-Gaussian clustering. *Biometrics 49*, 803–821.

Banjeree, T. and S. R. Paul (1999). An extension of Morel-Nagaraj's finite mixture distribution for modelling multinomial clustered data. *Biometrika 86*, 723–727.

Bar-Shalom, Y. and E. Tse (1975). Tracking in a cluttered environment with probabilistic data association. *Automatica 11*, 451–460.

Barndorff-Nielsen, O. E. (1978). *Information and Exponential Families in Statistical Theory*. Chichester: Wiley.

Bartholomew, D. J. (1980). Factor analysis for categorical data. *Journal of the Royal Statistical Society, Ser. B 42*, 293–321.

Baum, L. E. and T. Petrie (1966). Statistical inference for probabilistic functions of finite state Markov chains. *The Annals of Mathematical Statistics 37*, 1554–1563.

Baum, L. E., T. Petrie, G. Soules, and N. Weiss (1970). A maximization technique occurring in the statistical analysis of probabilistic functions of Markov chains. *The Annals of Mathematical Statistics 41*, 164–171.

Bauwens, L., C. S. Bos, H. K. van Dijk, and R. D. van Oest (2006). Adaptive radial-based direction sampling. *Journal of Econometrics*. Forthcoming.

Bauwens, L. and M. Lubrano (1998). Bayesian inference on GARCH models using the Gibbs sampler. *The Econometrics Journal 1*, C23–C46.

Bayes, T. (1763). An essay towards solving a problem in the doctrine of chances. Reprinted in *Biometrika 45*, 1958, 296–315.

Behboodian, J. (1970). On the modes of a mixture of two normal distributions. *Technometrics 12*, 131–139.

Bekaert, G. and C. R. Harvey (1995). Time-varying world market integration. *The Journal of Finance 50*, 403–444.

Bensmail, H. and G. Celeux (1996). Regularized Gaussian discriminant analysis through eigenvalue decomposition. *Journal of the American Statistical Association 91*, 1743–1748.

Bensmail, H., G. Celeux, A. E. Raftery, and C. P. Robert (1997). Inference in model-based cluster analysis. *Statistics and Computing 7*, 1–10.

Beran, R. (1977). Minimum Hellinger distance estimates for parametric models. *The Annals of Statistics 5*, 445–463.

Berger, J. O. (1985). *Statistical Decision Theory and Bayesian Analysis* (2 ed.). New York/Berlin/Heidelberg: Springer.

Berger, J. O. and M. Delampady (1987). Testing precise hypotheses. *Statistical Science 2*, 317–352.

Berger, J. O. and T. Sellke (1987). Testing a point null hypothesis: The irreconcilability of $p$ values and evidence. *Journal of the American Statistical Association 82*, 112–139.

Berkhof, J., I. van Mechelen, and A. Gelman (2003). A Bayesian approach to the selection and testing of mixture models. *Statistica Sinica 13*, 423–442.

Bernardinelli, L., D. Clayton, and C. Montomoli (1995). Bayesian estimates of disease maps: How important are priors? *Statistics in Medicine 14*, 2411–2431.

Bernardo, J. M. (1979). Reference posterior distributions for Bayesian inference. *Journal of the Royal Statistical Society, Ser. B 41*, 113–147.

Bernardo, J. M. and F. G. Girón (1988). A Bayesian analysis of simple mixture problems. In J. M. Bernardo, M. H. DeGroot, D. V. Lindley, and A. F. M. Smith (Eds.), *Bayesian Statistics 3*, pp. 67–78. New York: Clarendon.

Bernardo, J. M. and A. F. M. Smith (1994). *Bayesian Theory*. Chichester: Wiley.

Besag, J. (1974). Spatial interaction and the statistical analysis of lattice systems. *Journal of the Royal Statistical Society, Ser. B 36*, 192–236.

Besag, J. (1989). A candidate's formula: A curious result in Bayesian prediction. *Biometrika 76*, 183–183.

Bhar, R. and S. Hamori (2004). *Hidden Markov Models. Applications to Financial Economics*. Dordrecht: Kluwer Academic.

Bickel, P. J., Y. Ritov, and T. Rydén (1998). Asymptotic normality of the maximum likelhood estimator for general hidden Markov models. *The Annals of Statistics 26*, 1614–1635.

Biernacki, C., G. Celeux, and G. Govaert (2000). Assessing a mixture model for clustering with the integrated classification likelihood. *IEEE Transactions on Pattern Analysis and Machine Intelligence 22*, 719–725.

Biernacki, C., G. Celeux, and G. Govaert (2003). Choosing starting values for the EM algorithm for getting the highest likelihood in multivariate Gaussian mixture models. *Computational Statisics and Data Analysis 41*, 561–575.

Biernacki, C. and G. Govaert (1997). Using the classification likelihood to choose the number of clusters. *Computing Science and Statistics 29*, 451–457.

Billio, M. and A. Monfort (1998). Switching state-space models: Likelihood function, filtering and smoothing. *Journal of Statistical Planning and Inference 68*, 65–103.

Billio, M., A. Monfort, and C. P. Robert (1999). Bayesian estimation of switching ARMA models. *Journal of Econometrics 93*, 229–255.

Binder, D. A. (1978). Bayesian cluster analysis. *Biometrika 65*, 31–38.

Binder, D. A. (1981). Approximations to Bayesian clustering rules. *Biometrika 68*, 275–285.

Blackwell, D. and L. Koopmans (1957). On the identifiability problem for functions of finite Markov chains. *Annals of Mathematical Statistics 28*, 1011–1015.

Blom, H. and Y. Bar-Shalom (1988). The interacting multiple model algorithm for systems with Markovian switching coefficients. *IEEE Transactions on Automatic Control 33*, 780–783.

Bock, H. H. (1996). Probabilistic models in cluster analysis. *Computational Statisics and Data Analysis 23*, 5–28.

Böhning, D. (1998). Zero-inflated Poisson models and C.A.MAN: A tutorial collection of evidence. *Biometrical Journal 40*, 833–843.

Böhning, D. (2000). *Computer-Assisted Analysis of Mixtures and Applications.* London: Chapman & Hall.

Böhning, D., E. Dietz, and P. Schlattmann (1998). Recent developments in computer-assisted analysis of mixtures. *Biometrics 54*, 525–536.

Bollerslev, T. (1986). Generalized autoregressive conditional heteroskedasticity. *Journal of Econometrics 31*, 307–327.

Bollerslev, T., R. Y. Chou, and K. F. Kroner (1992). ARCH modelling in finance: A review of the theory and empirical evidence. *Journal of Econometrics 52*, 5–39.

Bottolo, L., G. Consonni, P. Dellaportas, and A. Lijoi (2003). Bayesian analysis of extreme values by mixture modeling. *Extremes 6*, 25–47.

Bougerol, P. and N. Picard (1992a). Stationarity of GARCH processes and of some nonnegative time series. *Journal of Econometrics 52*, 115–127.

Bougerol, P. and N. Picard (1992b). Strict stationarity of generalized autoregressive processes. *The Annals of Probability 20*, 1714–1730.

Box, G. E. P. and G. M. Jenkins (1970). *Time Series Analysis — Forecasting and Control.* San Francisco: Holden-Day.

Box, G. E. P. and G. C. Tiao (1968). A Bayesian approach to some outlier problems. *Biometrika 55*, 119–129.

Box, G. E. P. and G. C. Tiao (1973). *Bayesian Inference in Statistical Analysis.* Reading, MA: Addison-Wesley.

Bozdogan, H. (1987). Model selection and Akaike's information criterion (AIC): The generel theory and its analytical extensions. *Psychometrika 52*, 345–370.

Bozdogan, H. and S. L. Sclove (1984). Multi-sample cluster analysis using Akaike's information criterion. *Annals of the Institute of Statistical Mathematics 36*, 163–180.

Breslow, N. E. and D. G. Clayton (1993). Approximate inference in generalized linear mixed models. *Journal of the American Statistical Association 88*, 9–25.

Brockwell, P. J. and R. A. Davis (1991). *Time Series: Theory and Methods* (2 ed.). New York: Springer.

Brooks, S. P. (2001). On Bayesian analyses and finite mixtures for proportions. *Statistics and Computing 11*, 179–190.

Brooks, S. P., P. Giudici, and G. O. Roberts (2003). Efficient construction of reversible jump Markov chain Monte Carlo proposal distributions. *Journal of the Royal Statistical Society, Ser. B 65*, 3–55.

Brooks, S. P., B. J. T. Morgan, M. S. Ridout, and S. E. Pack (1997). Finite mixture models for proportions. *Biometrics 53*, 1097–1115.

Brusco, M. J. and J. D. Cradit (2001). A variable selection heuristic for $k$-means clustering. *Psychometrika 66*, 249–270.

Bryant, P. G. and J. A. Williamson (1978). Asymptotic behaviour of classification maximum likelihood estimates. *Biometrika 65*, 273–281.

Burmeister, E., K. D. Wall, and J. D. Hamilton (1986). Estimation of unobserved expected monthly inflation using Kalman filtering. *Journal of Business & Economic Statistics 4*, 147–160.

Cai, J. (1994). A Markov model of switching-regime ARCH. *Journal of Business & Economic Statistics 12*, 309–316.

Calinski, T. and J. Harabasz (1974). A dendrite method for cluster analysis. *Communications in Statistics 3*, 1–27.

Cameron, A. and P. Trivedi (1998). *Regression Analysis of Count Data.* Cambridge: Cambridge University Press.

Canova, F. (2004). Testing for convergence clubs in income per-capita: A predictive density approach. *International Economic Review 45*, 49–77.

Cappé, O., E. Moulines, and T. Rydén (2005). *Inference in Hidden Markov Models.* Springer Series in Statistics. New York: Springer.

Cappé, O., C. P. Robert, and T. Rydén (2003). Reversible jump, birth-and-death and more general continuous time Markov chain Monte Carlo samplers. *Journal of the Royal Statistical Society, Ser. B 65*, 679–700.

Carlin, B. P. and S. Chib (1995). Bayesian model choice via Markov chain Monte Carlo methods. *Journal of the Royal Statistical Society, Ser. B 57*, 473–484.

Carlin, B. P., N. G. Polson, and D. S. Stoffer (1992). A Monte Carlo approach to nonnormal and nonlinear state-space modeling. *Journal of the American Statistical Association 87*, 493–500.

Carreira-Perpiñán, M. A. and S. Renals (2000). Practical identifiability of finite mixtures of multivariate Bernoulli distributions. *Neural Computation 12*, 141–152.

Carreira-Perpiñán, M. A. and C. K. I. Williams (2003). On the number of modes in a Gaussian mixture. In L. D. Griffin and M. Lillholm (Eds.), *Scale-Space Methods in Computer Vision*, Number 2695 in Lecture Notes in Computer Science, pp. 625–640. Berlin/Heidelberg: Springer.

Carter, C. K. and R. Kohn (1994). On Gibbs sampling for state space models. *Biometrika 81*, 541–553.

Carter, C. K. and R. Kohn (1996). Markov chain Monte Carlo in conditionally Gaussian state space models. *Biometrika 83*, 589–601.

Casella, G. and R. L. Berger (2002). *Statistical Inference* (2 ed.). Pacific Grove, CA: Wadsworth Group.

Casella, G., M. Lavine, and C. P. Robert (2001). Explaining the perfect sampler. *The American Statistician 55*, 299–305.

Casella, G., K. L. Mengersen, C. P. Robert, and D. T. Titterington (2002). Perfect slice samplers for mixtures of distributions. *Journal of the Royal Statistical Society, Ser. B 64*, 777–790.

Casella, G., C. P. Robert, and M. T. Wells (2000). Mixture models, latent variables and partitioned importance sampling. Technical Report DT-2000-03, INSEE, CREST, Paris.

Catchpole, E. and B. J. T. Morgan (1997). Detecting parameter redundancy. *Biometrika 84*, 187–196.

Cecchetti, S. G., P. Lam, and N. C. Mark (1990). Mean reversion in equilibrium asset prices. *The American Economic Review 80*, 398–418.

Celeux, G. (1998). Bayesian inference for mixture: The label switching problem. In P. J. Green and R. Rayne (Eds.), *COMPSTAT 98*, pp. 227–232. Heidelberg: Physica.

Celeux, G. and J. Diebolt (1985). The SEM algorithm: A probabilistic teacher algorithm derived from the EM algorithm for the mixture problem. *Computational Statistics Quarterly 2*, 73–82.

Celeux, G. and G. Govaert (1991). Clustering criteria for discrete data and latent class models. *Journal of Classification 8*, 157–176.

Celeux, G. and G. Govaert (1992). A classification EM algorithm and two stochastic versions. *Computational Statisics and Data Analysis 14*, 315–332.

Celeux, G. and G. Govaert (1993). Comparison of the mixture and the classification maximum likelihood in cluster analysis. *Journal of Statistical Computation and Simulation 47*, 127–146.

Celeux, G. and G. Govaert (1995). Gaussian parsimonious clustering models. *Pattern Recognition 28*, 781–793.

Celeux, G., M. Hurn, and C. P. Robert (2000). Computational and inferential difficulties with mixture posterior distributions. *Journal of the American Statistical Association 95*, 957–970.

Celeux, G. and G. Soromenho (1996). An entropy criterion for assessing the number of clusters in a mixture model. *Journal of Classification 13*, 195–212.

Chan, K. S. and J. Ledolter (1995). Monte Carlo EM estimation for time series models involving counts. *Journal of the American Statistical Association 90*, 242–252.

Chandra, S. (1977). On the mixtures of probability distributions. *Scandinavian Journal of Statistics 4*, 105–112.

Chang, W. C. (1983). On using principal components before separating a mixture of two normal distributions. *Applied Statistics 32*, 267–275.

Charlier, C. V. L. and S. D. Wicksell (1924). On the dissection of frequency functions. *Arkiv för Matematik, Astronomi och Fysik 18*, 1–64.

Chen, C. and G. C. Tiao (1990). Random level-shift time series models, ARIMA approximations, and level-shift detection. *Journal of Business & Economic Statistics 8*, 83–97.

Chen, H., J. Chen, and J. D. Kalbfleisch (2001). A modified likelihood ratio test for homogeneity in finite mixture models. *Journal of the Royal Statistical Society, Ser. B 63*, 19–29.

Chen, H., J. Chen, and J. D. Kalbfleisch (2004). Testing for a finite mixture model with two components. *Journal of the Royal Statistical Society, Ser. B 66*, 95–115.

Chen, J. and J. D. Kalbfleisch (1996). Penalized minimum-distance estimates in finite mixture models. *The Canadian Journal of Statistics 24*, 167–175.

Chen, M.-H., Q.-M. Shao, and J. G. Ibrahim (2000). *Monte Carlo Methods in Bayesian Computation.* Springer Series in Statistics. New York: Springer.

Chen, R. and J. S. Liu (1996). Predictive updating methods with application to Bayesian classification. *Journal of the Royal Statistical Society, Ser. B 58*, 397–415.

Chib, S. (1995). Marginal likelihood from the Gibbs output. *Journal of the American Statistical Association 90*, 1313–1321.

Chib, S. (1996). Calculating posterior distributions and modal estimates in Markov mixture models. *Journal of Econometrics 75*, 79–97.

Chib, S. (1998). Estimation and comparison of multiple change-point models. *Journal of Econometrics 86*, 221–241.

Chib, S. (2001). Markov chain Monte Carlo methods: Computation and inference. In J. J. Heckman and E. Leamer (Eds.), *Handbook of Econometrics*, Volume 5, pp. 3569–3649. Amsterdam: North Holland.

Chib, S. and B. P. Carlin (1999). On MCMC sampling in hierarchical longitudinal models. *Statistics and Computing 9*, 17–26.

Chib, S. and E. Greenberg (1995). Understanding the Metropolis-Hastings algorithm. *Journal of the American Statistical Association 49*, 327–335.

Chib, S. and I. Jeliazkov (2001). Marginal likelihood from the Metropolis-Hastings output. *Journal of the American Statistical Association 96*, 270–281.

Chib, S., F. Nardari, and N. Shephard (2002). Markov chain Monte Carlo methods for stochastic volatility models. *Journal of Econometrics 108*, 281–316.

Chopin, N. (2002). A sequential particle filter method for static models. *Biometrika 89*, 539–551.

Chow, G. C. (1960). Tests of equality between sets of coefficients in two linear linear regressions (Ref: 74V42 p601-608). *Econometrica 28*, 591–605.

Clements, M. P. and H.-M. Krolzig (1998). A comparison of the forecast performance of Markov-switching and threshold autoregressive models of US GNP. *The Econometrics Journal 1*, 47–75.

Clogg, C. C. (1995). Latent class models. In G. Arminger, C. C. Clogg, and M. E. Sobel (Eds.), *Handbook of Statistical Modelling for the Social and Behavioral Sciences*, pp. 311–359. New York: Plenum.

Clogg, C. C. and L. A. Goodman (1984). Latent structure analysis of a set of multidimensional contincency tables. *Journal of the American Statistical Association 79*, 762–771.

Cohen, A. C. (1960). An extension of a truncated Poisson distribution. *Biometrics 16*, 446–450.

Cohen, A. C. (1966). A note on certain discrete mixed distributions. *Biometrics 22*, 566–572.

Congdon, P. (2001). *Bayesian Statistical Modelling.* Chichester: Wiley.

Congdon, P. (2005). *Bayes Models for Categorical Data.* Chichester: Wiley.

Copas, J. B. (1988). Binary regression models for contaminated data. *Journal of the Royal Statistical Society, Ser. B 50*, 225–265.

Cormack, R. M. (1971). A review of classification. *Journal of the Royal Statistical Society, Ser. A 134*, 321–367.

Cosslett, S. R. and L.-F. Lee (1985). Serial correlation in latent discrete variable models. *Journal of Econometrics 27*, 79–97.

Cox, D. R. (1981). Statistical analysis of time series: Some recent developments. *Scandinavian Journal of Statistics 8*, 93–115.

Crawford, S. L. (1994). An application of the Laplace method to finite mixture distributions. *Journal of the American Statistical Association 89*, 259–267.

Crowder, M. J. and D. J. Hand (1990). *Analysis of Repeated Measures.* London: Chapman & Hall.

Cruz-Medina, I. R., T. P. Hettmansperger, and H. Thomas (2004). Semiparametric mixture models and repeated measures: The multinomial cut point model. *Applied Statistics 53*, 463–474.

Cutler, A. and O. Cordero-Braña (1996). Minimum Hellinger distance estimation for finite mixture models. *Journal of the American Statistical Association 91*, 1716–1723.

Dacunha-Castelle, D. and E. Gassiat (1997). The estimation of the order of a mixture model. *Bernoulli 3*, 279–299.

Dalal, S. R. and W. J. Hall (1983). Approximating priors by mixtures of natural conjugate priors. *Journal of the Royal Statistical Society, Ser. B 45*, 278–286.

Dalrymple, M. L., I. L. Hudson, and R. P. K. Ford (2003). Finite mixture, zero-inflated Poisson and Hurdle models with application to SIDS. *Computational Statisics and Data Analysis 41*, 491–504.

Damien, P., J. Wakefield, and S. Walker (1999). Gibbs sampling for Bayesian non-conjugate and hierarchical models by using auxiliary variables. *Journal of the Royal Statistical Society, Ser. B 61*, 331–344.

Dasgupta, A. and A. E. Raftery (1998). Detecting features in spatial point processes with clutter via model-based clustering. *Journal of the American Statistical Association 93*, 294–302.

Davidian, M. and D. M. Giltinan (1998). *Nonlinear Models for Repeated Measurement Data*. London: Chapman & Hall.

Day, N. E. (1969). Estimating the components of a mixture of normal distributions. *Biometrika 56*, 463–474.

De Jong, P. and S. Chu-Chun-Lin (1994). Stationary and non-stationary state space models. *Journal of Time Series Analysis 15*, 151–166.

De Jong, P. and N. Shephard (1995). The simulation smoother for time series models. *Biometrika 82*, 339–350.

De Soete, G. and W. S. DeSarbo (1991). A latent class probit model for analyzing pick any/N data. *Journal of Classification 8*, 45–63.

Dellaportas, P., J. J. Forster, and I. Ntzoufras (2002). On Baysian model and variable selection using MCMC. *Statistics and Computing 12*, 27–36.

Dellaportas, P. and I. Papageorgiou (2006). Multivariate mixtures of normals with unknown number of components. *Statistics and Computing 16*, 57–68.

Delmar, P., S. Robin, D. Tronik-Le Roux, and J. J. Daudin (2005). Mixture model on the variance for the differential analysis of gene expression data. *Applied Statistics 54*, 31–50.

Dempster, A. P., N. M. Laird, and D. B. Rubin (1977). Maximum likelihood from incomplete data via the EM algorithm. *Journal of the Royal Statistical Society, Ser. B 39*, 1–37.

DeSarbo, W. S. and W. L. Cron (1988). A maximum likelihood methodology for clusterwise linear regression. *Journal of Classification 5*, 248–282.

DeSarbo, W. S., M. Wedel, M. Vriens, and V. Ramaswamy (1992). Latent class metric conjoint analysis. *Marketing Letters 3*, 273–288.

Dey, D. K., L. Kuo, and S. K. Sahu (1995). A Bayesian predictive approach to determining the number of components in a mixture distribution. *Statistics and Computing 5*, 297–305.

Diaconis, P. and D. Ylvisaker (1979). Conjugate priors for exponential families. *The Annals of Statistics 7*, 269–281.

Dias, J. G. and M. Wedel (2004). An empirical comparison of EM, SEM and MCMC performance for problematical Gaussian mixture likelihoods. *Statistics and Computing 14*, 323–332.

DiCiccio, T. J., R. E. Kass, A. Raftery, and L. Wasserman (1997). Computing Bayes factors by combining simulation and asymptotic approximations. *Journal of the American Statistical Association 92*, 903–915.

Dickey, D. A. and W. A. Fuller (1981). Likelihood ratio statistics for autoregressive time series with a unit root. *Econometrica 49*, 1057–1072.

Dickey, J. (1971). The weighted likelihood ratio, linear hypotheses on normal location parameters. *The Annals of Mathematical Statistics 42*, 204–23.

Diebold, F. X., J.-H. Lee, and G. C. Weinbach (1994). Regime switching with time-varying transition probabilities. In C. P. Hargreaves (Ed.), *Nonstationary Time Series Analysis and Cointegration*, pp. 283–302. Oxford: Oxford University Press.

Diebold, F. X. and G. D. Rudebusch (1996). Measuring business cycles: A modern perspective. *The Review of Economics and Statistics 78*, 67–77.

Diebolt, J. and C. P. Robert (1994). Estimation of finite mixture distributions through Bayesian sampling. *Journal of the Royal Statistical Society, Ser. B 56*, 363–375.

Diggle, P. J., P. Heagerty, K.-Y. Liang, and S. L. Zeger (2002). *Analysis of Longitudinal Data* (2 ed.). Oxford: Oxford University Press.

Douc, R., E. Moulines, and T. Rydén (2004). Asymptotic properties of the maximum likelihood estimator in autoregressive models with Markov regime. *The Annals of Statistics 32*, 2254–2304.

Doucet, A., N. De Freitas, and N. Gordon (2001). *Sequential Monte Carlo Methods in Practice.* New York: Springer.

Dueker, M. (1997). Markov switching in GARCH processes and mean-reverting stock-market volatility. *Journal of Business & Economic Statistics 15*, 26–34.

Durbin, J. and S. J. Koopman (2000). Time series analysis of non-Gaussian observations based on state space models from both classical and Bayesian perspectives. *Journal of the Royal Statistical Society, Ser. B 62*, 3–56.

Durbin, J. and S. J. Koopman (2001). *Time Series Analysis by State Space Methods.* Oxford: Oxford University Press.

Durbin, J. and S. J. Koopman (2002). A simple and efficient simulation smoother for state space time series analysis. *Biometrika 89*, 603–615.

Durland, J. M. and T. H. McCurdy (1994). Duration-dependent transitions in a Markov model of U.S. GNP growth. *Journal of Business & Economic Statistics 12*, 279–288.

Edwards, W., H. Lindman, and L. J. Savage (1963). Bayesian statistical inference for psychological research. *Psychological Review 70*, 193–242.

Efron, B. and C. Morris (1977). Stein's paradox in statistics. *Scientific American 236*, 119–127.

Engel, C. (1994). Can the Markov switching model forecast exchange rates? *Journal of International Economics 36*, 151–165.

Engel, C. and J. D. Hamilton (1990). Long swings in the Dollar: Are they in the data and do markets know it? *The American Economic Review 80*, 689–713.

Engel, C. and C.-J. Kim (1999). The long-run U.S./U.K. real exchange rate. *Journal of Money, Credit, and Banking 31*, 335–356.
Engle, R. F. (1982). Autoregressive conditional heteroscedasticity with estimates of the variance of United Kingdom inflation. *Econometrica 50*, 987–1007.
Engle, R. F. and T. Bollerslev (1986). Modelling the persistence of conditional variances. *Econometric Reviews 5*, 1–87.
Engle, R. F. and C. W. J. Granger (1987). Co-integration and error correction: Representation, estimation, and testing. *Econometrica 55*, 251–276.
Engle, R. F. and V. K. Ng (1995). Measuring and testing the impact of news on volatility. In R. Engle (Ed.), *ARCH: Selected Readings*, pp. 145–175. Oxford: Oxford University Press.
Engle, R. F. and M. Watson (1981). A one-factor multivariate time series model of metropolitan wage rates. *Journal of the American Statistical Association 76*, 774–781.
Escobar, M. D. and M. West (1995). Bayesian density estimation and inference using mixtures. *Journal of the American Statistical Association 90*, 577–588.
Escobar, M. D. and M. West (1998). Computing nonparametric hierarchical models. In D. Dey, P. Müller, and D. Sinha (Eds.), *Practical Nonparametric and Semiparametric Bayesian Statistics*, Number 133 in Lecture Notes in Statistics, pp. 1–22. Berlin: Springer.
Evans, M., Z. Gilula, and I. Guttman (1989). Latent class analysis of two-way contingency tables by Bayesian methods. *Biometrika 76*, 557–563.
Evans, M., I. Guttman, and I. Olkin (1992). Numerical aspects in estimating the parameters of a mixture of normal distributions. *Journal of Computational and Graphical Statistics 1*, 351–365.
Everitt, B. S. (1979). Unresolved problems in cluster analysis. *Biometrics 35*, 169–181.
Everitt, B. S. (1985). Mixture distributions. In S. Kotz and N. L. Johnson (Eds.), *Encyclopedia of Statistical Sciences*, Volume 5, pp. 559–569. New York: Wiley.
Everitt, B. S. (1988). A finite mixture model for the clustering of mixed mode data. *Statistics & Probability Letters 6*, 305–309.
Everitt, B. S. and D. J. Hand (1981). *Finite Mixture Distributions.* London: Chapman & Hall.
Everitt, B. S., S. Landau, and M. Leese (2001). *Cluster Analysis* (4 ed.). London: Edward Arnold.
Everitt, B. S. and C. Merette (1990). The clustering of mixed mode data: A comparison of possible approaches. *Journal of Applied Statistics 17*, 283–297.
Fahrmeir, L. and L. Knorr-Held (2000). Dynamic and semiparametric models. In M. Schimek (Ed.), *Smoothing and Regression: Approaches, Computation and Application*, pp. 417–452. New York: Wiley.

Fair, R. C. and D. M. Jaffee (1972). Methods of estimation for markets in disequilibrium. *Econometrica 40*, 497–514.

Falk, B. (1986). Further evidence on the asymmetric behaviour of economic time series over the business cycle. *Journal of Political Economy 94*, 1096–1109.

Fama, E. (1965). The behavior of stock market prices. *Journal of Business 38*, 34–105.

Fama, E. and M. Gibbons (1982). Inflation, real returns and capital investment. *Journal of Monetary Economics 9*, 297–323.

Farewell, V. T. (1982). The use of mixture models for the analysis of survival data with long-term survivors. *Biometrics 38*, 1041–1046.

Farewell, V. T. and D. A. Sprott (1988). The use of a mixture model in the analysis of count data. *Biometrics 44*, 1191–1194.

Fearnhead, P. and P. Clifford (2003). On-line inference for hidden Markov models via particle filters. *Journal of the Royal Statistical Society, Ser. B 65*, 887–899.

Feller, W. (1943). On a general class of contagious distributions. *Annals of Mathematical Statistics 14*, 389–400.

Feng, Z. and C. McCulloch (1996). Using bootstrap likelihood ratios in finite mixture models. *Journal of the Royal Statistical Society, Ser. B 58*, 609–617.

Ferguson, T. S. (1973). A Bayesian analysis of some nonparametric problems. *The Annals of Statistics 1*, 209–230.

Fernández, C. and P. J. Green (2002). Modeling spatially correlated data via mixtures: A Bayesian approach. *Journal of the Royal Statistical Society, Ser. B 64*, 805–826.

Filardo, A. J. (1994). Business-cycle phases and their transitional dynamics. *Journal of Business & Economic Statistics 12*, 299–308.

Filardo, A. J. and S. F. Gordon (1998). Business cycle duration. *Journal of Econometrics 85*, 99–123.

Finch, S. J., N. R. Mendell, and H. C. Thode (1989). Probabilistic measures of adequacy of a numerical search for global maximum. *Journal of the American Statistical Association 84*, 1020–1023.

Fokoué, E. and D. M. Titterington (2003). Mixture of factor analysers. Bayesian estimation and inference by stochastic simulation. *Machine Learning 50*, 73–94.

Follmann, D. A. and D. Lambert (1989). Generalizing logistic regression by nonparametric mixing. *Journal of the American Statistical Association 84*, 295–300.

Follmann, D. A. and D. Lambert (1991). Identifiability of finite mixtures of logistic regression models. *Journal of Statistical Planning and Inference 27*, 375–381.

Fong, W. (1997). Volatility persistence and switching ARCH in Japanese stock return. *Financial Engeneering and the Japanese Market 4*, 37–57.

Formann, A. K. (1982). Linear logistic latent class analysis. *Biometrical Journal 24*, 171–190.

Formann, A. K. (1992). Linear logistic latent class analysis for polytomous data. *Journal of the American Statistical Association 87*, 476–486.

Formann, A. K. (1993). Fixed-distance latent class models for the analysis of sets of two-way contingency tables. *Biometrics 49*, 511–521.

Formann, A. K. (1994a). Measurement errors in caries diagnosis: Some further latent class models. *Biometrics 50*, 865–871.

Formann, A. K. (1994b). Measuring change in latent subgroups using dichotomous data: Unconditional, conditional, and semiparametric maximum likelihood estimation. *Journal of the American Statistical Association 89*, 1027–1034.

Formann, A. K. and T. Kohlmann (1996). Latent class analysis in medical research. *Statistical Methods in Medical Research 5*, 179–211.

Fowlkes, E. B. (1979). Some methods for studying the mixture of two normal (lognormal) distributions. *Journal of the American Statistical Association 74*, 561–575.

Fowlkes, E. B., R. Gnanadesikan, and J. R. Kettering (1988). Variable selection in clustering. *Journal of Classification 5*, 205–228.

Fraley, C. and A. E. Raftery (1998). How many clusters? Which clustering method? Answers via model-based cluster analysis. *Computer Journal 41*, 578–588.

Fraley, C. and A. E. Raftery (2002). Model-based clustering, discriminant analysis, and density estimation. *Journal of the American Statistical Association 97*, 611–631.

Francq, C. and M. Roussignol (1997). On white noises driven by hidden Markov chains. *Journal of Time Series Analysis 18*, 553–578.

Francq, C. and M. Roussignol (1998). Ergodicity of autoregressive processes with Markov-switching and consistency of the maximimum-likelihood estimator. *Statistics 32*, 151–173.

Francq, C., M. Roussignol, and J. Zakoian (2001). Conditional heteroscedasticity driven by hidden Markov chains. *Journal of Time Series Analysis 22*, 197–220.

Francq, C. and J. Zakoian (1999). Linear-representations based estimation of switching-regime GARCH models. Proceedings of the NBER/NSF Time Series Conference.

Francq, C. and J. Zakoian (2001). Stationarity of multivariate Markov-switching ARMA models. *Journal of Econometrics 102*, 339–364.

Franses, P. H. and D. van Dijk (2000). *Nonlinear Time Series Models in Empirical Finance*. Cambridge: Cambridge University Press.

Friedman, H. P. and J. Rubin (1967). On some invariant criteria for grouping data. *Journal of the American Statistical Association 63*, 1159–1178.

Friedman, J. H. (1989). Regularized discriminant analysis. *Journal of the American Statistical Association 84*, 165–175.

Frühwirth, R. (1987). Application of Kalman filtering to track and vertex fitting. *Nuclear Instruments and Methods in Physics Research A262*, 444–450.

Frühwirth-Schnatter, S. (1994). Data augmentation and dynamic linear models. *Journal of Time Series Analysis 15*, 183–202.

Frühwirth-Schnatter, S. (1995). Bayesian model discrimination and Bayes factors for linear Gaussian state space models. *Journal of the Royal Statistical Society, Ser. B 57*, 237–246.

Frühwirth-Schnatter, S. (2001a). Fully Bayesian analysis of switching Gaussian state space models. *Annals of the Institute of Statistical Mathematics 53*, 31–49.

Frühwirth-Schnatter, S. (2001b). Markov chain Monte Carlo estimation of classical and dynamic switching and mixture models. *Journal of the American Statistical Association 96*, 194–209.

Frühwirth-Schnatter, S. (2004). Estimating marginal likelihoods for mixture and Markov switching models using bridge sampling techniques. *The Econometrics Journal 7*, 143–167.

Frühwirth-Schnatter, S. and R. Frühwirth (2006). Auxiliary mixture sampling with applications to logistic models. Research Report IFAS, http://www.ifas.jku.at/.

Frühwirth-Schnatter, S. and S. Kaufmann (2006a). How do changes in monetary policy affect bank lending? An analysis of Austrian bank data. *Journal of Applied Econometrics 21*, 275–305.

Frühwirth-Schnatter, S. and S. Kaufmann (2006b). Model-based clustering of multiple time series. Research Report IFAS, http://www.ifas.jku.at/.

Frühwirth-Schnatter, S. and T. Otter (1999). Conjoint-analysis using mixed effect models. In H. Friedl, A. Berghold, and G. Kauermann (Eds.), *Statistical Modelling. Proceedings of the Fourteenth International Workshop on Statistical Modelling*, pp. 181–191. Graz.

Frühwirth-Schnatter, S., R. Tüchler, and T. Otter (2004). Bayesian analysis of the heterogeneity model. *Journal of Business & Economic Statistics 22*, 2–15.

Frühwirth-Schnatter, S., R. Tüchler, and T. Otter (2005). Capturing unobserved consumer heterogeneity using the Bayesian heterogeneity model. In A. Taudes (Ed.), *Adaptive Information Systems and Modelling in Economics and Management Science*, pp. 57–70. Wien: Springer.

Frühwirth-Schnatter, S. and H. Wagner (2006). Auxiliary mixture sampling for parameter-driven models of time series of small counts with applications to state space modelling. Research Report IFAS 11, http://www.ifas.jku.at/.

Gamerman, D. (1997). *Markov Chain Monte Carlo. Stochastic Simulation for Bayesian Inference.* London: Chapman & Hall.

Gamerman, D. and A. F. M. Smith (1996). Bayesian analysis of longitudinal data studies. In J. M. Bernardo, J. O. Berger, A. P. Dawid, and A. F. M.

Smith (Eds.), *Bayesian Statistics 5*, pp. 587–597. Oxford: Oxford University Press.

Garcia, R. and P. Perron (1996). An analysis of real interest rate under regime shift. *The Review of Economics and Statistics 78*, 111–125.

Geisser, S. (1964). Estimation in the uniform covariance case. *Journal of the Royal Statistical Society, Ser. B 26*, 477–483.

Geisser, S. and J. Cornfield (1963). Posterior distributions for multivariate normal parameters. *Journal of the Royal Statistical Society, Ser. B 25*, 368–376.

Gelfand, A., S. Sahu, and B. Carlin (1995). Efficient parametrisations for normal linear mixed models. *Biometrika 82*, 479–488.

Gelfand, A. E. and D. K. Dey (1994). Bayesian model choice: Asymptotics and exact calculations. *Journal of the Royal Statistical Society, Ser. B 56*, 501–514.

Gelfand, A. E., D. K. Dey, and H. Chang (1992). Model determination using predictive distributions with implementation via sampling-based methods. In J. M. Bernardo, J. O. Berger, A. P. Dawid, and A. F. M. Smith (Eds.), *Bayesian Statistics 4*, pp. 147–167. Oxford: Oxford University Press.

Gelfand, A. E. and A. F. M. Smith (1990). Sampling-based approaches to calculating marginal densities. *Journal of the American Statistical Association 85*, 398–409.

Gelman, A., J. B. Carlin, H. S. Stern, and D. B. Rubin (2004). *Bayesian Data Analysis* (2 ed.). London: Chapman & Hall/CRC.

Gelman, A., X.-L. Meng, and H. Stern (1996). Posterior predictive assessment of model fitness via realized discrepancies (Disc: p760-807). *Statistica Sinica 6*, 733–760.

George, E. I. and R. McCulloch (1997). Approaches for Bayesian variable selection. *Statistica Sinica 7*, 339–373.

Gerlach, R., C. C. and R. Kohn (2000). Efficient Bayesian inference for dynamic mixture models. *Journal of the American Statistical Association 95*, 819–828.

Geweke, J. (1989). Bayesian inference in econometric models using Monte Carlo integration. *Econometrica 57*, 1317–1339.

Geweke, J. (1992). Evaluating the accuracy of sampling-based approaches to the calculation of posterior moments (Disc: p189-193). In J. M. Bernardo, J. O. Berger, A. P. Dawid, and A. F. M. Smith (Eds.), *Bayesian Statistics 4*, pp. 169–188. Oxford: Oxford University Press.

Geweke, J. (1993). Bayesian treatment of the independent Student-$t$ linear model. *Journal of Applied Econometrics 8(Supplement)*, 19–40.

Geweke, J. (1999). Using simulation methods for Bayesian econometric models: Inference, development, and communication (Disc: p75-126). *Econometric Reviews 18*, 1–73.

Geweke, J. and M. Keane (1999). Mixture of normals probit models. In C. Hsiao, M. H. Pesaran, K. L. Lahiri, and L. F. Lee (Eds.), *Analysis of*

*Panels and Limited Dependent Variables*, pp. 49–78. Cambridge: Cambridge University Press.

Geyer, C. and J. Møller (1994). Simulation procedures and likelihood inference for spatial point processes. *Scandinavian Journal of Statistics 21*, 359–373.

Ghosh, D. and A. M. Chinnaiyan (2002). Mixture modelling of gene expression data from microarray experiments. *Bioinformatics 18*, 275–286.

Gilula, Z. (1979). Singular value decomposition of probability matrices: Probabilistic aspects of latent dichotomous variables. *Biometrika 66*, 339–44.

Gnanadesikan, R., J. R. Kettering, and S. L. Tao (1995). Weighting and selection of variables for cluster analysis. *Journal of Classification 12*, 113–136.

Godsill, S. J. (1997). Bayesian enhancement of speech and audio signals which can be modelled as ARMA processes. *International Statistical Review 65*, 1–21.

Godsill, S. J. (2001). On the relation between MCMC model uncertainty methods. *Journal of Computational and Graphical Statistics 10*, 230–248.

Godsill, S. J. and P. Rayner (1998). Statistical reconstruction and analysis of autoregressive signals in impulsive noise using the Gibbs sampler. *IEEE Transactions on Speech and Audio Processing 6*, 352–372.

Goldfeld, S. and R. Quandt (1973). A Markov model for switching regression. *Journal of Econometrics 1*, 3–16.

Goodman, L. A. (1974a). The analysis of systems of quantitative variables when some of the variables are unobservable. Part I - a modified latent structure approach. *American Journal of Sociology 79*, 1179–1259.

Goodman, L. A. (1974b). Exploratory latent structure analysis using both identiable and unidentifiable models. *Biometrika 61*, 215–231.

Goodman, L. A. (1978). *Analyzing Qualitative/Categorical Data: Log-linear Models and Latent Structure Analysis.* Reading, MA: Addison-Wesley.

Goodwin, T. H. (1993). Business cycle analysis with a Markov switching model. *Journal of Business & Economic Statistics 11*, 331–339.

Gordon, A. (1999). *Classification* (2 ed.). London: Chapman & Hall.

Gordon, K. and A. F. M. Smith (1990). Modeling and monitoring biomedical time series. *Journal of the American Statistical Association 85*, 328–337.

Gosh, J. K. and P. K. Sen (1985). On the asymptotic performance of the log likelihood ratio statistic for the mixture model and related results. In L. LeCam and A. Olshen (Eds.), *Proceedings of the Berkeley Conference in Honor of Jerzy Neyman and Jack Kiefer*, Volume II, pp. 789–806. Monterey, CA: Wadsworth Advanced Books.

Granger, C. W. J. and D. Orr (1972). Infinite variance and research strategy in time series analysis. *Journal of the American Statistical Association 67*, 275–285.

Granger, C. W. J. and T. Teräsvirta (1993). *Modelling Nonlinear Economic Relationships.* Oxford: Oxford University Press.

Gray, S. F. (1996). Modeling the conditional distribution of interest rates as a regime switching process. *Journal of Financial Economics 42*, 27–62.

Green, P. J. (1995). Reversible jump Markov chain Monte Carlo computation and Bayesian model determination. *Biometrika 82*, 711–732.

Green, P. J. (2003). Trans-dimensional Markov chain Monte Carlo. In P. J. Green, N. L. Hjort, and S. Richardson (Eds.), *Highly Structured Stochastic Systems*, pp. 179–198. Oxford: Oxford University Press.

Green, P. J. and T. O'Hagan (2000). Model choice with MCMC on product spaces without using pseudo-priors. University of Nottingham Research Report.

Green, P. J. and S. Richardson (2001). Modelling heterogeneity with and without the Dirichlet process. *Scandinavian Journal of Statistics 28*, 355–375.

Green, P. J. and S. Richardson (2002). Hidden Markov models and disease mapping. *Journal of the American Statistical Association 97*, 1–16.

Grilli, V. and G. Kaminsky (1991). Nominal exchange rate regimes and the real exchange rate: Evidence from the United States and Great Britain, 1885-1986. *Journal of Monetary Economics 27*, 191–212.

Grim, J. and M. Haindl (2003). Texture modelling by discrete distribution mixtures. *Computational Statisics and Data Analysis 41*, 603–615.

Gruet, M.-A., A. Philippe, and C. P. Robert (1999). MCMC control spreadsheets for exponential mixture estimation. *Journal of Computational and Graphical Statistics 8*, 298–317.

Grün, B. (2002). Identifizierbarkeit von multinomialen Mischmodellen. Masters thesis, Technische Universität Wien, Vienna, Austria.

Grün, B. and F. Leisch (2004). Bootstrapping finite mixture models. In J. Antoch (Ed.), *COMPSTAT 2004. Proceedings in Computational Statistics*, pp. 1115–1122. Heidelberg: Physica-Verlag/Springer.

Guttman, I. (1973). Care and handling of univariate or multivariate outliers in detecting spuriousity – A Bayesian approach. *Technometrics 15*, 723–738.

Guttman, I., R. Dutter, and P. R. Freeman (1978). Care and handling of univariate outliers in the general linear model to detect spuriousity – A Bayesian approach. *Technometrics 20*, 187–193.

Hall, S. G., Z. Psaradakis, and M. Sola (1997). Cointegration and changes in regime: The Japanese consumption function. *Journal of Applied Econometrics 12*, 151–168.

Hall, S. G., Z. Psaradakis, and M. Sola (1999). Detecting periodically collapsing bubbles: A Markov-switching unit root test. *Journal of Applied Econometrics 14*, 143–154.

Hamilton, J. (1994a). State-space models. In R. Engle and D. McFadden (Eds.), *Handbook of Econometrics*, Volume 4, pp. 3039–3080. Amsterdam: North-Holland.

Hamilton, J. (1994b). *Time Series Analysis.* Princeton, NJ: Princeton University Press.

Hamilton, J. D. (1988). Rational expectations econometric analysis of changes in regime: An investigation on the term structure of interest rates. *Journal of Economic Dynamics and Control 12*, 385–423.

Hamilton, J. D. (1989). A new approach to the economic analysis of nonstationary time series and the business cycle. *Econometrica 57*, 357–384.

Hamilton, J. D. (1990). Analysis of time series subject to changes in regime. *Journal of Econometrics 45*, 39–70.

Hamilton, J. D. and G. Lin (1996). Stock market volatility and the business cycle. *Journal of Applied Econometrics 11*, 573–593.

Hamilton, J. D. and B. Raj (2002). New directions in business cycle research and financial analysis. *Empirical Economics 27*, 149–162.

Hamilton, J. D. and R. Susmel (1994). Autoregressive conditional heteroskedasticity and changes in regime. *Journal of Econometrics 64*, 307–333.

Han, C. and B. P. Carlin (2001). Markov chain Monte Carlo methods for computing Bayes factors: A comparative review. *Journal of the American Statistical Association 96*, 1122–1132.

Hannan, E. J. and M. Deistler (1988). *The Statistical Theory of Linear Systems.* New York: Wiley.

Hansen, B. (1992). The likelihood ratio test under non-standard conditions: Testing the Markov switching model of GNP. *Journal of Applied Econometrics 7*, S61 –S82.

Harrison, P. and C. Stevens (1976). Bayesian forecasting (with discussion). *Journal of the Royal Statistical Society, Ser. B 38*, 205–247.

Hartigan, J. A. (1975). *Clustering Algorithm.* Chichester: Wiley.

Hartigan, J. A. (1977). Distribution problems in clustering. In *Classification and Clustering*, pp. 45–72. San Diego: Academic.

Hartigan, J. A. (1985). A failure of likelihood ratio asymptotics for normal mixtures. In L. LeCam and A. Olshen (Eds.), *Proceeding of the Berkeley Conference in Honor of Jerzy Neyman and Jack Kiefer*, Volume II, pp. 807–810. Monterey, CA: Wadsworth Advanced Books.

Harvey, A., E. Ruiz, and E. Sentana (1992). Unobserved component time series models with ARCH disturbances. *Journal of Econometrics 52*, 129–157.

Harvey, A. C. (1989). *Forecasting, Structural Time Series Models and the Kalman Filter.* Cambridge: Cambridge University Press.

Harvey, A. C. (1993). *Time Series Models* (2 ed.). Harvester-Wheatsheaf.

Harvey, A. C. and C. Fernandes (1989). Time series models for count or qualitative observations. *Journal of Business & Economic Statistics 7*, 407–417.

Harvey, A. C., S. J. Koopman, and J. Penzer (1998). Messy time series: A unified approach. *Advances in Econometrics 13*, 103–143.

Harvey, A. C. and P. H. J. Todd (1983). Forecasting economic time series with structural and Box-Jenkins models: A case study. *Journal of Business & Economic Statistics 1*, 299–315.

Hasselblad, V. (1966). Estimation of parameters for a mixture of normal distributions. *Technometrics 8*, 431–444.

Hasselblad, V. (1969). Estimation of finite mixtures of distributions from the exponential family. *Journal of the American Statistical Association 64*, 1459–1471.

Hastie, T. and R. Tibshirani (1996). Discriminant analysis by Gaussian mixtures. *Journal of the Royal Statistical Society, Ser. B 58*, 155–176.

Hathaway, R. J. (1985). A constrained formulation of maximum-likelihood estimation for normal mixture distributions. *The Annals of Statistics 13*, 795–800.

Hathaway, R. J. (1986). Another interpretation of the EM algorithm for mixture distributions. *Statistics & Probability Letters 4*, 53–56.

Hawkins, D. S., D. M. Allen, and A. J. Stromberg (2001). Determining the number of components in mixtures of linear models. *Computational Statisics and Data Analysis 38*, 14–48.

Heckman, J. J., R. Robb, and J. R. Walker (1990). Testing the mixture of exponentials hypothesis and estimating the mixing distribution by the method of moments. *Journal of the American Statistical Association 85*, 582–589.

Heckman, J. J. and B. Singer (1984). A method for minimizing the impact of distributional assumptions in econometric models for duration data. *Econometrica 52*, 271–320.

Heller, A. (1965). On stochastic processes derived from Markov chains. *Annals of Mathematical Statistics 36*, 1286–1291.

Hennig, C. (2000). Identifiability of models for clusterwise linear regression. *Journal of Classification 17*, 273–296.

Hettmansperger, T. P. and H. Thomas (2000). Almost nonparametric inference for repeated measures in mixture models. *Journal of the Royal Statistical Society, Ser. B 62*, 811–825.

Hinton, G. E., P. Dayan, and M. Revow (1997). Modeling the manifolds of images of handwritten digits. *IEEE Transactions on Neural Networks 8*, 65–73.

Hobert, J. P. and G. Casella (1998). Functional compatibility, Markov chains, and Gibbs sampling with improper posteriors. *Journal of Computational and Graphical Statistics 7*, 42–60.

Hobert, J. P., C. P. Robert, and D. M. Titterington (1999). On perfect simulation for some mixtures of distributions. *Statistics and Computing 9*, 287–298.

Hoek, H., A. Lucas, and H. K. van Dijk (1995). Classical and Bayesian aspects of robust unit root inference. *Journal of Econometrics 69*, 27–59.

Holst, U., G. Lindgren, J. Holst, and M. Thuvesholmen (1994). Recursive estimation in switching autoregressions with a Markov regime. *Journal of Time Series Analysis 15*, 489–506.

Hosmer, D. W. (1973). A comparison of iterative maximum likelihood estimates of the parameters of a mixture of two normal distributions under three types of sample. *Biometrics 29*, 761–770.

Hosmer, D. W. (1974). Maximum likelihood estimates of the parameters of a mixture of two regression lines. *Communications in Statistics, Part A – Theory and Methods 3*, 995–1006.

Hunt, L. and M. Jorgensen (2003). Mixture model clustering for mixed data with missing information. *Computational Statisics and Data Analysis 41*, 429–440.

Hurn, M., A. Justel, and C. P. Robert (2003). Estimating mixtures of regressions. *Journal of Computational and Graphical Statistics 12*, 55–79.

Jacobs, R. A., M. I. Jordan, S. J. Nowlan, and G. E. Hinton (1991). Adaptive mixtures of local experts. *Neural Computation 3*, 79–87.

Jacobs, R. A., F. Peng, and M. A. Tanner (1997). A Bayesian approach to model selection in hierarchical mixtures-of-experts architectures. *Neural Networks 10*, 231–241.

Jacobs, R. A., M. A. Tanner, and F. Peng (1996). Bayesian inference for hierarchical mixtures-of-experts with applications to regression and classification. *Statistical Methods in Medical Research 5*, 375–390.

Jacquier, E., N. G. Polson, and P. E. Rossi (1994). Bayesian analysis of stochastic volatility models. *Journal of Business & Economic Statistics 12*, 371–417.

Jalali, A. and J. Pemberton (1995). Mixture models for time series. *Journal of Applied Probability 32*, 123–138.

Jansen, R. C. (1993). Maximum likelihood in a generalized linear finite mixture model by using the EM algorithm. *Biometrics 49*, 227–231.

Jazwinski, A. H. (1970). *Stochastic Processes and Filtering Theory.* New York: Academic.

Jeffreys, S. H. (1948). *Theory of Probability* (2 ed.). Oxford: Clarendon.

Jensen, J. L. and N. V. Petersen (1999). Asymptotic normality of the maximum likelihood estimator in state space models. *The Annals of Statistics 27*, 514–535.

Jiang, W. and M. A. Tanner (1999). On the identifiability of mixtures-of-experts. *Neural Network 12*, 1253–1258.

Johnson, N. L., S. Kotz, and N. Balakrishnan (1994). *Continuous Univariate Distributions* (2 ed.), Volume 1. New York: Wiley.

Johnson, N. L., S. Kotz, and N. Balakrishnan (1995). *Continuous Univariate Distributions* (2 ed.), Volume 2. New York: Wiley.

Johnson, N. L., S. Kotz, and A. Kemp (1993). *Univariate Discrete Distributions* (2 ed.). New York: Wiley.

Jordan, M. I. and R. A. Jacobs (1994). Hierarchical mixtures of experts and the EM algorithm. *Neural Computation 6*, 181–214.

Jorgensen, M. A. (2004). Using multinomial mixture models to cluster Internet traffic. *Australian & New Zealand Journal of Statistics 46*, 205–218.

Juang, B. H. and L. R. Rabiner (1985). Mixture autoregressive hidden Markov models for speech signals. *IEEE Transactions on Acoustics, Speech, and Signal Processing 33*, 1404–1413.

Justel, A. and D. Peña (1996). Gibbs sampling will fail in outlier problems with strong masking. *Journal of Computational and Graphical Statistics 5*, 176–189.

Justel, A. and D. Peña (2001). Bayesian unmasking in linear models. *Computational Statistics and Data Analysis 36*, 69–84.

Kadane, J. B. and N. Lazar (2004). Methods and criteria for model selection. *Journal of the American Statistical Association 99*, 279–290.

Kalman, R. E. (1960). A new approach to linear filtering and prediction problems. *Transactions ASME Journal of Basic Engeneering 82*, 35–45.

Kalman, R. E. (1961). New results in filtering and prediction theory. *Transactions ASME Journal of Basic Engeneering 81*, 95–108.

Kamakura, W. A. (1991). Estimating flexible distributions of ideal-points with external analysis of preferences. *Psychometrika 56*, 419–48.

Kamakura, W. A. and G. J. Russell (1989). A probabilistic choice model for market segmentation and elasticity structure. *Journal of Marketing Research 26*, 379–90.

Karawatzki, R., J. Leydold, and K. Pötzelberger (2005). Automatic Markov chain Monte Carlo procedures for sampling from multivariate distributions. Electronic publication available at `http://epub.wu-wien.ac.at`, Vienna University of Economics and Business Administration.

Karlin, S. and H. M. Taylor (1975). *A First Course in Stochastic Processes* (2 ed.). San Diego: Academic.

Karlis, D. and E. Xekalaki (1998). Minimum Hellinger distance estimation for Poisson mixtures. *Computational Statisics and Data Analysis 29*, 81–103.

Karlis, D. and E. Xekalaki (2001). Robust inference for finite Poisson mixtures. *Journal of Statistical Planning and Inference 93*, 93–115.

Karlis, D. and E. Xekalaki (2003). Choosing initial values for the EM algorithm for finite mixtures. *Computational Statisics and Data Analysis 41*, 577–590.

Karlis, D. and E. Xekalaki (2005). Mixed Poisson distributions. *International Statistical Review 73*, 35–58.

Karlsen, H. A. (1990). Existence of moments in a stationary stochastic difference equation. *Advances in Applied Probability 22*, 129–146.

Karlsen, H. A. and D. Tjøstheim (1990). Autoregressive segmentation of signal traces with application to geological dipmeter measurements. *IEEE Transactions on Geoscience and Remote Sensing 28*, 171–181.

Kashyap, R. L. (1970). Maximum likelihood identification of stochastic linear systems. *IEEE Transactions on Automatic Control 15*, 25–34.

Kass, R. E. and A. E. Raftery (1995). Bayes factors. *Journal of the American Statistical Association 90*, 773–795.

Kass, R. E., L. Tierney, and J. B. Kadane (1988). Asymptotics in Bayesian computation. In J. M. Bernardo, M. H. DeGroot, D. V. Lindley, and A. F. M. Smith (Eds.), *Bayesian Statistics 3*, pp. 261–278. New York: Clarendon.

Kass, R. E., L. Tierney, and J. B. Kadane (1990). The validity of posterior expansions based on Laplace's method. In S. Geisser, J. S. Hodges, S. J. Press, and A. Zellner (Eds.), *Bayesian and Likelihood Methods in Statistics and Econometrics: Essays in Honor of George A. Barnard*, pp. 473–488. New York: Elsevier.

Kass, R. E. and S. K. Vaidyanathan (1992). Approximate Bayes factors and orthogonal parameters, with application to testing equality of two binomial proportions. *Journal of the Royal Statistical Society, Ser. B 54*, 129–144.

Kaufman, L. and P. J. Rousseeuw (1990). *Finding Groups in Data: An Introduction to Cluster Analysis.* New York: Wiley.

Kaufmann, S. (2000). Measuring business cycles with a dynamic Markov switching factor model: An assessment using Bayesian simulation methods. *The Econometrics Journal 3*, 39–65.

Kaufmann, S. (2002). Is there an asymmetric effect of monetary policy over time. *Empirical Economics 27*, 277–297.

Kaufmann, S. and S. Frühwirth-Schnatter (2002). Bayesian analysis of switching ARCH models. *Journal of Time Series Analysis 23*, 425–458.

Keribin, C. (2000). Consistent estimation of the order of mixture models. *Sankhya A 62*, 49–66.

Kiefer, N. M. (1978). Discrete parameter variation: Efficient estimation of a switching regression model. *Econometrica 46*, 427–434.

Kiefer, N. M. and J. Wolfowitz (1956). Consistency of the maximum likelihood estimator in the presence of infinitely many incidental parameters. *Annals of Mathematical Statistics 27*, 887–906.

Kim, B. S. and B. H. Margolin (1992). Testing goodness of fit of a multinomial model against overdispersed alternatives. *Biometrics 48*, 711–719.

Kim, C.-J. (1993a). Sources of monetary growth uncertainty and economic activity: The time-varying parameter model with heteroscedastic disturbances. *The Review of Economics and Statistics 75*, 483–492.

Kim, C.-J. (1993b). Unobserved-component time series models with Markov-switching heteroscedasticity: Changes in regime and the link between inflation rates and inflation uncertainty. *Journal of Business & Economic Statistics 11*, 341–349.

Kim, C.-J. (1994). Dynamic linear models with Markov-switching. *Journal of Econometrics 60*, 1–22.

Kim, C.-J. and M.-J. Kim (1996). Transient fads and the crash of '87. *Journal of Applied Econometrics 11*, 41–58.

Kim, C.-J. and C. R. Nelson (1998). Business cycle turning points, a new coincident index, and tests of duration dependence based on a dynamic factor model with regime-switching. *The Review of Economics and Statistics 80*, 188–201.

Kim, C.-J. and C. R. Nelson (1999). *State-Space Models with Regime Switching: Classical and Gibbs-sampling Approaches with Applications.* Cambridge, MA: MIT Press.

Kim, C.-J. and C. R. Nelson (2001). A Bayesian approach to testing for Markov switching in univariate and dynamic factor models. *International Economic Review 42*, 989–1013.

Kim, I.-M. (1993c). A dynamic programming approach to the estimation of Markov switching regression models. *Journal of Statistical Computation and Simulation 45*, 61–76.

Kim, S., N. Shephard, and S. Chib (1998). Stochastic volatility: Likelihood inference and comparison with ARCH models. *Review of Economic Studies 65*, 361–393.

Kirnbauer, R., S. Schnatter, and D. Gutknecht (1987). Bayesian estimation of design floods under regional and subjective prior information. In R. Viertl (Ed.), *Probability and Bayesian Statistics*, pp. 285–294. New York: Plenum.

Kitagawa, G. (1981). A nonstationary time series model and its fitting by a recursive filter. *Journal of Time Series Analysis 2*, 103–116.

Kitagawa, G. and W. Gersch (1984). A smoothness priors state space modeling of time series with trend and seasonality. *Journal of the American Statistical Association 79*, 378–389.

Klaasen, F. (2002). Improving GARCH volatility forecasts with regime-switching GARCH. *Empirical Economics 27*, 363–394.

Kleibergen, F. and R. Paap (2002). Priors, posteriors and Bayes factors for a Bayesian analysis of cointegration. *Journal of Econometrics 111*, 223–249.

Kohn, R. and C. F. Ansley (1986). Estimation, prediction, and interpolation for ARIMA models with missing data. *Journal of the American Statistical Association 81*, 751–761.

Kohn, R. and C. F. Ansley (1987). A new algorithm for spline smoothing based on smoothing a stochastic process. *SIAM Journal on Scientific and Statistical Computing 8*, 33–48.

Kon, S. J. (1984). Models of stock returns – A comparison. *The Journal of Finance 39*, 147–165.

Koop, G. (2003). *Bayesian Econometrics.* Chichester: Wiley.

Koop, G. and H. K. van Dijk (2000). Testing for integration using evolving trend and seasonals models: A Bayesian approach. *Journal of Econometrics 97*, 261–291.

Koopman, S. J. (1993). Disturbance smoother for state space models. *Biometrika 80*, 117–126.

Koopman, S. J. (1997). Exact initial Kalman filtering and smoothing for nonstationary time series models. *Journal of the American Statistical Association 92*, 1630–1638.

Koopman, S. J. and J. Durbin (2000). Fast filtering and smoothing for multivariate state space models. *Journal of Time Series Analysis 21*, 281–296.

Koski, T. (2001). *Hidden Markov Models for Bioinformatics.* Dordrecht: Kluwer Academic.

Krishnamurthy, V. and T. Rydén (1998). Consistent estimation of linear and non-linear autoregressive models with Markov regime. *Journal of Time Series Analysis 19*, 291–307.

Krolzig, H.-M. (1997). *Markov-Switching Vector Autoregressions: Modelling, Statistical Inference, and Application to Business Cycle Analysis.* Lecture Notes in Economics and Mathematical Systems. New York/Berlin/Heidelberg: Springer.

Krolzig, H.-M. (2001). Business cycle measurement in the presence of structural change: International evidence. *International Journal of Forecasting 17*, 349–368.

Krolzig, H.-M. and M. Sensier (2000). A disaggregated Markov-switching model of the business cycle in UK manufacturing. *The Manchester School 68*, 442–460.

Krzanowski, W. J. and Y. T. Lai (1985). A criterion for determining the number of clusters in a data set. *Biometrics 44*, 23–34.

Kullback, S. and R. A. Leibler (1951). On information and sufficiency. *The Annals of Mathematical Statistics 22*, 79–86.

Kuttner, K. N. (1994). Estimating potential output as a latent variable. *Journal of Business & Economic Statistics 12*, 361–368.

Laird, N. M. and J. H. Ware (1982). Random-effects model for longitudinal data. *Biometrics 38*, 963–974.

Lam, P.-S. (1990). The Hamilton model with a general autoregressive component: Estimation and comparison with other models of economic time series. *Journal of Monetary Economics 26*, 409–432.

Lambert, D. (1992). Zero-inflated Poisson regression, with an application to defects in manufacturing. *Technometrics 34*, 1–14.

Lamoureux, C. G. and W. D. Lastrapes (1990). Persistence in variance, structural change and the GARCH model. *Journal of Business & Economic Statistics 8*, 225–234.

Lavine, M. and M. West (1992). A Bayesian method for classification and discrimination. *The Canadian Journal of Statistics 20*, 451–461.

Lawless, J. F. (1987). Negative binomial and mixed Poisson regression. *The Canadian Journal of Statistics 15*, 209–225.

Lawrance, A. J. and P. A. W. Lewis (1985). Modelling and residual analysis of nonlinear autoregressive time series in exponential variables (with discussions). *Journal of the Royal Statistical Society, Ser. B 47*, 165–202.

Lawrence, C. J. and W. J. Krzanowski (1996). Mixture separation for mixed-mode data. *Statistics and Computing 6*, 85–92.

Lazarsfeld, P. F. (1950). The logical and mathematical foundation of latent structure analysis. In S. A. Stouffer, L. Guttman, E. A. Suchman, P. F. Lazarsfeld, S. A. Star, and J. A. Clausen (Eds.), *Studies in Social Psycholgy in World War II, Vol. IV: Measurement and Prediction*, pp. 362–412. Princeton, NJ: Princeton University.

Lazarsfeld, P. F. and N. W. Henry (1968). *Latent Structure Analysis.* New York: Houghton Mifflin.

Le, N. D., B. G. Leroux, and M. L. Puterman (1992). Exact likelihood evaluation in a Markov mixture model for time series of seizure counts. *Biometrics 48*, 317–323.

Lehmann, E. L. (1983). *Theory of Point Estimation.* New York: Wiley.

Leisch, F. (2004). Exploring the structure of mixture model components. In J. Antoch (Ed.), *COMPSTAT 2004. Proceedings in Computational Statistics*, pp. 1405–1412. Heidelberg: Physica-Verlag/Springer.

Lenk, P. J. and W. S. DeSarbo (2000). Bayesian inference for finite mixtures of generalized linear models with random effects. *Psychometrika 65*, 93–119.

Leroux, B. G. (1992a). Consistent estimation of a mixing distribution. *The Annals of Statistics 20*, 1350–1360.

Leroux, B. G. (1992b). Maximum-likelihood estimation for hidden Markov models. *Stochastic Processes and Their Applications 40*, 127–143.

Leroux, B. G. and M. L. Puterman (1992). Maximum-penalized-likelihood estimation for independent and Markov-dependent mixture models. *Biometrics 48*, 545–558.

Levison, S. E., L. R. Rabiner, and M. M. Sondhi (1983). An introduction to the application of the theory of probabilistic functions of a Markov process to automatic speech recognition. *The Bell System Technical Journal 62*, 1035–1074.

Lewis, S. M. and A. E. Raftery (1997). Estimating Bayes factors via posterior simulation with the Laplace-Metropolis estimator. *Journal of the American Statistical Association 92*, 648–655.

Li, J. and R. M. Gray (2000). *Image Segmentation and Compression Using Hidden Markov Models.* Dordrecht: Kluwer Academic.

Li, J. Q. and A. R. Barron (2000). Mixture density estimation. Technical report, Department of Statistics, Yale University, New Haven, CT.

Li, L. A. and N. Sedransk (1988). Mixtures of distributions: A topological approach. *The Annals of Statistics 16*, 1623–1634.

Liang, Z., R. J. Jaszczak, and R. E. Coleman (1992). Parameter estimation of finite mixtures using the EM algorithm and information criteria with applications to medical image processing. *IEEE Transactions on Nuclear Science 39*, 1126–1133.

Liao, T. F. (2004). Estimating household structure in ancient China by using historical data: A latent class analysis of partially missing patterns. *Journal of the Royal Statistical Society, Ser. A 167*, 125–139.

Lindgren, G. (1978). Markov regime models for mixed distributions and switching regressions. *Scandinavian Journal of Statistics 5*, 81–91.

Lindley, D. V. (1957). A statistical paradoxon. *Biometrika 44*, 187–192.

Lindley, D. V. and A. F. M. Smith (1972). Bayes' estimates for the linear model. *Journal of the Royal Statistical Society, Ser. B 34*, 1–41.

Lindsay, B. G. (1989). Moment matrices: Applications in mixtures. *The Annals of Statistics 17*, 722–740.

Lindsay, B. G. (1994). Efficiency versus robustness: The case for minimum Hellinger distance and related methods. *The Annals of Statistics 22*, 1081–1114.

Lindsay, B. G. (1995). *Mixture Models: Theory, Geometry, and Applications.* Institute of Mathematical Statistics.

Lindsay, B. G. and P. Basak (1993). Multivariate normal mixtures: A fast consistent method of moments. *Journal of the American Statistical Association 88*, 468–476.

Lindsay, B. G. and K. Roeder (1992). Residual diagnostics for mixture models. *Journal of the American Statistical Association 87*, 785–794.

Liu, J. S. (1994). The collapsed Gibbs sampler in Bayesian computations with applications to a gene regulation problem. *Journal of the American Statistical Association 89*, 958–966.

Liu, J. S. (2001). *Monte Carlo Strategies in Scientific Computing*. Springer Series in Statistics. New York/Berlin/Heidelberg: Springer.

Liu, J. S., W. H. Wong, and A. Kong (1994). Covariance structure of the Gibbs sampler with applications to the comparisons of estimators and augmentation schemes. *Biometrika 81*, 27–40.

Liu, J. S., J. L. Zhang, M. J. Palumbo, and C. L. Lawrence (2003). Bayesian clustering with variable and transformation selection. In J. M. Bernardo, M. J. Bayarri, J. O. Berger, A. P. Dawid, D. Heckerman, A. F. M. Smith, and M. West (Eds.), *Bayesian Statistics 7*, pp. 249–275. Oxford: Oxford University Press.

Lopes, H. F., P. Müller, and G. L. Rosner (2003). Bayesian meta-analysis for longitudinal data models using multivariate mixture priors. *Biometrics 59*, 66–75.

Louis, T. A. (1982). Finding the observed information matrix when using the EM algorithm. *Journal of the Royal Statistical Society, Ser. B 44*, 226–233.

Lu, Z., Y. V. Hui, and A. H. Le (2003). Minimum Hellinger distance estimation for finite mixtures of Poisson regression models and its application. *Biometrics 59*, 1016–1026.

Luginbuhl, R. and A. de Vos (1999). Bayesian analysis of an unobserved-component time series model of GDP with Markov-switching and time-varying growths. *Journal of Business & Economic Statistics 17*, 456–465.

Lwin, T. and P. J. Martin (1989). Probits of mixtures. *Biometrics 45*, 721–732.

MacDonald, I. and W. Zucchini (1997). *Hidden Markov and Other Models for Discrete-valued Time Series*. Monographs on Statistics and Applied Probability. London: Chapman & Hall.

MacKay Altman, R. (2004). Assessing the goodness-of-fit of hidden Markov models. *Biometrics 60*, 444–450.

MacRae Keenan, D. (1982). A time series analysis of binary data. *Journal of the American Statistical Association 77*, 816–821.

Madigan, D. and J. York (1995). Bayesian graphical models for discrete data. *International Statistical Review 63*, 215–232.

Magill, D. T. (1965). Optimal adaptive estimation of sampled stochastic processes. *IEEE Transactions on Automatic Control 10*, 434–439.

Mandelbrot, B. (1963). The variation of certain speculative prices. *Journal of Business 36*, 394–419.

Manton, K. G., M. A. Woodbury, and E. Stallard (1981). A variance components approach to categorical data models with heterogeneous cell populations: Analysis of spatial gradients in lung cancer mortality rates in North Carolina counties. *Biometrics 37*, 259–269.

Marrs, A. D. (1998). An application of reversible-jump MCMC to multivariate spherical Gaussian mixtures. In M. I. Jordan, M. J. Kearns, and S. A. Solla (Eds.), *Advances in Neural Information Processing Systems*, Volume 10, pp. 577–583. Cambridge, MA: MIT Press.

Masreliez, C. J. (1975). Approximate non-Gaussian filtering with linear state and observation relations. *IEEE Transactions on Automatic Control 20*, 107–110.

Masreliez, C. J. and R. D. Martin (1975). Robust Bayesian estimation for the linear model and robustifying the Kalman filter. *IEEE Transactions on Automatic Control 20*, 361–371.

Mayer, J. (1999). Bayesian-type count data models with varying coefficients: estimation and testing against a parametric alternative. In H. Friedl, A. Berghold, and G. Kauermann (Eds.), *Statistical Modelling. Proceedings of the Fourteenth International Workshop on Statistical Modelling*, pp. 273–280. Graz.

McCullagh, P. and J. A. Nelder (1999). *Generalized Linear Models.* London: Chapman & Hall.

McCulloch, R. E. and P. E. Rossi (1992). Bayes factors for nonlinear hypotheses and likelihood distributions. *Biometrika 79*, 663–676.

McCulloch, R. E. and R. S. Tsay (1993). Bayesian inference and prediction for mean and variance shifts in autoregressive time series. *Journal of the American Statistical Association 88*, 968–978.

McCulloch, R. E. and R. S. Tsay (1994a). Bayesian inference of trend- and difference-stationarity. *Econometric Theory 10*, 596–608.

McCulloch, R. E. and R. S. Tsay (1994b). Statistical analysis of economic time series via Markov switching models. *Journal of Time Series Analysis 15*, 523–539.

McLachlan, G. J. and K. E. Basford (1988). *Mixture Models: Inference and Applications to Clustering.* New York/ Basel: Marcel Dekker.

McLachlan, G. J., R. W. Bean, and D. Peel (2002). A mixture-model based approach to the clustering of microarray expression data. *Bioinformatics 18*, 413–422.

McLachlan, G. J. and D. Peel (2000). *Finite Mixture Models.* Wiley Series in Probability and Statistics. New York: Wiley.

McLachlan, G. J., D. Peel, and R. W. Bean (2003). Modelling high-dimensional data by mixtures of factor analyzers. *Computational Statisics and Data Analysis 41*, 379–388.

McQueen, G. and S. Thorely (1991). Are stock returns predictable? A test using Markov chains. *The Journal of Finance 46*, 239–263.

Meinhold, R. J. and N. D. Singpurwalla (1983). Understanding the Kalman filter. *The American Statistician 37*, 123–127.

Meinhold, R. J. and N. D. Singpurwalla (1989). Robustification of Kalman filter models. *Journal of the American Statistical Association 84*, 479–486.

Meng, X.-L. (1994). Posterior predictive $p$-values. *The Annals of Statistics 22*, 1142–1160.

Meng, X.-L. (1997). The EM algorithm and medical studies: A historical link. *Statistical Methods in Medical Research 6*, 3–23.

Meng, X.-L. and D. B. Rubin (1991). Using EM to obtain asymptotic variance-covariance matrices: The SEM algorithm. *Journal of the American Statistical Association 86*, 899–909.

Meng, X.-L. and S. Schilling (2002). Warp bridge sampling. *Journal of Computational and Graphical Statistics 11*, 552–586.

Meng, X.-L. and D. Van Dyk (1997). The EM algorithm — An old folk-song sung to a fast new tune. *Journal of the Royal Statistical Society, Ser. B 59*, 511–567.

Meng, X.-L. and D. Van Dyk (1999). Seeking efficient data augmentation schemes via conditional and marginal data augmentation. *Biometrika 86*, 301–320.

Meng, X.-L. and W. H. Wong (1996). Simulating ratios of normalizing constants via a simple identity: A theoretical exploration. *Statistica Sinica 6*, 831–860.

Mengersen, K. L. and C. P. Robert (1996). Testing for mixtures: A Bayesian entropic approach. In J. M. Bernardo, J. O. Berger, A. P. Dawid, and A. F. M. Smith (Eds.), *Bayesian Statistics 5 — Proceedings of the Fifth Valencia International Meeting*, pp. 255–276. Oxford: Oxford University Press.

Milligan, G. W. and M. C. Cooper (1985). An examination of procedures for determining the number of clusters in a data set. *Psychometrika 50*, 159–179.

Miloslavsky, M. and M. J. van der Laan (2003). Fitting of mixtures with unspecified number of components using cross validation distance estimate. *Computational Statisics and Data Analysis 41*, 413–428.

Morbiducci, M., A. Nardi, and C. Rossi (2003). Classification of "cured" individuals in survival analysis: The mixture approach to the diagnostic-prognostic problem. *Computational Statisics and Data Analysis 41*, 515–529.

Morel, J. G. and N. K. Nagaraj (1993). A finite mixture distribution for modelling multinomial extra variation. *Biometrika 80*, 363–371.

Morris, C. N. (1983). Parametric empirical Bayes inference: Theory and applications. *Journal of the American Statistical Association 78*, 47–55.

Müller, P., A. Erkanli, and M. West (1996). Bayesian curve fitting using multivarite normal mixtures. *Biometrika 83*, 67–79.

Murtagh, F. and A. E. Raftery (1984). Fitting straight lines to point patterns. *Pattern Recognition 17*, 479–483.

Muthén, B. and K. Shedden (1999). Finite mixture modelling with mixture outcomes using the EM algorithm. *Biometrics 55*, 463–469.

Nahi, N. E. (1969). Optimal recursive estimation with uncertain observations. *IEEE Transactions on Information Theory 15*, 457–462.

Nakatsuma, T. (2000). Bayesian analysis of ARMA-GARCH models: A Markov chain sampling approach. *Journal of Econometrics 95*, 57–69.

Natarajan, R. and C. E. McCulloch (1998). Gibbs sampling with diffuse proper priors: A valid approach to data-driven inference? *Journal of Computational and Graphical Statistics 7*, 267–277.

Neal, R. N. (1998). Erroneous results in "Marginal likelihood from the Gibbs output.". http://www.cs.utoronto.ca/radford/radford@stat.utoronto.ca.

Neftçi, S. N. (1984). Are economic time series asymmetric over the business cycle? *Journal of Political Economy 92*, 307–328.

Nelder, J. A. and R. W. M. Wedderburn (1972). Generalized linear models. *Journal of the Royal Statistical Society, Ser. A 135*, 370–384.

Nelson, C. R. (1988). Spurious trend and cycle in the state space decomposition of a time series with a unit root. *Journal of Economic Dynamics and Control 12*, 475–488.

Nelson, C. R., J. Piger, and E. Zivot (2001). Markov regime switching and unit-root tests. *Journal of Business & Economic Statistics 19*, 404–415.

Nelson, C. R. and C. I. Plosser (1982). Trends and random walks in macroeconomic time series: Some evidence and implications. *Journal of Monetary Economics 10*, 139–162.

Nelson, D. B. (1990). Stationarity and persistence in the GARCH(1,1) model. *Econometric Theory 6*, 318–334.

Neuhaus, J. M., W. W. Hauck, and J. D. Kalbfleisch (1992). The effects of mixture distribution misspecification when fitting mixed-effects logistic models. *Biometrica 79*, 755–762.

Newcomb, S. (1886). A generalized theory of the combination of observations so as to obtain the best result. *American Journal of Mathematics 8*, 343–366.

Newton, M. A. and A. E. Raftery (1994). Approximate Bayesian inference with the weighted likelihood bootstrap (Disc: p26-48). *Journal of the Royal Statistical Society, Ser. B 56*, 3–26.

Nicholls, D. F. and A. R. Pagan (1985). Varying coefficient regression. In E. J. Hannan, P. R. Krishnaiah, and M. M. Rao (Eds.), *Handbook of Statistics*, Volume 5, Chapter 16, pp. 413–449. New York: Elsevier.

Nobile, A. (2004). On the posterior distribution of the number of components in a finite mixture. *The Annals of Statistics 32*, 2044–2073.

Nobile, A. and P. J. Green (2000). Bayesian analysis of factorial experiments by mixture modelling. *Biometrika 87*, 15–35.

Oberhofer, W. (1980). Die Nichtkonsistenz der ML Schätzer im Switching Regression Problem. *Metrika 27*, 1–13.

Omori, Y., S. Chib, N. Shephard, and J. Nakajima (2004). Stochastic volatility with leverage: Fast likelihood inference. Research Report.

Oskrochi, G. R. and R. B. Davies (1997). An EM-type algorithm for multivariate mixture models. *Statistics and Computing 7*, 145–151.

Otter, T., R. Tüchler, and S. Frühwirth-Schnatter (2002). Bayesian latent class metric conjoint analysis — A case study from the Austrian mineral water market. In O. Opitz and M. Schwaiger (Eds.), *Exploratory Data Analysis in Empirical Research*, Studies in Classification, Data Analysis and Knowledge Organization, pp. 157–169. New York/Berlin/Heidelberg: Springer.

Otter, T., R. Tüchler, and S. Frühwirth-Schnatter (2004). Capturing consumer heterogeneity in metric conjoint analysis using Bayesian mixture models. *International Journal of Marketing Research 21*, 285–297.

Paap, R. and H. van Dyck (2003). Bayes estimates of Markov trends in possibly cointegrated series: An application to U.S. consumption and income. *Journal of Business & Economic Statistics 21*, 547–563.

Pagan, A. R. and G. W. Schwert (1990). Alternative models for conditional stock volatility. *Journal of Econometrics 45*, 267–290.

Paul, S. R., K. Y. Liang, and S. G. Self (1989). On testing departure from the binomial and multinomial assumptions. *Biometrics 45*, 231–236.

Pauler, D. K., M. D. Escobar, J. A. Sweeney, and J. Greenhouse (1996). Mixture models for eye-tracking data: A case study. *Statistics in Medicine 15*, 1365–1376.

Peña, D. and I. Guttman (1988). Bayesian approach to robustifying the Kalman filter. In J. C. Spall (Ed.), *Bayesian Analysis of Time Series and Dynamic Linear Models*, pp. 227–253. New York: Marcel Dekker.

Peña, D. and I. Guttman (1993). Comparing probabilistic methods for outlier detection in linear models. *Biometrika 80*, 603–610.

Peña, D. and G. C. Tiao (1992). Bayesian robustness functions for linear models. In J. M. Bernardo, J. O. Berger, A. P. Dawid, and A. F. M. Smith (Eds.), *Bayesian Statistics 4. Proceedings of the Fourth Valencia International Meeting*, pp. 365–388. Oxford: Oxford University Press.

Pearson, K. (1894). Contributions to the mathematical theory of evolution. *Philosophical Transactions of the Royal Society of London A 185*, 71–110.

Pearson, K. (1915). On certain types of compound frequency distributions in which the components can be individually described by binomial series. *Biometrika 11*, 139–144.

Peel, D. and G. McLachlan (2000). Robust mixture modelling using the $t$ distribution. *Statistics and Computing 10*, 339–348.

Peng, F., R. A. Jacobs, and M. A. Tanner (1996). Bayesian inference in mixtures-of-experts and hierarchical mixtures-of-experts models with an application to speech recognition. *Journal of the American Statistical Association 91*, 953–960.

Peria, M. S. M. (2002). A regime-switching aproach to the study of speculative attacks: A focus on the EMS crisis. *Empirical Economics 27*, 299–334.

Perron, P. (1989). The great crash, the oil price shock, and the unit root hypothesis. *Econometrica 57*, 1361–1401.

Perron, P. (1990). Testing for a unit root in a time series with a changing mean. *Journal of Business & Economic Statistics 8*, 153–162.

Peskun, P. H. (1973). Optimum Monte Carlo sampling using Markov chains. *Biometrika 60*, 607–612.

Petrie, T. (1969). Probabilistic functions of finite state Markov chains. *The Annals of Mathematical Statistics 40*, 97–115.

Pfeiffer, R. and N. Jeffries (1999). A mixture model for the probability distribution of rain rate. In H. Friedl, A. Berghold, and G. Kauermann (Eds.), *Statistical Modelling. Proceedings of the Fourteenth International Workshop on Statistical Modelling*, pp. 610–615. Graz.

Phillips, D. B. and A. F. M. Smith (1996). Bayesian model comparison via jump diffusions. In W. Gilks, S. Richardson, and D. J. Spiegelhalter (Eds.), *Markov Chain Monte Carlo in Practice*, pp. 215–239. London: Chapman & Hall.

Phillips, K. L. (1991). A two-country model of stochastic output with changes in regime. *Journal of International Economics 31*, 121–142.

Pitt, M. K. and N. Shephard (1999). Analytic convergence rates and parameterization issues for the Gibbs sampler applied to state space models. *Journal of Time Series Analysis 20*, 63–85.

Ploberger, W., W. Kramer, and K. Kontrus (1989). A new test for structural stability in the linear regression model. *Journal of Econometrics 40*, 307–318.

Poskitt, D. S. and S.-H. Chung (1996). Markov chain models, time series analysis and extreme value theory. *Advances in Applied Probability 28*, 405–425.

Posse, C. (2001). Hierarchical model-based clustering for large data sets. *Journal of Computational and Graphical Statistics 10*, 464–486.

Pregibon, D. (1981). Logistic regression diagnostics. *The Annals of Statistics 9*, 705–724.

Press, S. J. (2003). *Subjective and Objective Bayesian Statistics* (2 ed.). New York: Wiley.

Psaradakis, Z. (2001). Markov-level shifts and the unit-root hypothesis. *The Econometrics Journal 4*, 225–241.

Psaradakis, Z. (2002). On the asymptotic behaviour of unit root tests in the presence of a Markov trend. *Statistics & Probability Letters 57*, 101–109.

Qian, W. and D. M. Titterington (1991). Estimation of parameters in hidden Markov models. *Philosophical Transactions of the Royal Society of London, Series A, Math and Physical Science 337*, 407–428.

Quandt, R. E. (1958). The estimation of the parameter of a linear regression system obeying two separate regimes. *Journal of the American Statistical Association 53*, 873–880.

Quandt, R. E. (1960). Tests of the hypothesis that a linear regression system obeys two separate regimes. *Journal of the American Statistical Association 55*, 324–330.

Quandt, R. E. (1972). A new approach to estimating switching regressions. *Journal of the American Statistical Association 67*, 306–310.

Quandt, R. E. and J. B. Ramsey (1978). Estimating mixtures of normal distributions and switching regressions. *Journal of the American Statistical Association 73*, 730–752.

Rabiner, L. R. (1989). A tutorial on hidden Markov models and selected applications in speech recognition. *Proceedings of the IEEE 77*, 257–286.

Rabiner, L. R. and B. H. Juang (1986). An introduction to hidden Markov models. *IEEE Transactions on Acoustics, Speech, and Signal Processing 34*, 4–16.

Raftery, A., D. Madigan, and J. Hoeting (1997). Bayesian model averaging for linear regression models. *Journal of the American Statistical Association 92*, 179–191.

Raftery, A. E. (1996a). Approximate Bayes factors and accounting for model uncertainty in generalised linear models. *Biometrika 83*, 251–266.

Raftery, A. E. (1996b). Hypothesis testing and model selection. In W. R. Gilks, S. Richardson, and D. J. Spiegelhalter (Eds.), *Markov Chain Monte Carlo in Practice*, pp. 163–188. London: Chapman & Hall.

Raj, B. (2002). Asymmetries of business cycle: The Markov switching approach. In A. Ullah, A. Wan, and A. Chaturvedi (Eds.), *Handbook of Applied Econometrics and Statistical Inference*, pp. 687–710. New York: Marcel Dekker.

Ramaswamy, V., E. W. Anderson, and W. S. DeSarbo (1994). Negative binomial models for count data. *Management Science 40*, 405–412.

Ramaswamy, V., W. S. DeSarbo, D. Reibstein, and W. T. Robinson (1993). An empirical pooling approach for estimating marketing mix elasticities with PIMS data. *Marketing Science 12*, 103–124.

Rao, C. (1948). The utilization of multiple measurements in problems of biological classification. *Journal of the Royal Statistical Society, Ser. B 10*, 159–203.

Rao, C. R. (1975). Simultaneous estimation of parameters in different linear models and applications to biometric problems. *Biometrics 31*, 545–554.

Ray, S. and B. Lindsay (2005). The topography of multivariate normal mixtures. *The Annals of Statistics 33*, 2042–2065.

Redner, R. A. and H. Walker (1984). Mixture densities, maximum likelihood and the EM algorithm. *SIAM Review 26*, 195–239.

Richardson, S. and P. J. Green (1997). On Bayesian analysis of mixtures with an unknown number of components. *Journal of the Royal Statistical Society, Ser. B 59*, 731–792.

Ripley, B. D. (1977). Modelling spatial patterns (C/R: P192-212). *Journal of the Royal Statistical Society, Ser. B 39*, 172–192.

Robert, C. P. (1996). Mixtures of distributions: Inference and estimation. In W. R. Gilks, S. Richardson, and D. J. Spiegelhalter (Eds.), *Markov Chain Monte Carlo in Practice*, pp. 441–464. London: Chapman & Hall.

Robert, C. P. and G. Casella (1999). *Monte Carlo Statistical Methods.* Springer Series in Statistics. New York/Berlin/Heidelberg: Springer.

Robert, C. P., G. Celeux, and J. Diebolt (1993). Bayesian estimation of hidden Markov chains: A stochastic implementation. *Statistics & Probability Letters 16*, 77–83.

Robert, C. P. and K. L. Mengersen (1999). Reparameterisation issues in mixture modelling and their bearing on MCMC algorithms. *Computational Statistics and Data Analysis 29*, 325–343.

Robert, C. P., T. Rydén, and D. M. Titterington (1999). Convergence controls for MCMC algorithms, with applications to hidden Markov chains. *Journal of Statistical Computation and Simulation 64*, 327–355.

Robert, C. P., T. Rydén, and D. M. Titterington (2000). Bayesian inference in hidden Markov models through the reversible jump Markov chain Monte Carlo method. *Journal of the Royal Statistical Society, Ser. B 62*, 57–75.

Robert, C. P. and D. M. Titterington (1998). Reparameterization strategies for hidden Markov models and Bayesian approaches to maximum likelihood estimation. *Statistics and Computing 8*, 145–158.

Robertson, C. A. and J. G. Fryer (1969). Some descriptive properties of normal mixtures. *Skand. Aktuarietidskr. 52*, 137–146.

Roeder, K. and L. Wasserman (1997a). Discussion of the paper by Richardson and Green "On Bayesian analysis of mixtures with an unknown number of components.". *Journal of the Royal Statistical Society, Ser. B 59*, 782.

Roeder, K. and L. Wasserman (1997b). Practical Bayesian density estimation using mixtures of normals. *Journal of the American Statistical Association 92*, 894–902.

Rossi, P. E., G. M. Allenby, and R. McCulloch (2005). *Bayesian Statistics and Marketing.* Chichester: Wiley.

Rothenberg, T. (1971). Identification in parametric models. *Econometrica 39*, 577–591.

Rubin, D. B. (1981). The Bayesian bootstrap. *The Annals of Statistics 9*, 130–134.

Rubin, D. B. (1984). Bayesianly justifiable and relevant frequency calculations for the applied statistician. *The Annals of Statistics 12*, 1151–1172.

Rue, H. (1995). New loss functions in Bayesian imaging. *Journal of the American Statistical Association 90*, 900–908.

Rydén, T., T. Teräsvirta, and S. Åsbrink (1998). Stylized facts of daily return series and the hidden Markov model. *Journal of Applied Econometrics 13*, 217–244.

Rydén, T. and D. Titterington (1998). Computational Bayesian analysis of hidden Markov models. *Journal of Computational and Graphical Statistics 7*, 194–211.

Sahu, S. K. and R. C. H. Cheng (2003). A fast distance based approach for determining the number of components in mixtures. *The Canadian Journal of Statistics 31*, 3–22.

Satagopan, J. M., M. A. Newton, and A. E. Raftery (2000). Easy estimation of normalizing constants and Bayes factors from posterior simulations: Stabilizing the harmonic mean estimator. Technical Report 382, Department of Statistics, University of Washington.

Scaccia, L. and P. J. Green (2003). Bayesian growth curves using normal mixtures with nonparametric weights. *Journal of Computational and Graphical Statistics 12*, 308–331.

Schall, R. (1991). Estimation in generalized linear models with random effects. *Biometrika 78*, 719–727.

Schaller, H. and S. van Norden (2002). Fads or bubbles? *Empirical Economics 27*, 335–362.

Schervish, M. J. and R. Tsay (1988). Bayesian modeling and forecasting in autoregressive models. In J. C. Spall (Ed.), *Bayesian Analysis of Time Series and Dynamic Linear Models*, pp. 23–53. New York: Marcel Dekker.

Schlattmann, P. and D. Böhning (1993). Mixture models and disease mapping. *Statistics in Medicine 12*, 1943–1950.

Schlattmann, P., E. Dietz, and D. Böhning (1996). Covariate adjusted mixture models and disease mapping with the program DismapWin. *Statistics in Medicine 15*, 919–929.

Schnatter, S. (1988a). Bayesian forecasting of time series by Gaussian sum approximation. In J. M. Bernardo, M. H. DeGroot, D. V. Lindley, and A. F. M. Smith (Eds.), *Bayesian Statistics 3*, pp. 757–764. London: Clarendon.

Schnatter, S. (1988b). *Dynamic Bayesian Models with Applications to Hydrological Short-Term Forecasting (in German)*. Ph. D. thesis, Vienna University of Technology, Austria.

Schnatter, S., D. Gutknecht, and R. Kirnbauer (1987). A Bayesian approach to estimating the parameters of a hydrological forecasting system. In R. Viertl (Ed.), *Probability and Bayesian Statistics*, pp. 415–422. New York: Plenum.

Schneider, W. (1988). Analytical uses of Kalman filtering in econometrics: A survey. *Statistical Papers 29*, 3–33.

Schwarz, G. (1978). Estimating the dimension of a model. *The Annals of Statistics 6*, 461–464.

Schweppe, F. C. (1965). Evaluation of likelihood functions for Gaussian signals. *IEEE Transactions on Information Theory 11*, 294–305.

Sclove, S. L. (1983). Time series segmentation: A model and a method. *Information Science 29*, 7–25.

Scott, A. J. and M. Symons (1971). Clustering methods based on likelihood ratio criteria. *Biometrics 27*, 387–397.

Scott, D. W. (1992). *Multivariate Density Estimation: Theory, Practice, and Visualization.* New York: Wiley.

Scott, D. W. (2004). Outlier detection and clustering by partial mixture modelling. In J. Antoch (Ed.), *COMPSTAT 2004. Proceedings in Computational Statistics*, pp. 453–464. Heidelberg: Physica-Verlag/Springer.

Scott, D. W. and W. F. Szewczyk (2001). From kernels to mixtures. *Technometrics 43*, 323–335.

Scott, S. L. (2002). Bayesian methods for hidden Markov models: Recursive computing in the 21st century. *Journal of the American Statistical Association 97*, 337–351.

Scott, S. L., G. M. James, and C. A. Sugar (2005). Hidden Markov models for longitudinal comparisons. *Journal of the American Statistical Association 100*, 359–369.

Shaked, M. (1980). On mixtures from exponential families. *Journal of the Royal Statistical Society, Ser. B 42*, 192–198.

Shephard, N. (1994). Partial non-Gaussian state space. *Biometrika 81*, 115–131.

Shephard, N. (1996). Statistical aspects of ARCH and stochastic volatility. In D. R. Cox, D. V. Hinkley, and O. E. Barndorff-Nielsen (Eds.), *Time Series Models: In Econometrics, Finance and Other Fields*, pp. 1–67. London: Chapman & Hall.

Shephard, N. and M. K. Pitt (1997). Likelihood analysis of non-Gaussian measurement time series. *Biometrika 84*, 653–667.

Shively, T. S. and R. Kohn (1997). A Bayesian approach to model selection in stochastic coefficient regression models and structural time series models. *Journal of Econometrics 76*, 39–52.

Shumway, R. H. and D. S. Stoffer (1982). An approach to time series smoothing and forecasting using the EM algorithm. *Journal of Time Series Analysis 3*, 253–264.

Shumway, R. H. and D. S. Stoffer (1991). Dynamic linear models with switching. *Journal of the American Statistical Association 86*, 763–769.

Shumway, R. H. and D. S. Stoffer (2000). *Time Series Analysis and Its Applications*. New York: Springer.

Sichel, D. E. (1994). Inventories and the three phases of the business cycle. *Journal of Business & Economic Statistics 12*, 269–277.

Silverman, B. W. (1999). *Density Estimation for Statistics and Data Analysis*. London: Chapman & Hall.

Simar, L. (1976). Maximum likelihood estimation of a compound Poisson process. *The Annals of Statistics 4*, 1200–1209.

Sims, C. A. (1980). Macroeconomics and reality. *Econometrica 48*, 1–48.

Sims, F. L. and D. G. Lainiotis (1969). Recursive algorithm for the calculation of the adaptive Kalman filter weighting coefficients. *IEEE Transactions on Automatic Control 14*, 215–218.

Slud, E. (1997). Testing for imperfect debugging in software reliability. *Scandinavian Journal of Statistics 24*, 555–572.

Smith, A. F. M. and G. O. Roberts (1993). Bayesian computation via the Gibbs sampler and related Markov chain Monte Carlo methods (Disc: p53-102). *Journal of the Royal Statistical Society, Ser. B 55*, 3–23.

Smith, A. F. M. and D. J. Spiegelhalter (1980). Bayes factors and choice criteria for linear models. *Journal of the Royal Statistical Society, Ser. B 42*, 213–220.

Smith, A. F. M. and M. West (1983). Monitoring renal transplants: An application of the multiprocess Kalman filter. *Biometrics 39*, 867–878.

Smith, D. R. (2002). Markov switching and stochastic volatility diffusion models of short-term interest rates. *Journal of Business & Economic Statistics 20*, 183–197.

Solka, J. L., E. J. Wegman, C. E. Priebe, W. L. Poston, and G. W. Rogers (1998). Mixture structure analysis using the Akaike information criterion and the bootstrap. *Statistics and Computing 8*, 177–188.

Sorenson, H. W. and D. L. Alspach (1971). Recursive Bayesian estimation using Gaussian sums. *Automatica 6*, 465–479.

Spiegelhalter, D. J., K. R. Abrams, and J. P. Myles (2003). *Bayesian Approaches to Clinical Trials and Health-Care Evaluation.* Chichester: Wiley.

Stephens, M. (1997a). *Bayesian Methods for Mixtures of Normal Distributions.* Ph. D. thesis, University of Oxford.

Stephens, M. (1997b). Discussion of the paper by Richardson and Green. *Journal of the Royal Statistical Society, Ser. B 59*, 768–769.

Stephens, M. (2000a). Bayesian analysis of mixture models with an unknown number of components – An alternative to reversible jump methods. *The Annals of Statistics 28*, 40–74.

Stephens, M. (2000b). Dealing with label switching in mixture models. *Journal of the Royal Statistical Society, Ser. B 62*, 795–809.

Stock, J. H. and M. W. Watson (2002). Macroeconomic forecasting using diffusion indexes. *Journal of Business & Economic Statistics 20*, 147–162.

Stone, M. (1974). Cross-validatory choice and assessment of statistical predictions. *Journal of the Royal Statistical Society, Ser. B 36*, 111–147.

Sugar, C. A. and G. M. James (2003). Finding the number of clusters in a data set: An information-theoretic approach. *Journal of the American Statistical Association 98*, 750–763.

Symons, M. J. (1981). Clustering criteria and multivariate normal mixtures. *Biometrics 37*, 35–43.

Tadesse, M. G., N. Sha, and M. Vannucci (2005). Bayesian variable selection in clustering high-dimensional data. *Journal of the American Statistical Association 100*, 602–617.

Tan, W. Y. and W. C. Chang (1972). Some comparisons of the method of moments and the method of maximum likelihood in estimating parameters of a mixture of two normal densities. *Journal of the American Statistical Association 67*, 702–708.

Tanner, M. A. and W. H. Wong (1987). The calculation of posterior distributions by data augmentation. *Journal of the American Statistical Association 82*, 528–540.

Taylor, J. M. G. (1995). Semi-parametric estimation in failure time mixture models. *Biometrics 51*, 899–907.

Teicher, H. (1960). On the mixture of distributions. *The Annals of Mathematical Statistics 31*, 55–73.

Teicher, H. (1961). Identifiability of mixtures. *The Annals of Mathematical Statistics 32*, 244–248.

Teicher, H. (1963). Identifiability of finite mixtures. *The Annals of Mathematical Statistics 34*, 1265–1269.

Teicher, H. (1967). Identifiability of mixtures of product measures. *The Annals of Mathematical Statistics 38*, 1300–1302.

Teräsvirta, T. and H. M. Anderson (1992). Characterizing nonlinearities in business cycles using smooth transition autoregressive models. *Journal of Applied Econometrics 7*, S119–S136.

Thompson, P. A. and R. B. Miller (1986). Sampling the future: A Bayesian approach to forecasting from univariate time series models. *Journal of Business & Economic Statistics 4*, 427–436.

Thyer, M. and G. Kuczera (2000). Modelling long-term persistence in hydro-climatic time series using a hidden state Markov model. *Water Resources Research 36*, 3301–3310.

Tibshirani, R., G. Walther, and T. Hastie (2001). Estimating the number of clusters in a data set via the gap statistic. *Journal of the Royal Statistical Society, Ser. B 63*, 411–423.

Tierney, L. (1994). Markov chain for exploring posterior distributions. *The Annals of Statistics 22*, 1701–1762.

Tierney, L., R. E. Kass, and J. B. Kadane (1989). Fully exponential Laplace approximations to expectations and variances of nonpositive functions. *Journal of the American Statistical Association 84*, 710–716.

Timmermann, A. (2000). Moments of Markov switching models. *Journal of Econometrics 96*, 75–111.

Tipping, M. E. and C. M. Bishop (1999). Mixtures of probabilistic principal component analysers. *Neural Computation 11*, 443–482.

Titterington, D. M. (1990). Some recent research in the analysis of mixture distributions. *Statistics 21*, 619–641.

Titterington, D. M., A. F. M. Smith, and U. E. Makov (1985). *Statistical Analysis of Finite Mixture Distributions.* Wiley Series in Probability and Statistics. New York: Wiley.

Tjøstheim, D. (1986). Some doubly stochastic time series models. *Journal of Time Series Analysis 7*, 51–72.

Tong, H. (1990). *Non-linear Time Series. A Dynamical System Approach.* Oxford: Oxford University Press.

Tsai, C. and L. Kurz (1983). An adaptive robustizing approach to Kalman filtering. *Automatica 19*, 279–288.

Tsay, R. S. (1987). Conditional heteroscedastic time series models. *Journal of the American Statistical Association 82*, 590–604.

Tsay, R. S. (1988). Outliers, level shifts, and variance changes in time series. *Journal of Forecasting 7*, 1–20.

Tucker, A. (1992). A reexamination of finite- and infinite-variance distributions as models of daily stock returns. *Journal of Business & Economic Statistics 10*, 73–81.

Tugnait, J. K. (1982). Detection and estimation for abruptly changing systems. *Automatica 18*, 607–615.

Tukey, J. W. (1960). A survey of sampling from contaminated distributions. In I. Olkin, S. Ghurye, W. Hoeffding, W. Madow, and H. Mann (Eds.), *Contributions to Probability and Statistics*, pp. 448–485. Stanford, CA: Stanford University Press.

Turner, C. M., R. Startz, and C. R. Nelson (1989). A Markov model of heteroscedasticity, risk, and learning in the stock market. *Journal of Financial Economics 25*, 3–22.

Utsugi, A. and T. Kumagai (2001). Bayesian analysis of mixtures of factor analyzers. *Neural Computation 13*, 993–1002.

van Dyk, D. and X.-L. Meng (2001). The art of data augmentation. *Journal of Computational and Graphical Statistics 10*, 1–50.

Verbeke, G. and E. Lesaffre (1996). A linear mixed-effects model with heterogeneity in the random-effects population. *Journal of the American Statistical Association 91*, 217–221.

Verbeke, G. and E. Lesaffre (1997). The effect of misspecifying the random-effects distribution in linear mixed models for longitudinal data. *Computational Statistics and Data Analysis 23*, 541–556.

Verbeke, G. and G. Molenberghs (2000). *Linear Mixed Models for Longitudinal Data.* Springer Series in Statistics. New York/Berlin/Heidelberg: Springer.

Verdinelli, I. and L. Wasserman (1991). Bayesian analysis of outlier problems using the Gibbs sampler. *Statistics and Computing 1*, 105–117.

Verdinelli, I. and L. Wasserman (1995). Computing Bayes factors using a generalization of the Savage-Dickey density ratio. *Journal of the American Statistical Association 90*, 614–618.

Viallefont, V., S. Richardson, and P. J. Green (2002). Bayesian analysis of Poisson mixtures. *Journal of Nonparametric Statistics 14*, 181–202.

Waagepetersen, R. and D. Sorensen (2001). A tutorial on reversible jump MCMC with a view toward applications in QTL-mapping. *International Statistical Review 69*, 49–61.

Wang, J. and E. Zivot (2000). A Bayesian time series model of multiple structural changes in level, trend, and variance. *Journal of Business & Economic Statistics 18*, 374–386.

Wang, P. and M. L. Puterman (1998). Mixed logistic regression models. *Journal of Agricultural, Biological, and Environmental Statistics 3*, 175–200.

Wang, P. and M. L. Puterman (1999). Markov Poisson regression models for discrete time series. Part I: Methodology. *Journal of Applied Statistics 26*, 855–869.

Wang, P., M. L. Puterman, I. Cockburn, and N. Le (1996). Mixed Poisson regression models with covariate dependent rates. *Biometrics 52*, 381–400.

Wasserman, L. (2000). Asymptotic inference for mixture models using data-dependent priors. *Journal of the Royal Statistical Society, Ser. B 62*, 159–180.

Watier, L., S. Richardson, and P. J. Green (1999). Using Gaussian mixtures with unknown number of components for mixed model estimation. In H. Friedl, A. Berghold, and G. Kauermann (Eds.), *Statistical Modelling. Proceedings of the Fourteenth International Workshop on Statistical Modelling*, pp. 394–401. Graz.

Watson, M. W. and R. F. Engle (1983). Alternative algorithms for the estimation of dynamic factor, mimic and varying coefficient regression models. *Journal of Econometrics 23*, 385–400.

Wecker, W. E. and C. Ansley (1983). The signal extraction approach to nonlinear regression and spline smoothing. *Journal of the American Statistical Association 78*, 81–89.

Wedel, M. and W. S. DeSarbo (1993a). A latent class binomial logit methodology for the analysis of paired comparison data. *Decision Sciences 24*, 1157–1170.

Wedel, M. and W. S. DeSarbo (1993b). A review of recent developments in latent class regression models. In R. P. Bagozzi (Ed.), *Advanced Methods of Marketing Research*, pp. 352–388. Oxford: Blackwell.

Wedel, M. and W. S. DeSarbo (1995). A mixture likelihood approach for generalized linear models. *Journal of Classification 12*, 21–55.

Wedel, M., W. S. DeSarbo, J. R. Bult, and V. Ramaswamy (1993). A latent class Poisson regression model for heterogeneous count data. *Journal of Applied Econometrics 8*, 397–411.

Wedel, M. and J.-B. Steenkamp (1991). A clusterwise regression method for simultaneous fuzzy market structuring and benefit segmentation. *Journal of Marketing Research 28*, 385–96.

West, M. (1981). Robust sequential approximate Bayesian estimation. *Journal of the Royal Statistical Society, Ser. B 43*, 157–166.

West, M. (1984). Outlier models and prior distribution in Bayesian linear regression. *Journal of the Royal Statistical Society, Ser. B 46*, 431–439.

West, M. (1985). Generalized linear models: Scale parameters, outlier accommodation and prior distributions. In J. M. Bernardo, M. H. DeGroot, D. V. Lindley, and A. F. M. Smith (Eds.), *Bayesian Statistics 2*, pp. 531–558. New York: Elsevier.

West, M. (1992). Modelling with mixtures. In J. M. Bernardo, J. O. Berger, A. P. Dawid, and A. F. M. Smith (Eds.), *Bayesian Statistics 4. Proceedings of the Fourth Valencia International Meeting*, pp. 503–524. Oxford: Oxford University Press.

West, M. (1993). Approximating posterior distributions by mixtures. *Journal of the Royal Statistical Society, Ser. B 55*, 409–422.

West, M. and P. J. Harrison (1997). *Bayesian Forecasting and Dynamic Models* (2 ed.). New York: Springer.

Whittaker, J. and S. Frühwirth-Schnatter (1994). A dynamic changepoint model for detecting the onset of growth in bacteriological infections. *Applied Statistics 43*, 625–640.

Wilks, S. S. (1938). The large sample distribution of the likelihood ratio for testing composite hypotheses. *The Annals of Mathematical Statistics 9*, 60–62.

Willse, A. and R. J. Boik (1999). Identifiable finite mixtures of location models for clustering mixed-mode data. *Statistics and Computing 9*, 111–121.

Windham, M. P. and A. Cutler (1992). Information ratios for validating mixture analyses. *Journal of the American Statistical Association 87*, 1188–1192.

Wolfe, J. H. (1970). Pattern clustering by multivariate mixture analysis. *Multivariate Behavioral Research 5*, 329–350.

Wong, C. S. and W. K. Li (2000). On a mixture autoregressive model. *Journal of the Royal Statistical Society, Ser. B 62*, 95–115.

Wong, C. S. and W. K. Li (2001). On a mixture autoregressive conditional heteroscedastic model. *Journal of the American Statistical Association 96*, 982–995.

Wu, C. F. J. (1983). On the convergence properties of the EM algorithm. *The Annals of Statistics 11*, 95–103.

Yakowitz, S. J. and J. D. Spragins (1968). On the identifiability of finite mixtures. *The Annals of Mathematical Statistics 39*, 209–214.

Yao, J. and J. Attali (2000). On stability of nonlinear AR processes with Markov switching. *Advances in Applied Probability 32*, 394–407.

Yau, K. K. W., A. H. Lee, and A. S. K. Ng (2003). Finite mixture regression model with random effects: Application to neonatal hospital lenght of stay. *Computational Statisics and Data Analysis 41*, 359–366.

Yeung, K. Y., C. Fraley, A. Murua, A. E. Raftery, and W. L. Ruzzo (2001). Model-based clustering and data transformations for gene expression data. *Bioinformatics 17*, 977–987.

Yeung, K. Y. and W. L. Ruzzo (2001). Principal component analysis for clustering gene expression data. *Bioinformatics 17*, 763–774.

Zeger, S. L. and B. Qaqish (1988). Markov regression models for time series: A quasi-likelihood approach. *Biometrics 44*, 1019–1031.

Zellner, A. (1971). *An Introduction to Bayesian Inference in Econometrics*. New York: Wiley.

Zhang, H. P. and K. Merikangas (2000). A frailty model of segregation analysis: Understanding the familial transmission of alcoholism. *Biometrics 56*, 815–823.

Zhu, H.-T. and H. Zhang (2004). Hypothesis testing in mixture regression models. *Journal of the Royal Statistical Society, Ser. B 66*, 3–16.

Zucchini, W. and P. Guttorp (1991). A hidden Markov model for space-time precipitation. *Water Resources Research 27*, 1917–1923.

# Index

# Springer Series in Statistics

(Continued from page ii)

*Heyde:* Quasi-Likelihood and its Application: A General Approach to Optimal Parameter Estimation.
*Huet/Bouvier/Poursat/Jolivet:* Statistical Tools for Nonlinear Regression: A Practical Guide with S-PLUS and R Examples, 2nd edition.
*Ibrahim/Chen/Sinha:* Bayesian Survival Analysis.
*Jolliffe:* Principal Component Analysis, 2nd edition.
*Knottnerus:* Sample Survey Theory: Some Pythagorean Perspectives.
*Küchler/Sørensen:* Exponential Families of Stochastic Processes.
*Kutoyants:* Statistical Inference for Ergodic Diffusion Processes.
*Lahiri:* Resampling Methods for Dependent Data.
*Le Cam:* Asymptotic Methods in Statistical Decision Theory
*Le Cam/Yang:* Asymptotics in Statistics: Some Basic Concepts, 2nd edition.
*Le/Zidek:* Statistical Analysis of Environmental Space-time Processes
*Liu:* Monte Carlo Strategies in Scientific Computing.
*Manski:* Partial Identification of Probability Distributions.
*Mielke/Berry:* Permutation Methods: A Distance Function Approach.
*Molenberghs/Verbeke:* Models for Discrete Longitudinal Data.
*Mukerjee/Wu:* A Modern Theory of Factorial Designs.
*Nelsen:* An Introduction to Copulas, 2nd edition.
*Pan/Fang:* Growth Curve Models and Statistical Diagnostics.
*Politis/Romano/Wolf:* Subsampling.
*Ramsay/Silverman:* Applied Functional Data Analysis: Methods and Case Studies.
*Ramsay/Silverman:* Functional Data Analysis, 2nd edition.
*Reinsel:* Elements of Multivariate Time Series Analysis, 2nd edition.
*Rosenbaum:* Observational Studies, 2nd edition.
*Rosenblatt:* Gaussian and Non-Gaussian Linear Time Series and Random Fields.
*Särndal/Swensson/Wretman:* Model Assisted Survey Sampling.
*Santner/Williams/Notz:* The Design and Analysis of Computer Experiments.
*Schervish:* Theory of Statistics
*Shaked/Shanthikumar:* Stochastic Orders.
*Shao/Tu:* The Jackknife and Bootstrap.
*Simonoff:* Smoothing Methods in Statistics
*Sprott:* Statistical Inference in Science.
*Stein:* Interpolation of Spatial Data: Some Theory for Kriging.
*Taniguchi/Kakizawa:* Asymptotic Theory for Statistical Inference for Time Series.
*Tanner:* Tools for Statistical Inference: Methods for the Exploration of Posterior Distributions and Likelihood Functions, 3rd edition.
*Tillé:* Sampling Algorithms.
*Tsaitis:* Semiparametric Theory and Missing Data.

*van der Laan/Robins:* Unified Methods for Censored Longitudinal Data and Causality.
*van der Vaart/Wellner:* Weak Convergence and Empirical Processes: With Applications to Statistics.
*Verbeke/Molenberghs:* Linear Mixed Models for Longitudinal Data.
*Weerahandi:* Exact Statistical Methods for Data Analysis.

## Semiparametric Theory and Missing Data

**Anastasios A. Tsiatis**

This book combines much of what is known in regard to the theory of estimation for semiparametric models with missing data in an organized and comprehensive manner. It starts with the study of semiparametric methods when there are no missing data. The description of the theory of estimation for semiparametric models is at a level that is both rigorous and intuitive, relying on geometric ideas to reinforce the intuition and understanding of the theory. These methods are then applied to problems with missing, censored, and coarsened data with the goal of deriving estimators that are as robust and efficient as possible.

2006. 395 p. (Springer Series in Statistics) Hardcover ISBN 0-387-32448-8

## Inference in Hidden Markov Models

**O. Cappé, E. Moulines, and T. Rydén**

This book is a comprehensive treatment of inference for hidden Markov models, including both algorithms and statistical theory. Topics range from filtering and smoothing of the hidden Markov chain to parameter estimation, Bayesian methods and estimation of the number of states.In a unified way the book covers both models with finite state spaces, which allow for exact algorithms for filtering, estimation etc. and models with continuous state spaces (also called state-space models) requiring approximate simulation-based algorithms that are also described in detail.

2005. 664 p. (Springer Series in Statistics) Hardcover ISBN 0-387-40264-0

## An Introduction to Bayesian Analysis

**Jayanta K. Ghosh, Mohan Delampady, and Tapas Samanta**

This is a graduate level textbook on Bayesian analysis blending modern Bayesian theory, methods, and applications. Starting from basic statistics, undergraduate calculus and linear algebra, ideas of both subjective and objective Bayesian analysis are developed to a level where real-life data can be analyzed using the current techniques of statistical computing. Advances in both low-dimensional and high-dimensional problems are covered, as well as important topics such as empirical Bayes and hierarchical Bayes methods and Markov chain Monte Carlo (MCMC) techniques.

2006. 365 p. (Springer Texts in Statistics) Hardcover ISBN 0-387-40084-2

**Easy Ways to Order►** Call: Toll-Free 1-800-SPRINGER • E-mail: orders-ny@springer.com • Write: Springer, Dept. S8113, PO Box 2485, Secaucus, NJ 07096-2485 • Visit: Your local scientific bookstore or urge your librarian to order.